Paul Price

COMPUTATIONAL TOXICOLOGY

BICENTENNIAL
1807 WILEY 2007
BICENTENNIAL

The Wiley Bicentennial–Knowledge for Generations

Each generation has its unique needs and aspirations. When Charles Wiley first opened his small printing shop in lower Manhattan in 1807, it was a generation of boundless potential searching for an identity. And we were there, helping to define a new American literary tradition. Over half a century later, in the midst of the Second Industrial Revolution, it was a generation focused on building the future. Once again, we were there, supplying the critical scientific, technical, and engineering knowledge that helped frame the world. Throughout the 20th Century, and into the new millennium, nations began to reach out beyond their own borders and a new international community was born. Wiley was there, expanding its operations around the world to enable a global exchange of ideas, opinions, and know-how.

For 200 years, Wiley has been an integral part of each generation's journey, enabling the flow of information and understanding necessary to meet their needs and fulfill their aspirations. Today, bold new technologies are changing the way we live and learn. Wiley will be there, providing you the must-have knowledge you need to imagine new worlds, new possibilities, and new opportunities.

Generations come and go, but you can always count on Wiley to provide you the knowledge you need, when and where you need it!

William J. Pesce
President and Chief Executive Officer

Peter Booth Wiley
Chairman of the Board

COMPUTATIONAL TOXICOLOGY

Risk Assessment for Pharmaceutical and Environmental Chemicals

Edited by

SEAN EKINS

WILEY-INTERSCIENCE
A JOHN WILEY & SONS, INC., PUBLICATION

Cover design/concept by Sean Ekins using images from Chapters 13, 16, and 19.

Published by John Wiley & Sons, Inc., Hoboken, New Jersey
Published simultaneously in Canada

Wiley Bicentennial logo: Richard J. Pacifico

Library of Congress Cataloging-in-Publication Data:

Computational toxicology : risk assessment for pharmaceutical and environmental chemicals / edited by Sean Ekins.
p. ; cm. – (Wiley series on technologies for the pharmaceutical industry)
Includes bibliographical references and index.
ISBN 978-0-470-04962-4 (cloth)
1. Toxicology – Mathematical models. 2. Toxicology – Computer simulation. 3. QSAR (Biochemistry) I. Ekins, Sean. II. Series.
[DNLM: 1. Toxicology – methods. 2. Computer Simulation. 3. Drug Toxicity. 4. Environmental Pollutants – toxicity. 5. Risk Assessment. QV 602 C738 2007]
RA1199.4.M37C66 2007
615.9001'5118–dc22

2006100242

Printed in the United States of America

10 9 8 7 6 5 4 3 2 1

To Maggie

It is very evident, that all other methods of improving medicine have been found ineffectual, by the stand it has been at these two or three thousand years; and that since of late mathematicians have set themselves to the study of it, men do already begin to talk intelligibly and comprehensibly, even about abstruse matters, that it may be hop'd in a short time, if those who are designed for this profession are early, while their minds and bodies are patient of labour and toil, initiated in the knowledge of numbers and geometry, that mathematical learning will be the distinguishing mark of a physician from a quack: and that he who wants this necessary qualification, will be as ridiculous as one without Greek or Latin.

Richard Mead
A mechanical account of poisons in several essays
2nd edition, London, 1708.

CONTENTS

SERIES INTRODUCTION

This book is the first in a series to be published by Wiley entitled *Technologies for the Pharmaceutical Industry*. The series aims to bring opinion leaders together to address important topics for the industry, from their implementation of technologies to current challenges. The pharmaceutical industry is at a critical juncture. It is pressured by patients on one side wanting effective treatments for diseases and governments trying to curtail health care spending while on the other side limited patent life and competition from generics all compound the issue. New technologies are one of the keys to maintaining competitiveness and minimizing the time for an idea coming from the bench to the bedside. Importantly these volumes will also describe how key technologies are likely to impact the direction in discovery and development for the future and will be accessible to readers both inside and outside the industry. Significant emphasis will also be put on the application rather than theory presented from both industrial and academic perspectives. At the time of going to press, two books are in preparation on in vitro–in vivo correlations, and biomarkers, with others in the pipeline. To ensure that the topics published are timely and relevant, an editorial board has been established and is listed in the front of this book. I gratefully acknowledge this team of scientists and those preparing the first volumes in the series for their time and willingness to assist me in this endeavor as we begin the series here.

PREFACE

It would have been unusual to mention the importance of mathematics to physicians in a book on poisons in the eighteenth century (Richard Mead, *A mechanical account of poisons in several essays*, second edition published in 1708), but nearly 300 years later mathematical and computational (in silico) methods are valuable assets for toxicology as they are in many other areas of science. From such a vantage point no one would have foreseen the broad impact and importance of toxicology itself, let alone its entwined relationship with pharmaceutical and environmental research. Now is the time for an assessment of the convergence of toxicology and computational methods in these areas and to outline where they will go in the future.

In pharmaceutical drug discovery and development, processes are in constant flux as new technologies are continually devised, tested, validated, and implemented. However, we have seen in the recent white paper from the US FDA on innovation stagnation in toxicology, that this is not always the case. Areas key to the overall development continue to use old technologies and processes and are not keeping pace with other developments in disparate fields of pharmaceutical research. This may be just the tip of the iceberg. If this industry is to improve its ability to rapidly identify and test therapeutics clinically with a high probability of success, it needs to discover and embrace new technologies early on.

Currently many companies, academics, regulatory authorities, and global organizations have or are evaluating the use of new predictive tools to improve human hazard assessment, (drug toxicity, P450 mediated drug metabolism etc.). For example the interaction of molecules can be predicted by using computer-based tools utilizing X-ray crystal structures, homology, receptor, pharmacophore, and QSAR models of human enzymes, transporters, nuclear receptors, ion channels, as well as other physicochemical properties and complex endpoints. In silico modeling for toxicology may therefore provide effective pre-screening for chemicals in pharmaceutical discovery and the chemical industry in general, and their effects on the environment may also be predicted. The criteria for the validation of computational toxicology models and other requirements for regulatory acceptance have not yet been widely discussed. This book addresses all of the above-mentioned areas and many more, presenting computational toxicology from an international and holistic perspective that differs significantly from recently published papers and books in the computational toxicology field.

The book is split into five key sections:

I. Introduction to Toxicology Methods
II. Computational Methods
III. Applying Computers to Toxicology Assessment: Pharmaceutical
IV. Applying Computers to Toxicology Assessment: Environmental
V. New Technologies for Toxicology: Future and Regulatory Perspectives

The book includes a comprehensive discussion on the state of the art of currently available molecular-modeling software for toxicology and their role in testing strategies for different types of toxicity when used alongside in vitro and in vivo models. The publication of this book comes at a critical time as we are now seeing REACH legislation coming into effect whose goal is to increase the amount of toxicological data required on tens of thousands of manufactured chemicals in order to predict the effect of chemicals on human health and their environmental impact. Naturally there has to be some means to prioritize in vitro and in vivo testing, and computational toxicology will be critical. The role of these computational approaches in addressing environmental and occupational toxicity is therefore covered broadly in this book, as well as new technologies and thoughts on the past, present, and future of computational toxicology and its applicability to chemical design. Each chapter is written by one or more leading expert in the field from industry, academe, or regulatory authorities, and each chapter has been edited to ensure consistency. Extensive use of explanatory figures is made, and all chapters include extensive key references for readers to delve deeper into topics at their own leisure.

This book is not aimed solely at laboratory toxicologists, as scientists of all disciplines in the pharmaceutical, chemical industries, and environmental sciences will find it of value. In particular, those researchers involved in ADMET, drug discovery, systems biology, and software development should benefit greatly from reading this book. The accessibility to the general reader with some scientific background should enable this volume to serve as an educational tool that inspires readers to pursue further the technologies presented. I hope you enjoy this book and benefit from the insights offered by the variety of contributing authors, as we take you on a tour of computational toxicology and go beyond—in silico.

ACKNOWLEDGMENTS

I am extremely grateful to Jonathan Rose at John Wiley who initiated this project and provided considerable assistance for initial chapter ideas and author suggestions. Thank you for getting me involved and allowing me to edit it. In addition I would like to thank all the team at Wiley for their assistance and in particular, Danielle Lacourciere for patiently putting the book together. My anonymous proposal reviewers are kindly acknowledged for their helpful suggestions, and along with other scientists who provided numerous ideas for additional authors, they greatly helped bring the book closer to its final format. Thank you!

I am immensely grateful to the many outstanding authors of the chapters for agreeing to contribute their valuable time, sharing their latest work and ideas, while patiently putting up with my editorial changes. This book represents their considerable talents. Although we have referenced many groups in these chapters, I acknowledge the many others that may have been omitted due to lack of space.

I would also like to take this opportunity to thank The Othmer Library at the Chemical Heritage Foundation in Philadelphia and, in particular, Ms. Ashley Augustyniak, Assistant Librarian, for providing access to a copy of the historic Mead text.

My studies in computational toxicology owe a great deal to collaboration with colleagues in both industry and academia, and several of these are contributors here. I acknowledge them all for letting me participate in stimulating science with them.

My parents and family have been incredibly supportive over what has been a tumultuous year. I dedicate this book to all my family in the United Kingdom and the United States, and to Maggie, in particular, for her continued steadfast support, valuable advice, and general encouragement to continue in the face of all adversity, this is for you.

Sean Ekins
Jenkintown, Pennsylvania
September 2006

CONTRIBUTORS

Daniel Acosta Jr., College of Pharmacy, University of Cincinnati, 3225 Eden Avenue, Cincinnati, OH 45267, USA. (daniel.acosta@uc.edu)

Alex M. Aronov, Vertex Pharmaceuticals Inc., 130 Waverly Street, Cambridge, MA 02139-4242, USA. (alex_aronov@vrtx.com)

Domenica Auteri, International Centre for Pesticides and Health Risk Prevention, Milano, Italy.

Giovanna Azimonti, International Centre for Pesticides and Health Risk Prevention, Milano, Italy.

Konstantin V. Balakin, ChemDiv, Inc. 11558 Sorrento Valley Road, Suite 5, San Diego, CA 92121, USA. (kvb@chemdiv.com)

Arianna Bassan, European Chemicals Bureau, Joint Research Centre, European Commission, Ispra, 21020 (VA), Italy.

Emilio Benfenati, Laboratory of Environmental Chemistry and Toxicology, Istituto di Ricerche Farmacologiche "Mario Negri," Milano, Italy. (benfenati@marionegri.it)

Frank E. Blaney, Computational, Analytical and Structural Sciences, GlaxoSmithKline Medicines Research, NFSP (North), Third Avenue, Harlow, Essex CM19 5AW, UK. (frank.e.blaney@gsk.com)

Alexander Breidenbach, Hoffmann-La Roche, PRBN-T, Bldg. 73/311B, CH-4070, Basel, Switzerland.

Curt M. Breneman, Department of Chemistry, Rensselaer Polytechnic Institute, 110 Eighth Street, Troy, NY 12180, USA.

Yu Zong Chen, Bioinformatics and Drug Design Group, Department of Computational Science, National University of Singapore, Blk SOC1, Level 7, 3 Science Drive 2, Singapore 117543. (yzchen@cz3.nus.edu.sg)

Peter Chiba, Institute of Medical Chemistry, Medical University Vienna, Waehringerstrasse 10, A-1090 Wien, Austria.

Alan B. Combs, College of Pharmacy, 2409 University Avenue, A2925 Austin, TX 78712, USA. (acombs@mail.utexas.edu)

Laura L. Custer, Drug Safety Evaluation, Bristol-Myers Squibb Pharmaceutical Research Institute, Syracuse, NY, USA. (laura.custer@bms.com)

J. Thom Deahl, Lilly Research Laboratories, Division of Eli Lilly and Company, Toxicology and Drug Disposition, Greenfield, IN 46140, USA.

James Devillers, CTIS, 3 Chemin de la Gravière, 69140 Rillieux La Pape, France. (j.devillers@ctis.fr)

Stephen K. Durham, Charles River Laboratories, 587 Dunn Circle, Sparks, NV 89431, USA. (stephen.durham@us.crl.com)

Gerhard F. Ecker, Emerging Field Pharmacoinformatics, Department of Medicinal Chemistry, University of Vienna, Althanstraße 14, A-1090 Wien, Austria. (gerhard.f.ecker@univie.ac.at)

Sean Ekins, ACT LLC, 1 Penn Plaza–36th Floor, New York, NY 10119, USA. (ekinssean@yahoo.com)

Mark J. Embrechts, Department of Decision Sciences and Engineering Systems, Rensselaer Polytechnic Institute, CII 5217, Troy, NY 12180, USA. (embrem@rpi.edu)

Jennifer M. Fostel, National Center for Toxicogenomics, National Institute of Environmental Health Sciences, PO Box 12233, MD F1-05, 111 Alexander Drive Research Triangle Park, NC 27709-2233, USA.

Christoph Funk, Hoffmann-La Roche, PRBN-T, Bldg. 73/311B, CH-4070, Basel, Switzerland.

Panos G. Georgopoulos, Department of Environmental and Occupational Medicine & Environmental and Occupational Health Sciences Institute, UMDNJ-RWJMS and Rutgers, the State University of New Jersey & Environmental Bioinformatics and Computational Toxicology Center (ebCTC), Piscataway, NJ 08854, USA.

Vijay K. Gombar, Lilly Research Laboratories, Division of Eli Lilly and Company, Lilly Corporate Center, Indianapolis, IN 46285, USA. (Gombar_Vijay_Kumar@Lilly.com)

Kam Jim, 5 Donald Avenue, Kendall Park, NJ 08824, USA. (kamcjim@gmail.com)

Dale E. Johnson, Emiliem, Inc., 6027 Christie Avenue, Emeryville, CA 94608, USA. (daleejohnson@sbcglobal.net)

Philip N. Judson, Judson Consulting Service, Heather Lea, Bland Hill, Norwood, Harrogate HG3 1TE, UK. (philip.judson@blubberhouses.net)

Stewart B. Kirton, NCE Discovery Ltd, 418 Science Park, Cambridge, CB24 0PZ, UK.

Alex Kiselyov, ChemDiv, Inc. 11558 Sorrento Valley Road, Suite 5, San Diego, CA 92121, USA.

Constantine Kreatsoulas, Merck Research Laboratories, Merck and Co. Inc., Rahway, NJ, USA.

Jason C. Lambert, Oak Ridge Institute for Science and Education, On assignment to the US Environmental Protection Agency, Cincinnati, OH, USA.

Edward L. LeCluyse, CellzDirect, 480 Hillsboro Street, Suite 130 Pittsboro, NC 27312, USA. (edl@cellzdirect.com)

Hu Li, Bioinformatics and Drug Design Group, Department of Computational Science, National University of Singapore, Blk SOC1, Level 7, 3 Science Drive 2, Singapore 117543.

Markus A. Lill, Institute of Molecular Pharmacy, Pharmacenter, University of Basel, Klingelbergstrasse 50, 4056 Basel, Switzerland and Biographics Laboratory 3R, Friendensgasse 35, 4056 Basel, Switzerland. (mlill@pharmacy.purdue.edu)

John C. Lipscomb, US Environmental Protection Agency, National Center for Environmental Assessment, 26 West Martin Luther King Drive (MS-190), Cincinnati, OH 45268, USA. (lipscomb.john@epa.gov)

Marco Lodi, Laboratory of Environmental Chemistry and Toxicology, Istituto di Ricerche Farmacologiche "Mario Negri," Milano, Italy.

Michael A. Lyons, Department of Environmental and Radiological Health Sciences Colorado State University, 1681 Campus Delivery, Fort Collins, CO 80523-1681, USA (lyonsm@lamar.ColoState.edu)

Brian E. Mattioni, Lilly Research Laboratories, Division of Eli Lilly and Company, Lilly Corporate Center, Indianapolis, IN 46285, USA.

Arthur N. Mayeno, Department of Environmental and Radiological Health Sciences Colorado State University, 1681 Campus Delivery, Fort Collins, CO 80523-1681, USA. (arthur.mayeno@colostate.edu)

B. Alex Merrick, National Center for Toxicogenomics, National Institute of Environmental Health Sciences, PO Box 12233, MD F1-05, 111 Alexander Drive Research Triangle Park, NC 27709-2233, USA.

John S. Mitcheson, Department of Cell Physiology and Pharmacology, University of Leicester, University Road, Leicester, LE1 7RH, UK.

Lutz Müller, Hoffmann-La Roche, PRBN-T, Bldg. 73/311B, CH-4070, Basel, Switzerland. (lutz.mueller@roche.com)

Moiz Mumtaz, Agency for Toxic Substances and Disease Registry, Atlanta, GA, USA.

Wolfgang Muster, Hoffmann-La Roche, PRBN-T, Bldg. 73/311B, CH-4070, Basel, Switzerland.

Tomoko Niwa, Discovery Research Laboratories, Nippon Shinyaku Co., Ltd. 14, Nishinosho-Monguchi-cho, Kisshoin, Minami-ku Kyoto, 601-8550, Japan. (t.niwa@po.nippon-shinyaku.co.jp)

Axel Pähler, Hoffmann-La Roche, PRBN-T, Bldg. 73/311B, CH-4070, Basel, Switzerland.

Brad Reisfeld, Department of Chemical and Biological Engineering and Department of Environmental and Radiological Health Sciences, Colorado State University, 1370, Campus Delivery, Fort Collins, CO 80523-1370, USA. (brad.reisfeld@colostate.edu)

Jim E. Riviere, Center for Chemical Toxicology Research and Pharmacokinetics Biomathematics Program, Carolina State University, Raleigh, NC, USA. (Jim_Riviere@ncsu.edu)

Amie D. Rodgers, Emiliem, Inc., 6027 Christie Avenue, Emeryville, CA 94608, USA.

J. Craig Rowlands, The Dow Chemical Company, Toxicology and Environmental Research and Consulting, 1803 Building, Midland, MI 48674, USA. (jcrowlands@dow.com)

Mike L. Shuler, Department of Biomedical Engineering, Cornell University, Ithaca, NY, USA. (mls50@cornell.edu)

Phillip J. Stansfeld, Department of Cell Physiology and Pharmacology, University of Leicester, University Road, Leicester, LE1 7RH, UK.

Sucha Sudarsanam, Emiliem, Inc., 6027 Christie Ave, Emeryville, CA 94608, USA.

Michael J. Sutcliffe, Manchester Interdisciplinary Biocentre & School of Chemical Engineering and Analytical Science, University of Manchester, 131 Princess Street, Manchester, M1 7ND, UK. (michael.sutcliffe@manchester.ac.uk)

Ben G. Tehan, Computational, Analytical and Structural Sciences, GlaxoSmithKline Medicines Research, NFSP (North), Third Avenue, Harlow, Essex CM19 5AW, UK.

Igor V. Tetko, Institute for Bioinformatics, GSF–National Research Centre for Environment and Health, Ingolstädter Landstraße 1, D-85764 Neuherberg, Germany. (itetko@vcclab.org)

Weida Tong, Center for Toxicoinformatics, US Food and Drug Administration–National Center for Toxicological Research (US FDA-NCTR), Jefferson, AR 72079, USA.

Shikha Varma-O'Brien, Accelrys, Inc., 10188 Telesis Court, Suite 100, San Diego CA, 92121, USA.

Angelo Vedani, Institute of Molecular Pharmacy, Pharmacenter, University of Basel, Klingelbergstrasse 50, 4056 Basel, Switzerland and Biographics Laboratory 3R, Friendensgasse 35, 4056 Basel, Switzerland. (angelo@biograf.ch)

Michael D. Waters, Integrated Laboratory Systems, Inc., PO Box 13501, Research Triangle Park, NC 27709, USA. (mwaters@ils-inc.com)

William J. Welsh, Department of Pharmacology, University of Medicine & Dentistry of New Jersey-Robert Wood Johnson Medical School (UMDNJ-RWJMS) & Environmental Bioinformatics and Computational Toxicology Center (ebCTC) & the Informatics Institute of UMDNJ, Piscataway, NJ 08854, USA. (welshwj@umdnj.edu)

Jean-Pierre Wery, INCAPS, 351 West 10th Street, Suite 350, Indianapolis, IN 46202, USA. (jwery@indianacaps.com)

Barbara A. Wetmore, National Center for Toxicogenomics, National Institute of Environmental Health Sciences, PO Box 12233, MD F1-05, 111 Alexander Drive Research Triangle Park, NC 27709-2233, USA.

Andrew P. Worth, European Chemicals Bureau, Joint Research Centre, European Commission, Ispra, 21020 (VA), Italy. (andrew.worth@jrc.it)

Hui Xu, Department of Biomedical Engineering, Cornell University, Ithaca, NY, USA. (hx28@cornell.edu)

Jinghai J. Xu, Pfizer Inc., Research Technology Center, 620 Memorial Drive, Rm. 367, Cambridge, MA 02139, USA. (jim.xu@pfizer.com)

Raymond S. H. Yang, Department of Environmental and Radiological Health Sciences, Colorado State University, 1681 Campus Delivery, Fort Collins, CO 80523-1681, USA. (raymond.yang@colostate.edu)

Chun Wei Yap, Bioinformatics and Drug Design Group, Department of Computational Science, National University of Singapore, Blk SOC1, Level 7, 3 Science Drive 2, Singapore 117543.

Katsumi Yoshida, Discovery Research Laboratories, Nippon Shinyaku Co., Ltd. 14, Nishinosho-Monguchi-cho, Kisshoin, Minami-ku Kyoto, 601-8550, Japan.

Craig Zwickl, Lilly Research Laboratories, Division of Eli Lilly and Company, Toxicology and Drug Disposition, Greenfield, IN 46140, USA.

PART I

INTRODUCTION TO TOXICOLOGY METHODS

1

AN INTRODUCTION TO TOXICOLOGY AND ITS METHODOLOGIES

ALAN B. COMBS AND DANIEL ACOSTA JR.

Contents

Computational Toxicology: Risk Assessment for Pharmaceutical and Environmental Chemicals, Edited by Sean Ekins

1.1 OVERVIEW

Toxicology in the broadest sense is the study of the adverse effects of drugs or chemicals on living systems. The questions asked by this discipline include what things are toxic, how and why toxicity is manifested, and how might toxicity be predicted, treated, or prevented. It is the purpose of this chapter to give a broad introduction to toxicology and to show how modern computational techniques are becoming so useful to the field.

1.2 WHERE AND WHY TOXICOLOGICAL KNOWLEDGE IS IMPORTANT

Our modern industrial society is highly dependent on chemical entities for its very existence. Useful chemicals cover the gamut from the building materials that make up our dwellings and machines, to the fertilizers and pesticides used in our production of food, to the chemicals used in our manufacture of electronics and communications. These chemicals include the drugs and materials used in medicine and health care. Many new biologically active and useful compounds result from the activity of our pharmaceutical industry in areas of biotechnology. The production of each of these materials leads to industrial waste and the potential for environmental pollution. Because we become exposed to all of these things, prudence and regulations dictate that their potentials for toxic risk must be determined. Toxicologists are involved in all facets of this risk evaluation. The purpose of this chapter is to introduce the endeavors used to evaluate risk in the field of toxicology and to indicate why instrumentation and computation are necessary.

1.3 INDISPENSABLE DISCIPLINES FOR THE SCIENCE OF TOXICOLOGY

It has long been a matter of honor and pride that pharmacologists and toxicologists must be highly conversant with so many different sciences. The disciplines needed for toxicology include many of the life sciences, mainly biology, zoology, botany, physiology, genetics, pharmacology, biochemistry, histology, and pathology. Analytically related methodologies used in toxicology include analytical chemistry, flow cytometry [1], the techniques and tools of modern genetics, and molecular biology. Statistics is involved in study design, data analysis, and interpretation. In effect, the topic of this book, the use of *computation* in the gathering of the massive amounts of data generated by modern toxicology, the documentation of these efforts, and the interpretation of the resulting data have become more and more essential and increasingly routine in toxicology.

Biology, zoology, and physiology predict the normal responses of living systems, whose deviations can help define the effects of toxic substances on these systems. Toxic effects may produce adverse changes at the biochemical, tissue, organ, and organism levels. Again, perturbations from normal function or anatomy can help define toxic effects. Histology is the study of normal microanatomy, and pathology describes what happens to these microstructures when they become injured by toxicants. Many different sophisticated analytical techniques are used in the most advanced studies.

1.4 SUBDISCIPLINES OF TOXICOLOGY

In its role of explaining, predicting, preventing, and treating the adverse effects of drugs and chemicals, toxicologists are working in many subdisciplines. They are involved in drug and chemical *safety screening* and in their *regulatory* counterparts, the EPA, FDA, and USDA in the United States. They are involved in occupational and industrial toxicology and in their own particular scientific and *regulatory* counterparts, NIOSH and OSHA. In addition there are people specializing in *forensic*, *veterinary*, and *clinical* toxicology. Finally there are scientists involved in mechanistic studies at all levels of the organism's organization.

The whole idea behind toxicological testing and safety screening is the potential benefit to humans and animals that will accrue. This entails defining the risk of exposure to drugs and chemicals, understanding the risk when it exists, and preventing the risk. This concept holds whether one is doing drug discovery, environmental, or regulatory toxicology.

As described, toxicologists in chemical and pharmaceutical industries work to define the risk associated with new drugs and chemicals. Such safety evaluation is part of the art and science of toxicology, though much of the methodology is codified in the law. Regulatory toxicologists acting for the general public welfare work to create and ensure adherence to safety regulations. The process of discovery is part collegial and part adversarial as investigators and regulators strive to fill their co-dependent function.

Forensic toxicology combines analytical chemistry, knowledge of toxicology, and detective work to determine the causes of those cases of poisoning that have become of interest to law enforcement or regulatory agencies. Veterinary toxicology and human clinical toxicology deal with the evaluation and treatment of poisoning.

1.5 TRADITIONAL TOOLS OF TOXICOLOGY

Epistemology is the undertaking of how we know what we know, or the study of knowledge and the basis of its validity. Toward this end in toxicology we bring all the tools of our science and our rationality. This entails the appropriate gathering of information and its proper evaluation and interpretation.

Properly designed and interpreted animal studies are the primary tools of safety screening and predictive toxicology. Among the basic techniques used by toxicologists are dose–response studies. Articulated first by Paracelsus [2] is the idea that it is the dose that makes the poison. All things are dangerous in large enough doses and all things are safe if exposure is small enough. Additionally the demonstration of a dose–response relationship between a tested substance and the effect it is suspected to produce provides strong evidence that a cause–effect relationship exists between them. As the basis for regulatory toxicology, the existence of a *threshold dose*, the dose below which no adverse effect occurs, provides the basis for recommended maximal exposures that are safe.

Many of the fields of study and types of toxicology are described below. These efforts are very broad and entail the use of many disciplines.

1.6 FIELDS OF EXPERTISE WITHIN TOXICOLOGY

Toxicology can be classified according to the effects on the organ systems damaged. Alternatively, it can be classified according to the mechanisms of toxicity. Almost anything that can go wrong with almost any tissue in the body will occur. Each of these areas comprises its own realm of toxicological expertise. We will first examine several examples of organ-based toxicity. In some cases there will be extensive overlap between categories. One of the important questions is why there frequently is specific target organ toxicity. We will examine some aspects of this question. Most frequently the answer relates to the specific biological characteristics of the tissues.

1.6.1 Chemical Carcinogenesis

Because of the intense public interest that exists in cancer prevention and the resulting political interest, chemical carcinogenesis is a gigantic and relatively well-funded field. Amazon.com (as of June 2006) lists nearly 80 books in print on the topic. Google.com lists over 200,000 hits on the topic. The National Library of Medicine's Medline lists nearly the same number of articles on the topic within the scientific or medical literature. Not only is the extent of effort indicated by these numbers, but also the diffuse, difficult, intractable, and fractal nature of the field.

Carcinogenesis is a multiple-step, progressive process. Causes of such damage can include alkylating agents, active oxygen species, and radiation. These causes of injury can also lead to genetic toxicology. Much of the current mechanistic interest is centered on dysregulation of cellular growth control mechanisms [3].

1.6.2 Genetic Toxicology

There is a great degree of overlap between the topics of genetic toxicity and chemical carcinogenesis. This is because many of the stepwise changes that occur during the development of neoplasia consist of somatic mutations that result in changes in growth regulation of the affected tissues.

The field more commonly thought of as *genetic toxicology* deals with changes in what might be termed *legacy genes*, those genes that are passed from one generation to the next. These changes occur as a result of unrepaired injury to the cellular DNA and the effects are almost always bad. The safety screening required for carcinogenesis in drug and chemical discovery is extensive [4].

1.6.3 Developmental Toxicology/Reproductive Toxicology

The topic of teratogenesis as a disruption of the control of embryological development is covered later in this chapter. Safety screening requires evaluation of developmental and reproductive toxicity of the compounds of interest [5].

1.6.4 Blood and Bone Marrow

Hematotoxicity is another area of active investigation. Benzene is an excellent example of the extremes that can be caused by substances that are toxic to the bone marrow. Chronic exposure to benzene can cause either leukemia, or bone marrow injury that can lead to aplastic anemia [6,7]. Agranulocytosis and aplastic anemia are infrequent but deadly toxic effects of several drugs. Anemias related to deficiencies of each of the formed elements of the blood also are known, and some of these are toxicological in origin. For example, thrombocytopenia is an established and potentially deadly adverse effect of heparin, though the etiology may be immunological [8].

1.6.5 The Immune System

The function of the immune system is to protect the internal environment of the body against external attack [9]. Because the nature of the attack can be so varied, bacterial, fungal, viral, and the presence of foreign proteins, the immune system has become one of great complexity. Antibody-mediated immunity and cellularly mediated immunity both exist, and the stimulus–response characteristics of this system and the necessary control mechanisms are also very complex.

Decreased immunological competency can lead to susceptibility to infections, and it can also lead to cells that lack the capacity to control their growth. In contrast, excess activity can cause the immune system to attack the host organism, itself. Both of these adverse effects can result from xenobiotic exposure.

Among the drugs that can decrease immunological competence are anti-inflammatory steroids, cyclosporine, and tacrolimus. Certain of these compounds are used to prevent transplant rejection, but they simultaneously carry the risk of allowing infection to occur. The aplastic anemia caused by the bone marrow toxicity of benzene was described above. Lead and chlorinated aryl hydrocarbons such as hexachlorobenzene also can cause bone marrow suppression.

Inappropriate immunological activation has been known for a long time. Anaphylaxis following sensitization is an example. Another example is that untreated beta-hemolytic streptococcus infections can lead to rheumatic fever and damage to the heart valves. This appears to occur because the streptococcus organism and our heart valves share a common antigen, and development of immunity against the former leads to damage to the latter.

Immunotoxicology is a discipline still in its infancy. Perhaps, this is most clearly bourn out by the recent experience in Great Britain in which a monoclonal antibody that was designed as an agonist to a receptor on T-lymphocytes was first given to six human volunteers. The dose given was much lower (500 times) than that which had been safe in animals. Nevertheless, the result was a massive release of cytokines leading to global organ failure. At this time all have survived the event, though it was not certain for some time that this would happen. This is an excellent example of species differences, and it is clear that much more work must be done to characterize human immunological responses when potential immunological stimulants are in the process of drug discovery [10,11].

1.6.6 The Liver

The liver has two main functions in the body [12]. The first is maintenance of internal nutritional homeostasis through facilitation of lipid absorption and intermediary metabolism. As described later, the large metabolic capacity of the liver renders it vulnerable to heavy metals through binding of the metals to and inactivation of electrophilic ligands.

The second function of the liver is to deal with various endogenous substrates and dietary xenobiotics through their metabolism and biliary excretion. Several toxicities are associated with disturbances of this function. One example is the oxidative dechlorination of carbon tetrachloride and other chlorinated hydrocarbons to free radical metabolites that bind to and destroy hepatic tissues. Another example of toxicity by metabolic activation occurs with acetaminophen. A trace metabolite of acetaminophen is a very reactive quinoneimine. Under normal usage of this analgesic, this metabolite is not a problem because it is inactivated by binding to reduced glutathione. However, when an overdose of acetaminophen is taken, the protective glutathione becomes depleted and the reactive metabolite covalently binds to and destroys hepatic parenchymal tissue.

Cirrhosis of the liver is one of the most well-known adverse effects of chronic alcohol abuse. The cholesterol-lowering, life-prolonging statin drugs must be monitored routinely for hepatotoxicity and rhabdomyolosis. A Google search on the terms "statins," "hepatotoxicity," and "review" produced over 22,000 hits indicating this is a very active field of interest.

1.6.7 The Kidney

For the same reasons as described for the liver, heavy metals and compounds converted to active metabolites can also be toxic to the kidney, which is very active metabolically [13,14]. With certain quinones, reduced glutathione can enhance toxicity, rather than being protective [15].

Gentamycin and other aminoglycoside antibiotics are toxic to the kidney. Use of these compounds necessitates repeated dosage adjustments according to drug blood levels.

1.6.8 The Respiratory System

As is the case with the skin, the lungs are in constant contact with the external environment [16]. Exposure to the toxins in cigarette smoke is one of the most common causes of congestive, obstructive damage in the respiratory system. Occupational exposure to asbestos and medically necessary exposure to drugs such as cyclophosphamide and carmustine can also cause lung injury. Inhalations of coal dust and cotton fibers are other occupational hazards to the lungs.

1.6.9 The Nervous System

The central nervous system is one of the most complex organs in living systems [17]. Neurotoxicity can be manifested rather globally, or very specifically, depending on the poison. One example of a very specific toxicity occurs with MPTP, a notorious meperidine analogue that can destroy the substantia nigra and leads to a very severe Parkinson-like syndrome. Another example of rather global neurotoxicity occurs with lead encephalopathy. Other metals can also be highly neurotoxic.

Developmental retardation occurs following exposure to metals, and this has been instrumental in decreasing the amount of lead used in gasoline and indoor paints. Maternal alcohol drinking during pregnancy can also cause developmental retardation manifested as fetal alcohol syndrome. Organophosphates can cause acute injury related to acetylcholine accumulation and certain ones such as triothocresylphosphate can cause delayed axonal degeneration. Picrotoxin, camphor, and strychnine are examples of powerful convulsants. Anesthetics and analgesics can lead to respiratory depression and hypoxia. Carbon monoxide and cyanide also cause general hypoxic damage to the brain and to other high oxygen demand organs of the body. The literature abounds with other examples of substances that are toxic to the nervous system.

1.6.10 Behavioral Toxicity

Many poisons can disturb mental and rational function leading to behavioral abnormalities. Psychototoxins include phencyclidine, LSD, and fungal toxins. Less commonly, stimulants such as cocaine and amphetamine can cause psychiatric problems. Psychiatric effects of high doses of corticosteroids have also been described. In addition to the developmental retardation, some investigators believe that cognitive impairment, hyperactivity, and perhaps even antisocial behavior may be caused by childhood lead exposure. Public discussion of these subtle toxic effects is highly politicized because childhood exposure to lead still occurs as a risk factor in slums and tenements.

1.6.11 Cardiotoxicity

Compared to many of the other organs, the heart must continuously maintain beating activity [18]. There is little energy storage capacity in the heart, which therefore must be producing the energy it uses in real time. Drugs that decrease the capacity of the heart to use substrate and generate ATP can be very harmful to the heart. Examples of toxicants believed to act by this mechanism include cyanide, glycolysis inhibitors such as emetine [19], and Krebs-cycle metabolism inhibitors such as the cardiotoxic anthracycline doxorubicin [18].

In addition to the necessity of continuous energy generation, the heart must maintain rhythmic function throughout its lifetime. Substances such as cocaine and cyclopropane that decrease the reuptake of norepinephrine after its release from noradrenergic neurons are prone to cause fatal arrhythmias. Additionally drugs that modify plasma membrane ion channel function can also cause arrhythmias. More recently cardiotoxicity from drugs that prolong the QT-interval has been reported. Such drugs include several antimicrobial agents, antidepressants, and anti-migraine agents. This broadly based toxicological effect has clear implications for the drug discovery process [20].

1.6.12 Dermal Toxicity

The skin is the primary organ of contact between the organism and its environment. There is extensive commercial interest in dermal toxicology and safety screening because of the many different products used topically for therapeutic and cosmetic purposes. Similar comments can be made about ocular products.

Some of the toxic effects to the skin are allergic in nature. The response to poison oak or poison ivy is an example. Corrosive injury to the skin can occur following contact with many household products. Cutaneous responses to certain drugs can include dangerous exfoliative dermatitis and the Stevens-Johnson syndrome [21,22].

1.6.13 The Reproductive Systems

In addition to chemical carcinogenesis, teratogenesis is a toxic effect that catches the public's attention. The public response to thalidomide was so great that it is still very difficult to get the drug approved for newer indications. Once we know what a toxic effect can be, toxicologists are quite effective in developing animal tests that screen for that effect. For example, the fetotoxic effects of compounds such as Accutane® and the angiotensin converting enzyme inhibitors are known from screening studies, and a large teratogenic disaster such as thalidomide should not happen, again. It is a commentary on human nature that the fetal alcohol syndrome still continues to occur.

1.6.14 Endocrine Systems

Toxic changes can be caused by endocrine agonists, antagonists, and disruptors. There are estrogen active compounds such as diethylstilbesterol and dioxin. Natural and synthetic thyroid antagonists such as propylthiouracil are known. Agonists and antagonists for adrenocortical hormones have been described. Oral contraceptives are risk factors for increased blood clotting and stroke. Estrogens are risk factors for breast and uterine cancers, and there is much interest in the associated risks from environmental estrogen pollutants (e.g., the REACH initiative).

1.7 IN VITRO METHODOLOGIES FOR FIELDS OF EXPERTISE WITHIN TOXICOLOGY

Biomedical and toxicological research and safety screening require the use of animals [23]. However, since the inception of the first animal welfare organizations, society's use of animals has been a matter of concern and controversy [24]. Because of this interest there has been much activity in the past few decades in finding alternative methods for doing research and screening. The

"Three Rs" of Russell and Burch [25], *replacement*, *reduction*, and *refinement*, have been the goals of much of this work. Many alternatives to animals have been suggested, and where the alternatives have been verified to be useful, it is appropriate that they be used. One of the best examples of replacement is the current use of Limulus (horseshoe crab) serum to detect the presence of the gram-negative organism endotoxins known as pyrogens [26]. Parenteral products must be sterile and pyrogen-free. Limulus serum is more sensitive to the presence of these harmful proteins than the rabbits that were previously used. Not many other alternatives have been so well verified, however.

The current all out attack mounted by animal activists on the societal use of animals is a matter of extreme concern to toxicology and the other biomedical sciences. Some of the best resources to inform ourselves and to counter these activists are the frequently asked questions (FAQ) detailed in the Animal Rights Myths FAQ [27].

Cell culture is one of the primary methods being studied for animal replacement. Primary cultures of heart cells, liver, keratinocytes, corneal cells, and many other tissues are actively being studied [28]. Much work has been done for decades in some cases to maintain the functions of the parent tissue close to those in vivo while the cells are in culture. Eventually most differentiated function of the cells is lost. The art is to maintain such function for as long as possible so that longer in vitro exposures can mimic in vivo dosing.

Propagated cell lines are also widely used. Such cells are immortalized by combination with neoplastic cells. One problem with these cell lines is that they frequently do not express any of the differentiated functions of the parent cells, and therefore do not provide tissue specific responses.

One of the failings of cell culture in predictive toxicology is that some examples of toxicity are multi-organ in nature. Methanol toxicity, for example, occurs when the methanol is oxidized in the liver to formate. The formate is transported by the blood to the retina and CNS where it produces its characteristic effects of blindness and brain damage. To model methanol toxicity in cell culture would require co-cultures of liver and retinal cells. Co-cultures are technically difficult, and it would be very difficult to predict which multiple cell types are needed in a co-culture to detect a previously unknown toxicity. It requires an intact organism to do this.

Innumerable cell lines are used in studies trying to understand intracellular messengers and control processes. Such models are particularly useful provided that the cells remain viable as almost any desired genetic alteration can be produced and studied.

1.8 MECHANISMS OF TOXIC INJURY

Although many different cells and tissues can be injured by toxicants, there are not many different fundamental mechanisms by which injury can occur. Each of these categories can be very broad, however. Mechanisms of injury include *ligand binding* by heavy metals, *covalent binding*, *oxidative stress* by active oxygen species, *antimetabolites*, the *extension of pharmacologic action*, *dysregulation of cellular signaling*, and a miscellaneous category, all of which are now described in a little more detail.

1.8.1 Ligand Binding by Heavy Metals

Electrophilic ligands such as sulfhydryls, amino groups, and hydroxyl groups are found at the active sites of many, if not most enzymes [29]. These ligands have a high affinity for many different metals (e.g., Hg, As, Pb, Sb), and the uptake of and binding of metals to these sites inactivates them. The effect that occurs depends on the location and function of the enzymes in the tissue involved. The more metabolically active a tissue is, the more it is likely to be adversely affected by metals. The liver, kidneys, gastrointestinal mucosa, and central nervous system are particularly vulnerable because they are so metabolically active. Teratology of the toxic metals is also an issue. The major differences between the metals are more a matter of pharmacokinetics than fundamental differences in mechanisms of toxicity.

The antidotes for heavy metals are called chelating agents, a picturesque term invoking an image of lobster claws (chelae) grabbing hold of the metal. Such drugs are rich sources of the ligands to which metals readily bind, and these drugs are able to compete effectively for the metal against the endogenous tissue ligands.

1.8.2 Covalent Binding to Biological Macromolecules

This toxic mechanism, which had its most active interest in the late 1970s and early 1980s, occurs when chemicals are metabolized to free radicals or other highly active molecules. These radicals then covalently alkylate nearby macromolecules. Such macromolecules can include proteins, cellular membranes (including plasma membranes, nuclear membranes, membranes of organelles, etc.), and genetic components such as DNA and RNA. If metabolically critical areas of these large molecules become covalently bound to a metabolic product, they may become inactivated. Specific toxic effects will depend on which biological macromolecules become inactivated. An example is the case of carbon tetrachloride that becomes oxidized by liver P450 enzymes to the trichloromethane free radical. This active alkylating agent attaches itself to nearby liver parenchyma, resulting in the classic liver toxicity described for carbon tetrachloride. Research interest in covalent binding as a mechanism of toxicity has decreased since the 1980s because it is very difficult to determine which of all the structures in the body that become alkylated is ultimately responsible for the toxicity produced.

1.8.3 Oxidative Stress

Oxidative stress is a general term for the excessive production of *active oxygen species* and the resulting biological responses [30]. The various oxygen species

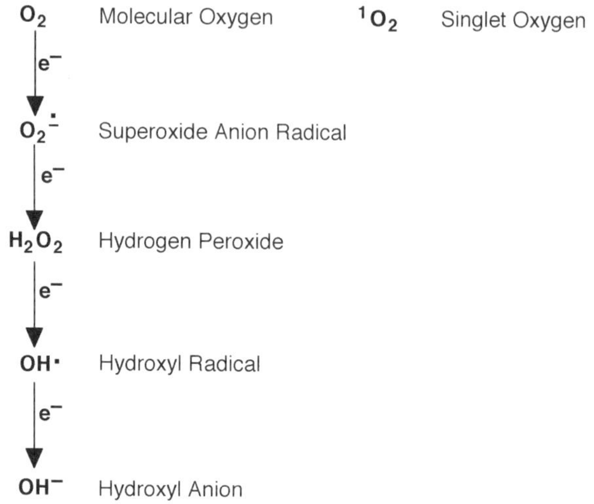

Figure 1.1 Various valence states of oxygen as a function of increasing single-electron reductions.

are depicted in the schematic of Figure 1.1 as a series of one-step electron reductions—starting with molecular oxygen and ending with the hydroxyl anion.

Active oxygen species are usually produced within cells as by-products of normal oxidative metabolism. The location of these processes can include the cytochromes P450, mitochondria, lysosomes, and peroxisomes. As was the case with damage by free radical metabolic remnants leading to covalent binding and injury, several of the active oxygen species can also cause damage to biological macromolecules.

In addition lipid peroxidation can result from action of active oxygen species. This leads to destruction of metabolically necessary lipid molecules and damage to the structural integrity of cellular membranes. Damage from oxidative stress can occur with excessive production of active oxygen species, inadequate protection against such species, or both. Examples of toxicity from active oxygen species include the pancreatic beta-cell destruction by alloxan, the neurotoxicity of 6-hydroxydopamine, the cardiotoxicity of the anthracycline antibiotics, and the pulmonary toxicity of the herbicide paraquat.

Singlet oxygen is a unique form of activated oxygen. It is most commonly involved in phototoxic reactions. Because of absorption of an energetic photon, one of the previously paired electrons of the molecule has been promoted to an orbital of higher energy. In the strictest sense it is not a free radical, but it can act as an active oxygen species. Certain compounds such as tetracyclines or amiodarone can trap photons upon exposure of the skin to the ultraviolet portion of the light spectrum. These photon-activated compounds can pass their energy to molecular oxygen, converting it to singlet oxygen. Singlet oxygen, in turn, passes its excess energy to dermal tissues, resulting in tissue damage and sunburn.

1.8.4 Antimetabolites

Antimetabolites compete with normal endogenous substrates and cause inhibition of the processes that require those substrates. Examples include purine and pyrimidine antagonists, which prevent nucleic acid replication and cellular division in cancer chemotherapy. Another example is methotrexate, which can inhibit folic acid metabolism.

1.8.5 Denaturing Agents

Denaturing agents can destroy the tertiary structure of proteins. Alcohol's antiseptic action results from denaturing of bacterial proteins. Corrosives can cause tissue damage upon accidental exposure.

1.8.6 Extension of Pharmacology

This is a broad category of toxic action in which exaggeration of the therapeutic effects of many drugs in overdose can lead to poisoning. For example, general anesthetics are also respiratory depressants, and too high concentrations can cause fatalities. Many antihypertensives cause potentially fatal vascular collapse and shock when taken in overdose. Overdoses of certain antiarrhythmic drugs can themselves cause fatal arrhythmias, actions that are related to their action on ion channels.

1.8.7 Dysregulation of Cell Signaling

This is one of the currently most active areas of toxicological research interest, and it is one with many different research thrusts. There is much current interest, for example, in the mechanisms of regulation of apoptosis, programmed cellular death. Production of apoptosis when it should not occur, or its lack when it normally should occur, can each be mechanisms of toxicity. Part of the adverse remodeling of cardiac and vascular tissue can occur because of hypertension-induced apoptosis of cardiac and vascular cells [31]. Apoptotic processes also have been implicated in alcoholic hepatotoxicity [32].

Excessive inflammatory responses may result from inappropriate cellular signaling. On the other hand, inflammation is normally a protective response, and its lack can lead to increased susceptibility to infections. Dysregulation of cellular division can lead to neoplasia or aplasia. Neoplastic changes reflect a dysregulation of cell growth, whether from failure of apoptosis or other mechanisms.

Embryological development is a highly conserved, highly regulated sequence of events in which many processes must be activated or deactivated in their proper sequence. We are just in the infancy of discovering what are the messages and messengers controlling development of the embryo into a fetus and eventual birth. Many substances perturb these processes and thus are fetotoxic teratogens. The most common human teratogen is alcohol, the use of which during pregnancy can cause the developmental retardation known as the *fetal alcohol syndrome* [33]. Among the many other effects it causes, angiotensin II is a growth regulator. Disturbing angiotensin II action by angiotensin-converting enzyme inhibition, or by angiotensin receptor block can be used therapeutically to reduce the inappropriate growth and remodeling that occurs in congestive heart failure and hypertension. However, these blocking actions also can result in severe fetotoxicity [34]. Numerous other examples of teratogens are known. Screening for these adverse effects is a necessary part of drug discovery.

1.8.8 Miscellaneous Other Mechanisms of Toxicity

Addition of a miscellaneous category to any list adds completeness. On the other hand, it is difficult to find toxic effects that do not fit into one or more of the previous categories. One example of such might be the necrosis of the mandible that appears to result from the use of bisphosphonates to prevent osteoporosis in postmenopausal women. The mechanism of this unexpected effect is not known, but the toxicity certainly has become of great concern to our dental colleagues [35].

1.9 COMPUTATION IN TOXICOLOGY

The primary advantage of the computer is to deal with work that is so large and so complex that it cannot otherwise readily be possible. One example of such a need is the highly complicated chemical/toxicological/biomedical literature that exists. A computer can search and mine the literature, and it can organize it into mutually relevant collections of articles. Data clustering is one example of such an intelligent organization of the literature [36]. Other examples include directory searches, keyword searches, and database searches. (See *types of search engines* [37].)

There are several different clustered search engines available. A case in point is Clusty.com's Vivisimo [38]. The default setting of this engine is to search the recent literature. Repeated searches, say at monthly intervals, enables one to keep up with topics of interest. A recent (June 2006) clustered search on "Computational Toxicology," the topic of this book, gave the results described in Table 1.1.

The relevance of these topics to computation and to toxicology is not trivial. Currently commercially available *gene microarrays* can characterize the expression of thousands of genes of several different species [39], and this

TABLE 1.1 Results of a Clustered Search on "Computational Toxicology" as Divided by the Vivisimo Search Engine into Topic Clusters

Microarray, gene expression profiling (13 articles)
Toxicogenomics (12 articles)
Protein, impact (13 articles)
Dose–response (11 articles)
Quantitative (8 articles)
Receptor, expression (7 articles)
Predict toxicity (7 articles)
Pharmacology and toxicology (6 articles)
Physiologically based pharmacokinetic (8 articles)
Properties (5 articles)
High-throughput data (3 articles)

Note: The search retrieved 105 articles. Less frequent items are not included above.

information has great potential in the drug discovery process [40]. Routines to interpret and correlate these findings are under active development, and the results are available on the internet through the NIH [41]. In the case of *toxicogenomics,* a subset of pharmacogenomics, several correlations between an individual's genome and susceptibility to particular toxicants are known. Databases exist and people are working to develop toxicogenomic in vitro procedures that might be useful early in drug discovery, or in predictive toxicology [42,43,44].

The *predictive toxicology* cluster provided a group of articles related to QSARs [45], bioinformatics [46], and expert systems [47]. Pharmacokinetic data acquisition and interpretation have been heavily intertwined with computation since the early days of the discipline. This hasn't changed with the more current field of *toxicokinetics*.

The *high-throughput data* cluster appears to be related to using computational and statistical techniques to separate useful data signals from large amounts of irrelevant noise (e.g., see [48]). This important endeavor is just in its infancy.

The connection between toxicology and *dose–response* relationships is several hundreds of years old [2,49]. In the pre-computational days these data were calculated by hand and nomogram [50]. The sheer labor involved has been greatly eased by computational techniques. Nevertheless, this author feels that working through such manual techniques at least once is very salutory for nacent pharmacologists and toxicologists.

Predictions of drug *receptor* interactions and related QSARs are useful for *predictive toxicology* and drug development. *Validation* of computer technology and predictions is another concern. Many of these topics are covered in the following chapters of this book.

REFERENCES

1. Osborne G, Gruninger S. <http://jcsmr.anu.edu.au/facslab/facshome.html> 2001.
2. Borzelleca JF. Profiles in toxicology—Paracelsus: Herald of modern toxicology. *Toxicol Sci* 2000;53:2–4.
3. Malarkey DE, Maronpot RR. Carcinogenesis. In: Wexler P, Anderson BD, de Peyster A, Gad SC, Hakkinen PJ, Kamrin MA, Locey BJ, Mehendale HM, Pope CN, Shugart LR, editors, *Encyclopedia of toxicology*, 2nd Ed., Vol. 1. Oxford: Elsevier Ltd., 2005. p. 445–66.
4. Gad SC. Genotoxicity. In: Gad SC, editor, *Drug safety evaluation*. New York: Wiley, 2002. p. 176–236.
5. Gad SC. Developmental and reproductive toxicity testing. In: Gad SC, editor, *Drug safety evaluation*, New York: Wiley, 2002. p. 258–96.
6. Anonymous. *Benzene material safety data sheet*. ScienceLab.com <http://www.sciencelab.com/xMSDS-Benzene-9927339>.
7. Bakhshi S, Baynes R. *Aplastic anemia*. eMedicine on-line 2006 <http://www.emedicine.com/med/topic162.htm>.
8. Sturtevant JM, Pillans PI, Mackenzie F, Gibbs HH. Heparin-induced thrombocytopenia: Recent experience in a large teaching hospital. *Intern Med J* 2006;36:431–6.
9. Haschek WM, Rousseaux CG, editors. *Immune system*. In: *Fundamentals of toxicologic pathology*. New York: Academic Press, 1998. p. 233–72.
10. Wood AJJ, Darbyshire J. Injury to research volunteers—The clinical-research nightmare. *N Eng J Med* 2006;354:1869–71.
11. Marshall E. Violent reaction to monoclonal antibody therapy remains a mystery. Science 2006;311:1688–9.
12. Haschek WM, Rousseaux CG, editors. Hepatobiliary system. In: *Fundamentals of toxicologic pathology*. New York: Academic Press, 1998. p. 127–51.
13. Haschek WM, Rousseaux CG, editors. The kidney. In: *Fundamentals of toxicologic pathology*. New York: Academic Press, 1998. p. 153–93.
14. Navarro VJ, Senior JR. Drug-related hepatotoxicity. *N Eng J Med* 2006;354:731–9.
15. Monks TJ, Lau SS. Toxicology of quinone-thioethers. *Crit Rev Toxicol* 1992;22:243–70.
16. Haschek WM, Rousseaux CG, editors. *Respiratory system*. In: *Fundamentals of toxicologic pathology*. New York: Academic Press, 1998. p. 91–126.
17. Haschek WM, Rousseaux CG, editors. Nervous system. In: *Fundamentals of toxicologic pathology*. New York: Academic Press, 1998. p. 355–93.
18. Combs AB, Ramos K, Acosta D. Cardiovascular toxicity. In: Hodgson E, Smart RC, editors, *Introduction to biochemical toxicology*, 3rd Ed. New York: Wiley-Interscience, 2001. p. 673–96.
19. Pan SJ, Combs AB. Emetine inhibits glycolysis in isolated, perfused rat hearts. *Cardiovasc Toxicol* 2003;03:311–18.
20. Fermini B, Fossa AA. The impact of drug-induced QT interval prolongation on drug discovery and development. *Nat Rev Drug Discov* 2003;2:439–47.

21. Roujeau JC, Stern RS. Severe adverse cutaneous reactions to drugs. *N Eng J Med* 1994;331:1272–85.
22. Parrillo SJ, Parrillo CV. *Stevens-Johnson syndrome*. eMedicine online 2006: <http://www.emedicine.com/emerg/topic555.htm>.
23. Cohen C. The case for the use of animals in biomedical research. *N Eng J Med* 1986;315:865–69.
24. Horton L. The enduring animal Issue. *J Natl Cancer Inst* 1989;81:736–43.
25. Anonymous. *The three "R's" of Russel & Burch*, 1959. 3R-INFO-BULLETIN 7—March 1996: <http://www.forschung3r.ch/de/publications/bu7.html>.
26. Anonymous. *Blue blood, The horseshoe crab, The history of Limulus and Endotoxin*. <http://www.mbl.edu/animals/Limulus/blood/bang.html>. Site verified as of June 2006.
27. O'Donnell K. *Animal rights myths*, FAQ v1.3 2000. <http://www.armyths.org/>. Site verified as of June 2006.
28. Gad SC. *In vitro toxicology*. London: Taylor and Francis, 2000.
29. Gossel TA, Bricker JD. Metals. In: *Principles of clinical toxicology*, 3rd Ed. New York: Raven Press, 1994, p. 179–214.
30. Kehrer JP. Free radicals as mediators of tissue injury and disease. *Critical Rev Toxicol* 1993;23:21–48.
31. Sabbah HN, Sharov VG. Apoptosis in heart failure. *Cardiovas Dis* 1998;40: 549–62.
32. Natori S, Rust C, Stadheim LM, Srinivasan A, Burgart LJ, Gores GJ. Hepatocyte apoptosis is a pathologic feature of human alcoholic hepatitis. *J Hepatol* 2001;34:248–53.
33. Ikonomidou C, Bittigau P, Ishimaru MJ, Wozniak DF, Koch C, Genz K, Price MT, Stefovska V, Horster F, Tenkova T, Dikranian K, Olney JW. Ethanol-induced apoptotic neurodegeneration and fetal alcohol syndrome. *Science* 2000;287: 1056–60.
34. Cooper WO, Hernandez-Diaz S, Arbogast PG, Dudley JA, Dyer S, Gideon PS, Hall K, Ray WA. Major congenital malformations after first-trimester exposure to ACE inhibitors. *N Engl J Med* 2006;354:2443–51.
35. Carreyrou J. Fosamax drug could become next merck woe. *Wall Street Journal Online*: April 12, 2006: p. B1 <http://online.wsj.com/article/SB114480509161723631.html?mod=djemHL>.
36. Anonymous. <http://en.wikipedia.org/wiki/Data_clustering>.
37. Anonymous. <http://dana.ucc.nau.edu/~pda/etc545/ews/su2.html>.
38. Anonymous. <http://clusty.com/>.
39. Liu ET, Karuturi KR. Microarrays and clinical investigations. *N Eng J Med* 2004;350:1595–7.
40. Evans WE, Guy RK. Gene expression as a drug discovery tool. *Nat Genet* 2004;36:214–5.
41. Anonymous. <http://www.ncbi.nlm.nih.gov/>.
42. Yang Y, Abel SJ, Ciurlionis R, Waring JF. Development of a toxicogenomics in vitro assay for the efficient characterization of compounds. *Pharmacogenomics* 2006;7:177–86.

43. Fielden MR, Kolaja KL. The state-of-the-art in predictive toxicogenomics. *Curr Opin Drug Discov Devel* 2006;9:84–91.
44. Borlak J, Drewes J, Hofmann K, Bosio A. Toxicogenomics applied to in vitro toxicology: A cDNA-array based gene expression and protein activity study in human hepatocyte cultures upon treatment with Aroclor 1254. *Toxicol In Vitro* 2005;20:736–47.
45. Salter AH, Nilsson KC. Informatics and multivariate analysis of toxicogenomics data. *Curr Opin Drug Discov Devel* 2003;6:117–22.
46. Benfenati E, Gini G. Computational predictive programs (expert systems) in toxicology, *Toxicology* 1997;119:213–25.
47. Nikolsky Y, Ekins S, Nikolskaya T, Bugrim A. A novel method for generation of signature networks as biomarkers from complex high throughput data. *Toxicol Lett* 2005;158:20–9.
48. Anonymous. <http://en.wikipedia.org/wiki/Paracelsus>.
49. Litchfield JT, Wilcoxon F. A simplified method of evaluating dose-effect experiments. *J Pharmacol Exp Ther* 1949;96:99–113.

2

IN VITRO TOXICOLOGY: BRINGING THE IN SILICO AND IN VIVO WORLDS CLOSER

Jinghai J. Xu

Contents

2.1 INTRODUCTION

At the turn of this century, an aerospace and a pharmaceutical executive found themselves seated next to each other at a social event. After exchanging some pleasantries, their conversation turned to their respective industry and applications of new technology. What single most transforming technology they found integral to the growth of these two different industries? Not

Computational Toxicology: Risk Assessment for Pharmaceutical and Environmental Chemicals, Edited by Sean Ekins

surprisingly, it is computer science and information technology. In silico calculations of aerodynamics are critical in the complex design of any new models of airplanes. Likewise 3D docking of drug molecules to their intended biological targets and pharmacokinetic/pharmacodynamic simulations are critical in discovering effective drug therapies. In each case computational tools have helped human experts see or visualize the light at the end of the long tunnel called product R&D.

Computer algorithms can also be useful to predict product failure. Actually the "dreaded analysis" of pointing to not the light but the destined abyss is just as important, if not more so, for both industries. Historically the pharmaceutical industry has a track record of failure or attrition rate that is more astronomical in comparison. More than 90% of the industry's "drug candidates" fail because of one reason or another [1]. If the industry can improve their failure analysis such that they can predict and avoid just 10% of the eventual failures, it can bring significant cost savings to the industry.

Among the various reasons for drug failures, lack of drug efficacy and presence of dose-limiting toxicity came at the top [1]. It therefore comes as no surprise that almost all major pharmaceutical companies have dedicated at least some resources to the better predictions of drug toxicity by developing, evaluating, and applying a variety of computational tools.

Yet developing in silico tools for the complex prediction of the in vivo biological world has turned out to be much more challenging than for the physical world of aerodynamics. Part of this is because in vivo biology has far too many interrelated parts, some of which have yet to be discovered. Another reason is that the prediction of any biological or toxicological property of a new chemical entity (NCE) is still a scientific hypothesis, meaning it still requires selected experimental verification. So what experimental systems can one use without resorting to the resource- and time-limiting in vivo animal or human testing? In vitro systems and hence in vitro toxicological approaches fit this need with unique capabilities, and they require much less time and cost compared to in vivo tests. It should be our collective vision that the integrated application of in vitro methods and models in the drug development process can bring the in silico and in vivo world closer for the ultimate benefit of improving human health and minimizing patient suffering.

Perhaps the first use of in silico toxicology is to demonstrate that it can predict in vitro toxicological outcomes. This is quite natural because only industrialized in vitro toxicology tests can provide the number of test results on such diverse chemical molecules needed to build a computational algorithm that has a broad coverage of chemical structures (e.g., [2]). These practices reinforced the model of the synergistic loop between the in silico and in vitro world, as depicted in Figure 2.1. In this loop the in silico and in vitro world enjoys a symbiotic relationship: in vitro assays provide a large compendium of data to train and validate in silico algorithms. In turn, once trained and validated, an in silico prediction can be used to predict the future experimental outcomes of a new chemical entity. For this relationship to be scientifi-

Figure 2.1 Synergistic relationship between the in silico and in vitro predictions of in vivo drug toxicity.

TABLE 2.1 Commonly Used In vitro Toxicological Assays in Pharmaceutical R&D Today

Toxicity	In vitro Assays
Torsade de Pointes	Ion channel inhibition, ion efflux, high-throughput patch clamp
Mutagenecity	Ames test, mini-Ames
Clastogenecity	In vitro micronucleus, in vitro chromosome aberration
General toxicity	Various cytotoxicity assays
Phototoxicity	Cytotoxicity assays performed under UV irradiation
Bioactivation	GSH adduct formation, covalent binding
Hepatotoxicity	Hepatocyte assays, mechanistic assays[a]
Other organ toxicity	Organ-specific mechanistic assays

[a]See Table 2.2.

cally sound, the in vitro toxicological assays need to have sufficient predictivity toward the in vivo outcome. The assays need to provide reproducible and consistent data over time, by different experimental operators, and by different laboratories. It is also critical that as new experimental data become available for NCEs, they are used to further refine computational models such that the in silico tools evolve with the chemical space that medicinal chemists are working in (or the biological space for biologics). Finally, a biological understanding of the underlying mechanisms of why certain in silico predictions fail is just as important as their successful usage in order to further improve computational tools (e.g., [3]).

2.2 IN VITRO TOXICOLOGICAL MODELS AND METHODS COMMONLY USED IN DRUG DISCOVERY

Through empirical trial and error, and decades of intense research, pharmaceutical scientists have developed and applied a plethora of in vitro toxicological assays in drug research and development. Table 2.1 summarizes the most commonly used in vitro toxicological assays in pharmaceutical R&D today.

2.2.1 Ion Channel Inhibition

In a recent publication cardiovascular ion channel inhibition was identified as the most frequently encountered small molecule drug liability [4]. Several ligand binding assays for the hERG (human ether-a-go-go-related gene) potassium channel have been developed to identify compounds that may have inhibitory activity and potential cardiotoxicity, especially Torsade de Pointes (see Chapters 13, 16, 19, and 20). The ligands used include [^{3}H]astemizole [5], [^{3}H]dofetilide [6], or other small molecule hERG ligands [7]. Alternatively, functional ion efflux across the biological membrane can be assayed by the nonradioactive Rb^+ flux assay [8]. Furthermore electrophysiological techniques such as high-throughput patch clamping has emerged as the whole-cell functional readout for predicting drug interaction potential with these membrane channels [9,10].

2.2.2 Ames

The Ames Salmonella mutagenicity assay (Salmonella test, Ames test) is used by practically every pharmaceutical company as an initial screen to determine the mutagenic potential of new chemicals and drugs (see Chapter 14). International guidelines have been developed for use by corporations and testing laboratories to ensure uniformity of testing procedures [11]. The test uses several histidine-dependent Salmonella strains each carrying different mutations in the histidine operon. When the Salmonella strains are grown on a minimal media agar plate containing a trace of histidine, only those bacteria that revert to histidine independence (either upon compound treatment or by natural conversion) are able to form colonies (reviewed by [12]). Recently several high-throughput variants of the original Ames assay have been developed to specifically increase the assay throughput and lower the compound requirement (e.g., [13,14]).

2.2.3 In vitro Micronucleus

The micronucleus (MN) test is a genotoxicity test relevant for the risk assessment of cancer-inducing ability of a new chemical entity [15]. It is currently used as a screening assay during the early stages of drug development by pharmaceutical companies to identify chemicals likely to produce positive outcomes in the classical in vitro chromosome aberration assay [16,17]. An MN or a small nucleus is formed during the metaphase/anaphase transition of mitosis (or cell division). An MN may contain a whole piece of chromosome that arises from a whole lagging chromosome (aneugenic event leading to chromosome loss), or a partial piece of a whole chromosome that arises from a broken chromosome (clastogenic event) that does not integrate into the daughter nuclei. The combination of the micronucleus test with fluorescence in situ hybridization (FISH) with a probe labeling the centromeric region of the chromosomes allows discrimination between micronuclei caused by aneugens (centromeric stain positive) and clastogens (centromeric stain negative) [18]. In addition the cells that have completed mitosis and nuclear division, as opposed to the cells that have not, can be distinguished by culturing the cells with cytochalasin B, an inhibitor of actins [19]. The undivided cells will have one large nucleus (mononucleated cells), and the divided cells will have two regular-sized nuclei (binucleated cells). Thus a valid MN positive cell is the presence of an MN in a binucleated cell [15].

Both the Ames and micronucleus tests have been used by pharmaceutical companies to identify the potential carcinogenicity hazards in the early stages of drug development, by utilizing automation techniques to increase assay throughput [20].

2.2.4 Cellular Toxicity

Several cell lines have been used to predict general cellular toxicity in drug discovery. The HepG2 cell line appeared to be slightly more sensitive among a panel of tumor cell lines in a multiparameter cytotoxicity assay [21,22]. However, these cells do not have normal levels of drug metabolizing enzymes, and may not demonstrate metabolism-mediated cellular toxicity. They may also fail to display specific target cell toxicity such as neurotoxicity as manifested by toxicity to neuronal cells (e.g., [23]). The THLE cell line is an SV40-immortalized human liver cell line that is nontumorigenic and exhibits certain differentiated human liver functions [24]. At least one pharmaceutical company is using this cell line in their discovery toxicology screening effort [25].

Recently multiparametric, live cell, prelethal cytotoxic assays for assessing compounds of potential human toxicity have been used to address some of the limitations of traditional in vitro methods. These limitations include the gross cytotoxic nature of the traditional readouts and an acute treatment regimen. Assays of this class were used to screen a library of drugs with varying degrees of toxicity. It was found that the sensitivity and specificity of these assays can meet the need for compound prioritization in drug discovery settings [26].

Many other cell lines have been used to address specific types of toxicity. Some of these have reached a level of validation that is sufficient for in vitro screening purposes. For example, the 3T3 NRU assay was regarded as an acceptable screen for hazard identification of potential phototoxicity [27,28]. This assay utilized the in vitro 3T3 cell line as a generic cell line and Neutral Red uptake as a cytotoxicity measurement. The compound in question is subjected to UV irradiation to assess the effect of UV rays on compound induced toxicity to 3T3 cells. Another example is the use of the MCF-7 cell proliferation assay as an in vitro screen for endocrine disrupters [29].

2.2.5 Bioactivation (Reactive Metabolite Formation)

It is well known that many toxic agents can be metabolized to reactive metabolites that can in turn react with glutathione, enzymes, nucleic acids, lipids, or proteins [30,31]. These reactive intermediates are electrophilic metabolites or free radicals that are generated during the metabolism of a broad range of chemical structures. Reactive metabolite formation is considered necessary but not sufficient in immune-mediated idiosyncratic drug reactions [32]. There are several rapid in vitro methods to detect and measure the generation of such reactive intermediates. For example, high-throughput assays for identifying pharmaceutical compounds that produce reactive metabolites have been developed. These methods involve incubating drug candidates with a liver microsomal drug metabolizing enzyme system in the presence of glutathione and detecting glutathione conjugates via mass spectrometry [33–36]. In a recent review biotransformation-related causes of toxicity were identified to account for 27% of all toxicities [4]. This should be an important area of in vitro and in silico toxicological efforts in the future, as the structural features prone to metabolic activation and the functional groups that can abolish such metabolic activation can be compiled and SAR predictions can also be potentially developed [37].

2.2.6 Hepatotoxicity and Other Organ-Specific Toxicity Assays

Hepatotoxicity accounts for 15% among the toxicity profiles of drug candidates, second only to cardiovascular toxicity [1]. A plethora of assays have been used by investigators to study organ-specific toxicity mechanisms including hepatotoxicity. This can be conducted in an ex vivo fashion using cells from a specific target organ, followed by in vitro studies where cells from a specific target organ are treated by the toxic agents, and a series of measurements are made depending on the mechanism of interest. This typically falls into the investigative toxicological studies as opposed to the toxicological screening approaches listed above. The rationale for such a distinction is due to cost versus benefit (the cost of running all possible mechanistic assays for many compounds, as opposed to the benefit of having high confidence that a subset of such mechanisms is actually causal to the observed in vivo side effects). As an NCE enters the body and interacts with a variety of biomolecules, the possible number of undesirable interactions is quite large. One needs to be careful in assigning a priori "a favorite mechanism" without the corroboration of any in vivo findings. That said, there are a limited number of key mechanisms involved in particular organ toxicity, and one can envision an in vitro "toolbox" approach to investigate a particular finding of organ toxicity. For example, even for complex organ toxicities such as hepatotoxicity, there is a panel of well-recognized in vitro assays for key liver injury mechanisms. They are summarized in Table 2.2. The toxicities include specific subtypes of liver injury such as steatosis (fatty liver), cholestasis

TABLE 2.2 Example "toolbox" of In vitro Assays Targeting Key Drug-Induced Liver Injury Mechanisms

Applications	Methodology	Example References
Multiparameter cytotoxicity	Combine several different readouts including multispectral cytometric analysis	[43]
Steatosis	Neutral lipid stain (e.g., Oil Red O)	[44]
Cholestasis	Uptake and efflux of taurocholate	[45]
Phospholipidosis	Phospholipid accumulation in cytoplasm or lysosomal stain	[43, 46]
Reactive metabolite	GSH adduct formation; GSH depletion (e.g., monochlorobimane)	[47, 48]
Oxidative stress	Redox sensitive dyes	[49–51]
Mitochondria damage	Mitochondria membrane potential dyes (e.g., TMRM)	[52]
Identify targets of toxicological importance	RNAi technology and/or specific inhibitors	[53–56]

(diminished bile flow), phospholipidosis (phospholipid accumulation), oxidative stress, and mitochondrial damage. Outside of the liver, a variety of other organ-specific toxicities are also amenable to in vitro toxicological testing approaches. These include haematotoxicity [38], skin sensitization [39], teratogenicity [40], neurotoxicity [41], and nephrotoxicity [42].

2.3 CONCLUSIONS

"Failure prediction" in pharmaceutical discovery and development is of critical importance in an industry that traditionally has a high product attrition rate. The successful prediction of drug toxicity relies heavily on the close synergy between computational and in vitro toxicological approaches. These predictions will undoubtedly require corroborations from selective in vivo experimentation. Table 2.3 gives an example of integrated applications of in silico, in vitro, and in vivo approaches to address toxicity issues in various stages of drug discovery. Depending on the stage and issues facing a drug discovery team, the approaches are likely to be different (Table 2.3). The combination of these perspectives breaks down any conceivable disciplinary boundaries, and requires an unprecedented integration of technology, computation, biology, and medicine. The successful drug R&D operations today are those that have fundamentally embedded in their culture the ability to (1) incorporate and integrate both data and predictions into practical decisions in their daily research process and (2) capture key institutional learnings in applying both in silico and in vitro tools toward predicting the ultimate in vivo outcomes of pharmaceuticals.

TABLE 2.3 Example of Integrated Applications of In silico and In vitro Approaches to Drug-Induced Toxicity in Drug Discovery and Early Development Process

	Target Identification	Lead Generation	Lead Optimization	Candidate Selection	Backup Seeking	Investigative Toxicology
Key decisions relating to toxicity	Are there any target-related toxicity issues?	Which lead library to focus on?	Which lead series to focus on?	Which development candidate to nominate?	Toxicity occurred, how to find a safer back up?	Toxicity occurred, why?
			Approach I			
	Use RNAi, transgenic mice to identify target-related toxicity	Rank order based on therapeutic index of well-characterized in vitro test	Rank order based on therapeutic index of better-characterized in vitro test	Check therapeutic index based on plasma concentration prediction	Develop plausible hypothesis that fits existing data	Develop plausible hypothesis that fits existing data
			Approach II			
	Pathway mining	In silico prediction (diversity set, target enriched chemical set)	Optimize efficacy, toxicity, and ADME	Optimize efficacy, toxicity, and ADME	Design in vitro/vivo studies to test hypothesis	Design in vitro/vivo studies to test hypothesis
			Approach III			
	Text mining of literature for all biological context for target	Use in silico model to identify potential issues	Gather data for in silico prediction (local model)	In silico prediction (local model)	Set up in vitro screen to select a safer compound	Predict relevance of such toxicity in human

REFERENCES

1. Schuster D, Laggner C, Langer T. Why drugs fail—A study on side effects in new chemical entities. *Curr Pharm Des* 2005;11:3545–59.
2. White AC, Mueller RA, Gallavan RH, Aaron S, Wilson AG. A multiple in silico program approach for the prediction of mutagenicity from chemical structure. *Mutat Res* 2003;539:77–89.
3. Snyder RD, Pearl GS, Mandakas G, Choy WN, Goodsaid F, Rosenblum IY. Assessment of the sensitivity of the computational programs DEREK, TOPKAT, and MCASE in the prediction of the genotoxicity of pharmaceutical molecules. *Environ Mol Mutagen* 2004;43:143–58.
4. Car BD. Enabling Technologies in Reducing Drug Attrition Due to Safety Failures. *Am Drug Discov* 2006;1:53–6.
5. Chiu PJ, Marcoe KF, Bounds SE, Lin CH, Feng JJ, Lin A, et al. Validation of a [3H]astemizole binding assay in HEK293 cells expressing HERG K+ channels. *J Pharmacol Sci* 2004;95:311–19.
6. Diaz GJ, Daniell K, Leitza ST, Martin RL, Su Z, McDermott JS, et al. The [3H]dofetilide binding assay is a predictive screening tool for hERG blockade and proarrhythmia: Comparison of intact cell and membrane preparations and effects of altering [K+]o. *J Pharmacol Toxicol Meth* 2004;50:187–99.
7. Raab CE, Butcher JW, Connolly TM, Karczewski J, Yu NX, Staskiewicz SJ, et al. Synthesis of the first sulfur-35-labeled hERG radioligand. *Bioorg Med Chem Lett* 2006;16:1692–5.
8. Rezazadeh S, Hesketh JC, Fedida D. Rb+ flux through hERG channels affects the potency of channel blocking drugs: correlation with data obtained using a high-throughput Rb+ efflux assay. *J Biomol Screen* 2004;9:588–97.
9. Guo L, Guthrie H. Automated electrophysiology in the preclinical evaluation of drugs for potential QT prolongation. *J Pharmacol Toxicol Meth* 2005;52:123–35.
10. Pugsley MK. Methodology used in safety pharmacology: Appraisal of the state-of-the-art, the regulatory issues and new directions. *J Pharmacol Toxicol Meth* 2005;52:1–5.
11. Kirkland D, Aardema M, Henderson L, Muller L. Evaluation of the ability of a battery of three in vitro genotoxicity tests to discriminate rodent carcinogens and non-carcinogens I. Sensitivity, specificity and relative predictivity. *Mutat Res* 2005;584:1–256.
12. Mortelmans K, Zeiger E. The Ames *Salmonella*/microsome mutagenicity assay. *Mutat Res* 2000;455:29–60.
13. Miller JE, Vlasakova K, Glaab WE, Skopek TR. A low volume, high-throughput forward mutation assay in *Salmonella typhimurium* based on fluorouracil resistance. *Mutat Res* 2005;578:210–24.
14. Flamand N, Meunier J, Meunier P, Agapakis-Causse C. Mini mutagenicity test: A miniaturized version of the Ames test used in a prescreening assay for point mutagenesis assessment. *Toxicol In Vitro* 2001;15:105–14.
15. Kirsch-Volders M, Sofuni T, Aardema M, Albertini S, Eastmond D, Fenech M, et al. Report from the in vitro micronucleus assay working group. *Mutat Res* 2003;540:153–63.

16. Garriott ML, Phelps JB, Hoffman WP. A protocol for the in vitro micronucleus test. I. Contributions to the development of a protocol suitable for regulatory submissions from an examination of 16 chemicals with different mechanisms of action and different levels of activity. *Mutat Res* 2002;517:123–34.
17. Phelps JB, Garriott ML, Hoffman WP. A protocol for the in vitro micronucleus test. II. Contributions to the validation of a protocol suitable for regulatory submissions from an examination of 10 chemicals with different mechanisms of action and different levels of activity. *Mutat Res* 2002;521:103–12.
18. Parry JM, Parry EM. The use of the in vitro micronucleus assay to detect and assess the aneugenic activity of chemicals. *Mutat Res* 2006;607:5–8.
19. Fenech M, Chang WP, Kirsch-Volders M, Holland N, Bonassi S, Zeiger E. HUMN project: detailed description of the scoring criteria for the cytokinesis-block micronucleus assay using isolated human lymphocyte cultures. *Mutat Res* 2003;534:65–75.
20. CEREP. Genetic Toxicity. <http://wwwcerepfr/Cerep/Users/pages/downloads/Documents/Marketing/Pharmacology%20&%20ADME/Application%20notes/GeneticToxicitypdf 2006>. Accessed on November 1, 2006.
21. Schoonen WG, de Roos JA, Westerink WM, Debiton E. Cytotoxic effects of 110 reference compounds on HepG2 cells and for 60 compounds on HeLa, ECC-1 and CHO cells. II. Mechanistic assays on NAD(P)H, ATP and DNA contents. *Toxicol In Vitro* 2005;19:491–503.
22. Schoonen WG, Westerink WM, de Roos JA, Debiton E. Cytotoxic effects of 100 reference compounds on Hep G2 and HeLa cells and of 60 compounds on ECC-1 and CHO cells. I. Mechanistic assays on ROS, glutathione depletion and calcein uptake. *Toxicol In Vitro* 2005;19:505–16.
23. Mannerstrom M, Toimela T, Ylikomi T, Tahti H. The combined use of human neural and liver cell lines and mouse hepatocytes improves the predictability of the neurotoxicity of selected drugs. *Toxicol Lett* 2006;165:195–202.
24. Tokiwa T, Yamazaki T, Xin W, Sugae N, Noguchi M, Enosawa S, et al. Differentiation potential of an immortalized non-tumorigenic human liver epithelial cell line as liver progenitor cells. *Cell Biol Int* 2006;30:992–98.
25. Dambach DM, Andrews BA, Moulin F. New technologies and screening strategies for hepatotoxicity: Use of in vitro models. *Toxicol Pathol* 2005;33:17–26.
26. O'Brien P, Haskins JR. In vitro cytotoxicity assessment. *Meth Mol Biol* 2006; 356:415–26.
27. Jones PA, King AV. High throughput screening (HTS) for phototoxicity hazard using the in vitro 3T3 neutral red uptake assay. *Toxicol In Vitro* 2003;17:703–8.
28. NICEATM I. *In vitro 3T3 NRU phototoxicity.* <http://iccvamniehsnihgov/methods/3t3_nruhtm>. Accessed November 1, 2006.
29. NICEATM. *NICEATM pre-screen evaluation of the in vitro endocrine disruptor assay* (Robotic MCF-7 cell proliferation assay of estrogenic activity). <http://iccvamniehsnihgov/methods/endodocs/CCiPrescreenEvalpdf>. Accessed November 1, 2006.
30. Pessayre D. Role of reactive metabolites in drug-induced hepatitis. *J Hepatol* 1995;23:S16–24.

31. Knowles SR, Uetrecht J, Shear NH. Idiosyncratic drug reactions: The reactive metabolite syndromes. *Lancet* 2000;356:1587–91.

32. Park BK, Naisbitt DJ, Gordon SF, Kitteringham NR, Pirmohamed M. Metabolic activation in drug allergies. *Toxicology* 2001;158:11–23.

33. Chen WG, Zhang C, Avery MJ, Fouda HG. Reactive metabolite screen for reducing candidate attrition in drug discovery. *Adv Exp Med Biol* 2001;500:521–4.

34. Pearson PG, Threadgill MD, Howald WN, Baillie TA. Applications of tandem mass spectrometry to the characterization of derivatized glutathione conjugates: Studies with *S*-(*N*-methylcarbamoyl)glutathione, a metabolite of the antineoplastic agent *N*-methylformamide. *Biomed Envir Mass Spectrom* 1988;16:51–6.

35. Haroldsen PE, Reilly MH, Hughes H, Gaskell SJ, Porter CJ. Characterization of glutathione conjugates by fast atom bombardment/tandem mass spectrometry. *Biomed Envir Mass Spectrom* 1988;15:615–21.

36. Samuel K, Yin W, Stearns RA, Tang YS, Chaudhary AG, Jewell JP, et al. Addressing the metabolic activation potential of new leads in drug discovery: a case study using ion trap mass spectrometry and tritium labeling techniques. *J Mass Spectrom* 2003;38:211–21.

37. Nelson SD. Structure toxicity relationships—How useful are they in predicting toxicities of new drugs? *Adv Exp Med Biol* 2001;500:33–43.

38. Casati S, Collotta A, Clothier R, Gribaldo L. Refinement of the colony-forming unit-megakaryocyte (CFU-MK) assay for its application to pharmaco-toxicological testing. *Toxicol In Vitro* 2003;17:69–75.

39. Kimber I, Pichowski JS, Betts CJ, Cumberbatch M, Basketter DA, Dearman RJ. Alternative approaches to the identification and characterization of chemical allergens. *Toxicol In Vitro* 2001;15:307–12.

40. Flick B, Klug S. Whole embryo culture: An important tool in developmental toxicology today. *Curr Pharm Des* 2006;12:1467–88.

41. Coecke S, Eskes C, Gartion J, Vliet EV, Kinsner A, Bogni A, et al. Metabolism-mediated neurotoxicity: the significance of genetically engineered cell lines and new three-dimensional cell cultures. *Altern Lab Anim* 2002;30(Suppl 2):115–18.

42. Prieto P. Barriers, nephrotoxicology and chronic testing in vitro. *Altern Lab Anim* 2002;30(Suppl 2):101–6.

43. O'Brien PJ, Irwin W, Diaz D, Howard-Cofield E, Krejsa CM, Slaughter MR, et al. High concordance of drug-induced human hepatotoxicity with in vitro cytotoxicity measured in a novel cell-based model using high content screening. *Arch Toxicol* 2006;80:580–604.

44. Amacher DE, Martin B-A. Tetracycline-induced steatosis in primary canine hepatocyte cultures. *Fund Appl Toxicol* 1997;40:256–63.

45. Kostrubsky SE, Strom SC, Kalgutkar AS, Kulkarni S, Atherton J, Mireles R, et al. Inhibition of hepatobiliary transport as a predictive method for clinical hepatotoxicity of nefazodone. *Toxicol Sci* 2006;90:451–9.

46. Gum RJ, Hickman D, Fagerland JA, Heindel MA, Gagne GD, Schmidt JM, et al. Analysis of two matrix metalloproteinase inhibitors and their metabolites for induction of phospholipidosis in rat and human hepatocytes. *Biochem Pharmacol* 2001;62:1661–73.

47. Thompson DC, Barhoumi R, Burghard RC. Comparative toxicity of eugenol and its quinone methide metabolite in cultured liver cells using kinetic fluorescence bioassays. *Toxicol Appl Pharmacol* 1998;149:55–63.

48. Lilius H, Haestbacka T, Isomaa B. A combination of fluorescent probes for evaluation of cytotoxicity and toxic mechanisms in isolated rainbow trout hepatocytes. *Toxicol In Vitro* 1996;10:341–8.

49. LeBel CP, Ischiropoulos H, Bondy SC. Evaluation of the probe 2′,7′-dichlorofluorescin as an indicator of reactive oxygen species formation and oxidative stress. *Chem Res Toxicol* 1992;5:227–31.

50. Lautraite S, Bigot-Lasserre D, Bars R, Carmichael N. Optimisation of cell-based assays for medium throughput screening of oxidative stress. *Toxicol In Vitro* 2003;17:207–20.

51. Wang H, Joseph JA. Quantifying cellular oxidative stress by dichlorofluorescein assay using microplate reader. *Free Radic Biol Med* 1999;27:612–16.

52. Haskins JR, Rowse P, Rahbari R, de la Iglesia FA. Thiazolidinedione toxicity to isolated hepatocytes revealed by coherent multiprobe fluorescence microscopy and correlated with multiparameter flow cytometry of peripheral leukocytes. *Arch Toxicol* 2001;75:425–38.

53. Tan NS, Michalik L, Desvergne B, Wahli W. Multiple expression control mechanisms of peroxisome proliferator-activated receptors and their target genes. *J Steroid Biochem Mol Biol* 2005;93:99–105.

54. Xu D, McCarty D, Fernandes A, Fisher M, Samulski RJ, Juliano RL. Delivery of MDR1 small interfering RNA by self-complementary recombinant adeno-associated virus vector. *Mol Ther* 2005;11:523–30.

55. Lee SH, Sinko PJ. siRNA—Getting the message out. *Eur J Pharm Sci* 2006; 27:401–10.

56. Pichler A, Zelcer N, Prior JL, Kuil AJ, Piwnica-Worms D. In vivo RNA interference-mediated ablation of MDR1 P-glycoprotein. *Clin Cancer Res* 2005;11:4487–94.

3

PHYSIOLOGICALLY BASED PHARMACOKINETIC AND PHARMACODYNAMIC MODELING

Brad Reisfeld, Arthur N. Mayeno, Michael A. Lyons, and Raymond S. H. Yang

Contents

Computational Toxicology: Risk Assessment for Pharmaceutical and Environmental Chemicals, Edited by Sean Ekins

3.1 INTRODUCTION

The adverse biological effects of toxic substances depend on the exposure concentration and the duration of exposure, as well as on the type of effect that is produced and the dose required to produce that effect. Thus, to fully understand and predict toxicity, we must consider both the pharmacokinetics (what the body does to the chemical) and pharmacodynamics (what the chemical does to the body). Pharmacokinetic (PK) models can quantitatively relate the external concentration of a toxicant in the environment to the internal dose of the toxicant in the target tissues of an exposed organism. The exposure concentration of a toxic substance is usually not the same as the concentration of the active form of the toxicant that reaches the target tissues following absorption, distribution, biotransformation, and elimination (ADME) of the parent toxicant. Pharmacodynamic (PD) models can be used to predict dose response; that is, given a chemical of interest, how the concentration at the site of action is related to some relevant toxicological endpoint. By coupling PK and PD models, it may be possible to predict the dose response of an organism to a chemical based on the exposure concentrations.

In this chapter we provide an overview of pharmacokinetics, focusing specifically on physiologically based pharmacokinetics (PBPK). Unlike classical compartmental pharmacokinetics, PBPK approaches incorporate the anatomical entities and physiological and biochemical processes of organisms, and are therefore able to make inter-species, inter-route, and/or inter-dose extrapolations useful for a variety of applications. After introducing concepts fundamental to this approach, we highlight the areas in toxicology where PBPK modeling is used, and present some limitations of this methodology. We then present several selected examples of the application of pharmacodynamics in computational toxicology. We conclude by describing future directions that may be taken to address certain weaknesses in these modeling areas.

3.2 PHYSIOLOGICALLY BASED PHARMACOKINETIC MODELING

3.2.1 Definitions

Pharmacokinetics describes in a quantitative manner the absorption, distribution, metabolism, and excretion of xenobiotics in living organisms. Pharmacokinetic *modeling* is a methodology for developing mathematical descriptions of these processes for the prediction of temporal variations in both total amounts, and tissue and organ concentrations of these chemicals. Two commonly used types of pharmacokinetic models are "classical" (or data based) and physiologically based. Classical pharmacokinetic models, routinely applied in the pharmaceutical industry [1], consider the organism as a single homogeneous (or a multicompartmental) system. The nature of the compartments chosen is determined by the type of equation selected to describe the data, regardless of the physiological characteristics of the organism. Since these "classical" models are not based on the true anatomy, physiology, and biochemistry of the species of interest, they cannot, in general, be used to produce reliable predictions outside the range of doses, dose routes, and species used in the studies on which they were based. Such extrapolation, which is essential in estimating the dose response of chemicals, can be performed more accurately using physiologically based pharmacokinetic (PBPK) modeling approaches.

PBPK models embody relevant biological and mechanistic information, enabling them to be used, with limited animal experimentation, for extrapolation of the kinetic behavior of chemicals from high dose to low dose, from one exposure route to another, and from test animal species to humans. Thus the power of PBPK modeling centers on its ability to perform these important extrapolations and to describe concentration–time profiles in individual tissues or organs and in the plasma or blood. These features are especially relevant when considering the wide variety of chemicals whose toxicities (or pharmaceutical efficacy) are closely related to their concentration in certain target tissues, rather than their concentration in plasma. In addition PBPK models are amenable to multilevel refinements of the basic transport and fate processes embodied in the compartments, allowing a model to evolve as requirements change and additional data and insights become available. These advantages, however, come at the expense of requiring a relatively large investment in resources to develop a validated PBPK model with well-founded parameter values.

In the following sections, we will describe the methodology of PBPK model construction, assumptions used in their derivation, and the utility of PBPK models for a variety of applications. In each section we present several representative and relevant references, as well as illustrative examples. In the sections to follow, we attempt neither to describe the history of PBPK modeling nor to create a *comprehensive* review of available references. For these topics the reader is referred to a recently published book from our laboratory [2] that contains a useful overview, introduction, and over 1000 pertinent references for further study.

3.2.2 Modeling Methodology

The fundamentals of PBPK modeling are straightforward: first identification of the principal organs or tissues involved in the ADME of the chemical of interest, and then creation of mathematical equations that describe these processes in each tissue compartment and that correlate the disposition of the chemical within and among these organs and tissues in an integrated and biologically plausible manner. Constructing a validated PBPK model consists of several steps: (1) defining a realistic model, (2) obtaining the necessary data, (3) performing simulations, and (4) validating and refining the model. Each of these steps will be described in some detail in the following sections.

Defining a Realistic Model A PBPK model is comprised of compartments, each of which represents a discrete tissue or groupings of tissues with similar properties with respect to the xenobiotics under consideration. Each compartment is parameterized with appropriate volume, blood flow, and biochemical and physicochemical parameters for metabolism and solubility of the chemicals (and metabolites) of interest. The compartments are interlinked by physiologically relevant fluid (typically blood) flows. Routes of exposure or dosing are included in their proper relationship to the overall physiology. For instance, orally absorbed compounds move through intestinal tissues and portal blood to the liver before moving to the mixed venous blood for distribution to the remainder of the body. Each compartment in the model is described by one or more differential or algebraic equations describing a chemical species mass-balance, germane biological processes, and/or appropriate property relationships. The set of equations is solved by analytical or numerical integration to simulate time-course concentrations of chemicals and their metabolites within each model compartment.

The Conceptual Model: Classification of Tissues into Compartments As was mentioned earlier, PBPK modeling is an approach for ADME prediction and analysis. A multitude of tissues, organs, and organ systems are involved in the various aspects of ADME (Table 3.1). However, despite all the organs and systems that control the ADME processes, it is not necessary to define every tissue as a separate compartment in a PBPK model. Kinetically similar tissues can be lumped together to simplify a model and allow more efficient computation. For example, it is a common practice to create two "lumped" compartments in a PBPK model, each of which comprises tissues with similar blood perfusion characteristics: rapidly (or richly) perfused (e.g., brain, lung, heart, kidney) and slowly (or poorly) perfused (e.g., muscle, skin, bone). Conversely, and despite the desire for model simplicity, there are a number of important reasons to split a tissue from a lumped compartment and treat it individually. Among these reasons are the following [2]: (1) the tissue is a target organ, (2) biotransformation occurs within the tissue, (3) the tissue is the site of some other clearance process, (4) the tissue affects pharmacokinetics in some way, and (5) data on the tissue are available for comparison with simulation results.

TABLE 3.1 ADME Considerations for PBPK Modeling

Process	Common Organs, Organ Systems, and Other Factors to Consider[a]
Absorption	GI tract, lungs, skin
Distribution	Storage in tissues (plasma proteins, liver and kidney, fat, bone), blood–brain barrier, passage across the placenta, membrane permeability
Metabolism	Liver, lungs, kidney, brain, phase I, phase II metabolism
Excretion	Urinary, fecal, exhalation, milk, sweat, saliva

[a]Not a comprehensive list; for example, metabolism can occur in the GI tract, and elimination of chemicals can occur through hair.

Example

As an example of a conceptual PBPK model, we consider the disposition of trichloroethylene (TCE) in a human exposed via both inhaled and oral routes. In this instance we are interested in developing a predictive model for the temporal biodistribution of TCE, especially in the liver compartment. We assume that time-course concentration data are available for this xenobiotic in the blood, fat, kidney, muscle, and liver. In general, the structure of a PBPK model is determined according to the exposure conditions, the pharmacokinetic characteristics of the chemical, and the available data. Specifically, since TCE is lipophilic and is relatively slowly metabolized in the liver, compartments representing the fat and liver are included in the model structure. To account for the routes of exposure, we include compartments for the lung (lung tissue and lung blood) and gastrointestinal (GI) tract. We expect that the other organs and tissues have no individually distinct impact on the pharmacokinetics; thus, they are lumped into slowly and rapidly perfused compartments. Accounting for the anatomical arrangement and interconnection of tissues, we arrive at the conceptual model for our system illustrated in [Figure 3.1].

In creating this, or any, PBPK model, we are mindful of Occam's razor and strive for a level of model complexity consistent with available data. If justified, further intricacy can be incorporated. For example, other PBPK models for TCE exposure have included compartments and processes in addition to those described above to better accommodate the objectives of the investigator, as well as the type and quantity of the available experimental data. The model created by Keys and coworkers [3] incorporated 10 compartments, including those for the brain, deep and shallow liver, kidney, heart, and spleen.

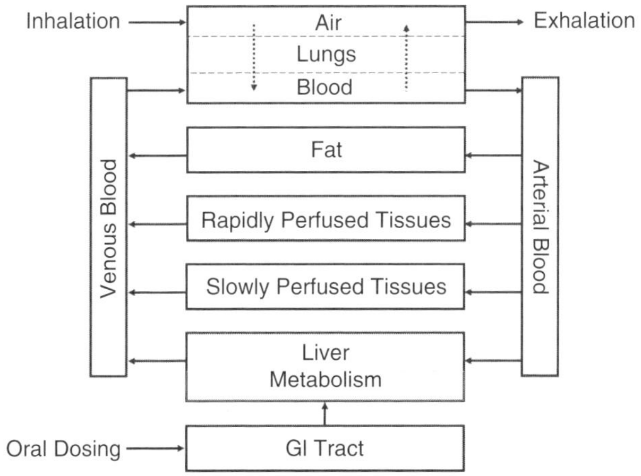

Figure 3.1 Schematic diagram of a PBPK model for trichloroethylene exposure.

The Mathematical Model Once a conceptual PBPK model has been created, with some knowledge of the chemical of interest, this representation can be translated into mathematical equations for use in predicting time-course disposition. The fundamental equations utilized arise from chemical species mass balances, which account for the rates at which molecules enter and leave each compartment, as well as other processes (e.g., rates of reactions) that produce or consume the chemical.

In the following development we assume that each of the compartments is homogeneous, that the complete oral dose in the GI compartment is absorbed, and that the chemical uptake in each tissue compartment is perfusion limited (i.e., the diffusion of the chemical into the tissue is "rapid" and the rate-limiting step is the blood perfusion rate). For a description of the equation structure appropriate for diffusion-limited analyses, see Krishnan and Andersen [4].

Example

We now recast the conceptual model depicted in Figure 3.1 into mathematical terms. The mass balance for TCE in the fat compartment can be expressed as

$$V_{fat}^{T} \cdot \frac{dC_{fat}^{T}}{dt} = Q_{fat} \cdot \left(C_{arterial}^{B} - C_{venous,fat}^{B}\right), \tag{3.1}$$

where V_{fat}^{T} and C_{fat}^{T} are the volume and concentration of the chemical in the fat compartment, respectively, Q_{fat} is the blood flow rate into fat, $C_{arterial}^{B}$ is the arterial blood concentration, and $C_{venous,fat}^{B}$ is the blood concentration of the chemical leaving the fat (venous blood concentration). In the concentration and volume terms presented herein, the superscripts T, B, and A denote tissue, blood, and air, respectively. Assuming the compartment is perfusion-limited, we can invoke the venous equilibrium assumption (i.e., the effluent venous blood concentration is in equilibrium with the tissue concentration), so

$$C_{venous,fat}^{B} = \frac{C_{fat}^{T}}{P_{fat:blood}},$$

where $P_{fat:blood}$ is the chemical-specific partition coefficient between the fat and blood, giving rise to the equation

$$V_{fat}^{T} \cdot \frac{dC_{fat}^{T}}{dt} = Q_{fat} \cdot \left(C_{arterial}^{B} - \frac{C_{fat}^{T}}{P_{fat:blood}}\right). \tag{3.2}$$

Mass-balance relationships completely analogous to equation (3.1) can be derived for the rapidly and slowly perfused tissue.

In the liver compartment, absorption from the GI compartment and metabolism should be considered. A mathematical representation incorporating these terms is

$$V_{liver}^{T} \cdot \frac{dC_{liver}^{T}}{dt} = Q_{liver} \cdot \left(C_{arterial}^{B} - \frac{C_{liver}^{T}}{P_{liver:blood}} \right) + A_{GI}(t) - M_{liver}(t), \quad (3.3)$$

where $A_{GI}(t)$ represents the rate of absorption from the GI compartment into the liver, and $M_{liver}(t)$ represents the rate of metabolism that results in a decrease in the amount of chemical in the liver. These terms are assumed to take the following forms:

$$A_{GI}(t) = K_{absorption} \cdot C_{GI}^{T}, \quad M_{liver}(t) = \frac{V_{max} \cdot \left(C_{liver}^{T} / P_{liver:blood} \right)}{K_m + \left(C_{liver}^{T} / P_{liver:blood} \right)}, \quad (3.4)$$

where $K_{absorption}$ is the rate constant for absorption, and V_{max} and K_m are parameters in the Michaelis-Menten equation, which is assumed to describe the kinetics of metabolism. The species mass balance for TCE in the GI lumen is simply

$$V_{GI}^{T} \cdot \frac{dC_{GI}^{T}}{dt} = -A_{GI}(t). \quad (3.5)$$

A mass balance on the venous blood gives a differential equation incorporating multiple terms arising from the inflows and outflows of TCE, that is,

$$V_{total}^{B} \cdot \frac{dC_{venous}^{B}}{dt} = \sum_{i} Q_i \cdot C_i^T - Q_{cardiac} \cdot C_{venous}^{B}, \quad (3.6)$$

where $Q_{cardiac}$ is the cardiac output, V_{total}^{B} is the total venous blood volume, and

$$\sum Q_i \cdot C_i^T = Q_{liver} \cdot \left(\frac{C_{liver}^{T}}{P_{liver:blood}} \right) + Q_{fat} \cdot \left(\frac{C_{fat}^{T}}{P_{fat:blood}} \right) + Q_{rp} \cdot \left(\frac{C_{rp}^{T}}{P_{rp:blood}} \right) + Q_{sp} \cdot \left(\frac{C_{sp}^{T}}{P_{sp:blood}} \right). \quad (3.7)$$

For the calculation of arterial blood concentration, $C_{arterial}^{B}$, it is assumed that steady state in the lung is quickly reached upon inhalation, that the exhaled concentration is in equilibrium with the concentration in the arterial blood, and that the chemical is absorbed only in the alveolar region. Consequently a species balance on TCE within the arterial blood gives

$$V^B_{arterial} \cdot \frac{dC^B_{arterial}}{dt} = Q_{cardiac} \cdot \left(C^B_{venous} - C^B_{arterial}\right) + Q_{alveolar} \cdot \left(C^A_{inhaled} - \frac{C^B_{arterial}}{P_{blood:air}}\right). \quad (3.8)$$

Here $Q_{alveolar}$ is pulmonary ventilation rate, $C^A_{inhaled}$ is the chemical concentration inhaled, and $P_{blood:air}$ is the partition coefficient between the blood and air, and $C^B_{arterial}/P_{blood:air}$ is the concentration exhaled.

The equations above, with appropriate parameter values and initial conditions, form a system that can be solved by appropriate integration methods to yield the time-course concentration values in all model compartments.

Obtaining the Necessary Data To produce a validated PBPK model, two basic types of data are necessary: parameter values and time-course concentration data. Parameter values are used in the mathematical formulation described above, while time-course data are used to critically evaluate and improve a model or to provide a means for deducing any missing parameter values (parameter identification).

Parameter Values Aside from the dependent and independent variables in the equations above, a variety of parameters must be specified. These include physiological parameters (e.g., ventilation rates, cardiac output, organ volumes and masses), physicochemical parameters (e.g., tissue partition coefficients, protein binding constants), and biochemical parameters (e.g., K_m and V_{max}).

Physiological parameters. Comprehensive sets of the physiological parameters for humans and common laboratory animals are available in the literature [5–7]. When information gaps exist, necessary values can often be obtained via experimentation or through allometric extrapolation, usually based on a power function of the body weight [8] (e.g., $X = \alpha W^{\beta}$, where X is the parameter of interest, W is body weight, and α, β are constants).

Physicochemical parameters. Partition coefficients are the most common type of physicochemical parameter in a PBPK model. Values for these quantities can be measured through experimental means (e.g., equilibrating tissue homogenates in a vial with an atmosphere containing the test chemical [9,10], or from ultrafiltration/equilibrium dialysis studies for nonvolatile chemicals), or through the use of quantitative structure activity/property relationships (QSA(P)Rs) [11].

Metabolic parameters. Constants for metabolism can be determined in vitro with tissue homogenates, microsomal preparations, and liver slices [12,13]. Another method for assessing metabolic parameters in vivo

TABLE 3.2 PBPK Model Parameters

Tissue (fraction of body weight)	
Liver	0.026
Rapidly perfused	0.05
Slowly perfused	0.62
Fat	0.19
Flows (Liter/h)	
Alveolar ventilation	12.6
Cardiac output	14.9
Tissue perfusion (fraction of cardiac output)	
Liver	0.26
Rapidly perfused	0.44
Slowly perfused	0.25
Fat	0.05
Partition coefficients	
Liver : blood	6.82
Rapidly perfused : blood	6.82
Slowly perfused : blood	2.35
Fat : blood	73.3
Blood : air	9.2
Metabolic constants	
V_{max} (mg/kg-h)	14.9
K_m (mg/Liter)	1.5
Oral uptake	
GI absorption (Liter/h)	1

Source: Values taken from [17, 128].

relies on closed chamber inhalation techniques in which a number of live animals are placed in a closed chamber to measure the rate of loss of chemical at a variety of chemical concentrations within the chamber [14,15].

To give the reader a sense of their magnitude, Table 3.2 lists typical parameters values that would be used in the PBPK model for TCE described earlier.

Time-Course Concentration Data To construct and validate a PBPK model, consistent in vivo pharmacokinetic data detailing blood and tissue time-course concentrations are essential. These data should include at least the following tissues and organs: blood (or plasma if blood-cell binding is not an issue), liver (organ of metabolism), kidney (representing rapidly perfused organs/tissues), muscle (representing slowly perfused organs/tissues), fat (for lipophilic compounds), and target organ(s) or tissue(s). Data taken using different doses, routes of dosing, and species will serve to enhance the veracity and realism of the constructed model. Whenever possible, a large enough quantity of data should be generated so that both a training set and validation set can be constructed.

Performing Simulations As illustrated earlier, the mathematical representation of a PBPK model generally comprises a system of coupled ordinary differential and algebraic equations. These equations are normally amenable to solution via numerical integration methods on the computer.

The available computer tools for PBPK modeling include general-purpose programming languages, special-purpose simulation software, and spreadsheets. A useful list of these tools, along with their developers/vendors, salient features, and application examples, has been compiled in a report on PBPK modeling [16]. Among the more popular examples in this list are MATLAB (MathWorks, Natick, MA), Berkeley Madonna (University of California at Berkeley, CA), SAAM II (University of Washington, Seattle, WA), SimuSolv (Dow Chemical Company, Midland, MI), and ACSL, ACSL Tox, and acslXtreme (Aegis Technologies Group, Huntsville, AL). Although the available software varies in flexibility and user-friendliness, any product selected for PBPK model construction and simulation should have stable, robust, and accurate (stiff and non-stiff) algorithms for integration; statistically driven methods for optimization and parameter identification; and capabilities for sensitivity analyses. Other useful features present in some analysis environments include a graphical user interface for "drag and drop" model construction, capabilities for Monte Carlo analysis, and functions to provide estimates of unknown physiological and physicochemical parameter values.

Regardless of the software chosen, a series of simulations should be performed over and beyond the set of conditions of interest and appropriate adjustments made to ensure that the results converge properly, all conservation of mass equations are satisfied, and the results are biologically plausible.

Validating and Refining the Model Once an initial set of simulations has been performed, validation against actual experimental data can be performed. During this phase, a priori simulations under a specific exposure scenario can be conducted and the simulation results compared with available experimental data. As an example, Figure 3.2 gives a comparison of experimental data and PBPK model simulations from Fisher and coworkers [17], who were interested in the disposition of TCE and its toxic metabolites. Parameters whose values were not available in the literature were adjusted to bring simulations in line with measured results over a series of different simulations.

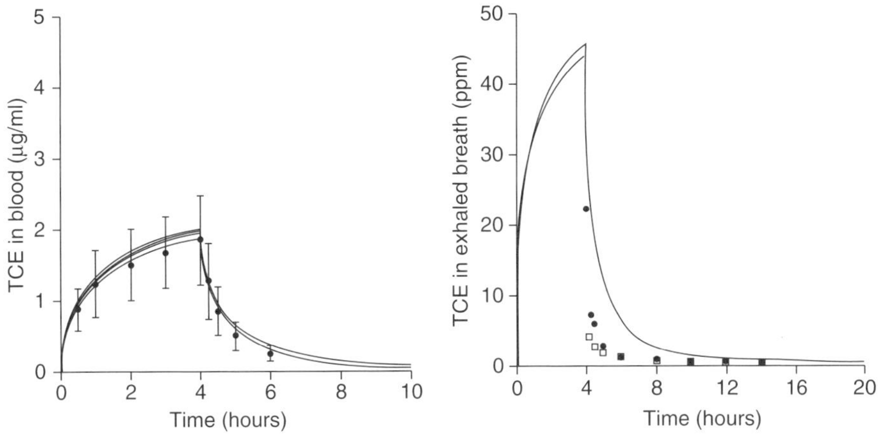

Figure 3.2 Biodistribution of TCE. (Reprinted from Fisher et al. [17] with permission from Elsevier.)

It is important to note that when performing parameter identification and assessing the validity of a model, it is crucial at each stage to ensure that the model structure and parameters are physiologically and biochemically realistic and appropriate. In addition, validation of the model with data sets other than the training set used to develop the model is essential. Once a PBPK model has been validated, a priori computer simulations may be performed for a given set of initial conditions, such as animal species of interest, dosing route, dosing levels, and regimen.

3.2.3 Extrapolation across Doses, Routes of Exposure, and Species

PBPK modeling can play an essential role in three common types of extrapolation used in classical toxicology: *dose to dose* (usually high dose in animals to low dose for realistic exposure scenarios), *route to route* (e.g., ingestion vs. inhalation), and *species to species* (animal or cell culture to human). Each of these types of extrapolation is described in some detail below.

Dose to Dose PBPK modeling permits reasonable extrapolation from one dose to another, if adequate information on physiology, physicochemical properties, and biochemistry is available. If the dynamic processes modeled by the PBPK approach are all directly proportional to administered concentrations, then the extrapolation can be relatively straightforward. However, this is not often the case, especially at higher doses, where saturation of metabolic or clearance processes can occur [14,19]. Further causes of nonlinearity of chemical kinetics include the induction and inhibition of metabolic enzymes [14]. Despite these difficulties successful applications of dose extrapolation using PBPK models for many chemicals have been published [20,21], and several recent studies have demonstrated the usefulness of PBPK modeling for the investigation of the chemical kinetics of lipophilic compounds at different dosage regimens [22,23].

Route to Route PBPK models have been used for route-to-route extrapolation for specific chemicals and systems, and have been shown to produce accurate predictions in many cases [24]. By assuming that the relationship between applied dose and tissue dose of the xenobiotic of interest is the same, regardless of the exposure route, route-to-route extrapolations may be performed by the addition of intake terms to the governing mass balance equations that represent each exposure pathway or mechanism. The uncertainty associated with this approach can arise from the first-pass effect as well as variations in rates and extent of absorption and metabolism from one route to another [25]. However, by accounting for these route-specific processes, PBPK models can be used to conduct route-to-route extrapolations [26].

Species to Species PBPK modeling is a highly appropriate approach for species-to-species extrapolation because all mammals have the same "compartment-scale" circulatory anatomy, and much is known about the comparative dimensions of their blood flow rates, organ volumes, and clearances. In order to conduct such an extrapolation, estimates of physiological parameters, partition coefficients, and metabolic rate constants must be obtained for the species of interest [14]. Although several methods to obtain these parameters were described earlier, it is worth considering this issue in the context of species-to-species extrapolation. It has been observed that many anatomical and physiological variables can be empirically correlated to the body mass of a species [27,28], and that the physiological function per unit of organ or body mass decreases as the size of the animal increases [29]. For those parameters used in the description of metabolism, the situation is generally much more complex. This is because there might be *qualitative* differences between species, such as the presence or absence of a given enzyme that would result in a (potentially dose-dependent) difference in metabolic capacity. Some correlations for metabolic rate constants in terms of animal body weight have been proposed for chemicals that have a high affinity for metabolizing enzymes [14], but a more generally applicable method has yet to be introduced.

3.2.4 Application to Chemical Mixture Toxicology

All organisms are exposed to multiple xenobiotics, through food, environmental contaminants, and drugs. Thus there is a need to evaluate the impact and risk associated with exposure, not to individual chemicals, but to *chemical mixtures*. Pioneering efforts in the use of PBPK modeling in the area of chemical mixture toxicology were lead by Krishnan and coworkers, whose early work concentrated on toxicological interactions and PBPK modeling between mixtures comprising two chemicals [30,31]. These investigators advanced the hypothesis that pharmacokinetic interactions of complex chemical mixtures, regardless of the number of components, can be predicted by utilizing *a combination of binary mixtures models* of the constituent chemicals [32,33]. Over the years these investigators analyzed increasingly complex chemical mixtures [31,32,34], and to date they have successfully carried out PBPK modeling on the pharmacokinetic interactions for mixtures involving up to five chemicals [32,33].

Applying the same approach created by Krishnan and coworkers, investigators in our laboratory used PBPK modeling to study the toxicological interactions of a ternary mixture of trichloroethylene, tetrachloroethylene, and 1,1,1-trichloroethane in rats and humans [35,36]. Later Dennison et al. [37,38] used an integrated PBPK modeling and lumping approach to characterize the pharmacokinetics of gasoline in rats. This complex mixture was represented by five individual chemicals and one lumped "pseudochemical." The PBPK model tracked selected target components (benzene, toluene, ethylbenzene, *o*-xylene, and *n*-hexane) and a combined chemical group representing all nontarget components.

For a more thorough discussion of the topic of the application of PBPK modeling to the analysis of chemical mixture toxicity and interactions, the reader is encouraged to consult a chapter on PBPK modeling of chemical mixtures [2].

3.2.5 Accounting for Population Variability

All living organisms share similarities in their structure and function, particularly in terms of DNA, RNA, and proteins. But even among members of a species, considerable variability exists in the features of the individual members. Variations in both physical and behavioral traits are especially evident among ourselves. The idea of certain shared characteristics of structure and function accompanied by variation in detail is contained within PBPK models applied to populations of animals. These models consist of an underlying dynamical structure based on general anatomical and physiological characteristics common to all members of the population, along with a set of parameters that distinguish the various subpopulations. Population variability is accounted for by the variability in the values of the parameters corresponding to each individual. An example of variability in body weight in a sample of human males and females throughout the life span is shown in Figure 3.3. This figure also illustrates a general feature of physiological and biochemical properties across age groups: the largest differences occur between young children/neonates and adults, more so than between young adults and the elderly [40,41]. In addition to differences in gross anatomical features, less apparent physiological or biochemical differences exist within populations that can significantly impact the ADME of xenobiotics. For example, interindividual differences in biotransformation among human populations with

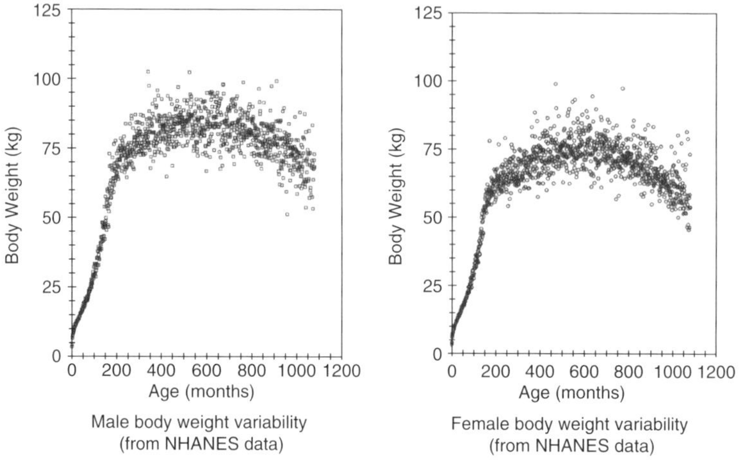

Figure 3.3 Human weight variability in the population. (Reprinted from [40].)

diverse genetics and lifestyles can lead to considerable variability in the metabolism and potential bioactivation of xenobiotics [42]. The topic of variability among various physiological and biochemical measures germane to PBPK modeling is detailed in a recent review by Clewell and coworkers [40].

Variability, which arises from heterogeneity of the population, is distinct from uncertainty; nevertheless, the two are usually discussed together [43,44]. Uncertainty arises from consideration of the accuracy and precision with which parameter measurements are made, and also from the approximate nature of the model itself. Uncertainty can be minimized through improved measurements and better model specification. Variability is intrinsic to the population under consideration and cannot be minimized. One consequence of this variability is that the exposure of a population to a certain amount of foreign chemical will result in a range of target tissue/organ concentrations, which in turn implies variation in dose–response relationships for this chemical. A proper accounting for population variability and uncertainty is relevant to the quantitative assessment of risk due to chemical exposure.

The simplest method of accounting for variability consists in identifying the parameters, experimental data, and model output with an average or reference individual [45]. One shortcoming of this approach is that although general model behavior is representative, much of the actual population may not be well described. Methods attempting to explicitly account for variability and uncertainty include Monte Carlo simulations [18,46,48], Bayesian population methods [49–51,53,56], the use of fuzzy sets [44,52], and other probability based methods [43]. Monte Carlo simulations, which model the parameter variability in terms of probability distributions, are the most common methods. Each individual is characterized by a set of parameters whose values are drawn from a random sampling of the associated probability distributions. The model is repeatedly evaluated with a sequence of randomly drawn parameter sets, producing a range of model output that can then be further statistically analyzed. Bayesian population methods can be viewed as an extension of Monte Carlo (MC) simulations [53]. They incorporate prior information regarding the parameter distributions along with experimental measurements to give an updated posterior distribution for the parameters via Bayes' theorem $p(\theta|y) \propto p(y|\theta)\mathrm{p}(\theta)$, where $p(\theta)$ is the prior parameter distribution, $p(y|\theta)$ is the probability of the data conditioned on the parameters, and $p(\theta|y)$ is the posterior probability distribution for the parameters conditioned on the data.

Population models describe the relationship between individuals and a population. Individual parameter sets are considered to arise from a joint population distribution described by a set of means and variances. The conditional dependencies among individual data sets, individual variables, and population variables can be represented by a graphical model, which can then be translated into the probability distributions in Bayes theorem. For most cases of practical interest, the posterior distribution is obtained via numerical simulation. It is also the case that the complexity of the posterior distribution for most PBPK models is such that standard MC sampling is inadequate leading instead to the use of Markov Chain Monte Carlo (MCMC) methods [54]. The difference between MC and MCMC simulation is that MC refers to a direct sampling of the distribution, while MCMC sampling is an iterative simulation in which draws of a random variable are made from a sequence of distributions which eventually converge to the target posterior distribution. Once convergence is obtained, the simulation is allowed to continue to run providing a set of values yielding a discrete approximation of the posterior distribution on which further statistical analysis can be performed. MCSim [55] is freely available software used for PBPK modeling, which includes tools for MCMC simulation. PBPK model calibration using Bayesian population modeling is well described in [49,53,56] with additional recent application to risk assessment [57].

An example in Figure 3.4 shows a distribution of cancer risk due to dichloromethane exposure that takes into account variability and uncertainty in the parameters of a physiologically based model as well as glutathione-*S*-transferase theta 1 (GSTT1) polymorphism [68]. Figures 3.3 and 3.4 depict variability between members of a population, referred to as inter-individual variability. Variability within an individual or subpopulation, referred to as intra-individual variability, may also be relevant depending on the focus of the investigation. Introducing a spatial and/or time dependence into the parameters for an individual or subpopulation can be used to describe this type of variability. An example of variability within an individual due to growth and aging, described as a life-stage model is presented by Clewell et al. [58]. Additional discussion of inter-individual and intra-individual variability can be found in the work of Banks and Potter [43].

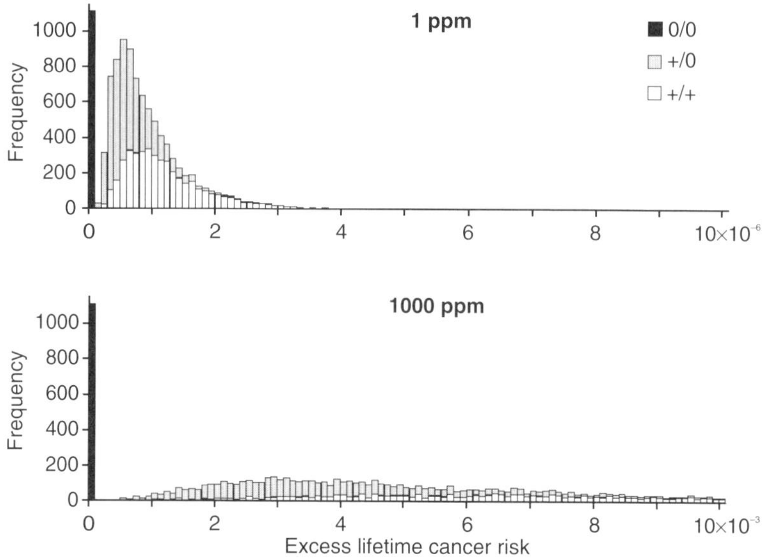

Figure 3.4 Distribution of cancer risk from dichloromethane exposure. (Reprinted from [68] with permission from Elsevier.)

3.2.6 Application to Risk Assessment

As shown previously, PBPK models allow the conversion of potential dose or exposure concentration to tissue dose, which can then be used for risk characterization purposes. The choice of an internal dose metric is based principally on an understanding of the mode of action of the chemical species of concern. The internal dose metric (sometimes called the biologically effective dose) is often used in place of the applied dose in quantitative dose–response assessments, in order to reduce the uncertainty inherent in using the applied dose to derive risk values.

PBPK modeling for risk assessment applies the major elements covered earlier in this chapter: interspecies extrapolation, route-to-route extrapolation, estimation of response from varying exposure conditions, estimation of human variability, and high-to-low dose extrapolation. One of the first PBPK models used for risk assessment extrapolations was developed for methylene chloride in the mid-1980s [45]; this model was used to predict tissue exposures to highly reactive metabolites from oxidative and conjugative pathways resulting from several different exposure scenarios for both mice and humans [59]. Since that time, PBPK models have been used for a wide variety of applications in risk assessment, including use in evaluating cancer risk [21,57,60–66], establishing biological exposure indexes (e.g., blood and urine concentrations) associated with industrial chemical exposures [67–69], reconstructing doses resulting from multi-route and multi-media (air, water, food, soil) exposures [70–72], developing metrics for the risk of exposure to chemical mixtures

[35,36,73], and appraising chemicals still in the developmental phase intended for therapeutic or nutritional purposes [74].

3.2.7 Limitations of PBPK Modeling

"All models are wrong; some are useful."—attributed to George Box.

As with any modeling approach, PBPK modeling has limitations, and each model has an appropriate range of applications. Extrapolation of model simulations beyond a designed range can yield serious errors in the predictions, and users of a model must be wary. Limitations in a model arise from diverse causes, such as the assumptions used in the model, insufficient understanding of the biological system simulated, the quality of data used for model parameterization, the hardware, software, numerical methods used, and other sources. It is beyond the scope of this section to cover each of these reasons in detail, and only a brief overview will be described here.

Assumptions are used to simplify the biological system to a mathematically and computationally tractable level. For any model the following questions must always be asked: What are the assumptions? and How valid are they? Moreover, a model is only as good as the parameters used. As published PBPK parameters are not always consistent [75], parameter values obtained from the literature must be used with care. Relatively recently PBPK model parameters have been compiled [76].

Before and, often, during PBPK model development, extensive in vitro and in vivo experimentation is required for model parameterization, for model validation, and when used for risk assessment, for elucidation of the mode(s)-of-action of toxicity. PBPK models can help in designing targeted experiments, to test hypotheses, and to minimize expensive experimentation, but they cannot replace experimentation.

When used for risk assessment, knowledge of the mechanism(s) by which a chemical exerts its toxic action is essential. Without a complete understanding of the mechanism of action of chemicals, prediction of toxicity is tenuous. In general, mechanisms of action cannot be gleaned from PBPK models, although such models can support or contradict hypothesized mechanisms [77]. An intriguing example is the toxicity of simple linear alkanes: to wit, *n*-pentane, *n*-hexane, and *n*-heptane. *n*-Hexane causes neurotoxicity in humans and animals [78] through the formation of the reactive γ-diketone metabolite 2,5-hexanedione, which is not formed from either *n*-pentane or *n*-heptane. Clearly, interpolation between pentane and heptane would fail to provide a good estimate of risk for *n*-hexane. Thus seemingly logical and rational interpolations or extrapolations, even within a single family of chemicals, could be specious. Through the use of PBPK modeling, unless *all* of the pertinent interactions of a chemical within an organism are known, it is impossible to know the risk of a new chemical entity. Moreover the effects of chronic (long-term) low-level exposure to a chemical are difficult, if not impossible, to assess based on simulations and animal testing, as studies in animals may not necessarily reflect the effects in humans. Current PBPK models cannot address this dilemma, since the mechanisms of action must be known beforehand and coded into the model. As the human population is diverse (e.g., young, elderly, obese, pregnant and/or nursing), investigators developing and using PBPK models for risk assessment must always be cognizant of these limitations, and one model is rarely appropriate for all individuals. For risk assessment of chemicals, the current way to address these uncertainties and variations is through the use of uncertainty or variability factors (UF) [79]. Recent efforts on Bayesian population PBPK modeling, as briefly outlined above, may offer a method to address this limitation.

For exposure to multiple chemicals, PBPK modeling is further limited as a predictive tool, as all interactions among the various chemicals within the organism must be known and incorporated into the model. Difficulties in multiple-chemical exposure are clearly illustrated in drug–drug and drug–food interactions, where one substance affects the pharmacokinetics or pharmacodynamics of another. To address the need for predictive capability for individual and mixtures of chemicals, new advances and approaches are required. Biochemical reaction network modeling is one nascent approach, as described in Section 3.4.

3.3 PHARMACODYNAMIC MODELING

As stated previously, the adverse biological effects of toxic substances depend on the exposure concentration and the duration of exposure, as well as on the type of effect that is produced and the dose required to produce that effect. Thus far we have focused on the use of PBPK modeling to simulate the distribution and clearance of xenobiotics, providing a biologically based method of translating an exposure dose of a parent chemical to a time-dependent target tissue dose of a toxicologically relevant chemical species. In this section we focus briefly on the description of pharmacodynamic models, which generally start with a tissue dose of a bioactive species, and predict a toxicological response.

Xenobiotics can exert effects on an organism via a number of different mechanisms, including cellular membrane disruption, chemical reactions, binding to receptors, and interactions with enzymes, structural and carrier proteins, nucleic acids, and ion channels. Corresponding adverse responses or endpoints can include inflammation, corrosive effects, sensitization, immune system compromise, neurological impairment, developmental effects, specific organ damage, and carcinogenesis. Owing to the diversity of these mechanisms and responses, there have been many pharmacodynamic modeling approaches taken to capture the salient biological phenomena. Therefore, rather than attempting to represent the few elements common to these approaches, we will look at the breadth of methodologies by briefly presenting selected pharmacodynamic models arising in both the pharmaceutical-toxicology and environmental-toxicology arenas.

3.3.1 Pharmacodynamic Studies of Drug–Drug Interactions

We limit our attention here to several models that attempt to describe adverse effects arising from drug–drug interactions, an increasingly important area in pharmaceutical toxicology. The models involved in these studies cover a wide spectrum of endpoints, including effects on the central nervous system, kidney, cardiovascular, as well as antimicrobial activities. We look briefly at the following distinctive types of pharmacodynamic models: sigmoid E_{max}, isobolographic, and response surface. In each case we briefly mention the purpose of the study and introduce the modeling approach without going into details of the outcome of the studies. The readers are encouraged to consult the original references for additional information.

Sigmoid E_{max} Model Jonkers and colleagues [80] studied the pharmacodynamics of racemic metoprolol, a cardioselective beta-blocker, and the active *S*-isomer in extensive metabolizers (EMs) and poor metabolizers (PMs). The drug effect studied was the antagonism by metoprolol of terbutaline-induced hypokalemia (abnormally low potassium concentration in the blood). The pharmacodynamic interaction was described by a sigmoidal function for competitive antagonism based on the earlier work of Holford and Sheiner [81]:

$$E = E_0 - \frac{(E_0 - E_{max}) \cdot C_e^n}{EC_{50}^n + \left[EC_{50}^n \cdot \left(C_{meto}/IC_{50}\right)\right] + C_e^n}, \tag{3.9}$$

where E_0 is the potassium concentration in the absence of terbutaline, E_{max} is the maximum effect of hypokalemia or potassium concentration, C_e is the effect compartment concentration of terbutaline, n is the factor expressing sigmoidicity of the concentration effect relationship, EC_{50} is the C_e that corresponds to an effect equal to the mean of the sum of E_0 and maximum effect, C_{meto} is the metoprolol concentration, and IC_{50} is the metoprolol concentration that corresponds with 50% maximum receptor occupancy.

The sigmoid E_{max} model formed the basis for a number of subsequent pharmacodynamic analyses of drug-drug interactions. For instance, Mandema et al. [82], used quantitative electroencephalographic effect measurements to study pharmacodynamic interactions among benzodiazepines in male Wistar rats. In a separate study, but utilizing the same pharmacodynamic endpoint in the same animal, these investigators explored the interactions of antispastic agents, racemic baclofen and its enantiomers, which selectively bind $GABA_b$ receptor sites [83]. Other drug interactions analyzed with similar pharmacodynamic modeling include alprazolam and caffeine [84] for central nervous system (CNS) effects, piperacillin/ciprofloxacin and piperacillin/tazobactam

[85] for antimicrobial combination activities, tiagabine and midazolam [86] for their antiepileptic effects, and irbesartan and hydrochlorothiazide [87] for their renal hypertensive effects.

Isobolographic Model Isobolographic analysis is a method to analyze dose–response curves (isobols) in a binary interaction study where the deviation from the line of additivity demonstrates antagonism or synergism. Levasseur et al. [88], in their studies on convulsant interactions between pefloxacin and theophylline in rats, developed a new approach for the isobolographic analysis of pharmacodynamic interactions. Their model for these interactions took the form of a quadratic equation for the combination index, *CI*, a measure indicative of whether the process is governed by additivity or synergism or antagonism:

$$\alpha \cdot R \cdot (1-R) \cdot CI^2 + CI - 1 = 0. \tag{3.10}$$

Here

$$R = \frac{C_1/\overline{IC}_1}{(C_1/\overline{IC}_1)+(C_2/\overline{IC}_2)}, \quad CI = C_1/\overline{IC}_1 + C_2/\overline{IC}_2, \tag{3.11}$$

and α is the interaction parameter. When α is positive, Loewe synergy is indicated; whereas a negative value reflects Loewe antagonism. When α is not significantly different from 0, the drug combination is Loewe additive. R represents the proportion of chemical 1, and C is the dose of drug in combination required to induce maximal seizures in rats. $\overline{IC}$ is the geometric mean dose of drug, which, when given alone, was required to induce maximal seizures. The subscripts 1 and 2 identify drug 1 (pefloxacin) and drug 2 (theophylline).

More recently Brochot et al. [89] reported an extension of the isobolographic approach to interaction studies for convulsant interaction among pefloxacin, norfloxacin, and theophylline in rats. Their contribution is unique in that they started out by explaining pharmacodynamic interactions for two drugs, but then extended the approach to derive an isobol for three drug interaction. In addition they included Bayesian analysis and developed a population model with Markov chain Monte Carlo methods.

Response Surface Model A dose–response surface is an extension of dose–response lines (isobols) to three dimensions. In this representation there can be a dose–response surface representing additivity and surfaces above and below suggesting deviation from additivity. Tam et al. [90] studied the combined pharmacodynamic interactions of two antimicrobial agents, meropenem and tobramycin. Total bacterial density data, expressed as CFU (colony forming units), were modeled using a three-dimensional surface. Effect summation was used as the definition of additivity (null interaction hypothesis) and the pharmacodynamic model was assumed to take the functional form

$$\log_{10} \text{CFU/ml} = Z_{intercept} - \left(E_{m-\max} \cdot \frac{C_m^{H_m}}{C_{50m}^{H_m} + C_m^{H_m}} \right) - \left(E_{t-\max} \cdot \frac{C_t^{H_t}}{C_{50t}^{H_t} + C_t^{H_t}} \right), \quad (3.12)$$

where $Z_{intercept}$ is the bacterial density at 24 hours in the absence of drug, $E_{m\text{-}max}$, $E_{t\text{-}max}$, are the maximal effect of meropenem or tobramycin, C_m, C_t, are the concentration of meropenem or tobramycin, H_m, H_t, are the sigmoidicity of meropenem or tobramycin, and C_{50m}, C_{50t}, are the concentration of meropenem or tobramycin needed to achieve 50% of the maximal effect.

The volume under the planes (VUP), of the observed and expected surfaces were computed by interpolation and integration, respectively. The interaction index was defined as $VUP_{observed}/VUP_{expected}$, which is used to assess synergism (interaction index <1) and antagonism (interaction index >1).

Hope and coworkers [91], in studying combination therapy of antifungal drugs in the treatment of invasive candidiasis, assessed pharmacodynamic interactions of amphotericin B deoxycholate and 5-fluorocytosine. The interaction model they used was based on the Greco model of drug interaction [92] represented by the equation

$$1 = \frac{D_{AmB}}{IC_{50,AmB} \cdot \left(\frac{E}{E_{con} - E} \right)^{1/m_{AmB}}} + \frac{D_{5FC}}{IC_{50,5FC} \cdot \left(\frac{E}{E_{con} - E} \right)^{1/m_{5FC}}} + \frac{\alpha \cdot D_{AmB} \cdot D_{5FC}}{IC_{50,AmB} \cdot IC_{50,5FC} \cdot \left(\frac{E}{E_{con} - E} \right)^{(1/2m_{AmB} + 1/2m_{5FC})}}, \quad (3.13)$$

where D_{AmB}, D_{5FC}, are the concentrations of amphotericin B deoxycholate and 5-fluorocytosine, respectively, that produce effect E; $IC_{50,AmB}$ is the AUC:MIC ratio for amphotericin B deoxycholate that produces 50% of the maximum effect, $IC_{50,5FC}$, is the fraction of the dosing interval that the serum concentration of 5-fluorocytosine is above the MIC that produces 50% of the maximum effect; m_{AmB}, m_{5FC}, are the respective slope parameters for the two drugs; and α is the interaction parameter.

On the right side of equation (3.13), the first two terms represent the additive effect and the third is the interaction term. Additivity, synergism, or antagonism are represented by $\alpha = 0$, $\alpha > 0$, and $\alpha < 0$, respectively.

3.3.2 Toxicodynamic Interactions between Kepone and Carbon Tetrachloride

Similar to PBPK, PBPD models are based on the physiological and other biological processes of the organism (e.g., chemical-receptor interactions), and they are used to predict the toxicological effects of chemicals. One of the earliest examples of PBPD modeling was an application to the, sometimes

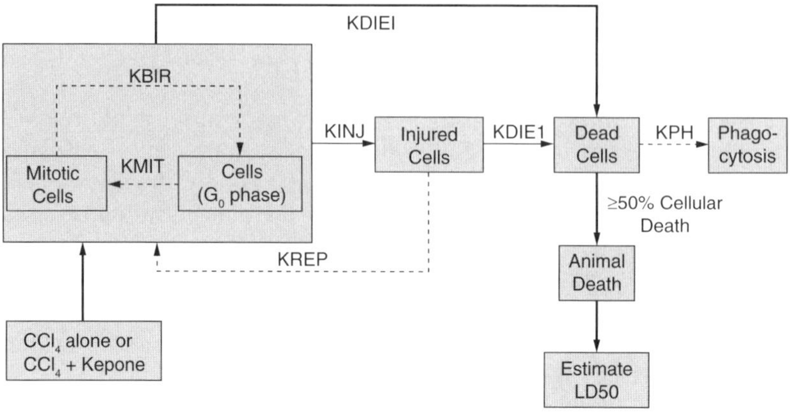

Figure 3.5 Conceptual model for CCl_4/Kepone interaction. KMIT is the rate constant for mitosis, KBIR is the rate constant for cell birth, KINJ is the rate constant for cell injury, KDIEI is the rate constant for general cell death, KDIE1 is the rate constant for cell death due to injury, and KPH is the rate constant for phagocytosis. (Adapted from [94]; reprinted with permission from Springer.)

lethal, toxicological interaction between Kepone (also known as chlordecone) and carbon tetrachloride (CCl_4). Briefly, CCl_4 is a well-known hepatotoxin. Following free radical formation through the cytochrome P450 enzyme system, the toxicity of CCl_4 is manifested through an accumulation of lipids (steatosis, fatty liver) and degenerative processes leading to liver cell death (necrosis). The toxicological interaction between Kepone and CCl_4 was elucidated to be the Kepone-mediated impairment of the liver's regeneration process. These mechanistic studies were summarized in a number of publications [93–95]. Based on the described mechanism of interaction, El-Masri et al. [96] constructed a PBPD model, a conceptual version of which is shown in Figure 3.5. A computer implementation of this model was capable of providing time-course simulations of mitotic, injured, and pyknotic (dead) cells after treatment with CCl_4 alone or with Kepone pretreatment. This implementation was further linked with Monte Carlo simulations to predict the acute lethality of CCl_4 alone and in combination with Kepone.

3.3.3 Toxicodynamic Interactions between Trichloroethylene and Dichloroethylene

A second application of PBPD modeling was the analysis of the toxicodynamic interactions between trichloroethylene (TCE) and 1,1-dichloroethylene (DCE) [97]. The interactions examined were related to the binding of these chemicals to, and depletion of, hepatic glutathione (GSH) in relation to the intrinsic hepatic GSH synthesis, a protective mechanism toward DCE toxicity. PBPK models for interactions leading to depletion of hepatic glutathione had been developed previously by several investigators [98,99]. Based, in part, on these results, a PBPD model was used to identify a critical time point at which hepatic GSH is at a minimum in response to both chemicals. Model-directed gas uptake experiments with DCE revealed that DCE was the only chemical capable of significantly depleting hepatic GSH. El-Masri and coworkers [97] extended these quantitative analyses to establish an "interaction threshold" between TCE and DCE.

3.3.4 Pharmacodynamic Modeling of the Clonal Growth of Initiated Liver Cells

A further example of the application of biologically based pharmacodynamic modeling is the simulation of clonal growth of initiated liver cells during carcinogenesis. The impetus of this research came from the desire to find a way to evaluate the carcinogenic potential of chemicals or chemical mixtures without utilizing resource-intensive chronic cancer bioassays [100–102]. Although most of the published work in the clonal growth modeling arena was focused on single chemicals [47,103,104], this approach was used for a chemical mixture study involving hexachlorobenzene and PCB-126 [23]. The experimental animal model used for this research development involved a modification of the medium-term initiation-promotion bioassay of Ito and coworkers [105,106]; this way to provide multiple time point data needed for pharmacodynamic analyses [23,104].

The pharmacodynamic model used was based on the two-stage model of carcinogenesis developed by Moolgavkar-Venzon-Knudson (MVK) [107,108]. In this model, a susceptible cell is at one of three states: normal, initiated, and malignant (Figure 3.6). The model allows the incorporation of relevant biological information such as the kinetics of tissue growth and differentiation and mutation rates. The growth of normal liver is described deterministically, whereas other cellular events use stochastic simulation. This approach facilitates description of complex biological process with time-dependent values.

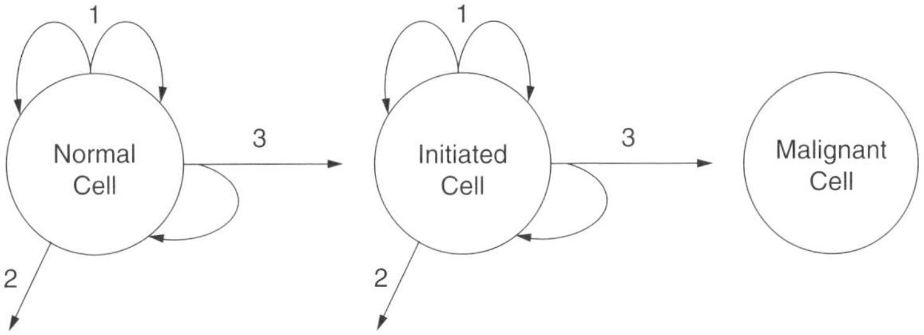

Figure 3.6 Conceptual two-stage carcinogenesis model. A normal or initiated cell is subjected to division into two susceptible cells without mutation (*event 1*), death/differentiation (*event 2*), division into two susceptible cells with one mutated (*event 3*), or no change.

BRN:
Predicts everything the chemical could turn into.

ıtical terms, the growth kinet-
rth rate, α, and death rate, β:

(3.14)

tocytes in the tissue. Initiated
rmal hepatocytes from dieth-
stochastic process that follows
initiated cells in a small time
ɔility and cell birth rate:

(3.15)

lls during a time step Δt, N is
the mutation probability per
genetic or epigenetic change
placental form) expression in

tions when applied to chemi-
and endogenous) chemical–
ɔach to address this issue is
ʒ and integrated models that
combine PBPK and BRN models. These approaches are described below.

3.4.1 Biochemical Reaction Network (BRN) Modeling

Biochemical reaction network modeling is an in silico approach being developed by Reisfeld, Mayeno, and Yang [102,109–113] that attempts to predict the biotransformation of one or a mixture of chemicals; more specifically, it predicts the chemical structures of all metabolites, the biotransformation pathways, and the interconnections among these pathways through metabolites in common. As such, it is a tool for predictive xenobiotic metabolomics. This modeling approach is an outgrowth of biochemistry and reaction network modeling that was historically used in chemical and petroleum engineering [114,115] and more recently in a limited context for biological applications [113,116]. BRN modeling is performed using a simulation tool called *BioTRaNS* (*B*iochemical *T*ool for *R*eaction *N*etwork *S*imulation). Once validated with experimental data, this methodology can predict and simulate in a quantitative and time-dependent manner the formation and disappearance of all metabolites, including potentially toxic reactive intermediates, such as been developed previously by several investigators [98,99]. Based, in part, on these results, a PBPD model was used to identify a critical time point at which hepatic GSH is at a minimum in response to both chemicals. Model-directed gas uptake experiments with DCE revealed that DCE was the only chemical capable of significantly depleting hepatic GSH. El-Masri and coworkers [97] extended these quantitative analyses to establish an "interaction threshold" between TCE and DCE.

3.3.4 Pharmacodynamic Modeling of the Clonal Growth of Initiated Liver Cells

A further example of the application of biologically based pharmacodynamic modeling is the simulation of clonal growth of initiated liver cells during carcinogenesis. The impetus of this research came from the desire to find a way to evaluate the carcinogenic potential of chemicals or chemical mixtures without utilizing resource-intensive chronic cancer bioassays [100–102]. Although most of the published work in the clonal growth modeling arena was focused on single chemicals [47,103,104], this approach was used for a chemical mixture study involving hexachlorobenzene and PCB-126 [23]. The experimental animal model used for this research development involved a modification of the medium-term initiation-promotion bioassay of Ito and coworkers [105,106]; this way to provide multiple time point data needed for pharmacodynamic analyses [23,104].

The pharmacodynamic model used was based on the two-stage model of carcinogenesis developed by Moolgavkar-Venzon-Knudson (MVK) [107,108]. In this model, a susceptible cell is at one of three states: normal, initiated, and malignant (Figure 3.6). The model allows the incorporation of relevant biological information such as the kinetics of tissue growth and differentiation and mutation rates. The growth of normal liver is described deterministically, whereas other cellular events use stochastic simulation. This approach facilitates description of complex biological process with time-dependent values.

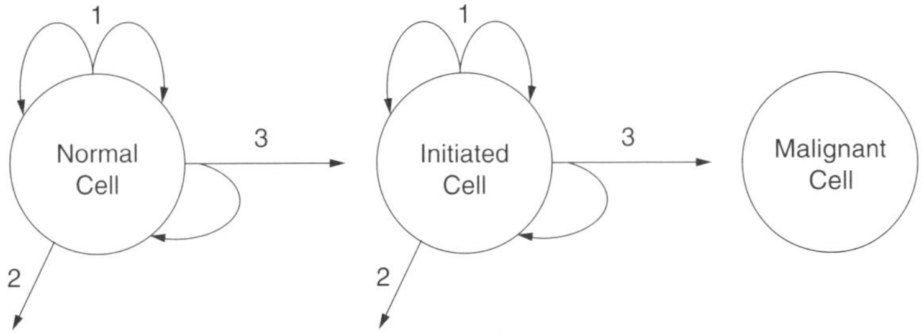

Figure 3.6 Conceptual two-stage carcinogenesis model. A normal or initiated cell is subjected to division into two susceptible cells without mutation (*event 1*), death/differentiation (*event 2*), division into two susceptible cells with one mutated (*event 3*), or no change.

To express the clonal growth model in mathematical terms, the growth kinetics of normal hepatocytes depend on the cell birth rate, α, and death rate, β:

$$\frac{dN}{dt} = N \cdot (\alpha - \beta), \tag{3.14}$$

where N is the number density of normal hepatocytes in the tissue. Initiated cells are produced through mutations in the normal hepatocytes from diethylnitrosamine treatment. This is assumed to be a stochastic process that follows a Poisson distribution. The expected number of initiated cells in a small time step is defined by a function of mutation probability and cell birth rate:

$$N_m = N \cdot \mu \cdot \alpha \cdot \Delta t. \tag{3.15}$$

Here N_m is the expected number of initiated cells during a time step Δt, N is the normal hepatocyte number density, and μ is the mutation probability per cell division. In this case mutation refers to any genetic or epigenetic change that leads to GST-P (glutathione *S*-transferase placental form) expression in the liver cells.

3.4 FUTURE DIRECTIONS

As discussed earlier, PBPK modeling has limitations when applied to chemical mixtures in which significant (xenobiotics and endogenous) chemical–chemical interactions are expected. An approach to address this issue is biochemical reaction network (BRN) modeling and integrated models that combine PBPK and BRN models. These approaches are described below.

3.4.1 Biochemical Reaction Network (BRN) Modeling

Biochemical reaction network modeling is an in silico approach being developed by Reisfeld, Mayeno, and Yang [102,109–113] that attempts to predict the biotransformation of one or a mixture of chemicals; more specifically, it predicts the chemical structures of all metabolites, the biotransformation pathways, and the interconnections among these pathways through metabolites in common. As such, it is a tool for predictive xenobiotic metabolomics. This modeling approach is an outgrowth of biochemistry and reaction network modeling that was historically used in chemical and petroleum engineering [114,115] and more recently in a limited context for biological applications [113,116]. BRN modeling is performed using a simulation tool called *BioTRaNS* (*B*iochemical *T*ool for *R*eaction *N*etwork *S*imulation). Once validated with experimental data, this methodology can predict and simulate in a quantitative and time-dependent manner the formation and disappearance of all metabolites, including potentially toxic reactive intermediates, such as

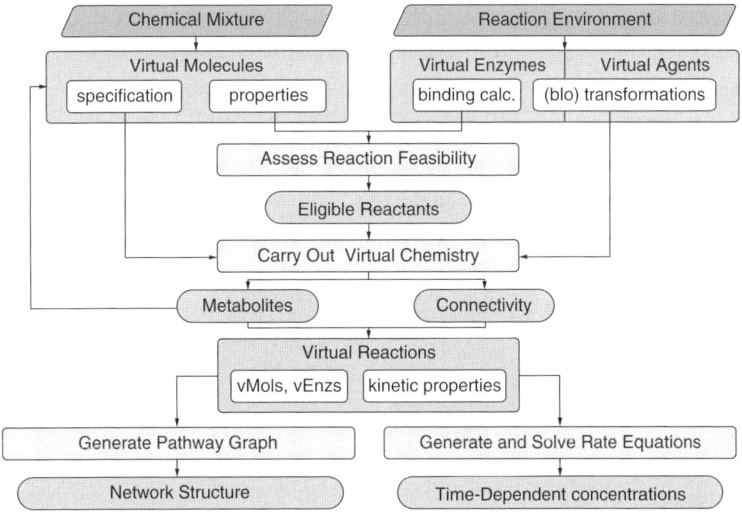

Figure 3.7 Information flow through the BioTRaNS. (Reprinted from [111] with permission from ACS.)

epoxides. This tool differs from most other computational metabolism methods [117–120] in that it is applicable for the prediction of metabolites of mixtures of chemicals, interlinks the pathways through shared metabolites, and is designed to be coupled with PBPK modeling approaches. Thus it can aid in predicting drug metabolism, toxicity, and understanding the modes of action of individual chemicals as well as chemical mixtures.

Modeling Framework Structure The conceptual flow of information through *BioTRaNS* is shown in Figure 3.7. A chemical mixture (concentration of each constituent in the mixture) and reaction environment (types and amounts of enzymes and other *agents*) are specified by the user. The chemical mixture is converted to a set of *virtual molecules* (vMols), each with its own specifications, namely a canonical simplified molecular line entry system (SMILES) format [121] and list of computed geometric, energetic, and physicochemical properties. In general, properties are retrieved from associated databases when available, or they can be computed as needed.

The reaction environment is translated into a set of *virtual enzymes* (vEnzs), representing the actual enzymes and *virtual agents* (vAgnts), representing nonenzymatic reactions. Each vEnz is endowed with (1) an appropriate binding calculator and (2) *transforms* governing the biotransformations that the vEnz can mediate.

The binding calculator computes the feasibility of a particular vMol binding to a vEnz, based on quantitative structure-activity relationships (or decision tree) derived from properties of known substrates for each enzyme. Transformations are stored as a list of SMIRKS (*SMI*LES *R*eaction *S*pecification) based representations (also called *transforms*) [122]. A unique feature of the methodology is the implementation of transforms that describe steps of proposed reaction mechanisms, including those for enzyme-mediated reactions. The use of mechanistic steps is key to automatically and accurately predicting metabolites (*vide infra* for an example). After the vMol is created, the binding calculator for each vEnz assesses binding feasibility; if feasible, the vMol becomes an eligible reactant. All eligible reactants then undergo appropriate virtual biotransformations, creating specific chemical reactions and converting the vMol into one or more metabolites, which also are checked for reaction feasibility. The information on the metabolites and the associated transform/agent interconnecting each substrate–product pair is converted to a *Virtual Reaction* (vRxn). Thus each vRxn comprises the vMols and vEnzs from which it was derived, as well as appropriate kinetic properties.

Based on the aggregate information in the vRxns, a reaction network is generated in the form of a mathematical graph [123], wherein the metabolites constitute the vertices or nodes, and the reactions form the edges. These network graphs can be depicted at different levels of detail, from highly detailed, showing steps of a reaction mechanism, to overview, showing only key metabolites.

Simultaneous to the graph creation, kinetic properties in each vRxn are used to create the appropriate reaction rate equations (ordinary differential equations, ODE). These properties include rate constants (e.g., Michaelis constant, K_m, and maximum velocity, V_{max}, for enzyme-catalyzed reactions, and k for nonenzymatic reactions), inhibitor constants, K_i; and modes of inhibition or allosterism. The total set of rate equations and specified initial conditions forms an initial value problem that is solved by a stiff ODE equation solver for the concentrations of all species as a function of time. The constituent transforms for the each virtual enzyme are compiled by carefully culling the literature for data on enzymes known to act on the chemicals and chemical metabolites of interest.

Application of BRNM *BioTRaNS* has been used to simulate the biotransformation reaction networks for selected individual and mixtures of volatile organic compounds (VOCs) (Figures 3.8 and 3.9) [111]. Figure 3.8*a* shows the *BioTRaNS*-generated biotransformation pathway of trichloroethylene (TCE) in the form of a graph. The interlinked biotransformation network of a mixture of four VOCs (TCE, tetrachloroethylene [Perc], methylchloroform [MC], and chloroform [$CHCl_3$]) is shown in Figure 3.8*b*; both graphs were generated using only selected vEnzs and vAgnts, giving a "high-level" or "overview" graph. Figure 3.9*a* shows the *BioTRaNS*-generated stepwise oxidation of TCE by CYP2E1, based on transforms representing each step of the proposed reaction mechanism [124]. The first step involves formation of an intermediate between the high-valent iron-oxo complex, $(FeO)^{3+}$, of the CYP heme [125] and the alkene, forming a carbocationic intermediate (Fe^{III}–O–C–C^+). Subsequent steps were performed by invoking other transforms. Finally, a more

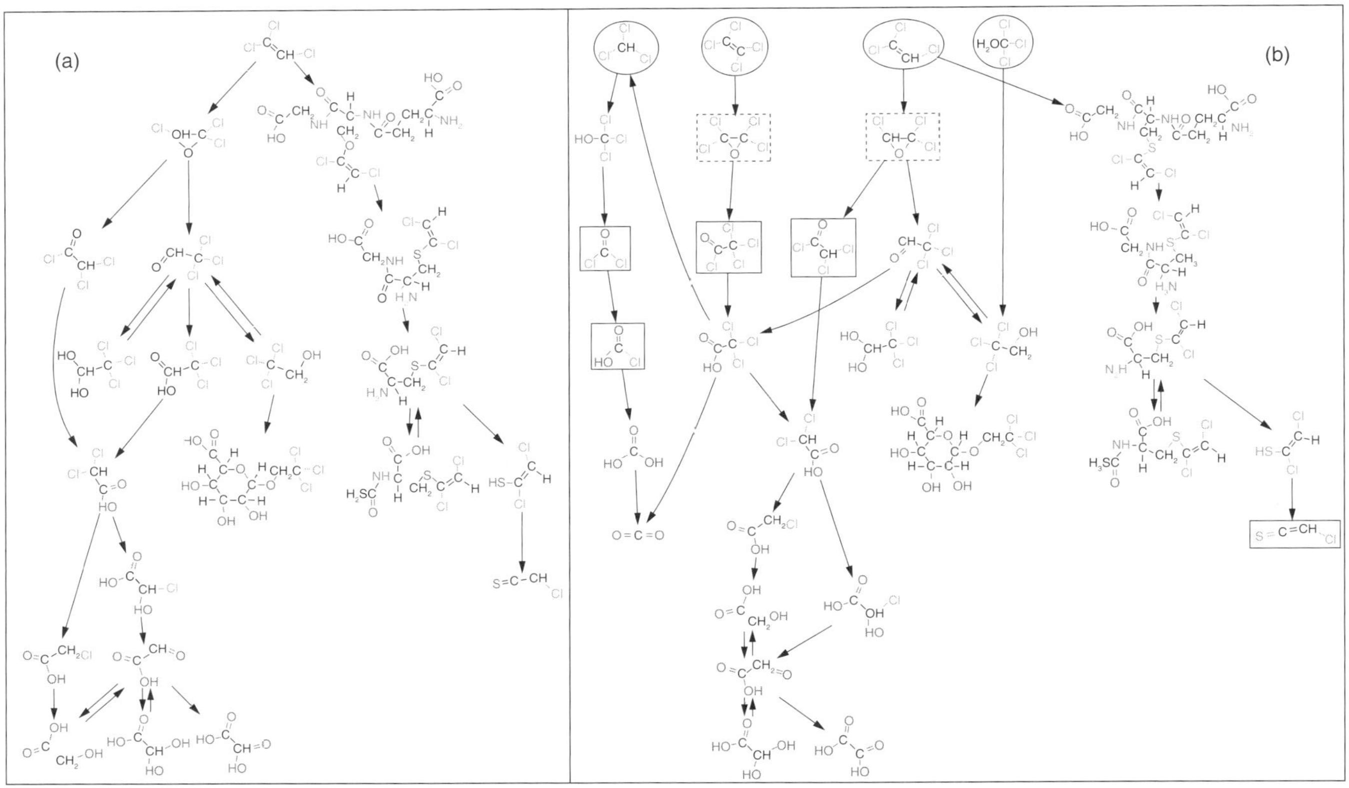

Figure 3.8 BioTRaNS-generated biotransformation pathways for (*a*) TCE and (*b*) a mixture of TCE, Perc, MC, and $CHCl_3$. For clarity, only selected agents (vEnzs and vAgnts) were used to generate these graphs. For (*b*), reactive metabolites are highlighted as follow: epoxides (brown, box, dashed); acid chlorides (orange, box, solid), thioketene (turquoise, box, solid); and starting chemicals (blue, ellipse, solid). (Reprinted from [111] with permission from ACS.) See color plates.

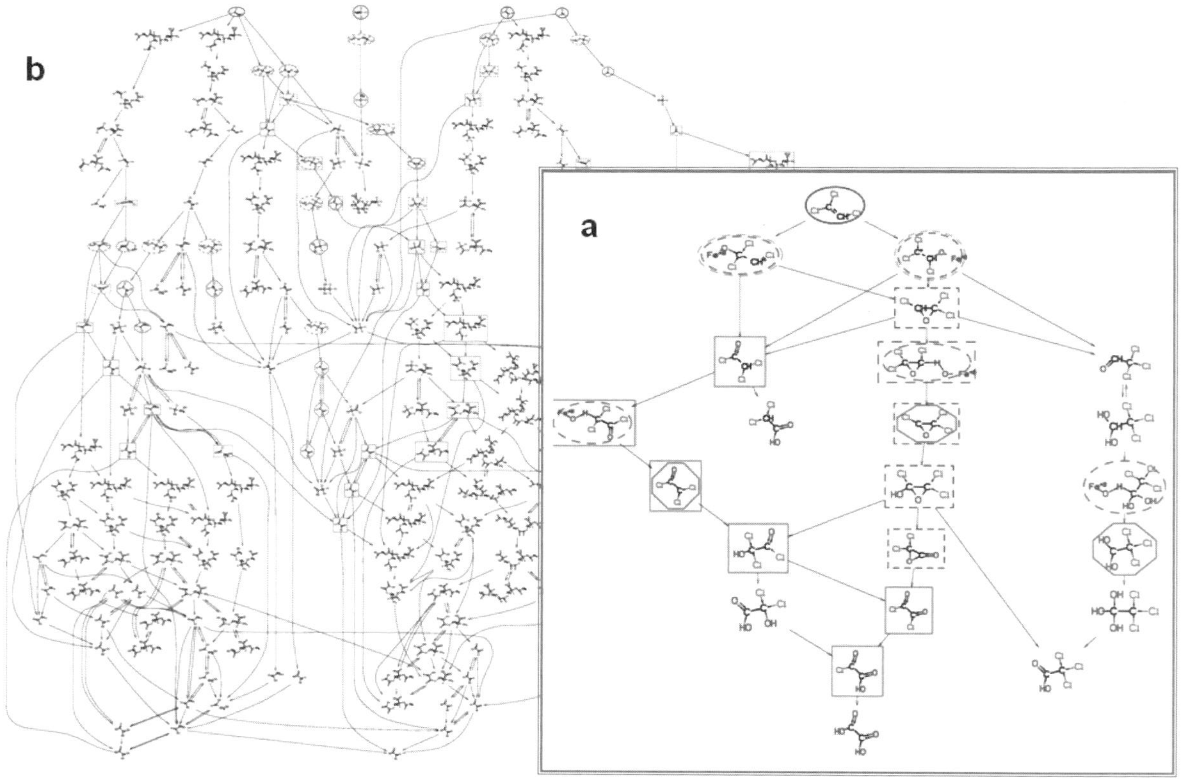

Figure 3.9 BioTRaNS-generated mechanism-based pathways for (*a*) the oxidation of TCE by CYP and (*b*) more comprehensive biotransformation pathways of TCE, Perc, MC, and $CHCl_3$. Chemicals are highlighted as follows: starting chemicals (blue, ellipse, solid); epoxides (brown, box, dashed); acid chlorides (orange, box, solid); radicals (red, octagon, solid); Fe-O-substrate complexes (magenta, ellipse, dashed), carbocations (salmon, ellipse, dashed). (Reprinted from [111] with permission from ACS.) See color plates.

complete depiction of the interconnected biotransformation pathways of all four VOCs, generated by *BioTRaNS*, is shown in Figure 3.9*b*. The complexity of the pathways is clearly illustrated despite the use of a noncomprehensive list of agents. Additional discussion on the figures and results is presented in the paper by Mayeno et al. [111].

With *BioTRaNS* the highlighting of various enclosing shapes and colors provides immediate visual feedback on species of interest, even in highly complex networks, and an investigator can easily locate highly reactive, and potentially toxic, species.

3.4.2 Integrated PBPK/BRN Modeling

PBPK modeling simulates ADME at the organ/tissue/organism level while BRN modeling focuses on biotransformations at the molecule level. Thus PBPK and BRN modeling each address different levels and facets within a biological system, and integrating them will provide a tool useful for studying various aspects of predictive toxicology. Specifically, while PBPK modeling provides time-dependent tissue concentrations of the parent chemicals, BRN modeling predicts metabolites within these tissues, with the number and types of metabolites dependent on the transforms employed. If metabolite-specific PBPK parameters are available in the database or can be estimated for the metabolites, these can be used to simulate the time-course dosimetry of predicted metabolites in tissues. Such multi-scale modeling reflects a "systems biology" approach, which attempts to study biology as whole "systems" [126]. As a first step in the development and validation of an integrated BRN/PBPK model, benzo[*a*]pyrene (BaP) was studied [113,127]. The results of these studies, which simulated rather well the metabolism of BaP and several BaP metabolites, demonstrate the potential of linking the predictive xenobiotic metabolomics capability of BRN and the ADME simulation capability of PBPK modeling approaches to yield a more comprehensive and predictive systems biology approach, as compared to each individual approach.

REFERENCES

1. van de Waterbeemd H, Gifford E. ADMET in silico modelling: Towards prediction paradise? *Nat Rev Drug Discov* 2003;2:192–204.
2. Reddy MB, Yang RSH, Clewell HJ, 3rd, Andersen ME. *Physiologically based pharmacokinetics modeling: Science and applications*. Hoboken, NJ: John Wiley & Sons, 2005.
3. Keys DA, Bruckner JV, Muralidhara S, Fisher JW. Tissue dosimetry expansion and cross-validation of rat and mouse physiologically based pharmacokinetic models for trichloroethylene. *Toxicol Sci* 2003;76:35–50.
4. Krishnan K, Andersen M. Physiologically based pharmacokinetic modeling in toxicology. In: Hayes AW, editor, *Principles and methods of toxicology*. New York: Raven Press, 1994. p. 149–88.

5. Brown RP, Delp MD, Lindstedt SL, Rhomberg LR, Beliles RP. Physiological parameter values for physiologically based pharmacokinetic models. *Toxicol Ind Health* 1997;13:407–84.
6. Davies B, Morris T. Physiological parameters in laboratory animals and humans. *Pharm Res* 1993;10:1093–5.
7. Price PS, Conolly RB, Chaisson CF, Gross EA, Young JS, Mathis ET, et al. Modeling interindividual variation in physiological factors used in PBPK models of humans. *Crit Rev Toxicol* 2003;33:469–503.
8. Lindstedt SL. Allometry: Body size constraints in animal design. In: *Pharmacokinetics in Risk Assessment: Drinking Water and Health*, Vol. 8. Washington, DC: National Academy Press, 1987; p. 65–79.
9. Sato A, Nakajima T. Partition coefficients of some aromatic hydrocarbons and ketones in water, blood and oil. *Br J Ind Med* 1979;36:231–4.
10. Gargas ML, Burgess RJ, Voisard DE, Cason GH, Andersen ME. Partition coefficients of low-molecular-weight volatile chemicals in various liquids and tissues. *Toxicol Appl Pharmacol* 1989;98:87–99.
11. Hansch C, Leo A, Hoekman DH. *Exploring QSAR*. Washington, DC: American Chemical Society, 1995.
12. Kim C, Manning RO, Brown RP, Bruckner JV. Use of the vial equilibration technique for determination of metabolic rate constants for dichloromethane. *Toxicol Appl Pharmacol* 1996;139:243–51.
13. Kedderis GL, Carfagna MA, Held SD, Batra R, Murphy JE, Gargas ML. Kinetic analysis of furan biotransformation by F-344 rats in vivo and in vitro. *Toxicol Appl Pharmacol* 1993;123:274–82.
14. Gargas ML, Andersen ME, Clewell HJ, 3rd. A physiologically based simulation approach for determining metabolic constants from gas uptake data. *Toxicol Appl Pharmacol* 1986;86:341–52.
15. Filser JG, Bolt HM. Pharmacokinetics of halogenated ethylenes in rats. *Arch Toxicol* 1979;42:123–36.
16. USEPA. Approaches for the application of physiologically based pharmacokinetic (PBPK) models and supporting data in risk assessment. Report number: EPA/600/R-05/043A. Washington, DC: US Environmental Protection Agency, 2005.
17. Fisher JW, Mahle D, Abbas R. A human physiologically based pharmacokinetic model for trichloroethylene and its metabolites, trichloroacetic acid and free trichloroethanol. *Toxicol Appl Pharmacol* 1998;152:339–59.
18. Portier CJ, Kaplan NL. Variability of safe dose estimates when using complicated models of the carcinogenic process. *Fundamentals and Applied Toxicology* 1989;13:533–44.
19. Licata AC, Dekant W, Smith CE, Borghoff SJ. A physiologically based pharmacokinetic model for methyl tert-butyl ether in humans: implementing sensitivity and variability analyses. *Toxicol Sci* 2001;62:191–204.
20. Reitz RH, McDougal JN, Himmelstein MW, Nolan RJ, Schumann AM. Physiologically based pharmacokinetic modeling with methylchloroform: implications for interspecies, high dose/low dose, and dose route extrapolations. *Toxicol Appl Pharmacol* 1988;95:185–99.
21. Clewell HJ 3rd, Gentry PR, Gearhart JM, Covington TR, Banton MI, Andersen ME. Development of a physiologically based pharmacokinetic model of isopropanol and its metabolite acetone. *Toxicol Sci* 2001;63:160–72.

22. Wang X, Santostefano MJ, DeVito MJ, Birnbaum LS. Extrapolation of a PBPK model for dioxins across dosage regimen, gender, strain, and species. *Toxicol Sci* 2000;56:49–60.
23. Lu Y, Lohitnavy M, Reddy MB, Lohitnavy O, Ashley A, Yang RSH. An updated physiologically based pharmacokinetic model for hexachlorobenzene: Incorporation of pathophysiological states following partial hepatectomy and hexachlorobenzene treatment. *Toxicol Sci* 2006;91:29–41.
24. Sarangapani R, Teeguarden J, Andersen ME, Reitz RH, Plotzke KP. Route-specific differences in distribution characteristics of octamethylcyclotetrasiloxane in rats: analysis using PBPK models. *Toxicol Sci* 2003;71:41–52.
25. Pauluhn J. Issues of dosimetry in inhalation toxicity. *Toxicol Lett* 2003; 140–1:229–38.
26. Clewell HJ, 3rd, Andersen ME. Physiologically-based pharmacokinetic modeling and bioactivation of xenobiotics. *Toxicol Ind Health* 1994;10:1–24.
27. Adolph EF. Quantitative relations in the physiological constitutions of mammals. *Science* 1949;109.
28. Dedrick R, Bischoff KB, Zaharko DS. Interspecies correlation of plasma concentration history of methotrexate (NSC-740). *Cancer Chemother Rep* 1970; 54:95–101.
29. Dedrick RL. Animal scale-up. *J Pharmacokinet Biopharm* 1973;1:435–61.
30. Pelekis M, Krishnan K. Assessing the relevance of rodent data on chemical interactions for health risk assessment purposes: A case study with dichloromethane-toluene mixture. *Regul Toxicol Pharmacol* 1997;25:79–86.
31. Beliveau M, Lipscomb J, Tardif R, Krishnan K. Quantitative structure-property relationships for interspecies extrapolation of the inhalation pharmacokinetics of organic chemicals. *Chem Res Toxicol* 2005;18:475–85.
32. Haddad S, Charest-Tardif G, Tardif R, Krishnan K. Validation of a physiological modeling framework for simulating the toxicokinetics of chemicals in mixtures. *Toxicol Appl Pharmacol* 2000;167:199–209.
33. Desiraju GR, Gopalakrishnan B, Jetti RKR, Nagaraju A, Raveendra D, Sarma JARP, et al. Computer-aided design of selective COX-2 inhibitors: Comparative molecular field analysis, comparative molecular similarity indices analyses, and docking studies of some 1,2-diarylimidazole derivatives. *J Med Chem* 2002;45:4847–57.
34. Haddad S, Tardif R, Charest-Tardif G, Krishnan K. Physiological modeling of the toxicokinetic interactions in a quaternary mixture of aromatic hydrocarbons. *Toxicol Appl Pharmacol* 1999;161:249–57.
35. Dobrev ID, Andersen ME, Yang RSH. Assessing interaction thresholds for trichloroethylene in combination with tetrachloroethylene and 1,1,1-trichloroethane using gas uptake studies and PBPK modeling. *Arch Toxicol* 2001;75:134–44.
36. Dobrev ID, Andersen ME, Yang RSH. In silico toxicology: Simulating interaction thresholds for human exposure to mixtures of trichloroethylene, tetrachloroethylene, and 1,1,1-trichloroethane. *Environ Health Perspect* 2002;110:1031–9.
37. Dennison JE. Physiologically-based pharmacokinetic modeling of simple and complex mixtures of gasoline and the gasoline components *n*-hexane, benzene, toluene, ethylbenzene, and xylene. Ph.D. dissertation. 2004, Colorado State University.

38. Dennison JE, Andersen ME, Yang RSH. Characterization of the pharmacokinetics of gasoline using PBPK modeling with a complex mixtures chemical lumping approach. *Inhal Toxicol* 2003;15:961–86.

39. Reddy MB, Yang RSH, Clewell HJ 3rd, Andersen ME. *Physiologically based pharmacokinetic modeling: Science and applications*. Hoboken, NJ: John Wiley & Sons, 2005.

40. Clewell HJ, Teeguarden J, McDonald T, Sarangapani R, Lawrence G, Covington T, et al. Review and evaluation of the potential impact of age- and gender-specific pharmacokinetic differences on tissue dosimetry. *Crit Rev Toxicol* 2002;32:329–89.

41. Ginsberg G, Hattis D, Sonawane B. Incorporating pharmacokinetic differences between children and adults in assessing children's risks to environmental toxicants. *Toxicol Appl Pharmacol* 2004;198:164–83.

42. Lipscomb JC, Teuschler LK, Swartout J, Popken D, Cox T, Kedderis GL. The impact of cytochrome P450 2E1-dependent metabolic variance on a risk-relevant pharmacokinetic outcome in humans. *Risk Anal* 2003;23:1221–38.

43. Banks HT, Potter LK. Probabilistic methods for addressing uncertainty and variability in biological models: Application to a toxicokinetic model. *Math Biosci* 2004;192:193–225.

44. Nestorov I. Modelling and simulation of variability and uncertainty in toxicokinetics and pharmacokinetics. *Toxicol Lett* 2001;120:411–20.

45. Andersen ME, Clewell HJ, 3rd, Gargas ML, Smith FA, Reitz RH. Physiologically based pharmacokinetics and the risk assessment process for methylene chloride. *Toxicol Appl Pharmacol* 1987;87:185–205.

46. Gentry PR, Hack CE, Haber L, Maier A, Clewell HJ, 3rd. An approach for the quantitative consideration of genetic polymorphism data in chemical risk assessment: Examples with warfarin and parathion. *Toxicol Sci* 2002;70:120–39.

47. Portier CJ, Kopp-Schneider A, Sherman CD. Calculating tumor incidence rates in stochastic models of carcinogenesis. *Math Biosci* 1996;135:129–46.

48. Thomas RS, Lytle WE, Keefe TJ, Constan AA, Yang RSH. Incorporating Monte Carlo simulation into physiologically based pharmacokinetic models using advanced continuous simulation language (ACSL): A computational method. *Fundam Appl Toxicol* 1996;31:19–28.

49. Bernillon P, Bois FY. Statistical issues in toxicokinetic modeling: A Bayesian perspective. *Environ Health Perspect* 2000;108(Suppl 5):883–93.

50. Gelman A, Bois FY, Jiang J. Physiological pharmacokinetic analysis using population modeling and informative prior distributions. *J. Am. Stat. Assoc.* 1996;91: 1400–12.

51. Jonsson F. Development of Bayesian population models. In: Marklund S, editor. *Physiologically Based Pharmacokinetic Modeling in Risk Assessment*.

52. Nestorov I, Gueorguieva I, Jones HM, Houston B, Rowland M. Incorporating measures of variability and uncertainty into the prediction of in vivo hepatic clearance from in vitro data. Drug Metab Dispos, 2002;30:276–82.

53. Jonsson F. Physiologically based pharmacokinetic modeling in risk assessment. In: *Development of Bayesian population methods*. Stockholm: Uppsala University, 2001.

54. Gelman A, Rubin DB. Markov chain Monte Carlo methods in biostatistics. *Stat Meth Med Res* 1996;5:339–55.

55. Bois FY, Maszle DR. MCSim: A Monte Carlo simulation program. *J Stat Software* 1997;2.

56. Gelman A, Bois FY, Jiang J. Physiological pharmacokinetic analysis using population modeling and informative prior distributions. *J Am Stat Assoc* 1996;91:1400–12.

57. Marino DJ, Clewell HJ, Gentry PR, Covington TR, Hack CE, David RM, et al. Revised assessment of cancer risk to dichloromethane: Part I. Bayesian PBPK and dose–response modeling in mice. *Regul Toxicol Pharmacol* 2006;45:44–54.

58. Clewell HJ, Gentry PR, Covington TR, Sarangapani R, Teeguarden JG. Evaluation of the potential impact of age- and gender-specific pharmacokinetic differences on tissue dosimetry. *Toxicol Sci* 2004;79:381–93.

59. Thomas RS, Rank DR, Penn SG, Zastrow GM, Hayes KR, Pande K, et al. Identification of toxicologically predictive gene sets using cDNA microarrays. *Mol Pharmacol* 2001;60:1189–94.

60. Andersen ME, Krishnan K. Physiologically based pharmacokinetics and cancer risk assessment. *Environ Health Perspect* 1994;102(Suppl 1):103–8.

61. Bogen KT, Gold LS. Trichloroethylene cancer risk: Simplified calculation of PBPK-based MCLs for cytotoxic end points. *Regul Toxicol Pharmacol* 1997;25:26–42.

62. Marino DJ, Clewell HJ, Gentry PR, Covington TR, Hack CE, David RM, et al. Revised assessment of cancer risk to dichloromethane: Part I. Bayesian PBPK and dose–response modeling in mice. *Regul Toxicol Pharmacol* 2006;45:44–54.

63. Maruyama W, Aoki Y. Estimated cancer risk of dioxins to humans using a bioassay and physiologically based pharmacokinetic model. *Toxicol Appl Pharmacol* 2006;214:188–98.

64. Sielken RL, Jr., Reitz RH, Hays SM. Using PBPK modeling and comprehensive realism methodology for the quantitative cancer risk assessment of butadiene. *Toxicology* 1996;113:231–7.

65. Krishnan K, Johanson G. Physiologically-based pharmacokinetic and toxicokinetic models in cancer risk assessment. *J Environ Sci Health C Environ Carcinog Ecotoxicol Rev* 2005;23:31–53.

66. David RM, Clewell HJ, Gentry PR, Covington TR, Morgott DA, Marino DJ. Revised assessment of cancer risk to dichloromethane: II. Application of probabilistic methods to cancer risk determinations. *Regul Toxicol Pharmacol* 2006;45:55–65.

67. Perbellini L, Mozzo P, Olivato D, Brugnone F. "Dynamic" biological exposure indexes for n-hexane and 2,5-hexanedione, suggested by a physiologically based pharmacokinetic model. *Am Ind Hyg Assoc J* 1990;51:356–62.

68. Jonsson F, Johanson G. The Bayesian population approach to physiological toxicokinetic models and example using the mcsim software. *Tox. Letters* 2003;138: 143–50.

69. Droz PO, Berode M, Jang JY. Biological monitoring of tetrahydrofuran: Contribution of a physiologically based pharmacokinetic model. *Am Ind Hyg Assoc J* 1999;60:243–8.

70. Roy A, Weisel CP, Lioy PJ, Georgopoulos PG. A distributed parameter physiologically–based pharmacokinetic model for dermal and inhalation exposure to volatile organic compounds. *Risk Anal* 1996;16:147–60.

71. Rao HV, Ginsberg GL. A physiologically-based pharmacokinetic model assessment of methyl *t*-butyl ether in groundwater for a bathing and showering determination. *Risk Anal* 1997;17:583–98.

72. Levesque B, Ayotte P, Tardif R, Ferron L, Gingras S, Schlouch E, et al. Cancer risk associated with household exposure to chloroform. *J Toxicol Environ Health A* 2002;65:489–502.

73. Yang RSH, el-Masri HA, Thomas RS, Constan AA. The use of physiologically-based pharmacokinetic/pharmacodynamic dosimetry models for chemical mixtures. *Toxicol Lett* 1995;82–83:497–504.

74. de Zwart LL, Haenen HE, Versantvoort CH, Wolterink G, van Engelen JG, Sips AJ. Role of biokinetics in risk assessment of drugs and chemicals in children. *Regul Toxicol Pharmacol* 2004;39:282–309.

75. Nestorov I. Whole body pharmacokinetic models. *Clin Pharmacokinetics* 2003;42:883–908.

76. Brown RP, Delp MD, Lindstedt SL, Rhomberg LR, Beliles RP. Physiological parameter values for physiologically based pharmacokinetic models. *Toxicol. Ind. Health* 1997;13:407–84.

77. Dixit R, Riviere J, Krishnan K, Andersen ME. Toxicokinetics and physiologically based toxicokinetics in toxicology and risk assessment. *J Toxicol Environ Health B—Crit Rev* 2003;6:1–40.

78. Agency for Toxic Substances and Disease Registry. *Toxicological profile for n-hexane.* Atlanta, GA: U.S. Department of Health and Human Services, 1999.

79. Dourson ML, Felter SP, Robinson D. Evolution of science-based uncertainty factors in noncancer risk assessment. *Regul Toxicol Pharmacol* 1996;24:108–20.

80. Jonkers RE, Koopmans RP, Portier EJ, van Boxtel CJ. Debrisoquine phenotype and the pharmacokinetics and beta-2 receptor pharmacodynamics of metoprolol and its enantiomers. *J Pharmacol Exp Ther* 1991;256:959–66.

81. Holford NH, Sheiner LB. Pharmacokinetic and pharmacodynamic modeling in vivo. *Crit Rev Bioeng* 1981;5:273–322.

82. Mandema JW, Kuck MT, Danhof M. In vivo modeling of the pharmacodynamic interaction between benzodiazepines which differ in intrinsic efficacy. *J Pharmacol Exp Ther* 1992;261:56–61.

83. Mandema JW, Heijligers-Feijen CD, Tukker E, De Boer AG, Danhof M. Modeling of the effect site equilibration kinetics and pharmacodynamics of racemic baclofen and its enantiomers using quantitative EEG effect measures. *J Pharmacol Exp Ther* 1992;261:88–95.

84. Edmond C, Michalek JE, Birnbaum LS, DeVito MJ. Comparison of the use of a physiologically based pharmacokinetic model and a classical pharmacokinetic

model for dioxin exposure assessments. *Environ Health Perspect* 2005;113:1666–8.

85. Strenkoski-Nix LC, Forrest A, Schentag JJ, Nix DE. Pharmacodynamic interactions of ciprofloxacin, piperacillin, and piperacillin/tazobactam in healthy volunteers. *J Clin Pharmacol* 1998;38:1063–71.
86. Jonker DM, Vermeij DA, Edelbroek PM, Voskuyl RA, Piotrovsky VK, Danhof M. Pharmacodynamic analysis of the interaction between tiagabine and midazolam with an allosteric model that incorporates signal transduction. *Epilepsia* 2003;44:329–38.
87. Huang XH, Qiu FR, Xie HT, Li J. Pharmacokinetic and pharmacodynamic interaction between irbesartan and hydrochlorothiazide in renal hypertensive dogs. *J Cardiovasc Pharmacol* 2005;46:863–9.
88. Levasseur LM, Delon A, Greco WR, Faury P, Bouquet S, Couet W. Development of a new quantitative approach for the isobolographic assessment of the convulsant interaction between pefloxacin and theophylline in rats. *Pharm Res* 1998;15:1069–76.
89. Brochot C, Marchand S, Couet W, Gelman A, Bois FY. Extension of the isobolographic approach to interactions studies between more than two drugs: Illustration with the convulsant interaction between pefloxacin, norfloxacin, and theophylline in rats. *J Pharm Sci* 2004;93:553–62.
90. Tam VH, Schilling AN, Lewis RE, Melnick DA, Boucher AN. Novel approach to characterization of combined pharmacodynamic effects of antimicrobial agents. *Antimicrob. Agents Chemother.* 2004;48:4315–21.
91. Hope WW, Warn PA, Sharp A, Reed P, Keevil B, Louie A, et al. Surface response modeling to examine the combination of amphotericin B deoxycholate and 5-fluorocytosine for treatment of invasive candidiasis. *J Infect Dis* 2005; 192:673–80.
92. Greco WR, Bravo G, Parsons JC. The search for synergy: A critical review from a response surface perspective. *Pharmacol Rev* 1995;47:331–85.
93. Mehendale HM. Potentiation of halomethane hepatotoxicity: chlordecone and carbon tetrachloride. *Fundam Appl Toxicol* 1984;4:295–308.
94. Mehendale HM. Role of hepatocellular regeneration and hepatolobular healing in the final outcome of liver injury. A two-stage model of toxicity. *Biochem Pharmacol* 1991;42:1155–62.
95. Mehendale HM. Mechanism of the interactive amplification of halomethane hepatotoxicity and lethality by other chemicals. In: Yang RSH, editor, *Toxicology of chemical mixtures: Case studies, mechanisms, and novel approaches.* San Diego, CA: Academic Press, 1994. p. 299–334.
96. El-Masri HA, Thomas RS, Sabados GR, Phillips JK, Constan AA, Benjamin SA, et al. Physiologically based pharmacokinetic/pharmacodynamic modeling of the toxicologic interaction between carbon tetrachloride and Kepone. *Arch Toxicol* 1996;70:704–13.
97. El-Masri HA, Constan AA, Ramsdell HS, Yang RSH. Physiologically based pharmacodynamic modeling of an interaction threshold between trichloroethylene and 1,1-dichloroethylene in Fischer 344 rats. *Toxicol Appl Pharmacol* 1996; 141:124–32.

98. D'Souza RW, Francis WR, Andersen ME. Physiological model for tissue glutathione depletion and increased resynthesis after ethylene dichloride exposure. *J Pharmacol Exp Ther* 1988;245:563–8.

99. Andersen ME, Green T, Frederick CB, Bogdanffy MS. Physiologically based pharmacokinetic (PBPK) models for nasal tissue dosimetry of organic esters: Assessing the state-of-knowledge and risk assessment applications with methyl methacrylate and vinyl acetate. *Regul Toxicol Pharmacol* 2002;36:234–45.

100. Yang RSH. Introduction to the toxicology of chemical mixtures. In: Yang RSH, editor, *Toxicology of chemical mixtures: Case studies, mechanisms, and novel approaches*. San Diego: Academic Press, 1994. p. 1–10.

101. Yang RSH. *Toxicologic interactions of chemical mixtures: Comprehensive toxicology*. Oxford: Elsevier Science Ltd., 1997. p. 189–203.

102. Yang RSH, El-Masri HA, Thomas RS, Dobrev ID, Dennison JE, Bae DS, et al. Chemical mixture toxicology: From descriptive to mechanistic, and going on to in silico toxicology. *Environ Toxicol Pharmacol* 2004;18:65–81.

103. Conolly RB, Andersen ME. Hepatic foci in rats after diethylnitrosamine initiation and 2,3,7,8-tetrachlorodibenzo-*p*-dioxin promotion: evaluation of a quantitative two-cell model and of CYP 1A1/1A2 as a dosimeter. *Toxicol Appl Pharmacol* 1997;146:281–93.

104. Thomas RS, Conolly RB, Gustafson DL, Long ME, Benjamin SA, Yang RSH. A physiologically based pharmacodynamic analysis of hepatic foci within a medium-term liver bioassay using pentachlorobenzene as a promoter and diethylnitrosamine as an initiator. *Toxicol Appl Pharmacol* 2000;166:128–37.

105. Ito N, Imaida K, Hasegawa R, Tsuda H. Rapid bioassay methods for carcinogens and modifiers of hepatocarcinogenesis. *Crit Rev Toxicol* 1989;19:385–415.

106. Ito N, Tatematsu M, Hasegawa RSH, Tsuda H. Medium-term bioassay system for detection of carcinogens and modifiers of hepatocarcinogenesis utilizing the GST-P positive liver cell focus as an endpoint marker. *Toxicol Pathol* 1989;17:630–41.

107. Moolgavkar SH, Luebeck G. Two-event model for carcinogenesis: Biological, mathematical, and statistical considerations. *Risk Anal* 1990;10:323–41.

108. Moolgavkar SH, Venzon DJ. Two-event model for carcinogenesis. *Math Biosci* 2000;47:55–77.

109. Reisfeld B, Reardon K, Yang RSH. A reaction network model for CYP2E1-mediated metabolism of toxicants. *Toxicol Sci* 2003;72:36–7.

110. Reisfeld B, Yang RSH. A reaction network model for CYP2E1-mediated metabolism of toxicant mixtures. *Environ Toxicol Pharmacol* 2004;18:173–9.

111. Mayeno AM, Yang RSH, Reisfeld B. Biochemical reaction network modeling: Predicting metabolism of organic chemical mixtures. *Environ Sci Technol* 2005;39:5363–71.

112. Reisfeld B, Mayeno AM, Yang RSH. Predictive metabolomics: The use of biochemical reaction network modeling for the analysis of toxicant metabolism. *Drug Metab Rev* 2004;36(supplement 1):9.

113. Liao KH, Dobrew ID, Dennison JE, Andersen ME, Reisfeld B, Reardon KF, et al. Application of biologically based computer modeling to simple or complex mixtures. *Environ. Health Persp.* 2002;110:957–63.

114. Kumar A, Campbell DM, Klein MT. Computer-assisted kinetic modeling: Interfacing structure and reaction network model builders. *Abstr Pap Am Chem S* 1997;214:82–PETR:Part 2.

115. Watson BA, Klein MT, Harding RH. Catalytic cracking of alkylbenzenes: Modeling the reaction pathways and mechanisms. *Appl Catal A-Gen* 1997;160:13–39.

116. Klein MT, Hou G, Quann RJ, Wei W, Liao KH, Yang RSH. BioMOL: a computer-assisted biological modeling tool for complex chemical mixtures and biological processes at the molecular level. *Environ Health Persp* 2002;110 (Suppl 6):1025–9.

117. Crivori P, Poggesi I. Computational approaches for predicting CYP-related metabolism properties in the screening of new drugs. *Eur J Med Chem* 2006;41: 795–808.

118. de Groot MJ. Designing better drugs: Predicting cytochrome P450 metabolism. *Drug Disc Today* 2006;11:601–6.

119. Ekins S, Andreyev S, Ryabov A, Kirillov E, Rakhmatulin EA, Sorokina S, et al. A combined approach to drug metabolism and toxicity assessment. *Drug Metab Dispos* 2006;34:495–503.

120. Kulkarni SA, Zhu J, Blechinger S. In silico techniques for the study and prediction of xenobiotic metabolism: A review. *Xenobiotica* 2005;35:955–73.

121. Weininger D. SMILES 1. Introduction and encoding rules. *J Chem Inf Comput Sci* 1988;28:31.

122. Daylight Chemical Information Systems Inc. Daylight Toolkits. Mission Viejo, CA: Daylight Chemical Information Systems, Inc.

123. Balaban AT. *Chemical applications of graph theory*. New York: Academic Press, 1976.

124. Miller RE, Guengerich FP. Oxidation of trichloroethylene by liver microsomal cytochrome P-450: Evidence for chlorine migration in a transition state not involving trichloroethylene oxide. *Biochemistry* 1982;21:1090–7.

125. Guengerich FP. Common and uncommon cytochrome P450 reactions related to metabolism and chemical toxicity. *Chem Res Toxicol* 2001;14:611–50.

126. Kitano H. Computational systems biology. *Nature* 2002;420:206–10.

127. Liao KH. Development and validation of a hybrid reaction network/ physiologically based pharmacokinetic model of benzo[a]pyrene and its metabolites. Ph.D. dissertation, 2004. Colorado State University.

128. Clewell HJ, 3rd, Gentry PR, Covington TR, Gearhart JM. Development of a physiologically based pharmacokinetic model of trichloroethylene and its metabolites for use in risk assessment. *Environ Health Perspect* 2000; 108(Suppl 2):283–305.

4

SPECIES DIFFERENCES IN RECEPTOR-MEDIATED GENE REGULATION

EDWARD L. LECLUYSE AND J. CRAIG ROWLANDS

Contents

4.1 INTRODUCTION TO RECEPTOR BIOLOGY

4.1.1 Hormone Receptors

Nuclear hormone receptors comprise a gene superfamily encoding for transcription factors that transfer endogenous (e.g., small, lipophilic hormones)

Computational Toxicology: Risk Assessment for Pharmaceutical and Environmental Chemicals, Edited by Sean Ekins

and exogenous (drugs, environmental compounds) stimuli into cellular responses by regulating the expression of their target genes. Regulation of gene expression at the transcriptional level by nuclear hormone receptors has an important role in both cellular developmental processes and the body's defensive systems, including the multiple phase I and phase II biotransformation pathways [1–4].

Comparative analysis of the structure of hormone nuclear receptors has revealed that they possess several independent but interacting functional modules. Two of the modulator domains are common structural motifs shared by all the known nuclear receptors. The highly conserved DNA-binding domain (DBD), characterized by two C4-type zinc fingers, links the receptor to the specific promoter regions of its target genes, termed hormone–response elements (HRE). As the most conserved domain and a signature of all nuclear receptors, the DBD can recognize the response elements that contain one or two consensus core half-sites related to the hexamer ACAACA (steroid receptors) or AGGTCA (estrogen receptors and other nuclear receptors) [5,6]. Different nuclear receptors bind to their response element as homodimers, heterodimers with RXR or as monomers. The HRE usually is composed of two hexameric half-sites that can be recognized as inverted, everted, or direct repeats with a 3 to 6 base-pair spacing.

In the past decade studies have demonstrated that specific hormone nuclear receptors play important roles in the induction of the enzymes involved in the biotransformation of drugs and other xenobiotics [4,7,8]. The induction profiles of specific cytochrome P450 subfamilies, such as CYP3A and CYP2C enzymes, are remarkably linked with the activation of specific hormone nuclear receptors. Compounds, such as hyperforin, clotrimazole, pregnenolone 16α-carbonitrile (PCN), and phenobarbital, are both CYP3A inducers and activators of the orphan nuclear receptor termed pregnane X receptor (PXR) (a.k.a., steroid and xenobiotic receptor [SXR]) [9,10]. This association of hormone nuclear receptors with particular CYPs not only shows the CYP subfamily specificity but also defines the species-specific differences in the CYP induction response [11]. Therefore hormone nuclear receptor binding and activation assays have been utilized as useful tools for screening drugs and chemicals in vitro for their potential to cause enzyme induction in vivo, and these data have been used for in silico models (see Chapters 12 and 17).

The recent advances in our knowledge and understanding of the species differences in the regulation of drug metabolizing enzymes by hormone nuclear receptors have been instrumental in allowing for better interspecies predictions of drug and chemical toxicity. Several hormone nuclear receptors, such as the PXR, constitutive androstane receptor (CAR), and peroxisome-proliferator activated receptor (PPAR), are especially emphasized because of their broad and overlapping xenobiotic specificity and their central role in regulating the biotransformation of both endogenous and exogenous substrates (Figure 4.1*a*). Moreover these receptors play complementary roles in the protection against xenobiotic exposure as their target genes represent redundant but distinct layers of defense (Figure 4.1*b*). More important, there are distinct

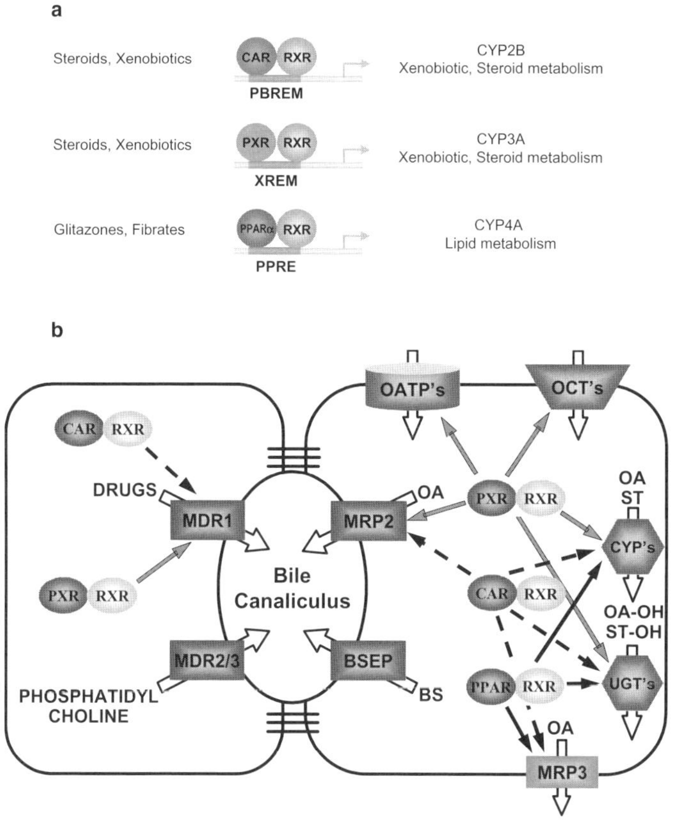

Figure 4.1 (*a*) Hormone nuclear receptors involved in the regulation of hepatic cytochrome P450 enzymes, phase II conjugation enzymes, and uptake/efflux transporters. (*b*) Integrated regulation of phase I and II enzymes as well as uptake and efflux proteins by nuclear receptors. (Modified from [98].)

differences among species in the ligand specificity and receptor response to xenobiotic exposure that must be considered when predicting drug-induced toxicity.

4.1.2 Aryl Hydrocarbon Receptor

The aryl hydrocarbon receptor (AHR) belongs to a family of environment-sensing proteins that share a common domain referred to as the PAS (Per, ARNT, Sim) homology domain of which it is the only member that is ligand-activated [12]. Also unique among the nuclear receptors, the AHR is a basic-helix-loop-helix (bHLH) protein much like the Myc-Mad-Max transcription factors. Thus the AHR shares common functional properties with the other receptors (e.g., ligand activation, nuclear translocation, dimerization, DNA binding, gene transactivation), but unlike the other receptors, the AHR belongs to very different classes of protein families (i.e., PAS and bHLH). Together with its heterodimeric bHLH-PAS protein partner ARNT (AH

Receptor Nuclear Translocator), the AHR regulates normal biochemical responses such as phase I (e.g., CYP1A) and phase II (e.g., UDPGT, GSTYa) enzymes, as well as the adverse health effects in animals associated with exposure to high levels of the ubiquitous environmental contaminant dioxin (2,3,7,8-tetrachlorodibenzo-*p*-dioxin, or TCDD) and other dioxin-like chemicals [13]. In addition chronically administered human medications such as omeprazole and sulindac are AHR activators inducing CYP1A, although no dioxin like toxicity has been associated with chronic use of these medications [14,15].

The precise physiological role of the AHR has not been fully elucidated, nor has a physiological ligand(s) yet been confirmed. However, a number of naturally occurring exogenous and endogenous chemicals have been identified as AHR ligands including indoles, heterocyclic aromatic amines, vitamin A derivatives, tryptophan derivatives, catechins, resveratrol, flavonoids, carotenoids, lipoxin A4, bilirubin and 7-ketocholesterol [16–22]. These other natural AHR ligands can exert significant AHR activity in a reporter gene assay that exceeds the AHR activity attributable to anthropomorphic halogenated aromatic hydrocarbons [23]. AHR activation appears necessary for the toxicity of dioxin-like chemicals, and the degree of biochemical and toxic responses induced by these chemicals in animals and cells is dependent on additional factors, including age, gender, strain, and species. In addition, gene array studies in wild-type and AHR knockout mice have reported that TCDD-activated AHR regulates approximately 450 genes with the AHR alone regulating nearly 400 genes [24]; differences in the AHR regulation of these genes are compounded by differences in ligands (e.g., dioxins, furans, PCBs, natural compounds) along with their dose, time exposure, and mixture composition.

4.2 SPECIES DIFFERENCES IN RECEPTOR ACTIVATION AND BIOLOGICAL RESPONSE

4.2.1 Aryl Hydrocarbon Receptor

TCDD, through activation of the AHR, induces a plethora of toxic effects in animals, which include thymic atrophy, a slow wasting syndrome, teratogenesis, reproductive/developmental impacts, and tumor promotion and cancer in laboratory animals [13]. Perhaps one of the best known yet least understood aspects of TCDD is the broad range in species sensitivity to TCDD-induced toxicity. For example, there is a 1000-fold difference in the LD_{50}s between hamsters and guinea pigs, and reproductive toxicities in mice and rats differ by approximately 200-fold, with less sensitivity observed in mice. The available epidemiological data in humans, however, indicates that we are relatively resistant or insensitive to TCDD-induced toxicity following accidental or occupational exposures, the exception being chloracne, a skin condition that develops after exposure to high levels of dioxin-like chemicals [25].

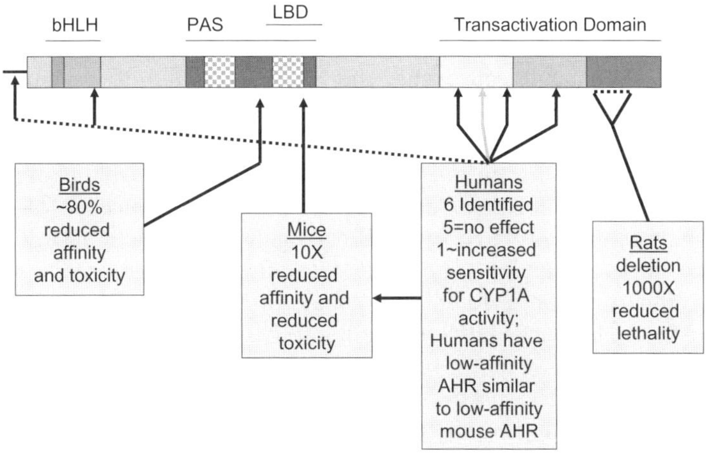

Figure 4.2 Structure of the AHR with polymorphisms and amino acid differences identified in different species. (See text for details.)

AHR molecular biology studies have identified regions for DNA binding (basic domain), protein dimerization (HLH/PAS), ligand binding (PAS), nuclear localization (NLS), nuclear export (NES) and transactivation (TAD) (Figure 4.2) [12,26,27]. The inactive cytoplasmic AHR is found in a complex with a dimer of 90 kD heat shock proteins (HSP90), the immunophilin-like protein XAP2, and a less well characterized 23 kD protein, p23. Upon ligand binding, the AHR sheds the cytosolic proteins, translocates to the nucleus and forms a heterdimer with ARNT. Dimerization with ARNT is required for formation a transcriptionally competant AHR complex that binds to genomic DNA at xenobiotic responsive elements (XRE, core recognition sequence), resulting in altered gene transcription. Gene expression changes occur through AHR-ARNT recruitment of transcriptional co-regulators (co-activators and co-repressors) [26], and similar to other nuclear receptors, the age-, gender-, species-, strain-, and tissue-specific selectivity of AHR ligands as receptor agonists and/or antagonists may be due, in part, to age-, gender-, species-, strain-, and tissue-dependent expression of co-regulatory proteins.

To the extent that the mechanisms for the differences in species and strain sensitivities are understood, the AHR appears to be central with important differences eminating from altered AHR ligand binding domains (LBDs) and transactivation domains (TAD). The reported TCDD binding affinities in mammals ranges considerably, but in general, the human receptor has a K_d of around 9.6 nM, which is approximately one-tenth compared with mice and many commonly used laboratory strains of rats such as the Sprague-Dawley [25]. The differences in ligand binding relate to a single amino acid difference in the human receptor LBD relative to the rodent receptor LBD [12]. A similar scenario has recently been reported in birds where differences in the LBD in terns causes an 80% decrease in receptor affinity and a nearly 250-fold lower sensitivity to TCDD-induced toxicity relative to chickens [28]. Alternatively, the resistance to TCDD toxicity in Han/Wistar (Kuopio) rats is not related to LBD changes, but rather the resistance is due to a small deletion in the transactivation domain (TAD) of the receptor that confers a significantly decreased sensitivity to TCDD-induced effects such as lethality, hepatotoxicity, and tumor promotion [29]. Given the differences in the rat receptors lie in the TAD, the molecular explanation for the differences in TCDD sensitivity are most likely related to AHR co-regulator interactions and resulting effects on gene expression. These rat data may have significance to understanding human responses to TCDD, since four of the five amino acid changes identified in the human AHR are found in the TAD and at least one of these sites reportedly alters receptor function increasing sensitivity for AHH induction [30]. No differences have been identified in the human AHR LBD indicating the lower binding affinity relative to rodents is stable across the human population (Figure 4.2).

The most thoroughly investigated biochemical response to AHR agonists is induction of cytochrome P450 1A1 (CYP1A1) mRNA and its associated enzyme activities, aryl hydrocarbon hydroxylase (AHH) and ethoxyresorufin-*O*-deethylase (EROD). Although induction of this gene and enzyme activity by TCDD is not considered by most to correlate with TCDD toxicity, the relative potency for induction of CYP1A1 by dioxin like chemicals is still used in determining the TCDD toxic equivalency of these chemicals by regulatory agencies (discussed further below) [31]. Significant species differences exist in the number of XRE's in the CYP1A1 gene with possibly six in the mouse gene, three in the rat gene and two in the human gene [12]. Moreover comparative analysis of XREs in human, mouse, and rat genomic sequences has identified that of the mouse–rat orthologous genes with a XRE between –1500 and +1500, only 37% had an equivalent human orthologue [32]. The decreased sensitivity of humans to induction of CYP1A by AHR ligands is illustrated in several studies conducted in primary human and rat hepatocytes and cell lines. The EC_{50}s for TCDD-induced CYP1A activities (EROD, AHH) in primary human hepatocytes and HepG2 cells was 8 to 20 times lower than in primary rat hepatocytes, and H4IIE cells [33–35]. Extensive dose–response studies have identified that TCDD-induced CYP1A1 and EROD in primary human hepatocytes is up to 50 times lower than in primary rat hepatocytes (Figure 4.3) [36]. Xu et al. [35] also noted species differences in TCDD-induced CYP1A mRNA expression with CYP1A1 predominantly induced in rat hepatocytes, whereas in human hepatocytes CYP1A2 was the predominantly induced mRNA. A recently developed humanized mouse model of the AHR, made by "knocking-in" the human AHR (hAHR) into an AHR knockout (AHRKO) mouse, has added evidence for the decreased sensitivity of humans to TCDD [37]. The hAHR mice treated with TCDD exhibited reduced induction of CYP1A and hydronephrosis, whereas the mice expressing the mouse AHR had higher induced levels of CYP1A and hydronephrosis and also cleft palate, the later not induced in the hAHR mice. Thus the hAHR appears to

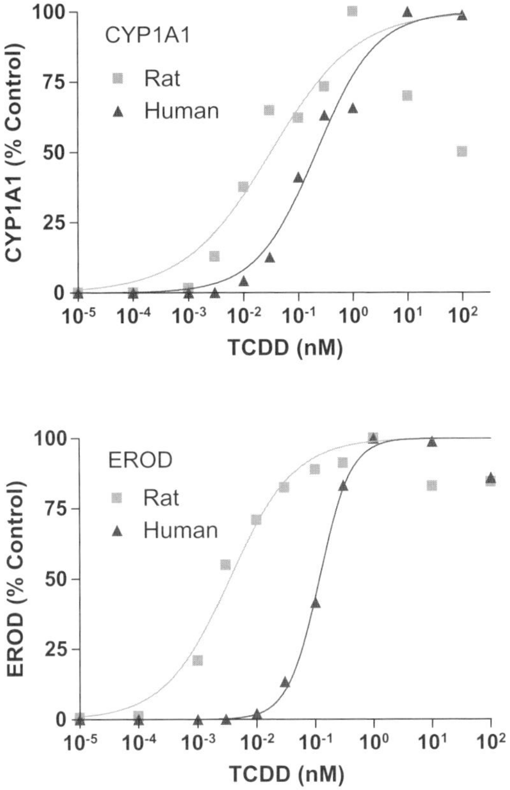

Figure 4.3 TCDD-Induced CYP1A1 mRNA and ethoxyresorufin *O*-dealkylation (EROD) in human and rat primary hepatocytes.

mediate a differential response to TCDD relative to the mAHR. In sum, comparative studies of human, rat and mouse cells, AHRs and AHR-regulated genes consistently report significantly lower sensitivity in humans compared with rodents. These studies also demonstrate the utility of the primary hepatocyte model and humanized mice for comparative species studies.

4.2.2 Peroxisome-Proliferator Activated Receptor

A peroxisome proliferator is a chemical that induces peroxisome proliferation in rodent liver and other tissues and includes a wide range of chemicals such as certain herbicides, plasticizers, drugs, and natural products [38,39]. The peroxisomes contain hydrogen peroxide and fatty acid oxidation systems important in lipid metabolism and activation of the peroxisome proliferator-activated receptor alpha (PPARα), is considered a key event in peroxisome proliferation in rodent hepatocytes [39]. A number of studies have identified

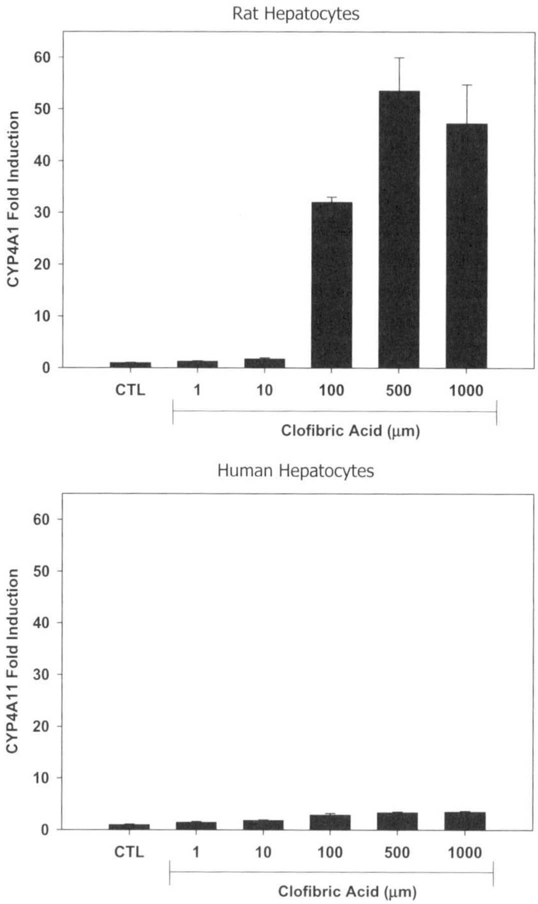

Figure 4.4 Species differences in clofibric acid-induced expression of CYP4A-specific activity (lauric acid 12-hydroxylation) represented as fold induction over control levels.

that rodents are more responsive than primates to peroxisome proliferators (PPs). Transient transfection studies comparing human and rodent PPARα have observed that target gene expression occurs with lower concentrations of agonists and to greater efficacies with rodent receptor than with human receptor [39]. For example, the EC_{50}s for MEHP activation of mouse and human PPARα are 0.6 and 3.12.mM, respectively [40], and induction of hepatic CYP4A by clofibric acid is more than 50-fold versus less than 5-fold in rats compared to humans, respectively (Figure 4.4).

Molecular explanations for the species differences in PP sensitivities include differences in PPAR expression levels and differences in PPAR transactivation capacities for target genes between species. The expression of PPARα in human hepatocytes was reported to be 10-fold lower than that in the rat or mouse [41,42], although others have not observed any differences in expression levels between species [43]. In addition truncated and polymorphic forms of PPARα have been reported in humans that may be less functional [39]. Studies have also indicated that the human PPARα does not bind to DNA response elements (PPREs) in target genes as efficiently as the rodent receptors and that there may be differences in the PPRE sequences between species that contribute to some of the differences in PPARα transactivation of target genes. The development of the humanized PPARα (hPPARα) has provided important evidence for the lack of sensitivity in human to the toxic effects of peroxisome proliferators. Although hPPARα mice had normal gene responses for mitochondrial and peroxisome lipid metabolizing enzymes, these mice did not develop hepatocellular proliferation and hepatomegally, nor did the hPPARα mice develop liver tumors following chronic treatment with Wy-14,643, a rodent peroxisome proliferating hepatocarcinogen [44,45]. Thus the prevailing evidence indicates that PPARα-mediated responses in humans occur differently and with less sensitivity than in rodents.

4.2.3 Pregnane X Receptor/Steroid and Xenobiotic Receptor

The full-length cDNA of mouse PXR was first cloned from the expressed sequence tags (ESTs) database derived from a mouse liver library by Kliewer and colleagues [9]. Soon, its homologous counterparts in human [46,47], rat [48,49], rabbit [49], pig, rhesus monkey, and dog [50] were identified. Because of the lack of a common nomenclature system, orthologous receptors from different species were given unrelated names. The human receptor has been referred to also as SXR, or pregnane-activated receptor (PAR), in some laboratories [46,47]. More recently a nomenclature system based on the cytochrome P450 superfamily designation scheme has been devised for the nuclear receptor superfamily. According to this system PXR has been classified as NR1I2, in which the gene family is indicated by an arabic numeral, the subfamily by a capital letter, and individual gene members by the second arabic numeral. For consistency within this review, we use "PXR" to refer to the human receptor, in preference to SXR or PAR. Studies conducted to determine the tissue distribution of PXR expression have demonstrated that all PXRs (mouse, human, rat, and rabbit) are predominantly expressed in liver and intestine, and, to a lesser extent, in the kidney and lung [51]. The tissue-specific pattern of PXR expression resembles that of CYP3A and other drug-metabolizing enzymes, suggesting that PXR may be of importance for the induced and constitutive expression of these enzymes.

There are several lines of evidence that strongly support PXR as the predominant regulator of CYP3A expression. As the most abundantly expressed CYP in human liver and intestine and capable of metabolizing a broad range of structurally diverse substrates, CYP3A4 can be induced by a variety of xenobiotics. Systematic deletion analysis of the promoter region of CYP3A has revealed PXR response elements in promoter regions of the rodent and human CYP3A genes as either a direct repeat of the half-site TGAACT spaced by 3 bp (DR3) [9], or an everted or inverted repeat of the TGAACT half-site spaced by 6 bp (ER6, and IR6) [47,52]. PXR can bind and transactivate these response elements after activation by a variety of CYP3A inducers, including the glucocorticoid receptor (GR) antagonist RU486, suggesting that the CYP3A gene is regulated by steroid and xenobiotics through a non-classical glucocorticoid receptor pathway [51,53]. These findings strongly support the hypothesis that PXR is the predominant regulator of the xenobiotic-responsive expression of the CYP3A gene.

Utilizing cell-based in vitro cotransfection assays, a large number of structurally diverse compounds have been identified as PXR activators [3]. As expected, most of these compounds had been shown previously to be CYP3A inducers (Figure 4.5). Important species-specific PXR activation profiles were

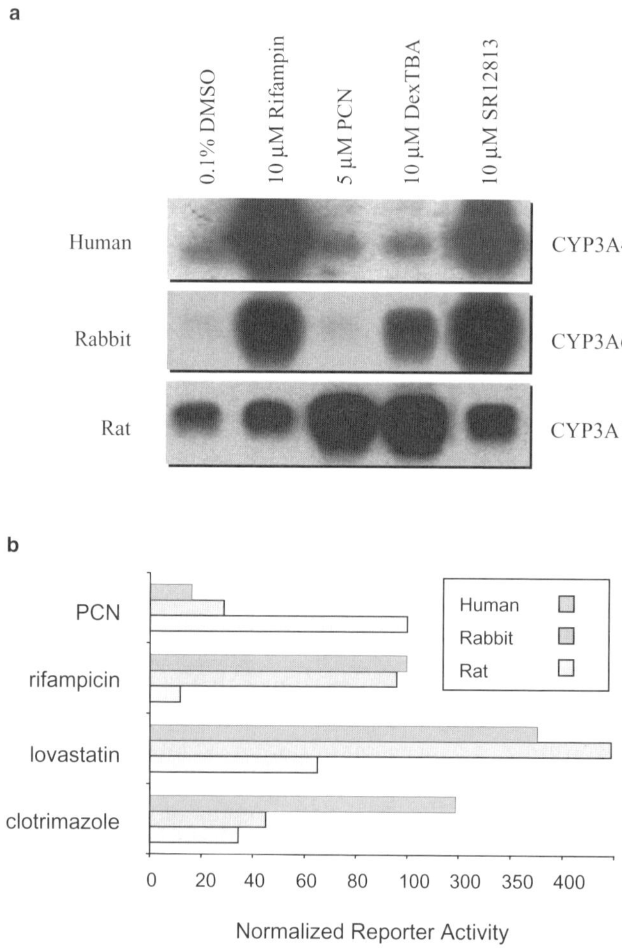

Figure 4.5 Species differences in chemical-induced expression of (*a*) CYP3A mRNA and (*b*) PXR reporter activity [8].

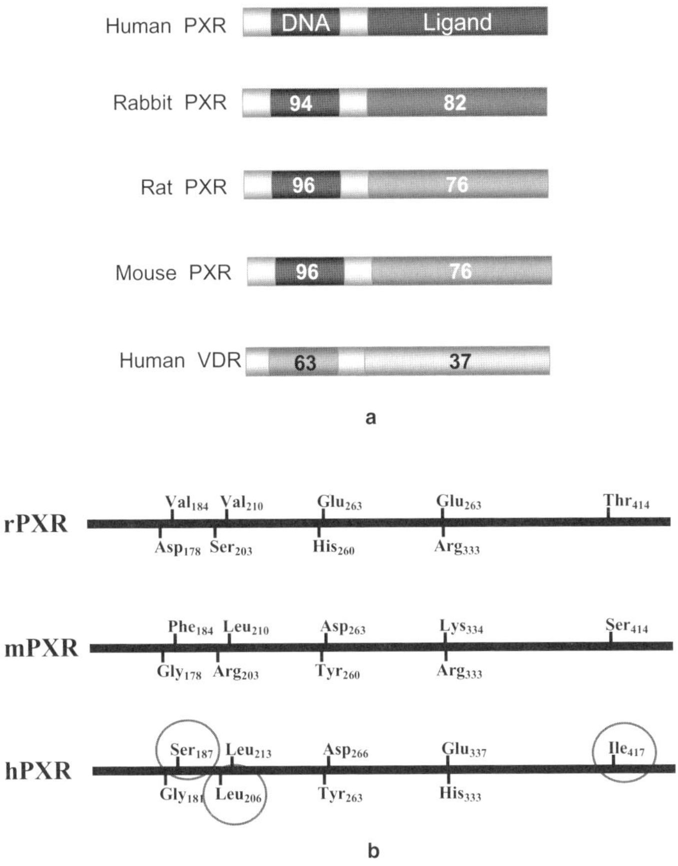

Figure 4.6 (*a*) Amino acid sequence homology (% human receptor) between the DNA- and ligand-binding domains of PXR from different species [8]. (*b*) Amino acid differences in the ligand-binding domain of rat, mouse, and human PXR [48]. Circles designate positions where major shifts in the physical-chemical properties occur.

observed also during this screening process. For example, the antibiotic rifampicin, an efficacious inducer of CYP3A expression in rabbit and human liver, but not in rat and mouse, was an efficacious activator of human and rabbit PXR but not rat or mouse. On the other hand, pregnenolone 16α-carbonitrile (PCN), a potent inducer of rat CYP3A but not of human or rabbit, was an efficacious activator of rat PXR but had little activity on the human and rabbit receptor (Figure 4.5) [49]. This selectivity in PXR activation reflects well the ability of compounds to induce CYP3A in different species.

Structural comparisons of PXR from different species demonstrated that there is more than 95% sequence homology in the DBD regions, but only 75% to 80% amino acid homology in the LBD (Figure 4.6*a*). This striking difference in the LBD sequence across species is unusual compared to the classic nuclear receptors, such as GR, which show nearly 90% sequence identity in the LBD region [2]. More recently the crystal structure of the LBD of human PXR was elucidated and showed the existence of an extensive hydrophobic ligand-binding cavity containing several polar residues, which allow hPXR to bind to a variety of structurally diverse ligands including small and large molecular weight compounds [54]. Furthermore mutagenesis analyses indicated that alteration of four amino acids in the ligand-binding pocket of mPXR to the corresponding hPXR amino acids (Arg203→Leu, Pro205→Ser, Gln404→His, and Gln407→Arg) was sufficient to switch the original PCN responsive mPXR to a human-like receptor, which could be activated effectively by SR12813 but no longer by PCN (Figure 4.6*b*). Using trans-species transfection assays, Xie et al. [11] demonstrated that co-transfection of hPXR into primary rat hepatocytes resulted in a significant induction of rat CYP3A23 reporter gene by compounds known to be human PXR activators and CYP3A4 inducers, including rifampicin, clotrimazole, phenobarbital, and RU486. More definitive proof for PXR regulation of CYP3A gene expression and species-specific induction was established by the generation of PXR-null mice, in which treatment with PCN fails to induce CYP3A gene expression [55,56]. By contrast, replacement of the mPXR with its human orthologue causes the xenobiotic responsiveness in this humanized mouse to be restored, but the response to xenobiotic stimulation matches that of humans [55]. Overall, these data indicate that PXR is the molecular basis for the species differences observed in the xenobiotic-responsive expression of CYP3A in mammalian liver. More important, the identification of new drugs likely to cause drug–drug interactions with CYP3A4 substrates has become attainable by utilizing PXR as a molecular tool for identifying efficacious activators.

Various hormone nuclear receptors are involved in the regulation of multiple CYPs by recognizing common response elements containing the half-site AGGTCA spaced by 3 to 6 base pairs (Figure 4.1*a*). The result is that individual response elements can be activated by more than a single nuclear receptor, which is often referred to as "cross-talk." Recent studies using electrophoretic mobility gel shift assays have shown that in vitro translated PXR can bind to the PB response element (PBREM containing both NR1 and NR2 sites) located in the 5′-flanking region of the CYP2B gene [57]. In addition co-transfection of hPXR with the natural promoter of mouse CYP2b10 linked to a reporter gene resulted in a significant induction of reporter activity in response to the known hPXR activators rifampicin and RU486 [11]. Studies in rodents have demonstrated that the rat CYP2B1 gene is inducible by PB in PXR abundant intestine [58]. Likewise dexamethasone (DEX), a ligand for mouse PXR but not an activator of mCAR, can efficiently induce CYP2b10 expression in mouse hepatocytes [59,60].

4.2.4 Constitutive Androstane Receptor

The hormone nuclear receptor CAR (NR1I3) was isolated through screening of a cDNA library with the nuclear receptor DBD based oligonucleotide as a probe [61]. The name CAR was originally referred to as *constitutively activated receptor*, since it forms a heterodimer with RXR that binds to retinoic acid response elements (RAREs) and transactivates target genes in the absence of ligands in transfection assays [61,62]. Belonging to the same subgroup of the orphan nuclear receptor superfamily as the farnesoid X receptor (FXR), the liver X receptor (LXR), and PXR, CAR is mainly expressed in liver with a minor presence in small intestine [61]. By screening several small hydrophobic compounds, two androstane metabolites, androstanol (5α-androstan-3α-ol) and androstenol (5α-androstan-16-en-3α-ol), were identified as endogenous CAR ligands. However, instead of activating CAR, both ligands act as antagonists by dissociating CAR from its coactivator and inhibiting the transactivation of CAR [63]. Thus CAR also was referred to as constitutive androstane receptor.

CYP2B, another important CYP subfamily, is effectively induced by phenobarbital (PB) in most mammalian species. The CYP2B gene and cDNA originally were sequenced and cloned in the 1980s [64]. However, the real breakthrough in the understanding of the mechanism of PB induction of this gene was with the identification of a 163 bp DNA sequence located between −2318 and −2155 bp upstream from the CYP2B2 encoding region, which was linked with PB responsiveness [65]. Furthermore consecutive deletion analysis of a similar 163 bp sequence in the mouse CYP2b10 indicated that a 51 bp minimum sequence was required for PB induction, and was termed the Phenobarbital–response enhancer module (PBREM) [66]. Later this PBREM was identified in rat, rabbit, and human CYP2B genes [67]. Sequence analysis of the PBREM revealed that this element was composed of two nuclear receptor-binding sites (NR1 and NR2) and a nuclear factor 1 (NF1) binding site. Both NR1 and NR2 are DR-4 motifs. NR1 is identical in rat and mice, while only a base-pair difference is observed in the human NR1. This conserved NR1 site is critical for conferring PB responsiveness [59,68]. The function of the NF1 site is unknown, but it is believed to be required for full PBREM activity.

Using a cell-based transfection assay, Negishi and colleagues screened a number of known nuclear receptors, such as RXR, CAR, LXR, thyroid receptor-α, hepatocyte nuclear factor 4 (HNF4), and chicken ovalbumin upstream promoter-transcription factor (COUP-TF), for their capacity to bind and transactivate the PBREM in reporter assays. The results indicated that CAR alone was able to stimulate the PBREM reporter gene expression [60]. Further NR1-affinity chromatography was used to purify the protein that bound to PBREM, and both binding assays and Western blot analysis proved that CAR was the protein mediating the PB induction response. Thereafter a group of structurally diverse compounds was identified to induce CYP2B through activation of CAR and thus was referred to as "phenobarbital-type" inducers. The most potent member from among these compounds is 1,4-*bis*[2-(3,5-dichlorpyridyloxy)]benzene (TCPOBOP), which was originally identified as a pesticide contaminant [69].

In contrast to PXR, CAR is located in the cytoplasm of hepatocytes in the absence of ligands, and it is translocated into the nucleus after treatment with PB-like CYP2B inducers [70]. It appears that nuclear accumulation of CAR is important and might be the first activation step in response to PB-type inducers. This translocation phenomenon is common among steroid receptors, such as GR, which are dissociated from their cytoplasmic complex with other proteins (e.g., heat shock proteins, immunophilins, and p23 proteins) and translocated into the nucleus upon binding to a ligand, such as DEX [71–73]. However, several lines of evidence indicate that ligand binding is not an absolute requirement for CAR translocation. First, in vitro binding assays showed that TCPOBOP can bind directly to mCAR but not hCAR, but in vivo transfection assays revealed that both hCAR and mCAR accumulate in the nucleus in mouse liver after TCPOBOP treatment [70]. Second, in the same in vitro binding assay, it was shown that PB does not bind directly to either mCAR or hCAR [3,74]. Finally, in HepG2 cells, transiently expressed CAR always accumulates in the nucleus even in the absence of CAR ligands or CYP2B inducers [60]. By contrast, transfected CAR behavior is more reflective of the in vivo condition in primary hepatocyte cultures. For example, in the absence of CAR ligands or CYP2B inducers, CAR is predominantly localized in the cell cytoplasm, while in the presence of CAR ligands or CYP2B inducers, marked activation of CAR is observed [75,76]. These results indicate that some unidentified cellular factor(s) plays an important role in the translocation and activation of CAR, and primary hepatocyte cultures may be an effective means to identify potential CAR activators.

CAR is the closest relative of PXR on the branch of the orphan nuclear receptor tree. Although they were originally recognized as the regulators of CYP2B and CYP3A, respectively, there is significant overlap in the inducers of these two gene subfamilies (Table 4.1). For example, PB can induce both

TABLE 4.1 Overlap of Nuclear Receptor Activation by Steroids and Xenobiotics

Ligand	Activated	Not Activated
RIF	hPXR, rbPXR	rPXR
PCN	Rat PXR	hPXR, rbPXR
PB	mCAR, rCAR	mPXR
	hCAR, hPXR	rPXR
Estrogen	mCAR	hCAR
TCPOBOP	mCAR, hPXR	hCAR, mPXR
Androstanol	mPXR, hPXR	Deactivates mCAR
Clotrimazole	hPXR	Deactivates hCAR
5β-Pregnane-3,20-dione	hCAR	
	hPXR, mPXR	
CITCO	hCAR	mCAR
Phenytoin	mCAR, hCAR	

Source: Modified from [99].

CYP2B and CYP3A, and rifampicin is a good inducer of both CYP3A4 and CYP2B6. However, no PBREM has been identified in the upstream region of the CYP3A4 gene, and the PXR response element (XREM) has not been identified in the CYP2B6 gene. This cross-talk may occur because both CAR and PXR recognize the other's response elements (especially, DR4 and DR3 elements) and trigger gene expression (CYP3A or CYP2B) upon activation by either common or selective ligands [77]. Using CV-1 cells and primary rat hepatocytes, Xie et al. [11] demonstrated that both CAR and PXR could regulate CYP3A and CYP2B gene expression upon activation by their specific ligands. Smirlis et al. [76] reported that when equimolar amounts of CAR and PXR expression vectors are co-transfected with a PBREM-reporter construct, a 60% decrease of the reporter gene expression was observed, suggested that the receptors could compete with each other for binding to the same response element.

Two issues that remain to be resolved are whether human CAR is regulated in the same fashion as the rodent CAR and whether CAR is the predominate regulator of CYP2B6 in human liver. At this point the conclusions regarding the role of CAR in the regulation of CYP2B or other genes are based almost entirely on rodent CAR, primarily mCAR [60,78]. Although human CAR exhibits some common characteristics with its rodent counterparts, such as undergoing nuclear translocation after PB treatment and binding to the PBREM, there are distinct differences between rodent and human CAR (Table 4.1). For example, TCPOBOP, the most potent mCAR ligand identified to date, cannot bind or activate either rat or human CAR. Conversely, CITCO can effectively bind and activate human CAR but does not activate mCAR [79]. Phenobarbital is a potent activator and inducer of rat CAR and CYP2B1, respectively; however, by comparison, it is a weak activator and inducer of human CAR and CYP2B6, respectively (Figure 4.7). Moreover phenobarbital is an efficacious inducer of human CYP3A4 but not rat CYP3A23, in part because it is an activator of human PXR. All known mCAR inhibitors, such as androstenol, progesterone, androgens, and CaMK inhibitors, do not inhibit hCAR activation. Overall, current evidence suggests that there are clear species-specific differences in CYP2B induction and CAR activation, which severely hampers the extrapolation of animal data to humans. For these reasons hCAR regulation by drugs and other xenobiotics has become a more complex, and yet urgent, issue that has only recently been resolved by the identification of the chemically responsive, splice variant of human CAR (hCAR3) [80,81].

Past studies have shown that the interspecies differences in CAR and PXR regulation of CYP's also happens at the level of cross-regulation. Generation of the transgenic PXR- and CAR-null mice has made it possible to specifically address the effects of these receptors on CYP regulation in vivo [55,82,83]. In PXR-null mice both PB and clotrimazole efficaciously induce CYP3A. Notably, PB elicited a stronger induction of CYP3A in the knockout animal than in their wild-type counterpart [11]. On the other hand, in CAR-null mice the

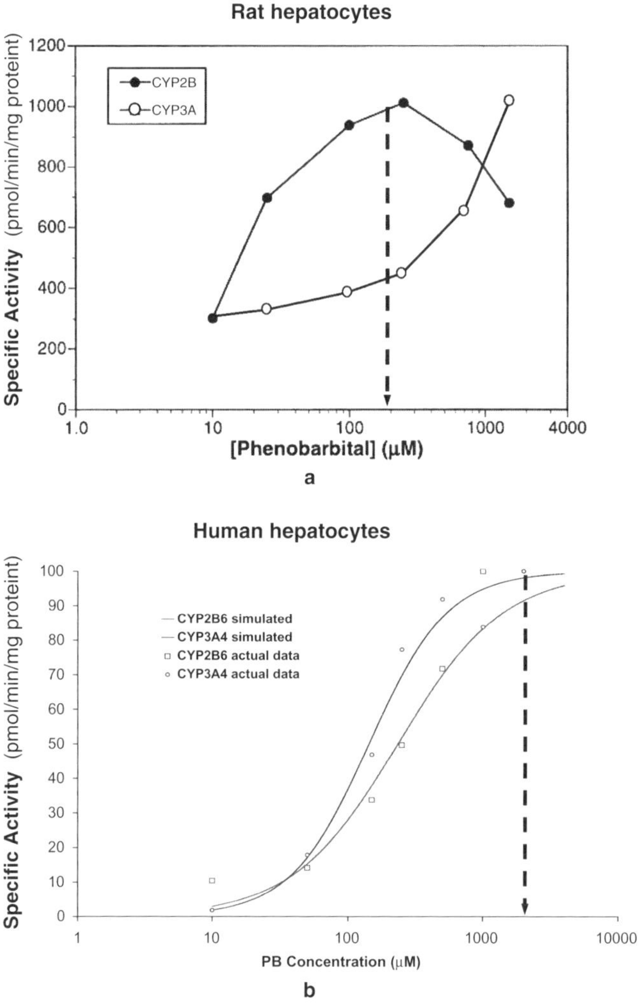

Figure 4.7 Differences in CYP2B and CYP3A induction by phenobarbital in rat and human hepatocytes. Arrows designate an approximate 10-fold difference between rat and human CYP2B in the concentration where optimal expression is observed.

strong induction of CYP2b10 gene expression by PB and TCPOBOP was totally absent [82]. Similarly, in obese Zucker rats that express extremely low levels of CAR but with normal levels of PXR expression, PB only poorly induces both CYP2B and 3A, while CYP3A was significantly induced by the PXR-specific activator PCN [84]. These studies using both transgenic animals and Zucker rats suggest that in mice and rats, CAR appears to cross-regulate both CYP3A and CYP2B, while PXR plays no role in the PB induction of CYP2B. On the contrary, studies have shown that most hPXR activators like rifampicin, clotrimazole, and PB could efficaciously induce human CYP2B6, and activate hPXR-mediated PBREM-tk-reporter gene expression [11,77].

Because only a limited number of chemicals activate both receptors, species differences in cross-talk between PXR and CAR result primarily from their differential binding to response elements within the CYP2B and CYP3A promoters. PXR binds to the DR4-type sequences in the rodent and human CYP2B PBREM as well as to the rat DR3 or human ER6 motifs in the CYP3A PXRE [9,47,52,57,81]. Conversely, CAR binds only the CYP3A direct repeat response elements (DR3) in addition to CYP2B promoter sequences [60,67,74,81]. hPXR and hCAR also share the ability to bind and activate the DR4-type NR3 motif in the CYP2B6 distal enhancer module and the DR3-type motif (dNR1) in the CYP3A4 distal enhancer module, but only PXR can bind to the ER6 motifs in the CYP3A4 promoter [10,57,81,85]. Several studies have demonstrated that the distal enhancer modules function cooperatively with the more proximal elements to elicit maximum induction of the respective genes [10,57,85]. As such, this would imply that activators of CAR would not induce CYP3A gene expression in humans to the same extent as PXR activators because of their inherent binding capacities for the individual response elements (Figure 4.8). These combined results indicate that cross-regulation of target genes by different nuclear receptors further broadens the

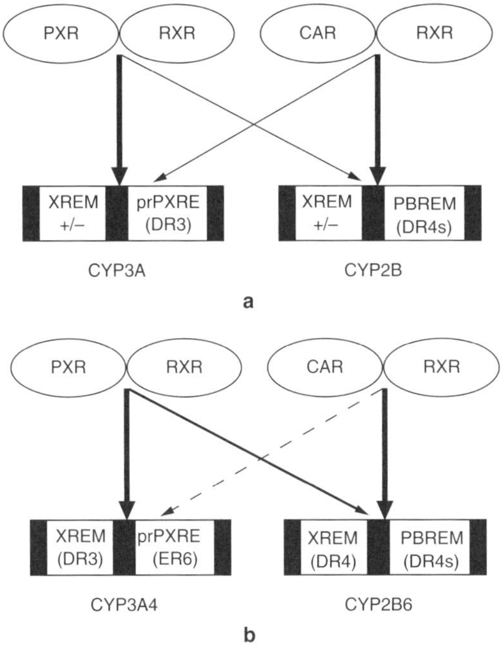

Figure 4.8 Species differences in CYP2B and CYP3A regulation by CAR and PXR. Rodent (*a*) versus human (*b*) models of cross-talk between PXR and CAR in the promoter regions of *CYP2B* and *CYP3A* genes [77].

possibilities for drug–drug interactions. At the same time, because of interspecies differences in the regulation of CYP2B and CYP3A between human and rodents, more studies with human relevant model systems should greatly benefit both our understanding of receptor regulation of CYPs and thus the development of safer drugs.

4.3 DISCUSSION AND CONCLUSIONS

4.3.1 Pharmaceutical Industry and FDA Perspective

According to the FDA, severe hepatotoxicity is one of the most common causes for pharmaceutical product recalls and labeling changes, and this raises the question of how effective nonclinical and clinical testing are in recognizing such toxicity. Although much attention has been focused on the predictivity of animal models for clinical findings in humans, clinical trials of pharmaceuticals are in fact a relatively poor source of information for ascertaining the probability that an event observed in an animal model is also observed in the human. Of critical importance to avoiding unanticipated toxicity in human is the need to understand why drugs that were judged safe to administer to humans on the basis of animal data sometimes cause unexpected toxicity or drug interactions. In other words, why do animal models in these cases fail to predict human hepatotoxicity? It is probably even more important to understand why these models sometimes fail to identify hepatotoxic potential in humans than to fully define the overall effectiveness of animal studies in predicting human outcome.

Drug–drug interactions are one of the major concerns for clinical practice and the pharmaceutical industry when several drugs are co-administered or during the development of a new drug. During the last 15 to 20 years drug interactions have received increasing attention because a large number of new drugs have been introduced into clinical practice for the treatment of several diseases, as cancer, HIV, and other opportunistic infections continue to involve multi-drug therapy. The majority of drug–drug interactions of clinical significance occur through induction or inhibition of cytochrome P450 enzymes. According to the new FDA draft guidance, a drug that induces a drug-metabolizing enzyme can increase the rate of metabolic clearance of a co-administered drug that is a substrate of the induced pathway [86]. A potential consequence of this type of drug–drug interaction is subtherapeutic blood concentrations. Drug–drug interactions due to induction of drug-metabolizing enzymes have greater clinical significance for those drugs that have a narrow therapeutic window, such as phenytoin, warfarin, and digoxin. Alternatively, the induced metabolic pathway could lead to increased formation of an active compound, resulting in an adverse event.

The classic example of this phenomenon is the bioactivation of acetaminophen to its reactive metabolite *N*-acetyl-benzoquinoneimine (NAPQI), which is a cytochrome P450-dependent oxidation step. Excessive formation of NAPQI in the liver can cause depletion of glutathione and may result in cell death and hepatotoxicity due to covalent binding to essential cellular macromolecules and/or other mechanisms such as oxidative stress. It has been shown in animal models that activators of PXR (PCN) or CAR (PB) can enhance the hepatotoxicity of APAP by the increased formation of NAPQI [87,88]. Notably, increased liver damage is not observed in CAR- or PXR-knockout animals, suggesting that receptor activation and enzyme induction play critical roles in the hepatotoxicity of APAP.

In humans, several hormone nuclear receptors that are activated by many drugs and steroids play major roles in the induced expression of CYPs and other drug-metabolizing enzymes that are involved in APAP elimination and toxicity. For example, PXR plays a key role in the regulation of CYP3A4, CYP2C9, CYP2B6, and UGT1A1 as well as transporters such as MDR1. Likewise CAR plays a key role in the regulation of the same enzymes and transport proteins; however, one important difference is that it does not up-regulate CYP3A genes as effectively as PXR [77]. Therefore, unlike in the rodent model, co-administration of APAP with ligands for CAR or PXR might lead to differential toxic side effects and liver damage in humans.

Chronic activation of PXR has been associated with disruption of homeostasis of endogenous substrates as well as other adverse physiological effects. For example, rifampicin is one of the most potent human PXR activators and inducers of CYP3A4, CYP2C9, and CYP2B6. Rifampicin can significantly impair the efficacy of a number of drugs that are substrates for these enzymes, such as tamoxifen, ifosfamide, paclitaxel, ethinyl estradiiol, warfarin, and cyclosporine A. In a double-blind crossover study, administration of rifampicin caused a 96% reduction in the AUC of midazolam in 10 healthy volunteers [89]. Terzolo et al. [90] reported that long-term treatment of tuberculosis by rifampicin could increase steroid clearance and lead to a misdiagnosis of Cushing's syndrome by interfering with the overnight dexamethasone test.

St. John's wort, a herbal remedy and a widely used antidepressant, has not been subjected to the rigorous clinical testing that most drug candidates currently in development receive. A number of case reports have demonstrated drug–drug interactions between St. John's wort and various drugs that are CYP3A4 substrates [91,92]. In women St. John's wort increased the clearance of oral contraceptives, which led to decreased circulating sex steroid levels and the loss of contraceptive efficacy [93]. Co-administration of St. John's wort also reduces the blood levels of HIV protease inhibitors and immunosuppressant drugs [94,95]. Recently Kliewer and colleagues demonstrated that these St. John's wort related drug interactions are due to activation of human PXR and subsequent induction of CYP3A4 expression [96]. Another example is paclitaxel, which is a widely used antineoplastic agent, an efficient human PXR activator, and a substrate of CYP3A4 and CYP2C8. The therapeutic efficacy of this drug might be limited by autoinduced metabolism. A paclitaxel analogue that has similar antineoplastic activity but does not activate PXR has been shown to have superior pharmacokinetic properties [7].

The co-administration of activators of individual or multiple nuclear receptors complicates further the prediction of drug interactions and toxicity in vivo. From the evidence thus far, it is becoming apparent that NR's have a primary function in the regulation of specific target genes, but they also appear to play a secondary or supporting role in the regulation of others. As such, there is much to be learned regarding the overlapping specificity of ligands and DNA response elements for individual receptors, as well as the manner in which they interact to affect gene expression in a synergistic or antagonistic fashion. Just as important, it is now understood that the molecular basis for the species differences observed in the xenobiotic-dependent induction of CYP3A genes can be simple mutations in the amino acid sequence of the ligand-binding domain of the receptor. Current induction predictions based on animal data are deficient inherently, and therefore new screening strategies combining reporter assays with expressed human receptors and primary cells of human origin are gaining widespread acceptance in the drug discovery and development process.

Although the role of PXR in the regulation of CYP3A4 and MDR1 has been firmly established, there are a number of unresolved issues remaining about the importance of other orphan receptors in drug-induced gene expression. Recent evidence suggests that PXR, CAR, and GR are the primary players involved in the regulation of a number of other phase I and II enzymes and transporters, whereas PXR appears to be the "master" regulator of cytochrome P450 expression in human liver. The role of the nuclear receptor CAR is ambiguous at this point in time, but it is certain that it is quite distinct from that of its rodent counterparts. The ability to identify the significance of CAR in human liver has been confounded historically by the lack of good human-relevant in vitro model systems. However, recent breakthroughs in the development of cell-based technologies and receptor-binding assays will enable investigators to determine whether CAR plays a dominant or supporting role in the drug-induced expression of hepatic enzymes and transporters.

The recent advances in orphan nuclear receptor biology have further expanded our ability to predict the potential pharmacological and toxicological properties of new drugs. Indeed great progress has been achieved in developing high-throughput screening of new chemical entities based on in vitro cell-based transfection assays for those likely to be involved in drug interactions with CYP3A4 substrates. More recently cellular-based assays have been employed to identify nuclear receptor agonists and antagonists as potential drug candidates. It is anticipated that these novel methods may lead to a more rational and molecular-based approach to developing drugs with enhanced therapeutic efficacy and improved safety profiles.

Finally, species differences in the induction of individual or multiple biotransformation and elimination pathways can lead to the production of different metabolite profiles in humans compared to animal models. The FDA considers that the quantitative and qualitative differences in metabolite profiles are important when comparing exposure and safety of a drug in a nonclinical species relative to humans during risk assessment. When the metabolic profile of a parent drug is similar qualitatively and quantitatively across species, it is generally assumed that potential clinical risks of the parent drug and its metabolites have been adequately characterized during standard nonclinical safety evaluations. However, because metabolic profiles and metabolite concentrations can vary across species, there may be cases when clinically relevant metabolites have not been identified or adequately evaluated during nonclinical safety studies. This may occur because the metabolite(s) being formed in humans are absent in the animal test species (unique human metabolite) or because the metabolite is present at much higher levels in humans (major metabolite) than in the species used during standard toxicity testing. Therefore understanding the species differences in the metabolic pathways involved in the clearance and toxicity of drugs, as well as how they are regulated by nuclear receptors, will continue to be a major challenge in the future.

4.3.2 Chemical Industry and EPA Perspective

Based on their common mechanism of action for AHR activation, regulatory agencies have developed the toxic equivalency factor (TEF) approach for risk assessment of dioxin-like chemicals (DLCs). The TEF approach involves assigning potency to DLCs relative to TCDD, meaning a DLC-specific TCDD relative potency (REP) (EC_{50}TCDD/EC_{50} DLC). Thus for complex mixtures of DLCs the toxic equivalent (TEQ) of the mixture is the sum of the concentration of each DLC multiplied by its TEF, a strictly additive approach that does not consider partial agonist and antagonist interactions for the AHR. A DLCs REP represents a point estimate selected from a range of potencies derived from both in vivo studies of toxicity and short-term in vitro cell bioassays of enzyme (e.g., CYP1A1, CYP1A2) and reporter gene (e.g., luciferase, CALUX) induction [97].

Although activation of the AHR by DLCs is a key event, mechanistic data indicate that AHR-mediated responses are not well conserved across species, with lower sensitivity in humans. A TEF value for a DLC based on rodent data may overestimate the potency of a DLC in humans, and this has not been considered in the current risk assessment of DLCs. Thus, the current TEF-Toxic Equivalency Quotient scheme tends to compound the conservative estimates of risk that exist within standard risk assessment approaches. Moreover mechanistic differences will now be considered by US EPA in the risk assessment of chemical carcinogens. The mechanistic data currently available for receptor-mediated DLCs and PPs clearly indicate that humans respond differently to these two classes of rodent carcinogens, and these data will need to be incorporated into cancer risk assessments for these chemicals. Full appreciation of the species differences in these receptor mechanisms will require continued development and refinement of models such as primary human cell cultures and human receptor expressing mice. However, in the end, the data generated from these models will no doubt markedly reduce the uncertainty associated with relying solely on animal data for human risk assessment of chemicals.

ACKNOWLEDGMENTS

The authors acknowledge the contribution of Dr. Richard Graham, GlaxoSmithKline, RTP, NC, in providing the rat and human CYP4A data for Figure 4.4.

REFERENCES

1. Kastner PM, Mark M, Chambon P. Nonsteroid nuclear receptors: What are genetic studies telling us about their role in real life? *Cell* 1995; 83(6):859–69.
2. Mangelsdorf DJ, Thummel C, Beato M, et al. The nuclear receptor superfamily: The second decade. *Cell* 1995;83:835–839.
3. Moore LB, Parks DJ, Jones SA, et al. Orphan nuclear receptors constitutive androstane receptor and pregnane X receptor share xenobiotic and steroid ligands. *J Biol Chem* 2000;275:15122–7.
4. Honkakoski P, Negishi M. Regulation of cytochrome P450 (CYP) genes by nuclear receptors. *Biochem J* 2000;347:321–37.
5. Cairns W, Cairns C, Pongratz I, et al. Assembly of a glucocorticoid receptor complex prior to DNA binding enhances its specific interaction with a glucocorticoid response element. *J Biol Chem* 1991;266:11221–6.
6. Lee MS, Kliewer SA, Provencal J, et al. Structure of the retinoid X receptor alpha DNA binding domain: A helix required for homodimeric DNA binding. *Science*. 1993;260:1117–21.
7. Synold TW, Dussault I, Forman BM. The orphan nuclear receptor SXR coordinately regulates drug metabolism and efflux. *Nat Med* 2001;7:584–90.
8. LeCluyse EL. Pregnane X receptor: molecular basis for species differences in CYP3A induction by xenobiotics. *Chem Biol Interact* 2001;134:283–9.
9. Kliewer SA, Moore JT, Wade L, et al. An orphan nuclear receptor activated by pregnanes defines a novel steroid signaling pathway. *Cell* 1998;92:73–82.
10. Goodwin B, Hodgson E, Liddle C. The orphan human pregnane X receptor mediates the transcriptional activation of CYP3A4 by rifampicin through a distal enhancer module. *Mol Pharmacol* 1999;56:1329–39.
11. Xie W, Barwick JL, Simon CM, et al. Reciprocal activation of xenobiotic response genes by nuclear receptors SXR/PXR and CAR. *Genes Dev* 2000;14:3014–23.
12. Rowlands JC, Gustafsson JA. Aryl hydrocarbon receptor-mediated signal transduction. *Crit Rev Toxicol* 1997;27:109–34.
13. Poland A, Knutson JC. 2,3,7,8-Tetrachlorodibenzo-*p*-dioxin and related halogenated aromatic hydrocarbons: Examination of the mechanism of toxicity. *An Rev Pharmacol Toxicol* 1982;22:517–54.

14. Backlund M, Weidolf L, Ingelman-Sundberg M. Structural and mechanistic aspects of transcriptional induction of cytochrome P450 1A1 by benzimidazole derivatives in rat hepatoma H4IIE cells. *Eur J Biochem/FEBS* 1999;26:66–71.
15. Ciolino HP, MacDonald CJ, Memon OS, Bass SE, Yeh GC. Sulindac regulates the aryl hydrocarbon receptor-mediated expression of phase 1 metabolic enzymes in vivo and in vitro. *Carcinogenesis* 2006;27:1586–92.
16. Bittinger MA, Nguyen LP, Bradfield CA. Aspartate aminotransferase generates proagonists of the aryl hydrocarbon receptor. *Mol Pharmacol* 2003;64:550–6.
17. Bjeldanes LF, Kim JY, Grose KR, Bartholomew JC, Bradfield CA. Aromatic hydrocarbon responsiveness-receptor agonists generated from indole-3-carbinol in vitro and in vivo: Comparisons with 2,3,7,8-tetrachlorodibenzo-*p*-dioxin. *Proc Natl Acad Sci USA* 1991;88:9543–7.
18. Denison MS, Heath-Pagliuso S. The Ah receptor: A regulator of the biochemical and toxicological actions of structurally diverse chemicals. *Bull Environ Contam Toxicol* 1998;61:557–68.
19. Denison MS, Nagy SR. Activation of the aryl hydrocarbon receptor by structurally diverse exogenous and endogenous chemicals. *An Rev Pharmacol Toxicol* 2003;43:309–34.
20. Oberg M, Bergander L, Hakansson H, Rannug U, Rannug A. Identification of the tryptophan photoproduct 6-formylindolo[3,2-b]carbazole, in cell culture medium, as a factor that controls the background aryl hydrocarbon receptor activity. *Toxicol Sci* 2005;85:935–43.
21. Tittlemier SA. Dietary exposure to a group of naturally produced organohalogens (halogenated dimethyl bipyrroles) via consumption of fish and seafood. *J Agric Food Chem* 2004;52:2010–5.
22. Zhang S, Qin C, Safe SH. Flavonoids as aryl hydrocarbon receptor agonists/antagonists: Effects of structure and cell context. *Environ Health Perspect* 2003;111:1877–82.
23. Connor K, Harris M, Edwards M, Budinsky R, Clark G, Chu A, Finley B, Rowlands J. The influence of diet on AH receptor activity in human blood measured with a cell-based bioassay: Evidence for naturally occurring AH receptor ligand activity in vivo. 2006; in press.
24. Tijet N, Boutros PC, Moffat ID, Okey AB, Tuomisto J, Pohjanvirta R. Aryl hydrocarbon receptor regulates distinct dioxin-dependent and dioxin-independent gene batteries. *Mol Pharmacol* 2006;69(1):140–53.
25. Connor KT, Aylward LL. Human response to dioxin: Aryl hydrocarbon receptor (AhR) molecular structure, function, and dose–response data for enzyme induction indicate an impaired human AhR. *J Toxicol Environ Health* 2006;9:147–71.
26. Hankinson O. Role of coactivators in transcriptional activation by the aryl hydrocarbon receptor. *Arch Biochem Biophys* 2005;433:379–86.
27. Ramadoss P, Perdew GH. The transactivation domain of the Ah receptor is a key determinant of cellular localization and ligand-independent nucleocytoplasmic shuttling properties. *Biochemistry* 2005;44:11148–59.
28. Karchner SI, Franks DG, Kennedy SW, Hahn ME. The molecular basis for differential dioxin sensitivity in birds: Role of the aryl hydrocarbon receptor. *Proc Natl Acad Sci USA* 2006;103:6252–7.

29. Okey AB, Franc MA, Moffat ID, Tijet N, Boutros PC, Korkalainen M, Tuomisto J, Pohjanvirta R. Toxicological implications of polymorphisms in receptors for xenobiotic chemicals: The case of the aryl hydrocarbon receptor. *Toxicol Appl Pharmacol* 2005;207:43–51.
30. Harper PA, Wong JY, Lam MS, Okey AB. Polymorphisms in the human AH receptor. *Chem Biol Inter* 2002;141:161–87.
31. Van den Berg M, Birnbaum LS, Denison M, De Vito M, Farland W, Feeley M, Fiedler H, Hakansson H, Hanberg A, Haws L, Rose M, Safe S, Schrenk D, Tohyama C, Tritscher A, Tuomisto J, Tysklind M, Walker N, Peterson RE. The 2005 World Health Organization reevaluation of human and mammalian toxic equivalency factors for dioxins and dioxin-like compounds. *Toxicol Sci* 2006;93:223–41.
32. Sun YV, Boverhof DR, Burgoon LD, Fielden MR, Zacharewski TR. Comparative analysis of dioxin response elements in human, mouse and rat genomic sequences. *Nucleic Acids Res* 2004;32:4512–23.
33. Schrenk D, Lipp HP, Wiesmuller T, Hagenmaier H, Bock KW. Assessment of biological activities of mixtures of polychlorinated dibenzo-*p*-dioxins: Comparison between defined mixtures and their constituents. *Arch Toxicol* 1991;65:114–8.
34. Wiebel FJ, Wegenke M, Kiefer F. Bioassay for determining 2,3,7,8-tetrachlorodibenzo-*p*-dioxin equivalents (TEs) in human hepatoma HepG2 cells. *Toxicol Lett* 1996;88:335–8.
35. Xu L, Li AP, Kaminski DL, Ruh MF. 2,3,7,8 Tetrachlorodibenzo-*p*-dioxin induction of cytochrome P4501A in cultured rat and human hepatocytes. *Chem Biol Interac* 2000;124:173–89.
36. Silkworth JB, Koganti A, Illouz K, Possolo A, Zhao M, Hamilton SB. Comparison of TCDD and PCB CYP1A induction sensitivities in fresh hepatocytes from human donors, sprague-dawley rats, and rhesus monkeys and HepG2 cells. *Toxicol Sci* 2005;87:508–19.
37. Moriguchi T, Motohashi H, Hosoya T, Nakajima O, Takahashi S, Ohsako S, Aoki Y, Nishimura N, Tohyama C, Fujii-Kuriyama Y, Yamamoto M. Distinct response to dioxin in an arylhydrocarbon receptor (AHR)-humanized mouse. *Proc Natl Acad Sci USA* 2003;100:5652–7.
38. Klaunig JE, Babich MA, Baetcke KP, Cook JC, Corton JC, David RM, DeLuca JG, Lai DY, McKee RH, Peters JM, Roberts RA, Fenner-Crisp PA. PPARalpha agonist-induced rodent tumors: Modes of action and human relevance. *Crit Rev Toxicol* 2003;33:655–780.
39. Rusyn I, Peters JM, Cunningham ML. Modes of action and species-specific effects of di-(2-ethylhexyl)phthalate in the liver. *Crit Rev Toxicol* 2006;36:459–79.
40. Hurst CH, Waxman DJ. Activation of PPARalpha and PPARgamma by environmental phthalate monoesters. *Toxicol Sci* 2003;74:297–308.
41. Palmer CN, Hsu MH, Griffin KJ, Raucy JL, Johnson EF. Peroxisome proliferator activated receptor-alpha expression in human liver. *Mol Pharmacol* 1998;53:14–22.
42. Sher T, Yi HF, McBride OW, Gonzalez FJ. cDNA cloning, chromosomal mapping, and functional characterization of the human peroxisome proliferator activated receptor. *Biochemistry* 1993;32:5598–604.

43. Walgren JE, Kurtz DT, McMillan JM. Expression of PPAR(alpha) in human hepatocytes and activation by trichloroacetate and dichloroacetate. *Res Commun Mol Pathol Pharmacol* 2000;108:116–32.
44. Cheung C, Akiyama TE, Ward JM, Nicol CJ, Feigenbaum L, Vinson C, Gonzalez FJ. Diminished hepatocellular proliferation in mice humanized for the nuclear receptor peroxisome proliferator-activated receptor alpha. *Cancer Res* 2004;64: 3849–54.
45. Morimura K, Cheung C, Ward JM, Reddy JK, Gonzalez FJ. Differential susceptibility of mice humanized for peroxisome proliferator-activated receptor alpha to Wy-14,643-induced liver tumorigenesis. *Carcinogenesis* 2006;27:1074–80.
46. Bertilsson G, Heidrich J, Svensson K, et al. Identification of a human nuclear receptor defines a new signaling pathway for CYP3A induction. *Proc Natl Acad Sci USA* 1998;95:12208–13.
47. Blumberg B, Sabbagh W, Jr, Juguilon H, et al. SXR, a novel steroid and xenobiotic sensing nuclear receptor. *Genes Dev* 1998;12:3195–205.
48. Zhang H, LeCluyse E, Liu L, et al. Rat pregnane X receptor: Molecular cloning, tissue distribution, and xenobiotic regulation. *Arch Biochem Biophys* 1999;368:14–22.
49. Jones SA, Moore LB, Shenk JL, et al. The pregnane X receptor: A promiscuous xenobiotic receptor that has diverged during evolution. *Mol Endocrinol* 2000;14:27–39.
50. Willson TM, Jones SA, Moore JT, et al. Chemical genomics: Functional analysis of orphan nuclear receptors in the regulation of bile acid metabolism. *Med Res Rev* 2001;21:513–22.
51. Moore JT, Kliewer SA. Use of the nuclear receptor PXR to predict drug interactions. *Toxicology* 2000;153:1–10.
52. Lehmann JM, McKee DD, Watson MA, et al. The human orphan nuclear receptor PXR is activated by compounds that regulate CYP3A4 gene expression and cause drug interactions. *J Clin Invest* 1998;102:1016–23.
53. Schuetz JD, Beach DL, Guzelian PS. Selective expression of cytochrome P450 CYP3A mRNAs in embryonic and adult human liver. *Pharmacogenetics* 1994;4:11–20.
54. Watkins RE, Wisely GB, Moore LB, et al. The human nuclear xenobiotic receptor PXR: Structural determinants of directed promiscuity. *Science* 2001;292:2329–33.
55. Xie W, Barwick JL, Downes M, et al. Humanized xenobiotic response in mice expressing nuclear receptor SXR. *Nature* 2000;406:435–39.
56. Staudinger JL, Goodwin B, Jones SA, et al. The nuclear receptor PXR is a lithocholic acid sensor that protects against liver toxicity. *Proc Natl Acad Sci USA* 2001;98:3369–74.
57. Goodwin B, Moore LB, Stoltz CM, et al. Regulation of the human CYP2B6 gene by the nuclear pregnane X receptor. *Mol Pharmacol* 2001;60:427–31.
58. Traber PG, Wang W, McDonnell M, et al. P450IIB gene expression in rat small intestine: Cloning of intestinal P450IIB1 mRNA using the polymerase chain reaction and transcriptional regulation of induction. *Mol Pharmacol* 1990;37:810–19.
59. Honkakoski P, Moore R, Washburn KA, et al. Activation by diverse xenochemicals of the 51-base pair phenobarbital-responsive enhancer module in the CYP2B10 gene. *Mol Pharmacol* 1998;53:597–601.

60. Honkakoski P, Zelko I, Sueyoshi T, et al. The nuclear orphan receptor CAR-retinoid X receptor heterodimer activates the phenobarbital-responsive enhancer module of the CYP2B gene. *Mol Cell Biol* 1998;18:5652–58.
61. Baes M, Gulick T, Choi HS, et al. A new orphan member of the nuclear hormone receptor superfamily that interacts with a subset of retinoic acid response elements. *Mol Cell Biol* 1994;14:1544–51.
62. Choi HS, Chung M, Tzameli I, et al. Differential transactivation by two isoforms of the orphan nuclear hormone receptor CAR. *J Biol Chem* 1997;272:23565–571.
63. Forman BM, Tzameli I, Choi HS, et al. Androstane metabolites bind to and deactivate the nuclear receptor CAR-beta. *Nature* 1998;395:612–15.
64. Mizukami Y, Sogawa K, Suwa Y, et al. Gene structure of a phenobarbital-inducible cytochrome P-450 in rat liver. *Proc Natl Acad Sci USA* 1983;80:3958–62.
65. Trottier E, Belzil A, Stoltz C, et al. Localization of a phenobarbital-responsive element (PBRE) in the 5′-flanking region of the rat CYP2B2 gene. *Gene* 1995;158:263–68.
66. Honkakoski P, Negishi M. Characterization of a phenobarbital-responsive enhancer module in mouse P450 Cyp2b10 gene. *J Biol Chem* 1997;272:14943–49.
67. Sueyoshi T, Kawamoto T, Zelko I, et al. The repressed nuclear receptor CAR responds to phenobarbital in activating the human CYP2B6 gene. *J Biol Chem* 1999;274:6043–46.
68. Ramsden R, Beck NB, Sommer KM, et al. Phenobarbital responsiveness conferred by the 5′-flanking region of the rat CYP2B2 gene in transgenic mice. *Gene* 1999;228:169–79.
69. Poland A, Mak I, Glover E, et al. 1,4-*Bis*[2-(3,5-dichloropyridyloxy)]benzene, a potent phenobarbital-like inducer of microsomal monooxygenase activity. *Mol Pharmacol* 1980;18:571–80.
70. Kawamoto T, Sueyoshi T, Zelko I, et al. Phenobarbital-responsive nuclear translocation of the receptor CAR in induction of the CYP2B gene. *Mol Cell Biol* 1999;19:6318–22.
71. Walker D, Htun H, Hager GL. Using inducible vectors to study intracellular trafficking of GFP-tagged steroid/nuclear receptors in living cells. *Methods* 1999;19:386–93.
72. Picard D, Yamamoto KR. Two signals mediate hormone-dependent nuclear localization of the glucocorticoid receptor. *EMBO J* 1987;6:3333–40.
73. Guiochon-Mantel A, Lescop P, Christin-Maitre S, et al. Nucleocytoplasmic shuttling of the progesterone receptor. *EMBO J.* 1991;10:3851–59.
74. Tzameli I, Pissios P, Schuetz EG, et al. The xenobiotic compound 1,4-*bis*[2-(3,5-di chloropyridyloxy)]benzene is an agonist ligand for the nuclear receptor CAR. *Mol Cell Biol* 2000;20:2951–58.
75. Muangmoonchai R, Smirlis D, Wong SC, et al. Xenobiotic induction of cytochrome P450 2B1 (CYP2B1) is mediated by the orphan nuclear receptor constitutive androstane receptor (CAR) and requires steroid co-activator 1 (SRC-1) and the transcription factor Sp1. *Biochem J* 2001;355:71–78.
76. Smirlis D, Muangmoonchai R, Edwards M, et al. Orphan receptor promiscuity in the induction of cytochromes P450 by xenobiotics. *J Biol Chem* 2001;276: 12822–6.

77. Faucette SR, Sueyoshi T, Smith CM, Negishi M, LeCluyse EL, Wang H. Differential regulation of hepatic CYP2B6 and CYP3A4 genes by constitutive androstane receptor but not pregnane X receptor. *J Pharmacol Exp Ther* 2006;317(3): 1200–09.

78. Kemper B. Regulation of cytochrome P450 gene transcription by phenobarbital. *Prog Nucleic Acid Res Mol Biol* 1998;61:23–64.

79. Maglich JM, Parks DJ, Moore LB, Collins JL, Goodwin B, Billin AN, et al. Identification of a novel human constitutive androstane receptor (CAR) agonist and its use in the identification of CAR target genes. *J Biol Chem* 2003;278:17277–83.

80. Auerbach SS, Stoner MA, Su S, Omiecinski CJ. Retinoid X receptor-α-dependent transactivation by a naturally occurring structural variant of human constitutive androstane receptor (NR1I3). *Mol Pharmacol* 2005;68:1239–53.

81. Faucette SR, Zhang TC, Moore R, Sueyoshi T, Omiecinski C, LeCluyse EL, et al. Relative activation of human pregnane X receptor versus constitutive androstane receptor defines distinct classes of CYP2B6 and CYP3A4 inducers. *J Pharmacol Exp Ther* 2007;320:72–80.

82. Wei P, Zhang J, Egan-Hafley M, et al. The nuclear receptor CAR mediates specific xenobiotic induction of drug metabolism. *Nature* 2000;407:920–3.

83. Ueda A, Hamadeh HK, Webb HK et al. Diverse roles of the nuclear orphan receptor CAR in regulating hepatic genes in response to phenobarbital. *Mol Pharmacol* 2002;61:1–6.

84. Yoshinari K, Sueyoshi T, Moore R, et al. Nuclear receptor CAR as a regulatory factor for the sexually dimorphic induction of CYB2B1 gene by phenobarbital in rat livers. *Mol Pharmacol* 2001;59:278–84.

85. Wang H, Faucette S, Sueyoshi T, Moore R, Ferguson S, Negishi M, et al. A novel distal enhancer module regulated by pregnane X receptor/constitutive androstane receptor is essential for the maximal induction of CYP2B6 gene expression. *J Biol Chem* 2003278:14146–52.

86. Food and Drug Administration. Draft guidance for industry on clinical lactation studies—Study design, data analysis, and recommendations for labeling. Department of Health and Human Services, Food and Drug Administration, Docket No. 2005D-0030, Sept. 2006.

87. Guo GL, Moffit JS, Nicol CJ, Ward JM, Aleksunes LA, Slitt AL, et al. Enhanced acetaminophen toxicity by activation of the pregnane X receptor. *Toxicol Sci* 2004;82:374–80.

88. Zhang J, Huang W, Chua SS, Wei P, Moore DD. Modulation of acetaminophen-induced hepatotoxicity by the xenobiotic receptor CAR. *Science* 2002;298:422–4.

89. Backman JT, Olkkola KT, Neuvonen PJ. Rifampin drastically reduces plasma concentrations and effects of oral midazolam. *Clin Pharmacol Ther* 1996;59:7–13.

90. Terzolo M, Borretta G, Ali A, et al. Misdiagnosis of Cushing's syndrome in a patient receiving rifampicin therapy for tuberculosis. *Horm Metab Res* 1995;27: 148–50.

91. Izzo AA, Ernst E. Interactions between herbal medicines and prescribed drugs: A systematic review. *Drugs* 2001;61:2163–75.

92. Fugh-Berman A, Ernst E. Herb–drug interactions: Review and assessment of report reliability. *Br J Clin Pharmacol* 2001;52:587–95.

93. Ernst E. Second thoughts about safety of St John's wort. *Lancet* 1999; 354:2014–6.
94. de Maat MM, Hoetelmans RM, Math t RA, et al. Drug interaction between St John's wort and nevirapine. *Aids* 2001;15:420–1.
95. Moschella C, Jaber BL. Interaction between cyclosporine and *Hypericum perforatum* (St. John's wort) after organ transplantation. *Am J Kidney Dis* 2001; 38:1105–7.
96. Moore L, Goodwin B, Jones SA, Wisely GB, Serabjit-Singh CJ, Willson TM, et al. St. John's wort induces hepatic drug metabolism through activation of the pregnane X receptor. *Proc Natl Acad Sci USA* 2000;97:7500–2.
97. Haws LC, Su SH, Harris M, Devito MJ, Walker NJ, Farland WH, Finley B, Birnbaum LS. Development of a refined database of mammalian relative potency estimates for dioxin-like compounds. *Toxicol Sci* 2006;89:4–30.
98. Kast HR, Goodwin B, Tarr PT, Jones SA, Anisfeld AM, Stoltz CM, et al. Regulation of multidrug resistance-associated protein 2 (ABCC2) by the nuclear receptors pregnane X receptor, farnesoid X-activated receptor, and constitutive androstane receptor. *J Biol Chem* 2002;277(4):2908–15.
99. Schuetz EG. Induction of cytochromes P450. *Curr Drug Metab* 2001;2(2):139–47.

5

TOXICOGENOMICS AND SYSTEMS TOXICOLOGY

MICHAEL D. WATERS, JENNIFER M. FOSTEL,
BARBARA A. WETMORE, AND B. ALEX MERRICK

Contents

Computational Toxicology: Risk Assessment for Pharmaceutical and Environmental Chemicals,
Edited by Sean Ekins

5.1 INTRODUCTION

Toxicology as the study of poisons is focused on the substances and exposures that cause adverse effects in living organisms. A critical part of this study is the empirical and contextual characterization of adverse effects at different levels of organization of the organism, ranging from animal health and function, organs, tissues, to cells, and intracellular and intercellular molecular systems. Thus studies in toxicology measure the effects of an agent on an organism's food consumption and digestion, on its body and organ weights, on microscopic histopathology, and on cell viability, immortalization, necrosis, and apoptosis [1].

Toxicology and, by extension toxicogenomics, are multidisciplinary in nature and have evolved significantly as relevant technologies have advanced. The invention of the microscope in the seventeenth century was the beginning of a long road of technological development to modern histology and pathology, which is an inherent component of toxicology. The intricacies of tissue organization and complex architecture of organs at a cellular level were made accessible by light microscopy, which then gave way to even further resolution with the advent of electron microscopy and more recently atomic force microscopy [2,3]. The 20th century saw the maturing of histology as a more refined tool for pathology and in a like manner has presaged the potential of toxicogenomics for the field of toxicology.

Similarly the modern achievements of sequencing whole genomes have been quickly followed by gene expression technologies that allow comprehensive queries of the transcriptome with regard to the refinement of traditional proteomics and to the creation of other -omic technologies. Toxicogenomics [4] evolved from the desire to characterize how genomes respond to environmental stressors or toxicants by combining genomewide mRNA expression profiling (transcriptomics) with global protein expression patterns (proteomics) that are interpreted by the use of bioinformatics to understand the role of gene–environment interactions in disease and dysfunction.

The increasing resolution of toxicogenomic analysis with oligonucleotide probe sets is becoming sufficiently refined for the production of single nucleotide polymorphism (SNP) genotyping arrays that can be used for cytochrome P450 genotyping and SNP mapping for identifying relevant gene loci for sensitivities or resistant responses [5–7]. The inherently reductive nature of toxicogenomic analysis down to the level of DNA, mRNA, and protein sequences is also being counterbalanced by a concerted attempt to reassemble these molecular pieces of information into pathways and networks that form the new field of systems toxicology [8–13].

The result of these concurrent reductive and assembly activities in gene expression information is a much greater depth of field now possible for examining toxicant responses. Toxicogenomics is leading the next revolution toward a better understanding of molecular pathology. Expectations are that toxicology and pathology of the future will couple three areas: traditional pathology and toxicology, differential protein and gene expression analysis, and systems biology [14]. The teaming of these technologies will extend the sensitivity of toxicity detection beyond what is currently achievable and may even uncover the earliest beginnings of acute toxicity onset or the molecular signatures of long-term toxicant exposure and disease [15,16].

The ability to discern mechanisms of toxicity as related to health issues is an important challenge facing scientists, public health decision-makers, and regulatory authorities, whose aim is to protect humans and the environment from exposures to hazardous drugs, chemicals, and environmental stressors (e.g., non-ionizing radiation). The problems of performing safety and risk assessments for drugs and chemicals and of identifying environmental factors involved in the etiology of human disease have long been formidable issues. Genomic technologies offer the potential to change the way in which toxicity and human health risk are assessed.

The rapid accumulation of genomic sequence data and associated gene and protein annotation has catalyzed the application of gene expression analysis to understand the modes of action of chemicals and other environmental stressors on biological systems (see Figure 5.1). These developments have

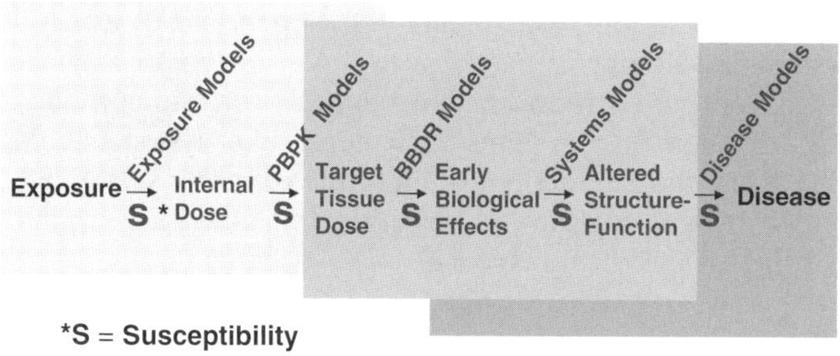

Figure 5.1 Role of genetic susceptibility and computational modeling on the continuum from exposure to disease outcome. The sequence of events between initial exposure and final disease outcome are shown from left to right. Following exposure, the body's ADME (absorption, distribution, metabolism, excretion) systems control local concentrations of a chemical stressor in various body compartments. The impact of genetics is felt in specific alleles encoding various transporters, xenobiotic metabolizing enzymes, etc. Mathematical models such as exposure models, PB/PK, and BBDR models can be used to approximate these processes. PB/PK models are a set of differential equations structured to provide a time course of a chemical's mass-balance disposition (wherein all inputs, outputs, and changes in total mass of the chemical are accounted for) in pre-selected anatomical compartments. BBDR models are dose–response models based on underlying biological processes. Once the target tissue is exposed to a local stressor, the cells response and adapt or undergo a toxic response; this process can be modeled with systems toxicology approaches. Finally, the disease outcome itself can be mimicked by genetic or chemically induced models of particular diseases, for instance, in the Zucker rat model of diabetes or streptozotocin-treated rat model.

facilitated the emergence of the field of toxicogenomics, which aims to study the response of a whole genome to toxicants or environmental stressors [17–28]. The related field of toxicoproteomics [29–31] is similarly defined with respect to the proteome, the protein subset of the genome. The emerging field of toxicometabonomics combines high-throughput intermediary metabolite or pharmacuetical metabolite profiling with computer-assisted pattern recognition approaches [32–34]. Global technologies, such as cDNA and oligonucleotide microarrays, protein chips, mass spectrometry, and nuclear magnetic resonance (NMR)-based molecular profiling, can simultaneously measure the expression of numerous genes, proteins, and metabolites, thus providing the potential to accelerate the discovery of toxicant pathways, modes-of-action, and specific chemical and drug targets. Toxicogenomics therefore combines toxicology with genetics, global -omics technologies (see Appendix 5.1), and appropriate pharmacological and toxicological models (Figure 5.1) to provide a comprehensive view of the function of the genetic and biochemical machinery of the cell.

This chapter explores the new field of toxicogenomics, delineates some of its research approaches and success stories, and describes the challenges it faces. It discusses how integrating data derived from transcriptomics, proteomics and metabonomics studies can contribute to the development of a toxicogenomics knowledgebase (Figure 5.2, Appendix 5.2) and to the evolution of systems toxicology as it relates to molecular expression profiling. In many ways current gene, protein, and metabolite expression profiles are simple "snapshots"; by contrast, systems toxicology, like systems biology [8,35], attempts to define the interactions of all of the elements in a given biological system, under stress or toxicant perturbation, to achieve a mechanistic understanding of the toxicological response.

5.2 TOXICOGENOMICS: AIMS AND METHODS

Toxicogenomics has three principal goals: to understand the relationship between environmental stress and human disease susceptibility (Figure 5.1), to identify useful biomarkers of disease and exposure to toxic substances, and to elucidate the molecular mechanisms of toxicity. A typical toxicogenomics study might involve an animal experiment with three treatment groups: high- and low-dose treatment groups and a vehicle control group that has received only the solvent used with the test agent. These groups will be observed at two to three points in time, with three to five animal subjects per group. In this respect a toxicogenomics investigation resembles a simple, acute toxicity study. Where the two approaches differ is in the scope of the response they each aim to detect and in the methods used. The highest dose regimen is intended to produce an overtly toxic response, which in a toxicogenomics study can be detected using the global measurement techniques described below (see also Appendix 5.1).

In a typical toxicogenomics experiment, lists of significantly differentially expressed genes are created for each biological sample [36]. Alternatively, profile analysis methods can be applied to dose- and time-course studies [36] to identify genes and gene profiles of interest. Then, with the aid of the relevant knowledge that is systematically extracted and assembled [37] through literature mining, comparative analysis, and iterative biological modeling of molecular expression datasets, it is possible to differentiate adaptive responses of biological systems from those changes (or biomarkers) associated with or precedent to clinical or visible adverse effects. Over the past 5 to 10 years the field of toxicogenomics has validated the concept of gene expression profiles as "signatures" of toxicant classes, disease subtypes, or other biological endpoints. These signatures have effectively directed the analytical search for predictive biomarkers of toxicant effects and contributed to the understanding of the dynamic alterations in molecular mechanisms associated with toxic and adaptive responses.

The experimental work involved in a toxicogenomics study and the amount of gene expression data generated are vast. To examine even only one tissue per animal in the study design described above requires 18 to 45 microarrays (more if technical replicates are used) and the attendant measurement of as many as 20,000 or more transcripts per array. In addition each animal will typically have treatment-associated data on total body and organ weight measurements, clinical chemistry measurements (often up to 25 parameters), and microscopic histopathology findings for several tissues [1]. The careful collection, management, and integration of these data, in the context of the experimental protocol, is essential for interpreting toxicological outcomes. Thus all data must be recorded in terms of dose, time, and severity of the toxicological and/or histopathological phenotype(s). The compilation of such experimental data, together with toxicoinformatics tools (see also Chapter 6) and computational modeling, will be important in obtaining a better understanding of toxicant-related diseases [22].

Toxicogenomics integrates the multiple data streams derived from transcriptomics, proteomics and metabonomics with traditional toxicological and histopathological endpoint evaluation (Figure 5.2). This integration has the potential to deepen our understanding of the relationship between toxicological outcomes and molecular genetics. Furthermore toxicology and toxicogenomics are progressively developing from studies done predominantly on individual chemicals and stressors into a knowledge-based science [26]. However, the evolution of a truly "predictive toxicology"—wherein knowledge of toxicogenomic responses of a prototypic agent in one species and strain is used to predict the mode-of-action of a similar agent in a related strain or another species—will require that the results of numerous toxicogenomics investigations across genotypes and species be assimilated into a multi-domain, multi-genome, knowledgebase (Figure 5.2, Appendix 5.2). This knowledgebase could be searchable by chemical formula/stressor-type, by gene/protein/metabolite molecular signature, or by phenotypic outcome, among other

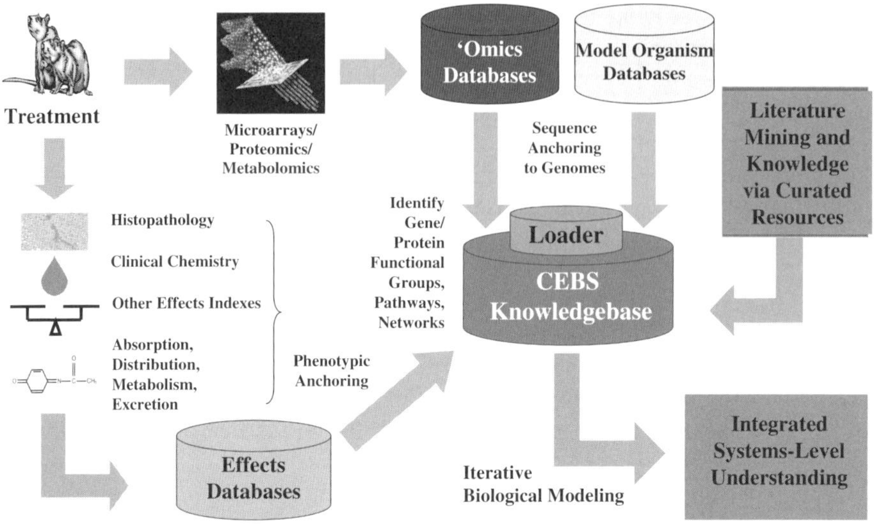

Figure 5.2 Framework for systems toxicology. The figure indicates the paths from the initial observation (rat in upper left) to an integrated toxicogenomics knowledgebase and thence to systems toxicology. The -omics data stream is shown by the clockwise path from rat to knowledgebase, and the traditional toxicology approach is shown in the counterclockwise path. The knowledgebase will integrate both data streams, along with literature knowledge, and, by virtue of iterative modeling, lead to a systems toxicology understanding. The framework involves "phenotypic anchoring" (to toxicological endpoints and study design information) and "sequence anchoring" (to genomes) of multidomain molecular expression datasets in the context of conventional indices of toxicology, and the iterative biological modeling of resulting data. See color plates.

entities, to find results analogous to those observed with a newly tested agent. It should capture current knowledge of chemical mode-of-action and facilitate the discovery of new modes. Toxicology will then have become an information science, and public health and risk assessment will be the beneficiaries.

5.3 TOXICOGENOMICS: EVOLUTION OF THE FIELD

Toxicogenomics evolved from early gene expression studies on the response of a biological system to a particular toxicant or panel of reference agents, toward more mature investigations that integrate several -omics domains with toxicology and pathology data (see Table 5.1). Exposure-specific and outcome-specific patterns of gene, protein, and metabolite profiles have been used to identify molecular changes that serve as biomarkers of toxicity [23,38–44] and provide insights into mechanisms of toxicity [45–55] and disease causation [56–60]. Critical to this evolution were extensive and ongoing genome sequencing and annotation efforts [61,62] and the ability to describe response profiles

TABLE 5.1 Scope and Evolution of Toxicogenomics

Study Aim	Key References
Toxicogenomics tools and model systems	Toxicogenomics began with "toxicology-specific" cDNA microarrays designed to measure the levels of acute phase and xenobiotic-metabolizing enzymes such as cytochrome P450s [4,199]. These were superseded as commercial platforms were developed for toxicologically important species such as rat. The armamentarium of pre-clinical gene expression platforms was completed with the canine microarray [62]. It is now possible to use commercial oligonucleotide microarrays (see Table 5.3) to measure expression responses in species ranging from nematode (*Caenorhabditis elegans*), to frog (*Xenopus laevis*) and zebrafish (*Danio rerio*) to rodents (rat and mouse), to nonhuman primates and humans. Toxicogenomics tools for sentinel aquatic species have been developed as well [97]. Later experiments began to focus on more challenging subjects such as subcellular organelles [200], nonstandard tissue such as saliva [118], less well characterized species [201], genetic models of diseases [202], and integration of data from different-omics disciplines [46,54,59,97,105,138]. Additionally comprehensive studies of yeast have become increasingly important [170,171,203].
Tissues used in toxicogenomics studies	Most toxicogenomics studies to date have involved hepatotoxicants [23,36, 38–44,47,49,50,52–55,57,59,60,204], as the liver is the primary source of xenobiotic metabolism and detoxification and because liver injury is the principal reason for withdrawal of new drugs from the market [205]. Toxicogenomics studies have also addressed nephrotoxicity [44,45,51], neurotoxicity [206,207], reproductive toxicity [48], as well as lung toxicity [39,56], skin toxicity [208], and cardiotoxicity [209].
Phenotypic anchoring	Phenotypic anchoring relates expression profiles to specific adverse effects defined by conventional measures of toxicity such as histopathology or clinical chemistry [22,26,70].
	Experiments have been designed to correlate expression patterns with disease pathologies such as necrosis, apoptosis, fibrosis, and inflammation [36,38,56,62,210]. Additionally phenotypic anchoring can be used to provide biological context for toxicogenomics observations made at subtoxic doses [41,53].
Classes of toxicants characterized	Studies have examined responses to toxicants with established mechanisms of toxicity [38,43,44,49,50,52,60,211], environmental toxicants [57,97,208,212], and exposures to suprapharmacological levels of drugs [39,41,46,47,53,54,59,204,209].

TABLE 5.1 (*Continued*)

Study Aim	Key References
Examples of toxicant or stressor mechanisms	Adverse effects of acetaminophen [41,46,54,204,213], estrogenic agents [48,214], oxidant stress [203,215], and peroxisome proliferators [23,42,44,50,52] are among the main toxicants being studied.
Importance of reporting husbandry and other technical details	Expression profiles are altered by experimental conditions, including the harvest method, the in vitro culture method, and the vehicle used to deliver an agent, time of day of sacrifice, and diet. Up to 9% of the transcripts in mouse liver fluctuated with circadian cycling [216]. These included genes controlling glucose metabolism and vesicle trafficking or cytoskeleton, as might be anticipated from changes in the diet of animals during the day and night. In addition transcript levels of Cyp17 and Cyp2a4, which are important for steroid synthesis, and Cyp2e1, which is important for detoxification of xenobiotics, fluctuated. These changes might be expected to impact the response to test agents and reflect a requirement to report the time of day of dosing and sacrifice, along with the diet, vehicle and harvest and culture methods, when summarizing or publishing results of toxicogenomics studies.
Commercial database resources for toxicogenomics profiles	Toxicogenomics studies for the purpose of developing commercial databases have been performed by both GeneLogic and Iconix, <http://www.genelogic.com> and <http://www.iconixpharm.com/>. These companies have each gathered data from several hundreds of samples produced from short-term exposures of agents at pharmacological and toxicological dose levels. Customers of either company can access the respective databases to classify the mode-of-action of novel agents of interest.
Integration of toxicogenomics efforts	Through such consortia as the ILSI Committee on the Application of Toxicogenomics to Risk Assessment [63,64], the Toxicogenomics Research Consortium (TRC), and the Consortium on Metabonomics and Toxicology (COMET) [65], the technical factors affecting data can be identified and overcome, approaches to data analysis and interpretation can be agreed upon, and high-quality public datasets prepared. The field of toxicoproteomics is currently not represented by a consortium (see Human Proteome Organization HUPO in Table 5.3), while the ILSI Genomics Committee and the TRC are working toxicogenomics consortia in transcriptomics, and COMET is a working toxicogenomics consortium in metabonomics.
Integration of data domains	By integration of data a more complete picture of the expression profiles associated with a particular treatment can be obtained, not only of what the cell is planning (transcriptomics) but what occurred in the proteome and metabonome [46,54,97,138,217].

Note: Peroxisome proliferators are compounds that induce increased numbers of peroxisomes—single-membrane cytoplasmic organelles that metabolize long-chain fatty acids.

in genetically and toxicologically important species such as mouse, rat, dog, and human. Another important contribution to toxicogenomics has been the formation of collaborative research consortia [63–65] that bring together scientists from regulatory agencies, industrial laboratories, and academic and governmental institutions to identify and address important issues for the field.

5.3.1 Profiles of Response to Toxicants

Nuwaysir et al. popularized the term "toxicogenomics" to describe the use of microarrays to measure the responses of toxicologically relevant genes and to identify selective sensitive biomarkers of toxicity [4]. The first published toxicogenomics study compared the gene expression profiles of human cells responding to the inflammatory agent lipopolysaccharide (LPS) or to mitogenic activation by phorbol myristate acetate (PMA) [66]. RNA samples isolated at various times following exposure showed the expected increases in cytokine, chemokine, and matrix metalloproteinase transcripts. Similar expression profiles were seen in synoviocytes and chondrocytes from a patient with rheumatoid arthritis, confirming the ability of the system to mimic the biological changes that occur during inflammatory disease. Subsequent studies extended this type of observation in other tissues and for a wide variety of toxicants, enabling the association of specific molecular profiles with specific toxicities [67].

5.4 PHENOTYPIC ANCHORING

Conventional toxicology has employed surrogate markers correlated with toxic responses to monitor adverse outcomes in inaccessible tissues [68]. For example, liver enzymes ALT (alanine aminotransferase) and AST (aspartate aminotransferase) are released following hepatic damage, and levels of these enzymes in serum correlate with histopathological changes in the liver [68,69]. These serum enzyme markers, in conjunction with histopathology, facilitate the "phenotypic anchoring" of molecular expression data [22,26,70]. "Phenotypic anchoring" is the process of determining the relationship between a particular expression profile and the pharmacological or toxicological phenotype of the organism for a particular exposure or dose and at a particular time [22]. The dose and time alone are often insufficient to define the toxicity experienced by an individual animal; thus another measure of toxicity is needed for full interpretation of the data obtained during a toxicogenomics study. Conversely, the phenotype alone may be insufficient to anchor the molecular profile, since an elevated value for serum ALT can be observed both before peak toxicity (as it rises) and after peak toxicity (as it returns to baseline). Anchoring the molecular expression profile in phenotype, dose, and time helps to define the sequence of key molecular events in the mode-of-action of a toxicant.

Phenotypic anchoring can also be used in conjunction with lower doses to classify agents and to explore the mechanisms of toxicity that occur before histopathological changes are seen. For example, transcriptional changes that occur following both low- and high-dose exposures of acetaminophen were identified, indicating that biological responses can be detected using transcriptome measurements before histopathological changes are easily detected [41]. Recent follow-up work shows the accumulation of nitrotyrosine and 8-hydroxy-deoxyguanosine adducts phenotypically anchors an oxidative stress gene expression signature observed with subtoxic dose of acetaminophen (APAP), lending support to the validity of gene expression studies as a sensitive and biologically meaningful endpoint in toxicology [71]. Additionally phenotypic anchoring can help to elucidate a toxicant's mechanism of action, for example, the transcriptional responses in a rat model to superpharmaceutical doses of WAY-144122 (a negative regulator of insulin) were observed before histopathological changes were seen in either liver or ovary, and reflected different mechanisms of toxicity in the two organs [53].

5.5 BIOMARKERS AND SIGNATURES

A biomarker is an objective measure that can indicate health, disease, pharmacologic response to therapy, or adverse response to toxicants [72,73]. Several reviews have been written about expression profiling and biomarker discovery [73–79]. Many classic biomarkers such as serum alanine aminotransferase, sorbitol dehydrogenase, and ornithine carbamyltransferase can almost function as singular indicators of specific types of liver pathology and dysfunction [80]. But at our current level of understanding for many complex diseases and toxicities, it is unlikely that any one readout will be a sufficient indicator. Multiple markers that function as a "molecular signature" or "metabolic fingerprint" are still being needed for classifying and better describing toxicant mechanisms of action [72,81,82].

To some degree, the wide use of cluster analysis in classifying toxicity and disease [83,84] seems to favor toxicity or disease characterization by transcript signature profile over a single biomarker. However, one appraisal of toxicogenomic studies outlined the following challenges that exist in developing such a profile: (1) it is difficult to generate a transcriptional profile that is truly predictive rather than diagnostic of an expected outcome, (2) few of the studied toxicants had a specific molecular target and unique mechanism of action that could provide a well defined response, and (3) it remains a challenge to distinguish between primary and secondary (cause or consequence) transcriptomic changes in toxicity among adaptive responses unrelated to toxicity [82]. In addition some of the other major factors that detract from robust DNA microarray toxicity signatures are variation in signal generational use of different transcriptomic platforms, variation in the number of arrayed genes and sequences, and inter-individual animal differences due to genetic or microenvironmental factors [82]. Such difficulties encountered in transcript profiling of toxicants in both experimental and preclinical studies will undoubtedly be a shared experience for proteomic analysis in similar settings. Alternatively, it might also be argued that internal redundancies in biochemical pathways and dilution effects by nonresponding cells can combine to either mask or dampen the expression of key, singular gene transcripts and proteins that are the rate-limiting step for toxicity in target cells during profiling studies.

Yet the number of citations in proteomics and toxicology to date suggest keen research and commercial interests in how to best use proteomic technologies for biomarker discovery. The terminology in biomarker research is also likely to continue expanding from biomarkers, signatures, and profiles to other terms like "barcode" [85] for simultaneous detection of multiple bioanalytes for a specific disease or "footprint" [86] from the signal transduction community to indicate proteins held in a partially activated state. Once a more sophisticated integration of proteomics technologies with toxicology occurs and biological responses brought on by subtle changes in activation status or post-translational modifications are more readily determined, the use of these terms will likely become more widespread. The use of transcriptomics in discovering individual biomarkers is still a valid and critical tool, particularly when conventional biomarkers are lacking. A class of lead compounds identified in a discovery program based on gamma secretase inhibition as therapy for Alzheimer's disease also had an undesirable effect of inhibiting cleavage by Notch1 of the Hes1 gene product, a process important for differentiation of intestinal epithelial cells. Through the use of gene expression profiling and subsequent protein analysis, Searfoss et al. [87] identified adipsin as a novel biomarker for this toxicity.

Carcinogenic potential is conventionally measured using a two-year study, incurring significant expense in both animals and human resources. It is therefore of great interest to identify biomarkers of carcinogenicity that can be detected in acute, short-term studies, and efforts toward this have been reported [36,40,58–60,88]. Biomarkers with clinical relevance have also been found using toxicogenomics approaches. For example, Petricoin et al. [89] found a set of protein markers that distinguished patients with high levels of prostate-specific antigen (PSA), a clinical marker correlated with prostate cancer, from those with low PSA levels and thus presumed to be healthy. In addition the marker set also correctly predicted 71% of patients with intermediate PSA levels. In a second example [90], three plasma biomarkers were discovered and then independently validated to detect early stage invasive epithelial ovarian cancer from healthy controls with high sensitivity and specificity compared to the traditional marker CA125 alone. These new biomarkers demonstrate the potential to improve the detection of early stage ovarian cancer using toxicoproteomics technologies.

The sequencing of the human genome has brought about a systematic means of viewing the molecular basis of disease through genes and their many levels of regulation. Genomics-based tools will be developed for diagnosis and prediction of disease onset or recurrence, personalized medicine, and assess treatment response [91]. Although in its beginnings, genomic biomarker research has the capability of providing high-definition viewing into pathophysiological processes and providing more precise predictors of outcome not previously possible with traditional biomarkers. Before genomic biomarkers can be regularly integrated into clinical practice, the necessary levels of evidence must be brought to bear to demonstrate analytical and clinical validity and ultimately their utility in improving patient care. Similar caveats apply prior to being embraced by the pharmaceutical and regulatory sectors for preclinical safety assessment.

5.6 PROTEOMIC ANALYSIS OF TOXIC SUBSTANCES

A previously published review on the role of proteomics in toxicology [74] covered the earlier toxicoproteomics literature, so the review in Table 5.1 will cover more recent citations with minimal overlap. Table 5.2 summarizes data from primary literature that reports on the effects of toxic agents on global protein expression. Entries are summarized by author, proteomic platform, chemicals or toxicants tested, the tissue analyzed, and a very brief statement of results and potential markers of toxic effect.

Of the 55 toxicology studies listed, 41 utilized the 2D gel-mass spectrometry (MS) platform, with 6 of these being DIGE-based 2D gels and the remainder using other forms of protein identification. Four (liquid chromatography) LC tandem-MS studies [92–95] and two RC-MS (surface-enhanced laser desorption ionization, SELDI) [96,97] studies are listed, as well as three reports using antibody arrays designed for phosphorylated proteins [98], cellular and signaling proteins [99], and cytokines/chemokines [100]. Most uses of proteomics involve protein separation from target organ homogenates, followed by protein identification from MALDI-MS and tandem MS from trypsinized proteins. A few studies examined protein changes in subcellular fractions or organelles. Liver was the principal organ studied by proteomics and the main biofluids in seven studies were plasma or serum and urine. Four reports were included on identification of chemical-adducted proteins in cells [95], tissue [101], or site-specific hemoglobin adducts by electrophilic chemicals [92,93]. There is substantial research interest in protein adducts, and here circulating red blood cells and serum proteins are frequent targets for special attention from researchers [102]. Only two studies examined posttranslationally modified proteins in a semi-global manner for nitrosylation [103] and phosphorylation [98] in response to toxicants or reactive intermediates.

Several studies reported both transcript and protein changes to toxicant exposure, indicated in Table 5.2 as "DNA microarray/proteomics" [45,54, 59,97,99,100,104,105]. These researchers found some disparities between transcriptomic and proteomics data (see section on transcriptomics and

TABLE 5.2 Proteomic Studies in Toxicology

Author	Platform	Chemical	Tissue	Results
[92]	LC-MS/MS	Epoxides	Human Hgb	Epoxide adduct sites IDed
[218]	2D gel	DEHP; decanoic acids, PPAR	Rat liver	Protein-mapping algorithm
[219]	2D-MS	4-Aminophenol D-serine	Rat plasma	↑ Plasma FAH in serum in renal toxicity
[219]	2D-MS	4-Aminophenol; D-, L-ser, *cis*/*tr*-Pt	Rat kidney; plasma	↑ Plasma kininogen in renal toxicity; plasma biomarkers
[107]	2D-MS	TCDD	Rat liver ER	NFkB proteins, IKBs, IKKs found in ER
[103]	2D-MS	$ONOO^-$	Human brain	6 targets of protein nitration IDed in Alzheimer brains
[220]	2D-MS	Gentamicin	Rat kidney	20 proteins IDed; mitochondria dysfunction in renal cortex
[45]	2D-MS	Puromycin	Rat urine	Proteomics/metabolomics to study renal glomerular toxicity
[108]	2D-MS	GT oligomers	Human	Basic isoform of eEF1A IDed in lymphocytes GT oligomer cytotoxicity complex
[96]	SELDI	TMPD	Rat urine	Parvalbumin-α biomarker of skeletal muscle toxicity
[221]	2D-MS	JP-8 jet fuel	Mouse lung	Proteins IDed during apoptosis and edema after exposure
[93]	LC-MS/MS	Acrylamides	Rat blood	Hgb adducted sites IDed
[104]	2D-MS	LPS	Human neutrophils in vitro–DNA array/proteomics	↑ Inflammamation signaling; cytoskeletal proteins IDed
[204]	2D-MS	APAP, AMAP	Mouse liver	35 proteins IDed; altered proteins are known targets for adducts
[94]	MudPIT	20 cytotoxic, nontoxic isomers	Human hepatocytes	↑ BMS-PTX-265 and BMS-PTX-837 proteins released in culture medium relate to toxicity
[105]	2D-MS	bromobenzene	Rat liver	DNA array/proteomics; IDed proteins infer degradation, oxidative stress from toxicity

TABLE 5.2 (*Continued*)

Author	Platform	Chemical	Tissue	Results
[98]	Anti-Phos Ab array	BCG mycobacterium infection of host	Monocyte THP-1 cells	Activation of SAP kinase cJun, ↓ PKC varε; ↑ α-adducin, GSK-3β by BCG infection
[100]	Ab array	Cardiotoxin cytokine	Mouse skeletal muscle	DNA array/proteomics; ↑ of osteopontin, C10/CCL6 with muscle injury
[97]	SELDI	Zinc	Trout, gill	DNA array/proteomics; proteins altered by SELDI; no protein ID
[59]	2D-MS	Oxazepam, Wyeth 14643	Mouse liver	DNA array/proteomics; subcellular fractions, protein IDs unique to each chemical
[222]	2D-MS	Microcystin	Mouse liver	PP1-NIPP1 complex ID as microcystin binding target
[99]	Ab array	Chromium VI	Rat lung	DNA array/proteomic; mRNA, protein expression coupled only during DNA damage and not homeostasis
[223]	2D-MS	MNNG	Human amnion FL cells	18 proteins IDed; Zn-finger family proteins altered
[224]	2D-MS	Acrolein and glycoaldehyde	CHO cells	ID of aldose reductase as protective from oxidative stress
[225]	2D-DIGE MS	Hydrazine	Rat liver	Lipid, Ca^{2+}, thyroid, stress pathways activated responses
[226]	2D-MS	Kainaic acid	Rat brain mitochondrial/cytosol fractions	Altered HSPs, α-internexin, cytoskeleton in cell death
[227]	2D-MS	Methyl-*tert* butyl ether	Pseudomonas KT2440	↑ AhpC, Sod-M, Sod-F, oxidative stress; ecotoxicology
[228]	2D-MS	CCl_4	Rat liver stellate cells	150 protein IDs, ↑ in calcyclin, calgizzarin, galectin-1
[101]	2D-MS	Quinones	Human bronchial epithelium	Protein adduct IDs including nucleophosmin, galectin-1, HSPs, others

[229]	2D-MS	Dopamine	MN9D neuronal cell line	↑ Calreticulin IDed with dopamine cell death
[230]	2D DIGE	DEHP	Mouse liver	59 protein IDs; lipid metabolism pathway, and others altered by DEHP
[231]	2D-MS	Methapyrilene cyproterone	Rat liver	Hepatotoxic drugs compared for protein signa-tures of toxicity
[232]	2D-MS	Nicotine	Rat blood	2D image protein-mapping proteins algorithm; fuzzy logic
[233]	2D DIGE MS	Compound A	Rat liver	Chemical liver steatosis study; ↑ Acetyl CoA pathway enzymes; ↓ sulfite oxidase
[109]	2D-MS	Microcystin-LR	Mel-7, J3 H4TG cell lines	3 protein adducts IDed: PP1, PP2A, and ATP-synthase-β that correlate to apoptosis
[234]	2D-MS	Daunarubicin	Human pancreas	12 protein IDs varied with carcinoma cells treatment
[95]	LC-MS/MS	Acrylonitrile	Rat liver	ID of Cys-186 adduct site on carbonic anhydrase III
[54]	2D DIGE MS	Acetaminophen	Mouse liver	DNA array/proteomics; ↓ Hsp 10,70; protein changes within 15 min of treatment
[235]	2D-PS	Carbamate	Planthopper	22 protein changes, ID of 15 insect of enzymes, structural proteins
[236]	2D-MS	Aroclor 1248,	Mussels	Protein changes differ by Cu, salinity treatment; ecotoxicology
[237]	2D gel	Estrogen, 4-nonylphenol	Zebrafish embryo	Image patterns differ by treatment; ecotoxicology
[238]	2D-PS	TCDD	Rat plasma	↑ GSH peroxidase, Ig's, cytokeratin 8
[239]	2D-MS	Fluvastatin	Rat liver	58 protein ID's; ↑ cholesterol pathway proteins; HMG CoA synthetase; IPDP-Δ-isomer

TABLE 5.2 (*Continued*)

Author	Platform	Chemical	Tissue	Results
[240]	2D-DIGE MS	Acetaminophen	Mouse liver	Proteins altered with treatment
[241]	2D-MS	Mutagens	Human fibroblasts	Stress-induced premature senescence
[110]	2D-MS	Ethanol, IBTP	Rat liver	Mitochondrial fraction; ↓ ALD deHase activity by EtOH; ↓ in mitochondrial matrix proteins
[242]	2D-MS	Cadmium	Yeast	↑ 54, ↓ 43 protein IDs with Cd
[243]	2D-MS	Ozone	Mouse lung	↑ CC16 and AOP2 isoforms are BALF fluid associated with pulmonary protection against ozone
[244]	2D-DIGE MS	Inhaled oxidants	Airway tract epithelium	Many differentially expressed proteins observed
[245]	2D-MS	J-8 jet fuel	Mouse lung	Cytosol fraction analyzed; protein changes consistent with lung injury
[246]	2D-MS	J-8 jet fuel	Mouse kidney	Cytosol fraction analyzed; protein changes show kidney injury
[212]	2D-MS	J-8 jet fuel	Rat testis	Organ fractions show 76 protein IDs
[246]	2D-MS	J-8 jet fuel	Rat liver,	3 protein IDs show consistent kidney changes in tissues
[247]	2D-MS	E2 and testosterone	Human neurons	E2 and testosterone induce Hsp70 to protect neurons from Aβ peptide

Note: Summaries of proteomics studies in toxicology are intended to briefly overview study details and results. Only abbreviations that are helpful in interpreting the summary notes are included. For further explanation, refer to the citation. Abbreviations are listed in order of appearance in the table. *Abbreviations*: ↑, ↓, increase, decrease; IDed, identified; IDs, identities; 2D-MS, two-dimensional gel electrophoresis—mass spectrometry; 2D PS, two-dimensional gel electrophoresis—protein sequencing; LC-MS/MS, liquid chromatography tandem mass spectrometry; Hgb, hemoglobin; D-, L-ser, D-serine, and L-serine; *cis*/*tr*-Pt, *cis*-platinum and *trans*-platinum; ER, endoplasmic reticulum; $ONOO^-$, peroxynitrite; DNA array/proteomics, transcriptomic and proteomic analysis in the same study; SELDI, retentate chromatography—mass spectrometry platform; PP1-NIPP1, protein phosphatase1-nuclear inhibitor of PP1; LPS, lipopolysaccharide; APAP/AMAP, acetaminophen/3-acetaminophenol; Ab array, antibody array; 2D-DIGE MS, two-dimensional gel differential gel electrophoresis; CCl_4, carbon tetrachloride; Compound A, unidentified pharmaceutical agent under study; Cu, copper; Cd, cadmium; E2, estrogen; Aβ peptide, amyloid beta peptide.

proteomics). A notable aspect of such proteomic studies accompanying DNA microarray data was that the numbers of proteins detected as altered and then identified were typically less than 100. The set of altered, identified proteins was uniformly far less than the number of altered transcripts found by DNA microarrays. However, it was encouraging to observe that several specific protein alterations from proteomic analysis appear to be linked to a toxicant effect and exposure, providing grounds for further testing and exploration as possible biomarkers [59,94,96,100,101,106–110]. The range of toxicant effects analyzed by proteomics from tissues, biofluids, subcellular fractions, and modified proteins by adduction or other posttranslational effects is impressively broad and promising for toxicology and pathology evaluations. The application of new proteomic platforms and an integration of proteomics with transcriptomic and metabolomic analysis in databases promises to greatly enhance the biomarker discovery capability for toxicoproteomics [111].

5.7 SERUM AND PLASMA PROTEOMES AND ACCESSIBLE BIOFLUIDS

Complete mapping of the human serum and plasma proteomes is underway. Serum and plasma are arguably the most informative, accessible biofluids of the body for biomarker development (Figure 5.3). Analysis of low concentration, bioinformative proteins can be greatly improved after immunoaffinity removal of abundant proteins such as albumin, IgG, IgA, transferrin, haptoglobin, α-1-antitrypsin, hemopexin, transthyretin, α-2-HS glycoprotein, α-1-acid glycoprotein, α-2-macroglobulin, and fibrinogen from human plasma [112]. Almost all cells are in contact with blood and often communicate through active endocrine secretion or passive release of peptides and proteins into the circulation (Figure 5.3).

Despite the enormous utility of clinical chemistry in diagnostics and medicine, remarkably little new biomarker development in serum has occurred in the last decade [113]. The commercial availability of "immunosubtraction" materials available for human sera and soon for rodent serum, and recent studies that survey the human serum and plasma proteomes (described below), will have major impacts on biomarker research. Three studies published over the past two years have attempted to comprehensively map the human plasma and serum proteomes defining them to 490 [114], 325 [115], and 341 [116] unique proteins with almost 3700 electrophoretically separable features [115]. The number of serum and plasma proteins is thought to be much higher than 500 proteins, but the exact number is unknown. The detection of nuclear, DNA binding and cytoplasmic, kinesin complex proteins in serum poses exciting possibilities about how the serum proteome reacts and adjusts during the course of disease and toxicity [29]. The complete mapping of the human proteome has taken on international importance as this is one of the prime goals of an international consortium called the Human Proteome Organization

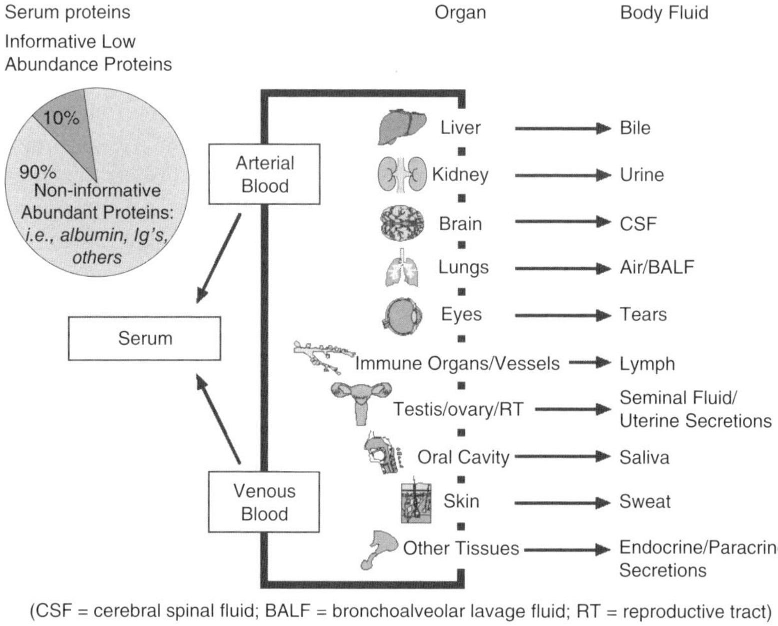

Figure 5.3 Protein profiling of serum and target organs: systems toxicology and pathology. Blood circulation in organs and tissues makes it an ideal body fluid to assess early overt injury and recovery periods of acute toxicity. More comprehensive profiling of the serum proteome can lead to improved markers of toxicity and disease, a better understanding of recovery and repair process, inter-organ and inter-tissue communication, and a more unifed, systems toxicology and pathology approach to temporal changes in the serum proteome and its relationship to human health in development, maturity, reproduction, and aging. Similar advances should occur in biomarker development after mapping the proteomes of tissue-specific biofluids and monitoring changes during toxicity and disease.

(HUPO) [117]. Cataloging of saliva [118], urine [119], cerebral spinal fluid [120,121] and the many other biofluids from various organs [122] (Figure 5.3) should lead to important advances in biomarker development. While there are no corresponding organizations dedicated to the proteomes of primates, mice, rats, and other experimental animals, such interests will evolve in time either as an outgrowth of HUPO or as separate entities.

5.8 TRANSCRIPTOMICS AND PROTEOMICS: DISAGREEMENT OR COMPLEMENTARITY?

An important issue that continues to arise in gene expression studies is the reportedly poor correlation between mRNA and protein levels reported for some studies that compare transcriptomic and proteomic datasets from the same experiment [9,123]. The correlation issue factors into how -omic integration efforts proceed at an experimental design level, at a database level, and eventually would affect choices of candidate protein biomarkers predicted from transcript changes in DNA microarrays, and vice versa.

Several studies in yeast have reported variable correlation between protein and transcript expression that ranges from a poor correlation in yeast, *Saccharomyces cerevisiae* [123–125], and lung adenocarcinoma [124], to a weakly positive correlation in yeast [126]. In the [126] study, a clustering approach took into account transcription and posttranscriptional control of expression for several proteins and arrived at a more positive correlation than previous reports. A positive correlation was noted for a limited set of proteins and transcripts in neuroblastoma cells during apoptosis [127] and a high correlation was noted for mRNAs and the corresponding 39 leukemia-relevant (AML and ALL) proteins in 113 patients [128]. In a more recent study, expression of 82 proteins identified from platelets were closely mirrored in the transcriptome [129]. In a study from chromium (VI)-treated rats, results from a 380 antibody array probing lung tissue suggested an uncoupling of transcriptome and proteome data under normal homeostatic conditions, but an establishment of parallel mRNA and protein expression when lung cells responded to acute genotoxic damage [99]. An interesting computational approach compared published literature knowledge with DNA microarray data of normal and malignant breast tissue and found correlation only among gene expression changes with a more than 10-fold difference [130]. A recent high impact study with the NCI-60 (60 human cancer cell lines) and controls demonstrated a high correlation of cell structure-related proteins between mRNA and protein levels across the NCI-60 cell line set, whereas non–cell-structure-related proteins were poorly correlated. The latter two studies suggest that highly expressed, abundant structural proteins appear to be most highly correlated at the transcript and protein levels. One important final study for consideration reports that proteomic mapping of mitochondria in the mouse produced a list of 591 mitochondrial proteins, including 163 proteins not previously associated with this organelle [131]. These researchers found protein expression data were largely concordant with large-scale surveys of RNA abundance and both measures indicated tissue-specific (brain, heart, kidney, and liver) differences in organelle composition. These researchers concluded that the combined proteomic and transcript analysis identified specific genes of biological interest, such as candidates for mtDNA repair enzymes, that offer new insight into the biogenesis and ancestry of mammalian mitochondria, and provide a framework for understanding the organelle's contribution to human disease [131].

The preceding studies suggest that some of the variability in correlating protein and transcript levels involve (1) sampling time after treatments (mRNA changes precede protein expression), (2) the high information density of DNA arrays and proteomic platform capacity for identified proteins (many transcripts compared to few proteins), (3) very different bioinformatic methods among studies for comparing correlated expression of protein and transcripts, and (4) differences among these studies in the particular cell or tissue sample, time of specific toxicant exposure, and experimental design. It is expected that levels of low concentration, short half-life, cell-specific regulatory proteins such as transcription factors or phosphorylated signaling intermediates are highly variable in their transcript and protein isoforms levels at any one point in time and cellular spatial compartment. The difficulty in correlating protein and transcript levels appears to escalate when considering entire transcriptomes and proteomes as encountered in reviewing published studies (above). However, the complementarity and sizable insight gained by performing proteomic and transcriptomic analysis as illustrated by gene expression studies of mouse mitochondria [131] seem highly justified by new knowledge gained. Further it is essential to eventually acquire transcriptomics, proteomics, and metabolomics data to construct a systems biology framework for toxicology and pathology. The variable nature of each experiment emphasizes the need to capture all pertinent experimental details in expression databases designed to accommodate and interrelate proteomics, transcriptomic, and metabolomic data.

5.9 CONSORTIA

The issues facing toxicogenomics are larger than can be solved by scientists independently, and the rapid advancement of the field requires common efforts toward data collection and comparison. Three main collaborative research consortia have been formed principally to standardize measurements and to guide the interpretation of toxicogenomics experiments. These groups of scientists from industry, government, and academic laboratories, as well as from regulatory agencies, were organized by research institutions around a relevant scientific question, such as methods and utility of applying toxicogenomics results to mechanism-based risk assessment.

The Health and Environmental Sciences Institute of the International Life Sciences Institute (ILSI/HESI) Genomics Committee, the first of these groups, began its work in 1999 and reported its main findings in 2004. These findings included the mechanisms of toxicity of several agents (hepatotoxicants clofibrate and methapyrilene [63], and the nephrotoxicants cisplatin [132], gentamicin and puromycin [133]), the successful applications of toxicogenomics to genotoxicity [88], and the establishment of a collaboration with the European Bioinformatics Institute (EBI) to develop a public toxicogenomics database [1]. The Toxicogenomics Research Consortium (TRC) of the National Institute of Environmental Health Sciences (NIEHS) National Center for Toxicogenomics is engaged in a project to standardize toxicogenomics investigations and to analyze environmental stress responses. In 2003 COMET (Consortium for Metabonomics Technology) reported its interim progress toward producing a metabonomics database containing studies of 80 agents [65]. Member laboratories reported data free of interlaboratory bias, suggesting that the

COMET standardized method was robust, and that findings obtained in different laboratories could be subjected to longitudinal data mining for patterns associated with various toxicity endpoints.

The ILSI Genomics Committee also found that microarray results from different laboratories and different platforms were comparable in identifying a common biological response profile, although the responses of individual genes contributing to the pattern differed between platforms [132,134–136]. This, together with the metabonomics reproducibility reported by COMET, is a critical finding that supports the use of public toxicogenomics databases for meaningful meta-analysis of results obtained in different laboratories. Although some researchers [20,137] are concerned that the capacity to assemble data on drug and toxicant effects using these technologies could result in inappropriate safety and risk decisions, collective efforts such as these will do much to help develop scientific consensus on the appropriate uses of gene expression data.

5.10 INTEGRATION OF DATA

A key objective in toxicogenomics is to integrate data from different studies and analytical platforms to produce a richer and biologically more refined understanding of the toxicological response of a cell, organ, or organism. For example, one would like to describe the interplay between protein function and gene expression, or between the activity of certain metabolizing enzymes and the excretion into serum or urine of populations of small metabolites. The integration of data from different domains such as proteomics and transcriptomics [54,97,138], or transcriptomics and metabonomics [46], has been reported. In these experiments, tissue samples derived from the same individual animals or from comparably treated animals were analyzed in parallel using different technologies. However, the data from different studies were integrated only after a short list of differentially responsive transcripts or protein spots had been obtained.

The experience gained from integrating global proteomics or metabonomics data, such as spot intensities from 2D gels or metabonomics fingerprint data from NMR, tells us that cluster or principle component analysis can be performed to obtain global signatures of molecular expression in much the same way as in transcriptomics analyses. If biological samples segregate into unique clusters that show similar expression characteristics, with additional effort the novel proteins or metabolites that are expressed in these samples can be discerned. Steps can then be taken to evaluate these proteins or metabolites as potential biomarkers and thus to determine the underlying toxicological response.

Although software is plentiful for managing expression profiling data at the laboratory level, there is a great need for public databases that combine profile data with associated biological, chemical, and toxicological endpoints [1]. Comparisons of gene, protein, and metabolite data in public databases can be a valuable tool and assist in global understanding of how biological systems function and respond to environmental stressors [65,139]. As these repositories are developed, experiments will be deposited from disparate sources, with different experimental designs and yet targeting the same toxicity endpoint or a similar class of toxicant. In these cases it will be important that the databases integrate data from related studies before performing data mining. To maximize the value of deposited datasets, the repositories must also be able to integrate data from different technological domains (see Appendixes 5.1, 5.2, and 5.3). Members of regulatory bodies are working with scientists from industrial, academic, and governmental laboratories participating in the ILSI Genomics Committee and Clinical Data Interchange Standards Consortium/Standards for Exchange of Nonclinical Data (CDISC/SEND) Consortia to develop standards for the exchange, analysis, and interpretation of transcriptomics data.

A proposal has been made to extend toxicogenomics and combine it with computational approaches such as physiologically based pharmacokinetic (PB/PK) and pharmacodynamic modeling [26]. PB/PK modeling can be used to derive quantitative estimates of the dose of the test agent or its metabolites that are present in the target tissue at any time after treatment, thereby allowing molecular expression profiles to be anchored to internal dose, as well as to the time of exposure and to the toxicant-induced phenotype. Relationships among gene, protein, and metabolite expression can then be described both as a function of the applied dose of an agent and the ensuing kinetic and dynamic dose–response behaviors that occur in various tissue compartments. Such models also must take into account the fact that the transcriptome, proteome, and metabolome are themselves dynamic systems, and are therefore subject to significant environmental influences, such as time of day and diet [140–142].

Despite the numerous successes of toxicogenomics in the context of toxicology, a poorly addressed but confounding issue pertinent to drug safety and human risk assessment is the impact of the individual genetic background on the response of the individual animal or human patient. The PharmGKB pharmacogenetics knowledgebase [143] cataloges the different human genetic backgrounds by their susceptibility to drug therapy. In addition the NIEHS Environmental Genome Project (EGP [24]) is identifying SNPs in genes that are important in environmental disease, detoxification, and repair. Linkages of toxicogenomics knowledgebases with those containing information about SNPs and human susceptibility will gradually lead to a more complete picture of the relevance of the responses and genotypes of surrogate animal species to human risk assessment.

5.11 CHALLENGES AND TECHNICAL CONSIDERATIONS

Predicting potential human health risks from chemical stressors incurs three main challenges: the diverse properties of thousands of chemicals and other stressors present in the environment, the time and dose parameters that define the relationship between exposure to a chemical and disease, and the genetic and experiential diversity of human populations and of organisms used as surrogates to determine the adverse effects of a toxicant. Figure 5.1 illustrates the effect of genetic susceptibility on the continuum from toxic exposure to disease outcome. Knowledge of this continuum, and the role that genetics has in it, can help us to understand environmentally induced diseases, assess risk, and make public health decisions. Associated with these challenges are others of a more technical nature; these pertain are specific to toxicogenomics studies and are described below.

Although genomewide alterations in mRNA, protein, or metabolite levels in tissue extracts clearly are useful in identifying "signature" gene changes, verifying that one or more gene products are involved in a toxic process depends on knowing the cell types in which the target-gene transcripts and products are located. Experimental methods, such as Northern or Western blotting, or real-time PCR, are typically used to verify the expression profile of a gene or to selectively analyze its expression as a function of toxicant dose or time of exposure. In situ hybridization, immunohistochemistry, and other techniques can be used to identify the cell types that express the gene(s).

The ability to focus molecular expression analysis on only a limited number of cell types depends on cell separation methods that minimize the opportunity for other cell types to contribute to gene expression in situ. Even the most carefully gathered biological samples contain many cell types, especially if the sample is from inflamed or necrotic tissue. More homogeneous samples are provided by laser capture microdissection (LCM), a method that isolates individual cells or sections of tissue from a fixed sample [144–148]. The use of LCM minimizes contributions by nontarget cell populations in comparisons of diseased and normal tissues, but also introduces handling and preparation steps that can affect detection accuracy.

Simultaneously with new technology that selectively samples cell populations must come the ability to reliably detect signals from increasingly smaller samples. For example, it will frequently be necessary to amplify mRNA signals from the same biological sample that was used for transcriptomics analysis. The need to detect weak signals or small but biologically important changes in expression levels remains, as toxicologists explore the initial steps in biological signaling cascades and compensatory processes. At present, cDNA microarray hybridization can detect strong signals within a mixed cell population in samples that are diluted by up to 20-fold [149]. Thus this technology is likely able to detect a strong signal from a population comprising 5–10% of the total tissue but might miss more subtle changes associated with signaling or other initial responses to a stressor. With LCM a relatively pure cell population could be sampled, so the technology is expected to detect much more subtle changes; for instance, responses seen only in a subpopulation or asynchronous responses occurring in 10% of the cells at the time of sampling. The ultimate goal would be the ability to quantify genomic changes occurring in a single cell.

Although mRNA analysis is a powerful tool for recognizing toxicant-induced effects, analyses of protein structure and modification and, more important, of global protein expression provide distinct advantages for understanding the functional state of the cell or tissue. Promising new methods are emerging, including the capacity to profile proteins with antibody arrays [150] and surface-enhanced laser desorption mass spectrometry (SELDI; see Appendix 5.1) [151,152]. Alterations in patterns of mRNA and protein expression in accessible tissues such as serum [30] may offer new insights into the function of genes in the context of toxicity and guide the search for protein biomarkers of toxicant exposure or predictive toxicity.

Whole genome queries by DNA array platforms are starting to address the complexity of gene expression in mammalian systems by incorporating experiments that include RNA splicing, inhibition and editing, and transcriptional silencing and modulation [153]. The perspective for quantifying and cataloging the human genome has changed markedly since the first maps were published [154,155]. Estimates have been placed for those portions of the human genome derived from alternative splicing of multiexon genes at 41–60%, for retrotransposition at 45%, for antisense transcription at 10–20% of genes, and for non-protein coding RNA at approximately 7% of full-length cDNAs [153].

Complete transcriptomic analysis with existing DNA array technology, although not routine, is becoming more practical with the use of tiled oligonucleotide microarrays [7]. However, generation of tissue-specific, normalized, and subtracted cDNA libraries has the potential to characterize the expression of rare transcriptional units not represented on available oligonucleotide DNA microarray gene chips [156]. Affymetrix initial sequence analysis of our murine cDNA clone collections showed that as much as 86%, 45%, and 30% of clones are not represented on the Affymetrix Mu11k, MG-U74, and MG-430 chip sets, respectively. A detailed study that compared EST sequences of a subtracted library generated from mouse retina to those of MG-430 consensus sequences was undertaken, using UniGene build 124 as the common reference. A set of 1111 nonredundant transcript regions, not represented on the commercial array, was identified from murine cDNA clone collections and subtracted cDNA libraries selected for expression of rare transcriptals units, including seven transcripts and splice variants with retina-specific expression [156].

The challenge can also be exemplified by the p53 gene family. The p53 tumor suppressor protein generally fits a linear expression pattern from gene, to transcript, to single protein; however, its two homologous family members, p63 and p73, have the peculiarity of sharing many splice variants that give rise to at least six major transcripts and subsequent proteins including TAp63/p73 α, β, and γ isoforms and ΔNTAp63/p73 α, β, and γ isoforms [157–159]. Even p53 splice variants and proteolytic cleavage products have been frequently observed in many different malignancies as well as developmental processes [158]. Despite the complex expression pattern of some genes and difficulties in distinguishing their variants with current microarray technologies, the current DNA microarray platforms for transcriptome assessments now provide an ever-improving coverage of gene expression.

Proteomic platforms do not yet deliver the same quantity or type of information as in the transcriptome. Indeed, for the human genome, the relationship between gene number and proteome size is far from simple when taking into account splice variants and pseudogenes [160]. One estimate has the number of protein-coding transcripts at about 100,000 but that the number might be higher [160]. Also, accounting for post-translational modifications of each protein would create many more structurally unique variants and hints at the enormous complexity of the human protein expression. Despite the vastness of the proteome, proteomics is creating revolutionary insights into protein expression of unicellular and higher order species that will reshape pathology, toxicology and many other disciplines.

First, the composition of subcellular structures such as the nucleolus [161], mitochondria [131], Golgi [162], and others are being comprehensively cataloged by new methods in proteomic analysis at a rapid rate and in a manner that will eventually apply to protein dysfunction underlying disease. Second, proteomes of important biological fluids such as serum and plasma, are planned for complete mapping, which will be crucial for biomarker development [113]. Third, development of global analysis methods for post-translational modifications, such as phosphorylation, glycosylation, and ubiquitination, will help elucidate the active forms of protein during homeostasis and disease. Fourth, identification of specific xenobiotic-protein adducts by proteomics will provide insights not only into acute cellular injury and necrosis [163] but also into immune system activation [164] and idiosyncratic responses to therapeutics and environmental chemicals [165]. Fifth, rapid, parallel, and nanovolume protein analysis of biofluids will be entering research and clinical realms ranging from therapeutic target discovery to diagnosis and to epidemiological investigations. These contributions of proteomics to toxicology and pathology represent a technological exploitation of the complex properties of proteins such as post-translational modifications (i.e., signal transduction) spatial location and structure (i.e., subcellular organelles) and functions (enzyme activities).

One of the great challenges in metabolomics—in using NMR spectroscopy, mass spectrometry, electrochemical detection, or other methods—is poor annotation (metabolite identification) of the observed signals. In a typical NMR study fewer than 30 metabolites of the estimated few hundred compounds within a spectrum will be unambiguously identified. This limits the metabolic information that can be extracted from the spectral data and restricts the mechanistic insight that could potentially be gained of the biological system under study. Furthermore peak annotation is a prerequisite for instrument-independent metabolic data, which is both the desired format for the construction of metabolic databases and facilitates the comparison of datasets regardless of the instrument on which they were measured. This problem must be addressed urgently if metabolomics is to fulfill its potential.

5.12 BIOINFORMATICS CHALLENGES

Full realization of the potential of molecular profiling in toxicogenomics requires a very substantial investment in bioinformatics in order to extract biological sense from the myriad of interrelated numerical molecular identifiers and their associated annotations. Advances in bioinformatics and mathematical modeling provide powerful approaches for identifying the patterns of biological response that are embedded in genomic datasets (Figure 5.4). However, facile interpretation of global molecular datasets derived from -omics technologies are currently constrained by the "bioinformatics bottleneck." Bioinformatics and -omics technologies must improve in gene, protein, and metabolite identification and annotation to open the field of toxicogenomics to high-throughput applications in drug development and toxicant evaluation. Several useful resources address the annotation problem by linking identifiers used in genomic databases at the National Center for Biotechnology Information (NCBI), European Molecular Biology Laboratory (EMBL), and DNA Data Bank of Japan (DDBJ) to other annotation resources (see Table 5.3). Critical to resolving annotation inconsistencies is the knowledge of the sequence of the actual nucleotide or protein that is used to query the genome.

The use of advanced bioinformatics tools to extract information from microarray results [166] is valuable only if the data employed by these tools have a high internal specificity and accuracy [1]. Additionally the interpretation of molecular expression profiles must emphasize both biological coherence and statistical validity when deriving knowledge from toxicogenomics experiments. This means that once a set of genes with altered expression is identified, the genes' biological functions must be ascertained. Mechanistic interpretation of transcript changes may be impeded by the nonstandard or

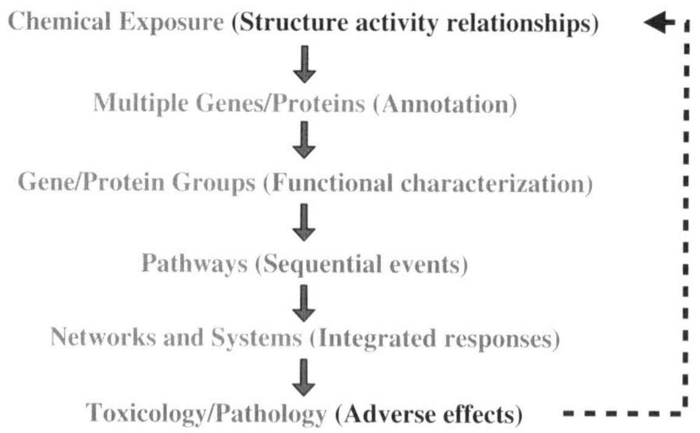

Figure 5.4 Bioinformatics challenges and biological complexity. The focus of bioinformatics (red) in interpreting molecular expression data depends on the level of biological complexity (blue)—here shown progressing from genes/proteins/metabolites to networks and systems. For toxicology/pathology the focus is on phenotypic anchors—observed biological responses that can be related to the chemical structure of the test agent or exposure.

TABLE 5.3 Online Links

Some annotation resources
Gene Ontology (GO): *http://www.geneontology.org*
GOStat: *http://gostat.wehi.edu.au* (Source: *http://source.stanford.edu*)
EASE: *http://david.niaid.nih.gov/david/*
MatchMiner: *http://discover.nci.nih.gov/matchminer/html/MatchMinerInteractiveLookup.jsp*
Commercial array-specific databases such as Affymetrix's NetAffx: *http://www.affymetrix.com/analysis/index.affx*
Multi-species commercial microarray platforms
Affymetrix: *http://www.affymetrix.com/index.affx*
Agilent: *http://www.chem.agilent.com/Scripts/PCol.asp?lPage=494*
Groups working toward standardization and exchange standards for toxicogenomics data
National Institute of Environmental Health Sciences: *http://www.niehs.nih.gov/cebs-df*
Toxicogenomics Research Consortium: *http://www.niehs.nih.gov/dert/trc/about.htm*
Food and Drug Administration (FDA), CDISC/SEND Consortium: *http://www.cdisc.org/*
EMBL-EBI European Bioinformatics Institute: *http://www.ebi.ac.uk/microarray/Projects/ilsi/index.html*
International Life Sciences Institute's Health and Environmental Sciences Institute: *http://hesi.ilsi.org/* and *http://hesi.ilsi.org/index.cfm?pubentityid=1* (Technical Committee on the Application of Genomics to Mechanism Based Risk Assessment)
Microarray Gene Expression Data (MGED) Society: *www.mged.org/*
MGED Toxicogenomics Working Group (MGED): *http://www.mged.org/*
Human Proteome Organization: *http://www.hupo.org/*
Protein Standard Initiative (PSI): *http://psidev.sourceforge.net/*
Protein interaction format [248]: *http://www.nature.com/cgi-taf/DynaPage.taf?file=/nbt/journal/v22/n2/abs/nbt926.html*
PEDRo prescribes standards for proteomic data for databases [249]: *http://psidev.sourceforge.net/*
MIAPE (Minimum Information about a Proteomics Experiment) [250]: *http://psidev.sourceforge.net/*
Metabolomics Standards Initiative (MSI) http://msi-workgroups.sourceforge.net/

imprecise annotation of a sequence element (i.e., gene). Without appropriate synonyms for gene names, the effectiveness of a literature search may be limited. Differences in annotation within and among different microarray platforms may hamper the comparison of results. Such inconsistency frequently arises from annotation resources using different lexicons, or from annotation information being compiled at different times.

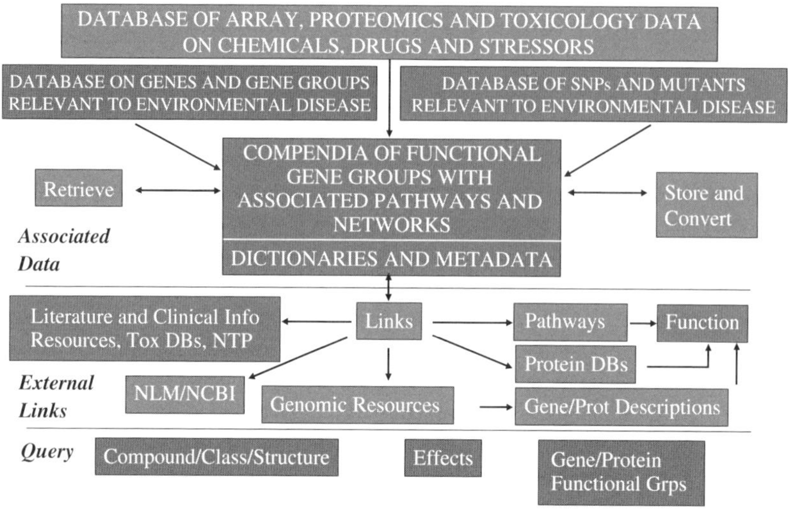

Figure 5.5 Conceptual framework for the development of the Chemical Effects in Biological Systems (CEBS) knowledgebase. CEBS knowledgebase is a cross-species reference toxicogenomics information system chemicals/stressors and their effects. The upper section indicates data associated in CEBS; the center section, external links from CEBS; and the lower section, sample query types that CEBS will support. The boxes in the upper section include primary data (blue), important genetic loci (red) and genetic markers such as SNPs (green). The tasks CEBS will carry out are shown with gray boxes. In the central section the links that are to databases are shown in gray, and the links to unstructured data are in green. Abbreviations: NTP, National Toxicology Program; NLM, National Library of Medicine; NCBI, National Center for Biotechnology Information. See color plates.

Additional bioinformatics and interpretive challenges arise at many levels of biological organization (Figure 5.5). Our current focus and level of understanding of the global molecular landscape encompasses only the lower levels of complexity (genes/proteins, gene/protein groups, functional pathways). The resolution of this knowledge might be termed linear toxicoinformatics, that is, the description of environmental stimuli and responses over dose and time following a toxicological stress. Toxicologists and risk assessors typically define a sequence of key events and linear modes-of-action for environmental chemicals and drugs [167–169]. By contrast, the networks and systems level of biological organization may demonstrate highly nonlinear cellular expression state changes in response to environmental stimuli [170,171]. Thus the statistical and bioinformatics-based separation of the complex adaptive, pharmacological, and toxicological responses of drugs, chemicals, and even dietary constituents will probably very much be a matter of degree. This reflects the kinetic and dynamic responses of specific tissues to toxicants as directed by the genome, the genetic heritage of the individual, and that individual's current and prior exposures.

5.13 SYSTEMS TOXICOLOGY

Ideker et al. [8] used the phrase "systems biology" to describe the integrated study of biological systems at the molecular level—involving perturbation of systems, monitoring molecular expression, integrating response data, and modeling the systems molecular structure and network function. Here we similarly use the phrase "systems toxicology" to describe the toxicogenomics evaluation of biological systems, involving perturbation by toxicants and stressors, monitoring molecular expression and conventional toxicological parameters, and iteratively integrating response data to model the toxicological system [26].

A number of approaches are being developed to model network behavior, with different assumptions, data requirements, and goals. However, it is not likely that toxicogenomics and systems toxicology models will be assembled exclusively from knowledge of cellular components, without equivalent knowledge of the response of these components to toxicants [170]. Thus the "stress testing" of the structural biology of the system and the capture of that data in the context of the functioning organism adapting, surviving, or succumbing to the stress will be required.

Development of a knowledgebase to accurately reflect network-level molecular expression and to facilitate a systems-level biological interpretation requires a new paradigm of data management, data integration, and computational modeling. A knowledgebase that fully embraces systems toxicology (see Appendix 5.2) will use precise sequence data to define macromolecules, interaction data based experimentally on co-localization, co-expression and analyses of protein–protein interactions, and functional and phenotypic data based on gene knockouts, knockins, and RNA-interference studies, besides studies of responses to chemical, physical, and biological stressors. These data will allow specific molecules to be accurately related to biological phenomena that reflect the normal as well as the stressed cell, tissue, organ, and organism. In the best of circumstances a systems toxicology approach will build a toxicogenomics understanding from global molecular expression changes that are informed by PBPK/PD modeling and biologically based dose–response (BBDR) modeling. The challenge in constructing a robust systems toxicology knowledgebase is formidable.

5.14 THE FUTURE OF TOXICOGENOMICS

New toxicogenomics methods have the power and potential to revolutionize toxicology. Technological innovations that are already in use permit RNA profiling of formalin-fixed tissues [172], potentially making archived tissues from generations of toxicological studies accessible to gene expression analysis. Methods to array hundreds of toxicologically relevant protein antibodies, and to profile hundreds of small molecules in high-throughput mode using GC/LC/MS, are in development.

Toxicoproteomics research is anticipated to lead to the identification, measurement, and evaluation of proteins and other biomarkers that are more accurate, sensitive and specific than those available now, and that might be targeted to particular human genetic subpopulations. Metabonomics research will help to identify small endogenous molecules as important in a sequence of key metabolic events; such "metabolite fingerprints" might then help to diagnose and define the ways in which specific chemicals, environmental exposures, or stressors cause disease. This, coupled with the ability to detect damage to particular organs by observing alterations in serum and urine components, is expected to lead to the more sensitive detection of exposure or risk factors [173]. Additional considerations in assessing the toxicogenomic response to environmental exposures are the individual genotype, lifestyle, age, and exposure history [141]. Toxicogenomics will help to ascertain the degree to which these factors influence the balance between healthy and disease states.

Toxicogenomics will increase the relevance of toxicology through the global observation of genomic responses with therapeutically and environmentally realistic dose regimens. It will help to delineate the mode-of-action of various classes of agents and the unique genetic attributes of certain species and population subgroups that render them susceptible to toxicants [25,168]. Studies on strains within a species that are sensitive or resistant to the chemical induction of specific disease phenotypes will be particularly valuable. Extending this thinking to the phylogenetic analysis of both core conserved biological processes [174] and to the toxicological responses seen in different species will provide additional comparative insight on genetic susceptibility and on probable disease outcomes.

The combined application of the -omics technologies will improve our overall understanding of mechanisms of toxicity and disease etiology as integrated toxicogenomics databases are developed more fully [25]. Data on gene/protein/metabolite changes collected in the context of dose, time, target tissue, and phenotypic severity across species from yeast to nematode to human will provide the comparative information needed to assess the genetic and molecular basis of gene–environment interactions. Toxicology will emerge as an information science that will facilitate scientific discovery across biological species, chemical classes, and disease outcomes [22]. Although there are large challenges in developing public toxicogenomic data repositories, the nucleotide sequence databases—GenBank, EMBL, and DDBJ—provide an excellent example of the benefit of sharing data with the larger scientific and medical community.

Concomitant with development of toxicogenomics databases must be the evolution of bioinformatics methods and data mining tools, and individuals trained to apply them [166]. We believe that a predictive systems toxicology will gradually evolve, aided by knowledge that is systematically generated [37] through literature mining [175,176], comparative analysis, and iterative biological modeling of molecular expression datasets over time. Given the vast numbers and diversity of drugs, chemicals, and environmental agents, and the diversity of species in which they act, we believe, however, that it is only through the development of a comprehensive and public knowledgebase that toxicology and environmental health can rapidly advance. The ultimate goal of the National Center for Toxicogenomics (NCT) is to create the Chemical Effects in Biological Systems (CEBS) knowledgebase [13,26], a public resource (see Appendix 5.2) that will enable health scientists and practitioners to understand and mitigate or prevent adverse environmental exposures and related diseases.

APPENDIX 5.1 DESCRIPTIONS OF SELECTED -OMICS TECHNOLOGIES

The terms transcriptomics, proteomics, and metabonomics or metabolomics refer to highly parallel analytical technologies wherein simultaneous measurements are made of expressed genes, proteins, or metabolites, respectively. These technologies are used to ascertain the function of the genome. Toxicogenomics makes use of all of these functional genomics technologies in the study of toxicology. The terms toxicoproteomics and toxicometabolomics are sometimes used in a technology-centric sense to discuss the response of the proteome or metabolome to toxicants. Toxicoproteomics, as part of the larger field of toxicogenomics, seeks to identify critical proteins and pathways in biological systems that are affected by and respond to adverse chemical and environmental exposures using global protein expression technologies.

Transcriptomics—DNA Microarray Hybridization and Analysis

Early gene-expression profiling experiments that were carried out for toxicogenomics studies employed cDNA microarrays [4]. Although this cDNA technology has been supplanted by synthetic short and long oligonucleotide microarrays, the technological concepts underlying the two approaches are largely analogous: cDNAs are derived from sequence-verified clones representing the 3′ ends of the genes, which are either spotted onto glass slides using a robotic arrayer or synthesized in situ. Each RNA sample is labeled with dye-conjugated dUTP by reverse transcription from an oligo dT primer. The fluorescently labeled cDNAs are then hybridized to the microarray, and the microarray is scanned using laser excitation of the fluorophores [36]. Raw pixel intensity images derived from the scanner are analyzed to locate targets on the array, measure local background for each target, and subtract it from the target intensity value. Prime discovery features of cDNA and oligonucleotide microarrays are the sequences representing EST's (expressed sequence tags), hypothetical proteins, homologues, orthologues, or genes of unknown function. ESTs or similar features are not currently represented on antibody arrays. When examining differential transcript profiles, however, many researchers usually focus more on annotated genes of known or homologous function that currently account for less than half the genes predicted for the human genome. Changes in expression of ESTs might be tracked in transcriptomic studies, but they are often not followed up in many gene expression experiments (except when used as part of discriminating signatures to classify disease of toxicity).

Proteomics

Global measurement of proteins and their many attributes in tissues and biofluids defines the field of proteomics. An established proteomics strategy [177] uses global protein stratification systems, such as polyacrylamide gel electrophoresis (PAGE), followed by protein identification by mass spectrometry (MS). Two-dimensional PAGE separation, by charge and by mass, can resolve thousands of proteins to near homogeneity. This separation is a necessary prerequisite to enzymatic digestion and MS identification, which requires unique peptide fingerprint masses or amino acid sequence tags. Where proteins are separated by liquid chromatography (LC) instead of PAGE, a new and promising platform involving multidimensional LC can be used to fractionate and reduce the complexity of the protein mixture before peptide sequencing by MS or tandem MS (LC/MS/MS). This approach is being augmented by SELDI-TOF MS (surface-enhanced laser desorption/ionization time-of-flight mass spectrometry), a method that results in the isolation of tens to hundreds of thousands of low molecular weight fragments representing a proteome.

The focus of proteomics can range from global protein analysis [178,179] via approaches such as "shotgun proteomics" [180,181] where researchers strive for maximum number of protein identifications, to a more discrete level of protein analysis often termed "targeted proteomics" [182]. In targeted proteomics, protein groupings or subproteomes [183] may be organelles [184], portions of structures like nuclear membrane proteins [185], signaling pathways [186], and protein complexes or protein families [187]. Inherent in this range of different biochemical perspectives is the assumption that protein structure and spatial location within cell(s) or tissue remain critical features in proteomic analysis. Other characteristics of proteins, such as function, protein–protein interactions, three-dimensional structure or specific post-translational modifications, have also helped define various subdisciplines of proteomics.

Metabolomics and Metabonomics

Quantitative analytical methods have been developed to identify metabolites in pathways or classes of compounds. This collective directed approach has been called metabolite profiling or metabolomics. Semiquantitative, NMR-based metabolic fingerprinting has also been applied to high abundance metabolites and has been termed "metabonomics" [188]. Peaks detected in NMR spectra carry information regarding the structure of the metabolites, whereas peaks detected by MS have associated molecular weights. In addition specific MS methods can be established to fragment the parent molecule, allowing metabolites to be identified through investigation of fragmentation patterns.

APPENDIX 5.2 CHEMICAL EFFECTS IN BIOLOGICAL SYSTEMS (CEBS) KNOWLEDGEBASE

To promote a systems biology approach to understanding the biological effects of environmental chemicals and stressors, the CEBS knowledgebase [13,26] is being developed to house data from many complex data streams in a manner that will allow extensive and complex queries from users. Unified data representation will occur through a systems biology object models (systems for managing diverse-omics and toxicology/pathology data formats) that incorporates current standards for data capture and exchange (CEBS SysBio-OM) [189] and CEBS SysTox-OM [190]. Data streams will include gene expression, protein expression, interaction, and changes in low molecular weight metabolite levels on agents studied, in addition to associated toxicology, histopathology, and pertinent literature [175].

The conceptual design framework for CEBS (see Figure 5.5) is based on functional genomics approaches that have been used successfully for analyzing yeast gene expression datasets [25,26,171]. Because CEBS will contain data on molecular expression, and associated chemical/stressor-induced effects in multiple species (e.g., from nematode to humans), it will be possible to derive functional pathway and network information based on cross-species homology. Genomic homology can be tapped within a knowledgebase such as CEBS to gain new understanding in toxicology as well as in basic biology and genetics.

CEBS will index and sequence-align to the respective genomes all datasets known to the knowledgebase. Thus changes or differences in the expression patterns of entire genomes at the levels of mRNA, protein, and metabolism can be determined. It will be possible to query CEBS globally, that is, to BLAST (Basic Local Alignment Search Tool) [191], the knowledgebase with a profile of interest and have it return information on similar profiles observed under defined experimental conditions of dose, time, and phenotype. CEBS will provide dynamic links to relevant sites such as genome browsers, animal model databases, genetic quantitative trait (QTL) and SNP susceptibility data, and PB/PK and BBDR modeling. Using search routines optimized for parsing known gene/protein groups onto toxicologically relevant pathways and networks, CEBS will automatically survey the literature and integrate this new knowledge with existing knowledgebase annotations. The current status of the CEBS infrastructure and that of other toxicogenomics databases is described in a recent review [1]. These repositories offer the regulatory community reference resources for comparison with toxicogenomics data submitted in the compound registration process [192]

Progress in the development of CEBS can be monitored at *http://cebs.niehs.nih.gov/* and *http://www.niehs.nih.gov/cebs-df*.

APPENDIX 5.3 DATABASES AND STANDARDS FOR EXCHANGE OF DATA

Databases

Public databases allow the scientific community to publish, share, and compare the data obtained from toxicology and toxicogenomics experiments (see Chapter 6). They are a resource for data mining, and for the discovery of novel genes/proteins through their coexpression with known molecules. They also help to identify and minimize the use of experimental practices that introduce undesirable variability into toxicogenomics datasets.

Guidelines

Public data repositories promote international database and data exchange standards [193–196] through guidelines developed by specific regulatory agencies. For example, the CDISC / SEND Data Consortium addresses the submission of toxicology study data (see Table 5.3). Minimum Information about a Microarray Experiment (MIAME) guidelines [197] specify sufficient and structured information to be recorded in order to correctly interpret and replicate microarray experiments or to retrieve and analyze the data from a public microarray database (such as ArrayExpress (Europe) [195], GEO (US) [193], or CIBEX (Japan) [198]). Similar guidelines describing what information should be included in a published set of toxicogenomics data are under development by the Microarray Gene Expression Data Society (MGED).

ACKNOWLEDGMENTS

We are indebted to the staff of the National Center for Toxicogenomics for their support and involvement with this work. This manuscript was derived from prior publications by the authors in *Nature Reviews Genetics* [12] and *Toxicological Pathology* [173]. This work was supported by the Intramural Research Program of the NIH, National Institute of Environmental Health Sciences and NIEHS contract number 273–02-C-0027.

REFERENCES

1. Mattes WB, Pettit SD, Sansone SA, Bushel PR, Waters MD. Database development in toxicogenomics: Issues and efforts. *Environ Health Perspect* 2004; 112(4):495–505.

2. Maunsbach A, Afzelium B. *Biomedical electron microscopy: illustrated methods and interpretations*. San Diego: Academic Press, 1999.
3. Braga PC, Ricci D. *Atomic force microscopy: Biomedical methods and applications*, Vol. 242. Totowa, NJ: Humana Press, 2004.
4. Nuwaysir EF, Bittner M, Trent J, Barrett JC, Afshari CA. Microarrays and toxicology: The advent of toxicogenomics. *Mol Carcinog* 1999;24(3):153–9.
5. Shi MM. Technologies for individual genotyping: Detection of genetic polymorphisms in drug targets and disease genes. *Am J Pharmacogenomics* 2002; 2(3):197–205.
6. Wen SY, Wang H, Sun OJ, Wang SQ. Rapid detection of the known SNPs of CYP2C9 using oligonucleotide microarray. *World J Gastroenterol* 2003;9(6): 1342–6.
7. Mandal MN, Heckenlively JR, Burch T, Chen L, Vasireddy V, Koenekoop RK, Sieving PA, Ayyagari R. Sequencing arrays for screening multiple genes associated with early-onset human retinal degenerations on a high-throughput platform. *Invest Ophthalmol Vis Sci* 2005;46(9):3355–62.
8. Ideker T, Galitski T, Hood L. A new approach to decoding life: Systems biology. *An Rev Genomics Hum Genet* 2001;2:343–72.
9. Ideker T, Thorsson V, Ranish JA, Christmas R, Buhler J, Eng JK, Bumgarner R, Goodlett DR, Aebersold R, Hood L. Integrated genomic and proteomic analyses of a systematically perturbed metabolic network. *Science* 2001;292(5518):929–34.
10. Tong W, Cao X, Harris S, Sun H, Fang H, Fuscoe J, Harris A, Hong H, Xie Q, Perkins R, Shi L, Casciano D. Arraytrack—Supporting toxicogenomic research at the U.S. Food and Drug Administration National Center for Toxicological Research. *Environ Health Perspect* 2003;111(15):1819–26.
11. Waters MD, Olden K, Tennant RW. Toxicogenomic approach for assessing toxicant-related disease. *Mutat Res* 2003;544(2–3):415–24.
12. Waters MD, Fostel JM. Toxicogenomics and systems toxicology: Aims and prospects. *Nat Rev Genet* 2004;5(12):936–48.
13. Waters M, Boorman G, Bushel P, Cunningham M, Irwin R, Merrick A, Olden K, Paules R, Selkirk J, Stasiewicz S, Weis B, Van Houten B, Walker N, Tennant R. Systems toxicology and the chemical effects in biological systems (CEBS) knowledge base. *EHP Toxicogenomics* 2003;111(1T):15–28.
14. Boorman GA, Anderson SP, Casey WM, Brown RH, Crosby LM, Gottschalk K, Easton M, Ni H, Morgan KT. Toxicogenomics, drug discovery, and the pathologist. *Toxicol Pathol* 2002;30(1):15–27.
15. Storck T, von Brevern MC, Behrens CK, Scheel J, Bach A. Transcriptomics in predictive toxicology. *Curr Opin Drug Discov Devel* 2002;5(1):90–7.
16. Suter L, Babiss LE, Wheeldon EB. Toxicogenomics in predictive toxicology in drug development. *Chem Biol* 2004;11(2):161–71.
17. Aardema MJ, MacGregor JT. Toxicology and genetic toxicology in the new era of “toxicogenomics”: Impact of “-omics” technologies. *Mutat Res* 2002;499(1): 13–25.
18. Afshari CA. Perspective: Microarray technology, seeing more than spots. *Endocrinology* 2002;143(6):1983–9.

19. Ulrich R, Friend SH. Toxicogenomics and drug discovery: Will new technologies help us produce better drugs? *Nat Rev Drug Discov* 2002;1(1):84–8.
20. Fielden MR, Zacharewski TR. Challenges and limitations of gene expression profiling in mechanistic and predictive toxicology. *Toxicol Sci* 2001;60(1):6–10.
21. Hamadeh HK, Amin RP, Paules RS, Afshari CA. An overview of toxicogenomics. *Curr Issues Mol Biol* 2002d;4(2):45–56.
22. Tennant RW. The National Center for Toxicogenomics: Using new technologies to inform mechanistic toxicology. *Environ Health Perspect* 2002;110(1):A8–10.
23. Thomas RS, Rank DR, Penn SG, Zastrow GM, Hayes KR, Pande K, Glover E, Silander T, Craven MW, Reddy JK, Jovanovich SB, Bradfield CA. Identification of toxicologically predictive gene sets using cDNA microarrays. *Mol Pharmacol* 2001;60(6):1189–94.
24. Olden K, Guthrie J. Genomics: Implications for toxicology. *Mutat Res* 2001;473(1): 3–10.
25. Waters MD, Olden K, Tennant RW. Toxicogenomic approach for assessing toxicant-related disease. *Mutat Res* 2003a;544(2–3):415–24.
26. Waters MD, Boorman G, Bushel P, Cunningham M, Irwin R, Merrick A, Olden K, Paules R, Selkirk J, Stasiewicz S, Weis B, Van Houten B, Walker N, Tennant R. Systems toxicology and the chemical effects in biological systems knowledge base. *Environ Health Perspect* 2003b;111(6):811–24.
27. Lobenhofer EK, Bushel PR, Afshari CA, Hamadeh HK. Progress in the application of DNA microarrays. *Environ Health Perspect* 2001;109(9):881–91.
28. Burchiel SW, Knall CM, Davis JW, II, Paules RS, Boggs SE, Afshari CA. Analysis of genetic and epigenetic mechanisms of toxicity: Potential roles of toxicogenomics and proteomics in toxicology. *Toxicol Sci* 2001;59(2):193–5.
29. Merrick BA, Tomer KB. Toxicoproteomics: A parallel approach to identifying biomarkers. *Environ Health Perspect* 2003;111(11):A578–9.
30. Petricoin EF, Rajapaske V, Herman EH, Arekani AM, Ross S, Johann D, Knapton A, Zhang J, Hitt BA, Conrads TP, Veenstra TD, Liotta LA, Sistare FD. Toxicoproteomics: Serum proteomic pattern diagnostics for early detection of drug induced cardiac toxicities and cardioprotection. *Toxicol Pathol* 2004;32 (Suppl 1):122–30.
31. Wilkins MR, Pasquali C, Appel RD, Ou K, Golaz O, Sanchez JC, Yan JX, Gooley AA, Hughes G, Humphery-Smith I, Williams KL, Hochstrasser DF. From proteins to proteomes: Large scale protein identification by two-dimensional electrophoresis and amino acid analysis. *Biotechnology (NY)* 1996;14(1):61–5.
32. Griffin JL. Understanding mouse models of disease through metabolomics. *Curr Opin Chem Biol* 2006;10(4):309–15.
33. Kasper P, Oliver G, Lima BS, Singer T, Tweats D. Joint EFPIA/CHMP SWP workshop: The emerging use of omic technologies for regulatory non-clinical safety testing. *Pharmacogenomics* 2005;6(2):181–4.
34. Lindon JC, Holmes E, Nicholson JK. Metabonomics techniques and applications to pharmaceutical research & development. *Pharm Res* 2006;23(6):1075–88.
35. Nurse P. Understanding cells. *Nat Biotechnol* 2003;424:883.

36. Hamadeh HK, Knight BL, Haugen AC, Sieber S, Amin RP, Bushel PR, Stoll R, Blanchard K, Jayadev S, Tennant RW, Cunningham ML, Afshari CA, Paules RS. Methapyrilene toxicity: Anchorage of pathologic observations to gene expression alterations. *Toxicol Pathol* 2002c;30:470–482.

37. Zweiger G. Knowledge discovery in gene-expression-microarray data: Mining the information output of the genome. *Trends Biotechnol* 1999;17(11):429–36.

38. Waring JF, Jolly RA, Ciurlionis R, Lum PY, Praestgaard JT, Morfitt DC, Buratto B, Roberts C, Schadt E, Ulrich RG. Clustering of hepatotoxins based on mechanism of toxicity using gene expression profiles. *Toxicol Appl Pharmacol* 2001; 175(1):28–42.

39. Mortuza GB, Neville WA, Delaney J, Waterfield CJ, Camilleri P. Characterisation of a potential biomarker of phospholipidosis from amiodarone-treated rats. *Biochim Biophys Acta* 2003;1631(2):136–46.

40. Kramer JA, Curtiss SW, Kolaja KL, Alden CL, Blomme EAG, Curtiss WC, Davila JC, Jackson CJ, Bunch RT. Acute molecular markers of rodent hepatic carcinogenesis identified by transcription profiling. *Chem Res Toxicol* 2004;17(4): 463–70.

41. Heinloth AN, Irwin RD, Boorman GA, Nettesheim P, Fannin RD, Sieber SO, Snell ML, Tucker CJ, Li L, Travlos GS, Vansant G, Blackshear PE, Tennant RW, Cunningham ML, Paules RS. Gene expression profiling of rat livers reveals indicators of potential adverse effects. *Toxicol Sci* 2004;80(1):193–202.

42. Hamadeh HK, Bushel PR, Jayadev S, DiSorbo O, Bennett L, Li L, Tennant R, Stoll R, Barrett JC, Paules RS, Blanchard K, Afshari CA. Prediction of compound signature using high density gene expression profiling. *Toxicol Sci* 2002b;67(2): 232–40.

43. Bulera SJ, Eddy SM, Ferguson E, Jatkoe TA, Reindel JF, Bleavins MR, De La Iglesia FA. RNA expression in the early characterization of hepatotoxicants in Wistar rats by high-density DNA microarrays. *Hepatology* 2001;33(5):1239–58.

44. Bartosiewicz MJ, Jenkins D, Penn S, Emery J, Buckpitt A. Unique gene expression patterns in liver and kidney associated with exposure to chemical toxicants. *J Pharmacol Exp Ther* 2001;297(3):895–905.

45. Cutler P, Bell DJ, Birrell HC, Connelly JC, Connor SC, Holmes E, Mitchell BC, Monte SY, Neville BA, Pickford R, Polley S, Schneider K, Skehel JM. An integrated proteomic approach to studying glomerular nephrotoxicity. *Electrophoresis* 1999;20(18):3647–58.

46. Coen M, Ruepp SU, Lindon JC, Nicholson JK, Pognan F, Lenz EM, Wilson ID. Integrated application of transcriptomics and metabonomics yields new insight into the toxicity due to paracetamol in the mouse. *J Pharm Biomed Anal* 2004;35(1):93–105.

47. Donald S, Verschoyle RD, Edwards R, Judah DJ, Davies R, Riley J, Dinsdale D, Lopez Lazaro L, Smith AG, Gant TW, Greaves P, Gescher AJ. Hepatobiliary damage and changes in hepatic gene expression caused by the antitumor drug ecteinascidin-743 (ET-743) in the female rat. *Cancer Res* 2002;62(15):4256–62.

48. Fertuck KC, Eckel JE, Gennings C, Zacharewski TR. Identification of temporal patterns of gene expression in the uteri of immature, ovariectomized mice following exposure to ethynylestradiol. *Physiol Genomics* 2003;15(2):127–41.

49. Fountoulakis M, de Vera MC, Crameri F, Boess F, Gasser R, Albertini S, Suter L. Modulation of gene and protein expression by carbon tetrachloride in the rat liver. *Toxicol Appl Pharmacol* 2002;183(1):71–80.

50. Hamadeh HK, Bushel PR, Jayadev S, Martin K, DiSorbo O, Sieber S, Bennett L, Tennant R, Stoll R, Barrett JC, Blanchard K, Paules RS, Afshari CA. Gene expression analysis reveals chemical-specific profiles. *Toxicol Sci* 2002a;67(2): 219–31.

51. Huang Q, Dunn RT, 2nd, Jayadev S, DiSorbo O, Pack FD, Farr SB, Stoll RE, Blanchard KT. Assessment of cisplatin-induced nephrotoxicity by microarray technology. *Toxicol Sci* 2001;63(2):196–207.

52. Kramer JA, Blomme EA, Bunch RT, Davila JC, Jackson CJ, Jones PF, Kolaja KL, Curtiss SW. Transcription profiling distinguishes dose-dependent effects in the livers of rats treated with clofibrate. *Toxicol Pathol* 2003;31(4):417–31.

53. Peterson RL, Casciotti L, Block L, Goad ME, Tong Z, Meehan JT, Jordan RA, Vinlove MP, Markiewicz VR, Weed CA, Dorner AJ. Mechanistic toxicogenomic analysis of WAY-144122 administration in Sprague-Dawley rats. *Toxicol Appl Pharmacol* 2004;196(1):80–94.

54. Ruepp SU, Tonge RP, Shaw J, Wallis N, Pognan F. Genomics and proteomics analysis of acetaminophen toxicity in mouse liver. *Toxicol Sci* 2002;65(1): 135–50.

55. Waring JF, Gum R, Morfitt D, Jolly RA, Ciurlionis R, Heindel M, Gallenberg L, Buratto B, Ulrich RG. Identifying toxic mechanisms using DNA microarrays: Evidence that an experimental inhibitor of cell adhesion molecule expression signals through the aryl hydrocarbon nuclear receptor. *Toxicology* 2002;181–182: 535–50.

56. Wagenaar GTM, ter Horst SAJ, van Gastelen MA, Leijser LM, Mauad T, van der Velden PA, de Heer E, Hiemstra PS, Poorthuis BJHM, Walther FJ. Gene expression profile and histopathology of experimental bronchopulmonary dysplasia induced by prolonged oxidative stress. *Free Rad Biol Med* 2004;36(6):782–801.

57. Lu T, Liu J, LeCluyse EL, Zhou YS, Cheng ML, Waalkes MP. Application of cDNA microarray to the study of arsenic-induced liver diseases in the population of Guizhou, China. *Toxicol Sci* 2001;59(1):185–92.

58. Hamadeh HK, Jayadev S, Gaillard ET, Huang Q, Stoll R, Blanchard K, Chou J, Tucker CJ, Collins J, Maronpot R, Bushel P, Afshari CA. Integration of clinical and gene expression endpoints to explore furan-mediated hepatotoxicity. *Mutat Res* 2004;549(1–2):169–83.

59. Iida M, Anna CH, Hartis J, Bruno M, Wetmore B, Dubin JR, Sieber S, Bennett L, Cunningham ML, Paules RS, Tomer KB, Houle CD, Merrick AB, Sills RC, Devereux TR. Changes in global gene and protein expression during early mouse liver carcinogenesis induced by non-genotoxic model carcinogens oxazepam and Wyeth-14,643. *Carcinogenesis* 2003;24(4):757–70.

60. Ellinger-Ziegelbauer H, Stuart B, Wahle B, Bomann W, Ahr HJ. Characteristic expression profiles induced by genotoxic carcinogens in rat liver. *Toxicol Sci* 2004;77(1):19–34.

61. Twigger S, Lu J, Shimoyama M, Chen D, Pasko D, Long H, Ginster J, Chen CF, Nigam R, Kwitek A, Eppig J, Maltais L, Maglott D, Schuler G, Jacob H,

Tonellato PJ. Rat Genome Database (RGD): Mapping disease onto the genome. *Nucleic Acids Res* 2002;30(1):125–8.

62. Higgins MA, Berridge BR, Mills BJ, Schultze AE, Gao H, Searfoss GH, Baker TK, Ryan TP. Gene expression analysis of the acute phase response using a canine microarray. *Toxicol Sci* 2003;74(2):470–84.
63. Ulrich RG, Rockett JC, Gibson GG, Pettit SD. Overview of an interlaboratory collaboration on evaluating the effects of model hepatotoxicants on hepatic gene expression. *Environ Health Perspect* 2004;112(4):423–7.
64. Pennie W, Pettit SD, Lord PG. Toxicogenomics in risk assessment: An overview of an HESI collaborative research program. *Environ Health Perspect* 2004;112(4): 417–9.
65. Lindon JC, Nicholson JK, Holmes E, Antti H, Bollard ME, Keun H, Beckonert O, Ebbels TM, Reily MD, Robertson D, Stevens GJ, Luke P, Breau AP, Cantor GH, Bible RH, Niederhauser U, Senn H, Schlotterbeck G, Sidelmann UG, Laursen SM, Tymiak A, Car BD, Lehman-McKeeman L, Colet JM, Loukaci A, Thomas C. Contemporary issues in toxicology the role of metabonomics in toxicology and its evaluation by the COMET project. *Toxicol Appl Pharmacol* 2003;187(3):137–46.
66. Heller RA, Schena M, Chai A, Shalon D, Bedilion T, Gilmore J, Woolley DE, Davis RW. Discovery and analysis of inflammatory disease-related genes using cDNA microarrays. *Proc Natl Acad Sci USA* 1997;94(6):2150–5.
67. Merrick BA, Bruno ME. Genomic and proteomic profiling for biomarkers and signature profiles of toxicity. *Curr Opin Mol Ther* 2004;6(6):600–7.
68. Loeb W, Quimby F, editors. *The clinical chemistry of laboratory animals*. Philadelphia: Taylor and Francis, 1999.
69. Travlos GS, Morris RW, Elwell MR, Duke A, Rosenblum S, Thompson MB. Frequency and relationships of clinical chemistry and liver and kidney histopathology findings in 13-week toxicity studies in rats. *Toxicology* 1996;107(1):17–29.
70. Paules R. Phenotypic anchoring: Linking cause and effect. *Environ Health Perspect.* 2003;111(6):A338–9.
71. Powell CL, Kosyk O, Ross PK, Schoonhoven R, Boysen G, Swenberg JA, Heinloth AN, Boorman GA, Cunningham ML, Paules RS, Rusyn I. Phenotypic anchoring of acetaminophen-induced oxidative stress with gene expression profiles in rat liver. *Toxicol Sci* 2006;93:213–22.
72. Guerreiro N, Staedtler F, Grenet O, Kehren J, Chibout SD. Toxicogenomics in drug development. *Toxicol Pathol* 2003;31(5):471–9.
73. Frank R, Hargreaves R. Clinical biomarkers in drug discovery and development. *Nat Rev Drug Discov* 2003;2(7):566–80.
74. Kennedy S. The role of proteomics in toxicology: Identification of biomarkers of toxicity by protein expression studies. *Biomarkers* 2002;7(4):269–90.
75. Gunn L, Smith MT. Emerging biomarker technologies. *IARC Sci Publ* 2004(157):437–50.
76. Hanash S. Disease proteomics. *Nature* 2003;422(6928):226–32.
77. Walgren JL, Thompson DC. Application of proteomic technologies in the drug development process. *Toxicol Lett* 2004;149(1–3):377–85.

78. Wallace KB, Hausner E, Herman E, Holt GD, MacGregor JT, Metz AL, Murphy E, Rosenblum IY, Sistare FD, York MJ. Serum troponins as biomarkers of drug-induced cardiac toxicity. *Toxicol Pathol* 2004;32(1):106–21.

79. Roberts R, Cain K, Coyle B, Freathy C, Leonard JF, Gautier JC. Early drug safety evaluation: Biomarkers, signatures, and fingerprints. *Drug Metab Rev* 2003; 35(4):269–75.

80. Amacher DE. A toxicologist's guide to biomarkers of hepatic response. *Hum Exp Toxicol* 2002;21(5):253–62.

81. Lindon JC. Biomarkers: Present concepts and future promise. *Preclinica* 2003; 1(5):221.

82. Bailey WJ, Ulrich R. Molecular profiling approaches for identifying novel biomarkers. *Expert Opin Drug Saf* 2004;3(2):137–51.

83. Bergmann S, Ihmels J, Barkai N. Iterative signature algorithm for the analysis of large-scale gene expression data. *Phys Rev E Stat Nonlin Soft Matter Phys* 2003;67(3 Pt 1):031902.

84. Abe T, Kanaya S, Kinouchi M, Ichiba Y, Kozuki T, Ikemura T. Informatics for unveiling hidden genome signatures. *Genome Res* 2003;13(4):693–702.

85. Choe LH, Dutt MJ, Relkin N, Lee KH. Studies of potential cerebrospinal fluid molecular markers for Alzheimer's disease. *Electrophoresis* 2002;23(14):2247–51.

86. Bratton SB, Cohen GM. Death receptors leave a caspase footprint that Smacs of XIAP. *Cell Death Differ* 2003;10(1):4–6.

87. Searfoss GH, Jordan WH, Calligaro DO, Galbreath EJ, Schirtzinger LM, Berridge BR, Gao H, Higgins MA, May PC, Ryan TP. Adipsin, a biomarker of gastrointestinal toxicity mediated by a functional gamma-secretase inhibitor. *J Biol Chem* 2003;278(46):46107–16.

88. Newton RK, Aardema M, Aubrecht J. The utility of DNA microarrays for characterizing genotoxicity. *Environ Health Perspect* 2004;112(4):420–2.

89. Petricoin EF, III. Ornstein DK, Paweletz CP, Ardekani A, Hackett PS, Hitt BA, Velassco A, Trucco C, Wiegand L, Wood K, Simone CB, Levine PJ, Linehan WM, Emmert-Buck MR, Steinberg SM, Kohn EC, Liotta LA. Serum proteomic patterns for detection of prostate cancer. *J Natl Cancer Inst* 2002;94(20):1576–8.

90. Zhang Z, Bast RC, Jr, Yu Y, Li J, Sokoll LJ, Rai AJ, Rosenzweig JM, Cameron B, Wang YY, Meng XY, Berchuck A, Van Haaften-Day C, Hacker NF, de Bruijn HW, van der Zee AG, Jacobs IJ, Fung ET, Chan DW. Three biomarkers identified from serum proteomic analysis for the detection of early stage ovarian cancer. *Cancer Res* 2004;64(16):5882–90.

91. Ginsburg GS, Haga SB. Translating genomic biomarkers into clinically useful diagnostics. *Expert Rev Mol Diagn* 2006;6(2):179–91.

92. Badghisi H, Liebler DC. Sequence mapping of epoxide adducts in human hemoglobin with LC-tandem MS and the SALSA algorithm. *Chem Res Toxicol* 2002; 15(6):799–805.

93. Fennell TR, Snyder RW, Krol WL, Sumner SC. Comparison of the hemoglobin adducts formed by administration of *N*-methylolacrylamide and acrylamide to rats. *Toxicol Sci* 2003;71(2):164–75.

94. Gao J, Ann Garulacan L, Storm SM, Hefta SA, Opiteck GJ, Lin JH, Moulin F, Dambach DM. Identification of in vitro protein biomarkers of idiosyncratic liver toxicity. *Toxicol In Vitro* 2004;18(4):533–41.
95. Nerland DE, Cai J, Benz FW. Selective covalent binding of acrylonitrile to Cys 186 in rat liver carbonic anhydrase III in vivo. *Chem Res Toxicol* 2003;16(5): 583–9.
96. Dare TO, Davies HA, Turton JA, Lomas L, Williams TC, York MJ. Application of surface-enhanced laser desorption/ionization technology to the detection and identification of urinary parvalbumin-alpha: A biomarker of compound-induced skeletal muscle toxicity in the rat. *Electrophoresis* 2002;23(18):3241–51.
97. Hogstrand C, Balesaria S, Glover CN. Application of genomics and proteomics for study of the integrated response to zinc exposure in a non-model fish species, the rainbow trout. *Compar Biochem Physiol Part B: Biochem Mol Biol* 2002; 133(4):523–35.
98. Hestvik AL, Hmama Z, Av-Gay Y. Kinome analysis of host response to mycobacterial infection: A novel technique in proteomics. *Infect Immun* 2003;71(10): 5514–22.
99. Izzotti A, Bagnasco M, Cartiglia C, Longobardi M, De Flora S. Proteomic analysis as related to transcriptome data in the lung of chromium(VI)-treated rats. *Int J Oncol* 2004;24(6):1513–22.
100. Hirata A, Masuda S, Tamura T, Kai K, Ojima K, Fukase A, Motoyoshi K, Kamakura K, Miyagoe-Suzuki Y, Takeda S. Expression profiling of cytokines and related genes in regenerating skeletal muscle after cardiotoxin injection: A role for osteopontin. *Am J Pathol* 2003;163(1):203–15.
101. Lame MW, Jones AD, Wilson DW, Segall HJ. Protein targets of 1,4-benzoquinone and 1,4-naphthoquinone in human bronchial epithelial cells. *Proteomics* 2003;3(4):479–95.
102. Sorensen M, Autrup H, Moller P, Hertel O, Jensen SS, Vinzents P, Knudsen LE, Loft S. Linking exposure to environmental pollutants with biological effects. *Mutat Res* 2003;544(2–3):255–71.
103. Castegna A, Thongboonkerd V, Klein JB, Lynn B, Markesbery WR, Butterfield DA. Proteomic identification of nitrated proteins in Alzheimer's disease brain. *J Neurochem* 2003;85(6):1394–401.
104. Fessler MB, Malcolm KC, Duncan MW, Worthen GS. A genomic and proteomic analysis of activation of the human neutrophil by lipopolysaccharide and its mediation by p38 mitogen-activated protein kinase. *J Biol Chem* 2002;277(35): 31291–302.
105. Heijne WH, Stierum RH, Slijper M, van Bladeren PJ, van Ommen B. Toxicogenomics of bromobenzene hepatotoxicity: A combined transcriptomics and proteomics approach. *Biochem Pharmacol* 2003;65(5):857–75.
106. Bandara LR, Kelly MD, Lock EA, Kennedy S. A correlation between a proteomic evaluation and conventional measurements in the assessment of renal proximal tubular toxicity. *Toxicol Sci* 2003;73(1):195–206.
107. Bruno ME, Borchers CH, Dial JM, Walker NJ, Hartis JE, Wetmore BA, Carl Barrett JC, Tomer KB, Merrick BA. Effects of TCDD upon IkappaB and IKK subunits localized in microsomes by proteomics. *Arch Biochem Biophys* 2002;406(2):153–64.

108. Dapas B, Tell G, Scaloni A, Pines A, Ferrara L, Quadrifoglio F, Scaggiante B. Identification of different isoforms of eEF1A in the nuclear fraction of human T-lymphoblastic cancer cell line specifically binding to aptameric cytotoxic GT oligomers. *Eur J Biochem* 2003;270(15):3251–62.

109. Mikhailov A, Harmala-Brasken AS, Hellman J, Meriluoto J, Eriksson JE. Identification of ATP-synthase as a novel intracellular target for microcystin-LR. *Chem Biol Interact* 2003;142(3):223–37.

110. Venkatraman A, Landar A, Davis AJ, Ulasova E, Page G, Murphy MP, Darley-Usmar V, Bailey SM. Oxidative modification of hepatic mitochondria protein thiols: Effect of chronic alcohol consumption. *Am J Physiol Gastrointest Liver Physiol* 2004;286(4):G521–7.

111. Patterson SD. Proteomics: Evolution of the technology. *Biotechniques* 2003; 35(3):440–4.

112. Pieper R, Su Q, Gatlin CL, Huang ST, Anderson NL, Steiner S. Multi-component immunoaffinity subtraction chromatography: An innovative step towards a comprehensive survey of the human plasma proteome. *Proteomics* 2003;3(4): 422–32.

113. Anderson NL, Anderson NG. The human plasma proteome: History, character, and diagnostic prospects. *Mol Cell Proteomics* 2002;1(11):845–67.

114. Adkins JN, Varnum SM, Auberry KJ, Moore RJ, Angell NH, Smith RD, Springer DL, Pounds JG. Toward a human blood serum proteome: Analysis by multidimensional separation coupled with mass spectrometry. *Mol Cell Proteomics* 2002;1(12):947–55.

115. Pieper R, Gatlin CL, Makusky AJ, Russo PS, Schatz CR, Miller SS, Su Q, McGrath AM, Estock MA, Parmar PP, Zhao M, Huang ST, Zhou J, Wang F, Esquer-Blasco R, Anderson NL, Taylor J, Steiner S. The human serum proteome: Display of nearly 3700 chromatographically separated protein spots on two-dimensional electrophoresis gels and identification of 325 distinct proteins. *Proteomics* 2003;3(7):1345–64.

116. Tirumalai RS, Chan KC, Prieto DA, Issaq HJ, Conrads TP, Veenstra TD. Characterization of the low molecular weight human serum proteome. *Mol Cell Proteomics* 2003;2(10):1096–103.

117. Merrick BA. The human proteome organization (HUPO) and environmental health. *EHP Toxicogenomics* 2003;111(1t):1–5.

118. Vitorino R, Lobo MJ, Ferrer-Correira AJ, Dubin JR, Tomer KB, Domingues PM, Amado FM. Identification of human whole saliva protein components using proteomics. *Proteomics* 2004;4(4):1109–15.

119. Pieper R, Gatlin CL, McGrath AM, Makusky AJ, Mondal M, Seonarain M, Field E, Schatz CR, Estock MA, Ahmed N, Anderson NG, Steiner S. Characterization of the human urinary proteome: A method for high-resolution display of urinary proteins on two-dimensional electrophoresis gels with a yield of nearly 1400 distinct protein spots. *Proteomics* 2004;4(4):1159–74.

120. Puchades M, Hansson SF, Nilsson CL, Andreasen N, Blennow K, Davidsson P. Proteomic studies of potential cerebrospinal fluid protein markers for Alzheimer's disease. *Brain Res Mol Brain Res* 2003;118(1–2):140–6.

121. Jiang L, Lindpaintner K, Li HF, Gu NF, Langen H, He L, Fountoulakis M. Proteomic analysis of the cerebrospinal fluid of patients with schizophrenia. *Amino Acids* 2003;25(1):49–57.

122. Kennedy S. Proteomic profiling from human samples: The body fluid alternative. *Toxicol Lett* 2001;120(1–3):379–84.

123. Gygi SP, Rist B, Gerber SA, Turecek F, Gelb MH, Aebersold R. Quantitative analysis of complex protein mixtures using isotope-coded affinity tags. *Nat Biotechnol* 1999;17(10):994–9.

124. Duan XJ, Xenarios I, Eisenberg D. Describing biological protein interactions in terms of protein states and state transitions: The LiveDIP database. *Mol Cell Proteomics* 2002;1(2):104–16.

125. Griffin TJ, Gygi SP, Ideker T, Rist B, Eng J, Hood L, Aebersold R. Complementary profiling of gene expression at the transcriptome and proteome levels in *Saccharomyces cerevisiae. Mol Cell Proteomics* 2002;1(4):323–33.

126. Washburn MP, Koller A, Oshiro G, Ulaszek RR, Plouffe D, Deciu C, Winzeler E, Yates JR, III. Protein pathway and complex clustering of correlated mRNA and protein expression analyses in *Saccharomyces cerevisiae. Proc Natl Acad Sci USA* 2003;100(6):3107–12.

127. Weinreb O, Mandel S, Youdim MB. cDNA gene expression profile homology of antioxidants and their antiapoptotic and proapoptotic activities in human neuroblastoma cells. *FASEB J* 2003;17(8):935–7.

128. Kern W, Kohlmann A, Wuchter C, Schnittger S, Schoch C, Mergenthaler S, Ratei R, Ludwig WD, Hiddemann W, Haferlach T. Correlation of protein expression and gene expression in acute leukemia. *Cytometry* 2003;55B(1):29–36.

129. McRedmond JP, Park SD, Reilly DF, Coppinger JA, Maguire PB, Shields DC, Fitzgerald DJ. Integration of proteomics and genomics in platelets: A profile of platelet proteins and platelet-specific genes. *Mol Cell Proteomics* 2004;3(2):133–44.

130. Hu Y, Hines LM, Weng H, Zuo D, Rivera M, Richardson A, LaBaer J. Analysis of genomic and proteomic data using advanced literature mining. *J Proteome Res* 2003;2(4):405–12.

131. Mootha VK, Bunkenborg J, Olsen JV, Hjerrild M, Wisniewski JR, Stahl E, Bolouri MS, Ray HN, Sihag S, Kamal M, Patterson N, Lander ES, Mann M. Integrated analysis of protein composition, tissue diversity, and gene regulation in mouse mitochondria. *Cell* 2003;115(5):629–40.

132. Thompson KL, Afshari CA, Amin RP, Bertram TA, Car B, Cunningham M, Kind Cl, Kramer JA, Lawton M, Mirsky M, Naciff JM, Oreffo V, Pine PS, Sistare FD. Identification of platform-independent gene expression markers of cisplatin nephrotoxicity. *Environ Health Perspect* 2004;112(4):488–94.

133. Kramer JA, Pettit SD, Amin RP, Bertram TA, Car B, Cunningham M, Curtiss SW, Davis JW, Kind C, Lawton M, Naciff JM, Oreffo V, Roman RJ, Sistare FD, Stevens J, Thompson K, Vickers AE, Wild S, Afshari CA. Overview on the application of transcription profiling using selected nephrotoxicants for toxicology assessment. *Environ Health Perspect* 2004;112(4):460–4.

134. Baker VA, Harries HM, Waring JF, Duggan CM, Ni HA, Jolly RA, Yoon LW, De Souza AT, Schmid JE, Brown RH, Ulrich RG, Rockett JC. Clofibrate-induced

gene expression changes in rat liver: A cross-laboratory analysis using membrane cDNA arrays. *Environ Health Perspect* 2004;112(4):428–38.

135. Chu TM, Deng S, Wolfinger R, Paules RS, Hamadeh HK. Cross-site comparison of gene expression data reveals high similarity. *Environ Health Perspect* 2004; 112(4):449–55.
136. Waring JF, Ulrich RG, Flint N, Morfitt D, Kalkuhl A, Staedtler F, Lawton M, Beekman JM, Suter L. Interlaboratory evaluation of rat hepatic gene expression changes induced by methapyrilene. *Environ Health Perspect* 2004;112(4):439–48.
137. Smith LL. Key challenges for toxicologists in the 21st century. *Trends Pharmacol Sci* 2001;22(6):281–5.
138. Juan HF, Lin JYC, Chang WH, Wu CY, Pan TL, Tseng MJ, Khoo KH, Chen ST. Biomic study of human meyloid leukemia cells differentiation to macrophages using DNA array, proteomic, and bioinformatic analytical methods. *Electrophoresis* 2002;23(15):2490–504.
139. Amin RP, Hamadeh HK, Bushel PR, Bennett L, Afshari CA, Paules RS. Genomic interrogation of mechanism(s) underlying cellular responses to toxicants. *Toxicology* 2002;181–82:555–63.
140. Kita Y, Shiozawa M, Jin W, Majewski RR, Besharse JC, Greene AS, Jacob HJ. Implications of circadian gene expression in kidney, liver and the effects of fasting on pharmacogenomic studies. *Pharmacogenetics* 2002;12(1):55–65.
141. Kaput J, Rodriguez RL. Nutritional genomics: The next frontier in the post-genomic era. *Physiol Genomics* 2004;16(2):166–77.
142. Kaput J. Diet-disease gene interactions. *Nutrition* 2004;20(1):26–31.
143. Klein TE, Altman RB. PharmGKB: The pharmacogenetics nad pharmacogenomics knowledge base. *Pharmacogenomics J.* 2004;4(1):1.
144. Wittliff JL, Erlander MG. Laser capture microdissection and its applications in genomics and proteomics. *Meth Enzymol* 2002;356:12–25.
145. Jain KK. Application of laser capture microdissection to proteomics. *Meth Enzymol* 2002;356:157–67.
146. Emmert-Buck MR, Bonner RF, Smith PD, Chuaqui RF, Zhuang Z, Goldstein SR, Weiss RA, Liotta LA. Laser capture microdissection. *Science* 1996;274(5289): 998–1001.
147. Bonner RF, Emmert-Buck M, Cole K, Pohida T, Chuaqui R, Goldstein S, Liotta LA. Laser capture microdissection: Molecular analysis of tissue. *Science* 1997;278(5342):1481,1483.
148. Karsten SL, Van Deerlin VM, Sabatti C, Gill LH, Geschwind DH. An evaluation of tyramide signal amplification and archived fixed and frozen tissue in microarray gene expression analysis. *Nucleic Acids Res* 2002;30(2):E4.
149. Hamadeh HK, Bushel P, Tucker CJ, Martin K, Paules R, Afshari CA. Detection of diluted gene expression alterations using cDNA microarrays. *Biotechniques* 2002e;32(2):322, 324, 326–9.
150. Huang RP. Detection of multiple proteins in an antibody-based protein microarray system. *J Immunol Meth* 2001;255(1–2):1–13.
151. Merchant M, Weinberger SR. Recent advancements in surface-enhanced laser desorption/ionization-time of flight-mass spectrometry. *Electrophoresis* 2000;21(6): 1164–77.

152. Liotta L, Petricoin E. Molecular profiling of human cancer. *Nat Rev Genet* 2000;1(1):48–56.

153. Herbert A. The four Rs of RNA-directed evolution. *Nat Genet* 2004;36(1):19–25.

154. Venter JC, Adams MD, Myers EW, Li PW, Mural RJ, Sutton GG, Smith HO, Yandell M, Evans CA, Holt RA, Gocayne JD, Amanatides P, Ballew RM, Huson DH, Wortman JR, Zhang Q, Kodira CD, Zheng XH, Chen L, Skupski M, Subramanian G, Thomas PD, Zhang J, Gabor Miklos GL, Nelson C, Broder S, Clark AG, Nadeau J, McKusick VA, Zinder N, Levine AJ, Roberts RJ, Simon M, Slayman C, Hunkapiller M, Bolanos R, Delcher A, Dew I, Fasulo D, Flanigan M, Florea L, Halpern A, Hannenhalli S, Kravitz S, Levy S, Mobarry C, Reinert K, Remington K, Abu-Threideh J, Beasley E, Biddick K, Bonazzi V, Brandon R, Cargill M, Chandramouliswaran I, Charlab R, Chaturvedi K, Deng Z, Di Francesco V, Dunn P, Eilbeck K, Evangelista C, Gabrielian AE, Gan W, Ge W, Gong F, Gu Z, Guan P, Heiman TJ, Higgins ME, Ji RR, Ke Z, Ketchum KA, Lai Z, Lei Y, Li Z, Li J, Liang Y, Lin X, Lu F, Merkulov GV, Milshina N, Moore HM, Naik AK, Narayan VA, Neelam B, Nusskern D, Rusch DB, Salzberg S, Shao W, Shue B, Sun J, Wang Z, Wang A, Wang X, Wang J, Wei M, Wides R, Xiao C, Yan C, et al. The sequence of the human genome. *Science* 2001;291(5507): 1304–51.

155. McPherson JD, Marra M, Hillier L, Waterston RH, Chinwalla A, Wallis J, Sekhon M, Wylie K, Mardis ER, Wilson RK, Fulton R, Kucaba TA, Wagner-McPherson C, Barbazuk WB, Gregory SG, Humphray SJ, French L, Evans RS, Bethel G, Whittaker A, Holden JL, McCann OT, Dunham A, Soderlund C, Scott CE, Bentley DR, Schuler G, Chen HC, Jang W, Green ED, Idol JR, Maduro VV, Montgomery KT, Lee E, Miller A, Emerling S, Kucherlapati, Gibbs R, Scherer S, Gorrell JH, Sodergren E, Clerc-Blankenburg K, Tabor P, Naylor S, Garcia D, de Jong PJ, Catanese JJ, Nowak N, Osoegawa K, Qin S, Rowen L, Madan A, Dors M, Hood L, Trask B, Friedman C, Massa H, Cheung VG, Kirsch IR, Reid T, Yonescu R, Weissenbach J, Bruls T, Heilig R, Branscomb E, Olsen A, Doggett N, Cheng JF, Hawkins T, Myers RM, Shang J, Ramirez L, Schmutz J, Velasquez O, Dixon K, Stone NE, Cox DR, Haussler D, Kent WJ, Furey T, Rogic S, Kennedy S, Jones S, Rosenthal A, Wen G, Schilhabel M, Gloeckner G, Nyakatura G, Siebert R, Schlegelberger B, Korenberg J, Chen XN, Fujiyama A, Hattori M, Toyoda A, Yada T, Park HS, Sakaki Y, Shimizu N, Asakawa S, et al. A physical map of the human genome. *Nature* 2001;409(6822):934–41.

156. Shearstone JR, Wang YE, Clement A, Allaire NE, Yang C, Worley DS, Carulli JP, Perrin S. Application of functional genomic technologies in a mouse model of retinal degeneration. *Genomics* 2005;85(3):309–21.

157. Benard J, Douc-Rasy S, Ahomadegbe JC. TP53 family members and human cancers. *Hum Mutat* 2003;21(3):182–91.

158. Courtois S, de Fromentel CC, Hainaut P. p53 protein variants: Structural and functional similarities with p63 and p73 isoforms. *Oncogene* 2004;23(3):631–8.

159. Demonacos C, La Thangue NB. Drug discovery and the p53 family. *Prog Cell Cycle Res* 2003;5:375–82.

160. Harrison PM, Kumar A, Lang N, Snyder M, Gerstein M. A question of size: The eukaryotic proteome and the problems in defining it. *Nucleic Acids Res* 2002; 30(5):1083–90.

161. Andersen JS, Lyon CE, Fox AH, Leung AK, Lam YW, Steen H, Mann M, Lamond AI. Directed proteomic analysis of the human nucleolus. *Curr Biol* 2002;12(1):1–11.

162. Wu CC, MacCoss MJ, Mardones G, Finnigan C, Mogelsvang S, Yates JR III, Howell KE. Organellar proteomics reveals Golgi arginine dimethylation. *Mol Biol Cell* 2004;15(6):2907–19.

163. Liu ZX, Kaplowitz N. Immune-mediated drug-induced liver disease. *Clin Liver Dis* 2002;6(3):467–86.

164. Ju C, Pohl LR. Immunohistochemical detection of protein adducts of 2,4-dinitrochlorobenzene in antigen presenting cells and lymphocytes after oral administration to mice: Lack of a role of Kupffer cells in oral tolerance. *Chem Res Toxicol* 2001;14(9):1209–17.

165. Ju C, Uetrecht JP. Mechanism of idiosyncratic drug reactions: Reactive metabolite formation, protein binding and the regulation of the immune system. *Curr Drug Metab* 2002;3(4):367–77.

166. Quackenbush J. Computational analysis of microarray data. *Nat Rev Genet* 2001;2(6):418–27.

167. Farland WH. The U.S. Environmental Protection Agency's Risk Assessment Guidelines: Current status and future directions. *Toxicol Ind Health* 1992;8(3): 205–12.

168. Farland WH. Cancer risk assessment: Evolution of the process. *Prev Med* 1996;25(1):24–5.

169. Larsen JC, Farland W, Winters D. Current risk assessment approaches in different countries. *Food Addit Contam* 2000;17(4):359–69.

170. Begley TJ, Rosenbach AS, Ideker T, Samson LD. Damage recovery pathways in *Saccharomyces cerevisiae* revealed by genomic phenotyping and interactome mapping. *Mol Cancer Res* 2002;1(2):103–12.

171. Hughes TR, Marton MJ, Jones AR, Roberts CJ, Stoughton R, Armour CD, Bennett HA, Coffey E, Dai H, He YD, Kidd MJ, King AM, Meyer MR, Slade D, Lum PY, Stepaniants SB, Shoemaker DD, Gachotte D, Chakraburtty K, Simon J, Bard M, Friend SH. Functional discovery via a compendium of expression profiles. *Cell* 2000;102(1):109–26.

172. Lewis F, Maughan NJ, Smith V, Hillan K, Quirke P. Unlocking the archive—Gene expression in paraffin-embedded tissue. *J Pathol* 2001;195(1):66–71.

173. Wetmore B, Merrick BA. Toxicoproteomics: Proteomics applied to toxicology and pathology. *Toxicol Pathol* 2004;32:619–42.

174. Stuart GW, Berry MW. A comprehensive whole genome bacterial phylogeny using correlated peptide motifs defined in a high dimensional vector space. *J Bioinform Comput Biol* 2003;1(3):475–93.

175. Chaussabel D, Sher A. Mining microarray expression data by literature profiling. *Genome Biol* 2002;3(10):RESEARCH0055.

176. Sluka JP. Extracting knowledge from genomic experiments by incorporating the biomedical literature. In: Lin SM, Johnson KF, editors, *Methods of microarray data analysis II* Boston: Kluwer Academics, 2002.

177. Patterson SD, Aebersold RH. Proteomics: The first decade and beyond. *Nat Genet* 2003;33(Suppl):311–23.

178. Davis TN. Protein localization in proteomics. *Curr Opin Chem Biol* 2004;8(1): 49–53.

179. Zhu H, Bilgin M, Snyder M. Proteomics. *An Rev Biochem* 2003;72:783–812.

180. Wolters DA, Washburn MP, Yates JR, III. An automated multidimensional protein identification technology for shotgun proteomics. *Anal Chem* 2001;73(23): 5683–90.

181. MacCoss MJ, McDonald WH, Saraf A, Sadygov R, Clark JM, Tasto JJ, Gould KL, Wolters D, Washburn M, Weiss A, Clark JI, Yates JR, III. Shotgun identification of protein modifications from protein complexes and lens tissue. *Proc Natl Acad Sci USA* 2002;99(12):7900–5.

182. Dongre AR, Opiteck G, Cosand WL, Hefta SA. Proteomics in the postgenome age. *Biopolymers* 2001;60(3):206–11.

183. Cordwell SJ, Nouwens AS, Verrills NM, Basseal DJ, Walsh BJ. Subproteomics based upon protein cellular location and relative solubilities in conjunction with composite two-dimensional electrophoresis gels. *Electrophoresis* 2000;21(6):1094–103.

184. Dreger M. Subcellular proteomics. *Mass Spectrom Rev* 2003;22(1):27–56.

185. Schirmer EC, Florens L, Guan T, Yates JR, III, Gerace L. Nuclear membrane proteins with potential disease links found by subtractive proteomics. *Science* 2003;301(5638):1380–2.

186. Ping P. Identification of novel signaling complexes by functional proteomics. *Circ Res* 2003;93(7):595–603.

187. Lee CL, Hsiao HH, Lin CW, Wu SP, Huang SY, Wu CY, Wang AH, Khoo KH. Strategic shotgun proteomics approach for efficient construction of an expression map of targeted protein families in hepatoma cell lines. *Proteomics* 2003;3(12): 2472–86.

188. Nicholson JK, Connelly J, Lindon JC, Holmes E. Metabonomics: A platform for studying drug toxicity and gene function. *Nat Rev Drug Discov* 2002;1(2): 153–61.

189. Xirasagar S, Gustafson S, Merrick BA, Tomer KB, Stasiewicz S, Chan DD, Yost KJ, III, Yates JR, III, Sumner S, Xiao N, Waters MD. CEBS object model for systems biology data, SysBio-OM. *Bioinformatics* 2004;20(13):2004–15.

190. Xirasagar S, Gustafson SF, Huang CC, Pan Q, Fostel J, Boyer P, Merrick BA, Tomer KB, Chan DD, Yost KJ, III, Choi D, Xiao N, Stasiewicz S, Bushel P, Waters MD. Chemical effects in biological systems (CEBS) object model for toxicology data, SysTox-OM: Design and application. *Bioinformatics* 2006;22(7): 874–82.

191. Altschul SF, Gish W, Miller W, Myers EW, Lipman DJ. Basic local alignment search tool. *J Mol Biol* 1990;215(3):403–10.

192. Petricoin EF, III, Hackett JL, Lesko LJ, Puri RK, Gutman SI, Chumakov K, Woodcock J, Feigal DW, Jr, Zoon KC, Sistare FD. Medical applications of microarray technologies: A regulatory science perspective. *Nat Genet* 2002;32(Suppl): 474–9.

193. Edgar R, Domrachev M, Lash AE. Gene expression omnibus: NCBI gene expression and hybridization array data repository. *Nucleic Acids Res* 2002;30(1): 207–10.

194. Mattingly CJ, Colby GT, Rosenstein MC, Forrest JN, Jr, Boyer JL. Promoting comparative molecular studies in environmental health research: An overview of the comparative toxicogenomics database (CTD). *Pharmacogenomics J* 2004;4(1): 5–8.

195. Brazma A, Parkinson H, Sarkans U, Shojatalab M, Vilo J, Abeygunawardena N, Holloway E, Kapushesky M, Kemmeren P, Lara GG, Oezcimen A, Rocca-Serra P, Sansone SA. ArrayExpress—A public repository for microarray gene expression data at the EBI. *Nucleic Acids Res* 2003;31(1):68–71.

196. Ball CA, Sherlock G, Parkinson H, Rocca-Sera P, Brooksbank C, Causton HC, Cavalieri D, Gaasterland T, Hingamp P, Holstege F, Ringwald M, Spellman P, Stoeckert CJ, Jr, Stewart JE, Taylor R, Brazma A, Quackenbush J. Standards for microarray data. *Science* 2002;298(5593):539.

197. Brazma A, Hingamp P, Quackenbush J, Sherlock G, Spellman P, Stoeckert C, Aach J, Ansorge W, Ball CA, Causton HC, Gaasterland T, Glenisson P, Holstege FC, Kim IF, Markowitz V, Matese JC, Parkinson H, Robinson A, Sarkans U, Schulze-Kremer S, Stewart J, Taylor R, Vilo J, Vingron M. Minimum information about a microarray experiment (MIAME)—Toward standards for microarray data. *Nat Genet* 2001;29(4):365–71.

198. Ikeo K, Ishi-i J, Tamura T, Gojobori T, Tateno Y. CIBEX: Center for information biology gene expression database. *CR Biol* 2003;326(10–11):1079–82.

199. Bartosiewicz M, Trounstine M, Barker D, Johnston R, Buckpitt A. Development of a toxicological gene array and quantitative assessment of this technology. *Arch Biochem Biophys* 2000;376(1):66–73.

200. Jiang XS, Zhou H, Zhang L, Sheng QH, Li SJ, Li L, Hao P, Li YX, Xia QC, Wu JR, Zeng R. A high-throughput approach for subcellular proteome: Identification of rat liver proteins using subcellular fractionation coupled with two-dimensional liquid chromatography tandem mass spectrometry and bioinformatic analysis. *Mol Cell Proteomics* 2004;3(5):441–55.

201. Talamo F, D'Ambrosio C, Arena S, Del Vecchio P, Ledda L, Zehender G, Ferrara L, Scaloni A. Proteins from bovine tissues and biological fluids: Defining a reference electrophoresis map for liver, kidney, muscle, plasma and red blood cells. *Proteomics* 2003;3(4):440–60.

202. Reddy PH, McWeeney S, Park BS, Manczak M, Gutala RV, Partovi D, Jung J, Yau V, Searles R, Mori M, Quinn J. Gene expression profiles of transcripts in amyloid precursor protein transgenic mice: Up-regulation of mitochondrial metabolism and apoptotic genes is an early cellular change in Alzheimer's disease. *Hum Mol Genet* 2004;13(12):1225–40.

203. Weiss A, Delproposto J, Giroux CN. High-throughput phenotypic profiling of gene-environment interactions by quantitative growth curve analysis in *Saccharomyces cerevisiae*. *Anal Biochem* 2004;327(1):23–34.

204. Fountoulakis M, Berndt P, Boelsterli UA, Crameri F, Winter M, Albertini S, Suter L. Two-dimensional database of mouse liver proteins: Changes in hepatic protein levels following treatment with acetaminophen or its nontoxic regioisomer 3-acetamidophenol. *Electrophoresis* 2000;21(11):2148–61.

205. Lee WM. Drug-induced hepatotoxicity. *N Engl J Med* 2003;349(5):474–85.

206. Xie T, Tong L, Barrett T, Yuan J, Hatzidimitriou G, McCann UD, Becker KG, Donovan DM, Ricaurte GA. Changes in gene expression linked to methamphetamine-induced dopaminergic neurotoxicity. *J Neurosci* 2002;22(1):274–83.
207. Dam K, Seidler FJ, Slotkin TA. Transcriptional biomarkers distinguish between vulnerable periods for developmental neurotoxicity of chlorpyrifos: Implications for toxicogenomics. *Brain Res Bull* 2003;59(4):261–5.
208. Hamadeh HK, Trouba KJ, Amin RP, Afshari CA, Germolec D. Coordination of altered DNA repair and damage pathways in arsenite-exposed keratinocytes. *Toxicol Sci* 2002;69(2):306–16.
209. Hu D, Cao K, Peterson-Wakeman R, Wang R. Altered profile of gene expression in rat hearts induced by chronic nicotine consumption. *Biochem Biophys Res Commun* 2002;297(4):729–36.
210. Waring JF, Ciurlionis R, Jolly RA, Heindel M, Ulrich RG. Microarray analysis of hepatotoxins in vitro reveals a correlation between gene expression profiles and mechanisms of toxicity. *Toxicol Lett* 2001;120(1–3):359–68.
211. Hamadeh HK, Bushel P, Paules R, Afshari CA. Discovery in toxicology: Mediation by gene expression array technology. *J Biochem Mol Toxicol* 2001;15(5): 231–42.
212. Witzmann FA, Bobb A, Briggs GB, Coppage HN, Hess RA, Li J, Pedrick NM, Ritchie GD, Iii JR, Still KR. Analysis of rat testicular protein expression following 91-day exposure to JP-8 jet fuel vapor. *Proteomics* 2003;3(6):1016–27.
213. Huang Q, Jin X, Gaillard ET, Knight BL, Pack FD, Stoltz JH, Jayadev S, Blanchard KT. Gene expression profiling reveals multiple toxicity endpoints induced by hepatotoxicants. *Mutat Res* 2004;549(1–2):147–67.
214. Adachi T, Koh KB, Tainaka H, Matsuno Y, Ono Y, Sakurai K, Fukata H, Iguchi T, Komiyama M, Mori C. Toxicogenomic difference between diethylstilbestrol and 17beta-estradiol in mouse testicular gene expression by neonatal exposure. *Mol Reprod Dev* 2004;67(1):19–25.
215. Nadadur SS, Schladweiler MC, Kodavanti UP. A pulmonary rat gene array for screening altered expression profiles in air pollutant-induced lung injury. *Inhal Toxicol* 2000;12(12):1239–54.
216. Akhtar RA, Reddy AB, Maywood ES, Clayton JD, King VM, Smith AG, Gant TW, Hastings MH, Kyriacou CP. Circadian cycling of the mouse liver transcriptome, as revealed by cDNA microarray, is driven by the suprachiasmatic nucleus. *Curr Biol* 2002;12(7):540–50.
217. Heijne WH, Slitt AL, Van Bladeren PJ, Groten JP, Klaassen CD, Stierum RH, Van Ommen B. Bromobenzene-induced hepatotoxicity at the transcriptome level. *Toxicol Sci* 2004;79(2):411–22.
218. Bajzer Z, Randic M, Plavsic D, Basak SC. Novel map descriptors for characterization of toxic effects in proteomics maps. *J Mol Graph Model* 2003;22(1):1–9.
219. Bandara LR, Kelly MD, Lock EA, Kennedy S. A potential biomarker of kidney damage identified by proteomics: Preliminary findings. *Biomarkers* 2003;8(3–4): 272–86.
220. Charlwood J, Skehel JM, King N, Camilleri P, Lord P, Bugelski P, Atif U. Proteomic analysis of rat kidney cortex following treatment with gentamicin. *J Proteome Res* 2002;1(1):73–82.

221. Drake MG, Witzmann FA, Hyde J, Witten ML. JP-8 jet fuel exposure alters protein expression in the lung. *Toxicology* 2003;191(2–3):199–210.

222. Imanishi S, Harada K. Proteomics approach on microcystin binding proteins in mouse liver for investigation of microcystin toxicity. *Toxicon* 2004;43(6): 651–9.

223. Jin J, Yang J, Gao Z, Yu Y. Proteomic analysis of cellular responses to low concentration *N*-methyl-*N′*-nitro-*N*-nitrosoguanidine in human amnion FL cells. *Environ Mol Mutagen* 2004;43(2):93–9.

224. Keightley JA, Shang L, Kinter M. Proteomic analysis of oxidative stres-resistant cells: A specific role for aldose reductase overexpression in cytoprotection. *Mol Cell Proteomics* 2004;3(2):167–75.

225. Kleno TG, Leonardsen LR, Kjeldal HO, Laursen SM, Jensen ON, Baunsgaard D. Mechanisms of hydrazine toxicity in rat liver investigated by proteomics and multivariate data analysis. *Proteomics* 2004;4(3):868–80.

226. Krapfenbauer K, Berger M, Lubec G, Fountoulakis M. Changes in the brain protein levels following administration of kainic acid. *Electrophoresis* 2001; 22(10):2086–91.

227. Krayl M, Benndorf D, Loffhagen N, Babel W. Use of proteomics and physiological characteristics to elucidate ecotoxic effects of methyl *tert*-butyl ether in *Pseudomonas putida* KT2440. *Proteomics* 2003;3(8):1544–52.

228. Kristensen DB, Kawada N, Imamura K, Miyamoto Y, Tateno C, Seki S, Kuroki T, Yoshizato K. Proteome analysis of rat hepatic stellate cells. *Hepatology* 2000;32(2):268–77.

229. Lee YM, Park SH, Chung KC, Oh YJ. Proteomic analysis reveals upregulation of calreticulin in murine dopaminergic neuronal cells after treatment with 6-hydroxydopamine. *Neurosci Lett* 2003;352(1):17–20.

230. Macdonald N, Chevalier S, Tonge R, Davison M, Rowlinson R, Young J, Rayner S, Roberts R. Quantitative proteomic analysis of mouse liver response to the peroxisome proliferator diethylhexylphthalate (DEHP). *Arch Toxicol* 2001;75(7): 415–24.

231. Man WJ, White IR, Bryant D, Bugelski P, Camilleri P, Cutler P, Heald G, Lord PG, Wood J, Kramer K. Protein expression analysis of drug-mediated hepatotoxicity in the Sprague-Dawley rat. *Proteomics* 2002;2(11):1577–85.

232. Marengo E, Robotti E, Gianotti V, Righetti PG, Cecconi D, Domenici E. A new integrated statistical approach to the diagnostic use of two-dimensional maps. *Electrophoresis* 2003;24(1–2):225–36.

233. Meneses-Lorente G, Guest PC, Lawrence J, Muniappa N, Knowles MR, Skynner HA, Salim K, Cristea I, Mortishire-Smith R, Gaskell SJ, Watt A. A proteomic investigation of drug-induced steatosis in rat liver. *Chem Res Toxicol* 2004;17(5): 605–12.

234. Moller A, Soldan M, Volker U, Maser E. Two-dimensional gel electrophoresis: A powerful method to elucidate cellular responses to toxic compounds. *Toxicology* 2001;160(1–3):129–38.

235. Sharma R, Komatsu S, Noda H. Proteomic analysis of brown planthopper: Application to the study of carbamate toxicity. *Insect Biochem Mol Biol* 2004;34(5): 425–32.

236. Shepard JL, Olsson B, Tedengren M, Bradley BP. Protein expression signatures identified in Mytilus edulis exposed to PCBs, copper and salinity stress. *Mar Environ Res* 2000;50(1–5):337–40.

237. Shrader EA, Henry TR, Greeley MS, Jr, Bradley BP. Proteomics in zebrafish exposed to endocrine disrupting chemicals. *Ecotoxicology* 2003;12(6):485–8.

238. Son WK, Lee DY, Lee SH, Joo WA, Kim CW. Analysis of proteins expressed in rat plasma exposed to dioxin using 2-dimensional gel electrophoresis. *Proteomics* 2003;3(12):2393–401.

239. Steiner S, Gatlin CL, Lennon JJ, McGrath AM, Seonarain MD, Makusky AJ, Aponte AM, Esquer-Blasco R, Anderson NL. Cholesterol biosynthesis regulation and protein changes in rat liver following treatment with fluvastatin. *Toxicol Lett* 2001;120(1–3):369–77.

240. Tonge R, Shaw J, Middleton B, Rowlinson R, Rayner S, Young J, Pognan F, Hawkins E, Currie I, Davison M. Validation and development of fluorescence two-dimensional differential gel electrophoresis proteomics technology. *Proteomics* 2001;1(3):377–96.

241. Toussaint O, Dumont P, Dierick JF, Pascal T, Frippiat C, Chainiaux F, Magalhaes JP, Eliaers F, Remacle J. Stress-induced premature senescence as alternative toxicological method for testing the long-term effects of molecules under development in the industry. *Biogerontology* 2000;1(2):179–83.

242. Vido K, Spector D, Lagniel G, Lopez S, Toledano MB, Labarre J. A proteome analysis of the cadmium response in Saccharomyces cerevisiae. *J Biol Chem* 2001;276(11):8469–74.

243. Wattiez R, Noel-Georis I, Cruyt C, Broeckaert F, Bernard A, Falmagne P. Susceptibility to oxidative stress: Proteomic analysis of bronchoalveolar lavage from ozone-sensitive and ozone-resistant strains of mice. *Proteomics* 2003;3(5): 658–65.

244. Wheelock AM, Zhang L, Tran MU, Morin D, Penn S, Buckpitt AR, Plopper CG. Isolation of rodent airway epithelial cell proteins facilitates in vivo proteomics studies of lung toxicity. *Am J Physiol Lung Cell Mol Physiol* 2004;286(2): L399–410.

245. Witzmann FA, Bauer MD, Fieno AM, Grant RA, Keough TW, Kornguth SE, Lacey MP, Siegel FL, Sun Y, Wright LS, Young RS, Witten ML. Proteomic analysis of simulated occupational jet fuel exposure in the lung. *Electrophoresis* 1999;20(18):3659–69.

246. Witzmann FA, Bauer MD, Fieno AM, Grant RA, Keough TW, Lacey MP, Sun Y, Witten ML, Young RS. Proteomic analysis of the renal effects of simulated occupational jet fuel exposure. *Electrophoresis* 2000;21(5):976–84.

247. Zhang Y, Champagne N, Beitel LK, Goodyer CG, Trifiro M, LeBlanc A. Estrogen and androgen protection of human neurons against intracellular amyloid beta1–42 toxicity through heat shock protein 70. *J Neurosci* 2004;24(23): 5315–21.

248. Hermjakob H, Montecchi-Palazzi L, Bader G, Wojcik J, Salwinski L, Ceol A, Moore S, Orchard S, Sarkans U, von Mering C, Roechert B, Poux S, Jung E, Mersch H, Kersey P, Lappe M, Li Y, Zeng R, Rana D, Nikolski M, Husi H, Brun C, Shanker K, Grant SG, Sander C, Bork P, Zhu W, Pandey A, Brazma A,

Jacq B, Vidal M, Sherman D, Legrain P, Cesareni G, Xenarios I, Eisenberg D, Steipe B, Hogue C, Apweiler R. The HUPO PSI's molecular interaction format—A community standard for the representation of protein interaction data. *Nat Biotechnol* 2004;22(2):177–83.

249. Jones A, Hunt E, Wastling J, Pizarro A, Stoeckert CJ, Jr. An object model and database for functional genomics. *Bioinformatics* 2004;20(10):1583–90.

250. Orchard S, Hermjakob H, Julian RK, Jr, Runte K, Sherman D, Wojcik J, Zhu W, Apweiler R. Common interchange standards for proteomics data: Public availability of tools and schema. *Proteomics* 2004;4(2):490–1.

PART II

COMPUTATIONAL METHODS

6

TOXICOINFORMATICS: AN INTRODUCTION

WILLIAM J. WELSH, WEIDA TONG, AND PANOS G. GEORGOPOULOS

Contents

6.1 INTRODUCTION

Toxicoinformatics is an emerging scientific discipline that provides computational approaches to support the elucidation of mechanisms of chemical

Computational Toxicology: Risk Assessment for Pharmaceutical and Environmental Chemicals, Edited by Sean Ekins

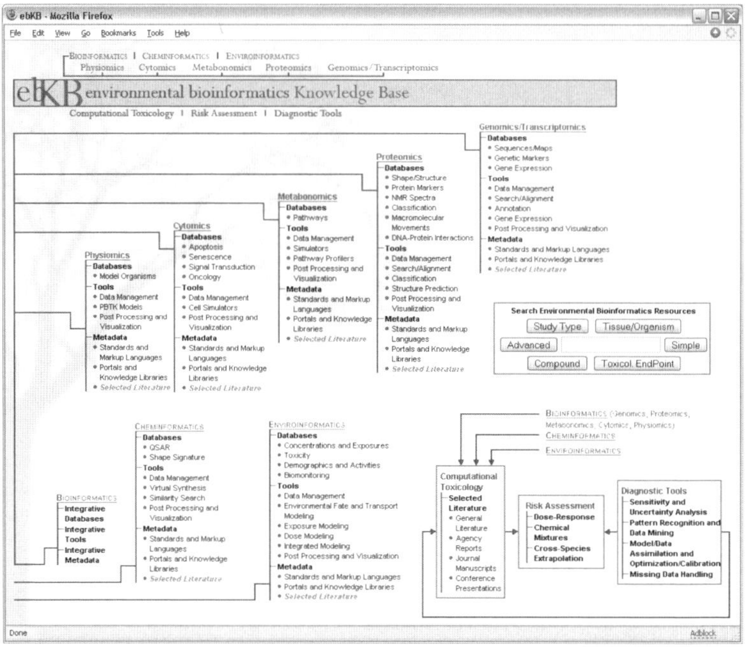

Figure 6.1 Evolving compendium of computational tools, databases, and literature information that make up the environmental bioinformatics Knowledge Base (ebKB).

toxicity. In its most general context toxicoinformatics refers to bioinformatic methods and computational tools (including databases and predictive models) that integrate knowledge and information from multiple scales of biological organization (molecular, cellular, organ, organism). This information is derived through the application of "genomic technologies" (genomics, transcriptomics, proteomics, metabolomics/metabonomics) as well as methods of cell biology (cytomics), and physiology (physiomics), in conjunction with computational chemistry (cheminformatics). The objective is to improve and quantify the understanding of biochemical mechanisms of toxicity (systems toxicology, computational toxicology). A wide spectrum of information covering this broad definition of toxicoinformatics can be found on the environmental bioinformatics Knowledge Base (ebKB) at www.ebkb.org. (Figure 6.1). In the more narrow context of toxicoinformatics which is adopted for the present introduction, the field can be viewed as a specialized area of cheminformatics that aims to relate molecular structure and properties to the potential toxicity of a chemical substance, using principles and methods of computational chemistry. To our best knowledge, the term "toxicoinformatics" was first coined in 2002 by the creation of the Center for Toxicoinformatics at the US Food and Drug Administration's National Center for Toxicological Research (US FDA-NCTR). The primary function of this program is to develop and implement toxicoinformatics approaches for "-omics" research and traditional toxicological studies at and beyond the US FDA-NCTR. Accordingly, this chapter reflects the authors' orientation in environmental science and regulatory affairs and their affiliation with the US EPA-supported Environmental Bioinformatics and Computational Toxicology Center (ebCTC: www.ebCTC.org). However, the topics addressed bear significance far beyond this single perspective.

The field of toxicoinformatics in the regulatory arena builds upon the tremendous progress achieved to date in predictive computational models and methods for hazard identification and risk assessment, particularly quantitative structure activity relationship (QSAR) models, pollution prevention strategies, and approaches to high-throughput screening [1–24]. A major concern about all predictive models (whether in vitro or computational) for regulatory purposes is the need to minimize, if not eliminate, uncertainty and false negatives [5–9,25,26]. New and more reliable computational strategies continue to be devised to meet the stringent requirements for regulatory applications. For instance, hierarchical frameworks (HFs) and consensus prediction models have recently attracted attention for their simplicity, scalability, and versatility. Because a vast numbers of chemicals may require screening, tiered HFs are best suited to this task. HFs organize component models, starting with the computationally fast but less accurate to the computationally intensive but more accurate. Consensus prediction modeling holds that the SAR analysis is encoded more effectively by multiple, independent models than by any single model. So, taken together, tiered HF and consensus prediction modeling architectures provide a versatile operational structure that is broadly capable of predicting chemical toxicity [2–9,11,12,19,20,26–29]. These novel approaches to toxicity prediction are described later in this chapter.

6.2 QUANTITATIVE STRUCTURE-ACTIVITY RELATIONSHIP (Q)SAR MODELS

Structure-activity relationships (SARs) and quantitative structure-activity relationships (QSARs), collectively known as (Q)SARs, are theoretical models that can be used to predict the physicochemical and biological properties of molecules. A SAR is a (qualitative) association between a chemical substructure and the potential of a chemical containing the substructure to exhibit a certain biological effect. A QSAR is a mathematical model that quantifies the relationship between the chemical's structure-related properties (descriptors) and normally a biological effect (e.g., toxicological endpoint) [2,5–10,12,30,31].

QSAR models have been widely used in the pharmaceutical industry, primarily for lead discovery and optimization (see Chapters 8–13). Likewise they have been employed in toxicology [32,33] and regulation [26,34,35] and have been particularly cost effective for prioritizing untested chemicals for more extensive and costly experimental evaluation. However, to ensure the proper use of QSAR models, it is important to recognize their inherent limitations [6,36–38]. Key regulatory uses of QSAR models include:

- For priority setting of chemicals
- For mechanistic information
- For grouping chemicals into categorical chemical families based on structure, activity/toxicity, mechanism, etc.
- For filling in data gaps in classification, labeling, and risk assessments
- For information on physicochemical properties, toxic potential and potency, environmental distribution and fate, and biokinetic processes

6.2.1 Construction of QSAR Models

Traditional QSAR modeling involves three basic steps: (1) collect or, if necessary, design a training set of chemicals, (2) choose descriptors that can properly relate chemical structure to the target property or activity, and (3) apply statistical regression methods that correlate changes in structure with changes in the target property. The final QSAR model is validated, *internally* by a procedure known as leave-n-out cross validation and *externally* by a test set of compounds with experimentally measured properties that were not used to build the QSAR model. Once validated, the QSAR model is considered useful for making predictions for untested chemicals generally within the "chemical space" of the training set and validation set.

These computational models must, however, undergo extensive validation in terms of the descriptors chosen to associate the target property and chemical structure, the applicable "chemical space" (applicability domain), and the quality of the data in the training sets [5–8,39]. Given the high-quality datasets, QSAR modeling methods can effectively predict discrete biological phenomenon at the molecular level in terms of interactions of chemicals with different classes of biological macromolecules (e.g., proteins).

6.2.2 Molecular Descriptors

The construction of QSAR models invariably requires the generation of molecular descriptors. For a single compound it is possible to calculate hundreds, and even thousands, of molecular descriptors directly from its molecular structure via a host of familiar open source and commercial application software packages. Molecular descriptors fall into several general categories (Table 6.1). In the present context we are interested in molecular descriptors that correlate with the physicochemical properties, biological activity, and toxicity of the target chemicals. Once appropriate molecular descriptors are chosen, QSAR models can be constructed to predict the target property of untested chemicals. Through QSAR models the specific molecular features

TABLE 6.1 Categories and Examples of Molecular Descriptors

Category	Requirements (Examples)
Constitutional	Molecular composition (M_w, number of atoms/bonds and of H-bond donors/acceptors)
Topological	2D structural formula (Kier-Hall indices, extent of branching)
Geometrical	3D structure of molecule (molecular volume, solvent accessible surface area, polar and nonpolar surface area)
Electrostatic	Charge distribution (atomic partial charges, electronegativities)
Quantum mechanical	Electronic structure (HOMO-LUMO energies, ionization potential)

that are associated with chemical toxicity can also offer insights into their modes-of-action. However, it must be emphasized that QSAR models provide a mathematical association or correlation, not a cause–effect relationship, between the target property and the descriptors.

6.2.3 Characteristics of QSAR Models

QSAR models can be broadly categorized into three types in terms of their construction and intended use: clustering, classification, and regression models. Clustering applies unsupervised learning techniques to explore data patterns on the basis of descriptors, without ascribing these patterns to specific categorical endpoints. In contrast, classification methods are supervised learning techniques that group chemicals into known categorical endpoints (e.g., highly active, moderately active, inactive). Regression models attempt to establish a quantitative relationship between structure and activity, through multivariate linear regression methods and, more recently, by way of machine learning approaches such as artifical neural networks (ANNs) (see also Chapter 8). In the interests of time, cost, and animal welfare, QSAR models will likely find extensive use in the assessment of high-production chemicals that depend on minimal animal testing [5–7,40,41]. QSAR models for computational toxicology will improve in their predictive ability as databases with reliable SAR data become more widely accessible [14,42–45].

Despite the wide acceptance of QSAR models within the regulatory community and beyond, QSAR models are not a panacea; they must be viewed realistically. QSAR models are an effective means for interpolation, not extrapolation. That is, it is risky and inadvisable to employ a QSAR model for predictions about chemicals that exceed the chemical space and range of biological activity of the training set. The single most common mistake in applying QSAR models would be to push them beyond these limits. It is not necessary for the chemicals in these various data sets to be congeneric, but they should share common structural features.

Many QSAR models are "global models," meaning they seek general trends that apply across the entire collection of chemicals in the dataset. In attempting to capture global associations and relationships, they sacrifice prediction accuracy for any single chemical for accuracy over the entire data set. Chemicals at the extremes of biological activity (i.e., weakly active) are especially troublesome. Lack of appreciation of these characteristics has led unfortunately to misuse, misunderstanding, and downright mistrust of QSAR models [2–12]. The disappointing results have often stemmed from unrealistic expectations. The simple truth is that QSAR models are, by nature, imprecise. Realization of this problem with QSAR models has inspired innovative solutions. Scientists have proposed novel schemes for addressing and resolving these and other issues surrounding QSAR models [1–24].

Obtaining a good quality QSAR model depends on many factors, such as the quality of biological data and the choice of descriptors and statistical methods. Given the technological advances and broader availability of various statistical methods and types of descriptors, it is now relatively easy and straightforward to develop a statistically sound model. However, methods for quantifying a QSAR model's quality and range of utility have not been addressed adequately, so this remains a challenge to those working in the field [9]. The utility of a QSAR model can be gauged by using three distinct criteria: (1) overall quality, (2) chance correlation, and (3) applicability domain. These concepts are described below.

The importance of validation has been generally acknowledged, and most QSAR models in the literature are validated either by cross validation or external test sets [13,46]. Model validation for classification models is typically specified by statistical quality measures of *overall quality* such as sensitivity, specificity, false positives, false negatives, and overall prediction. Unfortunately, it is often impossible to specify accuracy and prediction confidence for individual unknown chemicals, specifically those unknown chemicals with structures requiring the model to extend to, or beyond, the limits of chemistry space defined by the training set.

Assessing *chance correlation* is another measure of QSAR quality that is seldom reported with a QSAR model. Assessment of chance correlation intends to determine whether a valid model can be developed in the first place. This is usually accomplished by averaging many models where the activity values or classifications of the target property (e.g., acute toxicity) of a training set are randomly shuffled (scrambled or permuted). In general, skewed activity distributions and low signal-to-noise ratios can make training sets of small number of chemicals vulnerable to chance solutions. This effect becomes even more likely when a larger number of descriptors are available to build the QSAR model, a common occurrence today with the proliferation of software products for model development.

Differing from overall model validation, the assessment of *applicability domain* involves determining a model's confidence level in each prediction. Discussion in the literature regarding the accuracy or acceptability of QSAR models will often state the need to define a QSAR model's applicability domain. However, applicability domain means different things to different modelers, and such quantification can involve a wide assortment of statistical metrics [9]. Since a single simple definition is not feasible, applicability domain might best be viewed collectively as measures of confidence in *each prediction* when the overall quality of a model is acceptable. Model validation and applicability domain are separate assessments addressing model predictivity from distinct but related perspectives. Common practice is to develop a single QSAR model with a training set that is as large and structurally diverse as possible. This approach inherently results in an applicability domain that is difficult to determine and that is highly constrained by the training set. An alternative approach based on consensus of predictions of multiple models is described next for the case where the applicability domain is readily determinable.

External Validation versus Cross Validation A model fitted to the training set has minimal utility unless it can be generalized to predict unknown chemicals. Most experts in the QSAR field, as well as the present authors, concur that a model's predictive capability minimally needs to be demonstrated by some sort of cross validation or external validation procedure. Although both procedures share many common features, in principle, they are different in both ability and efficiency in assessing a model's overall prediction accuracy, applicability domain, and chance correlation during implementation.

When sufficient data are available, a fitted model should be validated by predicting chemicals not used in the training set but whose activities are known (the test set). This external validation method is analogous to a real-world application. However, external validation is not recommended unless the test set is sufficiently large and diverse, and encompasses the chemical space of the intended application. Using a small number of test set chemicals wastes valuable data that otherwise could improve the overall quality of a model. Design of the test set in terms of size and diversity, and most important, suitably for the intended application, is the major prerequisite for acceptable external validation. It is meaningless to design a test set without knowing how the QSAR model will be applied, which, unfortunately, accounts for most cases. Although the external validation method can discern the classes of chemicals that are not well predicted by the model, it generally provides only an overall assessment of a model with little indication of the prediction confidence for individual chemicals. In other words, external validation is of little value for assessing the applicability domain.

A common practice for defining a test set in external validation is to randomly select a portion of chemicals from a dataset. From this perspective, cross validation provides a similar measure of model performance for a given and fixed set of chemicals. In cross validation a fraction of chemicals in the training set are excluded, and then predicted by the model generated from the remaining chemicals. As each chemical is excluded one at a time, and the process is repeated for each chemical, this is known as leave-one-out cross validation. If the training set is divided into N groups with approximately equal numbers of chemicals, and the process is repeated for each group, it is called N-fold cross validation. The 10-fold cross validation procedure is commonly used to assess the predictive capability of a classification model. Cross validation results vary for each run due to random partitioning of the dataset, and thus it is recommended to repeat the cross validation process many times [47]. The average result of the multiple cross-validation runs provides an unbiased assessment of a model's predictivity. Compared with external validation, cross validation provides a systematic measurement of a model's performance without the loss of chemicals set aside for testing.

6.2.4 Consensus Modeling

Consensus modeling has been investigated for many years in the field of statistics as a means of combining multiple individual models to produce better single predictions [48]. A thorough review of this subject can be found in a number of papers [49–51]. The tacit assumption in consensus modeling is that the SAR relationship will be identified and encoded more effectively by multiple models than by a single model [52]. These benefits are realized only if the component models are independent; that is, they encode the complex relationship between chemical structure and biological activity in different ways. The ideal combined system should consist of a small number of accurate independent models. A significant added value of consensus modeling is that it lends itself to quantitative assessment of prediction confidence for individual chemicals, which is usually difficult to obtain from a single QSAR model [52].

Among the many approaches to consensus QSAR modeling, perhaps the simplest is to develop individual models using a subset of chemicals that are randomly selected from the entire original training dataset [53] (Figure 6.2). A recent study demonstrated that combining 10 models based on 10 datasets from a 10-fold cross validation procedure enhanced the overall performance of prediction [54]. Alternatively, the training set can be generated using more robust statistical "resampling" approaches, bagging [55] and boosting [56] (Figure 6.3). A drawback inherent to these techniques is that each component model is built from subsets of the entire training set, thereby forfeiting the full benefits otherwise attainable if each component model is built using the entire training set. Furthermore, since each chemical in a dataset encodes SAR information, reducing the number of chemicals in a training set for model construction will likely weaken the predictive accuracy of each individual model.

Alternatively, multiple models can be developed using different sets of descriptors [57]. One popular decision tree (DT) consensus method, called Random Forests, has recently demonstrated improved performance over bagging [58]. The DT method determines a chemical's activity through a

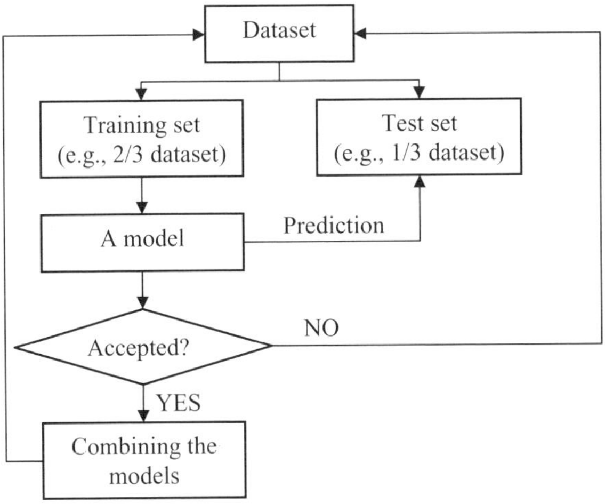

Figure 6.2 A resampling method. A dataset is first randomly divided into two sets, such as 2/3 for training and 1/3 for testing. A model developed with the training set is accepted if it gives satisfactory predictions for the testing set. A set of predictive models is generated by repeating the procedure, and the predictions of these models are then combined when predicting a new chemical.

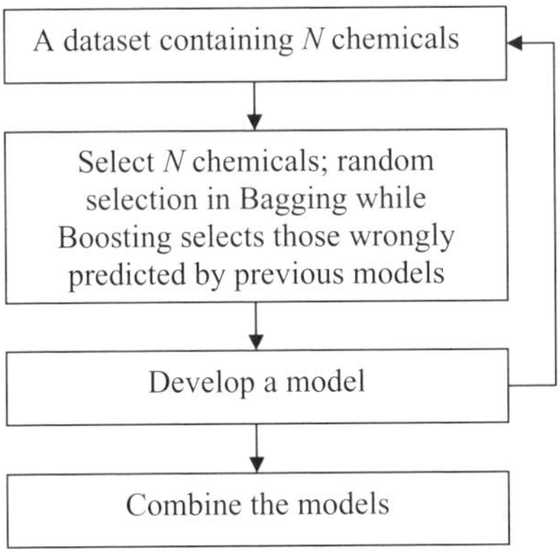

Figure 6.3 Consensus modeling based on bagging and boosting. Bagging is a "bootstrap" ensemble method by which each model is developed on a training set that is generated by randomly selecting chemicals from the original dataset. In the selection process some chemicals can be repeated more than once while others are left out so that the training set is the same size as the original dataset. In boosting, the training set for each model is also the same size as the original dataset. However, each training set is determined based on the performance of the earlier model/s. For the next training set, chemicals that were incorrectly predicted by the previous model are chosen more often than chemicals that were correctly predicted.

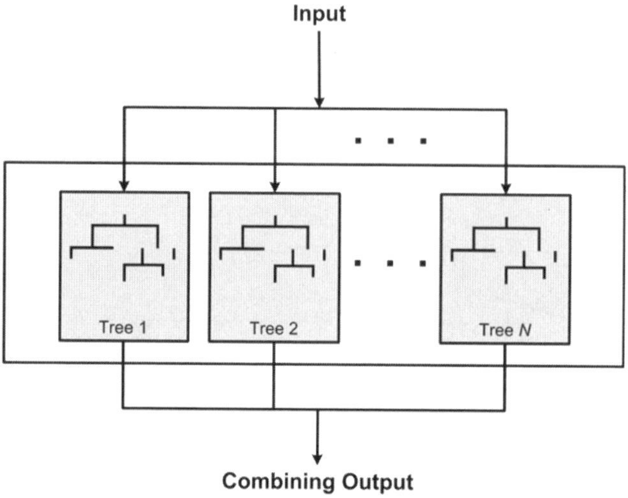

Figure 6.4 Illustration of the decision forest (DF). The individual trees are developed sequentially, and each tree uses a distinct set of descriptors. Classification (i.e., prediction) of an unknown chemical is based on the mean results of all trees.

series of rules based on selection of descriptors. These rules are operated by IF-THEN expressions and displayed as limbs in the form of a tree containing, in most cases, only binary branching. For example, a simple rule might be "IF molecular weight >300, THEN the chemical is active." The rules effectively interpret biological questions with respect to relationships with molecular descriptors. DTs offer speed of model development and prediction: a major attribute when screening large databases of chemical entities. In Random Forests, a small number (subset) of descriptors is randomly selected from the original descriptor pool in every split for growing a tree, and a descriptor in the subset giving the best split is chosen for splitting. Unfortunately, this method often requires a large number of individual trees (>400) for convergence.

6.2.5 The Decision Forest

The Decision Forest (DF) is a DT-based consensus modeling method that was conceived as a method for combining heterogeneous yet comparable trees that fully captures the association between molecular structure and biological activity (Figure 6.4) [59]. The heterogeneity requirement assures that each tree uniquely contributes to the combined prediction; whereas the quality comparability requirement assures that each tree equally contributes to the combined prediction.

Since a certain degree of noise is always present in biological data, optimizing a tree by increasing its complexity can actually reduce its predictive capability by overfitting to noise. The DF attempts to minimize the overfitting problem by building each tree using the entire training set but with a distinct set of descriptors. A pruning operation is employed whereby the descriptors are distributed among the trees such that individual models are uniformly but not overly predictive, thereby circumventing the overfitting problem.

Compared with other consensus modeling methods, the DF method provides numerous benefits: (1) in maximizing the differences among the trees by building each one from distinct descriptors, highly predictive forests are achieved using only a few trees; (2) DF-based models are entirely reproducible, hence the SAR relationships are constant and biological interpretation is unambiguous; and (3) since *all* chemicals in the training set are included in individual tree development, the SAR information in the original dataset is fully appreciated. Another notable feature of DF models is that they permit direct quantitative assessment of prediction confidence [60].

Hierarchical Frameworks for Toxicity Prediction In consensus modeling, individual models can be developed based on either the same or different methods. To cite a single example, a screening procedure for prioritizing potential environmental endocrine disruptors used a four-tiered hierarchical framework where each tier integrated different methods [61,62]. This strategy was adopted by the US FDA at the National Center for Toxicological Research (NCTR), in a cooperative program with the US EPA that led to development of the Endocrine Disruptor Knowledge Base (EDKB) [13,37,63–66]. A schematic of this four-tiered EDKB hierarchical framework is shown in Figure 6.5 [67]. The underlying motivation was to develop (Q)SAR models that support priority setting of tens of thousands of chemicals that will undergo assessment for potential as endocrine disrupting chemicals (EDCs). Starting with tier I, rejection filters were used to exclude chemicals deemed inactive. Tier II comprised 11 separate classifiers or models: three structural alerts, seven pharmacophores, and one decision tree. Tier III contained one or more QSAR-based

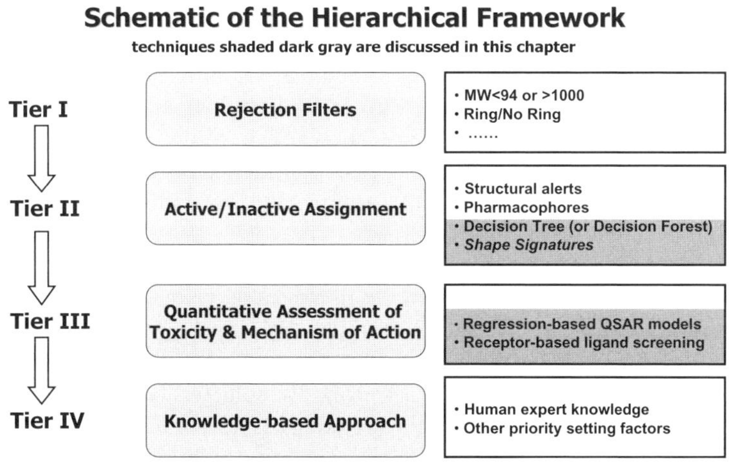

Figure 6.5 Schematic of the hierarchical framework, using the EDKB as an example.

models for quantitative prediction of activity, while tier IV provided for additional decision support such as input from human experts. The combined results dramatically reduced false negatives, which is crucial for regulatory applications and important for all applications. A variation of the HF approach is to develop individual models using the same type of computational methods, such as artificial neural networks [68–70] and DTs [71,72]. Combining models derived from the same method can reduce noise-induced error in individual models.

Like QSAR models, hierarchical schemes are ordinarily optimized in two steps: (1), calibration of the component model parameters to meet accepted criteria for performance (e.g., predictivity, number of false negatives) using a training set of chemicals; (2) validation of the scheme by assessing its ability to blind-predict test chemicals of known activity. It is presumed that the chemicals in the training and test sets share the same chemical space, range of activity, mode-of-action, and so on. The entire scheme and each component model are refined during validation.

Applicability Domain for DT-Based Models We describe applicability domain for QSAR models as being determined by two parameters: (1) *prediction confidence*, or the certainty of a prediction for an unknown chemical, and (2) *domain extrapolation*, or the prediction accuracy of an unknown chemical that lies beyond the chemical space of the training set [60]. Both parameters can be quantitatively estimated in the consensus tree approaches, where individual models are constructed as DTs. Taken together, prediction confidence and domain extrapolation assess the applicability domain of a model for each prediction.

For each tree in a consensus tree model the probability (0–1) for an unknown chemical to be active is taken as the percentage of active chemicals in the terminal node to which the chemical is assigned. For the consensus tree model the mean probability value for a chemical can be calculated by simply averaging the probabilities across all individual trees (or other combining methods such as voting). Chemicals that have a probability larger than 0.5 are designated *active*, whereas those that have a mean probability less than 0.5 are designated *inactive*. Importantly, this mean probability is also a measure of the confidence of each prediction. Larger probabilities approaching 1.0 indicate high confidence that the chemical is active while, conversely, smaller probabilities approaching 0.0 indicate high confidence the chemical is inactive. Probabilities near 0.5 are equivocal and indicate low confidence whether the prediction is active or inactive [52].

Most, if not all, QSAR methods require selection of relevant or informative descriptors before modeling is actually performed. This is necessary because the method could otherwise be more susceptible to the effects of noise. The a priori selection of descriptors, however, carries with it the additional risk of "selection bias" [73], when the descriptors are selected before the dataset is divided into the training and test sets (Figure 6.6A). Because of selection bias, both external validation and cross validation could significantly overstate pre-

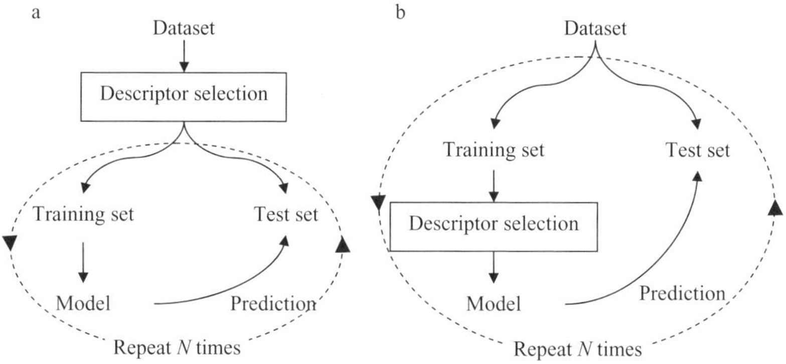

Figure 6.6 Two procedures for descriptor selection in validation processes (*a*) Descriptor selection occurs before dataset splitting (selection bias); (*b*) descriptor selection occurs after dataset splitting (correct procedure). The solid line illustrates the external validation process, and together the solid and dashed lines constitute the cross-validation process.

diction accuracy [74]. To avoid selection bias, the descriptor selection should be made after the dataset splitting (Figure 6.6B). It appears that this procedure is much easier to implement in external validation than in cross validation because the computational cost can be prohibitive for iterative descriptor selection during cross validation for many classification methods. However, the tree-based methods, including both DT and the consensus tree methods such as DF, hold the advantage of avoiding selection bias during cross validation, since the model is developed at each cycle by selecting descriptors from the entire descriptor pool. Cross validation thereby provides a realistic, unbiased assessment of the predictivity in a consensus tree model.

6.2.6 QSAR Models in Practical Scenarios

QSAR model validation mostly serves the purpose of demonstrating the overall prediction quality of the model. In practice, however, the way in which validation is performed largely depends on the model's intended use. If the model is to be applied to a known population of chemicals, regulatory acceptance of the model could depend entirely on the results of validation carried out that is specific to the particular chemical population. The model's validity can be demonstrated by comparing the predicted results with the experimental results on an external test set that is objectively selected from the application population Consequently the unbiased selection of an appropriate test set becomes an essential step in determining the validity of the model. The selected chemicals should represent the diversity of molecular structure and activity of the application population, and the selection process should provide statistically significant data to assess false positives and false negatives. A

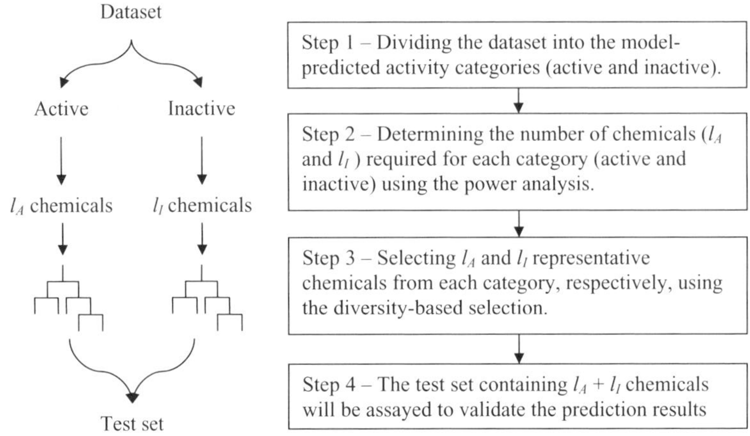

Figure 6.7 Schematic presentation of a model-driven selection method to construct a test set. In step 3, two diversity analysis methods can be used to select N chemicals from one category ($N = l_A$ for the active category and $N = l_I$ for the inactive category): (1) Group this category of chemicals into N clusters on the basis of their structural similarity using clustering methods; then select randomly one chemical from each cluster. (2) Group this category of chemicals into n clusters ($n < N$). Different number of chemicals are randomly selected from each cluster using a weighted factor, and the total number of chemicals will be N. Both approaches have been used in drug discovery for hit selection, and the weighted approach has proved to be more efficient.

model-driven selection method to determine a test set that meets the aforementioned criteria is depicted in Figure 6.7.

In many cases, however, an intended chemical application domain is broadly or vaguely defined, such as "the model will be used to predict estrogenic activity for the environmental chemicals." Often the chemical structure domain intended for application is not entirely known prior to either model development or validation. Validation for such models is best undertaken as part of a recursive process of incremental improvement, as in Figure 6.8 [75], where such a process was employed to predict so-called endocrine disrupting chemicals (EDCs). A QSAR model should be perceived as a "living model" that is successively challenged as new data become available, where upon incorrectly predicted chemicals are investigated to determine whether their inclusion in the training set will further improve the model's robustness and predictive capability. If so, the new chemicals will be assayed and incorporated into the model. Although this is a natural process with many benefits [46,75], it entails the drawback of being *reactive* in nature. While confidence in model prediction can grow over time as the training set expands, there is no quantitative measure of prediction accuracy available when the model is challenged by new untested chemicals.

Defining the training domain is the prerequisite for assessing the domain extrapolation. Commonly in QSAR modeling, a training domain is viewed as an N-dimensional space, where N is the number of descriptors in the model. We

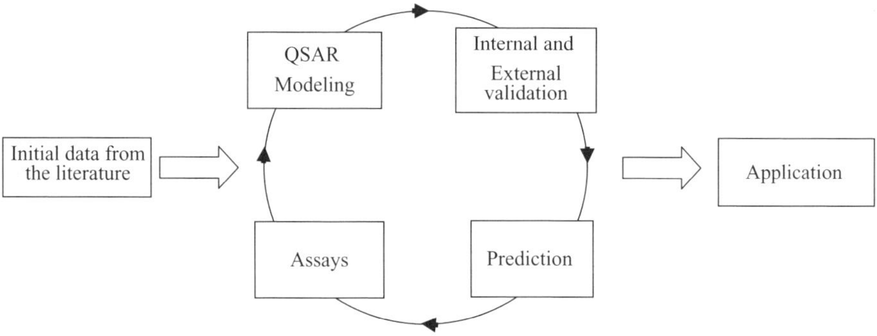

Figure 6.8 Depiction of the recursive process used in our laboratory to develop QSAR models for predicting estrogen receptor binding. The process starts with data from an initial set of chemicals from the literature QSAR modeling. These preliminary QSAR models are used prospectively to define a set of chemicals that will further improve the model's robustness and predictive capability. The new chemicals are assayed, and these data are then used to challenge and refine the QSAR models. Validation of the model is critical. The process emphasizes the living model concept.

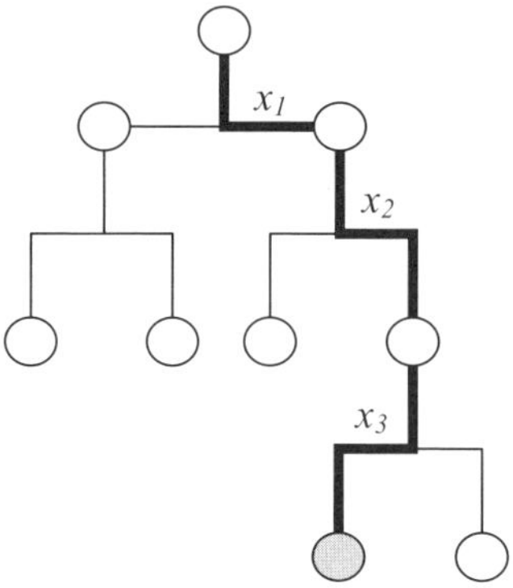

Figure 6.9 The focused domain representing the training domain of a tree. For an unknown chemical predicted by the tree, its classification is determined by a terminal node (e.g., dark circle) to which it belongs. There are three descriptors used in the path (bold line) from the root to the terminal node and the range of these three descriptors across all chemicals in the training set determines the training domain.

might rightly call this a global domain. In a tree, however, the classification of an unknown chemical is determined by only one terminal node that is descendent from the root node through a set of IF-THEN rules based on k descriptors x_i ($i = 1, \ldots, k$) (Figure 6.9). Analogous to the global domain, the training domain for DT and DF models can be defined as a k-dimensional space, called the focused domain. For the datasets we have evaluated, there was no appreciable difference in results for global domain compared to focused domain [60].

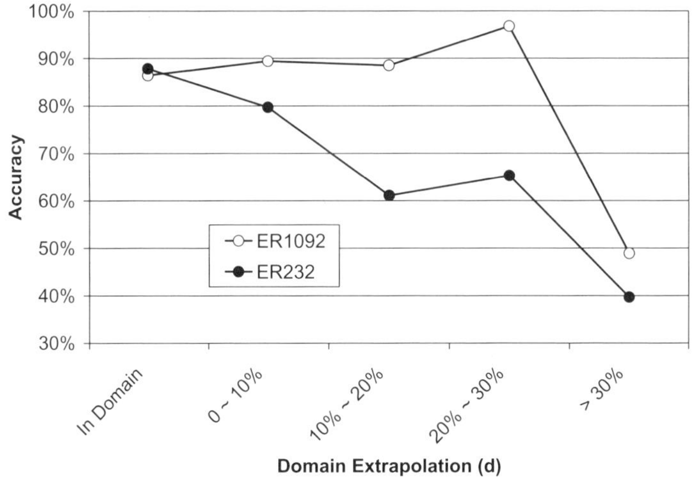

Figure 6.10 DF prediction accuracy versus domains extrapolation for ER232 and ER1092 based on 2000 runs of 10-fold cross validation. Domain extrapolation (*d*) for a chemical is defined as a percentage away from the focused domain as illustrated in Figure 6.8, while the prediction accuracy for the domain *d* is calculated by dividing correct predictions by total number of chemicals in this domain.

It is generally acknowledged that a model's prediction accuracy depends largely on the size and diversity of the training set. It is also widely acknowledged that models from a larger training set yield predictions with greater confidence, even for the chemicals that are less well represented by the training set. Despite the broad acceptance of these concepts, they have actually not been well tested in a quantitative sense. Recent studies offer compelling evidence that larger training sets provide predictions with both greater accuracy and confidence. When results from an evaluation of DF domain extrapolation for two training datasets, one containing 232 chemicals (ER232) and another containing 1092 chemicals (ER1092) with estrogen receptor binding activity, were plotted in Figure 6.10 [60] and compared for overall prediction accuracy for chemicals within the training domain, the accuracy for chemicals fell appreciably beyond the focused domain. The farther away the chemicals were from the training domain, the smaller was the prediction accuracy. A sharp drop in prediction accuracy was observed to occur at 10% extrapolation for ER232 but not until 30% extrapolation for ER1092. These results suggest that a DF model's applicability domain can be well determined by two inextricably linked concepts, prediction confidence and domain extrapolation. The results also imply that the model's applicability domain is predominantly determined by prediction confidence.

6.2.7 Assessment of Chance Correlation

Testing whether a fitted SAR model is a chance correlation is becoming increasingly imperative for smaller training data sets. Conditions with increasing numbers of descriptors, with increasing noise in biological data, and with an increasing skewed distribution of chemicals across activity categories, all increase the risk of obtaining a chance correlation lacking predictive value.

To assess the degree of chance correlation, the accepted practice is to first generate many pseudo-datasets (e.g., 2000 pseudo-datasets) using a randomization test where the activity classification is randomly scrambled across all chemicals in the training set. Next a 10-fold cross validation is applied on each of the pseudo-datasets. The null distribution, meaning the distribution of prediction accuracy for all pseudo-datasets, can then be compared with the distribution of multiple 10-fold cross validation results derived from the real dataset. The degree of chance correlation in the predictive model can be estimated from the overlap of the two distributions. A test for chance correlation of DF models to predict liver carcinogenicity was conducted based on four datasets. Although high cross-validation results were obtained for the models based on these four datasets, there was a significant overlap between the null and real distribution for each dataset, (Figure 6.11), indicating high degree of chance correlations for these models [60].

In another example, the null distribution was compared for 2000 pseudo-datasets with the real distribution generated from 2000 runs of 10-fold cross validation for ER232. The distribution of prediction accuracy of the real dataset centered around 82% while the pseudo-datasets were near 50% [60] (Figure 6.12). The distribution turned out to be much narrower for the real dataset than

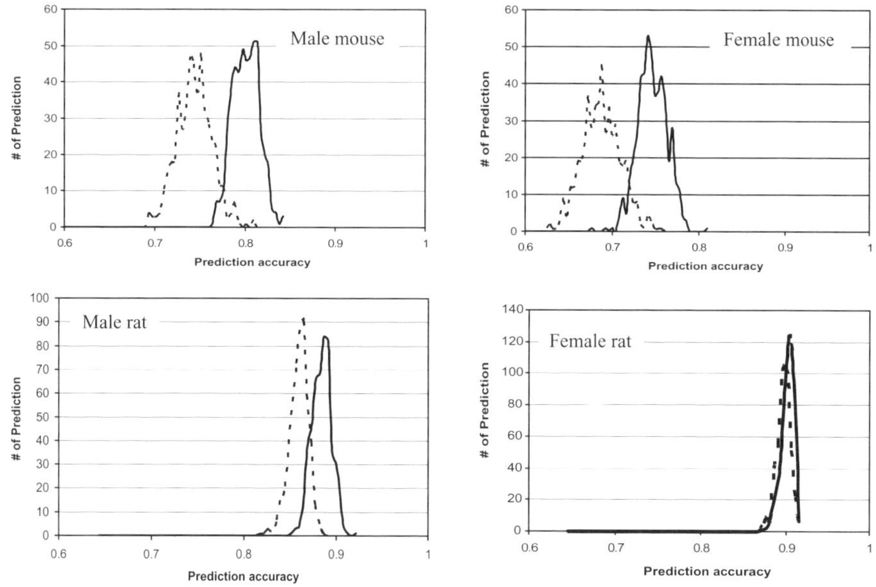

Figure 6.11 Assessment of the chance correlation in DF for four datasets. For each graph the null distribution (dashed) is generated from the results of 10-fold cross validation on 2000 pseudodatasets while the real distribution (solid) is derived from 2000 runs of 10-fold cross validation for the original dataset.

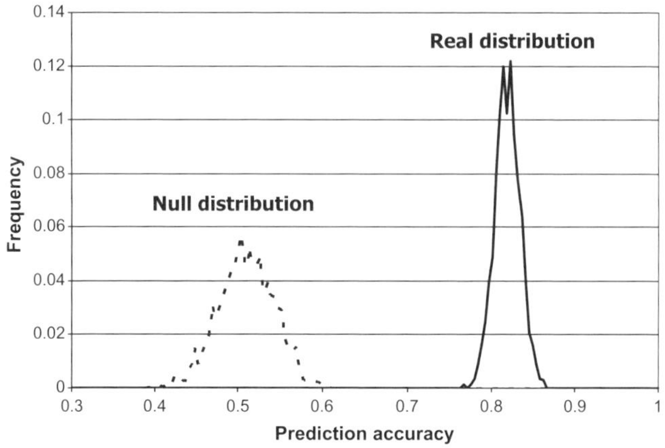

Figure 6.12 Assessment of the chance correlation in DF for ER232, an estrogenic dataset that contains 232 chemicals tested in an estrogen receptor binding assay.

for the pseudo-datasets, indicating that the training models generated from the cross-validation procedure for the real dataset give consistent and high prediction accuracy within corresponding test sets. In contrast, the prediction results of each pair of training and test sets in the 10-fold cross-validation process for the pseudo-datasets varied widely, implying a large variability of signal/noise ratio among these training models. Importantly, there was no overlap between two distributions, indicating that a statistically and biologically relevant DF model could be developed using the real dataset.

6.2.8 Final Reflections on QSAR Models

Every QSAR model will produce some degree of error, so it is advisable to recognize and appreciate the limitations of a model. Being able to quantitatively assess the accuracy limitation of each specific prediction allows selection of alternative methods, whether in silico or experimental, to supplement or supplant unreliable predictions, thus improving the value and utility of QSAR-based predictions. Assessing model limitations is a vital step toward the wider acceptance of QSAR models.

A model's limitations should be assessed from three different perspectives: (1) overall model predictivity (model validation), (2) individual prediction confidence (applicability domain), and (3) chance correlation. These attributes can be more readily assessed in the consensus tree modeling such as the DF method than in other QSAR methods. Using DF as an example, we have found the following:

- Combining multiple valid trees that use unique sets of descriptors into a single decision function produces a higher quality model than individual trees.

- The prediction confidence and domain extrapolation can be readily calculated and constitute the definition of the applicability domain for DF.
- Since the descriptor selection and model development are integrated, cross validation avoids descriptor selection bias and is more useful than the external validation in assessing a model's limitations.
- Performing many runs of cross validation is computationally inexpensive and ensures an unbiased assessment of a model's predictive capability, applicability domain, and potential chance correlation.

6.3 SHAPE SIGNATURES

We now introduce a new informatics tool called *Shape Signatures* that shows promise for drug discovery and computational toxicology. In contrast to QSAR and other descriptor-based approaches that compute hundreds of discrete descriptors for each molecule, the Shape Signature can be viewed as a very compact descriptor that encodes molecular shape and electrostatics in a single entity [76–78].

Shape Signatures employ a customized ray-tracing algorithm to explore the volume defined by the surface of a molecule, then uses the output to construct compact representations (i.e., signatures) of molecular shape, polarity, and other bio-relevant properties. The key steps of shape signature generation are shown in [76–78] Figure 6.13, where the surface encloses the molecule's solvent accessible volume (SAV). The Shape Signatures of different molecules can be compared by essentially subtracting their generated histograms. Among various metrics available, the most reliable is the simple linear metric.

$$\Delta = \sum_i \left| H_i^1 - H_i^2 \right|. \tag{6.1}$$

Shape Signatures have been shown to have utility for rapid screening and prioritizing of potential hazardous chemicals. For example, Shape Signatures

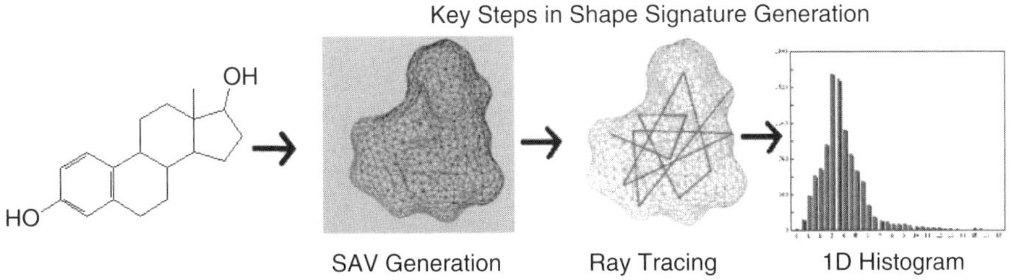

Figure 6.13 Key steps in generation 1-dimensional (1D) shape signature for 17β-estradiol.

can be used to identify molecules in a chemical library that are similar to a known toxicant. Alternatively, a query chemical could be compared to the Shape Signatures database of known and/or suspected toxicants. The algorithm is fast, physically intuitive, and accessible through a graphical user interface (GUI). Once the "signature" for a small-molecule compound (e.g., ligand) or for a receptor site (ligand-binding pocket) is generated, the compute time needed to compare two signatures is negligible. Unlike QSAR models and related methods, Shape Signatures obviates the need for model reformulation as new biological data become available. The method compares molecules based on the fundamental determinants of molecular recognition (e.g., shape, polarity); hence it can match substances that differ in chemical structure but share common properties like shape, H-bond patterns, and electrostatic features.

6.3.1 Shape Signature Database of PDB-Extracted Ligands

A particularly useful Shape Signatures database contains nearly 5000 ligands extracted from all high-quality crystal structures in the publicly available Protein Data Bank (PDB: http://www.rcsb.org/pdb). The organization of this Shape Signature database is illustrated in [76–78] Figure 6.14. The ligands are sorted alphabetically according to species (human, mouse, rat, etc.) and protein family as defined by CATH (http://www.cathdb.info/). The clear benefit of these structures is the knowledge of their receptor-bound conformation as well as the identity and crystal structure of the receptor itself. Many of the structures in the PDB are associated with environmentally induced human pathologies and xenobiotic toxicity pathways; hence a "hit" from a query compound can be traced to a wealth of information including the biologically active conformation, the target protein(s), and clues about toxicity pathways [79].

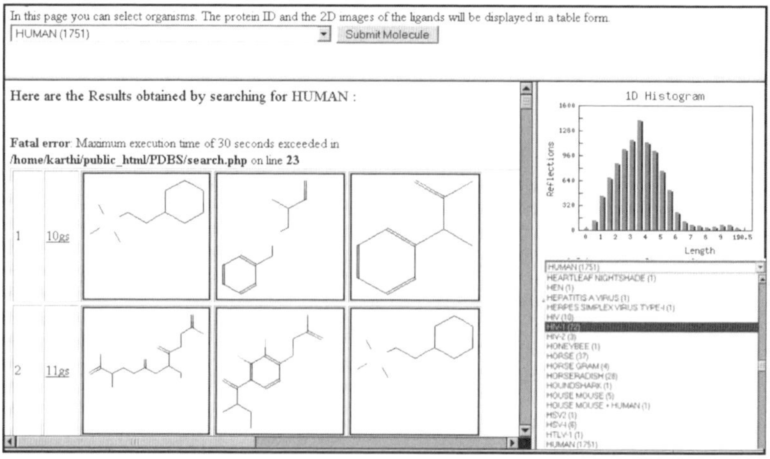

Figure 6.14 Snapshot of *Shape Signatures* database of PDB-extracted ligands.

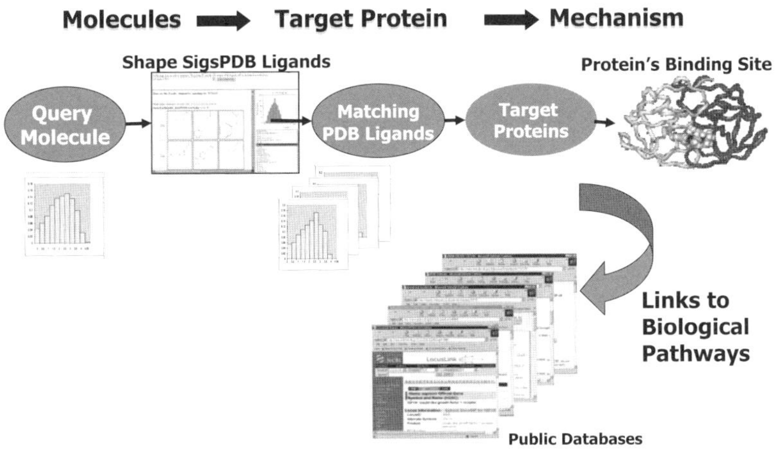

Figure 6.15 Depiction of the process by which a query molecule is used to search the PDB-based *Shape Signatures* database to discern potential protein targets and biological mechanisms.

This Shape Signatures database contains the 1D and 2D shape signatures of each ligand together with links to the source protein, quality of data (e.g., structural resolution), species, bibliographic references, and other biologically relevant information. This unique Shape Signatures database is expected to have broad utility, particularly for scientists engaged in risk assessment of real or potential chemical toxicants. Given the need to link adverse outcomes (e.g., reproductive or developmental changes, cancer) to initiating events through a cascade of biochemical and physiological changes that result from exposure to xenobiotics, the Shape Signatures database offers a linkage between a query chemical and potential receptor targets. This information may provide clues to upstream (initiating) events that trigger toxicity pathways (Figure 6.15) [1,80–83].

6.4 DATABASE SCREENING AND LIGAND-RECEPTOR DOCKING

Exploration of ligand-receptor interactions can provide a wealth of information useful in computational toxicology. First, calculated values of the ligand-receptor binding energy (*BE*) can be used to predict the relative binding affinity of untested small-molecule compounds. Second, these calculated *BE* values can be used as receptor-based molecular descriptors to produce much improved QSAR models. Third, key insights can be gained about ligand-receptor interactions and biological mechanisms of action (toxicity pathways). In many ways receptor-based approaches are highly complementary to ligand-based methods such as QSAR models.

The advent of fast computers (and computer networks) in recent years, coupled with the development of rapid ligand-receptor docking algorithms, has largely silenced past concerns about the speed of receptor-based approaches for processing large numbers of compounds. Values of *BE* are typically calculated as the difference in potential energy between the ligand-receptor complex ($E_{complex}$) and the ligand (E_{ligand}) and receptor ($E_{receptor}$) separately [i.e., $BE = E_{complex} - (E_{ligand} + E_{receptor})$]. Contributions for solvation effects can be included by employing explicit solvent (water) molecules or, more simply, by implicit solvation models. Solvation and entropy effects are fairly constant across a series of structurally related ligands and therefore can often be neglected as a first approximation.

Two recent studies demonstrate the utility of ligand-receptor docking approaches. In one study [84], a series of structural diverse estrogenic compounds, including steroids, phytoestrogens, and polychlorinated biphenyls of known activity, were docked to the crystal structure of estrogen receptor alpha (ERα) and an ERβ structural model constructed using homology modeling. In both cases a strong correlation ($R^2 > 0.82$) was found between values of *BE* and the relative binding affinity (RBA). Key insights were obtained about mechanisms of action and ligand selectivity. Compounds such as 16α-bromo-17β-estradiol and genistein exhibited selectivity for ERα and ERβ, respectively, as may have physiological implications given the known differential tissue distribution of these two ER isoforms. Calculated *BE* values reflected the observed degree of ERα/ERβ selectivity for this series of ligands, as indicated by the strong correlation ($R^2 = 0.76$) between the ratio of RBA values and the ratio of *BE* values for ERα/ERβ.

In another study [85], calculated BE values on a data set of 25 steroidal and nonsteroidal compounds for the rat androgen receptor (rAR) were plotted against the corresponding observed binding affinity (K_i). A strong correlation between the BE-predicted and experimental pK_i was found ($R^2 \cong 0.85$) (Figure 6.16) [85]. The BE values also reflected the observed loss in ligand specificity

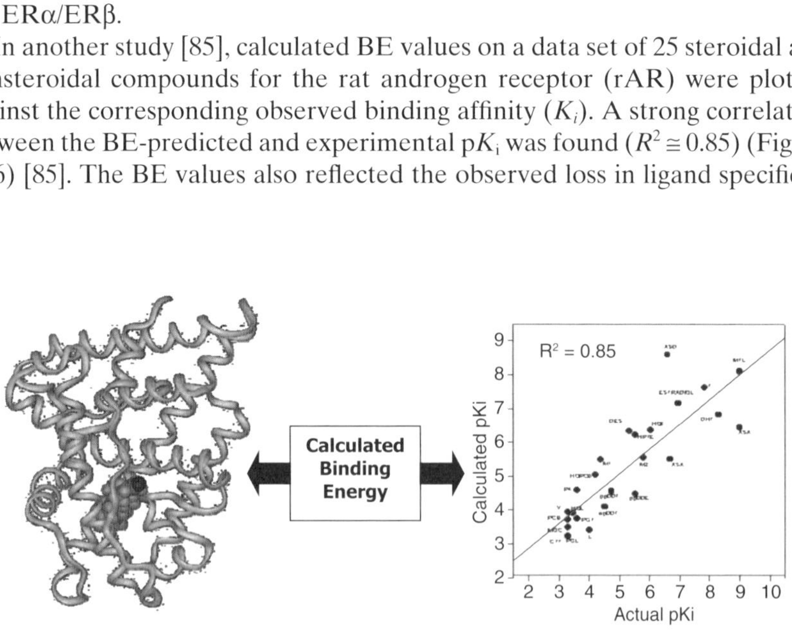

Figure 6.16 Binding energy of a docked ligand within receptor binding pocket (i.e., androgen receptor) is calculated and plotted versus the experimentally observed pKi.

between the wild-type (wt) AR and the T877A mutant AR associated with androgen-independent prostate cancer.

The studies described above revealed that BE values can be reliable predictors of ligand binding affinities for target proteins. The BE value can also serve as an additional descriptor to supplement QSAR models. A recent example demonstrated that the statistical quality of a 3D-QSAR model based on comparative molecular field analysis (CoMFA) improved dramatically—from $R^2 = 0.65$ to $R^2 = 0.93$—by inclusion of calculated BE values [86,87].

6.5 CLOSING REMARKS

While still in its early stages, the field of toxicoinformatics is flourishing. In addition to the US FDA-NCTR's Center for Toxicoinformatics (http://www.fda.gov/nctr/science/centers/toxicoinformatics/), the US EPA established a National Center for Computational Toxicology (http://www.epa.gov/comptox/) that coordinates and implements EPA's research in the field of computational toxicology. The EPA also recently awarded grants to establish two environmental bioinformatics research centers at the University of North Carolina at Chapel Hill and the University of Medicine and Dentistry of New Jersey (http://www.ebCTC.org). Similarly many other academic and governmental organizations have launched major initiatives in toxicoinformatics and computational toxicology. Practically all major biopharmaceutical companies have well-established predictive toxicology efforts in their drug discovery programs focused on ADME/Tox assessments. Another indicator is the proliferation of relevant databases (e.g., see http://www.ebkb.org and Chapter 5) and reviews, including the present book, that have appeared on the subject area in recent years [88–90]. If these developments meet forecasts of their potential growth and the impact of toxicoinformatics, the field will indelibly shape our perception of biology and toxicology.

ACKNOWLEDGMENT

The views presented in this article do not necessarily reflect those of the US Food and Drug Administration. Support for this work has been provided partially by the USEPA-funded Environmental Bioinformatics and Computational Toxicology Center (ebCTC), under STAR Grant number GAD R 832721-010. This work has not been reviewed by and does not represent the opinions of the funding agency.

REFERENCES

1. Benigni R, Richard AM. Quantitative structure-based modeling applied to characterization and prediction of chemical toxicity. *Methods* 1998;14:264–76.

2. Bradbury SP, Russom CL, Ankley GT, Schultz TW, Walker JD. Overview of data and conceptual approaches for derivation of quantitative structure-activity relationships for ecotoxicological effects of organic chemicals. *Environ Toxicol Chem* 2003;22:1789–98.
3. Browne LJ, L.Taylor L. Predictive chemoinformatics: Applications to the pharmaceutical industry. *Drug Discov World* 2002:2–8.
4. Comber MH, Walker JD, Watts C, Hermens J. Quantitative structure-activity relationships for predicting potential ecological hazard of organic chemicals for use in regulatory risk assessments. *Environ Toxicol Chem* 2003;22:1822–8.
5. Cronin MT, Dearden JC, Walker JD, Worth AP. Quantitative structure-activity relationships for human health effects: Commonalities with other endpoints. *Environ Toxicol Chem* 2003;22:1829–43.
6. Cronin MT, Jaworska JS, Walker JD, Comber MH, Watts CD, Worth AP. Use of QSARs in international decision-making frameworks to predict health effects of chemical substances. *Environ Health Perspect* 2003;111:1391–401.
7. Cronin MT, Walker JD, Jaworska JS, Comber MH, Watts CD, Worth AP. Use of QSARs in international decision-making frameworks to predict ecologic effects and environmental fate of chemical substances. *Environ Health Perspect* 2003; 111:1376–90.
8. Dearden JC. In silico prediction of drug toxicity. *J Comput Aided Mol Des* 2003;17:119–27.
9. Eriksson L, Jaworska J, Worth AP, Cronin MT, McDowell RM, Gramatica P. Methods for reliability and uncertainty assessment and for applicability evaluations of classification- and regression-based QSARs. *Environ Health Perspect* 2003;111:1361–75.
10. Fang H, Tong WD, Welsh WJ, Sheehan DM. QSAR models in receptor-mediated effects: the nuclear receptor superfamily. *J Mol Struct Theochem* 2003;622: 113–25.
11. Long A, Walker JD. Quantitative structure-activity relationships for predicting metabolism and modeling cytochrome p450 enzyme activities. *Environ Toxicol Chem* 2003;22:1894–9.
12. Patlewicz G, Rodford R, Walker JD. Quantitative structure-activity relationships for predicting mutagenicity and carcinogenicity. *Environ Toxicol Chem* 2003;22: 1885–93.
13. Perkins R, Fang H, Tong WD, Welsh WJ. Quantitative structure-activity relationship methods: Perspectives on drug discovery and toxicology. *Environ Toxicol Chem* 2003;22:1666–79.
14. Richard AM, Williams CR, Cariello NF. Improving structure-linked access to publicly available chemical toxicity information. *Curr Opin Drug Discov Devel* 2002;5:136–43.
15. Richardt AM, Benigni R. AI and SAR approaches for predicting chemical carcinogenicity: survey and status report. *SAR QSAR Environ Res* 2002;13:1–19.
16. Schmieder PK, Ankley G, Mekenyan O, Walker JD, Bradbury S. Quantitative structure-activity relationship models for prediction of estrogen receptor binding affinity of structurally diverse chemicals. *Environ Toxicol Chem* 2003;22:1844–54.

17. Soffers AE, Boersma MG, Vaes WH, Vervoort J, Tyrakowska B, Hermens JL, et al. Computer-modeling-based QSARs for analyzing experimental data on biotransformation and toxicity. *Toxicol In Vitro* 2001;15:539–51.
18. Tong W, Cao X, Harris S, Sun H, Fang H, Fuscoe J, et al. Arraytrack—supporting toxicogenomic research at the U.S. Food and Drug Administration National Center for Toxicological Research. *Environ Health Perspect* 2003;111:1819–26.
19. Walker JD, Rodford R, Patlewicz G. Quantitative structure-activity relationships for predicting percutaneous absorption rates. *Environ Toxicol Chem* 2003; 22:1870–84.
20. Walker JD, Enache M, Dearden JC. Quantitative cationic-activity relationships for predicting toxicity of metals. *Environ Toxicol Chem* 2003;22:1916–35.
21. Johnson DE, Smith DA, Park BK. Linking toxicity and chemistry: think globally, but act locally? *Curr Opin Drug Discov Devel* 2004;7:33–5.
22. Tong W, Fang H, Hong H, Xie Q, Perkins R, Sheehan DM. Receptor-mediated toxicity: QSARs for estrogen receptor binding and priority setting of potential estrogenic endocrine disruptors. In: Cronin MTD, Livingstone DJ, editors, *Predicting chemical toxicity and fate*. Boca Raton, FL: CRC Press, 2004.
23. Bradbury S, Kamenska V, Schmieder PK, Ankley G, Mekenyan O. A computationally based identification algorithm for estrogen receptor ligands: Part 1. Predicting hERalpha binding affinity. *Toxicol Sci* 2000;58:253–69.
24. Mekenyan O, Kamenska V, Schmieder PK, Ankley G, Bradbury S. A computationally based identification algorithm for estrogen receptor ligand: Part 2. Evaluation of hERalpha binding affinity model. *Toxicol Sci* 2000;58:270–81.
25. Liu J, Saavedra JE, Lu T, Song JG, Clark J, Waalkes MP, et al. O(2)-Vinyl 1-(pyrrolidin-1-yl)diazen-1-ium-1,2-diolate protection against D-galactosamine/endotoxin-induced hepatotoxicity in mice: Genomic analysis using microarrays. *J Pharmacol Exp Ther* 2002;300:18–25.
26. Russom CL, Breton RL, Walker JD, Bradbury SP. An overview of the use of quantitative structure-activity relationships for ranking and prioritizing large chemical inventories for environmental risk assessments. *Environ Toxicol Chem* 2003;22:1810–21.
27. Mazzatorta P, Benfenati E, Lorenzini P, Vighi M. QSAR in ecotoxicity: An overview of modern classification techniques. *J Chem Inf Comput Sci* 2004;44:105–12.
28. Knaak JB, Dary CC, Power F, Thompson CB, Blancato JN. Physicochemical and biological data for the development of predictive organophosphorus pesticide QSARs and PBPK/PD models for human risk assessment. *Crit Rev Toxicol* 2004; 34:143–207.
29. Ownby DR, Newman MC. Advances in quantitative ion character-activity relationships (QICARs): Using metal-ligand binding characteristics to predict metal toxicity. *Quant Struct-Act Rel* 2003;22:1–6.
30. McKinney JD, Richard A, Waller C, Newman MC, Gerberick F. The practice of structure activity relationships (SAR) in toxicology. *Toxicol Sci* 2000;56:8–17.
31. Siraki AG, Chevaldina T, Moridani MY, O'Brien PJ. Quantitative structure-toxicity relationships by accelerated cytotoxicity mechanism screening. *Curr Opin Drug Discov Devel* 2004;7:118–25.

32. Contrera JF, Matthews EJ, Kruhlak NL, Benz RD. Estimating the safe starting dose in phase I clinical trials and no observed effect level based on QSAR modeling of the human maximum recommended daily dose. *Regul Toxicol Pharmacol* 2004;40:185–206.
33. Lessigiarska I, Cronin MT, Worth AP, Dearden JC, Netzeva TI. QSARS for toxicity to the bacterium *Sinorhizobium meliloti*. *SAR QSAR Environ Res* 2004;15:169–90.
34. Walker JD, Carlsen L, Hulzebos E, Simon-Hettich B. Global government applications of analogues, SARs and QSARs to predict aquatic toxicity, chemical or physical properties, environmental fate parameters and health effects of organic chemicals. *SAR QSAR Environ Res* 2002;13:607–16.
35. Walker JD, Jaworska J, Comber MH, Schultz TW, Dearden JC. Guidelines for developing and using quantitative structure-activity relationships. *Environ Toxicol Chem* 2003;22:1653–65.
36. Tong W, Perkins R, Fang H, Hong H, Xie Q, Branham SW, et al. Development of quantitative structure-activity relationships (QSARs) and their use for priority setting in the testing strategy of endocrine disruptors. *Regul Res Perspec* 2002; 1:1–16.
37. Tong W, Fang H, Hong H, Xie Q, Perkins R, Anson JF, et al. Regulatory application of SAR/QSAR for priority setting of endocrine disruptors—A perspective. *Pure Appl Chem* 2003;75:2375–88.
38. Jaworska JS, Comber M, Auer C, Van Leeuwen CJ. Summary of a workshop on regulatory acceptance of (Q)SARs for human health and environmental endpoints. *Environ Health Perspect* 2003;111:1358–60.
39. Yu SJ, Keenan SM, Tong W, Welsh WJ. Influence of the structural diversity of data sets on the statistical quality of three-dimensional quantitative structure-activity relationship (3D-QSAR) models: predicting the estrogenic activity of xenoestrogens. *Chem Res Toxicol* 2002;15:1229–34.
40. Waller CL, Oprea TI, Chae K, Park HK, Korach KS, Laws SC, et al. Ligand-based identification of environmental estrogens. *Chem Res Toxicol* 1996;9:1240–8.
41. Xing L, Welsh WJ, Tong W, Perkins R, Sheehan DM. Comparison of estrogen receptor and subtypes based on comparative molecular field analysis (CoMFA). *SAR QSAR Environ Res* 1999;10:215–37.
42. Richard AM, Williams CR. Distributed structure-searchable toxicity (DSSTox) public database network: A proposal. *Mutat Res* 2002;499:27–52.
43. Richard AM. Commercial toxicology prediction systems: a regulatory perspective. *Toxicol Lett* 1998;102–103:611.
44. Julien E, Willhite CC, Richard AM, Desesso JM. Challenges in constructing statistically based structure-activity relationship models for developmental toxicity. *Birth Defects Res A Clin Mol Teratol* 2004;70:902–11.
45. Mattes WB, Pettit SD, Sansone SA, Bushel PR, Waters MD. Database development in toxicogenomics: Issues and efforts. *Environ Health Perspect* 2004;112: 495–505.
46. Tong W, Welsh WJ, Shi L, Fang H, Perkins R. Structure-activity relationship approaches and applications. *Environ Toxicol Chem* 2003;22:1680–95.
47. Shi LM, Fang H, Tong W, Wu J, Perkins R, Blair RM, et al. QSAR models using a large diverse set of estrogens. *J Chem Inf Comput Sci* 2001;41:186–95.

48. Bates JM, Granger CWJ. The combination of forecasts. *Oper Res Quart* 1969;20: 451–68.
49. Bunn DW. Expert use of forecasts: Bootstrapping and linear models. In: Wright G, Ayton P, editors, *Judgemental forecasting*. New York: Wiley, 1987. p. 229–41.
50. Bunn DW. Combining forecasts. *Eur J Oper Res* 1988;33:223–9.
51. Clemen RT. Combining forecasts: A review and annotated bibliography. *Int J Forecast* 1989;5:559–83.
52. Tong W, Hong H, Fang H, Xie Q, Perkins R. Decision forest: Combining the predictions of multiple independent decision tree model. *J Chem Inf Comput Sci* 2003;43:525–31.
53. Maclin R, Opitz D. An empirical evaluation of bagging and boosting. *Proceedings of the 14th National Conference on Artificial Intelligence*. Providence, RI: AAAI Press, 1997. p. 546–51.
54. Votano JR, Parham M, Hall LH, Kier LB, Oloff S, Tropsha A, et al. Three new consensus QSAR models for the prediction of Ames genotoxicity. *Mutagenesis* 2004;19:365–77.
55. Breiman L. Bagging predictors. *Mach Learn* 1996;24:123–40.
56. Freund Y, Schapire R. Experiments with a new boosting algorithm. In: Saitta L, editor. *Machine learning, Proceedings of the 13 International Conference (ICML '96)*. Bari, Italy: Morgan Kaufmann, 1996. p. 148–56.
57. Amit Y, Geman D. Shape quantization and recognition with randomized trees. *Neural Comput* 1997;9:1545–88.
58. Breiman L. *Random forests*. Berkeley, CA: Department of Statistics, University of California, 1999.
59. Tong W, Hong H, Fang H, Perkins R, Walker JD. From decision tree to decision forest – A novel chemometrics approach for structure activity relationship modeling. In: Barry Lavine and Curt Breneman, editors, Chemometrics and Chemoinformatics. American Chemical Society, Washington, DC, 2005. Chapter 12.
60. Weida Tong, Huixiao Hong, Qian Xie, Leming Shi, Hong Fang, Roger Perkins. Assessing QSAR limitations – A regulatory perspective. *Current Comp-Aided Drug Design* 2005;1:65–72.
61. Hong H, Tong W, Fang H, Shi L, Xie Q, Wu J, et al. Prediction of estrogen receptor binding for 58,000 chemicals using an integrated system of a tree-based model with structural alerts. *Environ Health Perspect* 2002;110:29–36.
62. Shi L, Tong W, Fang H, Xie Q, Hong H, Perkins R, et al. An integrated "4-phase" approach for setting endocrine disruption screening priorities—Phase I and II predictions of estrogen receptor binding affinity. *SAR QSAR Environ Res* 2002; 13:69–88.
63. Joseph P, Muchnok T, Ong T. Gene expression profile in BALB/c-3T3 cells transformed with beryllium sulfate. *Mol Carcinog* 2001;32:28–35.
64. Tong W, Harris S, Cao X, Fang H, Shi L, Sun H, et al. Development of public toxicogenomics software for microarray data management and analysis. *Mutat Res* 2002;549:241–53.
65. Shi LM, Tong W, Fang H, Perkins R, Wu J, Tu M, et al. An integrated "4-phase" approach for setting endocrine disruption screening priorities—Phase I and II predictions of estrogen receptor binding affinity. *Environ Res* 2002;13: 69–88.

66. Tong W, Lowis DR, Perkins R, Chen Y, Welsh WJ, Goddette DW, et al. Evaluation of quantitative structure-activity relationship methods for large-scale prediction of chemicals binding to the estrogen receptor. *J Chem Inf Comput Sci* 1998;38:669–77.
67. Walker JD, Fang H, Perkins R, Tong W. QSARs for endocrine disruption priority setting database 2: The integrated 4-phase model. *QSAR Comb Sci* 2003; 22:89–105.
68. Krogh A, Vedelsby J. Neural network ensembles, cross validation and active learning. In: Tesauro G, Touretzky D, Leen T, editors, *Advances in neural information processing systems*. Cambridge: MIT Press, 1995. p. 231–8.
69. Opitz D, Shavlik J. Actively searching for an effective neural-network ensemble. *Connect Sci* 1996;8:337–53.
70. Maclin R, Shavlik J. Combining the predictions of multiple classifiers: Using competitive learning to initialize neural networks. *Proceedings of the 14th International Joint Conference on Artificial Intelligence (IJCAI)*. Montréal, Québec, Canada: Morgan Kaufmann, 1995. p. 524–30.
71. Drucker H, Cortes C. Boosting decision trees. In: *Advances in neural information processing systems*. Cambridge: MIT Press, 1996. p. 479–85.
72. Quinlan JR. Bagging, boosting and C4.5. *Proceedings of the 13 National Conference on Artificial Intelligence*. Portland, OR: AAAI Press, 1996. p. 725–30.
73. Simon R, Radmacher MD, Dobbin K, McShane LM. Pitfalls in the use of DNA microarray data for diagnostic and prognostic classification. *J Natl Cancer Inst* 2003;95:14–8.
74. Ambroise C, McLachlan GJ. Selection bias in gene extraction on the basis of microarray gene-expression data. *Proc Natl Acad Sci USA* 2002;99:6562–6.
75. Tong W, Fang H, Hong H, Xie Q, Perkins R, Sheehan D. Receptor-mediated toxicity: QSARs for oestrogen receptor binding and priority setting of potential oestrogenic endocrine disruptors. In: Cronin MTD and Livingstone DJ, editors, *Predicting chemical toxicity and fate*. Boca Raton: CRC Press, 2004;285–314.
76. Meek PJ, Liu Z, Tian L, Wang CY, Welsh WJ, Zauhar RJ. Shape signatures: Speeding up computer aided drug discovery. *Drug Discov Today* 2006;11:895–904.
77. Nagarajan K, Zauhar R, Welsh WJ. Enrichment of ligands for the serotonin receptor using the shape signatures approach. *J Chem Inf Comput Sci* 2005;45:49–57.
78. Zauhar RJ, Moyna G, Tian L, Li Z, Welsh WJ. Shape signatures: A new approach to computer-aided ligand- and receptor-based drug design. *J Med Chem* 2003;46: 5674–90.
79. MacGregor JT. The future of regulatory toxicology: Impact of the biotechnology revolution. *Toxicol Sci* 2003;75:236–48.
80. Benigni R, Giuliani A. Putting the predictive toxicology challenge into perspective: Reflections on the results. *Bioinformatics* 2003;19:1194–200.
81. Benigni R, Passerini L. Carcinogenicity of the aromatic amines: from structure-activity relationships to mechanisms of action and risk assessment. *Mutat Res* 2002; 511:191–206.
82. Benigni R, Zito R. The second National Toxicology Program comparative exercise on the prediction of rodent carcinogenicity: Definitive results. *Mutat Res* 2004;566:49–63.

83. Benigni R, Zito R. Designing safer drugs: (Q)SAR-based identification of mutagens and carcinogens. *Curr Top Med Chem* 2003;3:1289–300.

84. DeLisle RK, Yu SJ, Welsh WJ. Homology modeling of the estrogen receptor subtype (ER-) and prediction of ligand binding affinities. *J Mol Graphics Model* 2001;20:155–67.

85. Ai N, DeLisle RK, Yu S, Welsh WJ. Computational models for predicting binding affinities of ligands for the wild-type androgen receptor and a mutated variant associated with prostate cancer. *Chem Res Toxicol* 2003;16:1652–60.

86. Jayatilleke P, Nair A, Zauhar R, Welsh WJ. Computational studies on HIV-1 protease inhibitors: Influence of calculated inhibitor-enzyme binding affinities on the statistical quality of 3D-QSAR Models. *J Med Chem* 2000;43:4446.

87. Nair A, Jayatilleke P, Wang X, Miertus S, Welsh WJ. Computational studies on tetrahydropyrimidine-2-one (THP) HIV-1 protease inhibitors: Improving 3D-QSAR comparative molecular field analysis (CoMFA) models by inclusion of calculated inhibitor- and receptor-based properties. *J Med Chem* 2000;45:973–83.

88. Richard AM. Future of toxicologys—Predictive toxicology: An expanded view of "chemical toxicity." *Chem Res Toxicol* 2006;19:1257–62.

89. Richard AM, Gold LS, Nicklaus MC. Chemical structure indexing of toxicity data on the Internet: Moving toward a flat world. *Curr Opin Drug Discov Devel* 2006;9:314–25.

90. Ekins S. Systems-ADME/Tox: Resources and network approaches. *J Pharmacol Toxicol Meth* 2006;53:38–66.

7

COMPUTATIONAL APPROACHES FOR ASSESSMENT OF TOXICITY: A HISTORICAL PERSPECTIVE AND CURRENT STATUS

VIJAY K. GOMBAR, BRIAN E. MATTIONI, CRAIG ZWICKL, AND J. THOM DEAHL

Contents

Computational Toxicology: Risk Assessment for Pharmaceutical and Environmental Chemicals, Edited by Sean Ekins

7.1 INTRODUCTION

As the world has become ever more industrialized, an alarmingly large number of chemicals have entered into commercial use. More than 100,000 chemical substances are listed in the European Inventory of Existing Chemical Substances [1] with more than 66,000 regulated under the Toxic Substances Control Act (TSCA) [2]. An additional 1000-plus new industrial chemicals are introduced annually into economic use [3,4]. Despite a multitude of testing initiatives such as the Genetox Program and the US National Toxicology Program, more than 99% of compounds lack a comprehensive toxicity evaluation.

Other enterprises dealing with extremely large numbers of compounds are the discovery departments of pharmaceutical companies. The synthesis and safety evaluation, among other tests of preclinical and clinical efficacy and pharmacokinetics, of these large libraries of compounds make up a significant portion of today's billion dollar-plus research and development cost to bring a new pharmaceutical to market. In the face of social and economic pressures to reduce the cost of drug discovery, the pharmaceutical industry is in search of inexpensive approaches for reliable assessment of the probability of success of late-stage compounds.

The US Environmental Protection Agency (EPA) Office of Pollution Prevention and Toxics (OPPT) has only a 90-day review period to issue a decision on Premanufacture Notices filed under the TSCA for any new compound to be manufactured or imported. The availability of inexpensive approaches for reliable assessment of hazard to environment and human health could be of immense value to OPPT. Likewise agencies such as the European Center for the Validation of Alternative Methods (ECVAM) aim to "promote the scientific and regulatory acceptance of alternative methods . . . which reduce, refine, or replace the use of laboratory animals" [5].

An ability to combine the speed and capacity of modern desktop computers in reliably predicting the toxicological properties of molecules, based on their chemical structure alone, holds tremendous promise for the aforementioned efforts, in particular, and for the field of predictive toxicology, in general. With such technology at hand, the pharmaceutical and chemical industry would position itself to only expend resources on those compounds that have the greatest likelihood of succeeding in the clinic and becoming a drug. An equally vital use of in silico toxicology tools that become positioned within a "virtual screening" paradigm is to reduce the number of compounds that must be synthesized before a decision can be made regarding the "drugability" of the chemical scaffold under study. Regulatory agencies could also benefit tremendously from the application of these tools, since they limit the use of experimental animals needed to effectively manage the safety of the environment and human health.

In this chapter we provide a historical perspective of the development of the field of computational toxicology. Beginning from the "similarity-based" grouping of elements into the periodic table, the chapter presents a chronology of developments from the simple observations of qualitative relations between structure and toxicity through LFER (linear free energy related) and QSAR (quantitative structure activity relationship) models, to the current artificial intelligence computational packages such as TOPKAT, HazardExpert, MultiCASE, DEREK, OncoLogic, ToxBoxes, KnowItAll, LeadScope, MDL QSAR, and CSGenoTox. We have made no attempt in this work to provide a comparative analysis of the different approaches and packages; so, readers are encouraged to deduce the benefits and utility of each of the approaches by reading the other contributions in this volume.

7.2 BIRTH OF COMPUTATIONAL TOXICOLOGY

Computational toxicology—the field of toxicity assessment through computer-assisted methods—is a relatively new area of applied science that combines key principles from chemistry, biology, and computer science in an attempt to discover those relationships between chemical structure and toxicological endpoints. These relationships, in turn, can be exploited for making predictions of the toxicity of untested compounds. Key conceptual breakthroughs began to be elucidated in the mid-nineteenth and early twentieth centuries with the discovery of a physical basis for chemical properties and for the interaction of molecules with biological systems. As early as 1858 Borodin, a Russian chemist and composer [6], while referring to compounds and their properties quoted, "... their toxicological properties and chemical makeup are closely related." However, it is the codification of the then-known elements into the Periodic Table in the late 1880s by Julius Lothar Meyer and Dmitriy Mendeleev [7] that can rightly be regarded as the beginning of similarity-based clustering of chemical elements. This grouping formed the basis for predictive chemistry and found use not only for predicting the reactive properties of the elements but also laid the groundwork for understanding, and ultimately for being able to predict, more complex physicochemical properties of combinations of elements. This work was quickly followed by the discovery of empirical correlations between toxicity and simple physicochemical properties of small sets of molecules. For example, in 1863, Cros observed an inverse relationship between mammalian alcohol toxicity and water solubility [8,9]; in 1869, Crum-Brown and Frazer [10] realized that the paralyzing properties of a set of quaternized strychnines depended on the nature of the quaternizing group, and Richardson [11] reported a relationship between the narcotic activity of alcohols and molecular weight. In 1893, Richet [12] observed a relationship between the toxicity of simple organic compounds and water solubility, and later, Meyer [13] and Overton [14] independently showed that the narcotic action of many compounds was dependent on their oil–water partition coefficient.

At the turn of the twentieth century, the advancement of the valence theory of chemical bond formation through electron transference by Gilbert Newton Lewis [15], and of the receptor theory based on the pioneering work of Langley [16], provided the conceptual framework for the physical basis of the interaction between xenobiotics and biological macromolecules. Valence theory established the hypothesis that only specific toxicants or drug molecules (or functional groups on these compounds) were capable of eliciting a specific biological response through selective physicochemical ligand-receptor interactions. A large volume of subsequent experimental work soon provided compelling evidence for a receptor-mediated mode of action for the majority of pharmaceuticals and toxicants, and conceptually established a rational basis for the structure-activity relationship (SAR). The thermodynamic basis of parametric procedures describing SAR, namely describing one property in terms of parameters that describe molecular structure, was provided by linear free energy relationships (LFER) developed in the 1930s by Hammet [17,18] through his pioneering work on quantification of the effects of substituents on ester hydrolysis. The use of steric and hydrophilicity descriptors decades later by Taft [19,20] and Hansch [21,22], respectively, not only confirmed the validity of LFERs, but also gave birth to the field of quantitative structure activity relationships (QSARs), the term "activity" referring to any molecular property of physicochemical, pharmacological, or toxicological interest. Around the time the concept of QSARs became firmly established in the literature, the availability and power of electronic computers were on the rise. Both academics (Klopman [23]) and entrepreneurs (Enslein [24], Daravas [25]) began to explore the roles of different approaches and descriptors of molecular structure by applying LFER principles to the toxicity data then available. This synthesis of chemical toxicology, statistical analysis, and computer technology spelled the birth of the field of computational toxicology.

Continued advances in sophisticated computational and statistical algorithms today give toxicologists access to a host of new approaches and tools to help them perform and manage SAR analyses for toxicologically important endpoints. The principal differences among the various computational methods arise from the way a molecule is quantified in terms of descriptors and how the relationship between these chemical descriptors and the toxicological endpoint of interest is established. For example, chemical structure descriptors range from measured or computed physical properties, such as logP (the logarithm of the *n*-octanol-water partition coefficient) (see Chapter 9), number of H-bond donors, and molecular weight, to all possible continuous pieces or "fragments" of molecules, to descriptors that represent the electronic configurations or surface maps of the whole or localized portions of the molecule. Similarly, a computational toxicology approach may be a rules-based expert system (see Chapter 18), namely based on the SAR expertise and experience of the scientific community, or a machine-learning system (see Chapter 8), in which the computer empirically determines the existence of a statistically significant relationship between one or more descriptors and the toxicological endpoint of interest, or a prepackaged suite of QSARs which are becoming more available commercially for a variety of toxicological endpoints. Brief historic accounts for each of these major approaches will be given in the sections that follow. Finally, just as there are a number of approaches used to define chemical descriptors and their relationship(s) with toxicological endpoints, there are a wide variety of software packages that specifically support each of the different approaches. Since there is not sufficient space to describe in this chapter all of the tools available, an effort will be made to discuss briefly in each section examples of several of the tools with which the authors have the most experience or are most familiar with in terms of the principles employed.

7.3 LINEAR FREE ENERGY RELATED (LFER) APPROACHES

The extension of the principles of LFER beyond substituent constants to different experimental and computed parameters of molecular structure led to the concept of QSAR [26–38], which was quickly extended from kinetic properties to toxicological endpoints. A QSTR (quantitative structure toxicity relationship) model relates the chemical structure of a compound to a toxicity metric (rodent median lethal dose, LD_{50}; fish median lethal concentration, LC_{50}; chronic lowest observed adverse effect level, LOAEL; mutagenicity, carcinogenicity, teratogenicity, etc.). The main underlying assumptions of QSTR methodology are (1) chemical structure is the primary determinant of observed toxicity response and (2) similar molecules possess similar properties. For developing a QSTR, each molecular structure contained within a set of molecules having known experimental toxicity values (i.e., a training set) is encoded with numerical values (the chemical descriptors), and a mathematical relationship between the descriptor set and numerically expressed toxicity value is generated by applying a suitable statistical method. Once a statistically sound model is generated, the model is validated to assess its ability to predict the toxicity of an independent set of compounds that are unknown to the training set (i.e., the test or prediction set).

Simple in concept, the formalism of QSTR allowed scientists to use different physicochemical properties as descriptors. For example, lipophilicity was first recognized by Meyer, Baum, and Overton [13,14,39] to be an important physicochemical parameter for predicting the toxic effects of chemicals. Meyer and his collaborator, Baum, proposed in 1899 that the ability of a compound to cause narcosis was dependent on its partitioning from water into a lipophilic environment to the site of action by demonstrating a correlation between the olive oil to water partition coefficient of a chemical and the minimum concentration required to retard the mobility of tadpoles. In the same year Overton independently proposed this same theory of narcosis and reiterated the importance of oil to water partitioning in narcotic events [14]. Later work by Lazarev also deserves mention as a continuation and expansion upon the initial findings of Meyer and Overton [40]. It is no surprise that the use of the octanol to water partition coefficient in many QSAR and pattern-recognition studies of toxicity has therefore become standard practice [41–43]. More detailed accounts of Meyer, Overton, and Lazarev's research can be found in the excellent comprehensive writings of Lipnick [40,44,45].

In 1962, the first formal QSAR paper was published by Hansch—considered by many as the father of QSAR—in which was emphasized the role of the octanol to water partition coefficient ($\log P$) [46]. Shortly thereafter, in 1964, publication of a method to compute a hydrophobic substituent constant π from $\log P$ values [47] led many investigators with access to toxicity data to

TABLE 7.1 Summary of QSARs for Toxicity Endpoints

Toxicity Endpoint	References
Mutagenicity	[75–121]
Carcinogenicity	[122–171]
Teratogenicity	[172–190]
Aquatic toxicity	[191–214]
Biodegradation	[215–247]
Bioconcentration	[248–272]
Bioaccumulation	[273–279]
Maximum tolerated dose	[280,281]
Acute toxicity	[282–288]
Cardiotoxicity	[289–304]

begin to explore relationships with π or $\log P$. So widespread were these investigations that it initially appeared that the goal of QSTR, especially in the field of environmental and aquatic toxicity, was to explore the relationship of toxicity endpoints to $\log P$ (or π). A series of early monographs attest to this as an area of focused effort at the time [48–51].

As the validity of Hansch's hypothesis was being confirmed with a variety of toxicity endpoints and experimentally determined substituent constants, other investigators with research interests aligned with the problem of identifying additional chemical structure-based descriptors were busy inventing ways to effectively describe a molecule's structure in terms of theoretically computed topological and geometrical properties and to apply more sophisticated statistical methods, such as partial least squares [52–55], artificial neural networks [56–61], recursive partitioning [62–66], and support vector machines [67–74] (see Chapter 8). The work and collaborative efforts among these investigators has today resulted in a wealth of QSARs now in existence for a variety of toxicity endpoints (Table 7.1) , collectively demonstrating the success of the various descriptors, statistical methods, and computational approaches that have been used to develop QSTRs for toxicity prediction.

7.3.1 TOPKAT

QSTRs were initially derived from small sets of congeneric molecules. With the formation of Health Designs, Inc. (HDi), a company founded by Enslein in 1978, the QSTR approach was applied to large heterogeneous sets of data with an intent to produce a commercial toxicology profiling package [24]. Small Business Innovation Research (SBIR) grants helped HDi to explore the use of predefined chemically and biologically relevant substructural fragments (then called "MOLSTAC keys") as structural descriptors for predicting the median lethal dose (LD_{50}) in the rat [305]. Having demonstrated the success of their approach with a variety of toxicity endpoints [306–308], in 1987 HDi launched the first QSTR-based toxicity prediction software application, TOPKAT—Toxicity Prediction by Komputer Assisted Technology—which is currently being marketed by Accelrys, Inc. The most recent version, TOPKAT 6.2, uses information-rich electrotopological, shape, and symmetry descriptors instead of a pre-defined set of substructures and allows assessment of 16 different toxicity endpoints (*http://www.accelrys.com/products/topkat/*) from pre-packaged, robust QSTRs, and provides the training databases from which the QSTRs were derived.

Before developing any QSTR, HDi needed to expend a tremendous amount of time and energy to diligently assess the quality of toxicity data that was then available from public sources, including the scientific literature. In the early to mid-1990s, easy access to the Internet made data sharing and downloading of large datasets commonplace, fostering a new round of activity in QSTR-based toxicity predictors.

7.3.2 ADMET Predictor

This package from SimulationsPlus allows prediction of many of the same toxicity endpoints available in TOPKAT but employs different modeling methodology. In addition, QSTRs for estrogen receptor modulation, maximum recommended therapeutic dose, carcinogenic potency, and cardiotoxicity (hERG-encoded K+ channel affinity) are available in ADMET Predictor (*http://www.simulationsplus.com/index.html*).

7.3.3 CSGenoTox

Like TOPKAT, CSGenoTox is a QSTR-based package. It is offered by ChemSilico for prediction of Ames mutagenicity (*http://www.chemsilico.com/*). CSGenoTox employs most of the descriptors and modeling methods used by TOPKAT, namely electrotopological state indexes, connectivity indexes, and shape indices, but it is based on a much larger data set for computing the mutagenicity index of a given test structure.

7.3.4 Admensa Interactive

Another QSAR-based commercial toxicity prediction system is Admensa Interactive (*http://www.inpharmatica.co.uk/AdmensaInteractive.htm*). Though primarily designed for ADME optimization, Admensa Interactive also allows for prediction of cardiotoxicity, similar to ADMET Predictor.

7.4 EXPERT SYSTEMS

Simply stated, expert systems are computer applications that carry out a degree of logical reasoning similar to those of human beings, making them ideal systems for making subject-matter expertise available to nonexperts (see

Chapter 18). Relying on mathematical and logic techniques, an expert system is comprised of two main components: (1) a knowledge base (KB); and (2) an inference engine (IE) [309]. The former holds subject-matter knowledge in the form of a collection of rules—formalized truths extracted from actual experience—and the latter draws appropriate deductions by logically compiling the set of rules triggered by input. Expert systems are important artificial intelligence systems, and their use is prevalent [310] in numerous areas, including medical diagnosis, credit authorization, genetic engineering, airline scheduling, analytical chemistry [311], chemical structure elucidation [312], computational toxicology, and many more.

In a computational toxicology expert system [313], structure-toxicity relationships recognized and trusted by professional toxicologists are stored as a computer-discernable collection of often largely nested IF-THEN constructs. When challenged to predict the toxicity of a compound, the expert system parses its structure and makes comparisons against the available KB of rules. The rules triggered by the compound's structure are collected and presented to the IE for qualitative deductive assessment of potential toxicity.

The need for expert systems in the complex field of toxicology has long been recognized because there is not enough toxicological expertise to satisfy the demand—certainly not where and when it is needed, that is, within regulatory agencies and drug discovery teams in the pharmaceutical industry. A predictive toxicology expert system can infuse a plethora of toxicology expertise in these bodies when they are faced with assessing toxicity profiles of compounds.

The KB of rules in a computational toxicology expert system is provided by the experience and expertise of the scientific community. As mentioned above, qualitative relationships between toxicological properties and chemical makeup have been known since the mid-1800s. More formal associations between physicochemical and toxicological properties, as reported by Crum-Brown and Fraser [10], Schmidt [314], Coulson [315], Pullman [316,317], and Nagata [318], make legitimate KB training input for an expert system, as do the established effects of small functional groups on carcinogenicity of congeneric molecules, such as methyl derivatives of benz[c]acridine, which tend to have higher carcinogenic activity than the methyl derivatives of benz[a]acridine [319]. As additional experimental and epidemiological toxicity data become available, toxicologists search for strong trends between toxic effects and molecular substructures and/or properties to expand the KB of rules. Three main commercially available expert systems of toxicology are available, namely HazardExpert [25], Oncologic [320], and DEREK [321]. These systems are briefly described below (see also Chapter 14).

7.4.1 HazardExpert

HazardExpert was the first computer-based toxicity prediction expert system developed by CompuDrug Chemistry Ltd. [25] in 1985. HazardExpert predicts a range of toxicity endpoints including irritation, neurotoxicity, immunotoxicity, teratogenicity, mutagenicity, and carcinogenicity. The KB of HazardExpert contains a Toxic Fragments Knowledge Base derived from literature on structure-toxicity relationships and from US EPA reports. The predictive rules in HazardExpert are based on the effects of these fragments on biological systems and incorporate expert judgment and fuzzy logic [322]. An important advantage that HazardExpert offers over other expert systems is that its predictions are attenuated by factors such as route of administration, dose level, and duration of exposure.

7.4.2 OncoLogic

This expert system was developed by LogiChem, Inc., under a cooperative agreement between the US EPA's Office of Pollution Prevention and Toxics (*http://www.epa.gov/oppt/cahp/actlocal/can.html*). Carcinogenicity is the only toxicity endpoint that OncoLogic predicts. Its rule base consists of structural alerts where the mechanism of carcinogenicity in humans and animals is well understood. A major advantage of OncoLogic is its ability to predict carcinogenicity not only of organic compounds but also polymers, metals, and fibers. Since the US EPA has obtained rights to OncoLogic, it may be made available for use free of charge in the future.

7.4.3 DEREK

Schering Agrochemical Company, the original creator of the DEREK (Deductive Estimation of Risk from Existing Knowledge) expert system [321], donated the system to its current not-for-profit distributor, Lhasa Limited (*www.lhasalimited.org*). Given the chemical structure of a compound, DEREK predicts a variety of toxicological hazards including genotoxicity, carcinogenicity, skin sensitization and irritation, based on the respective knowledge bases developed from the collective expertise of scientists representing over 20 member organizations. Due to differences in the number of compounds assayed and the level of analyses and understanding of the assay data, the KB for each property is generally powered by different numbers of structure-toxicity "rules" or structural "alerts." For instance, there are 96 rules for assessing genotoxicity but only 47 for prediction of carcinogenicity [323]. Likewise there are more than 70 rules for mutagenicity but far fewer for chromosomal aberration. Rules are generally derived from the organic chemistry mechanisms associated with a toxic response observed in a set of related compounds. Each rule in DEREK is named as a chemical class or substructure such as "aromatic hydroxylamine," "substituted pyrimidine or purine," and "epoxide," and each rule is numbered for quick reference and retrieval.

Being a rule-based system, DEREK outputs only a qualitative assessment of the potential toxicity associated with a chemical structure. Each prediction is assigned to one of the nine qualitative categories: Certain, Probable, Plausible, Equivocal, Doubted, Improbable, Impossible, Open, or Contradicted. Wherever applicable, the prediction is further justified with relevant literature references. Its performance has been tested on a number of independent data sets (see Chapters 14, 18, and 19).

7.5 MACHINE-LEARNING APPROACHES

In the mid-1940s the computer visionary Alan Turing first suggested that computers could provide knowledge by learning from data [324]. Machine-learning computational toxicology packages do just that: deduce empirical structure-toxicity relationship knowledge from molecular substructures by statistically determining the strength of an association with a given toxicity endpoint. When challenged to predict the toxicity of a compound, the structure-activity relationship and the strength of the relationship for each of the substructural features identified in the query compound are interpreted together to compute the probability of the compound being active (i.e., toxic). Any of a number of statistical methods, such as cluster analysis, discriminant analysis, nearest neighbor analysis, recursive partitioning, neural networks, and support vector machines (SVM), may be employed for creating prediction models under supervised or unsupervised conditions (see Chapter 8).

7.5.1 CASE/MultiCASE

Giles Klopman pioneered the use of machine-learning techniques to identify molecular fragments with a high probability of being associated with a given class of activity [325,326]. He conceived the outline of a machine-learning approach, called CASE (Computer Automated Structure Evaluation), during an airplane trip from Texas to Cleveland in 1981 after being inspired by a lecture that considered the feasibility of developing tools for chemists that could give them a head start on designing more environmentally friendly chemical products. In the knowledge extraction (i.e., learning) mode, when presented with a set of compounds and their associated toxicity data, CASE breaks down every molecule in terms of linear fragments containing two or more heavy atoms. If the prevalence of a fragment is not significantly different across toxic and nontoxic compounds, the fragment is considered irrelevant for the toxicity endpoint being analyzed. In contrast, and assuming a binomial distribution, if the frequency of a fragment is skewed either toward the active or nonactive set of compounds, the fragment is labeled respectively as a "biophore" or "biophobe." Based on the extent of statistical significance of the skew, each "biophore" and "biophobe" is assigned a probability value that expresses the strength of the predicted classification. In predictive mode, CASE evaluates the presence of biophores and biophobes in the input structures and then, based on the probability values associated with the biophores and biophobes identified, computes and returns the probability that the input structure belongs to the active (toxic) or inactive (nontoxic) class. A number of toxicity endpoints (Table 7.2) have been successfully analyzed using CASE methodology.

In 1992 the formalism developed for CASE was extended to a new product, MultiCASE (Multiple Computer Automated Structure Evaluation). MultiCASE introduced the concept of "modulators" of biophores and biophobes,

TABLE 7.2 Summary of Toxicity Endpoints Analyzed with CASE Methodology

Toxicity Endpoint	References
Genotoxicity	[327–336]
Carcinogenicity	[128,144,145,163]
Teratogenicity	[175,177,178,183]
Clinical hepatoxicity	[337]
Acute oral toxicity (maximum tolerated dose)	[338]
Biodegradibility	[215,226,227,231,233,235]

and this can be considered as a hierarchical CASE approach that combines machine learning with QSAR in making activity (toxicity) predictions [326]. Simply put, the MultiCASE algorithm first identifies the most significant biophore and compounds containing the biophore are separated out. The process is repeated iteratively to create groups of compounds corresponding to each identified biophore until all molecules are accounted for. Each biophore-specific set is then analyzed by the MultiCASE algorithm to identify neighboring fragments or functional groups capable of modulating the activity of the biophore, and a QSAR analysis is performed that assigns quantitative values to the categorized sets of molecules, typically in "CASE units"—10 for inactive (nontoxic) compounds and 30 and above for active (toxic) compounds.

Other enhancements in the MultiCASE approach include incorporation of molecular geometry and fragment environment in a molecule. For example, a distinction between *cis*- and *trans*-substituents in a fragment is recorded in a 7-bit geometry index. The environmental aspect of a fragment is included by utilizing a graph index that represents the geometric complexity of the molecule. For improved accuracy in toxicity prediction, MultiCASE also employs expanded and composite fragments, which are additional alerts embedded or extrapolated from the original biophores [326]. The MultiCASE methodology, available in the MCASE and MC4PC packages, is distributed by MultiCASE Inc. (*http://www.multicase.com/*).

7.5.2 Tox Boxes

A machine-learning approach for toxicity prediction has also been implemented in the Advanced Algorithm Builder (AAB) package from Pharma Algorithms (*http://www.ap-algorithms.com/*), founded in 1999 by Alanas Petrauskas. The fragment-based method in AAB, developed by the necessity to bring the in silico screening in parity with high-throughput screening (HTS) capabilities, expresses every structure in terms of generalized fragments constituting isolated carbons, functional groups, super fragments, and fragment interactions [339], and applies a classification SAR (C-SAR) approach to build a predictive model. The model is stored in AAB for predicting the toxicity of new compounds. Predictions are accompanied by 95% confidence intervals or probabilities, providing an indication of the reliability of the prediction. In addition the toxicophores are displayed to give insight as to which parts of the molecule are responsible for the toxic effect. The Tox Boxes application supports prepackaged toxicity prediction modules for prediction of acute toxicity [340], genotoxicity, and organ-specific health effects.

As per-unit cost of computer power and data storage decreases and the "-omics" data explosion continues [341,342], the application of machine-learning methods supporting data mining activities are also likely to proliferate [343]. Extensive application and validation of machine-learning packages is expected to be followed by the development of more sophisticated algorithms for generating more accurate predictions. In its current state most machine-learning methods view data as having a one-to-one relationship between a structure and property. New algorithms to explore one-structure-to-many-properties in parallel could lead to multivariate modeling and simultaneous optimization of some of the many interacting molecular properties that collectively determine whether a molecule is likely to be a good therapeutic candidate.

7.6 WEB-BASED TOXICITY PREDICTORS

Of late, a number of Web-based tools have appeared for assessment of small molecule toxicity. These tools are thin client-based and do not require installation of additional software on the user's computer (other than a Web browser). Some offer the service free of charge. For example, the lazar system (*http://www.predictive-toxicology.org/lazar/form.php*) offers prediction services for rodent carcinogenicity, Salmonella mutagenicity, fathead minnow LC_{50}, and human maximum recommended therapeutic dose at no cost [344]. Similarly the PreADMET site (*http://preadmet.bmdrc.org/preadmet/index.php*) offers predictions for mutagenicity and carcinogenicity, at no cost, simply by drawing the structure of a molecule. When using Web-based systems such as these, the user needs to keep in mind that the security of a proprietary structure *cannot* be guaranteed and is, in fact, breached once the structure or its codified equivalent leaves the protected confine of a secured intranet, so the user must exercise caution with these systems.

7.7 CONCLUSION

It should be obvious that all computational toxicology approaches require experimental data from which to learn. In our opinion, computational toxicology is most definitely not intended to replace the development and use of experimental, laboratory-based assay approaches. Quite the opposite—the approaches are intended to complement one another. The thoughtful application of computational toxicology tools can be used to rationalize the synthesis and safety testing of certain compounds, and this can help the pharmaceutical industry to cut costs and reduce cycle times associated with drug discovery. In addition, their prudent use by regulatory agencies to aid more rapid decision-making, to be more effective in hazard assessment, and to assist in managing the environmental and human health effects of chemicals can be leveraged to great advantage in making sound rational choices regarding society's use of particular chemical entities, both now and in the future.

REFERENCES

1. Vainio H, Coleman M, Wilbourn J. Carcinogenicity evaluations and ongoing studies: The IARC databases. *Environ Health Perspect* 1991;96:5–9.
2. Lynch DG, Tirado NF, Boethling RS, Huse GR, Thom GC. Performance of on-line chemical property estimation methods with TSCA premanufacture notice chemicals. *Ecotox Environ Safety* 1991;22:240–9.
3. Fishbein L. Critical elements in priority selections and ranking systems for risk assessment of chemicals. In: Mehlman MA, editor, *Advances in modern environmental toxicology: Safety evaluation; Toxicology, methods, concepts and risk assessment*, Vol. X. Princeton, NJ: Princeton Scientific Publications, 1987. p. 1–50.
4. Helma C, King RD, Kramer S, Srinivasan A. The predictive toxicology challenge 2000–2001. *Bioinform Appl Note* 2001;17:107–8.
5. Balls M, et al. Development and validation of non-animal tests and testing strategies: The identification of a coordinated response to the challenge and the opportunity presented by the sixth amendment to the Cosmetics Directive (76/768/EEC). ECVAN workshop report 7, *ATLA* 1995;23:398–409.
6. Borodin A. On the analogy of arsenic acid with phosphoric acid in chemical and toxicological behaviour. Doctorate dissertation. Medico-Surgical Academy, Russia, 1858.
7. Mendeleev D. The periodic law of the chemical elements. *J Chem Soc*. 1889;55: 634–56.
8. Borman S. New QSAR techniques eyed for environmental assessments. *Chem Eng News* 1990;68:20–3.
9. Cros AFA. *Action de l'alcohol amylique sur l'organisme*. University of Strasbourg, 1863.
10. Crum-Brown A, Fraser TR. On the connection between chemical constitution and physiological action. Part 1. On the physiological action of the salts of the smmonium bases, derived from Strychnia, Brucia, Thebaia, Codeia, Morphia, and Nicotia. *Trans R Soc Edinburgh* 1868–9;25:151–203.
11. Richardson BW. Lectures on experimental and practical medicine. Physiological research on alcohols. *Med Times Gaz* 1869;2:703–6.
12. Richet C. Note sur le Rapport Entre la Toxicite et les Propriretes Physiques des Corps. *Compt Rend Soc Biol* (*Paris*) 1893;45:775–6.
13. Meyer H. Zur theorie der alkoholnarkose. *Arch Exp Pathol Pharmakol* 1899;42: 109–18.
14. Overton E. Ueber die allgemeinen osmotischen Eigenschaften der Zelle, ihre vermutlichen Ursachen und ihre Bedeutung fur die Physiologie. *Vierteljahrsschr Naturforsch Ges Zurich* 1899;44:88–114.

15. Lewis GN. The atom and the molecule. *J Am Chem Soc* 1916;38:762–86.
16. Langley JN. On the contraction of muscle, chiefly in relation to the presence of "receptor" substances. Part IV. *J Physiol* 1904;33:374.
17. Hammett LP. Some relations between reaction rates and equilibrium constants. *Chem Rev* 1935;17:125–36.
18. Hammett LP. *Physical organic chemistry: Reaction rates, equilibria and mechanism*, 2nd edition. New York: McGraw-Hill, 1970.
19. Taft RW, Lewis IC. Evaluation of resonance effects on reactivity by application of the linear inductive energy relationship: V. Concerning a sR scale of resonance effects. *J Am Chem Soc* 1959;81:5343.
20. Taft RW Jr. Separation of polar, steric and resonance effects in reactivity. In: Newman MS, editor, *Steric effects in organic chemistry*. New York: Wiley, 1956. p. 556–675.
21. Hansch C, Maloney PP, Fujita T, Muir RM. Correlation of biological activity of phenoxyacetic acids with Hammett substituent constants and partition coefficients. *Nature* 1962;194:178–80.
22. Hansch C, Fujita T. β-σ-π analysis. A method for the correlation of biological activity and chemical structure. *J Am Chem Soc* 1964;86:1616.
23. Klopman G, Grinberg H, Hopfinger AJ. MINDO/3 calculations of the conformation and carcinogenicity of epoxy-metabolites of aromatic hydrocarbons: 7,8-Dihydroxy-9,10-oxy-7,8,9,10-tetrahydrobenzo(a)pyrene. *J Theor Biol* 1979 Aug 7;79(3):355–66.
24. Enslein K, Craig PN. A toxicity prediction system. *J Environ Toxicol* 1978;2: 115–21.
25. Smithing MP, Daravas F. HazardExpert: An expert system for predicting chemical toxicity. In: Finley, JW, Armstrong DJ, Robinson, SF, editors, *Food safety assessment.* Washington, DC: American Chemical Society, 1992. p. 191:200.
26. Bevan DR. QSAR and drug design. <http://www.netsci.org/Science/Compchem/feature12.html>.
27. Borman S. New QSAR techniques eyed for environmental assessments. *Chem Eng News* 1990;68:20–3.
28. Debnath AK. Quantitative structure-activity relationship (QSAR) paradigm—Hansch era to new millenium. *Mini Rev Med Chem* 2001;1:187–95.
29. Hansch C. Quantitative structure-activity relationships and the unnamed science. *Acc Chem Res* 1993;26:147–53.
30. Kubinyi H. From narcosis to hyperspace: The history of QSAR. *Quant Struct-Act Rel* 2002;21:348–56.
31. Martin YC. *Quantitative drug design: A critical introduction*. New York: Dekker, 1978.
32. McKinney JD, Richard A, Waller C, Newman MC, Gerberick F. The practice of structure activity relationships (SAR) in toxicology. *Toxicol Sci* 2000;56:8–17.
33. Purcell WP, Bass GE, Clayton JM. *Strategy of drug design: A molecular guide to biological activity.* New York: Wiley, 1973.

34. Rekker RF. The history of drug research: From Overton to Hansch. *Quant Struct-Act Rel* 1992;11:195–9.
35. Schultz TW, Cronin MTD, Walker JD, Aptula AO. Quantitative structure-activity relationships (QSARs) in toxicology: A historical perspective. *Theochem J Mol Struct* 2003;622:1–22.
36. Selassie CD. History of quantitative structure-activity relationships. In: Abraham DJ, editor, *Burger's medicinal chemistry and drug discovery*. Hoboken, NJ: Wiley, 2003. p. 1:48.
37. Tute MS. History and objectives of quantitative drug design. In: Ramsden CA, editor, *Quantitative drug design*. Vol. 4 of: Hansch C, Sammes PG, Taylor JB, editors, *Medicinal chemistry: the rational design, mechanistic study and therapeutic application of chemical compounds*. Oxford: Pergamon Press, 1990. p. 1:31.
38. Van de Waterbeemd H. The history of drug research: From Hansch to the present. *Quant Struct-Act Rel* 1992;11:200–4.
39. Baum F. *Arch Exp Pathol Pharmakol* (*Naunyn-Schmied*) 1899;42:119.
40. Lipnick RL, Filov VA. Nikolai Vasilyevich Lazarev, toxicologist and pharmacologist, comes in from the cold. *Trends Pharmacol Sci* 1992;13:56–60.
41. Leo A, Hansch C, Elkins D. Partition coefficients and their uses. *Chem Rev* 1971;71:525–616.
42. Leo AJ. Calculating log P_{oct} from structures. *Chem Rev* 1993;93:1281–306.
43. Rekker RF. *The hydrophobic fragmental constant.* Amsterdam: Elsevier, 1977.
44. Lipnick RL. Charles Ernest Overton: Narcosis studies and a contribution to general pharmacology. *Trends Pharmacol Sci* 1986;7:161.
45. Lipnick RL. Hans Horst Meyer and the lipoid theory of narcosis. *Trends Pharmacol Sci* 1989;10:265.
46. Hansch C, Maloney PP, Fujita T, Muir RM. Correlation of biological activity of phenoxyacetic acids with Hammett substituent constants and partition coefficients. *Nature* 1962;194:178–80.
47. Fujita T, Iwasa J, Hansch C. A new substituent constant, π, derived from partition coefficients. *J Am Chem Soc* 1964;86:51–75.
48. Kaiser KLE. *QSAR in environmental toxicology—II*. Dordrecht: Reidel, 1987.
49. Hermans JLM, Opperhuizen A. *QSAR in environmental toxicology—IV*. New York: Elsevier, 1991.
50. Karcher W, Devillers J. *Practical applications of QSAR in environmental chemistry and toxicology*. Dordrecht: Kluwer Academic, 1990.
51. Todeschini R, Consonni V. *Handbook of molecular descriptors*, Vol. 11. In: Mannhold R, Kubinyi H, Timmerman H, editors, *Methods and principles in medicinal chemistry*. Weinheim: Wiley-VCH, 2000.
52. Wold H. Estimation of principal components and related models by iterative least squares. In: Krishnaiah PR, editor, *Multivariate analysis*. New York: Academic Press, 1966. p. 391:420.
53. Wold H. Soft modelling with latent variables: The nonlinear iterative partial least squares approach. In: Gani J, editor, *Perspectives in probability and statistics: Papers in honour of M.S. Barlett*. London: Academic Press, 1975. p. 114–42.

54. Martens H, Jensen SA. Partial least Squares Regression: A new two-stage NIR calibration methods. In: Holas and Kratochvil, editors, *Proceedings of the World Cereal and Bread Congress*, Prague, June 1982. Amsterdam: Elsevier, 1982. p. 607–47.
55. Wold H. PLS regression. In: Johnson NL, Kotz S, editors, *Encyclopaedia of statistical sciences*, Vol 6. New York: Wiley, 1984. p. 581–91.
56. Bishop C. *Neural networks for pattern recognition*. Oxford: University Press, 1995.
57. Carling A. *Introducing neural networks*. Wilmslow, UK: Sigma Press, 1992.
58. Fausett L. *Fundamentals of neural networks*. New York: Prentice Hall, 1994.
59. Haykin S. *Neural networks: A comprehensive foundation*. New York: Macmillan, 1994.
60. Patterson D. *Artificial neural networks*. Singapore: Prentice Hall, 1996.
61. Ripley BD. *Pattern recognition and neural networks*. Cambridge: Cambridge University Press, 1996.
62. Nilakantan R, Nunn DS, Greenblatt L, Walker G, Haraki K, et al. A family of ring system-based structural fragments for use in structure-activity studies: Database mining and recursive partitioning. *J Chem Inf Model* 2006;46(3):1069–77.
63. Hawkins DM, Young SS, Rusinko A. Analysis of a large structure-activity data set using recursive partitioning. *Quant Struct-Act Rel* 1997;16(4):296–302.
64. Rusinko A, Farmen MW, Lambert CG, Brown PL, Young SS. Analysis of a large structure/biological activity data set using recursive partitioning. *J Chem Inf Comput Sci* 1999;39(6):1017–26.
65. Young SS, Gombar VK, Emptage MR, Cariello NF, Lambert C. Mixture deconvolution and analysis of Ames mutagenicity data. *Chemom Intell Lab Sys* 2002;60:5–11.
66. Blower P, Fligner M, Verducci J, Bjoraker J. On combining recursive partitioning and simulated annealing to detect groups of biologically active compounds. *J Chem Inf Comput Sci* 2002 Mar–Apr;42(2):393–404.
67. Steiner G, Suter L, Boess F, Gasser R, de Vera MC, et al. Discriminating different classes of toxicants by transcript profiling. *Environ Health Perspect* 2004;112:1236–48.
68. Yap CW, Cai CZ, Xue Y, Chen YZ. Prediction of torsade-causing potential of drugs by support vector machine approach. *Toxicol Sci* 2004;79:170–7.
69. Thukral SK, Nordone PJ, Hu R, Sullivan L, Galambos E, et al. Prediction of nephrotoxicant action and identification of candidate toxicity-related biomarkers. *Toxicol Pathol* 2005;33:343–55.
70. Yao XJ, Panaye A, Doucet JP, Chen HF, Zhang RS, et al. Comparative classification study of toxicity mechanisms using support vector machines and radial basis function neural networks. *Anal Chim Acta* 2005;535:259–73.
71. Panaye A, Fan BT, Doucet JP, Yao XJ, Zhang RS, et al. Quantitative structure-toxicity relationships (QSTRs): A comparative study of various nonlinear methods. General regression neural network, radial basis function neural network and support vector machine in predicting toxicity of nitro- and cyano- aromatics to *Tetrahymena pyriformis*. *SAR QSAR Environ Res* 2006;17:75–91.

72. Zhao CY, Zhang HX, Zhang XY, Liu MC, Hu ZD, et al. Application of support vector machine (SVM) for prediction toxic activity of different data sets. *Toxicology* 2006;217:105–19.
73. Liu HX, Yao XJ, Zhang RS, Liu MC, Hu ZD, et al. The accurate QSPR models to predict the bioconcentration factors of nonionic organic compounds based on the heuristic method and support vector machine. *Chemosphere* 2006;63:722–33.
74. Mazzatorta P, Cronin MTD, Benfenati E. A QSAR study of avian oral toxicity using support vector machines and genetic algorithms. *QSAR Comb Sci* 2006;25:616–28.
75. Votano JR, Parham M, Hall LH, Kier LB, Oloff S, et al. Three new consensus QSAR models for the prediction of Ames genotoxicity. *Mutagenesis* 2004;19:365–77.
76. Votano JR, Parham M, Hall LH, Kier LB. New predictors for several ADME/Tox properties: Aqueous solubility, human oral absorption, and Ames genotoxicity using topological descriptors. *Mol Divers* 2004;8:379–91.
77. Gombar VK, Emptage MR, Cariello NF, Lambert C. Mixture deconvolution and analysis of Ames mutagenicity data. *Chemom Intell Lab Sys* 2002;60(1–2):5–11.
78. Hall LH, Hall LM. QSAR modeling based on structure-information for properties of interest in human health. *SAR QSAR Environ Res*. 2005;16(1–2):13–41.
79. Kazius J, McGuire R, Bursi R. Derivation and validation of toxicophores for mutagenicity prediction. *J Med Chem* 2005;48(1):312–20.
80. Villemin D, Cherqaoui D, Cense JM. Neural networks studies—Quantitative structure-activity relationship of mutagenic aromatic nitro compounds. *J Chim Phys Phys Chim Biol* 1993;90(7–8):1505–19.
81. Debnath AK, Shusterman AJ, Decompadre RLL, Hansch C. The importance of the hydrophobic interaction in the mutagenicity of organic compounds. *Mutat Res* 1994;305(1):63–72.
82. Knize MG, Hatch FT, Tanga MJ, Lau EY, Colvin ME. A QSAR for the mutagenic potencies of twelve 2-amino-trimethylimidazopyridine isomers: Structural, quantum chemical, and hydropathic factors. *Environ Mol Mutagen*. 2006;47(2): 132–46.
83. Andrews LE, Bonin AM, Fransson LE, Gillson AME, Glover SA. The role of steric effects in the direct mutagenicity of *N*-acyloxy-*N*-alkoxyamides. *Mutat Res Genet Toxicol Environ Mutagen*. 2006;605(1–2):51–62.
84. Bhat KL, Hayik S, Sztandera L, Bock CW. Mutagenicity of aromatic and heteroaromatic amines and related compounds: A QSAR investigation. QSAR *Combin Sci*. 2005;24(7):831–43.
85. Snyder RD, Smith MD. Computational prediction of genotoxicity: Room for improvement. *Drug Discov Today* 2005;10(16):1119–24.
86. Wang XD, Lin ZF, Yin DQ, Liu SS. Wang LS. 2D/3D-QSAR comparative study on mutagenicity of nitroaromatics. *Sci China B—Chem*. 2005;48(3):246–52.
87. Maran U, Sild S. QSAR modeling of mutagenicity on non-congeneric sets of organic compounds. *Art Intell Meth Tools Sys Biol*. 2004;5:19–35.
88. Gonzalez MP, Moldes MDT, Fall Y, Dias LC, Helguera AM. A topological substructural approach to the mutagenic activity in dental monomers. 3. Heterogeneous set of compounds. *Polymer* 2005;46(8):2783–90.

89. Popelier PLA, Smith PJ, Chaudry UA. Quantitative structure-activity relationships of mutagenic activity from quantum topological descriptors: triazenes and halogenated hydroxyfuranones (mutagen-X) derivatives. *J Comput Aided Mol Des* 2004;18(11):709–18.
90. Winkler DA. Neural networks in ADME and toxicity prediction. *Drugs Future* 2004;29(10):1043–57.
91. Mekenyan O, Dimitrov S, Serafimova R, Thompson E, Kotov S, et al. Identification of the structural requirements for mutagenicity by incorporating molecular flexibility and metabolic activation of chemicals I: TA100 model. *Chem Res Toxicol* 2004;17(6):753–66.
92. Vracko M, Mills D, Basak SC. Structure-mutagenicity modelling using counter propagation neural networks. *Environ Toxicol Pharmacol* 2004;16(1–2):25–36.
93. Hawkins DM, Basak SC, Mills D. QSARs for chemical mutagens from structure: Ridge regression fitting and diagnostics. *Environ Toxicol Pharmacol* 2004; 16(1–2):37–44.
94. Vracko M, Szymoszek A, Barbieri P. Structure-mutagenicity study of 12 trimethylimidazopyridine isomers using orbital energies and "spectrum-like representation" as descriptors. *J Chem Inf Comput Sci* 2004;44(2):352–58.
95. Helguera AM, Gonzalez MP, Briones JR. TOPS-MODE approach to predict mutagenicity in dental monomers. *Polymer* 2004;45(6):2045–50.
96. Maran U, Sild S. QSAR modeling of genotoxicity on non-congeneric sets of organic compounds. *Artificial Intell Rev* 2003;20(1–2):13–38.
97. Sztandera L, Garg A, Hayik S, Bhat KL, Bock CW. Mutagenicity of aminoazo dyes and their reductive-cleavage metabolites: A QSAR/QPAR investigation. *Dyes Pigm* 2003;59(2):117–33.
98. Dearden JC. In silico prediction of drug toxicity. *J Comput Aided Mol Des* 2003;17(2):119–27.
99. Schultz TW, Cronin MTD, Walker JD, Aptula AO. Quantitative structure-activity relationships (QSARs) in toxicology: A historical perspective. *Theochem J Mol Struct* 2003;622(1–2):1–22.
100. Schultz TW, Cronin MTD, Netzeva TI. The present status of QSAR in toxicology. *Theochem J Mol Struct* 2003;622(1–2):23–38.
101. Greene N. Computer systems for the prediction of toxicity: An update. *Adv Drug Deliv Rev* 2002;54(3):417–31.
102. Cash GG. Prediction of the genotoxicity of aromatic and heteroaromatic amines using electrotopological state indices. *Mutat Res Genet Toxicol Environ Mutagen* 2001;491(1–2):31–7.
103. Benigni R, Giuliani A, Franke R, Gruska A. Quantitative structure-activity relationships of mutagenic and carcinogenic aromatic amines. *Chem Rev* 2000;100(10): 3697–714.
104. Karelson M, Sild S, Maran U. Non-linear QSAR treatment of genotoxicity. *Mol Simul* 2000;24(4–6):229–42.
105. Basak SC, Grunwald GD, Gute BD. Balasubramanian K. Opitz D. Use of statistical and neural net approaches in predicting toxicity of chemicals. *J Chem Inf Comput Sci* 2000;40(4):885–90.

106. Basak SC, Gute BD, Grunwald GD. Assessment of the mutagenicity of aromatic amines from theoretical structural, parameters: A hierarchical approach. *SAR QSAR Environ Res* 1999;10(2–3):117–29.

107. Compadre RL, Byrd C, Compadre CM. Comparative QSAR and 3-D-QSAR analysis of the mutagenicity of nitroaromatic compounds. *Comparative QSAR* 1998;111–36.

108. Baeten A, Tafazoli M, Kirsch-Volders M, Geerlings P. Use of the HSAB principle in quantitative structure-activity relationships in toxicological research: Application to the genotoxicity of chlorinated hydrocarbons. *Int J Quant Chem* 1999;74(3):351–5.

109. Maran U, Karelson M, Katritzky AR. A comprehensive QSAR treatment of the genotoxicity of heteroaromatic and aromatic amines. *Quant Struct-Act Rel* 1999;18(1):3–10.

110. Fan M, Byrd C, Compadre CM, Compadre RL. Comparison of CoMFA models for *Salmonella typhimurium* TA98, TA100, TA98 + S9 and TA100 + S9 mutagenicity of nitroaromatics. *SAR QSAR Environ Res* 1998;9(3–4):187.

111. Colvin ME, Seidl ET. Structural and quantum chemical factors affecting mutagenic potency of aminoimidazo-azaarenes. *Environ Mol Mutagen* 1996;27(4):314–30.

112. Hansch C, Hoekman D, Leo A, Zhang LT, Li P. The expanding role of quantitative structure-activity relationships (QSAR) in toxicology. Toxicol Lett 1995; 79(1–3):45–53.

113. Tuppurainen K. QSAR approach to molecular mutagenicity—A survey and a case study—mx compounds. *Theochem J Mol Struct* 1994;112(1):49–56.

114. Contrera JF, Matthews EJ, Kruhlak NL, Benz RD. In silico screening of chemicals for bacterial mutagenicity using electrotopological E-state indices and MDL QSAR software. *Regul Toxicol Pharmacol* 2005;43(3):313–23.

115. Anderson B, Mayo K, Bogaczyk S, Tunkel J. Predicting genotoxicity of aromatic and heteroaromatic amines using electrotopological state indices. *Mutat Res* 2005;585(1–2):170–83.

116. Mattioni BE, Kauffman GW, Jurs PC, Custer LL, Durham SK, et al. Predicting the genotoxicity of secondary and aromatic amines using data subsetting to generate a model ensemble. *J Chem Inf Comput Sci* 2003;43(3):949–63.

117. Lewis DF, Ioannides C, Parke DV. A quantitative structure-activity relationship (QSAR) study of mutagenicity in several series of organic chemicals likely to be activated by cytochrome P450 enzymes. *Teratogen Carcinogen Mutagen* 2003; (Suppl)1:187–93.

118. Livingstone DJ, Greenwood R, Rees R, Smith MD. Modelling mutagenicity using properties calculated by computational chemistry. *SAR QSAR Environ Res* 2002;13(1):21–33.

119. Tuppurainen K. Frontier orbital energies, hydrophobicity and steric factors as physical QSAR descriptors of molecular mutagenicity: A review with a case study: MX compounds. *Chemosphere* 1999;38(13):3015–30.

120. Ferguson LR, Denny WA. Potential antitumor agents. 33. Quantitative structure-activity relationships for mutagenic activity and antitumor activity of substituted 4′-(9-acridinylamino)methanesulfonanilide derivatives. *J Med Chem* 1980;23(3): 269–74.

121. Venger BH, Hansch C, Hatheway GJ, Amrein YU. Ames test of 1-(X-phenyl)-3,3-dialkyltriazenes: A quantitative structure-activity study. *J Med Chem* 1979; 22(5):473–6.
122. Wishnok JS, Archer MC, Edelman AS, Rand WM. Nitrosamine carcinogenicity: A quantitative Hansch-Taft structure-activity relationship. *Chem Biol Interact* 1978;20:43–54.
123. Chou JT, Jurs PC. Computer assisted structure-activity studies of chemical carcinogens: An *N*-nitroso compound data set. *J Med Chem* 1979;22:792–7.
124. Dunn WJ 3rd, Wold S. As assessment of carcinogenicity of *N*-nitroso compounds by the SIMCA method of pattern recognition. *J Chem Inf Comput Sci* 1981;21:8–13.
125. Niculescu-Duvaz I, Craescu T, Tugulea M, Croisy A, Jacquignon PC. A quantitative structure-activity analysis of the mutagenic and carcinogenic action of 43 structurally related heterocyclic compounds. *Carcinogenesis* 1981;2: 269–75.
126. Yuta K, Jurs PC. Computer-assisted structure-activity studies of chemical carcinogens: Aromatic amines. *J Med Chem* 1981;24:241–51.
127. Jurs PC, Hasan MN, Henry DR, Stouch TR, Whalen-Pedersen EK. Computer-assisted studies of molecular structure and carcinogenic activity. *Fundam Appl Toxicol* 1983;3:343–9.
128. Frierson MR, Klopman G, Rosenkranz HS. Structure-activity relationships (SARs) among mutagens and carcinogens: A review. *Environ Mutagen* 1986;8: 283–327.
129. Enslein K. An overview of structure-activity relationships as an alternative to testing in animals for carcinogenicity, mutagenicity, dermal and eye irritation, and acute oral toxicity. *Toxicol Ind Health* 1988;4:479–98.
130. Benigni R, Andreoli C, Giuliani A. Structure-activity studies of chemical carcinogens: Use of an electrophilic reactivity parameter in a new QSAR model. *Carcinogenesis* 1989;10:55–61.
131. Benigni R, Pellizzone G, Giuliani A. Comparison of different computerized classification methods for predicting carcinogenicity from short-term test results. *J Toxicol Environ* Health 1989;28:427–44.
132. Enslein K, Borgstedt HH. A QSAR model for the estimation of carcinogenicity: example application to an azo-dye. *Toxicol Lett* 1989;49:107–21.
133. Benigni R. QSAR prediction of rodent carcinogenicity for a set of chemicals currently bioassayed by the US National Toxicology Program. *Mutagenesis* 1991;6:423–5.
134. Enslein K, Gombar VK, Blake BW. International Commission for Protection against Environmental Mutagens and Carcinogens. Use of SAR in computer-assisted prediction of carcinogenicity and mutagenicity of chemicals by the TOPKAT program. *Mutat Res* 1994;305:47–61.
135. Moriguchi I. Development of quantitative structure-activity relationships and computer-aided drug design. *Yakugaku Zasshi* 1994;114:135–46.
136. Ashby J. International Commission for Protection against Environmental Mutagens and Carcinogens. Two million rodent carcinogens? The role of SAR and QSAR in their detection. *Mutat Res* 1994;305:3–12.

137. Villemin D, Cherqaoui D, Mesbah A. Predicting carcinogenicity of polycyclic aromatic hydrocarbons from back-propagation neural network. *J Chem Inf Comput Sci* 1994;34:1288–93.

138. Benigni R. Predicting chemical carcinogenesis in rodents: The state of the art in light of a comparative exercise. *Mutat Res* 1995;334:103–13.

139. Lewis DF, Parke DV. The genotoxicity of benzanthracenes: A quantitative structure-activity study. *Mutat Res* 1995;328:207–14.

140. Benigni R, Giuliani A. Quantitative structure-activity relationship (QSAR) studies of mutagens and carcinogens. *Med Res Rev* 1996;16:267–84.

141. Benigni R, Richards AM. QSARs of mutagens and carcinogens: Two case studies illustrating problems in the construction of models for noncongeneric chemicals. *Mutat Res* 1996;371:29–46.

142. King RD, Srinivasan A. Prediction of rodent carcinogenicity bioassays from molecular structure using inductive logic programming. *Environ Health Perspect* 1996;104:1031–40.

143. Moriguchi I, Hirano H, Hirono S. Prediction of the rodent carcinogenicity of organic compounds from their chemical structures using the FALS method. *Environ Health Perspect* 1996;104:1051–8.

144. Rosenkranz HS, Liu M, Cunningham A, Klopman G. Application of structural concepts to evaluate the potential carcinogenicity of natural products. *SAR QSAR Environ Res* 5(2):79–98, 1996.

145. Cunningham AR, Klopman G, Rosenkranz HS. Identification of structural features and associated mechanisms of action for carcinogens in rats. *Mutat Res* 1998;405:9–27.

146. Matthews EJ, Contrera JF. A new highly specific method for predicting the carcinogenic potential of pharmaceuticals in rodents using enhanced MCASE QSAR-ES software. *Regul Toxicol Pharm* 1998;28:242–64.

147. Gini G, Lorenzini M, Benfenati E, Grasso P, Bruschi M. Predictive carcinogenicity: A model for aromatic compounds, with nitrogen-containing substituents, based on molecular descriptors using an artificial neural network. *J Chem Inf Comput Sci* 1999;39:1076–80.

148. Bahler D, Stone B, Wellington C, Bristol DW. Symbolic, neural, and Bayesian machine learning models for predicting carcinogenicity of chemical compounds. *J Chem Inf Comput Sci* 2000;40:906–14.

149. Benigni R, Giuliani A, Franke R, Gruska A. Quantitative structure-activity relationships of mutagenic and carcinogenic aromatic amines. *Chem Rev* 2000; 100:3697–714.

150. Vracko M. A study of structure-carcinogenicity relationship for 86 compounds from NTP database using topological indices as descriptors. *SAR QSAR Environ Res* 2000;11:103–15.

151. Franke R, Gruska A, Giuliani A, Benigni R. Prediction of rodent carcinogenicity of aromatic amines: A quantitative structure-activity relationships model. *Carcinogenesis* 2001;22:1561–71.

152. Singh AK. Development of quantitative structure-activity relationship (QSAR) models for predicting risk of exposure from carcinogens in animals. *Cancer Invest* 2001;19:611–20.

153. Benigni R, Passerini L. Carcinogenicity of the aromatic amines: From structure-activity relationships to mechanisms of action and risk assessment. *Mutat Res* 2002;511:191–206.

154. Greene N. Computer systems for the prediction of toxicity: An update. *Adv Drug Deliver Rev* 2002;54:417–31.

155. Richards AM, Benigni R. AI and SAR approaches for predicting chemical carcinogenicity: Survey and status report. *SAR QSAR Environ Res* 2002;13:1–19.

156. Benigni R, Passerini L, Rodomonte A. Structure-activity relationships for the mutagenicity and carcinogenicity of simple and alpha-beta unsaturated aldehydes. *Environ Mol Mutagen* 2003;42:136–43.

157. Benigni R, Zito R. Designing safer drugs: (Q)SAR-based identification of mutagens and carcinogens. *Curr Top Med Chem* 2003;3:1289–300.

158. Contrera JF, Matthews EJ, Benz RD. Predicting the carcinogenicity potential of pharmaceuticals in rodents using molecular structural similarity and E-state indices. *Regul Toxicol Pharm* 2003;38:243–59.

159. Dearden JC. In silico prediction of drug toxicity. *J Comput Aid Mol Des* 2003;17:119–27.

160. Patlewicz G, Rodford R, Walker JD. Quantitative structure-activity relationships for predicting mutagenicity and carcinogenicity. *Environ Toxicol Chem* 2003;22: 1885–93.

161. Shen Q, Jiang JH, Shen GL, Yu RQ. Variable selection by an evolution algorithm using modified Cp based on MLR and PLS modeling: QSAR studies of carcinogenicity of aromatic amines. *Anal Bioanal Chem* 2003;375:248–54.

162. Benigni R. Chemical structure of mutagens and carcinogens and the relationship with biological activity. *J Exp Clin Canc Res* 2004;23:5–8.

163. Klopman G, Chakravarti SK, Zhu H, Ivanov JM, Saiakhov RD. ESP: A method to predict toxicity and pharmacological properties of chemicals using multiple MCASE databases. *J Chem Inf Comput Sci* 2004;44:704–15.

164. Sun H. Prediction of chemical carcinogenicity from molecular structure. *J Chem Inf Comput Sci* 2004;44:1506–14.

165. Contrera JF, MacLaughlin P, Hall LH, Kier LB. QSAR modeling of carcinogenic risk using discriminant analysis and topological molecular descriptors. *Curr Drug Discov Technol* 2005;2:55–67.

166. He L, Jurs PC, Kreatsoulas C, Custer LL, Durham SK, Pearl GM. Probabilistic neural network multiple classifier system for predicting the genotoxicity of quinolone and quinoline derivatives. *Chem Res Toxicol* 2005;18:428–40.

167. Helguera AM, Cabrera-Perez MA, Gonzalez MP, Ruiz RM, Gonzalez-Diaz H. A topological substructural approach applied to the computational prediction of rodent carcinogenicity. *Bioorg Med Chem* 2005;13:2477–88.

168. Lagunin AA, Dearden JC, Filimonov DA, Poroikov VV. Computer-aided rodent carcinogenicity prediction. *Mutat Res* 2005;586:138–46.

169. Luan F, Zhang R, Zhao C, Yao X, Liu M, et al. Classification of the carcinogenicity of *N*-nitroso compounds based on support vector machines and linear discriminant analysis. *Chem Res Toxicol* 2005;18:198–203.

170. Rallo R, Espinosa G, Giralt F. Using an ensemble of neural based QSARs for the prediction of toxicological properties of chemical contaminants. *Process Saf Environ* 2005;83:387–92.

171. Morales AH, Perez MA, Combes RD, Gonzalez MP. Quantitative structure-activity relationship for the computational prediction of nitrocompounds carcinogenicity. *Toxicology* 2006;220:51–62.

172. Kavlock RJ. Structure-activity relationships in the developmental toxicity of substituted phenols: In vivo effects. *Teratology* 1990;41:43–59.

173. Frierson MR, Mielach FA, Kochhar DM. Computer-automated structure evaluation (CASE) of retinoids in teratogenesis bioassays. *Fundam Appl Toxicol* 1990;14:408–28.

174. Dawson DA, Schultz TW, Baker LL, Mannar A. Structure-activity relationships for osteolathyrism: III. Substituted thiosemicarbazides. *J Appl Toxicol* 1990; 10:59–64.

175. Klopman G, Dimayuga ML. Computer automated structure evaluation (CASE) of the teratogenicity of retinoids with the aid of a novel geometry index. *J Comput Aided Mol Des* 1990;4:117–30.

176. Ridings JE, Manallack DT, Saunders MR, Baldwin JA, Livingstone DJ. Multivariate quantitative structure-toxicity relationships in a series of dopamine mimetics. *Toxicology* 1992;76:209–17.

177. Klopman G, Ptchelintsev D. Antifungal triazole alcohols: a comparative analysis of structure-activity, structure-teratogenicity and structure-therapeutic index relationships using the multiple computer-automated structure evaluation (MultiCASE) methodology. *J Comput Aided Mol Des* 1993;7:349–62.

178. Takihi N, Rosenkranz HS, Klopman G, Mattison DR. Structural determinants of developmental toxicity. *Risk Anal* 1994;14:649–57.

179. Hansch C, Telzer BR, Zhang LT. Comparative QSAR in toxicology—Examples from teratology and cancer chemotherapy of aniline mustards. *Crit Rev Toxicol* 1995;25:67–89.

180. Gombar VK, Enslein K, Blake BW. Assessment of developmental toxicity potential of chemicals by quantitative structure-toxicity relationship models. *Chemosphere* 1995;31:2499–510.

181. Dawson DA, Schultz TW, Hunter RS. Developmental toxicity of carboxylic acids to *Xenopus* embryos: A quantitative structure-activity relationship and computer-automated structure evaluation. *Teratogen Carcin Mut* 1996;16:109–24.

182. Richard AM, Hunter ES III. Quantitative structure-activity relationships for the developmental toxicity of haloacetic acids in mammalian whole embryo culture. *Teratology* 1996;53:352–60.

183. Ghanooni M, Mattison DR, Zhang YP, Macina OT, Rosenkranz HS, Klopman G. Structural determinants associated with risk of human developmental toxicity. *Am J Obstet Gynecol* 1997;176:799–805.

184. Pearl GM, Livingston-Carr S, Durham SK. Integration of computational analysis as a sentinel tool in toxicological assessments. *Curr Top Med Chem* 2001;1:247–55.

185. Devillers J, Chezeau A, Thybaud E, Rahmani R. QSAR modeling of the adult and developmental toxicity of glycols, glycol ethers and xylenes to *Hydra attenuata*. *SAR QSAR Environ Res* 2002;13:555–66.

186. Devillers J, Chezeau A, Thybaud E. PLS-QSAR of the adult and developmental toxicity of chemicals to *Hydra attenuata*. *SAR QSAR Environ Res* 2002;13:705–12.

187. Cronin MTD, Dearden JC, Walker JD, Worth AP. Quantitative structure-activity relationships for human health effects: Commonalities with other endpoints. *Environ Toxicol Chem* 2003;22:1829–43.

188. Schultz TW, Cronin MTD, Netzeva TI. The present status of QSAR in toxicology. *Theochem J Mol Struc* 2003;622:23–38.

189. Arena VC, Sussman NB, Mazumdar S, Yu S, Macina OT. The utility of structure-activity relationship (SAR) models for prediction and covariate selection in developmental toxicity: Comparative analysis of logistic regression and decision tree methods. *SAR QSAR Environ Res* 2004;15:1–18.

190. Matthews EJ, Kruhlak NL, Cimino MC, Benz RD, Contrera JF. An analysis of genetic toxicity, reproductive and developmental toxicity, and carcinogenicity data: II. Identification of genotoxicants, reprotoxicants, and carcinogens using in silico methods. *Regul Toxicol Pharm* 2006;44:97–110.

191. Dimitrov S, Koleva Y, Schultz TW, Walker JD, Mekenyan O. Interspecies quantitative structure-activity relationship model for aldehydes: Aquatic toxicity. *Environ Toxicol Chem* 2004;23(2):463–70.

192. Klopman G, Stuart SE. Multiple computer-automated structure evaluation study of aquatic toxicity. III. *Vibrio fischeri*. *Environ Toxicol Chem* 2003 Mar;22(3): 466–72.

193. Yan XF, Xiao HM, Ju XH, Gong XD. QSAR study of nitroaromatic compounds toxicity to the *Tetrahymena pyriformis*. *Acta Chim Sinica* 2006;64(5):375–80.

194. Netzeva TI, Schultz TW. QSARs for the aquatic toxicity of aromatic aldehydes from *Tetrahymena* data. *Chemosphere* 2005;61(11):1632–43.

195. Yan XF, Xiao HM, Ju XH, Gong XD. DFT study on the OSAR of nitroaromatic compound toxicity to the fathead minnow. *Chin J Chem* 2005;23(8):947–52.

196. Kamaya Y, Fukaya Y, Suzuki K. Acute toxicity of benzoic acids to the crustacean *Daphnia magna*. *Chemosphere* 2005;59(2):255–61.

197. Niculescu SP, Atkinson A, Hammond G, Lewis M. Using fragment chemistry data mining and probabilistic neural networks in screening chemicals for acute toxicity to the fathead minnow. *SAR QSAR Environ Res* 2004;15(4):293–309.

198. Netzeva TI, Schultz TW, Aptula AO, Cronin MTD. Partial least squares modelling of the acute toxicity of aliphatic compounds to *Tetrahymena pyriformis*. *SAR QSAR Environ Res* 2003;14(4):265–83.

199. Liu XH, Wang B, Huang Z, Han SK, Wang LS. Acute toxicity and quantitative structure-activity relationships of alpha-branched phenylsulfonyl acetates to *Daphnia magna*. *Chemosphere* 2003;50(3):403–8.

200. Papa E, Villa F, Gramatica P. Statistically validated QSARs, based on theoretical descriptors, for modeling aquatic toxicity of organic chemicals in *Pimephales promelas* (fathead minnow). *J Chem Inf Model* 2005;45(5):1256–66.

201. Oberg T. A QSAR for baseline toxicity: Validation, domain of application, and prediction. *Chem Res Toxicol* 2004;17(12):1630–7.

202. Tao S, Xi XH, Xu FL, Dawson R. A QSAR model for predicting toxicity (LC50) to rainbow trout. *Water Res* 2002;36(11):2926–30.

203. Martin TM, Young DM. Prediction of the acute toxicity (96-h LC50) of organic compounds to the fathead minnow (*Pimephales promelas*) using a group contribution method. *Chem Res Toxicol* 2001;14(10):1378–85.

204. Wang XD, Dong YY, Wang LS, Han SK. Acute toxicity of substituted phenols to *Rana japonica* tadpoles and mechanism-based quantitative structure-activity relationship (QSAR) study. *Chemosphere* 2001;44(3):447–55.

205. Uppgard L, Sjostrom M, Wold S. Multivariate quantitative structure-activity relationships for the aquatic toxicity of alkyl polyglucosides. *Tenside, Surfactants, Detergents* 2000;37(2):131–8.

206. Cash GG. Prediction of chemical toxicity to aquatic organisms—ECOSAR vs. Microtox(r) assay. *Environ Toxicol Water Qual* 1998;13(3):211–16.

207. Russom CL, Bradbury SP, Broderius SJ, Hammermeister DE, Drummond RA. Predicting modes of toxic action from chemical structure—Acute toxicity in the fathead minnow (*Pimephales promelas*). *Environ Toxicol Chem* 1997;16(5): 948–67.

208. Karabunarliev S, Mekenyan OG, Karcher W, Russom CL, Bradbury SP. Quantum-chemical descriptors for estimating the acute toxicity of electrophiles to the fathead minnow (*Pimephales promelas*)—An analysis based on molecular mechanisms. *Quant Struct-Act Rel* 1996;15(4):302–10.

209. Lindgren A, Sjostrom M, Wold S. QSAR modelling of the toxicity of some technical non-ionic surfactants towards fairy shrimps. *Quant Struct-Act Rel* 1996; 15(3):208–18.

210 Jackel H, Nendza M. Reactive substructures in the prediction of aquatic toxicity data. *Aquatic Toxicol* 1994;29(3–4):305–14.

211 Purdy R. The utility of computed superdelocalizability for predicting the LC_{50} values of epoxides to guppies. *Sci Total Environ* 1991;109–10:553–56.

212 Nendza M, Russom CL. QSAR modelling of the ERL-D fathead minnow acute toxicity database. *Xenobiotica* 1991;21(2):147–70.

213 Lipnick RL, Watson KR, Strausz AK. A QSAR study of the acute toxicity of some industrial organic chemicals to goldfish: Narcosis, electrophile and proelectrophile mechanisms. *Xenobiotica* 1987;17(8):1011–25.

214 Konemann H. Quantitative structure-activity relationships in fish toxicity studies: Part 1. relationship for 50 industrial pollutants. *Toxicology* 1981;19(3):209–21.

215 Klopman G, Tu M. Structure-biodegradability study and computer-automated prediction of aerobic biodegradation of chemicals. *Environ Toxicol Chem* 16(9):1829–35.

216 Roberts DW. Application of QSAR to biodegradation of linear alkylbenzene sulphonate (LAS) isomers and homologues. *Sci Total Environ* 1991 Dec; 109–10:301–6.

217 Degner P, Nendza M, Klein W. Predictive QSAR models for estimating biodegradation of aromatic compounds. *Sci Total Environ* 1991 Dec;109–10:253–9.

218 Peijnenburg WJGM. Structure-activity relationships for biodegradation: A critical review. *Pure Appl Chem* 1994;66(9):1931–41.

219 Dearden JC. Prediction of environmental toxicity and fate using quantitative structure-activity relationships (QSARs). *J Braz Chem Soc* 2002;13(6).

220. Loonen H, Lindgren F, Hansen B, Karcher W, Niemelä J, et al. Prediction of biodegradability from chemical structure: Modeling of ready biodegradation test data. *Environ Toxicol Chem* 1999;18(8):1763–8.

221. Gombar VK, Enslein K. Structure-biodegradability relationship model by discriminant analysis. In: Devillers J, Karcher W, editors, *Applied multivariate analysis in SAR and environmental studies.* Dordrecht: Kluwer Academic, 1991.

222. Sabljic A. Chemical topology and ecotoxicology. *Sci Total Environ* 1991;109–10:197–220.

223. Degner P, Nendza M, Klein W. Predictive QSAR models for estimating biodegradation of aromatic compounds. *Sci Total Environ* 1991;109–10:253–9.

224. Roberts DW. Application of QSAR to biodegradation of linear alkylbenzene sulphonate (LAS) isomers and homologues. *Sci Total Environ* 1991;109–10:301–6.

225. Devillers J. Neural modelling of the biodegradability of benzene derivatives. *SAR QSAR Environ Res* 1993;1:161–7.

226. Klopman G, Zhang Z, Woodgate SD, Rosenkranz HS. The structure-toxicity relationship challenge at hazardous waste sites. *Chemosphere* 1995;31:2511–9.

227. Klopman G, Zhang ZT, Balthasar DM, Rosenkranz HS. Computer-automated predictions of aerobic biodegradation of chemicals. *Environ Toxicol Chem* 1995;14:395–403.

228. Cowan CE, Federle TW, Larson RJ, Feijtel TC. Impact of biodegradation test methods on the development and applicability of biodegradation QSARs. *SAR QSAR Environ Res* 1996;5:37–49.

229. Okey RW, Stensel HD. A QSAR-based biodegradability model—A QSBR. *Water Res* 1996;30:2206–14.

230. Damborsky J, Schultz TW. Comparison of the QSAR models for toxicity and biodegradability of anilines and phenols. *Chemosphere* 1997;34:429–46.

231. Klopman G, Tu MH. Structure-biodegradability study and computer-automated prediction of aerobic biodegradation of chemicals. *Environ Toxicol Chem* 1997;16:1829–35.

232. Kompare B. Estimating environmental pollution by xenobiotic chemicals using QSAR (QSBR) models based on artificial intelligence. *Water Sci Technol* 1998;37:9–18.

233. Klopman G, Saiakhov R, Tu MH, Pusca F, Rorije E. Computer-assisted evaluation of anaerobic biodegradation products. *Pure Appl Chem* 1998;70:1385–94.

234. Damborsky J, Berglund A, Kuty M, Ansorgova A, Nagata Y, et al. Mechanism-based quantitative structure-biodegradability relationships for hydrolytic dehalogenation of chloro- and bromo- alkenes. *Quant Struct-Act Rel* 1998;17:450–8.

235. Rorije E, Peijnenburg WJGM, Klopman G. Structural requirements for anaerobic biodegradation of organic chemicals—A fragment model analysis. *Environ Toxicol Chem* 1998;17:1943–50.

236. Hornak V, Balaz S, Schaper KJ, Seydel JK. Multiple binding modes in 3D-QSAR—Microbial degradation of polychlorinated biphenyls. *Quant Struct-Act Rel* 1998;17:427–36.

237. Huuskonen J. Prediction of biodegradation from the atom-type electrotopological state indices. *Environ Toxicol Chem* 2001;20:2152–7.

238. Raymond JW, Rogers TN, Shonnard DR, Kline AA. A review of structure-based biodegradation estimation methods. *J Hazard Mater* 2001;84:189–215.

239. Cartwright HM. Investigation of structure-biodegradability relationships in polychlorinated biphenyls using self-organizing maps. *Neural Comput Appl* 2002;11:30–6.

240. Srikanth K, Debnath B, Jha T. QSAR study on adenosine kinase inhibition of pyrrolo[2,3-d]pyrimidine nucleoside analogues using the Hansch approach. *Bioorg Med Chem Lett* 2002;12:899–902.

241. Lindner AS, Whitfield C, Chen N, Semrau JD, Adriaens P. Quantitative structure-biodegradation relationships for ortho-substituted biphenyl compounds oxidized by *Methylosinus trichosporium* OB3b. *Environ Toxicol Chem* 2003;22: 2251–7.

242. Liu Y, Liu SS, Cui SH, Cai SX. A novel quantitative structure-biodegradability relationship (QSBR) of substituted benzenes based on MHDV descriptor. *J Chin Chem Soc* 2003;50:319–24.

243. Lu GH, Wang C, Yuan X, Zhao YH. QSBR study of substituted phenols and benzoic acids. *J Environ Sci* 2003;15:88–91.

244. Yang H, Jiang Z, Shi S. Anaerobic biodegradability of aliphatic compounds and their quantitative structure biodegradability relationships. *Sci Total Environ* 2004;322:209–19.

245. Liu SS, Yan DQ, Cui SH, Wang LS. VSMP for modeling the biodegradability of substituted benzenes based on electrotopological state indices for atom types. *Chin J Chem* 2005;23:622–6.

246. Kuanar M, Kuanar SK, Patel S, Mishra BK. QSAR studies on biological oxygen demand of alcohols. *Ind J Chem B* 2006;45:766–72.

247. Yang H, Jiang Z, Shi S. Aromatic compounds biodegradation under anaerobic conditions and their QSBR models. *Sci Total Environ* 2006;358:265–76.

248. Sabljic A. The prediction of fish bioconcentration factors of organic pollutants from the molecular connectivity model. *Zeitschrift für die Gesamte Hygiene und Ihre Grenzgebiete* 1987;33:493–6.

249. de Voogt P, Wegener JW, Klamer JC, van Zijl GA, Govers H. Prediction of environmental fate and effects of heteroaromatic polycyclic aromatics by QSARs: The position of *n*-octanol/water partition coefficients. *Biomed Environ Sci* 1988;1:194–209.

250. Geyer HJ, Scheunert I, Bruggemann R, Steinberg C, Korte F, et al. QSAR for organic chemical bioconcentration in *Daphnia*, algae, and mussels. *Sci Total Environ* 1991;109–10:387–94.

251. de Bruijn J, Hermens J. Qualitative and quantitative modelling of toxic effects of organophosphorous compounds to fish. *Sci Total Environ* 1991;109–10:441–55.

252. McCarty LS, Mackay D, Smith AD, Ozburn GW, Dixon DG. Interpreting aquatic toxicity QSARs: The significance of toxicant body residues at the pharmacologic endpoint. *Sci Total Environ* 1991;109–10:515–25.

253. Bintein S, Devillers J, Karcher W. Nonlinear dependence of fish bioconcentration on *n*-octanol/water partition coefficient. *SAR QSAR Environ Res* 1993;1:29–39.

254. Nendza M, Jackel H, Muller M, Giesrcuschel A, Klein W. Estimation of exposure and ecotoxicity related parameters by computer based structure-property and structure-activity relationships. *Toxicol Environ Chem* 1993;40:57–69.

255. Hermens J. Prediction of environmental toxicity based on structure-activity relationships using mechanistic information. *Sci Total Environ* 1995;171:235–42.

256. Todeschini R, Gramatica P. 3D-Modelling and prediction by WHIM descriptors: 6. Application of WHIM descriptors in QSAR studies. *Quant Struct-Act Rel* 1997;16:120–5.

257. Tao S, Hu H, Xu F, Dawson R, Li B, Cao J. QSAR modeling of bioconcentration factors in fish based on fragment constants and structural correction factors. *J Environ Sci Heal B* 2001;36:631–49.

258. Wei D, Zhang A, Wu C, Han S, Wang L. Progressive study and robustness test of QSAR model based on quantum chemical parameters for predicting BCF of selected polychlorinated organic compounds (PCOCs). *Chemosphere* 2001;44:1421–8.

259. Dearden JC. Prediction of environmental toxicity and fate using quantitative structure-activity relationships (QSARs). *J Braz Chem Soc* 2002;13:754–62.

260. Dimitrov SD, Mekenyan OG, Walker JD. Non-linear modeling of bioconcentration using partition coefficients for narcotic chemicals. *SAR QSAR Environ Res* 2002;13:177–84.

261. Nakai S, Saito S, Takeuchi M, Takimoto Y, Matsuo M. The inorganic and organic characters for predicting bioconcentration on wide variety of chemicals in fish. *SAR QSAR Environ Res* 2002;13:667–73.

262. Dimitrov SD, Dimitrova NC, Walker JD, Veith GD, Mekenyan OG. Predicting bioconcentration factors of highly hydrophobic chemicals: Effects of molecular size. *Pure Appl Chem* 2002;74:1823–30.

263. Dimitrov SD, Dimitrova NC, Walker JD, Veith GD, Mekenyan OG. Bioconcentration potential predictions based on molecular attributes—An early warning approach for chemicals found in humans, birds, fish and wildlife. *QSAR Comb Sci* 2003;22:58–68.

264. Fatemi MH, Jalali-Heravi M, Konuze E. Prediction of bioconcentration factor using genetic algorithm and artificial neural network. *Anal Chim Acta* 2003;486:101–8.

265. Gramatica P, Papa E. QSAR modeling of bioconcentration factor by theoretical molecular descriptors. *QSAR Comb Sci* 2003;22:374–85.

266. Khadikar PV, Singh S, Mandloi D, Joshi S, Bajaj AV. QSAR study on bioconcentration factor (BCF) of polyhalogenated biphenyls using the PI index. *Bioorg Med Chem* 2003;11:5045–50.

267. Sacan MT, Erdem SS, Ozpinar GA, Balcioglu IA. QSPR study on the bioconcentration factors of nonionic organic compounds in fish by characteristic root index and semiempirical molecular descriptors. *J Chem Inf Comp Sci* 2004;44:985–92.

268. Bermudez-Saldana JM, Escuder-Gilabert L, Medina-Hernandez MJ, Villanueva-Camanas RM, Sagrado S. Modelling bioconcentration of pesticides in fish using biopartitioning micellar chromatography. *J Chromatogr A* 2005;1063:153–60.

269. Gramatica P, Papa E. An update of the BCF QSAR model based on theoretical molecular descriptors. *QSAR Comb Sci* 2005;24:953–60.

270. Khadikar PV, Singh S, Mandloi D, Joshi S, Bajaj AV. QSAR study on bioconcentration factor (BCF) of polyhalogented biphenyls using the PI index. *Bioorg Med Chem* 2003 Nov 17;11(23):5045–50.

271. Tao S, Hu H, Xu F, Dawson R, Li B, Cao J. QSAR modeling of bioconcentration factors in fish based on fragment constants and structural correction factors. *J Environ Sci Health B* 2001 Sep;36(5):631–49.

272. Veith GD, DeFoe DL, Bergstedt BV. Measuring and estimating the bioconcentration factor in fish. *J Fish Res Bourd Can* 1979;36:1040–8.

273. Vanbavel B, Andersson P, Wingfors H, Ahgren J, Bergqvist PA, et al. Multivariate modeling of PCB bioaccumulation in three-spined stickleback (*Gasterosteus aculeatus*). *Environ Toxicol Chem* 1996;15:947–54.

274. MacDonald D, Breton R, Sutcliffe R, Walker J. Uses and limitations of quantitative structure-activity relationships (QSARs) to categorize substances on the Canadian domestic substance list as persistent and/or bioaccumulative, and inherently toxic to non-human organisms. *SAR QSAR Environ Res* 2002;13: 43–55.

275. Carlsen L, Walker JD. QSARs for prioritizing PBT substances to promote pollution prevention. *QSAR Comb Sci* 2003;22:49–57.

276. Arnot JA, Gobas FAPC. A generic QSAR for assessing the bioaccumulation potential of organic chemicals in aquatic food webs. *QSAR Comb Sci* 2003;22:337–45.

277. Dearden JC. QSAR modeling of bioaccumulation. In: Cronin MTD, Livingstone DJ, editors, *Predicting chemical toxicity and fate*. Boca Raton, FL: CRC Press, 2004. p. 333:55.

278. Mekenyan OG, Dimitrov SD, Pavlov TS, Veith GD. POPs: A QSAR system for developing categories for persistent, bioaccumulative and toxic chemicals and their metabolites. *SAR QSAR Environ Res* 2005;16:103–33.

279. Ivanciuc T, Ivanciuc O, Klein DJ. Modeling the bioconcentration factors and bioaccumulation factors of polychlorinated biphenyls with posetic quantitative super-structure/activity relationships (QSSAR). *Mol Divers* 2006;10:133–45.

280. Gombar VK, Enslein K, Hart JB, Blake BW, Borgstedt HH. Estimation of maximum tolerated dose for long-term bioassays from acute lethal dose and structure by QSAR. *Risk Anal* 1991 Sep;11(3):509–17.

281. Matthews EJ, Kruhlak NL, Benz RD, Contrera JF. Assessment of the health effects of chemicals in humans: I. QSAR estimation of the maximum recommended therapeutic dose (MRTD) and no effect level (NOEL) of organic chemicals based on clinical trial data. *Curr Drug Discov Technol* 2004 Jan;1(1):61–76.

282. Roy K, Ghosh G. QSTR with extended topochemical atom (ETA) indices. VI. Acute toxicity of benzene derivatives to tadpoles. *J Mol Model* 2006;12(3): 306–16.

283. Wang ZY, Han XY, Wang LS. Quantitative correlation of chromatographic retention and acute toxicity for alkyl(1-phenylsulfonyl) cycloalkane carboxylates and their structural parameters by DFT. *Chin J Struct Chem* 2005;24(7):851–7.

284. Devillers J. Prediction of mammalian toxicity of organophosphorus pesticides from QSTR modeling. *SAR QSAR Environ Res*. 2004;15(5–6):501–10.

285. Cronin MTD. Dearden JC. QSAR in toxicology: 2. Prediction of acute mammalian toxicity and interspecies correlations. *Quant Struct-Act Rel* 1995;14(2): 117–20.

286. Moore DR. Breton RL. MacDonald DB. A comparison of model performance for six quantitative structure-activity relationship packages that predict acute toxicity to fish. *Environ Toxicol Chem* 2003;22(8):1799–809.

287. Zmuidinavicius D, Japertas P, Petrauskas A, Didziapetris R. Progress in toxinformatics: The challenge of predicting acute toxicity. *Curr Topics Med Chem* 2003;3(11):1301–14.

288. Huuskonen J. QSAR modeling with the electrotopological state indices: Predicting the toxicity of organic chemicals. *Chemosphere* 2003;50(7):949–53.

289. Muzikant AL, Penland RC. Models for profiling the potential QT prolongation risk of drugs. *Curr Opin Drug Discover Devel* 2002;5:127–35.

290. Cavalli A, Poluzzi E, De Ponti F, Recanatini M. Toward a pharmacophore for drugs inducing the long QT syndrome: Insights from a CoMFA study of HERG K(+) channel blockers. *J Med Chem* 2002;45:3844–53.

291. Roche O, Trube G, Zuegge J, Pflimlin P, Alanine A, et al. A virtual screening method for prediction of the HERG potassium channel liability of compound libraries. *ChemBioChem* 2002;3:455–9.

292. Keseru GM. Prediction of hERG potassium channel affinity by traditional and hologram QSAR methods. *Bioorg Med Chem Lett* 2003;13:2773–5.

293. Pearlstein RA, Vaz RJ, Kang J, Chen XL, Preobrazhenskaya M, Shchekotikhin AE, Korolev AM, Lysenkova LN, Miroshnikova OV, Hendrix J, Rampe D. Characterization of HERG potassium channel inhibition using CoMSIA 3D QSAR and homology modeling approaches. *Bioorg Med Chem Lett* 2003;13: 1829–35.

294. Aptula AO, Cronin MTD. Prediction of hERG K+ blocking potency: Application of structural knowledge. *SAR QSAR Environ Res* 2004;15:399–411.

295. Fernandez D, Ghanta A, Kauffman GW, Sanguinetti MC. Physicochemical features of the HERG channel drug binding site. *J Biol Chem* 2004;279:10120–7.

296. Tobita M, Nishikawa T, Nagashima R. A discriminant model constructed by the support vector machine method for HERG potassium channel inhibitors. *Bioorg Med Chem Lett* 2005;15:2886–90.

297. Cianchetta G, Li Y, Kang J, Rampe D, Fravolini A, et al. Predictive models for hERG potassium channel blockers. *Bioorg Med Chem Lett* 2005;15:3637–42.

298. O'Brien SE, de Groot MJ. Greater than the sum of its parts: Combining models for useful ADMET prediction. *J Med Chem* 2005;48:1287–91.

299. Crumb WJ, Ekins S, Sarazan RD, Wikel JH, Wrighton SA, Carlson C, et al. Effects of antipsychotic drugs on I-to, I-Na, I-sus, I-K1, and hERG: QT prolongation, structure activity relationships, and network analysis. *Pharmaceut Res* 2006;23:1133–43.

300. Seierstad M, Agrafiotis DK. A QSAR model of HERG binding using a large, diverse, and internally consistent training set. *Chem Biol Drug Des* 2006;67: 284–96.

301. Yoshida K, Niwa T. Quantitative structure-activity relationship studies on inhibition of HERG potassium channels. *J Chem Inf Model* 2006;46:1371–8.

302. Coi A, Massarelli I, Murgia L, Saraceno M, Calderone V, et al. Prediction of hERG potassium channel affinity by the CODESSA approach. *Bioorg Med Chem* 2006;14:3153–9.

303. Song M, Clark M. Development and evaluation of an in silico model for hERG binding. *J Chem Inf Model* 2006;46:392–400.

304. Ekins S, Balakin KV, Savchuk N, Ivanenkov Y. Insights for human ether-a-go-go-related gene potassium channel inhibition using recursive partitioning and Kohonen and Sammon mapping techniques. *J Med Chem* 2006;49:5059–71.

305. Enslein K, Lander TR, Tomb ME, Craig PN. A predictive model for estimating rat oral LD_{50} values. In: *Benchmark papers in toxicology*, Vol 1. Princeton, NJ: Princeton Scientific Publishers, 1983.

306. Enslein K, Tuzzeo TM, Borgstedt HH, et al. Prediction of rat oral LD_{50} from *Daphnia magna* LC_{50} and chemical structure. In: Kaiser LE, editor, *Quantitative structure activity relationship (QSAR) in environmental toxicology: II. Proceedings of the 2nd International Workshop on QSAR in Environmental Toxicology*, McMaster University, Hamilton, Ontario, Canada, June 9–13, 1986. 1987. p. 91–106.

307. Enslein K, Blake BW, Tomb ME, Brgstedt HH. Prediction of Ames test results by structure-activity relationships. *In vitro Toxicol* 1986;1:33–44.

308. Enslein K, Tomb ME, Lander TR. Structure-activity models of biological oxygen demand. In: Kaiser, KLE, editor, *QSAR in environmental toxicology*. Dordrecht: Reidel, 1984. p. 89–109.

309. Luger GF. *Artificial intelligence: Structures and strategies for complex problem solving*, 5th Edition. Boston: Addison-Wesley, 2005.

310. Kahn J. From airports tarmacs to online job banks to medical labs, artificial intelligence is everywhere. *Wired* 2002;10 (3).

311. Zhu Q, Stillman MJ. Expert systems and analytical chemistry: Recent progress in the Acexpert Project. *J Chem Inf Comput Sci* 1996;36:497–509.

312. Huixiao H, Xinquan X. ESSESA: An expert system for structure elucidation from spectra: 5. Substructure constraints from analysis of first-order 1H-NMR spectra. *J Chem Inf Comput Sci* 1994;34:1259–66.

313. Benfenati E, Gini G. Computational predictive programs (expert systems) in toxicology. *Toxicology* 1997;119(3):213–25.

314. Schmidt O. Die Characterisierung der einfachen und Kerbs erzeugenden aromatischen Kohlenwasserstoffe durch die Dichteverteilung bestimmer Valenzelectronen. *Z Phys Chem* 1939;42:83–110.

315. Coulson CA. Electronic configuration and carcinogenesis. *Adv Cancer Res* 1953; 1:1–56.

316. Pullman A, Pullman B. Electronic structure and carcinogenic activity of aromatic molecules; new developments. *Adv Cancer Res* 1955;3:117–69.

317. Pullman B, Pullman A. Electron-donor or electron-acceptor properties and carcinogenic activity of organic molecules. *Nature* 1963;199:467–9.

318. Nagata C, Hukui K, Yonezava T, Tagashira Y. Electronic structure and carcinogenic activity of aromatic compounds. *Cancer Res* 1955;15:233–9.

319. Smith IA, Seybold PG. Substituent effects in chemical carcinogens: Methyl derivatives of the benzacridines. *J Hetero Chem* 1979;16:421–5.

320. Woo Y, Lai D, Argus M, Arcos J. Development of structure-activity relationship rules for predicting carcinogenic potential of chemicals. *Toxicol Lett* 1995;79: 219–28.

321. Sanderson DM, Earnshaw CG. Computer prediction of possible toxic action from chemical structure: The DEREK system. *Hum Exp Toxicol* 1991;10:261–73.

322. Dearden JC, et al. The development and validation of expert systems for predicting toxicity: The report and recommendations of an ECVAM/ECB workshop (ECVAM Workshop 24). *ATLA* 1997;25:223–52.

323. Crettaz P, Benigni R. Prediction of the rodent carcinogenicity of 60 pesticides by the DEREKfW expert system. *J Chem Inf Model* 2005;45:1864–73.

324. Turing AM. The automatic computing engine (ACE) report (1945). In: Davies DW, editor, *Com. Sci.* 57, National Physical Laboratory, Apr. 1972.

325. Klopman G. Artificial intelligence approach to structure-activity studies: Computer automated structure evaluation of biological activity of organic molecules. *J Am Chem Soc* 1984;106(24):7315–21.

326. Klopman G. MULTICASE. 1. A hierarchical computer automated structure evaluation program. *Quant Struct-Act Rel* 1992;11(2):176–84.

327. Klopman G, Zhu H, Fuller MA, Saiakhov RD. Searching for an enhanced predictive tool for mutagenicity. *SAR QSAR Environ Res* 2004;15(4):251–63.

328. Klopman G, Chakravarti SK, Harris N, Ivanov J, Saiakhov RD. In-silico screening of high production volume chemicals for mutagenicity using the MCASE QSAR expert system. *SAR QSAR Environ Res* 2003;14(2):165–80.

329. Rosenkranz HS, Mersch-Sundermann V, Klopman G. SOS chromotest and mutagenicity in *Salmonella*: Evidence for mechanistic differences *Mutat Res Fundam Mol* 1999;431:(1)31–38.

330. Zeiger E, Ashby J, Bakale G, Enslein K, Klopman G, Rosenkranz HS. Prediction of *Salmonella* mutagenicity. *Mutagenesis* 1996 Sep;11(5):471–84.

331. Klopman G, Rosenkranz HS. International Commission for Protection against Environmental Mutagens and Carcinogens: Approaches to SAR in carcinogenesis and mutagenesis. Prediction of carcinogenicity/mutagenicity using MULTI-CASE. *Mutat Res* 1994 Feb 1;305(1):33–46.

332. Klopman G, Rosenkranz HS. Approaches to SAR in carcinogenesis and mutagenesis—Prediction of carcinogenicity/mutagenicity using multi-case. *Mutat Res* 1994;305:(1)33–46.

333. Mersch-Sundermann V, Klopman G, Rosenkranz HS. Chemical structure and genotoxicity: Studies of the SOS chromotest. *Mutat Res* 1996;340(2–3):81–91.

334. Grant SG, Zhang YP, Klopman G, Rosenkranz HS. Modeling the mouse lymphoma forward mutational assay: The Gene-Tox program database. *Mutat Res-Genet Toxicol Environ Mutagen* 2000;465(1–2):201–29.

335. Henry B, Grant SG, Klopman G, Rosenkranz HS. Induction of forward mutations at the thymidine kinase locus of mouse lymphoma cells: Evidence for electrophilic and non-electrophilic mechanisms. *Mutat Res* 1998 Feb 2;397(2): 313–35.

336. Rosenkranz HS, Klopman G. Relationships between electronegativity and genotoxicity. *Mutat Res* 1995 May;328(2):215–27.

337. Matthews EJ, Kruhlak NL, Weaver JL, Benz RD, Contrera JF. Assessment of the health effects of chemicals in humans: II. Construction of an adverse effects database for QSAR modeling. *Curr Drug Discov Technol* 2004 Dec;1(4): 243–54.

338. Rosenkranz HS, Klopman G. Structural relationships between mutagenicity, maximum tolerated dose, and carcinogenicity in rodents. *Environ Mol Mutagen* 1993;21(2):193–206.

339. Japertas P, Didziapetris R, Petrauskas A. Fragmental methods in the analysis of biological activities of diverse compound sets. *Mini Rev Med Chem* 2003;3(8): 797–808.

340. Zmuidinavicius D, Japertas P, Petrauskas A, Didziapetris R. Progress in toxinformatics: The challenge of predicting acute toxicity. *Curr Top Med Chem.* 2003;3(11):1301–14.

341. Bicciato, S. Artificial neural network technologies to identify biomarkers for therapeutic intervention. *Curr Opin Mol Ther* 2004;6(6):616–23.

342. Islaih M, Li B, Kadura IA, Reid-Hubbard JL, Deahl JT, et al. Comparison of gene expression changes induced in mouse and human cells treated with direct-acting mutagens. *Environ Mol Mutagen* 2004;44(5):401–19.

343. Kramer S, Helma, C. Machine learning and data mining. In: Helma C, editor, *Predictive toxicology*. Philadelphia: Taylor and Francis, 2005. p. 223–54.

344. Helma C. Lazy structure-activity relationships (lazar) for the prediction of rodent carcinogenicity and *Salmonella* mutagenicity. *Mol Divers* 2006 May;10(2): 147–58.

8

CURRENT QSAR TECHNIQUES FOR TOXICOLOGY

YU ZONG CHEN, CHUN WEI YAP, AND HU LI

Contents

Computational Toxicology: Risk Assessment for Pharmaceutical and Environmental Chemicals,
Edited by Sean Ekins

8.1 INTRODUCTION

A quantitative structure-activity relationship (QSAR) is a mathematical model used to establish an approximate relationship between a biological property of a compound and its structure-derived physicochemical and structural features [1]. The two main objectives of QSAR are to allow prediction of the biological properties of chemically characterized compounds that are not yet biologically tested and to obtain information on the molecular characteristics of a compound that are important for the biological properties.

The process of development of a QSAR model starts with the collection of relevant biological data and the elimination of low-quality data that are likely to affect the quality of the model. The next step is the selection of representative compounds into a training set and a validation set to calibrate and evaluate the QSAR model respectively. Molecular descriptors are then computed for representing the physicochemical and structural properties of the compounds studied, and those that are redundant or contain little information are removed prior to the modelling process. A data analysis method, such as multiple linear regression or neural networks, is then used to develop a model that relates the biological property to the physicochemical and structural properties of the compounds. During the modeling process, optimization of the essential parameters of the data analysis methods and the selection of relevant descriptor subsets are conducted simultaneously. The optimum set of parameters and descriptor subset are used to construct a final QSAR model, which is subsequently subjected to evaluation by one or more of the various validation methods to ensure that the constructed model is valid and useful.

This chapter describes the algorithms of the various data analysis methods currently used for developing toxicological QSAR models. Data collection, data pre-processing, computation and selection of molecular descriptors, and model validation have been extensively reviewed elsewhere [2–11], so they are not described here. Freely available online software and commercial software available for constructing QSAR models of various toxicological properties prediction are also discussed.

8.2 DATA ANALYSIS METHODS FOR REGRESSION PROBLEMS

8.2.1 Multiple Linear Regression

Multiple linear regression (MLR) is one of the most common and simplest method for QSAR modeling. MLR has been used for the prediction of genotoxicity [12], *Daphnia magna* toxicity [13], photobacterium phosphoreum toxicity [14], aquatic toxicity [15], eye irritation [16], mutagenicity, and carcinogenicity [17]. An MLR model assumes that there is a linear relationship between the molecular descriptors of a compound, which is usually expressed as a feature vector **x** (with each descriptor as a component of this vector), and its target property, *y*. An MLR model can be described using the following equation:

$$\hat{y} = \beta_0 + \beta_1 X_1 + \beta_2 X_2 + \ldots + \beta_k X_k, \tag{8.1}$$

where $\{X1, \ldots, Xk\}$ are molecular descriptors, $\beta 0$ is the regression model constant, $\beta 1$ to βk are the coefficients corresponding to the descriptors $X1$ to Xk. The values for $\beta 0$ to βk are chosen by minimizing the sum of squares of the vertical distances of the points from the hyperplane so as to give the best prediction of *y* from **x**.

The advantages of MLR are that it is simple to use and the derived models are easy to interpret. The sign of the coefficients $\beta 1$ to βk shows whether the molecular descriptors contribute positively or negatively to the target property, and their magnitudes indicates the relative importance of the descriptors to the target property. However, MLR has some disadvantages. The molecular descriptors should be mathematically independent (orthogonal) of one another and the number of compounds in the training set should exceed the number of molecular descriptors by at least a factor of 5 [18]. Studies have shown that collinear descriptors may result in the coefficients $\beta 1$ to βk being larger than expected or have the wrong sign [11]. The assumption of a linear relationship between the molecular descriptor and target property may not be appropriate, especially for toxicological properties where multiple mechanisms may be involved in determining a particular toxicological property [2]. A variety of factors may interact in complex ways to affect the toxicological property of a compound. Therefore methods based only on linear relationships are not always the most efficient approach for constructing a QSAR model for toxicological prediction.

8.2.2 Partial Least Squares

Partial least squares (PLS) is similar to MLR in that it also assumes a linear relationship between a vector **x** and a target property *y*. However, it avoids the problems of collinear descriptors by calculating the principal components for the molecular descriptors and target property separately. The scores for the molecular descriptors are used as the feature vector **x** and are also used to predict the scores for the target property, which can in turn be used to predict *y*. An important consideration in PLS is the appropriate number of principal components to be used for the QSAR model. This is usually determined by using cross-validation methods like fivefold cross validation and leave-one-out. PLS has been applied to the prediction of carcinogenicity [19], fathead minnow toxicity [20], *Tetrahymena pyriformis* toxicity [21], mammalian toxicity [22], and *Daphnia magna* toxicity [23].

Comparative molecular field analysis (CoMFA) [24] is a popular 3D-QSAR technique that uses PLS as the data analysis method. In CoMFA, compounds are aligned to a common substructure, and the magnitudes of the steric and electrostatic fields of each compound are sampled at regular intervals and used as molecular descriptors. The number of molecular descriptors is usually much larger than the number of compounds used in the training set, and the descriptors may be highly correlated with one another. Thus PLS is a suitable data analysis method as it can reduce the number of descriptors to a few principal components. An advantage of CoMFA is that the derived QSAR models can be easily visualized using molecular graphics software, so they provide useful guides to medicinal chemists for designing new compounds without the toxicological property. CoMFA has been used in studies of genotoxicity [25], ecotoxicology [26], and mutagenicity [27] as well as many other areas.

8.2.3 Feedforward Backpropagation Neural Network

Feedforward backpropagation neural network (FFBPNN) is a form of artificial neural network that has two distinct phases: forward propagation of activation and backward propagation of error [28]. It is composed of an input layer, a variable number of hidden layers, and an output layer. The input and output layers contain neurons representing the molecular descriptors and target property, respectively. In a fully connected FFBPNN each neuron in the input layer sends its value to all neurons in the first hidden layer. Each neuron in the hidden layers receives inputs from all neurons in the previous layer and computes a weighted sum of the inputs. The neuron output is determined by passing the weighted sum through a transfer function, which is usually a linear or sigmoidal function. The single neuron in the output layer determines the predicted property value by computing a weighted sum of the outputs of all neurons in the last hidden layer. Weights for the connections between neurons in adjacent layers are initially randomly assigned. These weights are then refined via a backward propagation of the error process during training of the FFBPNN.

FFBPNN has been used for predictions of acute mammalian toxicity [29], fathead minnow toxicity[30], allergic contact dermatitis [31], and genotoxicity [32]. A difficulty in using FFBPNN is the design of an optimal architecture for a given problem. An undersized network prevents optimal learning of the relationship between the descriptors and target property while an oversized network has the danger of overfitting. The connection weights of FFBPNN are not easily interpreted, and thus it is difficult for medicinal chemists to optimize the structures of compounds from a FFPBNN model.

8.2.4 General Regression Neural Network

General regression neural network (GRNN) was introduced by Donald Specht in 1991 [33], and it has been successfully used in pharmacokinetic studies, including human intestinal absorption [34], blood–brain barrier prediction [35], human serum albumin binding [35], milk–plasma ratio [35], and drug clearance [36]. Recently it has been applied for the prediction of *Tetrahymena pyriformis* toxicity [37].

For GRNN the predicted value of the target property is the most probable value, which is given by

$$\hat{y} = \frac{\int_{-\infty}^{\infty} y f(\mathbf{x}, y)dy}{\int_{-\infty}^{\infty} f(\mathbf{x}, y)dy}, \tag{8.2}$$

where $f(\mathbf{x}, y)$ is the joint density. Joint density can be estimated by using Parzen's nonparametric estimator [38]:

$$g(x) = \frac{1}{n\sigma} \sum_{i=1}^{n} W\left(\frac{x - x_i}{\sigma}\right), \tag{8.3}$$

where n is the sample size, σ is a scaling parameter that defines the width of the bell curve that surrounds each compound, $W(d)$ is a weight function that has its largest value at $d = 0$, and $(x - x_i)$ is the distance between a given compound and a compound in the training set. The Parzen's nonparametric estimator was later expanded by Cacoullos [39] for the multivariate case:

$$g(x_1, \ldots, x_p) = \frac{1}{n\sigma_1 \ldots \sigma_p} \sum_{i=1}^{n} W\left(\frac{x_1 - x_{1,i}}{\sigma_1}, \ldots, \frac{x_p - x_{p,i}}{\sigma_p}\right). \tag{8.4}$$

The Gaussian function is frequently used as the weight function because it is well behaved, easily calculated, and satisfies the conditions required by Parzen's estimator. Thus the probability density function for the multivariate case becomes

$$g(\mathbf{x}) = \frac{1}{n} \sum_{i=1}^{n} \exp\left(-\sum_{j=1}^{p} \left(\frac{x_j - x_{j,i}}{\sigma_j}\right)^2\right). \tag{8.5}$$

To simplify the equation, a single σ that is common to all the descriptors (single-sigma model) can be used instead of an individual σ for each descriptor (multi-sigma model). Single-sigma models can be computed faster, and they produce reasonable models when all the descriptors are of approximately equal importance. However, multi-sigma models are more general than a single-sigma model, and they are useful when descriptors are of a different nature and importance [40].

Substituting Parzen's nonparametric estimator for $f(\mathbf{x}, y)$ and performing the integrations leads to the fundamental equation of GRNN:

$$\hat{y} = \frac{\sum_{i=1}^{n} y_i \exp(-D(\mathbf{x}, \mathbf{x}_i))}{\sum_{i=1}^{n} \exp(-D(\mathbf{x}, \mathbf{x}_i))}, \tag{8.6}$$

where

$$D(\mathbf{x}, \mathbf{x}_i) = \sum_{j=1}^{p} \left(\frac{x_j - x_{j,i}}{\sigma_j} \right)^2. \tag{8.7}$$

GRNN can be implemented as a neural network [40]. The network architecture of a GRNN is determined by the number of compounds and descriptors in the training set. There are four layers in a GRNN. The input layer provides input values to all neurons in the pattern layer, and it has as many neurons as the number of descriptors in the training set. The number of pattern neurons is determined by the total number of compounds in the training set. Each pattern neuron computes a distance measure between the input compound and the training compound represented by that neuron and then subjects the distance measure to the Parzen's nonparameteric estimator. The summation layer has two neurons that calculate the numerator and denominator of equation (8.6). The single neuron in the output layer then performs a division of the two summation neurons to obtain the predicted absorption, distribution, metabolism, excretion, and toxicity (ADME/Tox) values of the given compound.

8.3 DATA ANALYSIS METHODS FOR CLASSIFICATION PROBLEMS

8.3.1 Linear Discriminant Analysis

Linear discriminant analysis (LDA) [41] separates two data classes of feature vectors by constructing a hyperplane defined by a linear discriminant function:

$$L = \sum_{i}^{k} w_i x_i, \tag{8.8}$$

where L is the resultant classification score and w_i is the weight associated with the corresponding descriptor x_i. A positive or negative L value indicates that a feature vector **x** belongs to the positive or negative data class, respectively. In the area of toxicological prediction, LDA has been used for predicting *Tetrahymena pyriformis* toxicity [42], genotoxicity [43], and ecotoxicity [44].

8.3.2 Logistic Regression

Logistic regression (LR) [45] is based on the assumption that there is a logistic relationship between the probability of data class membership and one or more descriptors. The probability can be calculated by using

$$\hat{y} = \frac{1}{1 + e^{-(\beta 0 + \beta 1 X1 + \beta 2 X2 + \ldots + \beta k Xk)}}, \tag{8.9}$$

where $\{X1, \ldots, Xk\}$ are molecular descriptors, $\beta 0$ is the regression model constant, $\beta 1$ to βk is the coefficients corresponding to the descriptors $X1$ to Xk. $y > 0.5$ or $y < 0.5$ indicates that the vector **x** belongs to the positive or negative data class respectively. LR has been used for predicting *Hyalella azteca* toxicity [46], *Tetrahymena pyriformis* toxicity [42], and contact sensitization [47].

8.3.3 Decision Tree

A decision tree (DT) is a branch-test-based classifier [48]. DT modeling has been used for the prediction of ecotoxicity [44], mutagenicity [49], and genotoxicity [50]. A branch in a decision tree corresponds to a group of data classes, and a leaf represents a specific data class. A decision node specifies a test to be conducted on a single descriptor value, with one branch and its subsequent data classes as possible outcomes of the test. A given compound with vector **x** is classified by starting at the root of the tree and moving through the tree until a leaf is encountered. At each nonleaf decision node, a test is conducted and the classification process proceeds to the branch selected by the test. Upon reaching the destination leaf, the data class of the given compound is predicted to be that associated with the leaf.

The algorithm is a recursive greedy heuristic that selects descriptors for membership within the tree. It uses recursive partitioning to examine every descriptor of the compounds in the training set and ranks them according to their ability to partition the remaining compounds, thereby constructing a decision tree. Whether or not a descriptor is included within the tree is based on the value of its information gain. As a statistical property, information gain measures how well the descriptor separates training cases into subsets in which the data class is homogeneous. For descriptors with continuous values, a threshold value has to be established within each descriptor so that it could partition the training cases into subsets. These threshold values for each descriptor are established by rank ordering the values within each descriptor from lowest to highest and repeatedly calculating the information gain using the arithmetical midpoint between all successive values within the rank order. The midpoint value with the highest information gain is selected as the threshold value for the descriptor. That descriptor with the highest information gain (information being the most useful for classification) is then selected for inclusion in the DT. The algorithm continues to build the tree in this manner until it accounts for all training cases. Ties between descriptors that were equal in terms of information gain are broken randomly [51].

8.3.4 *k*-Nearest Neighbor

k-Nearest neighbor (*k*NN) is a basic instance-based method and was introduced by Evelyn Fix and Joseph L. Hodges, Jr. [52]. It has been used for the prediction of ecotoxicity [44] and genotoxicity [50]. *k*NN measures the Euclidean distance between a given compound with feature vector **x** and each compound in the training set with individual feature vector $\mathbf{x}_i$ [52,53]. The Euclidean distances for the feature vector pairs are calculated using the following formula:

$$D = \sqrt{\|\mathbf{x} - \mathbf{x}_i\|^2}. \tag{8.10}$$

A total of *k* number of training compounds nearest to the given compound is used to determine its data class:

$$y = \arg\max_{v \in V} \sum_{i=1}^{k} \delta(v, y_i), \tag{8.11}$$

where $\delta(a, b) = 1$ if $a = b$ and $\delta(a, b) = 0$ if $a \neq b$, argmax is the maximum of the function, V is a finite set of data classes. k is usually an odd number to prevent ambiguity in the estimation of y.

8.3.5 Probabilistic Neural Network

Probabilistic neural network (PNN) is similar to GRNN except that it is used for classification problems [54]. It has been used for pharmacodynamics [55], pharmacokinetics [34,56] studies and has recently been applied for genotoxicity [43,50,57] and torsade de pointes prediction [58]. PNN classifies compounds into their data class through the use of Bayes's optimal decision rule:

$$h_i c_i f_i(\mathbf{x}) > h_j c_j f_j(\mathbf{x}), \tag{8.12}$$

where h_i and h_j are the prior probabilities, c_i and c_j are the costs of misclassification and $f_i(\mathbf{x})$ and $f_j(\mathbf{x})$ are the probability density function for data class i and j, respectively. A given compound with vector **x** is classified into data class i if the product of all the three terms is greater for data class i than for any other data class j not equal to i. In most applications the prior probabilities and costs of misclassifications are treated as being equal. The probability density function for each data class for a univariate case can be estimated by the Parzen's nonparametric estimator (equation 8.3 or 8.4).

The network architecture of a PNN (Figure 8.1) is similar to that of a GRNN, except that its summation layer has a neuron for each data class and the neurons sum all the pattern neurons' output corresponding to members of that summation neuron's data class to obtain the estimated probability density function for that data class. The single neuron in the output layer then

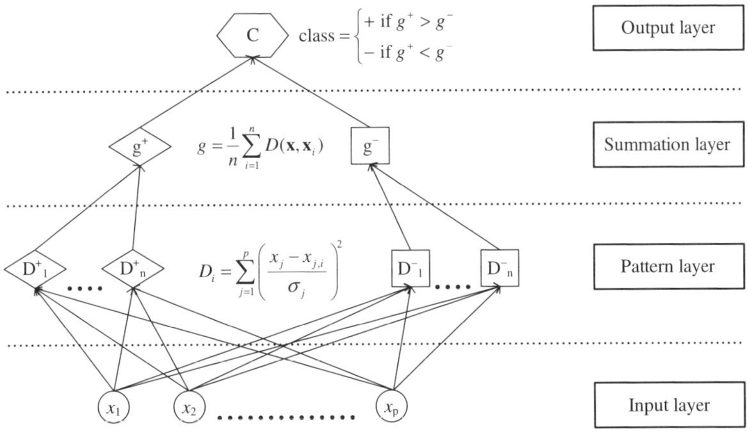

Figure 8.1 PNN architecture.

determines the final data class of the input compound after comparing all the probability density functions from the summation neurons and choosing the data class with the highest value for the probability density function.

8.3.6 Support Vector Machine

Support vector machine (SVM) is based on the structural risk minimization principle from statistical learning theory [59–61]. SVM has been used for prediction of torsade de pointes [58], genotoxicity [50], and *Tetrahymena pyriformis* prediction [37], as well as prediction of P450 isoenzyme substrates and inhibitors [62–64], P-glycoprotein substrates [56], blood–brain barrier penetration [65,66], and human intestinal absorption [67]. In linearly separable cases, SVM constructs a hyperplane that separates two data classes of compounds with a maximum margin. This is accomplished by finding another vector **w** and a parameter b that minimizes $||\mathbf{w}||^2$ and satisfies the following conditions:

$$\mathbf{w}\cdot\mathbf{x}_i + b \geq +1, \text{ for } y_i = +1 \quad \text{Class 1,} \tag{8.13}$$

$$\mathbf{w}\cdot\mathbf{x}_i + b \leq -1, \text{ for } y_i = -1 \quad \text{Class 2,} \tag{8.14}$$

where y_i is the data class index of compound i, $\mathbf{x}_i$ is a vector of molecular descriptors for compound i, **w** is a vector normal to the hyperplane, $|b|/||\mathbf{w}||$ is the perpendicular distance from the hyperplane to the origin, and $||\mathbf{w}||^2$ is the Euclidean norm of **w**. After the determination of **w** and b, a given compound with vector **x** can be classified by

$$\hat{y} = \text{sign}[(\mathbf{w}\cdot\mathbf{x}) + b]. \tag{8.15}$$

In nonlinearly separable cases, SVM maps the vectors into a higher dimensional feature space using a kernel function $K(\mathbf{x}_i, \mathbf{x}_j)$. The Gaussian radial basis function kernel has been extensively used in a number of different studies with good results [66,68,69]. Linear SVM is applied to this feature space and then the decision function is given by

$$\hat{y} = \text{sign}\left(\sum_{i=1}^{l} \alpha_i^0 y_i K(\mathbf{x}, \mathbf{x}_i) + b \right), \tag{8.16}$$

where l is the number of support vectors. The coefficients α_i^0 and b are determined by maximizing the Langrangian expression

$$\sum_{i=1}^{l} \alpha_i - \frac{1}{2} \sum_{i=1}^{l} \sum_{j=1}^{l} \alpha_i \alpha_j y_i y_j K(\mathbf{x}_i, \mathbf{x}_j) \tag{8.17}$$

under the following conditions:

$$0 \le \alpha_i \le C, \tag{8.18}$$

$$\sum_{i=1}^{l} \alpha_i y_i = 0, \tag{8.19}$$

where C is a penalty for training errors. A positive or negative value from equation (8.16) indicates that the compound with vector $\mathbf{x}$ belongs to the positive or negative data class.

The main advantage of SVM over other data analysis methods is its relatively low sensitivity to data overfitting, even with the use of a large number of redundant and overlapping molecular descriptors. This is due to its reliance on the structural risk minimization principle. Another advantage of SVM is the ability to calculate a reliability score, *R*-value, which provides a measure of the probability of a correct classification of a compound [70]. The *R*-value is computed by using the distance between the position of the compound and the hyperplane in the hyperspace. The expected classification accuracy for the compound can then be obtained from the *R*-value by using a chart which shows the statistical relationship between them. As with other methods, SVM requires a sufficient number of samples to develop a classification system and irrelevant molecular descriptors may reduce the prediction accuracies of the SVM classification systems.

8.4 AVAILABLE SOFTWARE FOR TOXICOLOGICAL PREDICTION

8.4.1 Software for QSAR Development

A number of commercial and free software are available for developing QSAR. Some of these software specialize in a particular data analysis method while others contain a number of data analysis methods. Two good resources for these software are The QSAR and Modelling Society (*http://www.qsar.org*) and Cheminformatics (*http://www.cheminformatics.org*) Web sites.

A disadvantage of the majority of these software is that they do not have the capability to calculate molecular descriptors. Thus additional software, such as DRAGON [71], Molconn-Z [72], or MODEL [73], are needed to enable toxicological QSAR models to be built.

C4.5 C4.5 (*http://www2.cs.uregina.ca/~dbd/cs831/notes/ml/dtrees/c4.5/tutorial.html*) is software to generate C4.5 decision trees, which are an extension of the basic ID3 algorithm [48]. C4.5 decision trees avoid overfitting of the data by determining how deeply to grow a decision tree, and they are capable of handling training sets that contain continuous descriptors with missing values. C4.5 decision trees also have reduced error pruning and have rule postpruning.

SVM^{light} SVM^{light} (*http://svmlight.joachims.org/*) is an implementation of SVM for classification and regression problems [74]. It has a fast optimization algorithm and can handle training sets with thousands of support vectors efficiently. The software assesses the generalization performance of a QSAR model by efficiently computing the leave-one-out estimates of the error rate, precision and recall. SVM^{light} provides all the standard kernel functions and allows the user to define additional kernels. Although the software does not have a graphical user interface, various extensions that provide a MATLAB interface and Java interface are available. The software is free for academic users.

ELECTRAS Electronic Data Analysis Service (ELECTRAS) (*http://www2.chemie.uni-erlangen.de/projects/eDAS/index.html*) is a Web-based application for developing QSAR models [75]. Currently it contains several data analysis methods like MLR, PLS, and FFBPNN and has a small module for calculating chemical descriptors such as total charges, topological autocorrelation, and 3D autocorrelation. The descriptors can be weighted, scaled and transformed by using various transformations such as Fourier transform and wavelet transform.

Weka Weka (*http://www.cs.waikato.ac.nz/ml/weka/*) is a collection of machine learning algorithms for data-mining tasks written in Java. It contains tools for data pre-processing, classification, regression, clustering, association rules, and visualization [76]. Weka is open source software issued under the GNU general public license. It is organized in a hierarchy of packages, and each package contains a collection of related classes. There are packages for core components, associations, attribute selection, classifiers, clustering, estimators, filters, experiments, and graphical user interface.

PHAKISO PHAKISO (*http://www.phakiso.com/*) is a Microsoft Windows software that uses YMLL, a machine-learning library. YMLL contains algorithms that are essential for performing QSAR experiments [77]. These include data analysis methods for both classification and regression problems, clustering algorithms, outlier detection algorithms, statistical molecular design algorithms, methods for dataset diversity measurement, descriptor scaling algorithms, descriptor selection algorithms, and model validation methods. In addition PHAKISO provides algorithms to automatically estimate the values of descriptors with missing values and performs principal component analysis of the training set. Both PHAKISO and YMLL are free for noncommercial uses.

NeuroSolutions NeuroSolutions (*http://www.nd.com/*) is a powerful commercial neural network modeling software that provides an icon-based graphical user interface and intuitive wizards to enable users to build and train neural networks easily [78]. It has a large selection of neural network architectures, which includes FFBPNN, GRNN, PNN, and SVM. A genetic algorithm is also provided to automatically optimize the settings of the neural networks.

MATLAB MATLAB (*http://www.mathworks.com/*) provides an interactive environment for algorithm development, data visualization, data analysis, and numeric computation. QSAR models can be easily built and tested using various add-on toolboxes like the Statistics Toolbox, Neural Network Toolbox, and PLS Toolbox [79]. The developed models and algorithms can be easily integrated with other applications or programming languages through specially provided functions. This enables QSAR models to be easily distributed as stand-alone programs or software modules.

Sybyl Sybyl (*http://www.tripos.com/*) is a collection of computational informatics software that provides comprehensive tools for molecular modeling, visualization of structures and associated data, annotation, and a wide range of force fields [80]. Sybyl has been used for ligand-based design, receptor-based design, structural biology, library design, and cheminformatics. It can be used to create QSAR models easily by way of the integrated CoMFA module, and different QSAR models can be compared statistically and visually.

DRAGON DRAGON (*http://www.talete.mi.it/dragon_exp.htm*) is a software program for the calculation of a large number of molecular descriptors [81]. Currently it can calculate 1664 molecular descriptors that can be classified into 20 classes, which include constitutional descriptors, topological descriptors, walk and path counts, connectivity indexes, information indexes, 2D autocorrelations, edge adjacency indexes, BCUT descriptors, topological charge indexes, eigenvalue-based indexes, Randic molecular profiles, geometrical descriptors, RDF descriptors, 3D-MoRSE descriptors, WHIM descriptors, GETAWAY descriptors, functional group counts, atom-centered fragments, charge descriptors, and molecular properties. This software also has a basic QSAR calculation capability.

Molconn-Z Molconn-Z (*http://www.edusoft-lc.com/molconn/*) is commercial software that calculates molecular connectivity, shape, and information indexes such as molecular connectivity chi indexes, kappa shape indexes, electrotopological state indexes, molecular connectivity difference chi indexes, topological indexes, counts of subgraphs, and vertex eccentricities [82]. The software is available for multiple platforms, which include UNIX/LINUX, Windows, and Mac OS-X.

MODEL Molecular Descriptor Lab (MODEL) (*http://jing.cz3.nus.edu.sg/cgi-bin/model/model.cgi*) is a free Web-based server for computing a comprehensive set of 3778 molecular descriptors, which can be divided into 6 classes: constitutional descriptors, electronic descriptors, physical chemistry properties, topological indexes, geometrical molecular descriptors, and quantum chemistry descriptors [73]. Compounds can be provided to the server in various molecular formats such as PDB, MDL, MOL2, and COR, and the computed molecular descriptors are displayed in a few seconds or less. Cross-links to the relevant sections of the reference manual page are also provided for some of the descriptors and descriptor classes.

8.4.2 Software for Specific Toxicological Properties

DEREK Deductive Estimation of Risk from Existing Knowledge (DEREK) (*http://www.lhasalimited.org/derek/index.html*) is a knowledge-based expert system (see also Chapter 18) that provides a high-throughput screen for genotoxicity/mutagenicity, carcinogenicity, skin sensitizers, and other potential toxicological hazards like irritancy and hepatotoxicity. The program applies QSAR and expert knowledge rules, such as chemical substructures that have been implicated in toxic effects, to predict the potential toxicity of a query compound. It also provides supporting evidence, such as comments, literature references, and toxicity data, for its prediction so that users can make their own judgments [83].

MCASE Multiple Computer Automated Structure Evaluation (MCASE) (*http://www.multicase.com*) automatically identifies biophores that are essential for activity from training sets of diverse compounds (see also Chapter 18). It then creates organized dictionaries of these biophores and develops local QSAR for the prediction of query compounds. Currently MCASE contains more than 180 models covering various areas of toxicology, such as acute toxicity in mammals, adverse effects, carcinogenicity, cytotoxicity, teratogenicity, ecotoxicity, mutagenicity, and skin/eye irritations [84].

TOPKAT Toxicity Prediction by Komputer Assisted Technology (TOPKAT) (*http://www.accelrys.com/products/topkat/index.html*) uses QSAR models for prediction of various toxicological properties such as mutagenicity, developmental toxicity potential, carcinogenicity, and skin/eye irritancy (see also Chapter 18). It employs the Optimum Prediction Space (OPS) technology to assess whether the query compound is well represented in its QSAR models and provides a confidence level on its prediction [85].

ECOSAE Ecological Structure Activity Relationships (ECOSAE) (*http://www.epa.gov/oppt/newchems/tools/21ecosar.htm*) predicts the toxicity of industrial chemicals to aquatic organisms such as fish, invertebrates, and algae, and estimates a compound's acute and chronic toxicity. Its QSAR database contains more than 100 models developed for 42 chemical classes [86].

HazardExpert HazardExpert (*http://www.compudrug.com/*) uses a rule-based system derived from toxic fragments to provide an estimation of toxic symptoms of compounds in human and animals (see also Chapter 18). It can make predictions for seven different toxicity classes, which include oncogenicity, mutagenicity, teratogenicity, membrane irritation, sensitivity, immunotoxicity, and neurotoxicity. ToxAlert is based on HazardExpert and can give additional information such as probability percentages for the different toxicity classes [87].

CSGenoTox CSGenoTox (*http://www.chemsilico.com/CS_prGT/GThome.html*) contains a single QSAR model that is based on topological structural descriptors and developed by using artificial neural networks. It can be used for the prediction of the mutagenic index of a query compound [88].

MetaDrug MetaDrug (*http://www.genego.com/about/products.shtml*) is a toxicogenomics platform for the prediction of the toxicity of novel compounds. It assesses toxicity by generating networks around proteins, genes, and the query compound. In addition it allows visualization of pre-clinical and clinical high-throughput date in the context of the complete biological system [89].

8.5 CONCLUSION

This chapter has introduced several data analysis methods that are commonly used for developing toxicological QSAR models. A list of the advantages and disadvantages of these methods are given in Table 8.1. Newer data analysis methods like SVM and GRNN are often applicable to a wide range of problems because they do not make any assumption about the type of relationship between the target property and the molecular descriptors. However, their models are more complex than those of traditional data analysis methods such as MLR and LR, and thus are more difficult to interpret. Hence an important

TABLE 8.1 Advantages and Disadvantages of Different Data Analysis Methods

Methods	Advantages	Disadvantages
MLR	Simple to use Models are easy to interpret	Molecular descriptors should be orthogonal to one another Number of compounds in the training set should exceed the number of molecular descriptors by at least a factor of 5 Assumes a linear relationship between target property and molecular descriptors
PLS	Able to model multiple target properties simultaneously Able to use collinear descriptors	Difficulty in interpreting scores for the molecular descriptors Assumes a linear relationship between target property and molecular descriptors
FFBPNN	Does not make any assumption of the type of relationship between target property and molecular descriptors	Models are difficult to interpret Difficult to design an optimal architecture Risk of overfitting
GRNN	Does not make any assumption of the type of relationship between target property and molecular descriptors Network architecture is simpler than FFBPNN Fast training time	Models are difficult to interpret Prediction speed may be slow with large training sets Does not extrapolate well
LDA	Simple to use Models are easy to interpret	Requires a balanced training set Binary classification only
LR	Provides a probability of target class membership	Models are difficult to interpret Assumes a logistic relationship between target property and molecular descriptors Binary classification only
DT	Does not make any assumption of the type of relationship between target property and molecular descriptors Models are easy to interpret Fast classification speed Multi-class classification	May have over fitting when training set is small and number of molecular descriptors is large Ranks molecular descriptors using information gain which may not be the best for some problems

TABLE 8.1 (*Continued*)

Methods	Advantages	Disadvantages
*k*NN	Docs not make any assumption of the type of relationship between target property and molecular descriptors Fast training time Multi-class classification	Models are difficult to interpret Classification speed may be slow with large training sets Classification is sensitive to the type of distance measures used
PNN	Does not make any assumption of the type of relationship between target property and molecular descriptors Fast training time Multi-class classification	Models are difficult to interpret Classification speed may be slow with large training sets
SVM	Does not make any assumption of the type of relationship between target property and molecular descriptors Low risk of over fitting Able to provide expected classification accuracies for individual compounds	Models are difficult to interpret Training speed may be slow with large training sets Predominantly binary classification only

Note: MLR: multiple linear regression; PLS: partial least squares; FFBPNN: feedforward back-propagation neural network; GRNN: general regression neural network; LDA: linear discriminant analysis; LR: logistic regression; DT: decision trees; *k*NN: *k*-nearest neighbor; PNN: probabilistic neural network; SVM: support vector machine.

area of research is the development of techniques to interpret these complex models. Current progress in this area includes the introduction of a weighting function to the molecular descriptors for SVM [90] and the use of a principal component analysis based method to perform a functional dependence study of a GRNN model [35].

Another important area of QSAR research is the determination of the reliability of QSAR models. Current research in this area includes the development of methods to define the applicability domain of a QSAR model [91] and to calculate the expected prediction accuracies for individual compounds [70].

The development of new data analysis methods is also an important area of QSAR research. Several methods have been developed in recent years, and these include kernel partial least squares (K-PLS) [92], robust continuum regression [93], local lazy regression [94], fuzzy interval number *k*-nearest neighbor (FINkNN) [95], and fast projection plane classifier (FPPC) [96]. These methods have been shown to be useful for the prediction of a wide variety of target properties, which include moisture, oil, protein and starch values of corn [92], output of a polymer processing plant [92], chaotic Mackey-Glass time-series [92], human signal detection performance monitoring [92], X-ray analysis of hydrometallugical solutions [93], activities of artemisinin analogues [94], activities of platelet-derived growth factor inhibitors [94], activities of dihydrofolate reductase inhibitors [94], sugar production [95], and classification of iris plants [96]. Thus they are potentially useful for toxicological prediction studies.

REFERENCES

1. Johnson MA, Maggiora GM. *Concepts and applications of molecular similarity*. New York: Wiley, 1990.
2. Cronin MTD, Schultz TW. Pitfalls in QSAR. *J Mol Struct Theochem* 2003; 622:39–51.
3. Susnow RG, Dixon SL. Use of robust classification techniques for the prediction of human cytochrome P450 2D6 inhibition. *J Chem Inf Comput Sci* 2003; 43:1308–15.
4. Wold S, Eriksson L. Statistical validation of QSAR results. In: van de Waterbeemd H, editor, *Chemometric methods in molecular design*. Weinheim: VCH, 1995. p. 309–18.
5. Gramatica P, Pilutti P, Papa E. Validated QSAR prediction of OH tropospheric degradation of VOCs: Splitting into training-test sets and consensus modeling. *J Chem Inf Comput Sci* 2004;44:1794–802.
6. Schultz TW, Netzeva TI, Cronin MTD. Selection of data sets for QSARs: analyses of *Tetrahymena* toxicity from aromatic compounds. *SAR QSAR Environ Res* 2003;14:59–81.
7. Rajer-Kanduc K, Zupan JM, N. Separation of data on the training and test set for modelling: A case study for modelling of five colour properties of a white pigment. *Chemom Intell Lab Sys* 2003;65:221–9.
8. Todeschini R, Consonni V. *Handbook of molecular descriptors*. Weinheim: Wiley-VCH, 2000.
9. Livingstone DJ. *Data analysis for chemists: Applications to QSAR and chemical product design*. Oxford: Oxford University Press, 1995.
10. Guyon I, Elisseeff A. An introduction to variable and feature selection. *J Mach Learn Res* 2003;3:1157–82.
11. Eriksson L, Jaworska J, Cronin M, Worth A, Gramatica P, McDowell R. Methods for reliability and uncertainty assessment and for applicability evaluations of classification- and regression-based QSARs. *Environ Health Perspect* 2003;111: 1361–75.
12. Cash GG. Prediction of the genotoxicity of aromatic and heteroaromatic amines using electrotopological state indices. *Mutat Res* 2001;491:31–7.
13. He YB, Wang LS, Liu ZT, Zhang Z. Acute toxicity of alkyl (1-phenylsulfonyl) cycloalkane-carboxylates to *Daphnia magna* and quantitative structure-activity relationships. *Chemosphere* 1995;31:2739–46.

14. Sixt S, Altschuh J, Brueggemann R. Quantitative structure-toxicity relationships for 80 chlorinated compounds using quantum chemical descriptors. *Chemosphere* 1995;30:2397–414.
15. Papa E, Villa F, Gramatica P. Statistically validated QSARs, based on theoretical descriptors, for modeling aquatic toxicity of organic chemicals in *Pimephales promelas* (fathead minnow). *J Chem Inf Model* 2005;45:1256–66.
16. Kulkarni AS, Hopfinger AJ. Membrane-interaction QSAR analysis: Application to the estimation of eye irritation by organic compounds. *Pharm Res* 1999; 16:1244–52.
17. Benigni R, Guiliani A, Franke R, Gruska A. Quantitative structure-activity relationships of mutagenic and carcinogenic aromatic amines. *Chem Rev* 2000; 100:3697–714.
18. Topliss JG, Edwards RP. Chance factors in studies of quantitative structure-activity relationships. *J Med Chem* 1979;22:1238–44.
19. Sun HM. Prediction of chemical carcinogenicity from molecular structure. *J Chem Inf Comput Sci* 2004;44:1506–14.
20. Öberg T. A QSAR for baseline toxicity: Validation, domain of application, and prediction. *Chem Res Toxicol* 2004;17:1630–7.
21. Cronin MTD, Schultz TW. Development of quantitative structure-activity relationships for the toxicity of aromatic compounds to *Tetrahymena pyriformis*: Comparative assessment of the methodologies. *Chem Res Toxicol* 2001;14:1284–95.
22. Devillers J. Prediction of mammalian toxicity of organophosphorus pesticides from QSTR modeling. *SAR QSAR Environ Res* 2004;15:501–10.
23. Lo Piparo E, Fratev F, Lemke F, Mazzatorta P, Smiesko M, Fritz JI, et al. QSAR models for *Daphnia magna* toxicity prediction of benzoxazinone allelochemicals and their transformation products. *J Agric Food Chem* 2006;54:1111–5.
24. Cramer RD, Patterson DE, Bunce JD. Recent advances in comparative molecular field analysis (CoMFA). *Prog Clin Biol Res* 1989;291:161–5.
25. Debnath AK, Hansch C, Kim KH, Martin YC. Mechanistic interpretation of the genotoxicity of nitrofurans (antibacterial agents) using quantitative structure-activity relationships and comparative molecular field analysis. *J Med Chem* 1993;36:1007–16.
26. Briens F, Bureau R, Rault S, Robba M. Applicability of CoMFA in ecotoxicology: A critical study on chlorophenols. *Ecotoxicol Environ Saf* 1995;31:37–48.
27. Fan M, Byrd C, Compadre CM, Compadre RL. Comparison of CoMFA models for Salmonella typhimurium TA98, TA100, TA98 + S9 and TA100 + S9 mutagenicity of nitroaromatics. *SAR QSAR Environ Res* 1998;9:187–215.
28. Wythoff BJ. Backpropagation neural networks. A tutorial. *Chemom Intell Lab Sys* 1993;18:115–55.
29. Eldred DV, Jurs PC. Prediction of acute mammalian toxicity of organophosphorus pesticide compounds from molecular structure. *SAR QSAR Environ Res* 1999; 10:75–99.
30. Eldred DV, Weikel CL, Jurs PC, Kaiser KLE. Prediction of Fathead Minnow Acute Toxicity of Organic Compounds from Molecular Structure. *Chem Res Toxicol* 1999;12:670–8.

31. Devillers J. A neural network SAR model for allergic contact dermatitis. *Toxicol Meth* 2000;10:181–93.
32. Maran U, Sild S. QSAR modeling of genotoxicity on non-congeneric sets of organic compounds. *Art Intell Rev* 2003;20:13–38.
33. Specht DF. A general regression neural network. *IEEE Trans* on *Neural Netw* 1991;2:568–76.
34. Niwa T. Using general regression and probabilistic neural networks to predict human intestinal absorption with topological descriptors derived from two-dimensional chemical structures. *J Chem Inf Comput Sci* 2003;43:113–9.
35. Yap CW, Chen YZ. Quantitative structure-pharmacokinetic relationships for drug distribution properties by using general regression neural network. *J Pharm Sci* 2005;94:153–68.
36. Yap CW, Li ZR, Chen YZ. Quantitative structure-pharmacokinetic relationships for drug clearance by using statistical learning methods. *J Mol Graph Mod* 2006;24:383–95.
37. Panaye A, Fan BT, Doucet JP, Yao XJ, Zhang RS, Liu MC, et al. Quantitative structure-toxicity relationships (QSTRs): A comparative study of various non linear methods. General regression neural network, radial basis function neural network and support vector machine in predicting toxicity of nitro- and cyano-aromatics to Tetrahymena pyriformis. *SAR QSAR Environ Res* 2006; 17:75–91.
38. Parzen E. On estimation of a probability density function and mode. *An Math Stat* 1962;33:1065–76.
39. Cacoullos T. Estimation of a multivariate density. *An I Stat Math* 1966; 18:179–89.
40. Masters T. *Advanced algorithms for neural networks: A C++ sourcebook*. New York: Wiley, 1995.
41. Huberty C. *Applied discriminant analysis*. New York: Wiley, 1994.
42. Schuurmann G, Aptula AO, Kuhne R, Ebert RU. Stepwise discrimination between four modes of toxic action of phenols in the *Tetrahymena pyriformis* assay. *Chem Res Toxicol* 2003;16:974–87.
43. Mosier PD, Jurs PC, Custer LL, Durham SK, Pearl GM. Predicting the genotoxicity of thiophene derivatives from molecular structure. *Chem Res Toxicol* 2003; 16:721–32.
44. Mazzatorta P, Benfenati E, Lorenzini P, Vighi M. QSAR in ecotoxicity: An overview of modern classification techniques. *J Chem Inf Comput Sci* 2004;44:105–12.
45. Hosmer DW, Lemeshow S. *Applied logistic regression*. New York: Wiley, 1989.
46. Lee JH, Landrum PF, Field LJ, Koh CH. Application of a sigmapolycyclic aromatic hydrocarbon model and a logistic regression model to sediment toxicity data based on a species-specific, water-only LC_{50} toxic unit for *Hyalella azteca*. *Environ Toxicol Chem* 2001;20:2102–13.
47. Fedorowicz A, Singh H, Soderholm S, Demchuk E. Structure-activity models for contact sensitization. *Chem Res Toxicol* 2005;18:954–69.
48. Quinlan JR. *C4.5: Programs for machine learning*. San Mateo, CA: Morgan Kaufmann, 1993.

49. Young SS, Gombara VK, Emptagea MR, Carielloa NF, Lambert C. Mixture deconvolution and analysis of Ames mutagenicity data. *Chemom Intell Lab Sys* 2002;60:5–11.
50. Li H, Ung CY, Yap CW, Xue Y, Li ZR, Cao ZW, et al. Prediction of genotoxicity of chemical compounds by statistical learning methods. *Chem Res Toxicol* 2005;18:1071–80.
51. Carnahan B, Meyer G, Kuntz L-A. Comparing statistical and machine learning classifiers: Alternatives for predictive modeling in human factors research. *Hum Factors* 2003;45:408–23.
52. Fix E, Hodges JL. Discriminatory analysis: Non-parametric discrimination: Consistency properties. *Int Stat Rev* 1989;57:238–47.
53. Johnson RA, Wichern DW. *Applied multivariate statistical analysis*. Englewood Cliffs, NJ: Prentice Hall, 1982.
54. Specht DF. Probabilistic neural networks. *Neural Netw* 1990;3:109–18.
55. Mosier PD, Jurs PC. QSAR/QSPR studies using probabilistic neural networks and generalized regression neural networks. *J Chem Inf Comput Sci* 2002;42: 1460–70.
56. Xue Y, Yap CW, Sun LZ, Cao ZW, Wang JF, Chen YZ. Prediction of *p*-glycoprotein substrates by support vector machine approach. *J Chem Inf Comput Sci* 2004;44:1497–505.
57. He L, Jurs PC, Custer LL, Durham SK, Pearl GM. Predicting the genotoxicity of polycyclic aromatic compounds from molecular structure with different classifiers. *Chem Res Toxicol* 2003;16:1567–80.
58. Yap CW, Cai CZ, Xue Y, Chen YZ. Prediction of torsade-causing potential of drugs by support vector machine approach. *Toxicol Sci* 2004;79:170–7.
59. Vapnik VN. *The nature of statistical learning theory*. New York: Springer, 1995.
60. Burges CJC. A tutorial on support vector machines for pattern recognition. *Data Min Knowl Disc* 1998;2:127–67.
61. Evgeniou T, Pontil M. Support vector machines: theory and applications. In: Paliouras G, Karkaletsis V, Spyropoulos CD, editors, *Machine learning and its applications*. Advanced lectures. New York: Springer, 2001. p. 249–57.
62. Kriegl JM, Arnhold T, Beck B, Fox T. Prediction of human cytochrome P450 inhibition using support vector machines. *QSAR Comb Sci* 2005;24: 491–502.
63. Kriegl JM, Arnhold T, Beck B, Fox T. A support vector machine approach to classify human cytochrome P450 3A4 inhibitors. *J Comput Aided Mol Des* 2005;19:189–201.
64. Yap CW, Chen YZ. Prediction of cytochrome P450 3A4, 2D6, and 2C9 inhibitors and substrates by using support vector machines. *J Chem Inf Model* 2005; 45:982–92.
65. Doniger S, Hofmann T, Yeh J. Predicting CNS permeability of drug molecules: comparison of neural network and support vector machine algorithms. *J Comp Biol* 2002;9:849–64.
66. Trotter MWB, Buxton BF, Holden SB. Support vector machines in combinatorial chemistry. *Meas Control* 2001;34:235–9.

67. Xue Y, Li ZR, Yap CW, Sun LZ, Chen X, Chen YZ. Effect of molecular descriptor feature selection in support vector machine classification of pharmacokinetic and toxicological properties of chemical agents. *J Chem Inf Comput Sci* 2004;44:1630–8.

68. Burbidge R, Trotter M, Buxton B, Holden S. Drug design by machine learning: support vector machines for pharmaceutical data analysis. *Comput Chem* 2001;26:5–14.

69. Czerminski R, Yasri A, Hartsough D. Use of support vector machine in pattern classification: Application to QSAR studies. *Quant Struct-Act Rel* 2001; 20:227–40.

70. Cai CZ, Han LY, Ji ZL, Chen X, Chen YZ. SVM-Prot: Web-based support vector machine software for functional classification of a protein from its primary sequence. *Nucleic Acids Res* 2003;31:3692–7.

71. Todeschini R, Consonni V, Mauri A, Pavan M. *DRAGON*. Milan: Talete SRL, 2005.

72. Hall LH, Kellogg GE, Haney DN. *Molconn-Z*. eduSoft, LC, 2002.

73. Li ZR, Han LY, Chen YZ. *MODEL—Molecular Descriptor Lab* <http://jing.cz3.nus.edu.sg/cgi-bin/model/model.cgi>. Singapore: Bioinformatics and Drug Design Group, 2005.

74. Joachims T. Making large-scale SVM learning practical. In: Schölkopf B, Burges CJC, Smola AJ, editors, *Advances in kernel methods: Support vector learning*. Cambridge: MIT Press, 1999. p. 169–84.

75. Gasteiger J, Burkard U, Lekishvili G. ELECTRAS – Electronic Data Analysis Service <http://www2.chemie.uni-erlangen.de/projects/eDAS/index.html>. 2003.

76. Witten IH, Frank E. *Data mining: Practical machine learning tools and techniques*. San Francisco: Morgan Kaufmann, 2005.

77. Yap CW. *PHAKISO* <http://www.phakiso.com>. 2006.

78. NeuroDimension. *NeuroSolutions 5.0* <http://www.nd.com/>. 2006.

79. MathWorks. *Matlab* <http://www.mathworks.com/>. 2006.

80. Tripos. *Sybyl* <http://www.tripos.com/>. 2006.

81. Talete. *DRAGON* <http://www.talete.mi.it/dragon_exp.htm>. 2006.

82. Haney DN. *Molconn-Z* <http://www.edusoft-lc.com/molconn/>. 2002.

83. Sanderson DM, Earnshaw CG. Computer prediction of possible toxic action from chemical structure: The DEREK system. *Hum Exp Toxicol* 1991;10:261–73.

84. Klopman G. MULTI-CASE: 1. A hierarchical computer automated structure evaluation program. *Quant Struct-Act Rel* 1992;11:176–84.

85. Accelrys. *TOPKAT* <http://www.accelrys.com/products/topkat/index.html>. 2005.

86. Nabholz JV, Cash GG. *ECOSAR* <http://www.epa.gov/oppt/newchems/tools/21ecosar.htm>. 2000.

87. Smithing MP, Darvas F. HazardExpert: An expert system for predicting chemical toxicity. In: Finlay JW, Robinson SF, Armstrong DJ, editors, *Food safety assessment*. Washington, DC: American Chemical Society, 1992. p. 191–200.

88. ChemSilico. CSGenoTox <http://www.chemsilico.com/CS_prGT/GThome.html>. 2003.

89. GeneGo. *MetaDrug* <http://www.genego.com/about/products.shtml>. 2006.

90. Chapelle O, Vapnik V, Bousquet O, Mukherjee S. Choosing multiple parameters for support vector machines. *Mach Learn* 2002;46:131–59.
91. Jaworska J, Nikolova-Jeliazkova N, Aldenberg T. QSAR applicability domain estimation by projection of the training set descriptor space: A review. *ATLA Altern Lab Anim* 2005;33:445–59.
92. Rosipal R, Trejo LJ. Kernel partial least squares regression in reproducing kernel Hilbert space. *J Mach Learn Res* 2001;2:97–123.
93. Serneels S, Filzmoser P, Croux C, Van Espen PJ. *Robust continuum regression. Chemom Intell Lab Sys* 2005;76:197–204.
94. Guha R, Dutta D, Jurs PC, Chen T. Local lazy regression: Making use of the neighborhood to improve QSAR predictions. *J Chem Inf Model* 2006;46:1836–47.
95. Petridis V, Kaburlasos VG. FINkNN: A fuzzy interval number k-nearest neighbor classifier for prediction of sugar production from populations of samples. *J Mach Learn Res* 2003;4:17–37.
96. Balthasar D, Priese L. Fast projection plane classifier. *Proceedings 16th International Conference on Pattern Recognition*. p. 200–3.

PART III

APPLYING COMPUTERS TO TOXICOLOGY ASSESSMENT: PHARMACEUTICAL

9

THE PREDICTION OF PHYSICOCHEMICAL PROPERTIES

Igor V. Tetko

Contents

Computational Toxicology: Risk Assessment for Pharmaceutical and Environmental Chemicals, Edited by Sean Ekins

9.1 INTRODUCTION

The physicochemical properties of compounds are extremely important in pharmaceutical and environmental studies. These properties determine the behavior of organic molecules in the environment as well as their biological and absorption, distribution, metabolism, excretion, and toxicity (ADME/T) properties as drugs.

There is a great gap in the number of chemical compounds that have measured physicochemical properties and those for which such information is needed. In the 1990s US EPA Office of Toxic Substances (OTS) listed approximately 70,000 industrial chemicals with approximately 1000 chemicals added each year [1]. The European Union has recently approved a new law, called REACH (Registration, Evaluation and Authorization of Chemicals) that will create a database of chemicals in EU use (see Chapter 25 and *http://ec.europa.eu/environment/chemicals/reach/reach_intro.htm*). This law will require testing for carcinogenic and mutagenic properties of all compounds produced in excess of 1 ton/per year and will lead to registration of more than 30,000 compounds. The total cost of tests required for the registration of compounds is estimated to be €5 billion over the next 11 years (*http://news.bbc.co.uk/2/hi/europe/4444550.stm*). The law considers the possibility of using QSAR and QSPR models in order to decrease dependence on animal tests.

While it is possible to experimentally measure physicochemical parameters for all chemicals registered in EPA OTS or REACH, the situation is much more complicated if such data are required for new, yet to be synthesized compounds. It can be a problem to predict values for molecules, for examples, from virtual computational libraries and/or during the early stages of drug or pesticide design. Already the size of virtual organic chemistry space accessible using currently known synthetic methods is estimated to be between 10^{20} and 10^{24} molecules [2]. The use of physicochemical filters could limit the experimental efforts to the most promising series of molecules and thus decrease the cost of drug discovery.

Consequently there is much interest in the development of reliable methods to predict physicochemical properties of molecules. This chapter presents methods for the prediction of log P, melting point, aqueous solubility, boiling point, and vapor pressure of chemical compounds.

The relative importance of each of these properties for pharmaceutical and environmental sciences is then judged by the publication rate. Figure 9.1 shows the number of publications for each of these properties in subject categories "pharmacology and pharmacy," "medicinal chemistry," and "environmental sciences" as defined by ISI Web of Knowledge (*http://www.isiknowledge.com*). Notice that the same properties are much differently represented in other sciences; for instance, in applied physics there are about 400 publications each year dealing with vapor pressure.

The most important property, as indicated by the number of publications, is definitely water solubility. The number of publications related to it has

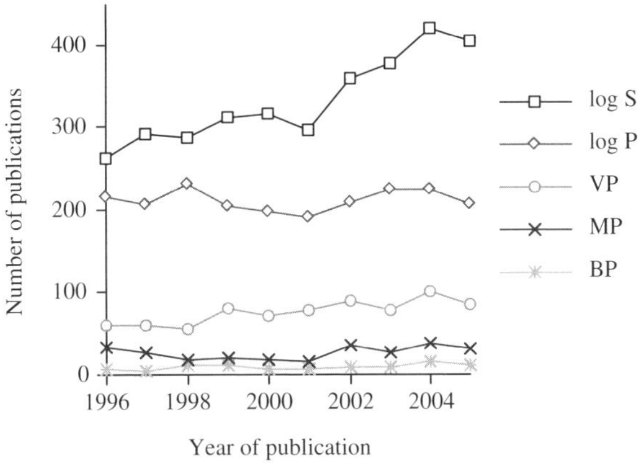

Figure 9.1 Number of publications dealing with physicochemical properties as keywords in environmental sciences, pharmacology and pharmacy, and medicinal chemistry. The following combinations of words: water/aqueous solubility, log *P*/lipophilicity, octanol/water, vapor/vapour pressure, melting point, boiling point were searched in the "topic category" of ISI Web of knowledge (*http://www.isiknowledge.com*).

almost doubled since 1996. There has also been an increase in the number of publications (from 60 to about 100) related to vapor pressure. While the other three properties have an approximately constant publication rate, there is a much larger number of publications related to log$P$ and lipophilicity compared to the melting and boiling point. Both boiling and melting points do not have widespread applications in pharmaceutical and environmental fields, and they mainly serve as supplementary properties to predict water solubility or vapor pressure.

A comprehensive review of methods used to predict each of these properties is not possible because of the size limitation of the chapter. We will mainly focus on properties in decreasing order of their importance, as identified by Figure 9.1. For each property we will start with a brief overview of the property followed by a discussion of the available data, the quality of the data and general overview of classes of models.

9.2 AQUEOUS SOLUBILITY, Log$S$

Aqueous solubility is one of the most important parameters for pharmaceutical and environmental studies. More than 130 years ago Benjamin Richardson showed that the toxicities of ethers and alcohols were inversely related to their water solubility [3].

The poor pharmacokinetic properties of compounds, particular due to problems with their solubility, constitute an important issue for the failure of candidate drugs in the later phases of clinical testing [4]. The famous "rule of five" [5] provides qualitative guidelines for compounds to be biologically available and thus to be soluble. Solubility in the range 2 ± 1 is indicated as a requirement for molecules to be bioavailable for agrochemicals [6].

It is possible to distinguish at least three more precise definitions of aqueous solubility:

- The *intrinsic solubility* S_i corresponds to the solubility of the neutral form of the compound.
- The *apparent solubility* at given pH, S_{pH}, is defined for compounds in ionized form. This kind of solubility is measured in pH-buffered solutions.
- The *solubility in pure water* corresponds to the *apparent solubility* of the compound in unbuffered water.

If the pK_a values of a monoprotolytic drug are known, the apparent solubility can be calculated from the intrinsic solubility using, for examples, the Henderson-Hasselbach equation [7]:

$$S_{pH} = S_i(1 + 10^{(\mathrm{pH}\text{-}\mathrm{p}Ka)\delta i}), \quad (9.1)$$

where $\delta_i = \{1, -1\}$ for the acid and basic group, respectively. Such equations should be combined for compounds that have multiple ionizable centers. There are several oversimplifications used in (9.1). For example, the equation assumes infinite solubility of ionized compounds, which is not the case. This problem can be approached by considering that the solubility range between the uncharged and the completely charged acids and bases is generally 4 and 3 log units, respectively [8]. However, experimental studies show that apparent solubility of some drug-like compounds had the ionization effects in the range from 1 to 6.5 log units [9], thus demonstrating large deviations from the proposed values. The equation does not take into consideration the dependency of solubility on the concentration and nature of the counter-ions, [10] aggregation, and salting out, namely factors that provide a considerable contribution to the apparent solubility [9]. Moreover, from a practical point of view, an increase of the apparent solubility does not always correlate with an increase in bioavailability due to interaction of drugs with the gastric or intestinal milieu because of common ion effects [11] or presence of surfactants such as bile salts [12].

The traditional way to measure aqueous solubility is by the shake-flask method, in which the compound is added in solvent and shaken until equilibrium is reached. This is, however, a time-consuming method. Another frequently used technique, the turbidimetric method, is based on aqueous titration of a DMSO solution of the compound under investigation. The solubility corresponds to the concentration at which precipitation occur. However, this method neglects the effect and particular kinetics of the solid state and provides a rough estimation of solubility [13]. A very good review of experimental approaches to measure aqueous solubility was recently published [8].

9.2.1 Data

The availability and the accuracy of publicly available aqueous solubility data are low. The largest datasets available in the field are the Physical Properties Database (PHYSPROP)[14] and AQUASOL databases (ca. 6000 compounds in each database). Both databases are available as commercial products. Recently the AQUASOL database was published [15]. The data entries in these databases are from the published literature and thus overlap. Moreover 25% of PHYSPROP solubility entries contain a reference to the AQUASOL database as a source of experimental data. In addition to these commercially available sets, there were several important studies that published relatively large sets of aqueous solubility data. A set of 1297 compounds was provided by Huuskonen [16] who collected data from PHYSPROP and AQUASOL databases. The dataset revised by Tetko [17] and Yan [18] is available for downloading at *http://www.vcclab.org/lab/alogps*. Three sets of 442 uncharged, 100 charged, and 50 zwitterionic compounds published in the *Journal of Medicinal Chemistry* from 1982 to 2002 were compiled by Lobell and Sivarajah [19]. Bergstrom et al. contributed a set of 85 molecules [20] that was measured in one laboratory under strict experimental conditions. However, a much larger amount of highly accurate experimental data is required to develop and to test aqueous solubility models.

There were several reports that estimated the accuracy of the aqueous solubility data. Analysis of experimental solubility values for 411 compounds provided an average standard deviation of 0.6 units [21]. In a more recent study [22] data for 1031 compounds that had at least two or more experimental values from the AQUASOL database [15] had a standard deviation of 0.5 log units for measurements at the same temperature. The large interlaboratory error masked the influence of temperature on the solubility. Differences as large as $\Delta T = 30°C$ did not increase the error. Thus the accuracy of models for diverse sets of compounds can be hardly below 0.5 to 0.6 log units.

9.2.2 Models

A number of methods to predict aqueous solubility of molecules use other experimental values, which can be more easily measured. The appearance of such methods is not surprising considering limited availability and low quality of experimental measurements for aqueous solubility. The melting point and lipophilicity are the two most frequently used parameters in such equations.

Irmann [23] was one of the first who used melting point (MP) in the equation

$$\log S = c + \sum n_i a_i + \sum n_i f_i + 0.0095(\text{MP} - 25), \tag{9.2}$$

where a_i and f_i are atomic and fragmental contributions, and c is a constant dependent on the compound type. The melting point was an important parameter in a number of other models [24–27]. Hansch et al. [28] correlated aqueous solubility of liquid nonelectrolytes with the octanol–water partition coefficient, $\log P$.

Yalkowsky [29,30] united both of these empiric relations and theoretically derived the general solubility equation (GSE). The derivation of the equation included three hypothetical steps. First, the analyzed crystal was heated until it melted. Next, the melted liquid was cooled to the water temperature, and eventually, the compound was dissolved in water. The equation was revised in 2001 by considering that the complete miscibility of a compound in octanol corresponds to a mole fraction of $X_0 = 0.5$ rather than to $X_0 = 1$ for each component [31]. The current GSE is written as

$$\log S = 0.5 - 0.01(\mathrm{MP} - 25) - \log P, \tag{9.3}$$

and for liquid compounds (i.e., MP < 25°C) this equation further simplifies to

$$\log S = 0.5 - \log P. \tag{9.4}$$

The charm of both equations is in their simplicity. In the case where the melting point is approximately constant for a particular scaffold of compounds, equation (9.3) can be further simplified by assuming constant MP to get the overall trend of changes of solubility. The chemists at Syngenta used a median melting point value of 125°C to apply the GSE in absence of experimental values [32].

A GSE-like equation

$$\log S = 0.69 - 0.96 \cdot \log P - 0.0031 \cdot \mathrm{MW} - 0.0092 \cdot (\mathrm{MP} - 25) + \sum f_i \tag{9.5}$$

that contains 15 correction factors f_i and molecular weight (MW) was derived by Meylan and Howard [33] with a database of 817 (RMSE = 0.62) compounds. Abraham and Le [34] used the linear free energy relationship (LFER) equation

$$\begin{aligned}\log S_w = 0.52 - R_2 + 0.77\pi_2^H + 2.17\sum \alpha_2^H + 4.24\sum \beta_2^H \\ - 3.36\sum \alpha_2^H \sum \beta_2^H - 3.99V_x\end{aligned} \tag{9.6}$$

to estimate the solubility of 659 solutes (standard error, SE = 0.56). A nice feature of this equation is the possibility to chemically interpret it. For example, positive coefficients for the solute hydrogen bond donor $\Sigma\alpha_2^H$, the acceptor $\Sigma\beta_2^H$, and the dipolarity π_2^H term help predict increased solubility. However, if the molecule is itself both a hydrogen-bond acid and a hydrogen-bond base, then intermolecular hydrogen-bond interactions, indicated by the $\Sigma\alpha_2^H\Sigma\beta_2^H$ term, will stabilize its crystal structure and thus decrease its solubility. The molar refraction R_2 refers to the tendency of a solute to interact with surrounding σ and π electrons while the volume of the solute V_x corresponds to the size of the solute and is proportional to the energy required to create a cavity in water. The same five descriptors (also known as Abraham's descriptors) can be also used to predict and interpret many other physicochemical properties.

There were several other methods that used experimentally derived descriptors to develop models by considering the energy change during the salvation process. For example, Ruelle and Kesserling used energy terms derived within mobile order (MOD) theory [35], while SPARC [36] calculated solubility based on summation over the free energy changes in a classical Boltzman model. Both methods can calculate some energy terms required for prediction of aqueous solubility in relation to simple physicochemical properties (e.g., boiling or melting point). In some studies the authors correlated aqueous solubility with other experimental properties of compounds such as free energy of solvation and vapor pressure of pure substances [37], and boiling point [38,39].

While these models have clear physical meaning and are easily interpretable, their disadvantage is the requirement to experimentally determine some descriptors. There are, however, possible solutions to this problem. Recently methods to calculate Abraham's descriptors directly from 2D molecular structure were proposed [40]. SPARC uses fragment contributions to predict solubility of new molecules. In this case, nevertheless, these methods can have the same problems as other fragmental methods.

In principle, it is possible to directly substitute the measured properties with the calculated ones. For example, Lobell and Sivarajah [19] confirmed the empirical observation of Syngenta [32] that MP is less important compared to log P for the prediction of aqueous solubility. Thus good results for the aqueous solubility can be calculated using just the lipophilicity of compounds. The model calculated using linear regression of lipophilicity values (predicted with AlogP98 program [41]),

$$\log\left(\frac{1}{S}\right) = 1.52 + 0.69\,\mathrm{A\log P98}, \tag{9.7}$$

provided the highest prediction accuracy (MAE = 0.66) for aqueous solubility of 442 predominately uncharged compounds and had higher accuracy than several in silico methods [19]. The equation was developed using a set of 202 uncharged British Biotech (BB) compounds for which a similar accuracy was calculated (MAE = 0.54). It is interesting that according to Lobell [42], another method, ALOGPS [43], "yields the most accurate prediction followed by AlogP98, AlogP and CLOGP with comparable accuracies" for 73 salt free BB-compounds with experimental $\log P$ values. The equation based on the ALOGPS method, however, provided worse results both for the training (MAE = 0.73) and test set (MAE = 0.90). Thus the observed results could be specific for the training/test sets analyzed by the authors. Indeed in another benchmarking study [44], equation 9.7 had one of the lowest prediction accuracies.

Methods Using 3D Descriptors Advances in quantum-chemical calculations and the increasing power of personal computers have made a great impact on the development of methods to predict aqueous solubility directly from the 3D structures of molecules. Monte Carlo statistical mechanics simulations by Jorgensen and Duffy [45] were used to predict the solubility of 150 compounds (MAE = 0.56) using the equation

$$\log S = 0.32 \cdot \mathrm{ESXL} + 0.65 \cdot \mathrm{HBAC} + 2.19 \cdot N_{amine} - 1.76 \cdot N_{nitro} - 162 \cdot \frac{\mathrm{HBAC} \cdot \mathrm{HBDN}^{1/2}}{\mathrm{SASA}} + 1.18, \tag{9.8}$$

where ESXL is the Lennard-Jones interaction energy, SASA is the solvent accessible surface area, HBDN and HBAC are the numbers of donor and acceptor hydrogen bonds and N_{amine} and N_{nitro} are the number of amino and nitro groups, respectively. To overcome limitations of time-consuming Monte Carlo calculations the authors also developed the QikProp solubility model [46]. The method requires 3D structures of molecules with optimized geometry, thus delegating a substantial optimization time to an external program. However, the accuracy of the method can be lower for complex, drug-like molecules. This is true if we consider computational difficulties to find the minimum energy of molecules and the optimization of molecules is normally performed in vacuum and not in water.

The COSMO-RS (continuum solvation model for real solvents) approach developed by Klamt considers interactions in a liquid system as contact interactions of the molecular surfaces [47]. The interaction energies of, for example, hydrogen bonding and electrostatic terms, are written as pairwise interactions of the respective polarization charge densities. The ensemble of interacting molecules is replaced by the corresponding system of surface segments, and the system is solved under the condition that all the surfaces of both solute and solution interact with each other. To overcome the speed limitations of the COSMO-RS, the authors developed a COSMOfrag method [48] that uses suitable fragments from a database of more than 40,000 pre-calculated molecules.

Considering the importance of the surface interactions of solute and solution, Bergstrom developed partitioned total surface area (PTSA) as an extension of PSA (polar surface area) [20]. PTSA calculates surface areas corresponding to particular types of atoms (i.e., sp, sp^2, and sp^3 hybridized carbon atoms and the hydrogen atoms bound to these carbon atoms). A comparison of the results calculated using 3D and 2D descriptors (calculated using Molconn-Z [49] and Selma [50]) favored the later set of descriptors [20]. The best results were calculated by combining both methods.

The previous considered methods usually depend on linear methods (MLR, PLS) to establish structure-solubility correlations for prediction of solubility of molecules. The work of Göller et al. [51] used a neural network ensemble to predict the apparent solubility of Bayer in-house organic compounds. The solubility was measured in buffer at pH 6.5, which mimics the medium in the human gastrointestinal tract. The authors used the calculated distribution coefficient log D (at several pH values), a number of 3D COSMO-derived parameters and some 2D descriptors. The final model was developed using 4806 compounds (RMSE = 0.72) and provided a similar accuracy (RMSE = 0.73) for the prediction of 7222 compounds that were not used to develop the model. The method, however, is quite slow, and it takes about 15 seconds to screen one molecule on an Intel Xeon 2.8 GHz CPU.

Methods Using 2D and 1D Descriptors A good number of articles on aqueous solubility used a nonlinear method of data analysis, in particular, for methods developed with 1D and 2D descriptors. Huuskonen [16] used E-state indexes [52,53] and several other topological indexes, with a total of 30 indexes, to develop his models. The predicted results for the 413 test set, SE = 0.71, calculated with MLRA were improved with a neural network, resulting in SE = 0.6. Tetko [17] noticed that E-state indexes represent a complete system of descriptors for molecules, and thus only these descriptors are sufficient to develop the aqueous solubility model. Indeed the model developed by the authors using exclusively E-state indexes provides similar results when compared to the model of Huuskonen [16]. Later on, the model was redeveloped using the Associative Neural Network (ASNN) method [54].

E-state indexes were used by Votano et al. [55] to develop two models (one for aromatic and one for nonaromatic compounds) for a dataset of 5694 molecules. A comparison of PLS, MLRA, and ANN models for their prediction of the test set molecules clearly indicated the advantage of the nonlinear methods. The neural networks calculated MAE = 0.62 and MAE = 0.56 for aromatic and nonaromatic sets, respectively. The second-best results for the same test sets, MAE = 0.76 and MAE = 0.66, were calculated using the MLRA model.

Interestingly in several studies (see also aforementioned work of Bergstrom [20]) the authors did not observe a significant improvement of solubility models when using 3D descriptors in addition or instead of 2D descriptors. For example, Cheng and Metz [56] did not report an increase in the performance of their model using 3D descriptors. Yan and Gasteiger applied neural networks to radial distribution function codes derived from 3D structure and some other descriptors [18] to analyze a dataset of 1293 compounds. In their later study, the same set was re-analyzed using 18 topological descriptors [57]. The neural networks trained with topological indexes provided similar prediction ability, RMSE = 0.52 as opposed to 0.59 log units for the test set, compared to the result using 3D descriptors. The higher performance when using 2D set of descriptors was further affirmed in another study by the same authors using a larger set of compounds [58].

In view of these studies Balakin et al. [22] concluded that there is no statistical support for the idea that 3D descriptors are more appropriate for prediction of aqueous solubility of chemicals compared to 2D or 1D methods. Because of the difficulties in generating and finding conformational minimum for 3D structures, one should initially try more simple indexes based on molecular topology. The big advantage of the later method is their speed: 2D methods can typically process tens of thousands of molecules per second, and that makes them very useful for screening of virtual libraries of compounds.

Figure 9.2 represents a plot of errors in the methods reviewed above as well as in those of several other works [22,44,59]. As the figure demonstrates, there is no apparent difference in the performance of methods developed with any particular group of descriptors, that is, quantum-chemical (median RMSE = 0.47) and topological (median RMSE = 0.53), or using physicochemical descriptors (median RMSE = 0.50). The median error of all methods (RMSE = 0.51) approximately corresponds to the experimental error of solubility measurements.

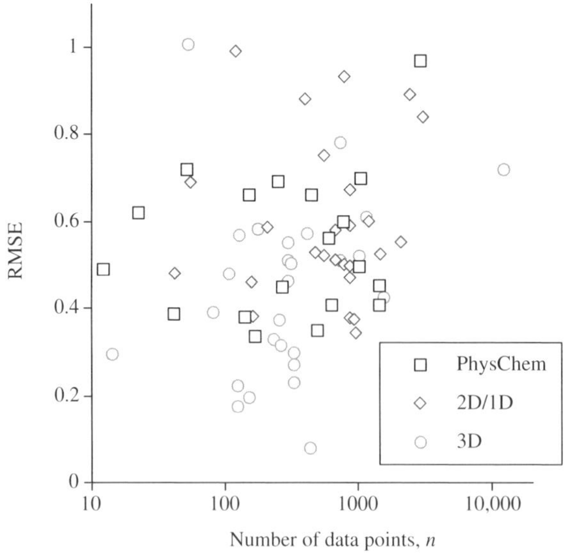

Figure 9.2 The root mean squared error (RMSE) of models for prediction of aqueous solubility of chemical compounds shown as a function of the number of molecules, n, used for model development and validation. The results of methods developed using quantum chemical (3D), topological descriptors (2D/1D), and methods based on other physicochemical descriptors (PhysChem) are shown.

9.3 APPROACHES TO PREDICT Log*P*

The octanol–water partition coefficient, log*P*, has a long history of applications in environmental and pharmaceutical studies. Already more than 100 years ago the activity of local anesthetics was described by Overton [60] and Meyer [61] in terms of oil–water partition coefficients. The log*P* was recognized as a standard property for description of the lipophilicity of chemicals due to the works of Corwin Hansch in the 1960s [62].

The partition coefficient is defined as the ratio of the concentration of a solute in the organic phase to its concentration in the water phase. This definition applies to neutral species. However, many pharmaceutical and environmental chemicals contain charged groups that could become ionized in water, thus contributing to the decrease of the lipophilicity of molecules. If one assumes that only the neutral form of a molecule will partition into the organic phase, then the observed log*D* can be related to the log*P* and the pK_a of the compound, at the pH of the measurement, by an equation such as that shown for monoprotic basic compounds below (compare to equation 9.1 for solubility):

$$\log D = \log P - \log\left(1 + 10^{pKa-pH}\right) \tag{9.9}$$

Notice that the assumption used in equation (9.9) is not valid in general. First, water dissolves in octanol, and thus charged compounds will partition in it too. Second, as in the case of aqueous solubility, the ionization of molecules depends on other parameters than just pH, such as the concentration and nature of the counter-ions [63].

While the theoretical background for log*D* calculation is quite complex, in practice, determination of this coefficient at a fixed pH is much simpler compared to the log*P* measurements, which usually require multiple experiments or extrapolations to the neutral state of the compound. Moreover, by using some specific ranges of pH (e.g., pH = 1–2 for stomach or neutral pH = 6.5 for jejunum), one can better simulate the medium in the gastrointestinal tract. Thus, not surprisingly, the experimental measurements in industry are mainly done for log*D*. At the same time, because of the complexity of effects determining distribution coefficients, one can expect the log*D* data measured at the same pH in different laboratories to be incompatible.

9.3.1 Available Experimental Data

The MedChem database [64] contains the largest publicly available collection of more than 60,000 measurements of log*P* and log*D* values, and it is available as a commercial product from BioByte Inc. (*http://www.biobyte.com*). This database provided experimental support for the development of the CLOGP program [65,66]. The PHYSPROP database [14] of Syracuse Research Inc. provides experimental log*P* values for 13,058 compounds. This database is publicly available at *http://esc.syrres.com/interkow/KowwinData.htm*.

Another publicly available database LOGKOW (*http://logkow.cisti.nrc.ca*) is supported by James Sangster. It provides online access to about 20,000 molecules, including $\log P$, $\log D$, and pK_a values. This database is updated quarterly and is the largest publicly available collection of $\log P/D$ values in the field. To estimate the experimental accuracy of the measurements, we selected 1106 molecules from LOGKOW that had at least three measurements with the same pH ($\Delta pH < 0.1$) and the same temperature ($\Delta T < 5°C$). There were in total 1312 such groups of values with average of five measurements per group. The interlaboratory variation $s = 0.45$ (MAE = 0.26) log units have a similar variation to that of the $\log S$ measurements.

The compounds both in the MedChem and LOGKOW databases provide recommended values that were selected according to the expert knowledge of the curators. The quality of the recommended data can be higher, and accuracy on the order of $s < 0.3$ log units can be expected. The KOWWIN database contains 9429 compounds that have experimental values in the MedChem database. The additional molecules in KOWWIN database are of similar quality to the MedChem database as shown by computational analysis [43].

9.3.2 Methods Using Experimental or Predicted Properties

Log P is already considered to be a simple property. So, contrary to aqueous solubility methods, there are not many methods trying to correlate it with even simpler properties.

However, experimental determination of $\log P$ does include correlation with such properties. For example, the ElogD method developed by Lombardo et al. [67] uses RP-HPLC retention data to determine octanol–water distribution coefficients at pH 7.4 for neutral and basic drugs in Pfizer. Moreover the authors used the same method to determine ElogD at pH 6.5, thus calculating important parameters for intestinal absorption [68].

The solute size (larger molecules favors octanol) together with solute hydrogen-bond basicity (favors water) were named as the main parameters of the LFER equation for prediction lipophilicity of chemicals [40,69]. The other methods from this group include SLIPPER [70], SPARC [36,71], as well as approaches developed within MOD theory [72,73].

9.3.3 Fragmental Methods

The $\log P$ is to a large extent an additive property, at least in comparison to the aqueous solubility. Not surprisingly, a large number of fragmental methods to predict this property have been published. The general equation for this group of methods can be represented as

$$\log P = a + \sum_{i=1}^{N} b_i G_i + \sum_{j=1}^{K} c_j F_j, \qquad (9.10)$$

where G_i is the number of occurrences of the group i, F_j are the correction factors, and a, b_i and c_j are the regression coefficients. A number of popular methods, such as ClogP [65,66] ACD/log P [74], KOWWIN [33,75], Σf-SYBYL [76,77] KlogP [78], HlogP [79], and AB/log P [80,81], use the fragmental representation of molecules to correlate activity and lipophilicity of molecules. These methods have been carefully reviewed and compared in previous publications [82–84].

Of course, the developed methods do not necessarily have to be based on a linear equation. For example, Artemenko et al. predicted lipophilicity using fragmental descriptors and artificial neural networks [85]. The use of linear equations nevertheless provides, an easier interpretation of calculated results that can be difficult when using nonlinear methods of data analysis.

The fragmental approaches can also use information about the 3D structure of molecules. A new version of the KlogP program [86] is based on fragmental indices but it benefits from knowledge of the 3D structure of molecules by means of steric hindrance indices, H, proposed by Cherkasov [87]. The modified equation

$$\log P = a + \sum_{i=1}^{N} b_i (1 - H_i) G_i \tag{9.11}$$

takes into account the hindrance H_i of the atoms in the fragment G_i as $H_i = \Sigma_{j \neq i}^{n} R_j^2 / A \cdot r_{ij}^2$, where A is a constant, R_j is the atomic radius of the jth atom, and r_{ij} is the distance between the ith and jth atoms [87]. The H_i index weights the contribution of different fragments according to their availability to solvent. The use of (9.11) instead of (9.10) remarkably decreased the standard deviation of the method from 1.08 to 0.78 log units for a test set of 137 drugs.

Problems with group contribution methods for the prediction of new compounds can be attributed to the presence of some groups that were not covered in the training set as well as the presence of nonlinear interactions between groups. To overcome this problem ClogP program (version 4.0 and higher) includes an algorithm for the *ab initio* calculation of the contribution of fragments if they are not found for the training set [65]. This method was claimed to provide an accuracy within 0.3 to 0.5 log units, but actually in another study many compounds with large prediction errors were found to contain the *ab initio* fragments [43].

9.3.4 Methods Based on Descriptors Calculated for the Whole Molecule

This group of methods includes methods that use the 3D structure representation of molecules, such as CLIP [88], QikProp [89], COSMOlogP [47,48], as well as methods based on topological indexes calculated for the whole molecules, such as AUTOLOGP [90], VLOGP [91,92], XLOGP [93], ALOGPS [43,94], or descriptors based on SMILES string analysis, LINGO method [95]. There have been a number of reviews that carefully described and compared these methods [76,82,84,96].

The main difference between this group of methods compared to fragmental ones is based on an assumption of absence of "missed" fragments, meaning the descriptors used in these algorithms are usually determined for any molecule and so these methods can generalize to new molecules. This assumption, however, does not always work. First, the performance of 3D methods critically depends on the success of correct conformation selection, and consequently on parameters of quantum-chemical or molecular mechanics methods. The difficulties with finding correct conformations increase with the size of a molecule. Second, even nominally the same descriptors can have very different values in training and test sets, and consequently the contributions of those descriptors to the target property can be different for training and test sets. Third, method development involves selection of a subset of significant descriptors. Unfortunately, some of the important descriptors can be underrepresented in the training set. Their elimination can hamper the prediction ability of models when the programs are used to predict compounds containing descriptors eliminated during variable selection.

For example, the authors noticed that molecules with the largest prediction errors usually contained descriptors missing in the training set, thus demonstrating a "missing fragment" problem for the ALOGPS method, [43] which is based on E-state descriptors [52,53]. To overcome this problem, the ALOGPS flags the molecules with missing fragments, this way warning the user about possible "nonreliability" of predictions. Later on, the ALOGPS was extended by the Associative Neural Network method [54] to provide "self-learning" of new data in a LIBRARY mode [94]. This approach can be useful for the inclusion of *in-house* data from pharmaceutical companies, but it cannot be used during the model development stage. The problems that this technique helps solve are missing fragments and the diversity of compounds in the training and test sets. The prediction efficiency of the LIBRARY mode was demonstrated for *in-house* data from BASF [54], AstraZeneca [97], and Pfizer [98] for both $\log P$ and $\log D$ data. Recently the authors demonstrated that the same methodology can be used to estimate the accuracy of model predictions [99]. It was shown that for over 50% of the *in-house* Pfizer compounds (characterized with property-based similarity >0.80) the method correctly predicted $\log P$ with an accuracy of 0.35 log units, which is similar to that of experimental measurements [99,100].

Figure 9.3 shows a plot of the errors from methods to predict lipophilicity of chemicals reviewed in this and in several other works [84,96] according to the same types of descriptors used in Figure 9.2. There are only four methods (see above) that were classified as those that use physicochemical descriptors. The median RMSE errors for 2D and 3D methods are 0.38 and 0.34 log units, respectively.

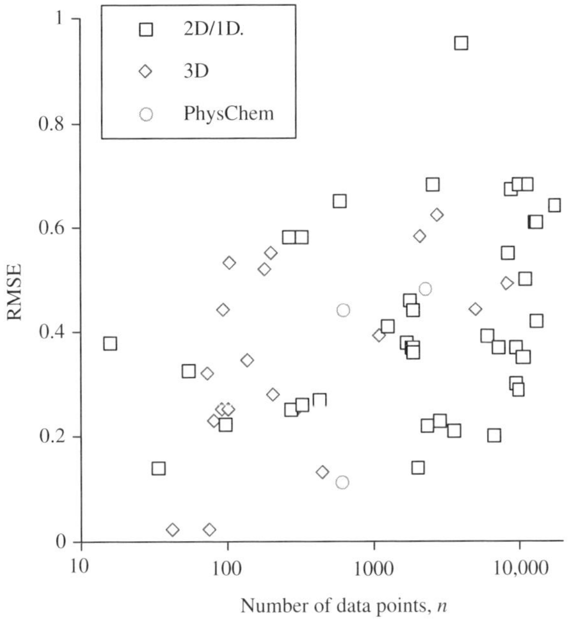

Figure 9.3 Root mean squared error (RMSE) of models for prediction of lipophilicity of chemical compounds shown as a function of the number of molecules, *n*, used for model development and validation. The results for methods developed using quantum chemical (3D), topological descriptors (2D/1D), and methods based on simpler physicochemical descriptors (PhysChem) are shown.

9.4 VAPOR PRESSURE (V_p)

Vapor pressure is used for estimating the volatility of chemical compounds and the environmental fate of chemicals. This coefficient is also important in pharmaceutical studies, for example, for development of perfumes and pressurized aerosols.

The PHYSPROP database [14] contains data for V_p of more than 2000 chemical compounds measured over the temperature range of 20° to 30°C. The NIST/TRC vapor pressure database (*http://www.nist.gov/srd/dblist.htm*) contains data for approximately 6000 pure compounds. Vapor pressure data are also available from a number of other databases, as reviewed elsewhere [101].

Vapor pressure is temperature dependent as demonstrated by the Clausius-Clapeyron equation

$$\ln \frac{V_{p1}}{V_{p2}} = -\frac{\Delta H_{vap}}{R(\frac{1}{T_1} - \frac{1}{T_2})}, \tag{9.12}$$

which can be rewritten as

$$\ln V_p = -\frac{\Delta H_{vap}}{RT} + c, \tag{9.13}$$

where ΔH_{vap} is the enthalpy of vaporization and c is a constant.

A thermodynamic equation was derived to predict vapor pressure by Mishra and Yalkowsky [102]. The authors integrated the Clausius-Clapeyron equation in a mental experiment considering the well-defined reversible processes: the heating of solids to melting point, melting, cooling (heating) of liquid to temperature of interest, heating to boiling point, evaporation, and cooling (heating) of gas to the temperature of interest. The resulting equation (also known as the Rankine-Kirchoff equation [103]) is written as

$$\begin{aligned}\ln V_p = &\frac{-\Delta S_{MP}(MP-T)}{RT} + \frac{\Delta C_{MP}}{R}\left[\frac{MP-T}{T} - \ln\frac{MP}{T}\right] \\ &- \frac{-\Delta S_{BP}(BP-T)}{RT} + \frac{\Delta C_{BP}}{R}\left[\frac{BP-T}{T} - \ln\frac{BP}{T}\right]\end{aligned} \tag{9.14}$$

where ΔS_{MP} and ΔS_{BP} are entropies of melting and boiling, respectively, and ΔC_{MP} and ΔC_{BP} are heat capacity change on boiling and melting, respectively.

By the further assumption S[102] on entropies and heat capacity, this equation can be simplified to include only four descriptors:

$$\begin{aligned}\ln V_p = &-\frac{(MP-T)(8.5-5.0\log\sigma+2.3\log\phi)}{T} \\ &-\frac{(BP-T)(10+0.08\log\phi)}{T} \\ &+\left(\frac{BP-T}{T}-\ln(BP/T)\right)(-6-0.9\log\phi),\end{aligned} \tag{9.15}$$

that is, MP, BP, and two molecular descriptors consisting of the rotational symmetry number, σ, and the conformational flexibility number, ϕ, of a molecule. While the later two numbers could be reasonably derived for simple hydrocarbons, their calculation for complex molecules is less obvious.

The Rankine-Kirchoff equation can be simplified further, as described by Mackay et al. [104]. The authors assumed that the change in both heat capacities is 0, so the entropy of vaporization is given by Trouton's rule,

$$\frac{\Delta S_{BP}}{R} = 10.75, \tag{9.16}$$

and the entropy of melting is given by Walden's rule,

$$\frac{\Delta S_{MP}}{R} = 6.75, \tag{9.17}$$

to

$$\ln V_P = \frac{[-10.75(BP-T)-6.75(MP-T)]}{T}, \tag{9.18}$$

which depends only on melting and boiling points. Equation (9.18) provides a lower accuracy (MAE = 0.47 vs. MAE = 0.18 for 72 organic compounds) compared to model of Mishra and Yalkowsky [102]. Although both equations (9.16) and (9.17) require experimental values and thus are difficult to use in practice, they are important for understanding the close interrelations among MP, BP, and vapor pressure.

Kühne et al. [105] used MP and 23 structural parameters to correlate the vapor pressure of 1838 hydrocarbons and halogenated hydrocarbons. The neural network model was built using 1200 compounds and provided a mean absolute error of 0.13 log units for the test set of 638 compounds. It should be noted that the authors predicted vapor pressure as a function of the temperature and total number of data points for model development was 8148. The MP and BP are also used in MPBPVP program developed by Syracuse Research Inc [101].

One of the first "group contribution" methods used for vapor pressure prediction was described by Ambrose and Sprake [106]. The authors used the group-contribution approach to accurately represent the temperature-dependent vapor pressures of 8 aliphatic alcohols. This study was followed by those of Macknick et al. [107] and Burkhard [108], who analyzed sets with up to several tens of hydrocarbons. Nevertheless, all these studies remain to some extent toy problems due to the limited diversity of compounds.

The Rankine-Kirchoff equation determines on the dependency of vapor pressure on temperature. Indeed it just contains terms proportional to $1/T$, $-\ln T$, and $-T$. Thus it does make more sense to explicitly derive a linear regression equation for V_p prediction that will also contain these dependencies. This was demonstrated by Jensen et al. [109] who predicted vapor pressure using the following equation:

$$\ln V_p = \sum N_k \left(\frac{A_{k,1} + A_{k,2}}{T - A_{k,3} \ln T - A_{k,4} T} \right) + \sum \sum N_{k,j} B_{k,j} + \sum N_k C_k - \ln D, \quad (9.19)$$

where $A_{k,1\text{–}4}$ are constant and independent of the structure of the compound for each group k and $B_{k,j}$ are structure-dependent contributions of each fragment within kth group, C_k are the residual group activity coefficients, and D is the fugacity coefficient. The groups were defined within UNIFAC (UNIversal quasichemical Functional group Activity Coefficients) approach. The $B_{k,j}$ are temperature dependent and are further parameterized as $B_{k,j} = B'_{k,j} + TB''_{k,j}$, where $B'_{k,j}$ and $B''_{k,j}$ are additional constants.

Tu [110] further simplified equation (9.19) by considering only the first term and combined all other terms in one correction term. He used only 42 UNIFAC groups for a relatively large set of 342 organic compounds with 5359 experimental vapor pressure values. The method was tested with 336 organic compounds and resulted in a mean absolute percentage error of 5%. The UNIFAC groups were also used in a number of other studies, for example, to estimate the vapor pressure of oxygen-containing compounds in the modeling of organic aerosols [111].

A feature of this group of methods was an attempt by the authors to structurally restrict the explored temperature dependency of vapor pressure to the theoretically derived dependency. This motivation is clearly justified by the nonlinear dependency of vapor pressure from temperature, which cannot be easily captured with the pure linear regression approach. The use of nonlinear methods, however, can solve this problem. For example, neural networks predicted saturated vapor pressure for 352 hydrocarbons with an RMSE = 0.12 compared to an RMSE = 0.25 using a linear method with the same descriptors [112].

A temperature-dependent quantum-chemical model based on a set of 7681 measurements for 2349 molecules selected from the Beilstein database (*http://www.beilstein.com/*) was developed by Chalk et al. [113]. A set of 27 descriptors representing geometrical, electrotopological, and electrostatic interactions was selected by the authors. The mean absolute error of 0.21 to 0.29 was calculated following cross-validation and validation sets. Moreover the authors further validated their approach by calculating the experimental enthalpy of vaporization from a model prediction using equation (9.12) as well as predicting the boiling point. The results were in good agreement with experiment (i.e., calculated MAE = 19 K for boiling point was similar to MAE = 13.6 calculated in their boiling point model [114]). A number of models that were developed for vapor pressure prediction are reviewed elsewhere [59,101,115].

9.5 BOILING POINT (BP)

The normal boiling point corresponds to the temperature at which a substance has a vapor pressure of 760 mm Hg. Thus methods to predict vapor pressure at ambient temperature can be also used to calculate the normal boiling point, as described by Chalk et al. [113] in their model. The main factors determining the boiling point are the size of the molecule (required to create a cavity in the water) and the intermolecular interactions in the liquid that are controlled by polar and hydrogen forces and molecular flexibility. The entropy of boiling is generally assumed to be constant according to Trouton's rule (equation 9.16). Similar to the $\log P$ partition coefficient, there is an approximately additive effect of different groups at the boiling point. Therefore the boiling point is a simpler property to predict compared with the vapor pressure or the melting point.

More than 6000 compounds with a boiling point values at 760 mm Hg or 1 atm have been collected by ACDlabs (*http://www.acdlabs.com/products/phys_chem_lab/bp/*). The same number of compounds, with BP measured at different pressures are available in the PHYSPROP database [14]. CHEMEXPER (*http://www.chemexper.com*) has experimental BP data for more than 7500 compounds. The experimental BP data are also provided by ChemFinder (*http://www.chemfinder.com*) and ChemIDplus (*http://chem.sis.nlm.nih.gov/*

chemidplus/). A comprehensive overview of other experimental resources for BP can be found elsewhere [101]. The boiling point data are expected to be highly accurate (in the range of 1–2 degrees), apart from compounds with very high boiling points that can decompose at high temperatures. Nevertheless, the literature data contain serious discrepancies, even for alkanes; for instance, the boiling point of cyclooctane had 23 reported BP values in the range from 120.3° to 156°C in the Beilstein database, as noted elsewhere [116]. Such discrepancies presumably can be attributed to the imprecise control of atmospheric pressure. If the BP is measured at pressures that differ from 1 atm, the values should be extrapolated to 1 atm using the Clausius-Clapeyron equation or some of its simplifications. This was done in the work of Stein and Brown [117] who found that extrapolations from two measured values to the same pressure of 1 atm depended on the selected equations. In accuracy, the extrapolations had MAE in the range of 12 to 18 K. These values were used as approximate estimates of the limit of the accuracy of boiling point data in very diverse datasets.

The boiling point was a test bed for many topological indexes. The saturated acyclic hydrocarbons (the alkanes) were the easiest compounds to model, starting from the pioneering work of Wiener [118] who introduced the path number w (the Wiener index) as the sum of the distances between any carbon atoms in a molecule. A lot of studies have been performed since to successfully model boiling points in homologous and congeneric series of compounds [116,119–121]. Rücker and Rücker [116] carefully collected and verified experimental data for 531 saturated hydrocarbons (up to $n = 10$) and tested about 20 topological descriptors used for a subseries of these compounds in 20 previous studies. They concluded that there was no descriptor or combination of descriptors that can adequately describe the BP in all or most of the various samples with a precision of a few degrees. At the same time they also demonstrated that the use of incomplete or biased data (due to the presence of multiple BP values in the literature) could produce chance correlations, as can be sometimes found in the literature.

The prediction of BP for noncongeneric diverse sets of compounds is more interesting and is still a challenging problem. Despite the wide availability of reasonably high-quality experimental BP data, there are just a few methods developed using thousands of molecules.

One of the most important contributions to this field was provided by Stein and Brown [117], who extended Joback and Reid's group contribution method [122]. The authors calculated 85 group contributions and used a nonlinear quadratic equation to correctly predict BP at high temperatures. The authors calculated MAE = 15.5 K for a set of 4426 compounds and predicted another independent set of 6584 compounds with MAE = 20.4 K. The method was further parameterized by Syracuse Research Inc using 6484 compounds and was used in the MPBPVP program [101].

Two interesting approaches to predict BP using quantum-chemical methods and neural networks were provided by the group of Clark et al. [114,123]. In both studies the authors used 6629 experimental boiling points that were separated into training (6000) and test (629) sets. Following generation of the 3D structure of molecules using CORINA, the molecules were optimized with AM1 and PM3 semi-empirical quantum-chemical methods. In the first study the final models had 18 parameters and included electrostatic (mean, max, and variance of positive/negative electrostatic potential, etc.) and structural (sum of nitrogens, oxygens, halogens, surface area, etc.) variables. The AM1-based approach calculated MAE = 13 K for the test set [114] that was 1 to 2 K lower compared to the results using the PM3 method. In the second study the authors attempted to eliminate element-specific descriptors by generating an orthogonal set of descriptors calculated at the molecular surface. These types of descriptors may allow better generalization for the prediction of new molecules. The selected set included just 13 descriptors and calculated a similar performance to the previous results. Unfortunately, both these studies did not compare models using 1D or 2D descriptors, which could have been very interesting.

9.6 MELTING POINT (MP)

The melting point is dramatically affected by crystal packing and energies that contribute to it, including ionic, polar, and hydrogen bonding forces. Thus, according to Gavezzotti [124], MP remains one of the most difficult crystal properties to predict. The 3D structure of molecules plays a crucial role for determining the MP. A practical example is given by anthracene, which melts at 215°C, while its asymmetric isomer phenanthrene melts at 100°C (Figure 9.4) [14,115]. Considering the difficulties to predict the MP and the crucial role of this property in aqueous solubility and vapor pressure, it is not surprising that quite a few methods to predict both these properties used MP as an external parameter, as mentioned above. Recently an increasing interest in MP prediction has appeared due to development of ionic liquids and green chemistry [125].

The amount of experimental data for MP is large, and it is not a major limiting factor for modeling, as is the case with aqueous solubility or vapor pressure. Indeed, when submitting a new article for publication, the major publishers require the authors to provide spectral and MP data of molecules synthesized. However, the quality of MP data may significantly vary depending on sample preparation and its purity. Moreover the same substance can be in a crystal or amorphic state that can dramatically change its MP and solubility.

The PHYSPROP database [14] contains data for more than 10,000 compounds. Some of the values are given as inequalities or ranges. The CHEMEXPER (*http://www.chemexper.com*) has data for more than 20,000 compounds. Another large collection of MP data is provided by the Beilstein database. The prediction of MP has a long history, starting from Mills [126] who derived QSPR models for homologous series and calculated very low errors (<1°C)

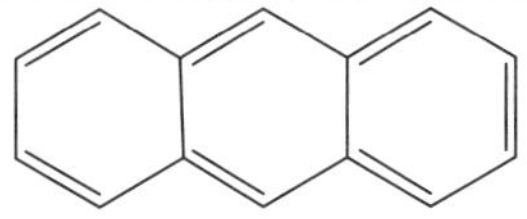	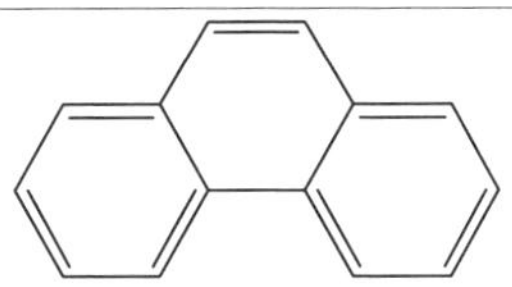
Anthracene [120-12-7]	Phenanthrene [85-01-8]
Melting point: 215°C	Melting point: 99.2°C
Boiling point: 339.9°C	Boiling point: 340°C
log S: -6.61 (0.043 mg/l)	log S: -5.19 (1.15 mg/l)
log P: 4.45	log P: 4.46
Vapor pressure: $6.53*10^{6}$ mm Hg (T=25°C)	Vapor pressure: $1.2*10^{-3}$ mm Hg (T=25°C)

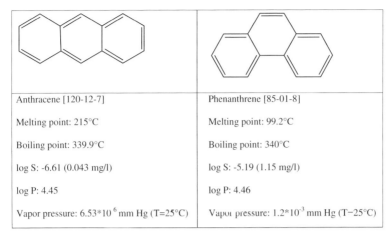

Figure 9.4 Physicochemical properties of anthracene and phenanthrene according to the PHYSPROP database [14]. The large discrepancy in MP of both these compounds, ΔMP = 115°C, explains their differences in aqueous solubility and vapor pressures. The difference in solubility of molecules calculated using GSE (equation 9.3) $\Delta \log S = 0.01*115 = 1.15$ approximately corresponds to the experimentally observed value $\Delta \log S = 1.42$.

for melting and boiling points as function of the number of methylene groups, x, in the hydrocarbon chain as

$$t(°\mathrm{C}) = \frac{\beta x}{1+\gamma x}, \tag{9.20}$$

where β and γ are constants for a given class [115].

Since that work a lot of studies were performed using mainly homologous series of compounds [115,127–130]. We will concentrate mainly on models developed with a large number of molecules (>1000).

Syracuse Research Inc predict MP in their MPBPVP program [101] as a weighted average of two methods: Gold and Ogle [131] and the group contribution method of Joback and Reid [122]. The Gold and Ogle [131] approach calculates MP as a simple correlation with boiling point

$$\mathrm{MP} = 0.5839\ \mathrm{BP}. \tag{9.21}$$

The Joback method was modified in the MPBPVP program to include the same 85 groups used by the Stein and Brown [117] method for BP calculation. The MPBPVP program has a calculated MAE = 44 K for a set of 2045 compounds.

The group contribution method was also used by Tu and Wu [132] who used 15 group contribution and 6 structural parameters to calculate an impressive accuracy (8.2%) for 1310 compounds.

A diverse set of 4173 compounds was used by Karthikeyan et al. [133] to derive their models with a large number of 2D and 3D descriptors (all calculated by MOE following 3D-structure generation of Concord [134]) and an artificial neural network. The authors found that 2D descriptors provided better prediction accuracy (RMSE = 48–50°C) compared to models developed using 3D indexes (RMSE = 55–56°C) for both training and test sets. The use of a combined 2D and 3D dataset did not improve the results.

Another study comparing 2D and 3D descriptors to predict the MP of ionic liquids of 717 bromides of nitrogen-containing organic cations was done by Varnek et al. [135]. Depending on the sets of molecules used, the authors calculate RMSE in range from 26° to 49°C. The higher errors corresponded to more diverse sets of molecules. The authors found that Dragon descriptors [120] calculated using online E-Dragon software [136] (3D structures as calculated by Corina [137]) provided better results compared to the models calculated using E-state indexes ($p < 0.05$ according to Wilcoxon matched-pairs signed-ranks test). The differences in results of Varnek et al. [135] compared to Karthikeyan et al. [133] can be explained by a lower structural diversity of compounds used in Varnek et al. [135] and/or use of different 3D structure generation software and descriptors.

Prediction ability (SE = 48.9°C) was calculated in the study of Clark [138], who used 261 overlapping 2D structure-based descriptors and the largest dataset of 5598 molecules from the PHYSPROP database. In addition to simple 2D descriptors, such as atoms, secondary, and tertiary carbons, the author used rings, fused ring (including attached heteroatoms), and other drug-like motifs. The use of large functional overlapping descriptors could help one better understand the influence of each fragment on activity as well as distinguish molecules with the same components but substituted in different positions.

9.7 APPLICABILITY DOMAIN OF MODELS

Each year an increasing number of computational methods dealing with the development of physicochemical models are published. For example, recent reviews [22,84] indicate that as many as 50 articles concerning prediction of lipophilicity and aqueous solubility of molecules were published in 2005. This is about a fivefold increase compared to 1995. However, the prediction accuracy obtained from "blind testing" of the methods using, for example, proprietary datasets of pharmaceutical companies remains disappointingly low [97,98,139,140]. A similar problem of low prediction accuracy of models can be expected when using QSAR approaches to predict properties of molecules within the REACH regulatory framework in the European Union. The main reason for failure of such models is due to an implicit assumption that both training and test molecules cover the same chemical space. However, this is very often not the case and molecules in both sets can be very different. Since the contribution of descriptors to a target property can be highly nonlinear, the structure–property correlations calculated from the training set will not be valid for the test set.

Several methods considered in this article, such as the quantum-chemical approaches developed by Clark et al. [51,113,114] or the ALOGPS program [99], provide an estimation of the accuracy of model predictions. Other approaches to estimate accuracy of model predictions were reviewed elsewhere [99,141].

The calculation of the confidence of predictions will be definitely one of the most important features of physicochemical property prediction in the future, and this field is quickly developing now.

9.8 BENCHMARKING OF METHODS

The benchmarking of methods for physicochemical property predictions is a topic of considerable interest, particularly for the comparison of commercial programs. There were a number of studies and reviews that addressed this problem for prediction of lipophilicity [82–84] aqueous solubility [19,44] as well as other physicochemical property prediction methods [115]. Indeed, since the final purpose of any model development is to predict new data, the benchmarking of different models allows one to compare the advantages of different methodologies.

Unfortunately, there is a huge difference in the accuracy of prediction depending on whether the analyzed compound (or similar compounds) was used in the training set or if it was not used. For example, ALOGPS [43] and KOWWIN [75] provided high accuracy (RMSE = 0.44–0.46) compared to CLOGP [65] results (RMSE = 0.62) for prediction of compounds that were missed in the BioByte database (the so-called nova dataset). However, when the ALOGPS program was redeveloped using only a subset of molecules from the BioByte dataset, its prediction accuracy (RMSE = 0.57) for the nova dataset was similar to that of the CLOGP.

The LIBRARY mode [54,94] of the ALOGPS program makes it possible to add additional compounds to the training set without retraining the model. Thus it provides an easy way to demonstrate the influence of the training set size on the accuracy of prediction. When applied to the Lobell and Sivarajah [19] dataset, the LIBRARY mode had a twofold lower prediction error for prediction of aqueous solubility compared to the blind prediction of the same compounds [22]. In a number of other studies a 2- to 10-fold increase in the prediction accuracy of the ALOGPS was demonstrated following extension of the training set in the LIBRARY mode [84,94,97,98].

These studies clearly demonstrate that the similarity of molecules in the training and test sets is one of the crucial contributing factors for benchmarking of methods. Thus any benchmarking study is meaningless until there is available information on the composition of the training sets of benchmarked programs. Unfortunately, this is unlikely occur for commercial programs.

It is expected that comparing differences of estimated and calculated experimental errors is an alternative and presumably better means of comparison of the performance of the methods. Those methods that fail to correctly estimate their errors (or do not provide them at all) should receive a low rating.

9.9 PROGRAMS FOR DATA ANALYSIS

A number of methods to predict physicochemical properties of compounds are available as public or commercial products. Table 9.1 lists some of the programs and their providers. Many programs have been developed for prediction of lipophilicity and aqueous solubility of chemicals. A more detailed overview of Internet-based tools that can be used for physicochemical property prediction can be also found elsewhere [142].

9.10 CONCLUSIONS

In recent years large number of methods to predict physicochemical properties were developed. These methods have been applied to speed up development of drugs and pesticides. They have also been used in environmental studies, particularly, as parameters in models to estimate toxicity and hazardous properties of chemical compounds.

Nevertheless, the development of production methods remains a challenging problem. One of the main difficulties limiting progress in this field is the restricted amount of experimental data available for model development. Moreover the data that are usually available for modeling are frequently incompatible due to the differing experimental protocols used. A larger amount of data is available in pharmaceutical companies, but these data are usually not publicly available for external use and model development.

There are several emerging technologies that may change the situation [136] (see Chapter twenty-four). First, there has been considerable progress in automation of physicochemical property measurements. The development of HTS ADME/T technologies can dramatically decrease the cost of experimental measurements and thus facilitate their increased availability for modeling. These technologies, such as that of the turbidimetric method for aqueous solubility determination, have already produced thousands of data points. Second, because of the expense of drug failures [143], there is motivation for large pharmaceutical companies to release some of their data in order to promote development of new predictive technologies. Since the success of the drug industry depends on maintaining privacy concerning molecular structures, the development of approaches to enable release of data without the underlying molecular structures is actively being explored in the field [144,145]. Third, the implementation of REACH and the PubChem initiative of the National Institute of Health [146] will soon provide large amounts of

TABLE 9.1 Programs to Predict Physicochemical Properties of Molecules

Property	Product Name	Descriptors	Method	Supplier	References
			Commercial		
Log P,[a] log S[a]	AB/log P/S	Fragmental	Linear	www.ap-algorithms.com	[81]
Log P, log S, BP, V_p	ACDlabs log P/S	Fragmental	Linear	www.acdlabs.com	[74]
Log P, log S	TSAR, VLOGP, CERIUS[2]	Several models		www.accelrys.com	[91]
Log P		Fragmental	Linear	www.biobyte.com	[64,65]
Log P, log S	CS log P, CS log S	Topological descriptors	Neural networks	www.chemsilico.com	[55]
Log P, log S	Pallas	Fragmental	Linear	www.compudrug.com	
Log P, log S	S + log P/S	Topological	Neural networks	www.simulations-plus.com	
Log P, log S	SLIPPER	Properties	Linear	www.timtec.net	[70]
Log P,[a] log S	COSMOTherm, COSMOFrag	Quantum-chemical	Linear	www.cosmologic.de	[47]
Log P, log S	QikProp	Quantum-chemical	Linear	www.schrodinger.com	[46]
Log P, log S, V_p, BP	SPARC	Quantum chemical, fragmental	Linear	ibmlc2.chem.uga.edu/sparc	[36,71]

TABLE 9.1 (*Continued*)

Property	Product Name	Descriptors	Method	Supplier	References
			Free		
Log P, log S	CHEMICALC-2	Atomic values	Linear	www.osc.edu/ccl/qcpe	[26]
Log P,[a] log S, MP, BP, V_p	KOWWIN, WSKOWWIN, MPBPVP	Fragmental	Linear	www.epa.gov/opptintr/exposure/docs/episuite.htm	[33,101]
Log P[a]	XLOGP	Atomic values	Linear	mdl.ipc.pku.edu.cn/drug_design/work/xlogp.html	[93]
			Via the Web		
Log P, log S	Osiris	Atomic values	Linear	organic-chemistry.org	
Log P,[a] log S[a]	ALOGPS	Topological descriptors	Neural networks	www.vcclab.org	[17,43]
Log P[a], log S[a]	IA_log P, IA_log S	Topological descriptors	Neural networks	www.logp.com	
Log P[a]	miLogP	Group contributions	Linear	www.molinspiration.com	
Log P,[a] log S[a]	Qlog P, Qlog S	Quantum-chemical		q-pharm.com	
Log P, log S, BP, MP		Topological descriptors	Neural networks	preadme.bmdrc.org	

[a]These models are also available for online comparison at the Virtual Computational Chemistry Laboratory site [147] (*http://www.vcclab.org*).

high-quality data for public access. The development of the Semantic Web (*http://www.w3.org/2001/sw*) has opened up ways of publishing scientific articles and, by including comprehensive experimental results in a structured format, should improve automatic verification (when submitting a publication) and collection of information for use in future models.

The availability of data will dramatically transform the field and boost development of new, reliable methods for physicochemical property predictions. The development of methods to estimate the accuracy of a prediction and the applicability domain of models will make it possible to obtain more confident results on their wider use in environmental and pharmaceutical studies.

REFERENCES

1. Karickhoff SW, Carreira LA, Melton C, McDaniel VK, Vellino AN, Nute DE. Computer prediction of chemical reactivity—The ultimate SAR. In: US Environmental Protection Agency CfERI, editor, *Environmental research brief EPA/600/M-89/017*. Cincinnati, OH, 1989.
2. Ertl P. Cheminformatics analysis of organic substituents: Identification of the most common substituents, calculation of substituent properties, and automatic identification of drug-like bioisosteric groups. *J Chem Inf Comput Sci* 2003;43:374–80.
3. Richardson BJ. Lectures on experimental and practical medicine. *Med Times Gaz* 1868;2:703.
4. DiMasi JA, Hansen RW, Grabowski HG. The price of innovation: New estimates of drug development costs. *J Health Econ* 2003;22:151–85.
5. Lipinski CA, Lombardo F, Dominy BW, Feeney PJ. Experimental and computational approaches to estimate solubility and permeability in drug discovery and development settings. *Adv Drug Deliv Rev* 2001;46:3–26.
6. Clarke ED, Delaney JS. Physical and molecular properties of agrochemicals: An analysis of screen inputs, hits, leads, and products. *Chimia* 2003;57:731–4.
7. Hasselbalch KA. Die Berechnung der Wasserstoffzahl des Blutes aus der freien und gebunden Kohlensäure desselben, und die Sauerstoffbindung des Blutes als Funktion der Wasserstoffzahl. *Die Biochem Z* 1916;78.
8. Avdeef A. *Absorption and drug development: Solubility, permeability and charge state*. Hoboken, NJ: Wiley-Interscience, 2003.
9. Bergstrom CA, Luthman K, Artursson P. Accuracy of calculated pH-dependent aqueous drug solubility. *Eur J Pharm Sci* 2004;22:387–98.
10. Agharkar S, Lindenbaum S, Higuchi T. Enhancement of solubility of drug salts by hydrophilic counterions: Properties of organic salts of an antimalarial drug. *J Pharm Sci* 1976;65:747–9.
11. Crowley PJ, Martini LG. Physicochemical approaches to enhancing oral absorption. *Pharm Technol Eur* 2004;16:18–27.

12. Glomme A, März A, Dressman JB. Predicting the Intestinal Solubility of Porly Soluble Drugs. In: Testa B, Krämer SD, Wunderli-Allenspach H, Folkers G, editors, *Pharmacokinetics profiling in drug research*. Zürich and Weinheim: Verlag Helvetica Chimica Acta and Wiley-VCH, 2006. p. 259–80.
13. Avdeef A, Testa B. Physicochemical profiling in drug research: A brief survey of the state-of-the-art of experimental techniques. *Cell Mol Life Sci* 2002;59:1681–9.
14. The Physical Properties Database (PHYSPROP) is a trademark of Syracuse Research Corporation <www.syrres.com>.
15. Yalkowsky SH, He Y. *Handbook of aqueous solubility data*. Boca Raton: CRC Press, 2003.
16. Huuskonen JJ, Livingstone DJ, Tetko IV. Neural network modeling for estimation of partition coefficient based on atom-type electrotopological state indices. *J Chem Inf Comput Sci* 2000;40:947–55.
17. Tetko IV, Tanchuk VY, Kasheva TN, Villa AE. Estimation of aqueous solubility of chemical compounds using E-state indices. *J Chem Inf Comput Sci* 2001;41: 1488–93.
18. Yan A, Gasteiger J. Prediction of aqueous solubility of organic compounds based on a 3D structure representation. *J Chem Inf Comput Sci* 2003;43:429–34.
19. Lobell M, Sivarajah V. In silico prediction of aqueous solubility, human plasma protein binding and volume of distribution of compounds from calculated p*K*a and AlogP98 values. *Mol Divers* 2003;7:69–87.
20. Bergstrom CA, Wassvik CM, Norinder U, Luthman K, Artursson P. Global and local computational models for aqueous solubility prediction of drug-like molecules. *J Chem Inf Comput Sci* 2004;44:1477–88.
21. Katritzky AR, Wang YL, Sild S, Tamm T, Karelson M. QSPR studies on vapor pressure, aqueous solubility, and the prediction of water–air partition coefficients. *J Chem Inf Comput Sci* 1998;38:720–5.
22. Balakin KV, Savchuk NP, Tetko IV. In silico approaches to prediction of aqueous and DMSO solubility of drug-like compounds: trends, problems and solutions. *Curr Med Chem* 2006;13:223–41.
23. Irmann F. A Simple correlation between water solubility and structure of hydrocarbons and halohydrocarbons. *Chem Ing Tech* 1965;37:789–98.
24. Wakita K, Yoshimoto M, Miyamoto S, Watanabe H. A method for calculation of aqueous solubility of organic compounds using new fragment solubility constants. *Chem Pharm Bull* 1986;34:4663–81.
25. Nirmalakhandan NN, Speece RE. Prediction of aqueous solubility of organic chemicals based on molecular structure: 2. Application to PNAs, PCBs, PCDDs, etc. *Environ Sci Technol* 1989;23:708–13.
26. Suzuki T. Development of an automatic estimation system for both the partition coefficient and aqueous solubility. *J Comput Aided Mol Des* 1991;5:149–66.
27. Kühne R, Ebert R-U, Kleint F, Scmidt G, Schuurmann G. Group contribution methods to estimate water solubility of organic chemicals. *Chemosphere* 1995; 30:2061–77.
28. Hansch C, Quinlan JE, Lawrence GL. The linear free-energy relationship between partition coefficients and the aqueous solubility of organic liquids. *J Org Chem* 1968;33:347–50.

29. Yalkowsky SH, Valvani SC. Solubility and partitioning: I. Solubility of non-electrolytes in water. *J Pharm Sci* 1980;69:912–22.
30. Yalkowsky SH, Valvani SC, Roseman TJ. Solubility and partitioning: VI. Octanol solubility and octanol-water partition coefficients. *J Pharm Sci* 1983;72:866–70.
31. Jain N, Yalkowsky SH. Estimation of the aqueous solubility: I. Application to organic nonelectrolytes. *J Pharm Sci* 2001;90:234–52.
32. Delaney JS. Predicting aqueous solubility from structure. *Drug Discov Today* 2005;10:289–95.
33. Meylan WM, Howard PH. Estimating log P with atom/fragments and water solubility with log P. *Perspect Drug Discov Des* 2000;19:67–84.
34. Abraham MH, Le J. The correlation and prediction of the solubility of compounds in water using an amended solvation energy relationship. *J Pharm Sci* 1999;88:868–80.
35. Ruelle P, Kesselring UW. The hydrophobic effect. 2. Relative importance of the hydrophobic effect on the solubility of hydrophobes and pharmaceuticals in H-bonded solvents. *J Pharm Sci* 1998;87:998–1014.
36. Hilal SH, Karickhoff SW, Carreira LA. Prediction of the solubility, activity coefficient and liquid/liquid partition coefficient of organic compounds. *QSAR Combin Sci* 2004;23:709–20.
37. Thompson JD, Cramer CJ, Truhlar DG. Predicting aqueous solubilities from aqueous free energies of solvation and experimental or calculated vapor pressures of pure substances. *J Chem Phys* 2003;119:1661–70.
38. Miller MM, Ghodbane S, Wasik SP, Tewari YB, Martire DE. Aqueous solubilities, octanol/water partition coefficients, and entropies of melting of chlorinated benzenes and biphenyls. *J Chem Eng Data* 1984;29:184–90.
39. Yaws CL, Xiang P, Xiaoyin L. Water solubility data for 151 hydrocarbons. *Chem Eng* 1993;100:108.
40. Platts JA, Abraham MH, Butina D, Hersey A. Estimation of molecular linear free energy relationship descriptors by a group contribution approach: 2. Prediction of partition coefficients. *J Chem Inf Comput Sci* 2000;40:71–80.
41. Ghose AK, Viswanadhan VN, Wendoloski JJ. Prediction of hydrophobic (lipophilic) properties of small organic molecules using fragmental methods: An analysis of ALOGP and CLOGP methods. *J Phys Chem A* 1998;102:3762–72.
42. Lobell M. Advances in the in-silico prediction of aqueous solubility from structure. Cerius2 User Group Meeting. Cerep, Paris, 2001.
43. Tetko IV, Tanchuk VY, Villa AE. Prediction of *n*-octanol/water partition coefficients from PHYSPROP database using artificial neural networks and E-state indices. *J Chem Inf Comput Sci* 2001;41:1407–21.
44. Dearden JC. In silico prediction of aqueous solubility. *Expert Opin Drug Discov* 2006;1:31–52.
45. Jorgensen WL, Duffy EM. Prediction of drug solubility from Monte Carlo simulations. *Bioorg Med Chem Lett* 2000;10:1155–8.
46. Jorgensen WL, Duffy EM. Prediction of drug solubility from structure. *Adv Drug Deliv Rev* 2002;54:355–66.
47. Klamt A, Eckert F, Hornig M, Beck ME, Burger T. Prediction of aqueous solubility of drugs and pesticides with COSMO-RS. *J Comput Chem* 2002;23:275–81.

48. Hornig M, Klamt A. COSMOfrag: A novel tool for high-throughput ADME property prediction and similarity screening based on quantum chemistry. *J Chem Inf Model* 2005;45:1169–77.
49. Molconn-Z. Quincy, MA: Hall Associates Consulting.
50. Selma. Mölndal: AstraZeneca in house software package.
51. Goller AH, Hennemann M, Keldenich J, Clark T. In silico prediction of buffer solubility based on quantum-mechanical and HQSAR- and topology-based descriptors. *J Chem Inf Model* 2006;46:648–58.
52. Kier LB, Hall LH. *Molecular structure description: The electrotopological state.* London: Academic Press, 1999.
53. Kier LB, Hall LH. An electrotopological-state index for atoms in molecules. *Pharml Res* 1990;7:801–7.
54. Tetko IV. Neural network studies: 4. Introduction to associative neural networks. *J Chem Inf Comput Sci* 2002;42:717–28.
55. Votano JR, Parham M, Hall LH, Kier LB, Hall LM. Prediction of aqueous solubility based on large datasets using several QSPR models utilizing topological structure representation. *Chem Biodiver* 2004;1:1829–41.
56. Cheng A, Merz KM, Jr. Prediction of aqueous solubility of a diverse set of compounds using quantitative structure-property relationships. *J Med Chem* 2003;46:3572–80.
57. Yan AX, Gasteiger J. Prediction of aqueous solubility of organic compounds by topological descriptors. *QSAR Combin Sci* 2003;22:821–9.
58. Yan A, Gasteiger J, Krug M, Anzali S. Linear and nonlinear functions on modeling of aqueous solubility of organic compounds by two structure representation methods. *J Comput Aided Mol Des* 2004;18:75–87.
59. Taskinen J, Yliruusi J. Prediction of physicochemical properties based on neural network modelling. *Adv Drug Deliv Rev* 2003;55:1163–83.
60. Overton E. Über die osmotischen Eigenschaften der Zelle in ihrer Bedeutung für die Toxikologie und Pharmakologie. *Z Phys Chem* 1897;22:189–209.
61. Meyer H. Lipoidtheorie der Narkose. *Arch Exp Path Pharm* 1899;42:109–18.
62. Hansch C, Maloney PP, Fujita T, Muir RM. *Nature* 1962;194:178–80.
63. Wang PH, Lien EJ. Effects of different buffer species on partition coefficients of drugs used in quantitative structure-activity relationships. *J Pharm Sci* 1980; 69:662–8.
64. Hansch C, Leo A, Hoekman D. *Hydrophobic, electronic, and steric constants.* Washington, DC: American Chemical Society, 1995.
65. Leo AJ, Hoekman D. Calculating log P(oct) with no missing fragments: The problem of estimating new interaction parameters. *Perspect Drug Discov Des* 2000;18:19–38.
66. Leo AJ. Calculating $\log P_{\text{oct}}$ from structures. *Chem Rev* 1993;93:1281–306.
67. Lombardo F, Shalaeva MY, Tupper KA, Gao F. ElogD(oct): A tool for lipophilicity determination in drug discovery. 2. Basic and neutral compounds. *J Med Chem* 2001;44:2490–7.
68. Yoshida F, Topliss JG. QSAR model for drug human oral bioavailability. *J Med Chem* 2000;43:2575–85.

69. Abraham MH, Chadha HS, Whiting GS, Mitchell RC. Hydrogen bonding: 32. An analysis of water-octanol and water-alkane partitioning and the delta log*P* parameter of seiler. *J Pharm Sci* 1994;83:1085–100.

70. Raevsky OA, Trepalin SV, Trepalina HP, Gerasimenko VA, Raevskaja OE. SLIPPER-2001—Software for predicting molecular properties on the basis of physicochemical descriptors and structural similarity. *J Chem Inf Comput Sci* 2002;42:540–9.

71. Hilal SH, Karickhoff SW, Carreira LA. Prediction of the vapor pressure boiling point, heat of vaporization and diffusion coefficient of organic compounds. *QSAR Combin Sci* 2003;22:565–74.

72. Ruelle P. Universal model based on the mobile order and disorder theory for predicting lipophilicity and partition coefficients in all mutually immiscible two-phase liquid systems. *J Chem Inf Comput Sci* 2000;40:681–700.

73. Ruelle P. The *n*-octanol and *n*-hexane/water partition coefficient of environmentally relevant chemicals predicted from the mobile order and disorder (MOD) thermodynamics. *Chemosphere* 2000;40:457–512.

74. Petrauskas AA, Kolovanov EA. ACD/log*P* method description. *Perspect Drug Discov Des* 2000;19:99–116.

75. Meylan WM, Howard PH. Atom/fragment contribution method for estimating octanol-water partition coefficients. *J Pharm Sci* 1995;84:83–92.

76. Mannhold R, Rekker RF. The hydrophobic fragmental constant approach for calculating log*P* in octanol/water and aliphatic hydrocarbon/water systems. *Perspect Drug Discov Des* 2000;18:1–18.

77. Mannhold R, Rekker RF, Dross K, Bijloo G, de Vries G. The lipophilic behaviour of organic compounds: 1. An updating of the hydrophobic fragmental constant approach. *Quant Struc-Act Rel* 1998;17:517–36.

78. Klopman G, Li J-Y, Wang S, Dimayuga M. Computer automated log*P* calculations based on an extended group contribution approach. *J Chem Inf Comput Sci* 1994;34:752–81.

79. Viswanadhan VN, Ghose AK, Wendoloski JJ. Estimating aqueous solvation and lipophilicity of small organic molecules: A comparative overview of atom/group contribution methods. *Perspect Drug Discov Des* 2000;19:85–98.

80. Japertas P, Didziapetris R, Petrauskas A. Fragmental methods in the design of new compounds: Applications of the Advanced Algorithm Builder. *Quant Struct-Act Rel* 2002;21:23–37.

81. Japertas P, Didziapetris R, Petrauskas A. Fragmental methods in the analysis of biological activities of diverse compound sets. *Mini Rev Med Chem* 2003;3: 797–808.

82. Mannhold R, van de Waterbeemd H. Substructure and whole molecule approaches for calculating log*P*. *J Comput Aided Mol Des* 2001;15:337–54.

83. Mannhold R, Petrauskas A. Substructure versus whole-molecule approaches for calculating log*P*. *QSAR Combin Sci* 2003;22:466–75.

84. Tetko IV, Livingstone DJ. Rule-based systems to predict lipophilicity. In: Testa B, van de Waterbeemd H, editors, *Comprehensive medicinal chemistry: II. In silico tools in ADMET*. Amsterdam: Elsevier, 2006;5: pp. 649–68.

85. Artemenko NV, Palyulin VA, Zefirov NS. Neural-network model of the lipophilicity of organic compounds based on fragment descriptors. *Doklady Chem* 2002;383:114–6.

86. Zhu H, Sedykh A, Chakravarti SK, Klopman G. A new group contribution approach to the calculation of log*P*. *Curr Comput Aided Drug Des* 2005;1:3–9.

87. Cherkasov A, Jonsson M. Substituent effects on thermochemical properties of free radicals: New substituent scales for C-centered radicals. *J Chem Inf Comput Sci* 1998;38:1151–6.

88. Carrupt P-A, Gaillard P, Billois F, Weber P, Testa B, Meyer C, et al. The molecular lipophilicity potential (MLP): A new tool for log*P* calculations and docking, and in comparative molecular field analysis (CoMFA). In: Pliska V, Testa B, van de Waterbeemd H, editors, *Lipophilicity in drug action and toxicology*. Weinheim: VCH, 1996. p. 195–217.

89. Duffy EM, Jorgensen WL. Prediction of properties from simulations: Free energies of solvation in hexadecane, octanol, and water. *J Am Chem Soc* 2000;122:2878–88.

90. Devillers J, Domine D, Guillon C, Bintein S, Karcher W. Prediction of Partition coefficient ($\log P_{oct}$) using autocorrelation descriptors. *SAR QSAR Environ Res* 1997;7:151–72.

91. Gombar VK, Enslein K. Assessment of *n*-octanol/water partition coefficient: When is the assessment reliable? *J Chem Inf Comput Sci* 1996;36:1127–34.

92. Gombar VK. Reliable assessment of log*P* of compounds of pharmaceutical relevance. *SAR QSAR Environ Res* 1999;10:371–80.

93. Wang RX, Gao Y, Lai LH. Calculating partition coefficient by atom-additive method. *Perspect Drug Discov Des* 2000;19:47–66.

94. Tetko IV, Tanchuk VY. Application of associative neural networks for prediction of lipophilicity in ALOGPS 2.1 program. *J Chem Inf Comput Sci* 2002;42: 1136–45.

95. Vidal D, Thormann M, Pons M. LINGO, an efficient holographic text based method to calculate biophysical properties and intermolecular similarities. *J Chem Inf Model* 2005;45:386–93.

96. Tetko IV, Poda GI. Property-based log*P* calculations. In: Mannhold R, editor, *Drug properties: Measurement and computation*. Weinheim: Wiley-VCH, 2007.

97. Tetko IV, Bruneau P. Application of ALOGPS to predict 1-octanol/water distribution coefficients, log*P*, and log*D*, of AstraZeneca in-house database. *J Pharm Sci* 2004;93:3103–10.

98. Tetko IV, Poda GI. Application of ALOGPS 2.1 to predict log*D* distribution coefficient for Pfizer proprietary compounds. *J Med Chem* 2004;47:5601–4.

99. Tetko IV, Bruneau P, Mewes HW, Rohrer DC, Poda GI. Can we estimate the accuracy of ADME-Tox predictions? *Drug Discov Today* 2006;11:700–7.

100. Tetko IV, Bruneau P, Mewes HW, Rohrer DC, Poda GI. Can we estimate the accuracy of ADMET predictions? 232th ACS National Meeting. San Francisco, CA, 2006.

101. Boethling RS, Howard PH, Meylan WM. Finding and estimating chemical property data for environmental assessment. *Environ Toxicol Chem* 2004;23: 2290–308.

102. Mishra DS, Yalkowsky SH. Estimation of vapor pressure of some organic compounds. *Ind Eng Chem Res* 1991;30:1609–12.
103. Bondi A. *Physical properties of molecular crystals, liquids, and gases*. New York: Wiley, 1968.
104. Mackay D, Bobra AM, Chan D, Shiu WY. Vapor pressure correlations for low volatility environmental chemicals. *Environ Sci Technol* 1982;16:645–9.
105. Kühne R, Ebert RU, Schuurmann G. Estimation of vapour pressures for hydrocarbons and halogenated hydrocarbons from chemical structure by a neural network. *Chemosphere* 1997;34:671–86.
106. Ambrose D, Sprake CHS. Thermodynamic properties of organic oxygen compounds XXV: Vapour pressures and normal boiling temperatures of aliphatic alcohols. *J Chem Thermodynamics* 1970;2:631–45.
107. Macknick AB, Prausnitz JM. Vapor pressures of heavy liquid hydrocarbons by a group-contribution method. *Ind Eng Chem Fundam* 1979;18:348–51.
108. Burkhard LP. Estimation of vapor pressures for halogenated aromatic hydrocarbons by a group-contribution method. *Ind Eng Chem Fundam* 1985;24: 119–20.
109. Jensen T, Fredenslund A, Rasmussen P. Pure-component vapor pressures using UNIFAC group contribution. *Ind Eng Chem Fundam* 1981;20:239–46.
110. Tu CH. Group-contribution method for the estimation of vapor pressures. *Fluid Phase Equilibria* 1994;99:105–20.
111. Asher WE, Pankow JF, Erdakos GB, Seinfeld JH. Estimating the vapor pressures of multi-functional oxygen-containing organic compounds using group contribution methods. *Atmos Environ* 2002;36:1483–98.
112. Artemenko NV, Baskin II, Palyulin VA, Zefirov NS. Artificial neural network and fragmental approach in prediction of physicochemical properties of organic compounds. *Russ Chem Bull* 2003;52:20–9.
113. Chalk AJ, Beck B, Clark T. A temperature-dependent quantum mechanical/neural net model for vapor pressure. *J Chem Inf Comput Sci* 2001;41:1053–9.
114. Chalk AJ, Beck B, Clark T. A quantum mechanical/neural net model for boiling points with error estimation. *J Chem Inf Comput Sci* 2001;41:457–62.
115. Dearden JC. Quantitative structure-property relationships for prediction of boiling point, vapor pressure, and melting point. *Environ Toxicol Chem* 2003; 22:1696–709.
116. Rucker G, Rucker C. On topological indices, boiling points, and cycloalkanes. *J Chem Inf Comput Sci* 1999;39:788–802.
117. Stein SE, Brown RL. Estimation of normal boiling points from group contributions. *J Chem Inf Comput Sci* 1994;34:581–7.
118. Wiener H. Structural determination of paraffin boiling points. *J Am Chem Soc* 1947;69:17–20.
119. Katritzky AR, Lobanov VS, Karelson M. Normal boiling points for organic compounds: Correlation and prediction by a quantitative structure-property relationship. *J Chem Inf Comput Sci* 1998;38:28–41.
120. Todeschini R, Consonni V. *Handbook of molecular descriptors*. Weinheim: Wiley-VCH, 2000.

121. Devillers J, Balaban AT. *Topological indices and related descriptors in QSAR and QSPR*. Philadelphia: Gordon and Breach, 1999.

122. Joback KG, Reid RC. Estimation of pure-component properties from group contributions. *Chem Eng Commun* 1987;57:233–43.

123. Ehresmann B, de Groot MJ, Alex A, Clark T. New molecular descriptors based on local properties at the molecular surface and a boiling-point model derived from them. *J Chem Inf Comput Sci* 2004;44:658–68.

124. Gavezzotti A. Are crystal-structures predictable? *Acc Chem Res* 1994;27: 309–14.

125. Endres F, Zein El Abedin S. Air and water stable ionic liquids in physical chemistry. *Phys Chem Chem Phys* 2006;8:2101–16.

126. Mills EJ. On melting point and boiling point as related to composition. *Phil Mag* 1884;17:173–87.

127. Horvath AL. *Molecular design: Chemical structure generation from the properties of pure organic compounds*. Amsterdam: Elsevier, 1992.

128. Dearden JC. The QSAR prediction of melting point, a property of environmental relevance. *Sci Total Environ* 1991;109–10:59–68.

129. Baum EJ. *Chemical property estimation: Theory and application*. Boca Raton, FL: CRC Press, 1998.

130. Tesconi M, Yalkowsky SH. Melting point. In: Boethling RS, Mackay D, editors, *Handbook of property estimation methods for chemicals*. Boca Raton, FL: Lewis, 2000. p. 2–27.

131. Gold PI, Ogle GJ. Estimating thermophysical properties of liquids: Part 4. Boiling, freezing and triple-point temperatures. *Chem Eng* 1969;76.

132. Tu CH, Wu YS. Group-contribution estimation of normal freezing points of organic compounds. *J Chin Inst Chem Eng* 1996;27:323–8.

133. Karthikeyan M, Glen RC, Bender A. General melting point prediction based on a diverse compound data set and artificial neural networks. *J Chem Inf Model* 2005;45:581–90.

134. Pearlman RS. CONCORD: Rapid generation of high quality approximate 3D molecular structures. *Chem Des Autom News* 1987;2:5–7.

135. Varnek A, Kireeva N, Tetko IV, Baskin II, Solov'ev VP. Exhaustive QSPR studies of large diverse set of ionic liquids: how accurately can we predict the melting point? *J Chem Inf Mod* 2007.

136. Tetko IV. Computing chemistry on the Web. *Drug Discov Today* 2005;10:1497–500.

137. Sadowski J, Gasteiger J, Klebe G. Comparison of automatic three-dimensional model builders using 639 X-ray structures. *J Chem Inf Comput Sci* 1994;34: 1000–8.

138. Clark M. Generalized fragment-substructure based property prediction method. *J Chem Inf Model* 2005;45:30–8.

139. Morris JJ, Bruneau PP. Prediction of physicochemical properties. In: Bohm HJ, Schneider G, editors, *Virtual screening for bioactive molecules*. Weinheim: Wiley-VCH, 2000. p. 33–58.

140. Walker MJ. Training ACD/log *P* with experimental data. *QSAR Combin Sci* 2004; 23:515–20.

141. Netzeva TI, Worth A, Aldenberg T, Benigni R, Cronin MT, Gramatica P, et al. Current status of methods for defining the applicability domain of (quantitative) structure-activity relationships. The report and recommendations of ECVAM Workshop 52. *Altern Lab Anim* 2005;33:155–73.

142. Tetko IV. The WWW as a tool to obtain molecular parameters. *Mini Rev Med Chem* 2003;3:809–20.

143. Landers P. Cost of developing a new drug increases to about $1.7 billion. *Wall Street Journal*, 2003. p. B4.

144. Wilson EK. Is safe exchange of data possible? *Chem Eng News* 2005;83:24–9.

145. Tetko IV, Abagyan R, Oprea TI. Surrogate data—A secure way to share corporate data. *J Comput Aided Mol Des* 2005;19:749–64.

146. Morrissey SR. NIH initiatives target chemistry. *Chem Eng News* 2005;83:23–4.

147. Tetko IV, Gasteiger J, Todeschini R, Mauri A, Livingstone D, Ertl P, et al. Virtual computational chemistry laboratory—Design and description. *J Comput Aided Mol Des* 2005;19:453–63.

10

APPLICATIONS OF QSAR TO ENZYMES INVOLVED IN TOXICOLOGY

SEAN EKINS

Contents

10.1 INTRODUCTION

Xenobiotics can be absorbed across the cellular barriers and may be biologically active and possibly toxic to the cell. Metabolism of these molecules by enzymes to hydrophilic metabolites is a prerequisite for their eventual elimination from the body. However, in some cases bioactivation may also occur, and such metabolites may be toxic. Xenobiotic metabolism has therefore been widely studied since the early 1800s [1]. The parent molecule and the products of metabolic pathways may also be involved in drug interactions where they

Computational Toxicology: Risk Assessment for Pharmaceutical and Environmental Chemicals, Edited by Sean Ekins

interfere with metabolism of endogenous or other co-administered compounds. Such drug–drug interactions, or other adverse drug reactions, can have potentially fatal consequences for the patient or be very costly for health care providers [2–6]. Research in metabolism is aimed at answering the questions about how the molecule is metabolized, which enzymes are involved in metabolism, what are the site/s of metabolism, what are the resulting metabolite/s, what is the rate of metabolism [7], and whether the metabolites are biologically active or potentially toxic [8]. There are also such questions as whether there are species differences in metabolism due to differences in the substrate specificity of the homologous enzymes involved and could this knowledge aid in extrapolation from preclinical species to human. As there have been increases in the throughput of experimental in vitro systems using various enzymes, datasets are being generated that can be used for computational models. Since the majority of xenobiotics undergo phase I metabolism via the cytochrome P450 (P450) enzymes primarily in liver, intestine, and kidney [9], this family of enzymes has been well studied. These enzymes have further been extensively modeled by various computational and structural methods. Enzymes are also expressed extrahepatically and demonstrate considerable activity in organs such as the intestine and kidney [9,10] (Table 10.1; also see Chapters 15–18). Secondary or tertiary metabolism may also occur via phase II reactions including glucuronidation, sulfation, and other phase II biotransformations (Table 10.1) [11]. However, the role of phase II enzymes in drug metabolism has not been studied to the same extent as the P450s [11], and this is reflected in the smaller number of computational models available. The prediction of the potential metabolites via the many endogenous enzymes in human is desirable but also extremely complex (Chapters 3 and 18). Computational approaches have only been applied to a fraction of the known enzymes involved in human drug metabolism (Table 10.1). The aim of this chapter is to focus on quantitative structure activity relationships (QSAR) methods (see Chapter 8) that have been used for modeling enzymes involved in human metabolism.

10.2 QSAR FOR ENZYMES

For a long time in the absence of X-ray crystal structures for many P450s, computational models provided considerable insights. A recent exhaustive review of computational methods for prediction of P450 metabolism and inhibition across all species has documented how these approaches have been used over nearly 20 years, and the reader is encouraged to refer to this [12].

10.2.1 2D and 3D-QSAR

Quantitative structure metabolism relationships (QSMR) were pioneered by Hansch and coworkers [13–16] using very small sets of similar molecules and

TABLE 10.1 Examples of Human Enzymes Involved in Drug Metabolism with Sources of Computational Data

Enzyme	Cellular Location	Reaction	Cofactor	Phase	QSAR	Pharmacophore
Cytochrome P450	Microsomal	Oxidation and reaction	NADPH	1	[13–16] [17–22, 26–28, 56,58, 59,87, 88]	[24,25,42–53]
Flavin containing monooxygenase	Microsomal	Oxidation	NADPH	1	[89]	
Monoamine oxidases	Mitochondrial	Oxidation	FAD	1	[90–94]	[95]
Aromatases	Mitochondrial	Oxidation	FAD	1	[96–101]	[101,102]
Esterases (Carboxylesterase, pseudocholinesterase, paraoxonase)	Microsomal, lysosomes and cytosolic	Hydrolysis		1	[103–106]	
Epoxide hydrolases	Microsomal and cytosolic	Hydration		1	[107,108]	[109]
Alcohol dehydrogenase	Cytosol, blood, microsomes	Carbonyl reduction	NAD^+	1	[110]	[111]
NAD(P)H-quinone oxidoreductase	Cytosol, microsomes	Quinone reduction	NAD[P]H	1	[112,113]	
Dihydropyrimidine Dehydrogenase	Cytosol	Reduction	NADPH	1		

TABLE 10.1 (*Continued*)

Enzyme	Cellular Location	Reaction	Cofactor	Phase	QSAR	Pharmacophore
Dihydrodiol dehydrogenase	Cytosol	Oxidation	NADP(H)	1		
Molybdenum hydroxylases	Cytosol	Oxidation	O_2 or NAD^+	1		
Polyamine oxidase	Cytosol	Oxidation	FAD	1		
Diamine oxidase	Cytosol	Oxidation	FAD	1		
Prostaglandin H-synthase	Microsomal	Oxidation		1	[114]	
UDP-glucuronosyltransferases	Microsomal	Glucuronidation	UDPGA	2	[32–39, 115]	[33,34,36]
Sulfotransferases	Cytosolic	Sulfation	PAPS	2	[40,115, 116]	
Phenyl-*O*-methyltransferase	Microsomal	*O*-Methylation	*S*-Adenosylmethionine	2		
Catechol *O*-methyltransferase	Cytosolic and microsomal	*O*-Methylation	*S*-Adenosylmethionine	2	[115–118]	
Thiopurine methyltransferase	Cytoplasm	*S*-Methylation	*S*-Adenosylmethionine	2	[119]	
N-acetyl transferases	Cytosolic	Acetylation	Acetyl-coenzyme A	2	[120]	
Amino acid conjugation	Mitochondria, microsomal	Conjugation	Acyl-CoA, serine, proline			
Glutathione *S*-transferases	Microsomal and cytosolic	Glutathione conjugation	Glutathione	2	[90]	

Source: Modified from [10].

Note: Abbreviations NADPH, β-nicotinamide adenine dinucleotide phosphate reduced from; NAD, β-nicotinamide adenine dinucleotide; FAD, flavin adenine dinucleotide; PAPS, 3′-phosphoadenosine 5′-phosphosulfate; UDPGA, uridine diphosphate-glucuronic acid.

a few molecular descriptors. Later Lewis and coworkers [17–22] provided many QSAR studies for the human P450s that resulted in a decision tree for classifying human P450 substrates [18]. Lipophilicity expressed as log *P* or molecular refractivity were the first important molecular properties related to enzyme substrate binding. These were followed by steric, electronic, and molecular shape properties that were also found to be important for enzyme binding and transformation. Conversely, metabolite release likely requires the opposite properties to binding [7]. QSAR models have been constructed for virtually all major human P450 enzymes using in vitro inhibition data or substrate data derived with human liver microsomal or expressed enzymes. As more computationally complex and graphically intensive software tools became available in the late 1980s and in the 1990s, this resulted in increased levels of QSAR analysis applied to drug-metabolizing enzymes. Software that has been used for 3D-QSAR includes Catalyst (Accelrys, San Diego, CA), DISCO, CoMFA, ALMOND (Tripos Associates, St. Louis, MO), and GOLPE (Multivariate Infometric Analysis, S.r.l., Perugia), have all been described in detail [23]. CoMFA was used to describe key molecular features of ligands for human CYP1A2 [24] and CYP2C9 [25], and more recently for CYP2C9 ligands [26]. These methods introduced researchers to computational models for enzymes important for human drug metabolism, but the datsets used were small and likely capture only a very small portion of chemical space. Many of the early models rarely used test sets of molecules to evaluate their predictive capability.

More recently other QSAR methods have been used to generate predictions. Kohonen maps have been useful for differentiating high- and low-affinity CYP3A4 substrates [27], while neural networks have been used to predict *N*-dealkylation rates for CYP3A4 and CYP2D6 substrates [28]. NMR T1 relaxation data have been proposed as descriptors for use in QSAR models to describe molecules in the P450 enzyme's binding site [29].

In terms of human drug metabolism there has been only very limited QSAR modeling of phase II enzymes. For example, small lipophillic molecules can also undergo glucuronidation, which is a further important route for drug clearance [30]. These membrane-bound enzymes have not been crystallized. Several QSAR models have been described (Table 10.1), including a study that described the glucuronidation of 4-substituted phenols by the human recombinant UGT1A6 and UGT1A9 enzymes [31]. A genetic algorithm and a range of molecular surface and atomic descriptors enabled one of the first computational attempts to predict the K_m for these enzymes [31]. More recently other QSAR algorithm methods such as pharmacophores and support vector machines have been used with quantum chemical and 2D descriptors for UGTs [32–39]. The datasets are limited in terms of structural diversity compared with the P450 models, but this situation is likely to improve as more data are generated. A further class of conjugating enzymes are the sulphotransferases, which, in contrast to UGTs, have been crystallized [40,41], and a QSAR method has been used to predict substrate affinity to SULT1A3 [40].

10.2.2 Pharmacophores

Computational pharmacophore models have been widely applied to predicting metabolism and interactions with P450s. A pharmacophore represents the key features present in ligands likely important for a biological response. The ligand molecular features are generally translated into spheres onto which molecule structures themselves can be mapped in 3D space [23]. Many pharmacophores have been generated for P450s [42], including CYP1A2, CYP2A6, CYP2B6, CYP2C9, CYP2D6, CYP3A4, CYP3A5, and CYP3A7 [24,43–53]. The CYP3A enzymes are perhaps the most important human drug-metabolizing enzymes [54], with a very broad substrate specificity metabolizing a very large proportion of marketed drugs. Computational pharmacophores for CYP3A4 have been developed for substrates [55] and inhibitors [51,55,56] using an array of kinetic constants (K_m, $K_{i(apparent)}$, and IC_{50}) [42]. The more recent development of benzbromarone analogues that are CYP2C9 [57,58] and 2C19 [59] inhibitors with K_i values in the nM range has enabled the further extension of the pharmacophore /3D-QSAR models that this group has been developing and refining over many years [26]. These analogues pointed to a role for hydrophobic interactions that was addressed in an earlier pharmacophore study [49]. The incorporation of such high-affinity inhibitors will improve the model statistics and be useful for searching databases for other inhibitors consisting of the same pharmacophore.

The computational pharmacophore approach has been used to develop a model for the features of molecules that increase their own metabolism (autoactivators) via CYP3A4 [55]. As such, nonhyperbolic kinetics have been reported for numerous P450s [60–62] and UGTs [11,63]. This may be important, however, the in vivo relevance is unclear. Several CYP2B6 reactions have also been reported to display atypical enzyme kinetics with recombinant enzyme [61], yet there has not been any development of a pharmacophore for CYP2B6 autoactivators. There have been several CYP2B6 substrate pharmacophores, one of which [45] may have some similarity to the CYP3A4 autoactivator model, which also has three hydrophobic features around a hydrogen-bond acceptor. This may represent a common autoactivator motif. The pharmacophore approach has further been used with heteroactivators (a molecule increases the metabolism of another molecule that is metabolized by the same enzyme) of CYP3A4 and CYP2C9 metabolism [64,65]. These pharmacophores define the key features necessary for autoactivation and heteroactivation and possibly could be used to relate to a specific binding site/s in the respective enzyme.

Pharmacophores have been applied to various human enzymes involved in glucuronidation using a custom metabolism pharmacophore feature [33,34,36]. This way it was possible to derive pharmacophore models for UDPGT1A4 [36], UDPGT1A1 [34,35], and others [39]. This work has been reviewed in detail by Miners et al. [37]. Pharmacophores assume a similar binding mode and interaction with the protein, which are unlikely to be the situation in enzymes like P450s as they generally do not indicate reactivity (although some pharmacophores have included features for the site of metabolism [66]). Pharmacophores are in many ways simplistic representations for enzymes, yet they have been useful in delineating the shapes and distances between key features in molecules likely to be substrates or inhibitors, allowing researchers to visualize features in molecules that can be avoided or enable rapid database searching for molecules.

10.2.3 Electronic Models

Molecular models accounting for electronic effects of ligands for P450-mediated metabolism have been published [25,67,68]. These methods generally depend on the calculation of ground state energies and in some cases have combined aliphatic and aromatic oxidation reactions. With steric and orientation terms, predictions have been generated for metabolic regioselectivities of enzymes, in general [67,69], or for specific enzymes such as CYP2E1 [70] and CYP3A4 [68]. In the case of CYP3A4 a partial least squares method was trained with AM1 calculated hydrogen abstraction energy data to rapidly speed up the prediction of these values for molecules. The combination of electronic methods with steric and orientation terms has also been described to limit overfitting of the training data and improve predictions [71]. An electronic model has been developed for hydrogen abstraction for a series of steroidal androgens [72]. Electronic methods have to date been much less widely applied than QSMR and QSAR methods likely due to their slow calculation speeds, and there have been no comparisons of predictions from electronic models and other QSMR. There is certainly a need for the further development of these technologies alongside the other methods described here.

10.2.4 Hybrid Methods

The history of methods used for the computational prediction of human drug metabolism includes several different approaches such as databases, QSMR/QSAR, pharmacophores, rule-based approaches, electronic models, homology models, and crystal structures with docking approaches (see Chapters 15–18). These techniques, when used individually, have different levels of success. They could be combined to improve predictions, perhaps by deploying the most appropriate models based on the molecule neighborhood for the test molecule compared to training set/s. P450-substrate/inhibitor recognition interactions have been studied extensively and have generally shown the importance of hydrophobic, hydrogen bonding, and ionizable features for both substrates based on K_m data and inhibitors using K_i, IC_{50}, and percent inhibition data [42]. Molecular models that account for electronic effects of ligands for P450-mediated metabolism have also been produced [25,67], and these have combined aliphatic and aromatic oxidation reactions to generate predictions for metabolic regioselectivities. The combination of approaches may also balance the strengths and weaknesses of each approach, and hence the introduction of hybrid methods is ongoing. For example, a recent technique called MetaSite (Molecular Discovery, Middlesex, UK) generates GRID field descriptors for determining energetically favorable binding sites on molecules of known structure using crystal structures or homology models for the P450 enzymes and the interaction energy descriptors for the molecules evaluated as substrates [73]. A reactivity component is also used in the MetaSite calculation to produce a probability for an atom to be metabolized. This approach has been applied with AT receptor antagonists predicting the site of metabolism for CYP2C9 and CYP3A4 [73]. A recent study has expanded the application to CYP2D6, CYP2C19, and CYP1A2 with 75–86% correct predictions [74]. MetaSite has been compared with docking of ligands into crystal structures and homology models of CYP3A4 and MetaSite had 78% overall success compared with 57% for docking [75]. MetaSite has a limitation of not being able to predict the absolute or relative amounts of the major and minor metabolites as well as the rate of metabolite formation for a molecule. A second hybrid method MetaDrug™ includes a manually annotated Oracle™ database of human drug metabolism information comprised of xenobiotic reactions, enzyme substrates, and enzyme inhibitors with kinetic data. The MetaDrug™ database has been used to predict some of the major metabolic pathways and identify the involvement of P450s by way of multiple QSAR models enabling the prediction of affinity and rate of metabolism for numerous enzymes [56,76–78]. The user can also upload their own QSAR or QSMR data into the software to offer a further level of utility. Steps have been undertaken to provide more confidence in predictions, since the QSAR methods also provide Tanimoto similarity as a measure of similarity to training set molecules. Structural alerts for likely reactive metabolites [79–81] are also provided. MetaDrug™ has been tested with 66 molecules and captured approximately 79% of first-pass metabolites [82], which is very similar to the success rates for MetaSite. The results of the QSAR model predictions can be visualized as nodes on a network diagram, representing a novel graphical method for presenting predicted drug–drug interactions. Since this method combines various methods, for example, QSAR and rules-based methods, the user should be aware of their individual limitations.

10.3 LIMITATIONS

A drug-metabolizing enzyme may produce several products from a single substrate, and these metabolites may in turn be substrates for other enzymes or for close family members. Predicting the major metabolites is therefore challenging. The substrate promiscuity of enzymes such as CYP3A4 [83] known to metabolize molecules at multiple positions also severely complicates predictions for metabolite formation. This provides further weight for combining different computational methods to predict affinity for enzymes to improve their overall accuracy.

Methods to reliably predict the further effects of metabolites on the complete biological system are needed to aid in the selection of molecules to be synthesized and tested in vitro or in vivo, and thus to ultimately provide feedback on any likely toxicity [84]. Graphical representations of this type of information are also important for the biologist and chemist to understand what new research directions they should take.

The limitations of QSAR for enzymes are related to the fact that the experimental measurement of kinetic parameters is inherently prone to errors. Kinetic constants for the same compound vary substantially among studies, depending on the enzyme source (recombinant enzyme, purified enzyme, subcellular fraction, etc.) or experimental conditions. Reported V_{max} values for the same compound can vary by 2 to 3 orders of magnitude, seriously impacting regression-based QSAR modeling. Therefore much larger, consistent datasets for each enzyme will be required to increase the predictive scope of such models.

The number of QSAR approaches that have been applied to the P450s (CYP2C9, CYP2D6, and CYP3A4 predominantly) is quite large; nevertheless, because of the availability of only very small datasets for other enzymes, these have been less well studied with QSAR methods (Table 10.1). This is similar to the case for drug transporters in which many QSAR methods have been used with P-glycoprotein resulting in many different datasets (see Chapter 11). This is also similar with ion channels for which there are numerous QSAR for hERG, and only several hundred molecules with useful data (see Chapter 13) while other ion channels (besides the calcium channel) have been less widely modeled. We may also be seeing the same with the receptors as PXR has been the subject of several QSAR models (see Chapters 12 and 17). Interestingly all of these proteins have been the focus of QSAR modeling and are relatively promiscuous in their ability to bind a large array of primarily hydrophobic molecules. This represents a challenge for both the utility of the final models and the confidence with which they can be applied to different molecular series that are not represented in the training sets. There is certainly scope in both generating new datasets for enzymes relevant to toxicology that have not been crystallized yet and that have not been used for QSAR modeling such as flavin containing monooxygenase and thiopurine methyltransferase (Table 10.1).

10.4 CONCLUSIONS

QSAR and pharmacophore models for enzymes represent an approach for rapidly screening databases of molecules from companies or vendors that can then be selected and tested in vitro as a rapid approach for finding new inhibitors or substrates. This approach has also been used with transporters (see Chapter 11) [85,86]. With numerous crystal structures for mammalian P450s becoming available in recent years (Chapter 17), QSAR models may be used in addition to docking and other computational methods for both prediction and ultimately validation.

REFERENCES

1. Leibman KC. Drug metabolism: Prospects for the future. *Drug Metab Rev* 1979;10:299–309.
2. Doucet J, Chassagne P, Trivalle C, Landrin I, Pauty MD, Kadri N, et al. Drug–drug interactions related to hospital admissions in older adults: A prospective study of 1000 patients. *JAGS* 1996;44:944–8.
3. Yee JL, Hasson NK, Schreiber DH. Drug-related emergency department visits in an elderly veteran population. *An Pharmacother* 2005;39:1990–4.
4. Pezalla E. Preventing adverse drug reactions in the general population. *Manag Care Interface* 2005;18:49–52.
5. Zhan C, Correa-de-Araujo R, Bierman AS, Sangl J, Miller MR, Wickizer SW, et al. Suboptimal prescribing in elderly outpatients: Potentially harmful drug-drug and drug-disease combinations. *J Am Geriatr Soc* 2005;53:262–7.
6. Klarin I, Wimo A, Fastbom J. The association of inappropriate drug use with hospitalisation and mortality: A population-based study of the very old. *Drugs Aging* 2005;22:69–82.
7. Austel V, Kutter E. Absorption, distribution, and metabolism of drugs. In: Topliss JG, editor, *Quantitative structure-activity relationships of drugs*. New York: Academic Press, 1983. p. 437–96.
8. Smith DA, Obach RS. Seeing through the mist: Abundance versus percentage. Commentary on metabolites in safety testing. *Drug Metab Dispos* 2005;33: 1409–17.
9. Paine MF, Khalighi M, Fisher JM, Shen DD, Kunze KL, Marsh CL, et al. Characterization of interintestinal and intraintestinal variations in human CYP3A-dependent metabolism. *J Pharm Exp Ther* 1997;283:1552–62.
10. Ekins S. Computer methods for predicting drug metabolism. In: Ekins S, editor, *Computer applications in pharmaceutical research and development*. Hoboken, NJ: Wiley, 2006. p. 445–68.
11. Williams JA, Hyland R, Jones BC, Smith DA, Hurst S, Goosen TC, et al. Drug–drug interactions for UDP-glucuronosyltransferase substrates: A pharmacokinetic explanation for typically observed low exposure (AUCi/AUC) ratios. *Drug Metab Dispos* 2004;32:1201–8.
12. de Graaf C, Vermeulen NP, Feenstra KA. Cytochrome P450 in silico: An integrative modeling approach. *J Med Chem* 2005;48:2725–55.
13. Hansch C. Quantitative relationships between lipophilic character and drug metabolism. *Drug Metab Rev* 1972;1:1–14.
14. Hansch C. The QSAR paradigm in the design of less toxic molecules. *Drug Metab Rev* 1984;15:1279–94.
15. Hansch C, Lien EJ, Helmer F. Structure-activity correlations in the metabolism of drugs. *Arch Biochem Biophys* 1968;128:319–30.

16. Hansch C, Zhang L. Quantitative structure-activity relationships of cytochrome P-450. *Drug Metab Rev* 1993;25:1–48.

17. Lewis DFV. Quantitative structure activity relationships in substrates, inducers, and inhibitors of cytochrome P4501 (CYP1). *Drug Metab Rev* 1997;29:589–650.

18. Lewis DFV. On the recognition of mammalian microsomal cytochrome P450 substrates and their characteristics. *Biochem Pharmacol* 2000;60:293–306.

19. Lewis DFV. Structural characteristics of human P450s involved in drug metabolism: QSARs and lipophilicity profiles. *Toxicology* 2000;144:197–203.

20. Lewis DFV, Eddershaw PJ, Dickins M, Tarbit MH, Goldfarb PS. Structural determinants of cytochrome P450 substrate specificity, binding affinity and catalytic rate. *Chem Bio Interact* 1998;115:175–99.

21. Lewis DFV, Eddershaw PJ, Dickins M, Tarbit MH, Goldfarb PS. Erratum to structural determinants of cytochrome P450 substrate specificity, binding affinity and catalytic rate. *Chem Biol Interact* 1999;117:187.

22. Lewis DF, Jacobs MN, Dickins M. Compound lipophilicity for substrate binding to human P450s in drug metabolism. *Drug Disc Today* 2004;9:530–7.

23. Ekins S, Swaan PW. Development of computational models for enzymes, transporters, channels and receptors relevant to ADME/TOX. *Rev Comp Chem* 2004;20:333–415.

24. Fuhr U, Strobl G, Manaut F, Anders E-M, Sorgel F, Lopez-de-brinas E, et al. Quinolone antibacterial agents: relationship between structure and in vitro inhibition of human cytochrome P450 isoform CYP1A2. *Mol Pharmacol* 1993;43: 191–9.

25. Jones JP, Korzekwa KR. *Predicting the rates and regioselectivity of reactions mediated by the P450 superfamily.* New York: Academic Press, 1996.

26. Locuson CW, Wahlstrom JL. Three-dimensional quantitative structure-activity relationship analysis of cytochromes p450: Effect of incorporating higher-affinity ligands and potential new applications. *Drug Metab Dispos* 2005;33:873–8.

27. Balakin KV, Ekins S, Bugrim A, Ivanenkov YA, Korolev D, Nikolsky Y, et al. Kohonen maps for prediction of binding to human cytochrome P450 3A4. *Drug Metab Dispos* 2004;32:1183–9.

28. Balakin KV, Ekins S, Bugrim A, Ivanenkov YA, Korolev D, Nikolsky Y, et al. Quantitative structure-metabolism relationship modeling of the metabolic *N*-dealkylation rates. *Drug Metab Dispos* 2004;32:1111–20.

29. Yao H, Costache AD, Sem DS. Chemical proteomic tool for ligand mapping of CYP antitargets: An NMR-compatible 3D QSAR descriptor in the heme-based coordinate system. *J Chem Inf Comput Sci* 2004;44:1456–65.

30. Tukey RH, Strassburg CP. Human UDP-glucuronosyltransferases: Metabolism, expression, and disease. *An Rev Pharmacol Toxicol* 2000;40:581–616.

31. Ethell BT, Ekins S, Wang J, Burchell B. Quantitative structure activity relationships for the glucuronidation of simple phenols by expressed human UGT1A6 and UGT1A9. *Drug Metab Dispos* 2002;30:734–8.

32. Sorich MJ, McKinnon RA, Miners JO, Winkler DA, Smith PA. Rapid prediction of chemical metabolism by human UDP-glucuronosyltransferase isoforms using quantum chemical descriptors derived with the electronegativity equalization method. *J Med Chem* 2004;47:5311–17.

33. Smith PA, Sorich M, McKinnon R, Miners JO. QSAR and pharmacophore modelling approaches for the prediction of UDP-glucuronosyltransferase substrate selectivity and binding. *Pharmacologist* 2002;44(Suppl).
34. Sorich M, Smith PA, McKinnon RA, Miners JO. Pharmacophore and quantitative structure activity relationship modelling of UDP-glucuronosyltransferase 1A1 (UGT1A1) substrates. *Pharmacogenetics* 2002;12:635–45.
35. Smith PA, Sorich MJ, McKinnon RA, Miners JO. In silico insights: Chemical and structural characteristics associated with uridine diphosphate-glucuronosyltransferase substrate selectivity. *Clin Exp Pharmacol Physiol* 2003;30:836–40.
36. Smith PA, Sorich MJ, McKinnon RA, Miners JO. Pharmacophore and quantitative structure-activity relationship modeling: Complementary approaches for the rationalization and prediction of UDP-glucuronosyltransferase 1A4 substrate selectivity. *J Med Chem* 2003;46:1617–26.
37. Miners JO, Smith PA, Sorich MJ, McKinnon RA, Mackenzie PI. Predicting human drug glucuronidation parameters: Application of in vitro and in silico modeling approaches. *An Rev Pharmacol Toxicol* 2004;44:1–25.
38. Smith PA, Sorich MJ, Low LS, McKinnon RA, Miners JO. Towards integrated ADME prediction: Past, present and future directions for modelling metabolism by UDP-glucuronosyltransferases. *J Mol Graph Model* 2004;22:507–17.
39. Sorich MJ, Miners JO, McKinnon RA, Smith PA. Multiple pharmacophores for the investigation of human UDP-glucuronosyltransferase isoform substrate selectivity. *Mol Pharmacol* 2004;65:301–8.
40. Dajani R, Cleasby A, Neu M, Wonacott AJ, Jhoti H, Hood AM, et al. X-ray crystal structure of human dopamine sulfotransferase, SULT1A3. *J Biol Chem* 1999;53:37862–8.
41. Gamage NU, Duggleby RG, Barnett AC, Tresillian M, Latham CF, Liyou NE, et al. Structure of a human carcinogen-converting enzyme, SULT1A1. *J Biol Chem* 2003;278:7655–62.
42. Ekins S, de Groot M, Jones JP. Pharmacophore and three dimensional quantitative structure activity relationship methods for modeling cytochrome P450 active sites. *Drug Metab Dispos* 2001;29:936–44.
43. Ekins S, VandenBranden M, Ring BJ, Wrighton SA. Examination of purported probes of human CYP2B6. *Pharmacogenetics* 1997;7:165–79.
44. Ekins S, VandenBranden M, Ring BJ, Gillespie JS, Yang TJ, Gelboin HV, et al. Further characterization of the expression and catalytic activity of human CYP2B6. *J Pharm Exp Ther* 1998;286:1253–9.
45. Ekins S, Bravi G, Ring BJ, Gillespie TA, Gillespie JS, VandenBranden M, et al. Three dimensional-quantitative structure activity relationship (3D-QSAR) analyses of substrates for CYP2B6. *J Pharm Exp Ther* 1999;288:21–9.
46. Ekins S, Bravi G, Wikel JH, Wrighton SA. Three dimensional quantitative structure activty relationship (3D-QSAR) analysis of CYP3A4 substrates. *J Pharm Exp Ther* 1999;291:424–33.
47. Ekins S, Bravi G, Binkley S, Gillespie JS, Ring BJ, Wikel JH, et al. Three and four dimensional-quantitative structure activity relationship (3D/4D-QSAR) analyses of CYP2D6 inhibitors. *Pharmacogenetics* 1999;9:477–89.

48. Ekins S, Bravi G, Binkley S, Gillespie JS, Ring BJ, Wikel JH, et al. Three and four dimensional-quantitative structure activity relationship analyses of CYP3A4 inhibitors. *J Pharm Exp Ther* 1999;290:429–38.
49. Ekins S, Bravi G, Binkley S, Gillespie JS, Ring BJ, Wikel JH, et al. Three and four dimensional-quantitative structure activity relationship (3D/4D-QSAR) analyses of CYP2C9 inhibitors. *Drug Metab Dispos* 2000;28:994–1002.
50. Snyder R, Sangar R, Wang J, Ekins S. Three dimensional quantitative structure activity relationship for CYP2D6 substrates. *Quant Struct-Act Rel* 2002;21: 357–68.
51. Ekins S, Stresser DM, Williams JA. In vitro and pharmacophore insights into CYP3A enzymes. *Trends Pharmacol Sci* 2003;24:191–6.
52. Ekins S, Wrighton SA. Application of in silico approaches to predicting drug–drug interactions: A commentary. *J Pharm Tox Meth* 2001;44:1–5.
53. Asikainen A, Tarhanen J, Poso A, Pasanen M, Alhava E, Juvonen RO. Predictive value of comparative molecular field analysis modelling of naphthalene inhibition of human CYP2A6 and mouse CYP2A5 enzymes. *Toxicol In Vitro* 2003;17: 449–55.
54. Wrighton SA, Schuetz EG, Thummel KE, Shen DD, Korzekwa KR, Watkins PB. The human CYP3A subfamily: Practical considerations. *Drug Metab Rev* 2000; 32:339–61.
55. Ekins S, Wrighton SA. The role of CYP2B6 in human xenobiotic metabolism. *Drug Metab Rev* 1999;31:719–54.
56. Ekins S, Berbaum J, Harrison RK. Generation and validation of rapid computational filters for CYP2D6 and CYP3A4. *Drug Metab Dispos* 2003;31:1077–80.
57. Locuson CW, 2nd, Wahlstrom JL, Rock DA, Rock DA, Jones JP. A new class of CYP2C9 inhibitors: Probing 2C9 specificity with high-affinity benzbromarone derivatives. *Drug Metab Dispos* 2003;31:967–71.
58. Locuson CW II, Rock DA, Jones JP. Quantitative binding models for CYP2C9 based on benzbromarone analogues. *Biochemistry* 2004;43:6948–58.
59. Locuson CW, 2nd, Suzuki H, Rettie AE, Jones JP. Charge and substituent effects on affinity and metabolism of benzbromarone-based CYP2C19 inhibitors. *J Med Chem* 2004;47:6768–76.
60. Korzekwa KR, Krishnamachary N, Shou M, Ogai A, Parise RA, Rettie AE, et al. Evaluation of atypical cytochrome P450 kinetics with two-substrate-models: Evidence that multiple substrates can simultaneously bind to cytochrome P450 active sites. *Biochemistry* 1998;37:4137–47.
61. Ekins S, Ring BJ, Binkley SN, Hall SD, Wrighton SA. Autoactivation and activation of cytochrome P450s. *Int J Clin Pharmacol Ther* 1998;36:642–51.
62. Atkins WM. Non–Michaelis-Menten kinetics in cytochrome P450-catalyzed reactions. *An Rev Pharmacol Toxicol* 2005;45:291–310.
63. Bauman JN, Goosen TC, Tugnait M, Peterkin V, Hurst SI, Menning LC, et al. Udp-glucuronosyltransferase 2b7 is the major enzyme responsible for gemcabene glucuronidation in human liver microsomes. *Drug Metab Dispos* 2005;33: 1349–54.
64. Egnell AC, Houston JB, Boyer CS. Predictive models of CYP3A4 heteroactivation: In vitro–in vivo scaling and pharmacophore modelling. *J Pharmacol Exp Ther* 2005;312:926–37.

65. Egnell AC, Eriksson C, Albertson N, Houston B, Boyer S. Generation and evaluation of a CYP2C9 heteroactivation pharmacophore. *J Pharmacol Exp Ther* 2003;307:878–87.
66. Wang Q, Halpert JR. Combined three-dimensional quantitative structure-activity relationship analysis of cytochrome P450 2B6 substrates and protein homology modeling. *Drug Metab Dispos* 2002;30:86–95.
67. Jones JP, Mysinger M, Korzekwa KR. Computational models for cytochrome P450: A predictive electronic model for aromatic oxidation and hydrogen abstraction. *Drug Metab Dispos* 2002;30:7–12.
68. Singh SB, Shen LQ, Walker MJ, Sheridan RP. A model for likely sites of CYP3A4-mediated metabolism on drug-like molecules. *J Med Chem* 2003;46:1330–6.
69. Csanady GA, Laib JG. Metabolic transformation of halogenated and other alkenes- a theoretical approach: Estimation of metabolic reactivities for in vivo conditions. *Toxicology* 1995;75:217–23.
70. Yin H, Anders MW, Korzekwa KR, Higgins L, Thummel KE, Kharasch ED, et al. Designing safer chemicals: Predicting the rates of metabolism of halogenated alkanes. *Proc Natl Acad Sci USA* 1995;92:11076–80.
71. Korzekwa K, Ewing TJ, Kocher JP, Carlson TJ. Models for cytochrome P450-mediated metabolism. In: Borchardt RT, Kerns EH, Lipinski CA, Thakker DR, Wang B, editors, *Pharmaceutical profiling in drug discovery for lead selection.* AAPS Press, 2004. Arlington, VA. p. 69–80.
72. Bursi R, de Gooyer ME, Grootenhuis A, Jacobs PL, van der Louw J, Leysen D. (Q)SAR study on the metabolic stability of steroidal androgens. *J Mol Graph Model* 2001;19:552–6.
73. Berellini G, Cruciani G, Mannhold R. Pharmacophore, drug metabolism, and pharmacokinetics models on non-peptide AT1, AT2, and AT1/AT2 angiotensin II receptor antagonists. *J Med Chem* 2005;48:4389–99.
74. Cruciani G, Carosati E, De Boeck B, Ethirajulu K, Mackie C, Howe T, et al. MetaSite: Understanding metabolism in human cytochromes from the perspective of the chemist. *J Med Chem* 2005;48:6970–9.
75. Zhou D, Afzelius L, Grimm SW, Andersson TB, Zauhar RJ, Zamora I. Comparison of methods for the prediction of the metabolic sites for CYP3A4-mediated metabolic reactions. *Drug Metab Dispos* 2006;34:976–83.
76. Korolev D, Balakin KV, Nikolsky Y, Kirillov E, Ivanenkov YA, Savchuk NP, et al. Modeling of human cytochrome P450-mediated drug metabolism using unsupervised machine learning approach. *J Med Chem* 2003;46:3631–43.
77. Young SS, Ekins S, Lambert C. So many targets, so many compounds, but so few resources. *Curr Drug Disc* 2002;Dec:17–22.
78. Young SS, Gombar VK, Emptage MR, Cariello NF, Lambert C. Mixture deconvolution and analysis of Ames mutagenicity data. *Chemom Intell Lab Sys* 2002;60: 5–11.
79. Li AP. A review of the common properties of drugs with idiosyncratic hepatotoxicity and the "multiple determinant hypothesis" for the manifestation of idiosyncratic drug toxicity. *Chem Biol Interact* 2002;142:7–23.
80. Williams DP, Park BK. Idiosyncratic toxicity: The role of toxicophores and bioactivation. *Drug Discov Today* 2003;8:1044–50.

81. Uetrecht J. Screening for the potential of a drug candidate to cause idiosyncratic drug reactions. *Drug Discov Today* 2003;8:832–7.

82. Ekins S, Andreyev S, Ryabov A, Kirillov E, Rakhmatulin EA, Sorokina S, et al. A combined approach to drug metabolism and toxicity assessment. *Drug Metab Dispos* 2006;34:495–503.

83. Ekins S. Predicting undesirable drug interactions with promiscuous proteins in silico. *Drug Discov Today* 2004;9:276–85.

84. Ekins S, Giroux C. Computers and systems biology for pharmaceutical research and development. In: Ekins S, editor, *Computer applications in pharmaceutical research and development*. Hoboken, NJ: Wiley, 2006. p. 139–65.

85. Chang C, Ekins S, Bahadduri P, Swaan PW. Pharmacophore-based discovery of ligands for drug transporters. *Adv Drug Del Rev* 2006;58(12–13):1431–30.

86. Chang C, Bahadduri PM, Polli JE, Swaan PW, Ekins S. Rapid identification of *P*-glycoprotein substrates and inhibitors. *Drug Metab Dispos* 2006;34:1976–84.

87. Lill MA, Dobler M, Vedani A. Prediction of small-molecule binding to cytochrome P450 3A4: Flexible docking combined with multidimensional QSAR. *ChemMedChem* 2006;6:73–81.

88. Mao B, Gozalbes R, Barbosa F, Migeon J, Merrick S, Kamm K, et al. QSAR modeling of in vitro inhibition of cytochrome P450 3A4. *J Chem Inf Model* 2006;46:2125–34.

89. Kim YM, Ziegler DM. Size limits of thiocarbamides accepted as substrates by human flavin-containing monooxygenase 1. *Drug Metab Dispos* 2000;28:1003–6.

90. Soffers AEMF, Ploeman JHTM, Moonen MJH, Wobbes T, van Ommen B, Vervoort J, et al. Regioselectivity and quantitative structure-activity relationships for the conjugation of a series of fluoronitrobenzenes by purified glutathione *S*-transferase enzymes from rat and man. *Chem Res Toxicol* 1996;9:638–46.

91. Medvedev AE, Veselovsky AV, Shvedov VI, Tikhonova OV, Moskvitina TA, Fedotova OA, et al. Inhibition of monoamine oxidase by pirlindole analogues: 3D-QSAR and CoMFA analysis. *J Chem Inf Comput Sci* 1998;38:1137–44.

92. Miller JR, Edmondson DE. Structure-activity relationships in the oxidation of para-substituted benzylamine analogues by recombinant human liver monoamine oxidase A. *Biochemistry* 1999;38:13670–83.

93. Chimenti F, Bolasco A, Manna F, Secci D, Chimenti P, Granese A, et al. Synthesis, biological evaluation and 3D-QSAR of 1,3,5-trisubstituted-4,5-dihydro-(1H)-pyrazole derivatives as potent and highly selective monoamine oxidase A inhibitors. *Curr Med Chem* 2006;13:1411–28.

94. Carrieri A, Carotti A, Barreca ML, Altomare C. Binding models of reversible inhibitors to type-B monoamine oxidase. *J Comput Aided Mol Des* 2002;16: 769–78.

95. Ekins S, Waller CL, Swaan PW, Cruciani G, Wrighton SA, Wikel JH. Progress in predicting human ADME parameters in silico. *J Pharmacol Toxicol Meth* 2000;44:251–72.

96. Cavalli A, Greco G, Novellino E, Recanatini M. Linking CoMFA and protein homology models of enzyme-inhibitor interactions: An application to non-steroidal aromatase inhibitors. *Bioorg Med Chem* 2000;8:2771–80.

97. Leonetti F, Favia A, Rao A, Aliano R, Paluszcak A, Hartmann RW, et al. Design, synthesis, and 3D QSAR of novel potent and selective aromatase inhibitors. *J Med Chem* 2004;47:6792–803.

98. Nagy PI, Tokarski J, Hopfinger AJ. Molecular shape and QSAR analyses of a famiy of substituted dichlorodiphenyl aromatase inhibitors. *J Chem Inf Comput Sci* 1994;34:1190–7.

99. Recanatini M. Comparative molecular field analysis of nonsteroidal aromatase inhibitors related to fadrozole. *J Comp Aided Mol Des* 1996;10:74–82.

100. Recanatini M, Cavalli A. Comparative molecular field analysis of non-steroidal aromatase inhibitors: An extended model for two different structural classes. *Bioorg Med Chem* 1998;6:377–88.

101. Sonnet P, Dallemagne P, Guillon J, Enguehard C, Stiebing S, Tanguy J, et al. New aromatase inhibitors, synthesis and biological activity of aryl-substituted pyrrolizine and indolizine derivatives. *Bioorg Med Chem* 2000;8:945–55.

102. Schuster D, Laggner C, Steindl TM, Palusczak A, Hartmann RW, Langer T. Pharmacophore modeling and in silico screening for new P450 19 (aromatase) inhibitors. *J Chem Inf Model* 2006;46:1301–11.

103. Wadkins RM, Hyatt JL, Yoon KJ, Morton CL, Lee RE, Damodaran K, et al. Discovery of novel selective inhibitors of human intestinal carboxylesterase for the amelioration of irinotecan-induced diarrhea: Synthesis, quantitative structure-activity relationship analysis, and biological activity. *Mol Pharmacol* 2004;65: 1336–43.

104. Wadkins RM, Hyatt JL, Wei X, Yoon KJ, Wierdl M, Edwards CC, et al. Identification and characterization of novel benzil (diphenylethane-1,2-dione) analogues as inhibitors of mammalian carboxylesterases. *J Med Chem* 2005;48: 2906–15.

105. Knaak JB, Dary CC, Power F, Thompson CB, Blancato JN. Physicochemical and biological data for the development of predictive organophosphorus pesticide QSARs and PBPK/PD models for human risk assessment. *Crit Rev Toxicol* 2004;34:143–207.

106. Buchwald P, Bodor N. Quantitative structure-metabolism relationships: Steric and non steric effects in the enzymatic hydrolysis of noncongener carboxylic esters. *J Med Chem* 1999;42:5160–8.

107. McElroy NR, Jurs PC, Morisseau C, Hammock BD. QSAR and classification of murine and human epoxide hydrolase inhibition by urea-like compounds. *J Med Chem* 2003;46:1066–80.

108. Nakagawa Y, Wheelock CE, Morisseau C, Goodrow MH, Hammock BG, Hammock BD. 3D-QSAR analysis of inhibition of murine soluble epoxide hydrolase (MsEH) by benzoylureas, arylureas, and their analogues. *Bioorg Med Chem* 2000;8:2663–73.

109. Kim EJ, Kim KS, Shin WH. Electrophysiological safety of DW-286a, a novel fluoroquinolone antibiotic agent. *Hum Exp Toxicol* 2005;24:19–25.

110. Hansch C, Klein T, McClarin J, Langridge R, Cornell NW. A quantitative structure-activity relationship and molecular graphics analysis of hydrophobic effects in the interactions of inhibitors with alcohol dehydrogenase. *J Med Chem* 1986;29:615–20.

111. Kutsenko AS, Kuznetsov DA, Poroikov VV, Tumanian VG. Mapping of active site of alcohol dehydrogenase with low-molecular ligands. *Bioorg Khim* 2000;26: 179–86.

112. Skibo EB, Xing C, Dorr RT. Aziridinyl quinone antitumor agents based on indoles and cyclopent[b]indoles: Structure-activity relationships for cytotoxicity and antitumor activity. *J Med Chem* 2001;44:3545–62.

113. Khadikar P, Jaiswal M, Gupta M, Mandloi D, Sisodia RS. QSAR studies on 1,2-dithiole-3-thiones: Modeling of lipophilicity, quinone reductase specific activity, and production of growth hormone. *Bioorg Med Chem Lett* 2005;15:1249–55.

114. Pouplana R, Lozano JJ, Perez C, Ruiz J. Structure-based QSAR study on differential inhibition of human prostaglandin endoperoxide H synthase-2 (COX-2) by nonsteroidal anti-inflammatory drugs. *J Comput Aided Mol Des* 2002;16: 683–709.

115. Taskinen J, Ethell BT, Pihlavisto P, Hood AM, Burchell B, Coughtric MW. Conjugation of catechols by recombinant human sulfotransferases, UDP-glucuronosyltransferases, and soluble catechol *O*-methyltransferase: structure-conjugation relationships and predictive models. *Drug Metab Dispos* 2003;31: 1187–97.

116. Sipila J, Taskinen J. CoMFA modeling of human catechol *O*-methyltransferase enzyme kinetics. *J Chem Inf Comput Sci* 2004;44:97–104.

117. Lautala P, Ulmanen I, Taskinen J. Molecular mechanisms controlling the rate and specificity of catechol *O*-methylation by human soluble catechol *O*-methyltransferase. *Mol Pharmacol* 2001;59:393–402.

118. Chen D, Wang CY, Lambert JD, Ai N, Welsh WJ, Yang CS. Inhibition of human liver catechol-*O*-methyltransferase by tea catechins and their metabolites: Structure-activity relationship and molecular-modeling studies. *Biochem Pharmacol* 2005;69:1523–31.

119. Ames MM, Selassie CD, Woodson LC, Van Loon JA, Hansch C, Weinshilboum RM. Thiopurine methyltransferase: Structure-activity relationships for benzoic acid inhibitors and thiophenol substrates. *J Med Chem* 1986;29:354–8.

120. Mesangeau C, Yous S, Chavatte P, Ferry G, Audinot V, Boutin JA, et al. Design, synthesis and in vitro evaluation of novel benzo[b]thiophene derivatives as serotonin *N*-acetyltransferase (AANAT) inhibitors. *J Enzyme Inhib Med Chem* 2003;18:119–25.

11

QSAR STUDIES ON DRUG TRANSPORTERS INVOLVED IN TOXICOLOGY

GERHARD F. ECKER AND PETER CHIBA

Contents

11.1 INTRODUCTION

Almost 35% of all compounds in the drug development pipeline fail because of improper ADME behavior and toxicity. Thus in silico models and algorithms addressing bioavailability, phase I and phase II metabolism, blood–

Computational Toxicology: Risk Assessment for Pharmaceutical and Environmental Chemicals, Edited by Sean Ekins

brain barrier permeation, and toxicity are more or less routinely applied to further shape both the screening collection and combinatorial libraries. However, bioavailability and toxicity are a multifactorial issue, being mainly influenced by interactions with so-called nontarget proteins. Some of the key nontarget proteins responsible for poor ADME properties and/or high toxicity are the human ether-a-go-go-related gene (*hERG*) potassium channel (see Chapters 13, 16, 19, and 27), the cytochrome P450 enzyme complex (CYPs) (see Chapters 10 and 16), and multidrug efflux transporters such as the P-glycoprotein (ABCB1). Over the past decade a number of drug transporters have been identified as being expressed at various physiological barriers. There is increasing recognition that these transporters have a major role in drug absorption, disposition, toxicity, and efficacy [1]. Thus unexpected drug–drug interactions, such as those between digoxin and atorvastatin [2] or digoxin and St. John's wort [3], have been observed. Furthermore some of these transporters show considerable overlap with CYPs in their substrate profiles [4], and this necessitates a systems-oriented approach when predicting ADME parameters of a given compound.

11.2 THE PROBLEM OF MULTISPECIFICITY

With a deeper understanding of the processes involved in ADMET, the notion of avoiding interaction with drug transporters has gained increasing importance. Several key ABC-proteins identified so far in the ADMET cascade appear promiscuous in their binding interactions. Additionally they are membrane-bound proteins that up to now have resisted elucidation of their structure in high resolution. The inherent promiscuity of ABC-transporters accompanied by the limited knowledge on the molecular basis of this multispecificity makes it difficult to apply traditional molecular modeling methods, in predicting ligand-binding conformation, and thus special attention is required of the medicinal chemist. At the molecular level, promiscuity can have several fundamental causes. These include binding sites (or "binding zones") accommodating more than one ligand, multiple separate (maybe in part overlapping) binding sites and high protein flexibility (see Chapter 27). The current armory of methods used in the computational drug design field is only in part suited to deal with complex phenomena. Traditional QSAR methods require distinct ligand-binding conformations and in this context are only suited for homologous series of compounds.

There have been considerable modeling efforts to target promiscuous proteins, especially using pharmacophore modeling and machine-learning approaches. However, when the 3D-pharmacophore models are compared, almost no overlap can be identified. Thus a general applicable pharmacophoric pattern unifying the currently available hypotheses is still missing. Nevertheless, from data on X-ray structures of the proton motif force driven transporter AcrB in *E. coli*, this might even be an impossible task. AcrB shows extremely broad substrate specificity, ranging from most of the currently used antibiotics, disinfectants, dyes, and detergents up to simple solvents. Koshland Jr. et al. obtained crystallographic structures of this bacterial pump with four diverse ligands [5]. The structures show three ligands binding simultaneously to the extremely large central cavity of 5000 Å^3. Binding is mainly driven by hydrophobic, aromatic, and van der Waals interactions and utilizes different subsets of AcrB residues. Additionally the bound ligands stabilize the binding of each other via intermolecular interactions. Thus *the* universal pharmacophore for ligands of promiscuous drug efflux pumps simply might not exist.

11.3 QSAR APPROACHES TO DESIGN INHIBITORS OF P-GLYCOPROTEIN (ABCB1)

In 1976 Victor Ling and coworkers identified P-glycoprotein (ABCB1) as being responsible for reduced drug accumulation in multidrug resistant chinese hamster ovary cells [6]. In tumor cells, ABCB1 functions as a membrane-bound, ATP-dependent efflux pump extruding a wide variety of functionally and structurally diverse natural product toxins out of mammalian cells [7]. Overexpression of the protein leads to multiresistance to cytotoxic agents, which mainly is observed in tumor therapy. Only five years later Tsuruo et al. found verapamil to be able to block P-gp mediated transport [8]. This activity gives rise to a restoration of sensitivity in multidrug-resistant cells to chemotherapeutic agents and thus represents a versatile way of overcoming drug resistance. Since this study a wide variety of structurally and functionally diverse compounds have been identified, and several are in clinical studies reaching up to phase III [9,10].

In lead optimization programs, numerous QSAR studies on structurally homologous series of compounds have been performed. Especially verapamil analogues, triazines, acridonecarboxamides, phenothiazines, thioxanthenes, flavones, dihydropyridines, propafenones, and cyclosporine derivatives have been extensively studied, and the results are summarized in several excellent reviews [11,12]. The studies show that the arrangement of H-bond acceptors and H-bond acceptor strength, the distance between aromatic moieties, and H-bond acceptors as well as global physicochemical parameters, such as lipophilicity and molar refractivity, are correlated with P-glycoprotein inhibitory activity. Systematic quantitative structure–activity relationship studies have been performed, mainly on phenothiazines and propafenones [13]. The latter have been carried out using Hansch and Free-Wilson analyses [14], hologram QSAR, CoMFA, and CoMSIA studies [15], as well as nonlinear methods [16], and similarity-based approaches [17] (see Chapter 8). Hansch analyses typically give excellent correlations between lipophilicity and pIC_{50} values within structurally homologous series of compounds (Figures 11.1 and 11.2). This further supports the hypothesis that the interaction with the protein takes place within the membrane bilayer and that lipophilicity of the com-

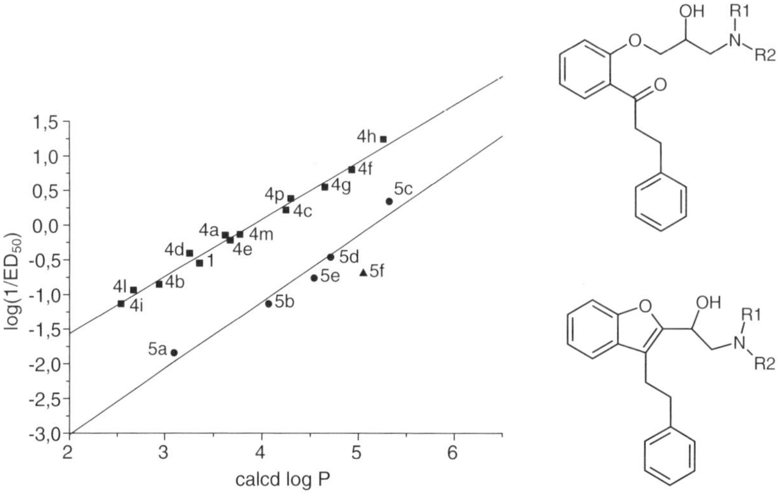

Figure 11.1 Plot of calculated log$P$ values against log(1/EC_{50}) values for a series of propafenones (■) and analogous benzofuranes (•).

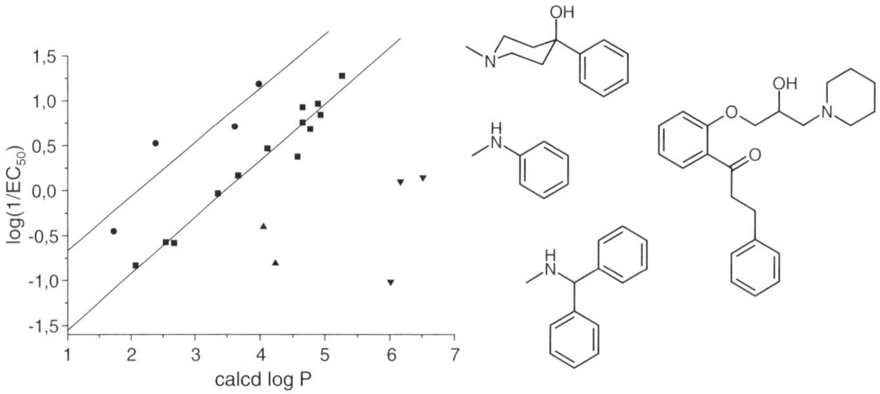

Figure 11.2 Plot of calculated log$P$ values against log(1/EC_{50}) values for a series of propafenones modified at the nitrogen atom; (■) *N*-alkyl derivatives, (▲) *N*-aryl derivatives, (▼) *N*-diphenylalkyl derivatives, and (•) 4-hydroxy-4-phenylpiperidine derivatives.

pounds triggers their concentration at the binding site. Different intercepts of correlation lines and outliers point to altered pharmacophoric patterns. This is especially exemplified in Figure 11.1 showing the log$P$/pIC_{50} correlation for series of propafenones and analogous conformationally restricted benzofuranes [18].

Within both series an excellent correlation is obtained, whereby benzofuranes generally show a lower log potency/logP ratio than propafenones.

Thus for an equi-lipophilic pair of compounds the benzofurane is generally an order of magnitude less active than the corresponding propafenone. This is further supported by equation (11.1), which shows a coefficient of –1.16 for the indicator variable that encodes for the benzofurane scaffold (I_{bf}):

$$\log\left(\frac{1}{EC_{50}}\right) = 0.86\log P - 1.16 I_{bf} - 3.33. \tag{11.1}$$

This relative loss of activity is probably due to the transformation of the H-bond accepting C=O group to a C=C double bond rather than the increased rigidity of the molecule. The latter seems unlikely because several highly rigid compounds, such as steroids, flavones, and benzopyranones, still show P-gp inhibitory activity. Furthermore, decreasing the H-bond acceptor strength of this group by modification to $-OCH_3$ and –OH, or increasing it via electron-donating groups attached to the aromatic ring in para-position to the C=O group, accordingly influences the pharmacological activity of the compounds [19].

Figure 11.2 illustrates the opposite case where a series of 4-hydroxy-4-phenylpiperidines exhibit higher $pIC_{50}/\log P$ ratios than expected by their lipophilicity [20]. This way the –OH group utilizes an additional H-bond interaction with the protein. Preliminary docking studies obtained in our group support this hypothesis showing that the 4-hydroxy4-phenyl-piperidine moiety is fixed in a network of H-bonds made by amino acids Y307, S766, and T769. Figure 11.2 also shows modifications in close proximity to the basic nitrogen atom that can influence pharmacological activity independently of lipophilicity. Introduction of large groups such as diphenylalkyl- gives rise to a dramatic relative loss of activity (relative to the $\log P$ of the moiety!). Systematic variation of the H-bond acceptor strength of the nitrogen atom has revealed that H-bond acceptor strength in this region is quantitatively correlated to P-gp inhibitory activity [21]. Thus anilines, amides, and even esters exhibit pharmacological activity, and this rules out the hypothesis that the nitrogen atom interacts in positively charged form. The fundamental importance of H-bond acceptors has already been pointed out by several studies by Anna Seelig's group [22]. She described a scheme in which the spatial distance between two H-bond acceptors is either 2.5 or 4.6 Å. To the latter case, a third H-bond acceptor could be located in between the two primary electron-donating groups.

2D-QSAR studies have been complemented by a technique called hologram QSAR. Each input molecule is dissected into all possible fragments of a pre-defined length (usually 4–7 atoms). A unique integer identifier is generated for each of the resulting types of fragments, and these integer identifiers are hashed into an array of a user-defined length (usually 23–353 bins) that represents the molecular hologram of the respective compound. These fragments are subject to PLS analysis to identify which of them have a statistically significant influence on pharmacological activity. For a set of propafenones, positive influence was found for the basic nitrogen atom and the phenyl ring (the importance increases when electron donating substituents are attached). Highly negative influence was found for diphenylalkyl- groups in close vicinity to the nitrogen atom, as further evidence of steric hindrance [15].

More proof on the limitations of sterically demanding substituents in close vicinity to the nitrogen atom has been obtained by CoMFA and CoMSIA analyses. These grid-based 3D-QSAR methods are well suited for structural analysis on a 3D basis. Analysis of a set of 131 propafenone-type compounds revealed an important unfavorable steric interaction with compounds possessing two bulky substituents (e.g., diphenylmethyl) in close vicinity to the nitrogen atom. A favorable steric interaction was also observed in the region of the phenyl ring of the phenylpropionyl-moiety (i.e., more bulky substituents should improve activity). In case of electrostatic interactions, both the carbonyl oxygen and the propanolamine nitrogen atom are important for high activity. Favorable hydrophobic interactions were identified along the propanolamine chain and in the vicinity of the phenyl-ring of the arylpiperazine moiety [15]. This space-directed character of lipophilicity was first demonstrated by Pajeva and Wiese for a series of phenothiazines and thioxanthenes [23] and a subset of our propafenone-based library [24]. Since steric and electrostatic fields could not satisfactorily explain the variance in the data set, Pajeva and Wiese added HINT-derived hydrophobic fields to the CoMFA input matrix, which remarkably improved the quality of the 3D-QSAR models. Very recently the same authors extended their studies to a series of 32 tariquidar analogues. The best models performed with q^2 values from 0.66 to 0.75 and confirmed H-bond acceptor, steric, and hydrophobic fields to be the top-contributing properties [25]. Additionally Free-Wilson analysis of the data set revealed that a bulky aromatic ring system bearing a heteroatom in position 3 is beneficial for high activity.

One of the big disadvantages of field-based 3D-QSAR methods is the need for a proper alignment of the molecules. This can be overcome by using descriptors derived from molecular interaction fields, such as VolSurf or GRIND. These are alignment-free and thus allow the analysis of structurally diverse compound sets. Also a model based on VolSurf descriptors has been successfully applied to identify new inhibitors in a virtual screening protocol [26]. Although these approaches do not enable one to rationalize the ligand–protein interaction, they represent versatile tools for prefiltering large combinatorial libraries for compounds with P-gp activity.

Even with all the information from QSAR studies taken together, the consensus picture remains general. Strong inhibitors are characterized by high lipophilicity (and/or molar refractivity) and possess at least two H-bond acceptors. Other features, such as H-bond donors, may act as additional interaction points. Furthermore some steric constraints seem to apply in the vicinity of pharmacophoric structures. These basic requirements are supported by various pharmacophore modeling studies. Pharmacophoric features, such as hydrophobic, aromatic, H-bond acceptor, H-bond donor, and positive and negative charges are assigned to the respective substructures of the molecules. Subsequently a small training set of highly active, structurally diverse molecules is selected and a multi-conformation database is generated. This database is searched for a consensus alignment of pharmacophoric features and respective arrangements are retrieved. Using the genetic algorithm based similarity program GASP, Pajeva and Wiese derived a general pharmacophore model for P-gp modulators using a diverse set of compounds binding to the verapamil site [27]. The final model comprized two hydrophobic planes, three H-bond acceptors, and one H-bond donor. Garrigues et al. calculated the intramolecular distribution of polar and hydrophobic surfaces of a set of structurally diverse P-gp ligands and used the respective fields for superposition of the molecules. This led to the identification of two different, but partially overlapping binding pharmacophores [28]. Penzotti et al. utilized pharmacophore sampling with more than three million pharmacophores. The top 100 models, denoted as pharmacophore ensemble, contained 53 four-point pharmacophores, 39 three-point pharmacophores, and 8 two-point pharmacophores [29]. Roughly half of the models include an H-bond acceptor, an H-bond donor, and hydrophobic areas.

The most comprehensive studies were performed by Ekins and coworkers [30,31]. They used training sets of different kinds, such as inhibitors of digoxin transport, inhibitors of vinblastine binding, inhibitors of vinblastine accumulation, and inhibition of calcein accumulation). Interestingly all four models retrieved differences in both the number and type of features involved and in the special arrangement of these features. Furthermore three out of four models were comprised of four hydrophobic/aromatic features and at most one H-bond acceptor feature. The calcein accumulation model had two hydrophobic features, one H-bond acceptor and one H-bond donor. The consensus model, which correctly ranked the four data sets, consisted of one H-bond acceptor, one aromatic feature, and two hydrophobic features. This further supports the hypothesis that toxins may bind to P-gp at different but overlapping sites. We used a CATALYST model based on propafenone-type inhibitors for an in silico approach to identify new inhibitors of P-gp. The training set comprized 27 propafenone-type inhibitors of daunorubicin efflux, and the model derived included one H-bond acceptor, two aromatic features, one hydrophobic area, and one positively charged group (Figure 11.3). The model was validated with an additional 81 compounds from our in-house data set and subsequently used to screen the World Drug Index. Out of the 32 structurally diverse hits retrieved, 9 compounds have already been described as P-gp inhibitors [32]. Thus it is probable that the other compounds selected also bind to P-gp.

A completely different approach to model P-gp is the use of algorithms based on machine learning (see Chapter 8). Wang et al. utilized Bayesian-regularized neural networks to establish a model for a set of 57 flavonoids binding to the C-terminal nucleotide-binding domain. Descriptors used comprised molecular connectivity indexes and electrotopological state values. The Bayesian-regularized network performed slightly better than an analogous backpropagation network and by far better than PLS [33]. We used both

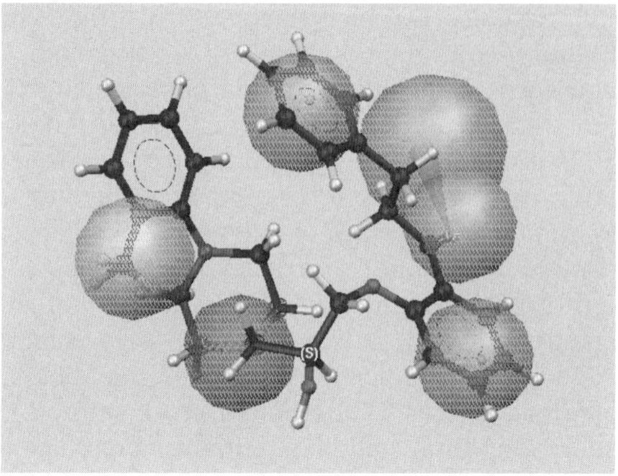

Figure 11.3 P-glycoprotein pharmacophore model derived from propafenones used for screening of the world drug index. Blue: hydrophobic/aromatic; green: hydrophobic; brown: H-bond acceptor; red: positive ionisable. See color plates.

supervised and unsupervised learning routines to create predictive models for P-gp inhibitors. Thus a feedforward backpropagation network was trained to predict IC_{50} values of a series of propafenone-type derivatives. The final model obtained used log P values as well as several Free-Wilson type indicator variables, denoting the presence or absence of distinct substructures as the input vector, and this model outperformed the analogous linear combined Hansch–Free-Wilson model [16]. In silico screening of a small virtual library retrieved several compounds predicted to be active in the nanomolar range. Synthesis and pharmacological testing of selected derivatives proved the validity of the neural network model [34]. A completely different approach was used for identification of structurally new scaffolds. A set of 131 propafenone-type P-gp inhibitors was projected onto a self-organizing map. From a set of roughly 30 2D-autocorrelation vectors, a good separation between actives and inactives could be obtained. Subsequently the size of the map was enlarged, and the propafenones were merged with the SPECS compound library (134,000 compounds). After the training procedure was repeated, all SPECS compounds were retrieved, which localized close to the most active propafenones 1 and 2 (Figure 11.4) [35]. All seven hits retrieved show completely different scaffolds than the training set compounds. Pharmacological testing revealed that two compounds showed inhibitory activity with IC_{50}-values in the submicromolar range, which definitely renders them new lead compounds for P-gp.

11.4 OTHER ABC-TRANSPORTERS

Within the last decade, inhibitors of the MDR-related proteins ABCC1 (MRP1) and ABCC2 (MRP2), the breast cancer resistance protein ABCG2

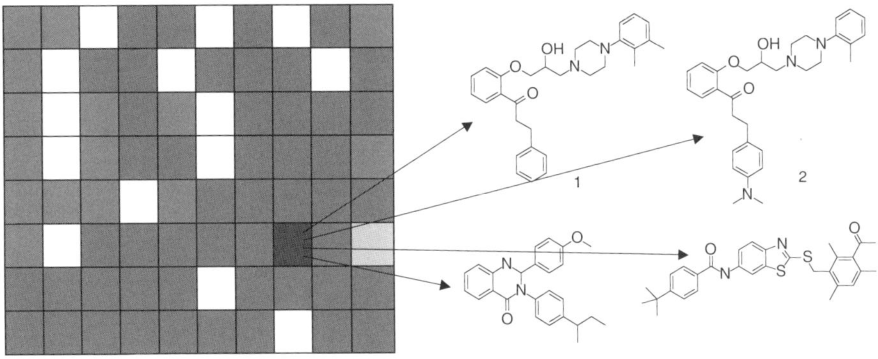

Figure 11.4 Self-organizing map used for screening the SPECS compound library.

(BCRP), and the sister of P-gp ABCB11 (SPGP, BSEP) have been published [36]. Further ABC-proteins capable of transporting drugs comprise ABCC3 (MRP3), ABCC4 (MRP4), ABCC5 (MRP5), and ABCA2 [37]. These proteins are of increasing interest as targets, and the above-mentioned computational methods might also be applied to these transporters both for identification of inhibitors and for selectivity profiling. However, currently only few in vitro data are available for these transporters, and QSAR studies with adequate validation sets are therefore rather rare.

11.4.1 ABCG2 (Breast Cancer Resistance Protein, BCRP, MXR)

Analogous to ABCB1, ABCG2 has a broad, partly overlapping and diverse substrate specificity comparable to the former transporter. It is mainly expressed in the small intestine, placenta, the blood–brain barrier, and the liver, transporting mitoxantrone, methotrexate, camptothecins (topotecan, irinotecan), anthracyclines, etoposide, and flavonoids [38,39]. The latter have also served as lead structures for development of ABCG2 inhibitors. Zhang et al. selected a panel of 25 flavonoids covering 5 different structural subclasses in order to identify structural features important for ABCG2 inhibitory activity. Results showed that the presence of a 2,3-double bond in ring C, ring B attached at position 2, hydroxylation at position 5, lack of an OH group at position 3 and hydrophobic substituents at positions 6, 7, 8, or 4′ are prerequisites for strong interaction with ABCG2 [40]. Subsequent QSAR analysis using calculated log P values, molecular connectivity indexes, kappa shape indexes, electrotopological state indexes, information indexes, subgraph count indexes, molecular polarizability, weight, and volume as the input vector and multiple linear regression analysis coupled with a genetic algorithm gave a model with good predictive power ($q^2 = 0.78$). Descriptors remaining as statistically significant in the final equation are log P, count of all =C-groups and the moment of the displacement between the center of mass and the center of dipole along the inertial

Y-axis. Boumendjel et al. linked piperazines and phenylalkylamines to benzopyranones in order to obtain new inhibitors of ABCG2 [41]. The most active compounds shared several structural features, such as an alkylpiperazine moiety or methoxyphenylalkylamino groups with the highly active ABCG2 inhibitors imatinib (STI 571) and the natural product fumitremorgin C (FTC). The latter served also as starting point for synthesis of a series of 42 structural analogous indolyl diketopiperazines. SAR studies demonstrated that lipophilic side chains in position 3 are important for high inhibition activity [42]. Also within a series of propafenone analogues lipophilicity was shown to be highly predictive for ABCG2 inhibitory potency. Both QSAR studies using a set of 10 ADME-related descriptors and qualitative pharmacophore feature modeling revealed that hydrophobicity, number of rotable bonds and number of H-bond acceptors are key features both for activity and selectivity toward ABCB1 [43]. Results further indicate that for the class of propafenones, ABCG2 is more tolerant for structural modification than ABCB1. Selectivity is therefore mainly determined by the distinct QSAR pattern with respect to ABCB1 rather than a specific interaction with ABCG2.

11.4.2 ABCC1 and ABCC2 (Multidrug Resistance Related Proteins 1 and 2, MRP1 and MRP2)

Multidrug resistance protein 1 (MRP1, ABCC1) is a high-affinity transporter of leukotriene C_4. Additionally it confers resistance toward vinca alkaloids, anthracyclines, epipodophyllotoxins, mitoxanthrone, and methotrexate, but not toward taxanes and bisantrene [44]. In contrast to ABCB1, ABCC1 functions mainly as a (co)transporter of amphipathic organic anions. It transports hydrophobic drugs that are conjugated or complexed to the anionic tripeptide glutathione (GSH), to glucuronic acid, or to sulphate [45]. As in the case of ABCB1 and ABCG2, a lot of structurally and functionally diverse inhibitors for ABCC1 have been identified and are summarized in a recent review [46]. These comprise verapamil, flavonoids, raloxifene, isoxazoles, quinazolinones, quinolines, pyrrolopyrimidines, and peptides. Lather and Madan recently presented a topological model for the classification of a series of 82 pyrrolopyrimidines into active/inactive. An overall classification accuracy of 88% could be achieved based solely the Wiener's index descriptor. Threshold values used were below 3190 for inactives and 4395 to 5223 for actives [47]. For the group of flavonoids, QSAR-studies for both ABCC1 and ABCC2 have been performed. Results demonstrate three structural characteristics to be most important for ABCC1 inhibition: the total number of methoxy-groups, the number of OH-groups, and the dihedral angle between ring B and ring C. ABCC2-inhibitory potency was investigated in parallel. ABCC2 (cMOAT) has been characterized as an organic anion transporter with a broad range of substrates such as methotrexate and drugs conjugated to glutathione [48]. For flavonoid-type inhibitors of ABCC2, the presence of a flavanol B-ring pyrogallol group seems to be a critical structural characteristic. Only robinetin and myricetin were able to inhibit the activity by more than 50%. All other flavonoids did not reach 50% ABCC2 inhibition at concentrations up to 50 μm [49]. For a series of methotrexate analogues, the octanol–water partition coefficient, hydrophobicity, and negative charge were identified as important features for high affinity to rat ABCC2. Furthermore the addition of a benzoyl ornithine group at a distance of 9.3 Å from the negatively ionizable center gave rise to a 40-fold increase in affinity. These findings were supported by a pharmacophore model, which consists of two hydrophobic features, a negative ionizable feature, and two aromatic rings. The latter are consistent with the possibility for hydrophobic and/or π–π interactions at the highly conserved tryptophan residue Trp1254 in TM17 [50]. Iwase and Hirono [51] used 3D-QSAR based receptor mapping of a series of 16 structurally diverse ABCC2 ligands in order to identify key functional groups for ligand binding. Molecular dynamics based generation of conformers, superposition using the SUPERPOSE program [51], and subsequent CoMFA analysis gave a statistically significant model with a predictive power of $q^2 = 0.59$. This model comprised two hydrophobic and two electrostatically positive sites as primary binding sites [52]. The model also suggested that secondary binding sites, which correspond to specific contour levels in the CoMFA contour map, are important in explaining the broad substrate specificity of ABCC2.

11.4.3 ABCB11 (Bile Salt Export Pump, BSEP)

ABCB11 mainly eliminates bile salts from liver cells, and it thereby may be involved in several liver diseases. Hirano et al. used plasma vesicles prepared from insect cells to assess the ABCB11 inhibitory potency of a set of 40 structurally diverse compounds. The authors identified a set of chemical fragmentation codes generated with Markush TOPFRAG that are statistically significant and linked to the ABCB11 interaction [53]. Examples are the descriptors M132 (ring-linking group containing one C atom), H181 (one amine bonded to aliphatic C), and ESTR (one ester group bonded to heterocyclic C via C=O).

11.4.4 OATP1 (Organic Anion Transporting Polypeptide 1)

The organic anion transporting polypetides mediate sodium-independent transport of a diverse array of molecules that are mostly negatively charged as well as steroid conjugates and xenobiotics. They have been implicated in drug–drug interactions such as the interaction between cerivastatin and cyclosporine A [54]. Using a set of 12 structurally diverse compounds, Chang et al. created a chemical function based pharmacophore feature model for the human OATP1B1. Although data were limited and had to be taken from different laboratories using multiple cell types and species (rat and human), the authors successfully derived a model with two H-bond acceptors and three hydrophobes [55].

11.5 PREDICTING SUBSTRATE PROPERTIES—THE ANTITARGET CONCEPT

Despite its overexpression in a wide variety of tumor cells, P-glycoprotein is known to be constitutively expressed in several organs, such as kidney, liver, and intestine and at the blood–brain barrier (BBB). Thus P-gp substrates show poor oral absorption, enhanced renal and biliary excretion, and usually do not enter the brain [56]. This renders P-gp an antitarget, and at least medium-throughput systems have been developed to address the P-gp substrate properties of compounds of interest. These systems mostly rely on transport studies through a monolayer of P-gp expressing Caco-2 [57] or MDCK cells [58]. Especially in the case of brain targeting, it is of vital interest to avoid interaction with P-gp. For these reasons, for the past few years, the focus has shifted from inhibitor design to the development of in silico screening tools for P-gp substrates. Methods applied to in silico screening span the whole range of classification algorithms utilizing decision trees, discriminant analysis, self-organizing maps, and even support vector machines. Seelig proposed a general recognition pattern for P-gp substrates based on a set of structural elements related to H-bond acceptor characteristics (see Section 11.3) [59]. Unfortunately this classification is based on the analysis of a data set of 100 compounds, and no attempts were made to check the predictive performance of this simple and general rule. In analogy to Lipinski's rule of five, Didziapetris et al. used a set of 220 compounds to introduce the "rule of fours": compounds with the number of N and O atoms ≥8, molecular weight >400, and acid pK_a <4 are likely to be P-gp substrates, whereas compounds with the number of N and O atoms ≤4, molecular weight <400, and base $pK_a < 8$ are likely to be nonsubstrates [60].

Gombar et al. applied a two-group linear discriminant model based on 27 electrotopological state values to classify a set of 95 compounds into P-gp substrates and nonsubstrates [61]. The model performs with a sensitivity of 100% (ability to correctly identify substrates) and a specificity of 90.6% (ability to correctly identify nonsubstrates). For a test set of 58 compounds, a total accuracy of 86.2% was obtained. Furthermore a striking relationship between the molecular E-State (MolES) and the P-gp substrate property was found that represents the molecular bulk of a compound. Compounds with MolES greater than 110 are predominantly substrates (18/19; 95%) and those with MolES less than 49 are nonsubstrates (11/13; 84.6%). This so-called Gombar-Polli rule may serve as a fast preliminary estimate for substrate properties. However, it should be noted that out of 98 compounds only 29 (30%) were subject to these values; all other compounds had MolES values between 49 and 110 and were likely to be poorly categorized by this simple classifier. Cabrera et al. pursued a topological substructural approach for the prediction of P-gp substrates. A linear discriminant model classified 163 compounds with an accuracy of 81% based on standard bond distance, polarizability, and the Gasteiger-Marsilli atomic charge [62]. Furthermore the predictive potential of the TOPS-MODE approach was demonstrated for a set of 6-fluoroquinolones not covered by the original training set.

11.6 NOVEL METHODS

One of the problems when applying classical QSAR techniques is the right choice of the method and the descriptor combination. To overcome this issue, which normally is pursued on a trial and error basis, two approaches have recently been introduced: Cartmell et al. presented an automated QSPR through a competitive workflow [63] and the group of Tropsha introduced the combinatorial QSAR approach [64]. The competitive workflow technique allows exhaustive exploration of descriptor and model space, automated model validation, and continuous updating. This is achieved via a workflow comprising six-agent calculating descriptors, one agent for feature selection and seven agents for model-building utilizing, for example, multiple linear regression, neural networks, PLS, regression trees, and genetic algorithms. To date, the best models obtained for a set of 184 P-gp substrates and nonsubstrates was comparable with the support vector machine model published by Xue [71] (see below). In the combinatorial QSAR approach, all possible descriptors are combined with all possible methods in a combinatorial way. Thus the authors used the data set of Penzotti et al. and calculated molecular connectivity indexes, atom pair descriptors, VolSurf descriptors, and MOE-descriptors. These input matrices were then analyzed with the k-nearest neighbor classification, decision tree, binary QSAR, and support vector machines, respectively [65]. The best model obtained used VolSurf descriptors and a support vector machine based classifier and showed an overall accuracy of 94% for the training set of 94% and of 0.81 for the test set. VolSurf descriptors were also applied previously by Crivori et al. [66] to a P-gp data set. They developed both a model discriminating between substrates and nonsubstrates and a model that classifies P-gp substrates having poor inhibitory activity versus inhibitors showing no evidence of significant transport [66]. The latter was achieved via a partial least squares discriminant (PLSD) analysis using GRIND-descriptors. These methods allowed the authors to identifying key pharmacophoric features for substrates and inhibitors. The most important descriptors for P-gp substrates were mainly related to H-bonding properties and comprise intense O–O (regions around H-bond donor groups; distance 6.5 Å), N1–N1 (regions around H-bond acceptor groups; distance 15 Å) and mixed O–N1 (distance 12.5 Å) interaction energies. GRIND-based 3D-QSAR was also pursued by Cianchetta et al. to derive a hypothesis for P-gp substrate recognition [67]. From a set of 129 compounds the authors derived a pharmacophore hypothesis that contains the following recognition elements: two hydrophobic groups 16.5 Å apart, two H-bond acceptor groups 11.5 Å apart, and the size of the molecule (21.5 Å between the two edges of the molecule).

Penzotti et al. used an ensemble of pharmacophore models that discriminated between P-gp substrates and nonsubstrates as described earlier [68]. The model correctly classified 50–60% of the substrates and 80% of the nonsubstrates. Very recently the group of Ekins published a series of pharmacophore models for rapid identification of P-gp substrates and inhibitors [69]. They used a combination of their already established CATALYST models for substrates and inhibitors and generated one additional inhibitor model. All three models were used for in silico screening of an in house database of approximately 600 frequently prescribed drugs. A selected subset of predicted positives was subject to pharmacological testing, which proved the applicability of the models.

As already outlined for P-gp inhibitors, self-organizing maps are a versatile tool for classifying actives and inactives. Thus Yang and coworker developed a self-organizing map to separate P-gp substrates from inhibitors on the basis of a set of molecular connectivity indices and electrotopological state descriptors. The average accuracy of classification obtained was 82.3%. Comparison with feedforward backpropagation neural networks showed the superiority of the SOM method [70]. Very recently Xue et al. reported the application of support vector machines (SVM) for prediction of P-gp substrates [71]. They used a set of 201 compounds comprising 116 substrates and 85 nonsubstrates. On the basis of 159 molecular descriptors, the SVM yielded a prediction accuracy of 81% for substrates and of 79% for nonsubstrates. These values are slightly higher than those obtained with other classification methods, such as *k*-nearest neighbors, probabilistic neural networks and decision trees.

Adenot and Lahana used discriminant analysis and PLS-DA to simultaneously model the passive diffusion component of BBB permeation and potential physicochemical requirements for P-gp substrates [72]. They could demonstrate that a set of relatively simple descriptors (number of heteroatoms, number of H-bond donors and acceptors, number of halogen atoms, etc.) are sufficient to discriminate between potential CNS and non-CNS drugs including P-gp substrates. An identical data set was used by the group of Arodz to demonstrate the applicability of their new classification method, called random feature subset boosting, to linear discriminant analysis [73]. Via the introduction of ensembles of linear discriminant models the analysis of more complex problems, such as those involving multiple mechanisms of action, seems possible. The ensemble model gave slightly better accuracy than the SVM model and enabled the identification of 10 descriptors with the highest importance factors. However, data sets commonly used in P-gp substrate studies are rather small and very often also inconsistent. When analysing six publications dealing with P-gp substrate/nonsubstrate classification, we identified 50 compounds (out of 326) as being differently classified in the literature. Especially when it comes to the molecular level, the classification between substrates and inhibitors is rather fuzzy, so it is important to recognize this problem. At the macroscopic level, compounds showing net transport are considered substrates, and compounds blocking this transport are classified as inhibitors. At the mechanistic level, it has been shown that some inhibitors also increase ATP-activity of P-gp and thus might be transported too [74]. Because of their high lipophilicity such molecules rapidly diffuse back into the membrane and block the pump by keeping it occupied. Undoubtedly, much larger data sets are needed to carefully address these details and to serve as basis for further computational studies.

11.7 CONCLUSIONS AND OUTLOOK

The success of traditional QSAR-methods, such as Hansch-analysis and CoMFA, heavily relies on the basic assumption that all compounds used bind to the same site and in the same mode to the target protein. In the case of polyspecific drug transport pumps such as those described in this chapter, there is experimental evidence that drug-binding occurs at the interface of the two transmembrane domains and therefore the binding cavity is rather large, accommodating simultaneously up to three ligands in the case of some transporters. Thus conventional QSAR methods fail to decipher clear and distinct ligand–protein interaction patterns when structurally diverse compound set are used. Success stories published so far mainly rely on VolSurf/GRIND descriptors, pharmacophore models, and machine-learning methods. The latter two approaches were also successfully applied for in silico screening of medium to large compound libraries in order to identify structurally new molecular scaffolds as ligands for P-gp. New approaches such as SVM and similarity-based descriptors may pave the way for the establishment of rapid in silico filters that can be routinely applied in the early drug discovery phase. This will have special importance in the field of predicting substrate properties of ABC-transporters, as these are increasingly considered antitargets in the pharmaceutical industry.

P-glycoprotein, the paradigm transporter for the whole class of drug efflux pumps, has been known for over 30 years. However, both the molecular basis of the drug–protein interaction and the mechanism of transport remain elusive. Recent X-ray structures of analogous bacterial transporters as well as combined photoaffinity labeling/protein homology modeling approaches have started to shed some light on the molecular basis of polyspecificity. Further in silico and in vitro studies and additional X-ray structures are expected soon and will help to solve this amazingly complex biological puzzle.

REFERENCES

1. Endres CJ, Hsiao P, Chung FS, Unadkat JD. The role of transporters in drug interactions. *Eur J Pharm Sci* 2006;27:501–17.
2. Boyd RA, Stern RH, Stewart BH, Wu X, Reyner EL, Zegarac EA, Randinitis EJ, Whitfield L. Atorvastatin coadministration may increase digoxin concentrations

by inhibition of intestinal *P*-glycoprotein-mediated secretion. *J Clin Pharmacol* 2000;40:91–8.

3. Johne A, Brockmoller J, Bauer S, Maurer A, Langheinrich M, Roots I. Pharmacokinetic interaction of digoxin with an herbal extract from St John's wort (*Hypericum perforatum*). *Clin Pharmacol Ther* 1999;66:338–45.
4. Fromm MF. Importance of *P*-glycoprotein at blood-tissue barriers. *Trends in Pharmacol Sci* 2004;25:423–9.
5. Yu EW, McDermott G, Zgurskaya HI, Nikaido H, Koshland DE Jr. Structural basis of multiple drug-binding capacity of the AcrB multidrug efflux pump. *Science* 2003;300:976–80.
6. Juliano RL, Ling V. A surface glycoprotein modulating drug permeability in Chinese hamster ovary cell mutants. *Biochim Biophys Acta* 1976;455:152–62.
7. Gottesman MM, Pastan I. Biochemistry of multidrug resistance mediated by multidrug transporter. *An Rev Biochem* 1993;62:385–427.
8. Tsuruo T, Iida H, Tsukagoshi S, Sakurai Y. Overcoming of vincristine resistance in P388 leukemia in vivo and in vitro through enhanced cytotoxicity on vincristine and vinblastine by verapamil. *Cancer Res* 1981;41:1967–72.
9. Leonard GD, Polgar O, Bates SE. ABC transporters and inhibitors: new targets, new agents. *Curr Opin Investig Drugs* 2002;3:1652–9.
10. Szakacs G, Paterson JK, Ludwig JA, Booth-Genthe C, Gottesman MM. Targeting multidrug resistance in cancer. *Nat Rev Drug Discov* 2006;5:219–34.
11. Raub TJ. P-glycoprotein recognition of substrates and circumvention through rational drug design. *Mol Pharm* 2006;1:3–25.
12. Pleban K, Ecker GF. Inhibitors of *P*-glycoprotein—Lead identification and optimisation. *Minirev Med Chem* 2005;5:153–63.
13. Wiese M, Pajeva IK. Structure-activity relationships of multidrug resistance reversers. *Curr Med Chem* 2001;8:685–713.
14. Tmej C, Chiba P, Huber M, Richter E, Hitzler M, Schaper KJ, Ecker G. A combined Hansch/free-Wilson approach as predictive tool in QSAR studies on propafenone-type modulators of multidrug resistance. *Arch Pharm (Weinheim)* 1998;331:233–40.
15. Kaiser D, Smiesko M, Kopp S, Chiba P, Ecker GF. Interaction field based and hologram based QSAR analysis of propafenone-type modulators of multidrug resistance. *Med Chem* 2005;1:431–44.
16. Tmej C, Chiba P, Schaper KJ, Ecker G, Fleischhacker W. Artificial neural networks as versatile tools for prediction of MDR-modulatory activity. *Adv Exp Med Biol* 1999;457:95–105.
17. Klein C, Kaiser D, Kopp S, Chiba P, Ecker GF. Similarity based SAR (SIBAR) as tool for early ADME profiling. *J Comput Aided Mol Des* 2002;16:785–93.
18. Ecker G, Chiba P, Hitzler M, Schmid D, Visser K, Cordes HP, Csöllei J, Seydel JK, Schaper KJ. Structure-activity relationship studies on benzofurane analogs of propafenone-type modulators of tumor cell multidrug resistance. *J Med Chem* 1996;39:4767–74.
19. Chiba P, Ecker G, Schmid D, Drach J, Tell B, Goldenberg S, Gekeler V. Structural requirements for activity of propafenone type modulators in PGP-mediated multidrug resistance. *Mol Pharmacol* 1996;49:1122–30.

20. Chiba P, Hitzler M, Richter E, Huber M, Tmej C, Ecker G. Studies on propafenone-type modulators of multidrug resistance: III. Variations on the nitrogen. *Quant Struct-Act Rel* 1997;16:361–6.
21. Ecker G, Huber M, Schmid D, Chiba GF. The importance of a nitrogen atom in modulators of multidrug resistance. *Mol Pharmacol* 1999;56:791–6.
22. Seelig A, Gatlik-Landwojtowicz E. Inhibitors of multidrug efflux transporters: Their membrane and protein interactions. *Minirev Med Chem* 2005;5:135–51.
23. Pajeva IK, Wiese M. Molecular modelling of phenothiazines and related drugs as multidrug resistance modifiers: A comparative molecular field analysis study. *J Med Chem* 1998;41:1815–26.
24. Pajeva IK, Wiese M. A comparative molecular field analysis of propafenone-type modulators of cancer multidrug resistance. *Quant Struct-Act Rel* 1998;17:301–12.
25. Globisch C, Pajeva IK, Wiese M. Structure-activity relationships of a series of tariquidar analogs as multidrug resistance modulators. *Bioorg Med Chem* 2006; 14:1588–98.
26. Kaiser D, Bohl M, Kopp S, Chiba P, Ecker GF. A Volsurf model of propafenone-type inhibitors of *P*-glycoprotein for in silico screening of the Leadquest compound library. *Drugs Future* 2004;29 (Suppl A):119.
27. Pajeva IK, Wiese M. Pharmacophore model of drugs involved in *P*-glycoprotein multidrug resistance: Explanation of structural variety (hypothesis). *J Med Chem* 2002;45:5671–86.
28. Garrigues A, Loiseau N, Delaforge M, Ferte J, Garrigos M, Andre F, Orlowski S. Characterization of two pharmacophores on the multidrug transporter *P*-glycoprotein. *Mol Pharmacol* 2002;62:1288–98.
29. Penzotti JE, Lamb ML, Evensen E, Grootenhuis PD. A computational ensemble pharmacophore model for identifying substrates of *P*-glycoprotein. *J Med Chem* 2002;45:1737–40.
30. Ekins S, Kim RB, Leake BF, Dantzig AH, Schuetz E, Lan LB, et al. Application of three-dimensional quantitative structure-activity relationships of *P*-glycoprotein inhibitors and substrates. *Mol Pharmacol* 2002;61:974–81.
31. Ekins S, Kim RB, Leake BF, Dantzig AH, Schuetz EG, Lan LB, et al. Three-dimensional quantitiative structure-activity relationships of inhibitors of *P*-glycoprotein. *Mol Pharmacol* 2002;61:964–73.
32. Langer T, Eder M, Hoffmann RD, Chiba P, Ecker GF. Lead identification for modulators of multidrug resistance based on in silico screening with a pharmacophoric feature model. *Arch Pharm (Weinheim)* 2004;337:317–27.
33. Wang YH, Li Y, Yang SL, Yang L. An in silico approach for screening flavonoids as *P*-glycoprotein inhibitors based on a Bayesian-regularized neural network. *J Comput Aided Mol Des* 2005;19:137–47.
34. Kaiser D, Tmej C, Chiba P, Schaper KJ, Ecker G. Artificial neural networks in drug design: II. Influence of learning rate and momentum factor on the predictive ability. *Sci Pharm* 2000;68:57–64.
35. Kaiser D, Terfloth L, Kopp S, Chiba P, Gasteiger J, Ecker GF. Artificial neural networks for identification of new modulators of multidrug resistance. In: Sener E, Yalcin I, editors, *QSAR & molecular modelling in rational design of bioactive molecules*. Computer aided Drug Design & Development Society in Turkey, Antrara, Turkey, 2006, p. 274–7.

36. Chiba P, Ecker GF. Inhibitors of ABC-type drug efflux pumps—An overview on the actual patent situation. *Expert Opin Ther Patents* 2004;14:499–508.
37. Gottesman MM, Fojo T, Bates SE. Multidrug resistance in cancer: Role of ATP-dependent transporters. *Nat Rev Canc* 2002;2:48–58.
38. van Herwaarden AE, Schinkel AH. The function of breast cancer resistance protein in epithelial barriers, stem cells and milk secretion of drugs and xenotoxins. *Trends in Pharmacol Sci* 2006;27:10–16.
39. Mao Q, Unadkat JD. Role of breast cancer resistance protein (ABCG2) in drug transport. *AAPS J* 2005;7:E118–33.
40. Zhang S, Yang X, Coburn RA, Morris ME. Structure activity relationships and quantitative structure activity relationships for the flavonoid-mediated inhibition of breast cancer resistance protein. *Biochem Pharmacol* 2005;70:627–39.
41. Boumendjel A, Nicolle E, Moraux T, Gerby B, Blanc M, Ronot X, Boutonnat J. Piperazinobenzopyranones and phenylalkylaminobenzopyranones: Potent inhibitors of breast cancer resistance protein (ABCG2). *J Med Chem* 2005;48:7275–81.
42. van Loevezijn A, Allen JD, Schinkel AH, Koomen GJ. Inhibiton of BCRP-mediated drug efflux by fumitremorgin-type indolyl dikeotpiperazines. *Bioorg Med Chem Lett* 2001;11:29–32.
43. Cramer J, Kopp S, Bates SE, Chiba P, Ecker GF. Multispecificity of drug transporters: Probing inhibitor selectivity for the human drug efflux transporters ABCB1 and ABCG2. Submitted.
44. Schinkel AH, Jonker JW. Mammalian drug efflux transporters of the ATP binding cassette (ABC) family: An overview. *Adv Drug Deliv Rev* 2003;55:3–29.
45. Deeley RG, Cole SPC. Substrate recognition and transport by multidrug resistance protein 1 (ABCC1). *FEBS Lett* 2006;580:1103–11.
46. Boumendjel A, Baubichon-Cortay H, Trompier D, Perrotton T, Di Pietro A. Anticancer multidrug resistance mediated by MRP1: Recent advances in the discovery of reversal agents. *Med Res Rev* 2005;25:453–72.
47. Lather V, Madan AK. Topological model for the prediction of MRP1 inhibitory activity of pyrrolopyrimidines and templates derived from pyrrolopyrimidine. *Bioorg Med Chem Lett* 2005;15:4967–72.
48. Faber KN, Muller M, Jansen PL. Drug transport proteins in the liver. *Adv Drug Deliv Rev* 2003;55:107–24.
49. van Zanden JJ, Wortelboer HM, Bijlsma S, Punt A, Usta M, van Bladeren PJ, et al. Quantitative structure activity relationship studies on the flavonoid mediated inhibition of multidrug resistance proteins 1 and 2. *Biochem Pharmacol* 2005; 69:699–708.
50. Ng C, Xiao YD, Lum BL, Han YH. Quantitative structure-activity relationships of methotrexate and methotrexate analogues transported by the rat multispecific resistance-associated protein 2 (rMrp2). *Eur J Pharm Sci* 2005;26:405–13.
51. Iwase K, Hirono S. Estimation of active comformation of drugs by a new molecular superposing procedure. *J Comput Aided Mol Des* 1999;13:305–15.
52. Hirono S, Nakagome I, Imai R, Maeda K, Kusuhara H, Sugiyama Y. Estimation of the three-dimensional pharmacophore of ligands for rat multidrug-resistance associated protein 2 using ligand-based drug design techniques. *Pharm Res* 2005; 22:260–9.

53. Hirano H, Kurata A, Onishi Y, Sakurai A, Saito H, Nakagawa H, et al. High-speed screening and QSAR analysis of human ATP-binding cassette transporter ABCB11 (bile salt export pump) to predict drug-induced intrahepatic cholestasis. *Mol Pharm* 2006;3:252–65.

54. Kim RB. Organic anion-transporting polypeptide (OATP) transporter family and drug disposition. *Eur J Clin Investig* 2003;33:1–5.

55. Chang C, Pang KS, Swaan PW, Ekins S. Comparative pharmacophore modelling of organic anion transporting polypeptides: A meta-analysis of rat Oatp1a1 and human OATP1B1. *J Pharmacol Exp Ther* 2005;314:533–41.

56. Chan LM, Lowes S, Hirst BH. The ABCs of drug transport in intestine and liver: Efflux proteins limiting drug absorption and bioavailability. *Eur J Pharm Sci* 2004;21:25–51.

57. Delie F, Rubas WA. A human colonic cell line sharing similarities with enterocytes as a model to examine oral absorption: Advantages and limitations of the Caco-2 model. *Crit Rev Ther Drug Carrier Sys* 1997;14:221–86.

58. Irvine JD, Takahashi L, Lockhart K, Cheong J, Tolan JW, Selick HE, Grove JR. MDCK (Madin-Darby canine kidney) cells: A tool for membrane permeability screening. *J Pharm Sci* 1999;88:28–33.

59. Seelig A. A general pattern for substrate recognition by *P*-glycoprotein. *Eur J Biochem* 1998;251:252–61.

60. Didziapetris R, Japertas P, Avdeef A, Petrauskas A. classification analysis of *P*-glycoprotein substrate specificity. *J Drug Target* 2003;11:391–406.

61. Gombar VK, Polli JW, Humphreys JE, Wring SA, Serabjit-Singh CS. Predicting *P*-glycoprotein substrates by a quantitative structure-activity relationship model. *J Pharm Sci* 2004;93:957–68.

62. Cabrera MA, Gonzalez I, Fernandez C, Navarro C, Bermejo M. A topological substructural approach for the prediction of *P*-glycoprotein substrates. *J Pharm Sci* 2006;95:589–606.

63. Cartmell J, Enoch S, Krstajic D, Leahy DE. Automated QSPR through competitive workflow. *J Comput Aided Mol Des* 2005;19:821–33.

64. Kovatcheva A, Golbraikh A, Oloff S, Xiao YD, Zheng W, Wolschann P, et al. A combinatorial QSAR of Ambergris fragrance compounds. *J Chem Inf Comput Sci* 2004;44:582–95.

65. De Cerqueira Lima P, Golbraikh A, Oloff S, Xiao Y, Tropsha A. Combinatorial QSAR modeling of *P*-glycoprotein substrates. *J Chem Inf Model* 2006;46:1245–54.

66. Crivori P, Reinach B, Pezzetta D, Poggesi I. Computational models for identifying potential *P*-glycoprotein substrates and inhibitors. *Mol Pharm* 2006;3:33–44.

67. Cianchetta G, Singleton RW, Zhang M, Wildgoose M, Giesing D, Fravolini A, et al. A pharmacophore hypothesis for *P*-glycoprotein substrate recognition using GRIND-based 3D-QSAR. *J Med Chem* 2005;48:2927–35.

68. Penzotti JE, Lamb ML, Evensen E, Grootenhuis PDJ. A computational ensemble pharmacophore model for identifying substrates of *P*-glycoprotein. *J Med Chem* 2002;45:1737–40.

69. Chang C, Bahadduri PM, Polli JE, Swaan PW, Ekins S. Rapid identification of *P*-glycoprotein substrates and inhibitors. *Drug Metab Dispos* 2006;34:1976–84.

70. Wang YH, Li Y, Yang SL, Yang L. Classification of substrates and inhibitors of *P*-glycoprotein using unsupervised machine learning approach. *J Chem Inf Model* 2005;45:750–7.

71. Xue Y, Yap CW, Sun LZ, Cao ZW, Wang JF, Chen YZ. Prediction of *P*-glycoprotein substrates by a support vector machine approach. *J Chem Inf Comput Sci* 2004;44:1497–505.

72. Adenot M, Lahana R. Blood-brain barrier permeation models: Discrimination between potential CNS and non-CNS drugs including *P*-glycoprotein substrates. *J Chem Inf Comput Sci* 2004;44:239–48.

73. Arodz T, Yuen DA, Dudek AZ. Ensemble of linear models for predicting drug properties. *J Chem Inf Model* 2006;46:416–23.

74. Schmid D, Ecker G, Richter E, Hitzler M, Chiba P. Structure-activity relationship studies of propafenone analogs based on *P*-glycoprotein ATPase activity measurements. *Biochem Pharmacol* 1999;58:1447–56.

12

COMPUTATIONAL MODELING OF RECEPTOR-MEDIATED TOXICITY

MARKUS A. LILL AND ANGELO VEDANI

Contents

12.1 INTRODUCTION

Receptor proteins initiate a cellular response upon binding of endogenous ligand molecules, such as hormones and neurotransmitters [1,2,3,4]. Ligand-induced modifications of the physicochemical properties or conformational changes of the receptor can trigger transcription processes or signal-transduction cascades. This may result in physiological changes that

Computational Toxicology: Risk Assessment for Pharmaceutical and Environmental Chemicals, Edited by Sean Ekins

constitute the biological actions of the ligand molecules. Drugs and environmental chemicals may share pharmacophoric properties similar to the endogenous molecule, thus simulating the biochemical effect activated by the very receptor [5]. Molecules can also trigger the inverse physiological result, agonism or antagonism, or function as co-repressors or co-activators [6]. Computational approaches have not yet coped with the latter successfully.

Different endpoints, ranging from hormone-dependent cancer [7], developmental toxicity [8], to neurotoxicity [9], are triggered by the binding of ligands to specific receptors. While, in principle, most receptors might be capable of mediating toxic effects, specific proteins, such as the estrogen receptor (ER), androgen receptor (AR), aryl hydrocarbon receptor (AhR), peroxisome proliferator activated receptor (PPAR), and pregnane xenobiotic receptor (PXR/SXR), are currently receiving most attention in biomedical research. It should be noted that this focus does not reflect the actual contributions of the various proteins to receptor-mediated toxicity. Receptors currently recognized as relevant to the toxicity of drugs and chemicals include proteins that are targets for natural and human-made environmental chemicals (ER, AR, thyroid receptors TR, AhR), those that are involved in the regulation of drug metabolism and consequently in drug–drug interactions (AhR, PXR/SXR, CAR), those that are involved in metabolic diseases (PPAR, farnesoid X receptor FXR, liver X receptors LXR), and the retinoid X receptor (RXR), which can form heterodimers with other receptors (including TR, PXR, CAR, PPAR, FXR, and LXR).

12.2 RECEPTORS INVOLVED IN TOXICITY OF ENVIRONMENTAL CHEMICALS

Since the publication of Rachel Carson's *Silent Spring* [10], there has been increasing awareness and scientific evidence that human-made chemicals in the environment can exert deleterious effects in humans and animals by interfering with the endocrine system. These compounds, referred to as endocrine-disrupting chemicals (EDCs), can modulate the endocrine system and potentially cause adverse effects by a number of different mechanisms. The modes of action include receptor-mediated responses, inhibition of hormone synthesis, metabolism [11,12,13], and transport [14], as well as activation of receptors through processes such as receptor phosphorylation. The generally accepted theory for receptor-mediated toxicity includes binding of a ligand (e.g., a hormone) to its receptor, typically a nuclear receptor, transport of the ligand-bound receptor to the nucleus, followed by a complex series of events that leads to changes in the gene-expression profile [15]. This change in gene expression is thought to be an early but critical step in altering the regulation of biological functions, including cell proliferation and differentiation, both essential for normal development and function of multiple organ systems.

12.2.1 Estrogen Receptors

Estrogens are a group of steroid compounds that function as the primary female sex hormone. They promote the development of female secondary sex characteristics, such as breasts and are involved in regulating the menstrual cycle. The estrogen receptor, a member of the class of nuclear receptors, is an intracellular protein that binds estrogens such as estradiol, the main endogenous human estrogen. There are two estrogen receptor subtypes, ER_α and ER_β, both of which have a DNA-binding domain and can function as transcription factors. Although they show significant sequence homology in their DNA- and ligand-binding domains, they exhibit profound differences in their tissue distribution patterns, ligand selectivity (but on the order of only one log unit for most compounds) and transcriptional properties (agonism/antagonism) [16,17]. Consequently they may have distinct biological roles [18, 19–21].

A variety of compounds in the environment have been shown to display agonistic or antagonistic activity toward the ER, including natural products and synthetic compounds [22–28]. The concern over xenobiotics binding to the ER has created a need to screen and monitor compounds that can modulate endocrine effects. This has been underscored by US legislation in 1995 and 1996 by mandating that chemicals and formulations must be screened for potential estrogenic activity before they are manufactured or used in certain processes [29,30].

Reflecting the availability of pertinent biological data, the largest number of computational studies in the literature is associated with the binding of molecules to the ER. These studies can be classified as lead-optimization projects aiming to develop high-affinity compounds and projects that predict the toxic potential of structurally diverse sets of environmental chemicals. While the former studies often cope with selectivity issues toward ER_α and ER_β, most of the latter models are limited to the ER_α. In the following, we will focus on computational models associated with the prediction of toxicity. As comprehensive reviews of computational modeling of ER in the field of toxicity exist [31,32], we will not aim to discuss every approach in detail. Instead, we will focus on practical aspects relevant to toxicity: First, what is the diversity of the data set modeled and the associated applicability domain, and second, how is the model validated? The latter affects the model's reliability to predict the toxicity of new or hypothetical ligand molecules. Consequently we will mainly discuss approaches that fulfill at least one of two criteria: structural diversity of ligand data and validation via a test set. As demonstrated by Golbraikh and Tropsha [33], leave-one-out cross validation is, in general, not sufficient for validating a model.

To deal with the wealth of structure classes of compounds that bind to ER, two approaches seem to be logical (Figure 12.1): either a QSAR model is derived for each class separately, resulting in a conglomeration of equations, or a single computational model is used to cover a diversity of compound classes.

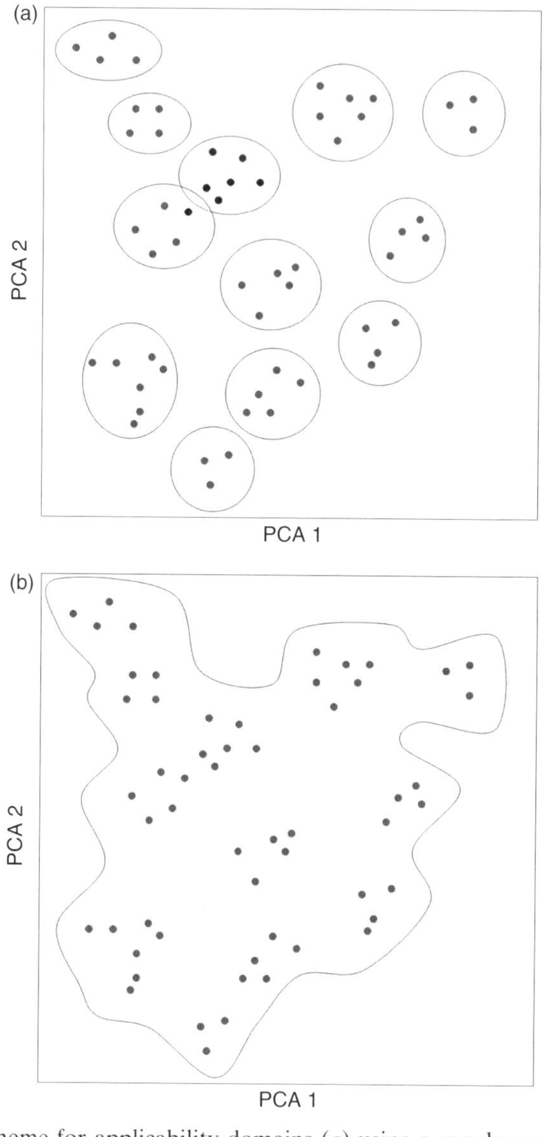

Figure 12.1 Scheme for applicability domains (*a*) using a conglomeration of QSAR models for each compound class separately and (*b*) using a single QSAR model covering all compound classes (PCA refers to principal component axis).

The first approach mentioned was chosen in the C-QSAR database [34] using Hansch-type QSAR equations [35]. Its advantage is that these models are easy to apply and interpret. However, each model has a limited applicability domain (Figure 12.1); thus only toxicity predictions for compounds belonging to these specific classes are suitable. Unfortunately, no analysis of the predictive power of a test set has been published to date.

Alternatively, various attempts were made to derive computational models for a large and diverse set of molecules ranging from 2D-QSAR, 3D-QSAR, multidimensional QSAR, to structure-based simulations and combinations of docking and QSAR.

Gao et al. [36] used binary QSAR based on topological descriptors and indicator variables (including one for the phenolic hydroxyl group) to derive a classification model that separates "active" from "inactive" compounds. The model was trained on 410 diverse molecules, and it demonstrated its predictive power on a test set of 53 randomly selected molecules from which 94% were correctly classified. The biological data were selected from four different laboratories, so there might be some inconsistency with respect to the classification of the model.

Similarly Zheng and Tropsha [37] used topological descriptors (molecular-connectivity indices [38–40] and atom pairs [41]) in combination with an automated variable-selection method based on the *k*-nearest neighbor principle (*k*NN QSAR). A diverse set of 58 ER ligands was used for training with a leave-one-out cross-validated r^2 reaching 0.77, with the largest individual deviation being two log units. The statistical significance of the model was tested using a randomization of the biological activity data. Again, no external validation of the test set was performed. Asikainen et al. [42] extended the method to consensus *k*NN QSAR and applied it to predict binding affinities to the estrogen receptor of four different species: human (ER_α and ER_β), calf, mouse, and rat. The biological data was retrieved from the EKDB database [43]—thus measured in different laboratories with varying assays—resulting in 245 structurally diverse compounds, 61 for human, 53 for calf, 68 for mouse, and 130 for rat ER. From DRAGON 3.0 there were 1497 descriptors calculated [44], and annealing was simulated as the variable selection method [37] was applied to obtain 50 locally optimized models containing 250 descriptors. The average over all 50 models yielded the consensus QSAR. Leave-one-out crossvalidation and randomization tests were applied to the QSAR models for internal validation, and separate training and test sets were used for external validation. The internal predictive power of the consensus models was demonstrated by cross-validated correlation coefficients ranging from 0.69 (human ER_β data) to 0.79 (hER_α). The external predictive power was demonotrated by predictive r^2 scores varying from 0.62 (hER_β) to 0.77 (calf and mouse data).

Marini et al. [45] applied various multivariate modeling techniques using a pool of 280 different descriptors (constitutional, topological, geometrical, electrostatic, and quantum-chemical) on a biologically consistent data set [46] of 132 structurally diverse compounds. To reduce the wealth of descriptors, a genetic algorithm was applied as the variable-selection method to the neural-network models. The variable selection was on the basis of a genetic algorithm, and it was controlled by an internal test set of compounds extracted from the data set available. The best model was obtained by a nonlinear neural network model based on an error backpropagation algorithm that resulted in an r^2 of 0.92 and a leave-one-out q^2 of 0.71. An external test set was not employed. Qualitatively the descriptors that dominantly correlate with the binding affinity in all models are the hydrophobicity of the compounds, parameters for shape, and flexibility and minimal and maximal partial charge.

Brown et al. [47] applied a novel type of geometric fingerprinting (Fingal3D) to a data set of 58 estrogen agonists from eight diverse structure classes. Using all 58 molecules for training yielded a model with an r^2 of 0.887 and a sevenfold cross-validated q^2 of 0.771. In this approach the similarity between r^2 and q^2 was statistically superior to all other tested methods, including CoMFA, HQSAR, FRED/SKEYS, Fingal2D, Dragon2D, and Dragon3D. Different test protocols were set up to challenge the predictive power of the models: First, eight models were generated, and of these, seven classes were used for training one class as an alternating test set. The predictive r^2 values for the test sets vary dramatically between 0.02 and 0.94, with no clear superiority of Fingal3D compared to the other approaches. Second, five models with randomly assigned test sets in a splitting ratio of 40 to 18 demonstrated a statistically slightly better performance compared to Dragon2D and Dragon3D in both q^2 (Fingal3D: 0.79) and predictive r^2 (0.64).

Mekenyan et al. [48–50] made use of their COmmon REactivity PAttern (COREPA) concept—a pattern-recognition method identifying the common stereoelectronic (reactivity) pattern of structurally diverse chemicals exerting similar biological effects via the same mode of action. To elucidate these patterns, the conformational flexibility of the molecules was examined. The areas in the multidimensional descriptor space were found to be most populated by the conformers of the biologically active molecules and least populated by inactive ones. Relative to 17β-estradiol, 151 compounds with measured human ER_{α}, mouse uterine, rat uterine, and MCF7 cell relative binding affinities (RBAs) were used to generate a model ranking the compounds with RBA of >150%, 150–10%, 10–1%, and 1–0.1%. The model was validated by screening of the EU chemical inventories for potentially active ER ligands and subsequent experimental testing of selected chemicals with reasonable agreement for many compounds.

Various CoMFA [51] analyses have been performed with different data sets. Waller [52], for example, used a data set of 58 estrogen agonists from eight diverse structure classes (see also Brown [47] above). The model with the highest LOO q^2 value of 0.59 was obtained by a combination of steric, electrostatic and hydrophobic HINT [53] fields. A validation test predicting each of the eight subclasses via a model trained by the remaining seven classes lead to averaged absolute errors between one and two log units with a largest individual deviation of 3.4 log units.

A study on the issue of ER_{α}–ER_{β} selectivity was performed by Demyttenaere-Kovatcheva et al. [54]. They derived CoMFA models for 104 benzoxazole and benzisoxazole derivatives. The models displayed high r^2 values for the 72 training compounds (ER_{α}: 0.91; ER_{β}: 0.95) but moderate cross-validated q^2 of 0.60 and 0.40, not including 6 and 5 outliers for ER_α and ER_β, respectively. From the 32 test molecules 6 and 8 compounds were outside one log unit in the RBA.

Beside HQSAR, Coleman et al. [55] used CoMFA and CoMSIA on a data set of 25 Bisphenol A derivatives with different measures of in vitro hormone activity (estrogen receptor binding, reporter gene induction, and cell proliferation). The models are statistically similar; the q^2 values range from 0.513 to 0.617, which is significantly lower than the r^2 scores, which vary between 0.774 and 0.999—an effect, possibly caused by overfitting. For the CoMFA and CoMSIA model four compounds were used for testing, yielding accurate prediction with a maximum deviation of 0.76 log units from the experimental value for binding affinity.

Shi et al. [56] evaluated HQSAR and CoMFA on a diverse set of 130 compounds, and the binding affinities of these compounds were measured in the same laboratory. These data act as a kind of benchmark for various QSAR methodologies. While the HQSAR model yielded an r^2 value of 0.76 and a LOO q^2 of 0.59, the CoMFA model showed superior behavior with an r^2 = 0.91 and a q^2 = 0.66. Adding a phenol indicator as descriptor improved the q^2 to 0.71. Leave-*n*-out cross validation was performed with up to 50% of the chemicals left out. The average q^2 over 100 different cross validations remained significant and stayed above 0.57. Finally the models were tested by the test sets of Kuiper at el. [57] and of Waller et al. [52]. As the biological data were measured in different laboratories, a significant variability in the absolute activity value of a chemical was observed. Consequently the activity values of the two data sets were normalized to the data of the training set: the data of each test set was first correlated with the training data on the basis of the shared compounds in both data sets, and then the activity value for each compound not in the training set was normalized to the data of the training set on the basis of the correlation equation. The r^2 for the test set of Kuiper and Waller were 0.71 and 0.62, respectively, with a maximum deviation of 1.9 and 1.6 log units in RBA.

Analyzing the contour maps of various CoMFA studies [52,56,58,59–61], the data from X-ray structures [62], and qualitative structure-activity data [63] gives a consistent picture for structural prerequisites of compounds binding to the ER (Figure 12.2): a negative partial charge vicinal to the 3–OH and 17β–OH of estradiol favors binding, corresponding to H-bonding abilities with Glu353/Arg394 and His524, where the 3-position H-bonding ability of phenols seems to be a significant requirement for ER binding. Furthermore the hydrophobic centers mimick steric 7α- and 11β-substituents, and their exact position in space favors binding to ER. Steric intolerance in the vicinity of the steroid A ring was interpreted to indicate that this region of the receptor's binding pocket exhibits a preference for planar rings. These observations suggest that a phenolic ring is likely a common structural feature associated with tight binding to the receptor. Where a direct comparison can be made, potent estrogens tend to be more hydrophobic in nature.

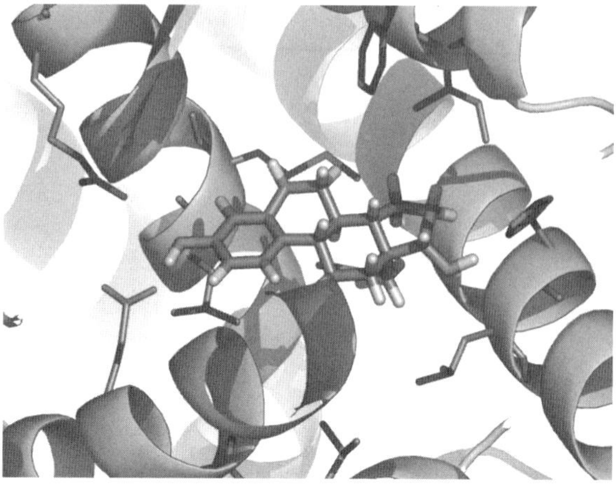

Figure 12.2 Estradiol binding to the ER as obtained by X-ray crystallography. The image was created with PyMol. (Delano, WL, *The PyMol user's manual*. San Cartos, CA: Delano Scientific, 2002.) See color plates.

The knowledge derived from this analysis was integrated in a four-phase hierarchical system for priority setting of EDCs [64,65] at the FDA's National Center for Toxicological Research (NCTR) [66]. The system is focused on minimizing possible false negatives, since it is essential for regulatory purposes to not ignore any potential harmful compound. For this purpose a progressive protocol is used where the four phases work in a hierarchical way to incrementally reduce the size of a dataset with increasing precision of prediction. In phase I, two rejection filters are used to eliminate those chemicals extremely unlikely to bind to ER with high confidence. If the molecular weight is less than 94 or more than 1000 and no ring structure is present in the compound, the compound is removed from the pipeline. According to an analysis of around 2000 compounds [65], those with the aforementioned properties will not bind to ER with significant affinity. In phase II, the chemicals passing through phase I are tested in 11 models composed of three 2D-structural alerts, seven 3D pharmacophores, and one classification model used to discriminate active from inactive chemicals. A chemical predicted to be active by any of these models is subsequently evaluated in phase III, where the previously described CoMFA model (see above) is used to more accurately predict binding affinities for the chemicals. Chemicals with higher predicted binding affinity are given higher priority for further evaluation in phase IV: a rule-based system incorporating accumulated human knowledge and expertise as well as information about production volume, environmental fate, and other factors that are employed to make a final decision on prioritizing experimental risk assessment.

At NCTR an alternative priority-setting model, called Decision Forest [67,68], was developed. In this model the results of multiple heterogeneous but statistically comparable decision trees are combined to produce a consensus prediction. An extensive cross-validation protocol (2000 runs with 10-fold cross validation) was elaborated to assess the prediction confidence and extrapolation sensitivity of the model for predicting unknown chemicals. The models consistently showed a poor accuracy (accuracy level [67] ≈ 0.64) for chemicals within the domain of low confidence (confidence level [67] <0.4), whereas accuracy was inversely proportional to the degree of domain extrapolation in the high-confidence domain and was on average around 0.86 for the high-confidence region.

Recently numerous studies were reported that explicitly take experimental X-ray data for the ER into account. Jacobson et al. [69] evaluated seven scoring functions (two scoring functions implemented in ICM [70,71] and five scoring functions as implemented in CScore [72]: FlexX-score [73], DOCK score [74], PMF score [75], GOLD score [76], and ChemScore [77]) based on ICM docking results of both mimics and toxins binding to ER_{α}. The scoring values were stored in a seven-dimensional vector for each ligand and were subjected to three different multivariate statistical methods: PLS discriminant analysis [78], rule-based methods [79], and Bayesian classification [80]. Rule-based classification methods showed the best performance (with respect to accuracy, precision, recall, and enrichment) and were also superior compared to classical consensus scoring and the best single scoring functions. For validation a test set was used based on a 2:1 splitting of the data. Out of 36 toxic compounds 29 were identified as toxic, and only two nontoxic compounds were predicted to be toxic. It should be emphasized that the authors were developing this method to optimize the lead-finding process and not toxicity predictions. Consequently they accepted a compromise, removing as many nonbinders as possible without losing many true binders. For regulatory purposes the method can also be tuned to identify as many toxic compounds as possible.

Another study on the same data set was performed by Prathipati et al. [81]. They derived binary QSAR [82] models using LUDI [83,84] and MOE [85] scoring functions and obtained discriminative ability comparable with the models derived by Jacobsson [69].

Akahori et al. [86] performed docking with the program ADAM [87] on four different molecule classes binding to ER_{α}. The docked configurations were minimized with BRUTO, keeping the protein side-chains flexible, and were energetically evaluated based on extended AMBER [88] force-field parameters for ligand and protein, respectively. Individual energy contributions (electrostatic and nonpolar protein–ligand interaction energies, desolvation energy of ligand and receptor) and terms for the lost degrees-of-freedom of the ligand conformational changes and the desolvation number (difference in the solvation numbers of the functional groups in the ligand and receptor surface between the unbound and bound forms) were used as independent variables in a discriminant analysis to separate binders from nonbinders. Additionally the same data were used for a multiple linear regression (MLR) analysis to quantify the affinities of the binding molecules. The analysis was performed for the four compound classes separately, and no validation with a test set was performed. The LOO cross-validation results display false negative rates, of 8%, 8%, 22%, and 8% for alkylphenols, phthalates, diphenylethanes, and benzophenones, respectively, where most of the false negative diphenylethanes were weak binders and were located by the discriminant analysis at the borderline between binders and nonbinders. The MLR analysis yielded models with q^2 values of 0.75 and 0.74 for the alkylphenols and diphenylethanes, respectively, whereas they were negative for phthalates and benzophenones due to a rather narrow range in affinity of about two orders of magnitude.

Two other studies used a combination of docking and QSAR for predicting binding affinities to ER. Sippl et al. [89] used AutoDock 3.0 [90] to dock 30 structurally diverse ER agonists and re-ranked the best 20 configurations utilizing the Yeti force field [91]. The energetically lowest conformation of each ligand was then deployed for input in the 3D-QSAR GRID/GOLPE [92], yielding a high correlation between predicted and experimental binding affinity in a fivefold cross validation q^2 of 0.90. The model was challenged with an external test set of 36 compounds structurally diverse when compared with the training set; their binding affinities were, however, measured at different laboratories and yielded a predictive r^2 value of 0.66. Three compounds were excluded from the prediction, as AutoDock was not able to identify any low-energy conformation without changing the orientation of several amino acids, which was not implemented in the version of AutoDock (see also Figure 12.4 for the androgen receptor). A direct correlation between AutoDock/Yeti scoring values and experimental binding affinities as well as using a ligand-based alignment utilizing FlexS [93], yielded models with very low if any predictive power (predictive $r^2 < 0.2$).

Vedani et al. [94] recognized the importance of protein flexibility for compounds binding to ER. They developed an automated docking protocol (Yeti 7.0 [95]), including side-chain flexibility of the protein and "on-the-fly" solvation to identify feasible binding modes. As several ligand orientations with similar scoring values were identified for most ligands (Figure 12.3), multidimensional QSAR concepts, Quasar [96,97,98], and Raptor [99,100,101] were deployed to allow for the induced-protein fit and an ensemble of different ligand configurations as input. The models yielded a cross-validated q^2 of 0.903 and 0.902 for 88 training molecules and a predictive r^2 of 0.885 and 0.846 for 18 test compounds with Quasar and Raptor, respectively.

In summary, a large variety of methods have been deployed to predict ER binding of possible endocrine-disrupting chemicals. Consistently measured data for a large inventory of structurally diverse compounds are available [102,103] that will make extensive validation of the methods in future applications possible and also mandatory before helping regulatory bodies in the risk-assessment process. This validation should also include the prediction of affinities for compound classes that are not part of the training process.

(a)

(b)

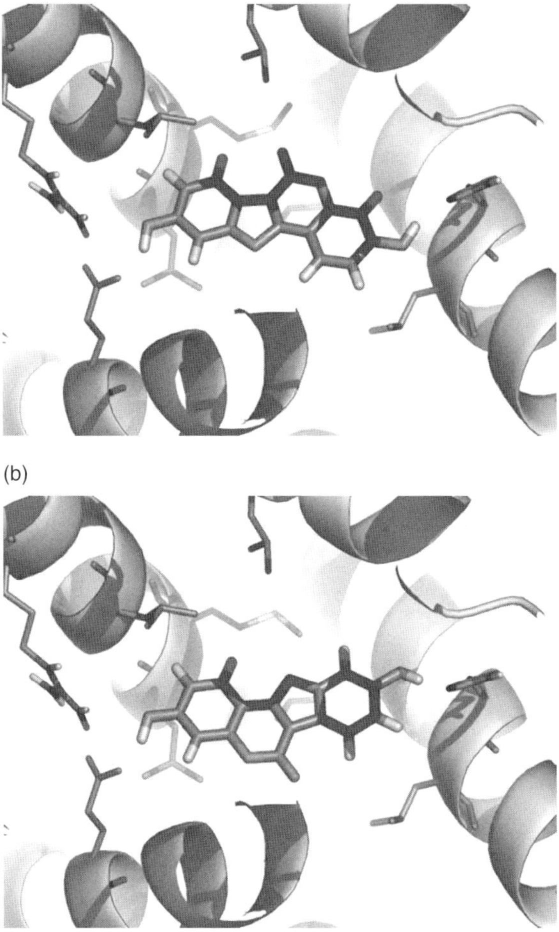

Figure 12.3 Four orientations representing a low-energy state of coumestrol in the binding pocket of the estrogen receptor identified by docking simulations with Yeti 7.0 [96]. Image was created with PyMol. See color plates.

12.2.2 Androgen Receptor

Androgens and the AR play an essential role in the growth of normal prostate. Moreover they are involved in the development of prostate cancer [104,105], representing the most common male malignancy in the United States. Many environmental chemicals, such as flavones or kepone, bind to the AR and act as AR antagonists [106,107,108], influencing the balance of the endocrine system [109,110,111]. It has been further noticed that estrogenic compounds can act as anti-androgens [112,113], while only few molecules have been identified to possess AR agonistic properties [107]. From data on animal species

(c)

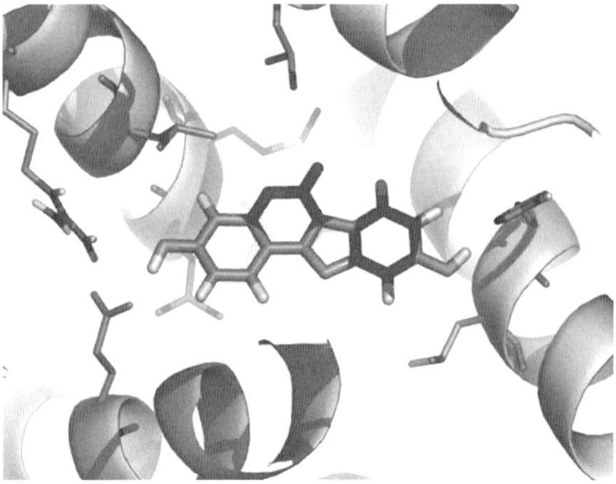

(d)

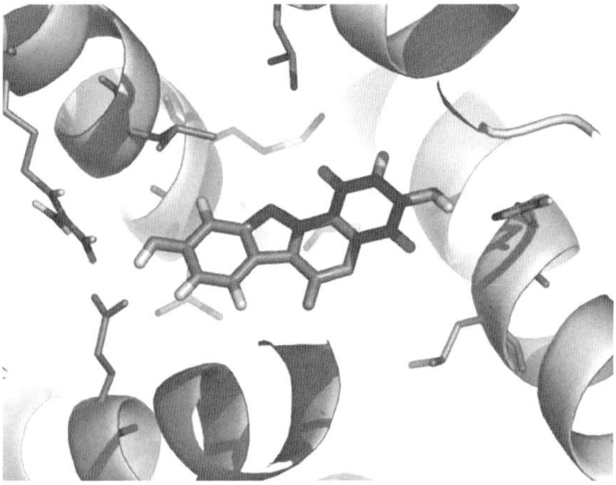

Figure 12.3 (*Continued*)

where, for example, activation of an androgen-regulated gene in fish has been observed [114,115], it is reasonable to think that future studies will uncover other androgenic compounds that are emitted into the environment with the potential to cause reproductive disturbances.

Only few computational studies have attempted to quantify the binding affinity, and thus estimate a toxic potential, of compounds. Only recently high-quality data of large sets of compounds have become available [106,107,108]. Serafimova et al. [116] applied the COREPA approach on a training set consisting of 28 steroidal and nonsteroidal ligands binding to the AR, where interatomic distances between nucleophilic sites and their charges were correlated with the binding affinities toward the AR.

Zhao et al. [117] applied a heuristic method to select five variables that represent the essential determinants of the molecular interaction out of a pool of 381 constitutional, geometrical, topological, electrostatic, and quantum-chemical descriptors. Three methods, multiple linear regression, radical basis function neural network, and support vector machine (SVM), were used to construct QSARs and to predict the binding of 146 structurally diverse chemicals to the AR. The statistically best model was obtained using an SVM yielding a LOO q^2 of 0.76 for 118 training compounds and a RMS deviation of 0.59 log units for the test set comprising 28 molecules.

Several CoMFA and CoMSIA studies were performed, both for optimizing selective androgen receptor modulators and for toxicity predictions. An early AR CoMFA model was reported by Waller et al. [118]. It was based on 28 structurally diverse chemicals from which 21 served as training set and yielded a cross-validated q^2 value of 0.792. From the seven test compounds six were predicted within one order of magnitude to the experiment, the remaining deviated by 1.8 log units in K_i from the experimental value.

Hong et al. [119] used CoMFA on a large and diverse training set of 146 compounds. The alignment was performed by optimizing the alignment of manually specified pharmacophores (overlap of ring structures and hydrogen-bond acceptors, maximizing steric overlap, etc.) of every compound class with the steroid R1881. The orientation of this molecule was specified based on X-ray crystallographic information [120]. The CoMFA simulation yielded an r^2 value of 0.90 and a LOO q^2 value of 0.57. The model was also tested on an external validation set of eight compounds; seven were predicted within one log unit to that of the experimental value, one deviated 1.5 log units from the experiment. The binding characteristics derived from the CoMFA contour map are in agreement with these observed in a human AR crystal structure: Regions that display favorable binding contributions of charged groups around the 3-keto and 17β-OH groups indicate a positive contribution to the affinity by forming hydrogen bonds, for example, between 17β-OH and Asn705 with Thr877, as well as between the 3-keto group with Arg752 and Gln 711 (see Figure 12.4).

Söderholm et al. [121] performed CoMSIA simulations on 70 structurally diverse AR binding compounds whose binding affinity was measured in two different laboratories, and the dataset contained analogues of flutamide, nilutamide, and bicalutamide. The binding mode of each ligand to AR was determined using docking with GOLD [76]. As the top-scored poses did not produce a consistent alignment of the pharmacophore features of the molecules, manual selection of docking poses was included into the alignment generation. Based on the alignment, CoMSIA produced a statistically significant model (LOO q^2 = 0.66, fivefold cross validation q^2 = 0.57) and showed a good predictive power for a test set of nine compounds (predictive r^2 = 0.80). The areas of the 3D-QSAR model showing favorable and unfavorable binding contributions of the structural features of the ligands fit with the chemical environment observed in the X-ray structure of the AR ligand-binding domain [122].

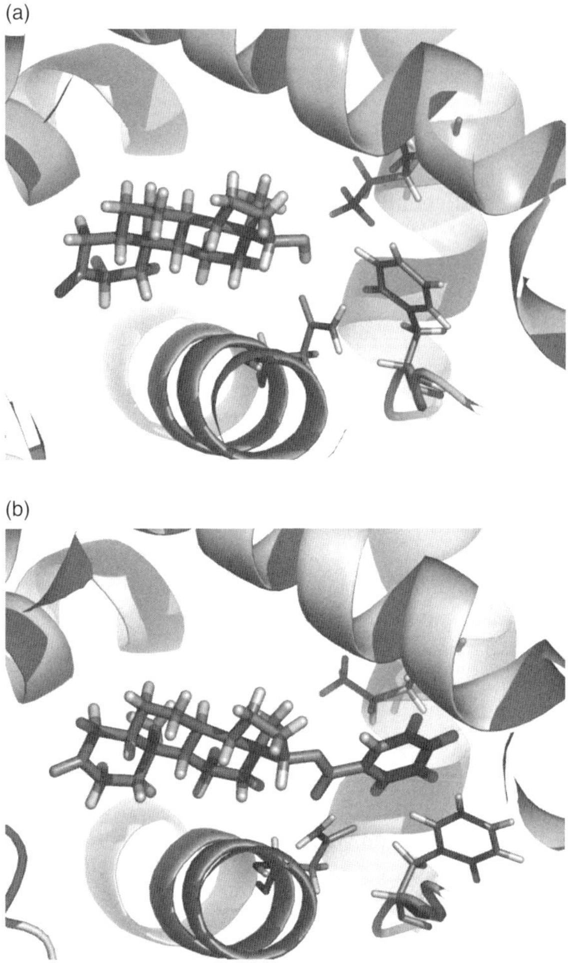

Figure 12.4 (*a*) Dihydrotestosterone (DHT) and DHT benzoate binding to the androgen receptor. Local induced fit is necessary to accommodate the additional volume of the benzoate group. The image was created with PyMol. See color plates.

Bohl et al. [123] docked six lead compounds with FlexX into the ligand-binding domain of a model of hAR based on homology to the human progesterone receptor. Subsequently they aligned 116 molecules, including hydroxyflutamide analogues, bicalutamide analogues, tricyclic quinolinones, and steroids, to the lead compounds structurally most similar to the very molecule. The CoMFA model based on this alignment resulted in r^2 of 0.95. Although the cross-validated q^2 dropped to a value of 0.593, the model seems to have a good predictive ability when validated by a test set ($r^2 = 0.95$) of 10 compounds.

Lill et al. [124] used flexible docking (software Yeti [95]) and linear-interaction analysis [125] to identify the bioactive conformations and orientations of 119 structurally diverse compounds. They demonstrated the relevance of induced protein fit upon ligand binding to identify the correct binding modes (Figure 12.4). Based on the superposition of the ligands in their identified poses, they then performed multidimensional QSAR using Raptor on 88 training compounds that converged at a cross-validated r^2 of 0.86. The model was validated by a test set of 26 ligands, yielding a predictive r^2 of 0.79, and it was further tested using five compounds from two compound classes that were structurally diverse compared to the training and test set. These compounds were all predicted within a factor of 4.5 of the experimental affinity.

12.2.3 Thyroid Receptors

Thyroid hormones exert profound effects on growth, development, and homeostasis in mammals [126,127,128, 129]. They regulate important genes in intestinal, skeletal, and cardiac muscles as well as in the liver and the central nervous system. In addition they influence the metabolic rate, cholesterol and triglyceride levels, heart rate, and an overall sense of well-being. Two major subtypes exist for the thyroid hormone receptor—TR_{α} and TR_{β}—encoded by two different genes. Differential RNA processing results in the formation of at least two isoforms each: $TR_{\alpha1}$, $TR_{\beta1}$, and $TR_{\beta2}$ bind thyroid hormones and act as ligand-regulated transcription factors; $TR_{\alpha2}$ species is prevalent in the pituitary and other part of the central nervous system. It has been suggested that most effects of thyroid hormones on heart rate and rhythm are mediated by the $TR_{\alpha1}$ species. On the other hand, most actions of hormones in the various tissues are predominantly mediated through the β subtypes. Thyroid hormones are primarily used for the treatment of hypothyroidism; other applications include obesity, hyperlipidemia, depression, and osteoporosis. The first pharmacological attempts to utilize thyroid hormones to treat these disorders were limited by the manifestation of hypothyroidism and cardiovascular toxicity [130].

A first attempt to quantitatively study thyromimetic activity was published by Bruice and coworkers in 1956 who derived equations relating thyroxine-like activity in amphibia and mammals [131]. Two decades later Kubinyi and Kehrhahn put forward a series of equations considering thyroxine-like activity of thyroxine analogues in the rat, as an example of a mixed approach to quantitative structure-activity relationships based on the Hansch and Free-Wilson analysis [132,133]. Following this in 1977, Dietrich and coworkers published a comprehensive account, describing the extension of these studies using modified equations and parameters and their application to 44 thyromimetics, in which they achieved a good agreement between calculated and observed binding affinities ($r = 0.90$) [134]. In 1989, Leeson and coworkers published a QSAR study of substituent effects on cardiac-sparing thyroid hormone analogues [135]. The underlying data sets used in the previous studies were quite congeneric, and a test set was not used in this case.

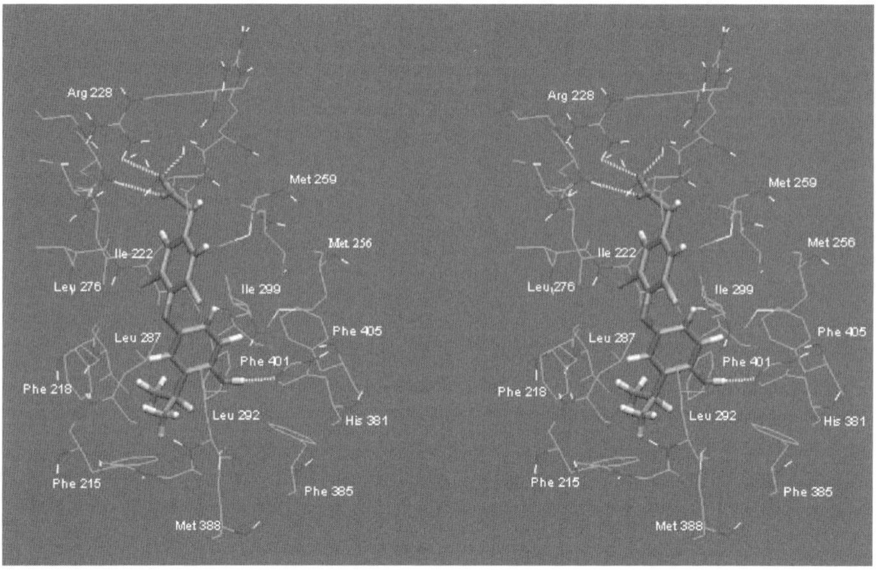

Figure 12.5 Stereo view of 3,5-dichloro-3′-isopropyl-thyronine bound to the thyroid hormone receptor α. Details of the binding pocket; dashed lines indicate stabilizing ligand-protein hydrogen bonds. See color plates.

In 1998, McKinney and Waller analyzed the triggering of thyromimetic action by environmental chemicals in a broader mechanistic context and suggested the beneficial use of 3D-QSAR [136]. The possibility to generate a QSAR in the presence of structural information of the thyroid receptor α and β became possible in 2003, when Ye and coworkers at KaroBio made their crystallographic studies public [130]. Figure 12.5 shows details of small-molecule agonist binding to TR_α at atomic resolution (after molecular-mechanical optimization).

In 2005, Zumstein established the first quantitative structure-activity relationship based on the experimental structures of TR_α and TR_β [137]. Following this, Vedani and coworkers validated a comprehensive QSAR model for TR_α and TR_β by means of consensus scoring (using the software Quasar [94,97] and Raptor [99]) [138]. The simulations were based on 82 agonists/antagonists, provided as a 4D data set with up to four conformations/orientations per ligand. With the Quasar simulation, the surrogate for the thyroid receptor α converged at a cross-validated r^2 of 0.846 for the 64 training compounds; the simulation yielded a predictive r^2 of 0.812 for the 18 test compounds. The corresponding values for the thyroid receptor β were $q^2 = 0.823$ and $p^2 = 0.665$, respectively.

The corresponding Raptor simulations converged at a q^2 of 0.919 and 0.909 for TR_α and TR_β, respectively, for the 64 training compounds; the simulation reached a predictive r^2 of 0.814 and 0.796 for TR_α and TR_β, respectively for

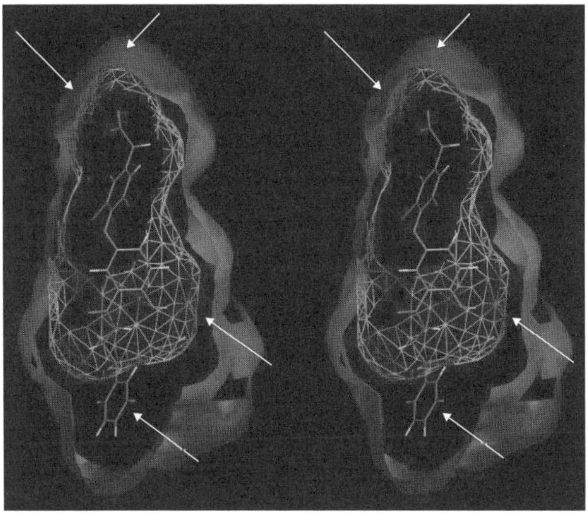

Figure 12.6 Stereo view of the Raptor surrogate for the thyroid receptor β with the largest ligand of the training set depicted. The front section has been clipped to display inner (wireframe) and outer shells (smooth surface). Areas colored in brown represent hydrophobic properties; areas in red correspond to H-bond acceptors, areas in blue to H-bond donors and green reflects H-bond flip-flops. See color plates.

the 18 test compounds. Comparison of the mapped properties in the Raptor model (Figure 12.6) with the amino acid residues lining the binding pocket at the true biological receptor demonstrates that the algorithm is well capable of identifying the appropriate interaction properties. The key amino acid residues Arg228 (inner shell: blue arrow in Figure 12.4), Arg262 (outer shell: white arrow in Fig 12.6), and His381 (inner and outer shell: red arrow in Figure 12.6) are perfectly mapped. In addition the two-shell concept allows one to realistically mimic the induced fit occurring when ligands with bulky substituents at the phenolic moiety bind. At the true biological receptor this refers to a larger movement of residues Phe215, Phe218, Met388, and Phe401 enlarging the hydrophobic pocket at the lower end of the binding site (gray arrow in Figure 12.6). To probe the sensitivity of these models toward the biological data, we have conducted a series of scramble tests for each of the receptor subtypes and obtained negative predictive r^2 values. This demonstrates that both surrogates are sufficiently sensitive toward the biological data, as well as demons trades the robustness of the approach [137].

Finally, a consensus analysis using the two approaches (Raptor and Quasar) was achieved as, on average, the calculated activities of the training set differ by a factor 2.2 in K_i and those of the test set by a factor 2.8 when predicted by Quasar and Raptor, respectively. It is important to note this size of deviation corresponds to the uncertainties in the underlying biological assays. Employing two different concepts for the prediction of binding affinities should be considered clearly advantageous, particularly as Quasar and Raptor are based on fundamentally different scoring functions.

12.2.4 Aryl Hydrocarbon Receptor

Although drugs such as acetaminophen bind to AhR, the majority of AhR agonists or antagonists are environmental chemicals. Polychlorinated dibenzodioxins such as 2,3,7,8-tetrachlorodibenzo-*p*-dioxin (TCDD), dibenzofurans, biphenyls, and a number of other chemicals are widespread pollutants in aquatic ecosystems. These compounds cause a high reproductive and developmental toxicity, which is mediated via binding to the AhR. Thus they pose a serious threat to many populations of mammals, birds, and fish. Various adverse effects—including structural malformations, reduced fertility, tumor promotion, immunotoxicity, and skin disorders like chloracne—have been observed [139].

AhR agonists are efficient inducers of cytochrome P450 (CYP) 1A1 and 1A2 genes. These enzymes catalyze certain carcinogens, for example, polyaromatic hydrocarbons (PAHs), into their ultimate carcinogenic and reprotoxic forms. By inducing CYP1A, AhR agonists may consequently enhance toxicity through the reactive intermediates formed from these molecules in the environment. Since many CYP1A-metabolized molecules, such as benzo(a)pyrene, are also AhR agonists, they can enhance their own metabolic activation. Moreover toxicity from CYP1A induction maybe caused by oxidative stress.

There is a wealth of QSAR studies on the AhR. Among the earliest accounts is that of Bandiera and coworkers who, in 1983, formulated a regression equation on 13 4′-substituted tetrachlorobiphenyls (PCBs) [140]. In 1986, they extended their calculations to include a series of 16 polyhalogenated dibenzofurans (PCDFs) [141], and in 1990, added 13 dibenzodioxins (PCDDs) [142].

In 1991, Rannug and coworkers published results from the first comprehensive study using artificial intelligence (using the software CASE), which included 136 polyhalogenated dibenzodioxins, dibenzofurans, biphenyls, and polyaromatic hydrocarbons (PAHs) [143]. They employed both a training and test set and achieved a good correlation between the predicted activity and the toxicity classes. The activating fragments from PAHs and heterocyclic compounds were observed to differ from those found in halogenated compounds such as dibenzo-*p*-dioxins, dibenzofurans, and biphenyls, suggesting the presence of two different recognition sites involved in AhR binding.

Mekenyan et al. developed an algorithm that was not restricted to minimum-energy structures and applied it to a series of PCBs, PCDFs, and PCDDs. The resulting QSAR models were robust and showed good utility across multiple classes of halogenated aromatic compounds [144]. In 1997, So and Karplus [145] applied molecular similarity matrices and genetic neural networks on a series of 78 polyhalogenated aromatic compounds (which were previously studied by Waller and McKinney [146] and Wagener et al. [147]) and obtained a cross-validated r^2 of 0.72 to 0.85. Waller and McKinney [146] applied the

CoMFA approach to rat SAR data for different classes of ligands (dibenzo-*p*-dioxins, dibenzofurans, and biphenyls). This study suggests that the basic tendency of lateral chlorine substitutions is to increase binding affinity whereas nonlateral groups decrease binding affinity.

Mhin et al. [148] used a molecular quadrupole-moment parameter to describe the electrostatic interaction of polychlorinated dibenzo-*p*-dioxins at the Ah receptor active site. Their analysis suggests that the most potent molecules have a high polarizability across the lateral direction.

QSAR methods have also been applied specifically to AhR agonists. Lewis et al. [149] demonstrated in their study the importance of planarity and molecular length, along with other parameters such as the frontier orbital HOMO and lipophilicity for AhR agonists. Mankowski and Ekins [150] have generated a Catalyst HIPHOP pharmacophore based on four ligands binding in the nanomolar range: indirubin, indigo, ITE, and TCDD. Two different planar pharmacophores were generated, one containing a hydrogen bond acceptor and two hydrophobic groups, and the other two ring aromatic features. Both pharmacophores have hydrophobic features, confirming their importance as suggested by earlier studies. This model also agrees with a homology model for the ligand-binding domain of mouse AhR based on bacterial oxygen sensing protein FixL. The high level of structural similarity with other PAS domains for hERG and PYP proteins [151] suggests that Gln377 can form favorable electrostatic interactions with chlorine atoms on TCDD and that Phe345 can have hydrophobic interactions with the planar ring system.

Vedani and coworkers published a QSAR study including 121 PCDDs, PCDFs, PCBs, and PAHs whereby a multiple representation of binding modes (4D-QSAR) yielded superior results: for the 91 molecules in the training set, a cross-validated r^2 of 0.816 was reached and the 30 compounds defining the test set were predicted with an r^2 of 0.763 (using the software Quasar) [152,153]. Because of the symmetric core topology and the absence of structural information on the true biological receptor, the compounds were represented in up to four different orientations and were thus used to compose a 4D data set. A scramble test yielded a $q^2 = -0.129$ and a predictive $r^2 = -0.540$. The model is depicted in Figure 12.7. A simulation on an extended data set of 105 training and 35 test compounds, including a series of 19 aza-PAHs, yielded a cross-validated r^2 of 0.782 and a predictive r^2 of 0.766.

Multimode ligand binding was also used for the CoMFA study of Lukacova and Balaz on a set of 34 compounds [154]. The 12 test compounds were predicted with a RMS deviation of 1.0 log units from the experimental values.

Various density–functional theory based descriptors were probed with success by Arulmozhiraja and Morita [155]. In 2005, Hirokawa and coworkers employed Hartree-Fock theory, which identifies the polarization as a key parameter for QSAR on AhR binding [156]. Wang et al. extended their work to include polybrominated compounds and reached a cross-validated r^2 of 0.580 and 0.680 using CoMFA and CoMSIA, respectively [157]. Zheng and coworkers [158] employed radial basis function neural networks and obtained

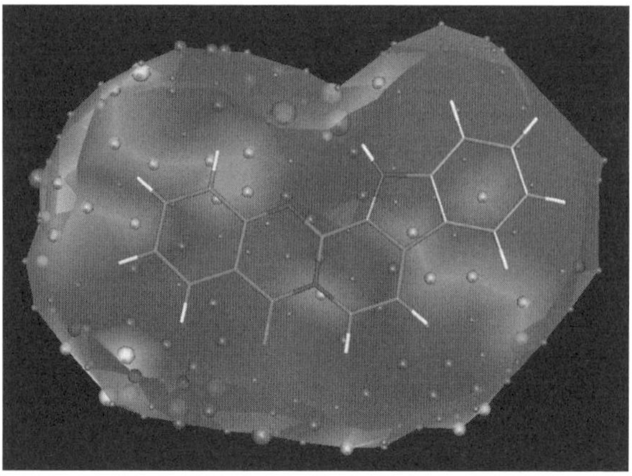

Figure 12.7 AhR surrogate with a bound aza-PAH as generated by Quasar. For clarity, the front section has been clipped. Areas colored in gray/brown represent hydrophobic properties; areas in green H-bond donor functions; areas in yellow indicate H-bond acceptors and purple domains correspond to H-bond flip-flops. No salt bridges are observed in this model as the ligands lack any charged groups. See color plates.

a q^2 of 0.882. In 2006, Sovadinova and coworkers established a QSAR that relates AhR receptor-mediated activity of PAHs and cytotoxicity [159]. At the 2006 ECVAM workshop on toxicity modeling, Benfenati reported a comparison of CoMFA, HQSAR, and VolSurf on a series of 93 AhR ligands that converged at a cross-validated r^2 of 0.91/0.85/0.79 for the 84 training compounds. The simulations yielded a predictive r^2 of 0.86/0.78/0.71 for the nine compounds defining the test set [160]. These results suggest that many QSAR methods are applicable to the analysis of AhR.

12.3 RECEPTORS INVOLVED IN DRUG METABOLISM AND DRUG–DRUG INTERACTIONS

Adverse drug–drug interactions may occur when a drug alters the pharmacokinetic properties of another, co-administered drug. This effect is most often caused by drug-induced induction or inhibition of CYP enzymes. Inhibition of CYP enzymes may increase the concentration of the co-administered drugs and can have a significant effect on their toxicity. Induction of specific P450 enzymes can alter the metabolic profile of a drug by increasing metabolic rates or by creating an alternate pathway of metabolism with profound effects on its toxicity profile. Drugs such as cisapride, terfenadine, and mibefradil are examples of drugs that have been withdrawn from the market because of adverse drug–drug interactions associated with CYP induction or inhibition.

The mechanisms of induction or inhibition of the induction of these CYPs are primarily due to activation or inhibition of ligand-activated transcription factors such as the AhR for the CYP1A family, CAR for the CYP2B family, and PXR/SXR, GR, and VDR for the CYP3A family. As there is, however, some considerable degree of cross talk between these receptors and other proteins, the process of induction is not a single receptor–protein relationship. As a comprehensive review on predicting CYP induction has recently been published [150], we will not discuss every computational model in detail.

12.3.1 Pregnane X Receptor/Steroid and Xenobiotic Receptor

The PXR/SXR regulates the transcription of CYP3A genes [161,162,163]. It binds structurally quite diverse compounds, such as bile acids [164], statins [165], HIV protease inhibitors [166], calcium-channel modulators [167], steroids [168], plasticizers [169], and organochlorine pesticides [170] ranging in molecular weight from 232 Da (phenobarbital) up to 854 Da (paclitaxel). In addition PXR/SXR is involved in the regulation of various other genes, such as CYP2B6 [171] and CYP2C8/9 [172] as well as drug transporters [172].

PXR/SXR is also thought to be responsible for switching on genes in response to environmental pollutants that are potential teratogens [173]. Being able to predict this potential activity is of definite value for developmental toxicology of environmental chemicals.

Three-dimensional structures of PXR [174,175,176,177] obtained by X-ray crystallography display a large hydrophobic ligand-binding domain with a few polar residues and initially three distinct binding modes of the co-crystallized ligand SR12813. With co-activator binding, a single site is apparent. The structures further reveal that significant flexibility of the topological elements lining the binding site enable the binding of structurally diverse ligands (Figure 12.8): movement of two loops adjacent to the binding cavity increases the volume of the binding site and allows for binding of hyperforin, which makes additional hydrophobic interactions, such as with Ile414.

Despite the flexibility of the PXR ligand-binding domain, pharmacophore models were derived to attempt to understand the key features for ligand binding to PXR. From on EC_{50} data for 12 molecules, Ekins and Erickson [178] generated a pharmacophore using Catalyst software. The study implies that four hydrophobic features and at least one hydrogen-bonding group may be required for a ligand to bind to PXR. These features are consistent with one of the earlier experimentally determined binding modes of SR12813 at PXR. The pharmacophore was also used to predict the binding affinity for 28 test molecules known to be PXR ligands. The pharmacophore distinguishes the most potent activators of PXR, such as ecteinascidin, troglitazone, nifedipine, and dexamethasone-*t*-butylacetate, from poor activators, such as scopoletin and kaempferol. Schuster and Langer [179] used Catalyst software to derive pharmacophore models for PXR activation. A pharmacophore hypothesis manually annotated based on the X-ray structure of PXR with the ligand SR12813, and several ligand-based ones were compared in order to identify ligand receptor interactions essential for receptor activation. The results sug-

(a)

(b)

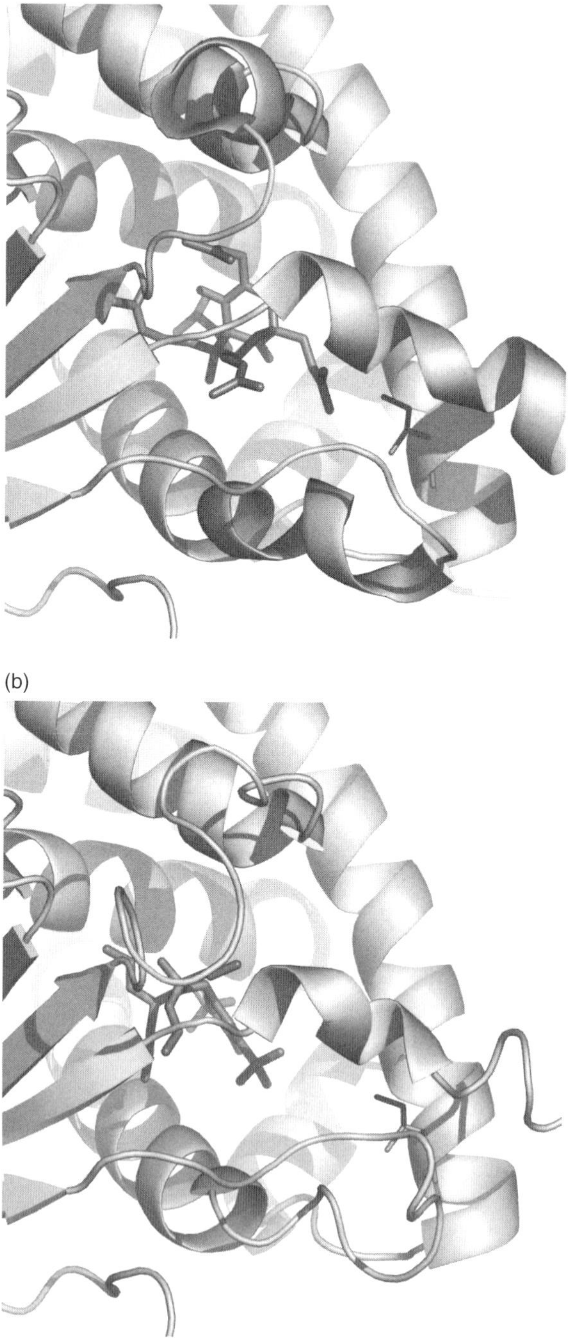

Figure 12.8 (*a*) SR12813 and (*b*) hyperforin binding to the PXR as obtained by X-ray crystallography. When hyperforin binds, two loops move out to provide additional space and a hydrophobic interaction is observed with Ile414. The image was created with PyMol. See color plates.

gested that hydrogen bonding to Gln285 is necessary for PXR activation, while a second hydrogen bond to His407 could be identified for most ligands. Further hydrophobic interactions contribute to ligand affinity, where highly active compounds share up to five hydrophobic features that allow the ligand to occupy the predominantly hydrophobic binding pocket. When testing the model with 2361 commercial drugs, they obtained a low rate of 1.6% for PXR-activating compounds, agreeing with the assumption that noticeable binding to PXR would induce significant side effects.

Jacobs [180] reported a QSAR model ($r^2 = 0.83$) based on PXR relative induction data of 33 structurally diverse compounds using VolSurf descriptors. Favorable for binding were regions accessible by an amide probe and an optimal distance for hydrogen bonding with three of the five polar amino acid residues (Ser208, Ser247, His407, Arg410, and Gln285) lining the PXR ligand-binding domain (Figure 12.8).

12.3.2 Constitutive Androstane Receptor

Although it appears to be much less promiscuous than PXR and the binding site is smaller [181], CAR also responds to structurally diverse xenobiotics, such as clotrimazole, the progesterone metabolite 5α-pregnane-3,20-dione or androstanol, and controls expression of CYP genes (CYP2B6 and CYP3A4). CAR is activated by various environmental chemicals, such as the pesticide TCPOBOP (1,4-*bis*[2-(3,5-dichloropyridyloxy)]benzene), methoxychlor, and PCB, and can thus also be responsible for endocrine-disrupting effects. It should be noted that profound species differences in ligand selectivity exist for both PXR and CAR [182,183,184].

Prior to the publication of the first X-ray structures of CAR [185,186,187] some pharmacophore models had been generated. Based on small numbers of molecules suggested to be mouse CAR activators, a study of Ekins et al. [188] suggested a relatively planar pharmacophore model with two hydrophobic groups and one hydrogen-bond acceptor relevant for binding. The study takes into account the finding that CAR has high affinity for the rigid repressors and androstane metabolites that fit well within the confines of this model [189] and suggests that CAR accommodates less flexibility in the ligands involved in activation compared to PXR/SXR.

Another pharmacophore model put forward by Ekins et al. [188] based on the alignment of clotrimazole, androstanol, and 5a-pregnane-3,20-dione consisted of three hydrophobic elements and one hydrogen bond acceptor, indicating a smaller binding pocket when compared with PXR, which is agreement with information from X-ray crystallography [181]. Steric overlap was observed between the potent ligand 5α-pregnane-3,20 dione and the low-affiniy PXR ligands lithocholic acid and ketolithocholic acid, suggesting that a compound could therefore potentially regulate both CYP3A4 and CYP2B6 via different receptors to differing extents simultaneously.

12.3.3 Glucocorticoid Receptor

Glucocorticoids such as dexamethasone may induce basal CYP3A4 expression by binding to GR [190]; they are also capable of up-regulating PXR, CAR, and RXR expression and increasing CYP3A4 expression [191,192,193] along with it. To a lesser extent glucocorticoids binding to GR may induce CYP3A5 and CYP3A7 genes.

Several computational studies on rather small datasets have been reported. The key features suggested by these studies are in agreement with the information gained from X-ray structures [194,195]. These structures show that apart from hydrophobic contacts, an extensive hydrogen bonding network is responsible for ligand recognition, and that a side pocket in the binding site accommodating the bulky substituents at the 17α position of glucocorticoids might be responsible for selectivity against the mineralcorticoid receptor.

Lewis et al. [196] performed a study on 14 steroidal compounds using structural descriptors. A consecutive study by the same group [197] using homology modeling and QSAR suggested that besides hydrogen bonding molecular planarity and rectangularity is responsible for induction.

Mankowski and Ekins [150] derived a Catalyst model based on nine molecules with a good agreement to the experiment for the training set and a pharmacophore consisting of one hydrogen-bond donor, a hydrophobic feature, and a aromatic ring feature.

12.4 CONCLUSIONS

A wealth of computational models for receptors involved in mediating toxicity of environmental chemicals (predominantly endocrine-disrupting effects) as well as for receptors involved in drug metabolism and adverse drug–drug interactions have been developed. Because of the availability of consistent biological data, the ER and AhR received most attention in the past. For toxicological purposes the computational models and the underlying data were focused on chemicals for the prediction of toxicity of environmental chemicals, such as for ER, AR, and AhR, while they were considering more drug-like compounds for toxicity prediction associated with receptors involved in drug metabolism and drug–drug interactions, such as PXR and CAR. However, as discussed in the previous sections, both chemicals and drugs can bind to receptors like AhR, PXR, and CAR and activate transcription of CYP genes. Therefore the separation into toxicity triggered by environmental xenobiotic chemicals and drug interactions remains doubtful. Further investigations would seem to be necessary showing how computational models for PXR based predominantly on drug-like molecules might be applicable to environmental chemicals due to possibly nonoverlapping applicability domains.

The prediction of activity, agonism versus antagonism, adds an additional layer of complexity to the problem of toxicity classification of compounds.

Furthermore the binding of different co-activators influences the type of activity for several compounds. Although very critical for toxicity predictions, extensive and sufficient modeling studies on this subject are currently lacking.

That still many of the models discussed in this account are not sufficiently validated against external test sets might undermine an otherwise solid concept. Particularly for regulatory purposes, this issue is critical, but also to provide confidence in prediction and toxicity warnings when used in a drug-development process.

Therefore the development of more diverse and thoroughly validated models for many of these receptors will remain an important scientific task. Such data should mainly become available with high-throughput screening approaches.

REFERENCES

1. Tsai MJ, Omalley BW. Molecular mechanisms of action of steroid/thyroid receptor superfamily members. *An Rev Biochem* 1994;63:451–86.
2. Ribeiro RCJ, Kushner PJ, Baxter JD. The nuclear hormone-receptor gene superfamily. *An Rev Med* 1995;46:443–53.
3. Moras D, Gronemeyer H. The nuclear receptor ligand-binding domain: Structure and function. *Curr Opin Cell Biol* 1998;10:384–91.
4. Katzenellenbogen JA, Katzenellenbogen BS. Nuclear hormone receptors: Ligand-activated regulators of transcription and diverse cell responses. *Chem Biol* 1996;3:529–36.
5. Gray LE, Kelce WR, Wiese T, Tyl R, Gaido K, Cook J, et al. Endocrine screening methods workshop report: Detection of estrogenic and androgenic hormonal and antihormonal activity for chemicals that act via receptor or steroidogenic enzyme mechanisms. *Reprod Toxicol* 1997;11:719–50.
6. Weatherman RV, Fletterick RJ, Scanlan TS. Nuclear-receptor ligands and ligand-binding domains. *An Rev Biochem* 1999;68:559–81.
7. Damstra T, Barlow S, Bergman A, Kavlock R, van der Kraak G, editors. *Global assessment of the state of the science of endocrine disruptors*. Geneva: World Health Organization, 2002, ch. 5.4, and references therein.
8. Damstra T, Barlow S, Bergman A, Kavlock R, van der Kraak G, editors. *Global assessment of the state of the science of endocrine disruptors*. Geneva: World Health Organization, 2002, ch. 5.1, and references therein.
9. Damstra T, Barlow S, Bergman A, Kavlock R, van der Kraak G, editors. *Global assessment of the state of the science of endocrine disruptors*. Geneva: World Health Organization, 2002, chs. 3.14 and 5.2, and references therein.
10. Carson R. *Silent spring*. Boston: Houghton Mifflin, 2002.
11. Schurmeyer T, Nieschlag E. Effect of ketoconazole and other imidazole fungicides on testosterone biosynthesis. *Acta Endocrinol* 1984;105:275–80.
12. Pepper G, Brenner S, Gabrilove J. Ketoconazole use in the treatment of ovarian hyperandrogenism. *Fertil Steril* 1990;54:38–44.

13. Williams DR, Fisher MJ, Rees HH. Characterization of ecdysteroid 26-hydroxylase: An enzyme involved in molting hormone inactivation. *Arch Biochem Biophys* 2000;376:389–98.
14. Sheehan DM, Young M. Diethylstilbestrol and estradiol binding to serum-albumin and pregnancy plasma of rat and human. *Endocrinology* 1979;104:1442–46.
15. Birnbaum LS. Endocrine effects of prenatal exposure to PCBs, dioxins and other xenobiotics: Implications for policy and research. *Environ Health Perspect* 1994;102:676–9.
16. Kuiper GGJM, Gustafsson J-A. The novel estrogen receptor-β subtype: Potential role in the cell- and promoterspecific actions of estrogens and anti-estrogens. *FEBS Lett* 1997;410:87–90.
17. Kuiper GGJM, Carlsson B, Grandien K, Enmark E, Haeggblad J, et al. Comparison of the ligand binding specificity and transcript tissue distribution of estrogen receptors α and β. *Endocrinology* 1997;138:863–70.
18. Hewitt SC, Couse JF, Korach KS. Estrogen receptor transcription and transactivation. Estrogen receptor knockout mice: What their phenotypes reveal about mechanisms of estrogen action. *Breast Cancer Res* 2000;2:345–52.
19. Couse JF, Korach KS. Estrogen receptor null mice: What have we learned and where will they lead us? *Endocr Rev* 1999;20:358–417.
20. Dupont S, Krust A, Gansmuller A, Dierich A, Chambon P, et al. Effect of single and compound knockouts of estrogen receptors α (ER_{α}) and β (ER_{β}) on mouse reproductive phenotypes. *Development* 2000;127:4277–91.
21. Barkhem T, Carlsson B, Nilsson Y, Enmark E, Gustafsson J, Nilsson S. Differential response of estrogen receptor alpha and estrogen receptor beta to partial estrogen agonists/antagonists. *Mol Pharmacol* 1998;54:105–12.
22. Dibb S. Swimming in a sea of estrogens, chemical hormone disrupters. *Ecologist* 1995;25:27–31.
23. McLachlan JA, Arnold SF. Environmental estrogens. *Am Sci* 1996;84:452–61.
24. Guillette LJ, Crain DA, Rooney AA, Pickford DB. Organization versus activation: The role of endocrine disrupting contaminants EDCs during embryonic development in wildlife. *Environ Health Perspect* 1995;103:157–64.
25. Colborn T. Environmental estrogens: Health implications for humans and wildlife. *Environ Health Perspect* 1995;103:135–36.
26. Feldman D, Krishnan A. Estrogens in unexpected places: Possible implications for researchers and consumers. *Environ Health Perspect* 1995;103:129–33.
27. Hoare SA, Jobling S, Parker MG, Sumpter JP, White R. Environmental persistent alkylphenolic compounds are estrogenic. *Endocrinology* 1994;135:175–82.
28. Korach KS, Levy LA, Sarver PJ. Estrogen receptor stereochemistry: Receptor binding and hormonal responses. *J Steroid Biochem* 1987;27:281–90.
29. Safe Drinking Water Act Amendment; Public Law 104-182 (Section 136); US Environmental Protection Agency, 1996 <http://www.epa.gov.safewater/sdwa/index.html>.
30. Food Quality Protection Act; Public Law 104-170 (Section 408); US Food and Drug Administration, 1996 <http://www.fda.gov/opacom/laws/foodqual/fqpatoc.htm>.

31. Fang H, Tong W, Welsh WJ, Sheehan, DM. QSAR models in receptor-mediated effects: the nuclear receptor superfamily. *J Molec Structure* 2003;622:113–25.
32. Schmieder PK, Ankley G, Mekenyan O, Walker JD. Quantitative structure-activity relationship models for prediction of estrogen receptor binding affinity of structurally diverse chemicals. *Environ Toxicol Chem* 2003;22:1844–54.
33. Golbraikh A, Tropsha A. Beware of q^2! *J Mol Graphics Model* 2002;20:269–76.
34. BioByte Corp. 201 W. 4th St., #204 Claremont, CA 91711-4707.
35. Gao H, Katzenellenbogen JA, Garg R, Hansch C. Comparative QSAR Analysis of estrogen receptor ligands. *Chem Rev* 1999;99:723–44.
36. Gao H, Williams C, Labute P, Bajorath J. Binary quantitative structure-activity relationship (QSAR) analysis of estrogen receptor ligands. *J Chem Inf Comput Sci* 1999;39:164–8.
37. Zheng W, Tropsha A. Novel variable selection quantitative structure-property relationship approach based on the *k*-nearest-neighbor principle. *J Chem Inf Comput Sci* 2000;40:185–94.
38. Kier LB, Hall LH. *Molecular connectivity in chemistry and drug research*. New York: Academic Press, 1976.
39. Kier LB, Hall LH. *Molecular connectivity in structure-activity analysis*. Chichester: Research Studies Press, 1986.
40. Hall LH, Kier LB. The molecular connectivity chi indexes and kappa shape indexes in structure-property modeling. In: Lipkowitz KB, Boyd DB, editors, *Reviews in computational chemistry II*. Weinheim: VCH, 1991;367–422.
41. Carhart RE, Smith DH, Venkataraghavan R. Atom pairs as molecular features in structure-activity studies: Definition and applications. *J Chem Inf Comput Sci* 1985;25:64–73.
42. Asikainen A, Ruuskanen J, Tuppurainen K. Consensus kNN QSAR: A versatile method for predicting the estrogenic activity of organic compounds in silico: A comparative study with five estrogen receptors and a large, diverse set of ligands. *Environ Sci Technol* 2004;38:6724–9.
43. <http://edkb.fda.gov/databasedoor.html>.
44. Todeschini R, Consonni V, Pavan M. DRAGON. Software for the calculation of molecular descriptors, Version 3.0.
45. Marini F, Roncaglioni A, Novic M. Variable selection and interpretation in structure-affinity correlation modeling of estrogen receptor binders. *J Chem Inf Model* 2005;45:1507–19.
46. Blair RM, Fang H, Branham WS, Hass BS, Dial SL, Moland CL, et al. The estrogen receptor relative binding affinities of 188 natural and xenochemicals: structural diversity of ligands. *Toxicol Sci* 2000;54:138–53.
47. Brown N, McKay B, Gasteiger J. Fingal: A novel approach to geometric fingerprinting and a comparative study of its application to 3D-QSAR modelling. *QSAR Comb Sci* 2005;24:480–4.
48. Mekenyan O, Ivanov J, Karabunarliev S, Bradbury S, Ankley G, Karcher W. A computationally-based hazard identification algorithm that incorporates ligand flexibility: 1. Identification of potential androgen receptor ligands. *Environ Sci Technol* 1997;31:3702–11.

49. Mekenyan O, Nikolova N, Karabunarliev S, Bradbury S, Ankley G, Hansen, B. New developments in a hazard identification algorithm for hormone receptor ligands. *Quant Struct-Act Rel* 1999;18:139–53.
50. Mekenyan O, Kamenska V, Serafimova R, Poellinger L, Brouwer A, Walker J. Development and validation of an average mammalian estrogen receptor-based QSAR model. *SAR QSAR Environ Res* 2002;13:579–95.
51. Cramer R III, Patterson DE, Bunce JD. Comparative molecular field analysis (CoMFA): 1. Effect of shape on binding of steroids to carrier proteins. *J Am Chem Soc* 1988;110:5959–67.
52. Waller CL, Oprea TI, Chae K, Park H-E, Korach KS, Laws SC, et al. ligand-based identification of environmental estrogens. *Chem Res Toxicol* 1996;9:1240–8.
53. Kellogg GE, Semus SF, Abraham DJ. HINT: A new method of empirical hydrophobic field calculation for CoMFA. *J Comput Aided Mol Des* 1991;5:545–52.
54. Demyttenaere-Kovatcheva A, Cronin MTD, Benfenati E, Roncaglioni A, LoPiparo E. Identification of the structural requirements of the receptor-binding affinity of diphenolic azoles to estrogen receptors α and β by three-dimensional quantitative structure-activity relationship and structure-activity relationship analysis. *J Med Chem* 2005;48:7628–36.
55. Coleman KP, Toscano WA Jr, Wiese TE. QSAR models of the in vitro estrogen activity of bisphenol A analogs. *QSAR Comb Sci* 2003;22:78–88.
56. Shi LM, Fang H, Tong W, Wu J, Perkins R, Blair RM, et al. QSAR models using a large diverse set of estrogens. *J Chem Inf Comput Sci* 2001;41:186–95.
57. Kuiper GG, Lemmen JG, Carlsson B, Corton JC, Safe SH, van der Saag PT, et al. Interaction of estrogenic chemicals and phytoestrogens with estrogen receptor beta. *Endocrinology* 1998;139:4252–63.
58. Wiese TE, Polin LA, Palomino E, Brooks SC. Induction of the estrogen specific mitogenic response of MCF-7 cells by selected analogues of estradiol-17 beta: A 3D QSAR study. *J Med Chem* 1997;40:3659–69.
59. Tong W, Perkins R, Xing L, Welsh WJ, Sheehan DM. QSAR models for binding of estrogenic compounds to estrogen receptor alpha and beta subtypes. *Endocrinology* 1997;138:4022–5.
60. Sadler BR, Cho SJ, Ishaq KS, Chae K, Korach KS. Three-dimensional quantitative structure-activity relationship study of nonsteroidal estrogen receptor ligands using the comparative molecular field analysis/cross-validated r^2-guided region selection approach. *J Med Chem* 1998;41:2261–7.
61. Gantchev TG, Ali H, van Lier JE. Quantitative structure-activity relationships/comparative molecular field analysis (QSAR/CoMFA) for receptor-binding properties of halogenated estradiol derivatives. *J Med Chem* 1994;37:4164–76.
62. <http://www.rcsb.org/pdb>.
63. Fang H, Tong W, Shi LM, Blair R, Perkins R, Branham W, et al. Structure-activity relationships for a large diverse set of natural, synthetic, and environmental estrogens. *Chem Res Toxicol* 2001;14:280–94.
64. Shi LM, Tong W, Fang H, Perkins R, Wu J, Tu M, et al. An integrated "4-phase" approach for setting endocrine disruption screening priorities—Phase I and II predictions of estrogen receptor binding affinity. *SAR QSAR Environ Res* 2002;13:69–88.

65. Hong H, Tong W, Fang H, Shi LM, Xie Q, Wu J, et al. Prediction of estrogen receptor binding for 58,000 chemicals using an integrated system of a tree-based model with structural alerts. *Environ Health Perspect* 2002;110:29–36.

66. <http://www.fda.gov/nctr>.

67. Tong W, Hong H, Fang H, Xie Q, Perkins R. Decision forest: Combining the predictions of multiple independent decision tree models. *J Chem Inf Comput Sci* 2003;43:525–31.

68. Tong W, Xie Q, Hong H, Shi LM, Fang H, Perkins R. Assessment of prediction confidence and domain extrapolation of two structure-activity relationship models for predicting estrogen receptor binding activity. *Environ Health Perspect* 2004;112:1249–54.

69. Jacobsson M, Lidén P, Stjernschantz E, Boström H, Norinder U. Improving structure-based virtual screening by multivariate analysis of scoring data. *J Med Chem* 2003;46:5781–9.

70. Abagyan R, Totrov M. *ICM* online manual <http://www.molsoft.com>.

71. Abagyan R, Totrov M. Biased probability Monte Carlo conformational searches and electrostatic calculations for peptides and proteins. *J Mol Biol* 1994;235: 983–1002.

72. Clark RD, Strizhev A, Leonard JM, Blake JF, Matthew JB. Consensus scoring for ligand-protein interactions. *J Mol Graphics Model* 2002;20:281–95.

73. Rarey M, Kramer B, Lengauer T, Klebe G. A fast flexible docking method using an incremental construction algorithm. *J Mol Biol* 1996;261:470–89.

74. Kuntz ID, Blaney JM, Oatley SJ, Langridge R, Ferrin TE. A geometric approach to macromolecule-ligand interactions. *J Mol Biol* 1982;161:269–88.

75. Muegge I, Martin YC. A general and fast scoring function for protein-ligand interactions: A simplified potential approach. *J Med Chem* 1999;42:791–804.

76. Jones G, Willett P, Glen RC, Leach AR, Taylor R. Development and validation of a genetic algorithm for flexible docking. *J Mol Biol* 1997;267:727–48.

77. Eldridge MD, Murray CW, Auton TR, Paolini GV, Mee RP. Empirical scoring functions: I. The development of a fast empirical scoring function to estimate the binding affinity of ligands in receptor complexes. *J Comput Aided Mol Des* 1997;11:425–45.

78. Wold S, Johansson E, Cocchi M. PLS-partial least-squares projections to latent structures. In *3D QSAR in Drug Design*. Leiden: ESCOM, 1993. p. 523–50.

79. *Rule Discovery System (RDS) 0.8* <http://www.compumine.com>; Compumine AB.

80. Domingos P, Pazzani M. On the optimality of the simple Bayesian classifier under zero-one loss. *Mach Learn* 1997;29:103–30.

81. Prathipati P, Saxena AK. Evaluation of binary qsar models derived from ludi and moe scoring functions for structure based virtual screening. *J Chem Inf Model* 2006;46:39–51.

82. Labute P. Binary QSAR: A new method for the determination of quantitative structure-activity relationships. In Altman RB, Dunker AK, Hunter L, Klein TE, Lauderdale K, editors, *Proceedings of the Pacific Symposium on Biocomputing '99*. River Edge, NJ: World Scientific, p. 444–55.

83. Bohm HJ. The development of a simple empirical scoring function to estimate the binding constant for a protein-ligand complex of known three-dimensional structure. *J Comput Aided Mol Des* 1994;8:243–56.

84. Bohm HJ. Prediction of binding constants of protein ligands: a fast method for the prioritization of hits obtained from de novo design or 3D database search programs. *J Comput Aided Mol Des* 1998;12:309–23.

85. Chemical Computing Group, MOE, Quebec, Canada, 2004.

86. Akahori Y, Nakai M, Yakabe Y, Takatsuki M, Mizutani M, Matsuo M, et al. Two-step models to predict binding affinity of chemicals to the human estrogen receptor *a* by three-dimensional quantitative structure-activity relationships (3D-QSARs) using receptor-ligand docking simulation. *SAR QSAR Environ Res* 2005;16:323–37.

87. Institute of Medicinal Molecular Design Inc. Japan <http://www.immd.co.jp/en/product_2.html>.

88. Cornell WD, Cieplak P, Bayly CI, Gould IR, Merz KM Jr, Ferguson DM, et al. A second generation force field for the simulation of proteins, nucleic acids and organic molecules. *J Am Chem Soc* 1995;117:5179–97.

89. Sippl W. Binding affinity prediction of novel estrogen receptor ligands using receptor-based 3-D QSAR methods. *Bioorg Med Chem* 2002;10:3741–55.

90. Rao MS, Olson AJ. Modelling of factor Xa-inhibitor complexes: A computational flexible docking approach. *Prot Struct Funct Gen* 1999;34:173–83.

91. Vedani A, Huhta DW. A new force field for modeling metalloproteins. *J Am Chem Soc* 1990;112:4759–67.

92. Baroni M, Costantino G, Cruciani G, Riganelli D, Valigi R, Clementi S. Generating optimal linear PLS estimations (GOLPE): An advanced chemometric tool for handling 3D-QSAR problems. *Quant Struct-Act Rel* 1993;12:9–20.

93. Lemmen C, Lengauer T, Klebe G. FLEXS: A method for fast flexible ligand superposition. *J Med Chem* 1998;41:4502–20.

94. Vedani A, Dobler M, Lill MA. Combining protein modeling and 6D-QSAR. simulating the binding of structurally diverse ligands to the estrogen receptor. *J Med Chem* 2005;48:3700–3.

95. <http://www.biograf.ch/downloads/yeti.pdf>.

96. Vedani A, Briem H, Dobler M, Dollinger K, McMasters DR. Multiple conformation and protonation-state representation in 4D-QSAR: The neurokinin-1 receptor system. *J Med Chem* 2000;43:4416–27.

97. Vedani A, Dobler M. 5D-QSAR: The key for simulating induced fit? *J Med Chem* 2002;45:2139–49.

98. <http://www.biograf.ch/downloads/quasar.pdf>.

99. Lill MA, Vedani A, Dobler M. *Raptor*—Combining dual-shell representation, induced-fit simulation and hydrophobicity scoring in receptor modeling: Application towards the simulation of structurally diverse ligand sets. *J Med Chem* 2004;47:6174–86.

100. Lill MA, Dobler M, Vedani A. Prediction of small-molecule binding to cytochrome P450 3A4: Flexible docking combined with multidimensional QSAR. *ChemMedChem*. 2006;1:73–81.

101. <http://www.biograf.ch/downloads/raptor.pdf>.

102. U.S. Food and Drug Administration <http://edkb.fda.gov>.

103. Chemicals Evaluation and Research Institute, Japan, 1-4-25 Koraku, Bunkyo-ku, Tokyo 112-0004.

104. Heinlein CA, Chang CS. Androgen receptor in prostate cancer. *Endocr Rev* 2004;25:276–308.

105. Berry SJ, Coffey DS, Walsh PC, Ewing LL. The development of human benign prostatic hyperplasia with age. *J Urol* 1984;132:474–9.

106. Fang H, Tong W, Branham WS, Moland CL, Dial SL, Hong H, et al. Study of 202 natural, synthetic, and environmental chemicals for binding to the androgen receptor. *Chem Res Toxicol* 2003;16:1338–58.

107. Araki N, Ohno K, Nakai M, Takeyoshi M, Iida M. Screening for androgen receptor activities in 253 industrial chemicals by in vitro reporter gene assays using AR-EcoScreen cells. *Toxicol In Vitro* 2005;19:831 42.

108. Kojima H, Katsura E, Takeuchi S, Niiyama K, Kobayashi K. Screening for estrogen and androgen receptor activities in 200 pesticides by in vitro reporter gene assays using Chinese hamster ovary cells. *Environ Health Perspect* 2005;112: 524–31.

109. Carlsen E, Giwercman A, Keiding N, Skakkebaek NE. Evidence for decreasing quality of semen during past 50 years. *Br Med J* 1992;305:609–13.

110. Kavlock RJ, Daston GP, DeRosa C, Fenner-Crisp P, Gray LE, Kaattari S. Research needs for the risk assessment of health and environmental effects of endocrine disruptors: A report of the U.S. EPA-sponsored workshop. *Environ Health Perspect* 1996;104:715–40.

111. Damstra T, Barlow S, Bergman A, Kavlock R, van der Kraak G, editors. *Global assessment of the state of the science of endocrine disruptors*. Geneva: World Health Organization, 2002, ch. 3.12.

112. Toft G, Edwards TM, Baatrup E, Guillette LJ Jr. Disturbed sexual characteristics in male mosquitofish (*Gambusia holbrooki*) from a lake contaminated with endocrine disruptors. *Environ Health Perspect* 2003;111:695–701.

113. Kelce WR, Wilson EM. Environmental antiandrogens: Developmental effects, molecular mechanisms, and clinical implications. *J Mol Med* 1997;75:198–207.

114. Jones I, Lindberg C, Jakobsson S, Hellqvist A, Hellman U, Borg B, et al. Molecular cloning and characterization of spiggin: An androgen-regulated extraorganismal adhesive with structural similarities to von willebrand factor-related proteins. *J Biol Chem* 2001;276:17857–63.

115. Katsiadaki I, Scott AP, Hurst MR, Matthiessen P, Mayer I. Detection of environmental androgens: A novel method based on enzyme-linked immunosorbent assay of spiggin, the stickleback (*Gasterosteus aculeatus*) glue protein. *Environ Toxicol Chem* 2002;21:1946–54.

116. Serafimova R, Walker J, Mekenyan O. Androgen receptor binding affinity of pesticide "active" formulation ingredients: QSAR evaluation by COREPA method. *SAR QSAR Environ Res* 2002;13:127–34.

117. Zhao CY, Zhang RS, Zhang HX, Xue CX, Liu HX, Liu MC, et al. QSAR study of natural, synthetic and environmental endocrine disrupting compounds for binding to the androgen receptor. *SAR QSAR Environ Res* 2005;16:349–67.

118. Waller CL, Juma BW, Gray LE Jr, Kelce WR Three-dimensional quantitative structure-activity relationships for androgen receptor ligands. *Toxic Appl Pharm* 1996;137:219–27.

119. Hong H, Fang H, Xie Q, Perkins R, Sheehan DM, Tong W. Comparative molecular field analysis (CoMFA) model using a large diverse set of natural, synthetic and environmental chemicals for binding to the androgen receptor. *SAR QSAR Environ Res* 2003;14:373–88.

120. Matias PM, Donner P, Coelho R, Thomaz M, Peixoto C, Macedo S, et al. Structural evidence for ligand specificity in the binding domain of the human androgen receptor: Implications for pathogenic gene mutations. *J Biol Chem* 2000;275: 26164–71.

121. Söderholm AA, Lehtovuori PT, Nyrönen TH. Three-dimensional structure-activity relationships of nonsteroidal ligands in complex with androgen receptor ligand-binding domain. *J Med Chem* 2005;48:917–25.

122. Sack JS, Kish KF, Wang C, Attar RM, Kiefer SE, et al. Crystallographic structures of the ligand-binding domains of the androgen receptor and its T877A mutant complexed with the natural agonist dihydrotestosterone. *Proc Natl Acad Sci USA* 2001;98:4904–9.

123. Bohl CE, Chang C, Mohler ML, Chen J, Miller DD, Swaan PW, et al. A ligand-based approach to identify quantitative structure-activity relationships for the androgen receptor. *J Med Chem* 2004;47:3765–76.

124. Lill MA, Winiger F, Vedani A, Ernst B. Impact of induced fit on ligand binding to the androgen receptor: A multidimensional QSAR study to predict endocrine-disrupting effects of environmental chemicals. *J Med Chem* 2005;48:5666–74.

125. Hansson T, Marelius J, Aqvist J. Ligand binding affinity prediction by linear interaction energy methods. *J Comput Aided Mol Des* 1998;12:27–35.

126. Lazar MA. Thyroid hormone receptors: Multiple forms, multiple possibilities. *Endocrine Rev* 1993;14:184–93.

127. Kavlock RJ, Daston GP, DeRosa C, Fenner-Crisp P, Gray LE, Kaattari S, et al. Research needs for the risk assessment of health and environmental effects of endocrine disruptors: A report of the U.S. EPA-sponsored workshop. *Environ Health Perspect* 1996;104:715–40.

128. Colborn T, von Saal FS, Soto AM. Developmental effects of endocrine-disrupting chemicals in wildlife and humans. *Environ Health Perspect* 1993;101:378–84.

129. Carlsen E, Giwercman A, Keiding N, Skakkebaek NE. Evidence for decreasing quality of semen during past fifty years. *Br Med J* 1992;305:609–13.

130. Ye L, Li YL, Mellstrøm K, Mellin C, Bladh L-G, Koehler KF, et al. Thyroid receptor ligands: 1. Agonists selective for the thyroid receptor β_1. *J Med Chem* 2003;46:1580–8.

131. Bruice TC, Kharash N, Winzler RJ. A correlation of thyroxine-like activity and chemical structure. *Arch Biochem Biophys* 1956;62:305–17.

132. Kubinyi H, Kehrhahn OH. Quantitative structure-activity relationships: 1. The modified free-Wilson approach. *J Med Chem* 1976;19:578–86.

133. Kubinyi H. Quantitative structure-activity relationships: 1. The modified free-Wilson approach. 2. A mixed approach, based on Hansch and Free-Wilson analysis. *J Med Chem* 1976;19:587–600.

134. Dietrich SW, Bolger MB, Kollman PA, Jorgensen EC. Thyroxine analogues: 23. Quantitative structure-correlation studies of in vivo and in vitro thyromimetic activities. *J Med Chem* 1977;20:863–80.

135. Leeson PD, Emmet JC, Shah VP, Showell GA, Novelli R, Prain HD, et al. Selective thyromimetics: Cardiac-sparing thyroid hormone analogs containing 3′-arylmethyl substituents. *J Med Chem* 1989;32:320–36.

136. McKinney JD, Waller CL. Molecular determinants of hormone mimicry: Halogenated aromatic hydrocarbon environmental agents. *Toxicol Environ Health B Crit Rev* 1998;1:27–58.

137. Zumstein M. Development of a quantitative structure-activity relationship for the α/β specificity of the human thyroid hormone receptor. M.Sc. thesis. University of Basel, Switzerland, 2005.

138. Vedani A, Zumstein M, Lill MA, Ernst B. Simulating α/β specificity at the thyroid receptor: Consensus scoring in multidimensional QSAR. *ChemMedChem*, 2007;2:78–87.

139. Damstra T, Barlow S, Bergman A, Kavlock R, van der Kraak G, editors. *Global assessment of the state of the science of endocrine disruptors*. Geneva: World Health Organization, 2002, ch. 3.12.5, and references therein.

140. Bandiera S, Sawyer TW, Campbell MA, Fujita T, Safe S. Competitive binding to the cytosolic 2,3,7,8-tetrachlorodibenzo-*p*-dioxin receptor: Effects of structure on the affinities of substituted halogenated biphenyls—A QSAR analysis. *Biochem Pharmacol* 1983;32:3803–13.

141. Denomme MA, Homonko K, Fujita T, Sawyer T, Safe S. Substituted polychlorinated dibenzofuran receptor-binding affinities and hydrocarbon hydroxylase induction potenties—A QSAR analysis. *Chem Biol Inter* 1986;57:175–87.

142. Golas CL, Prokipcak RD, Okey AB, Manchester DK, Safe S, Fujita T. Competitive binding of 7-substituted-2,3-dichlorodibenzo-*p*-dioxins with human Ah receptor—A QSAR analysis. *Biochem Pharmacol* 1990;40:737–41.

143. Rannug U, Sjogren M, Rannun A, Gillner M, Toftgard R, Gustafsson JA, et al. Use of artificial intelligence in structure-affinity correlations of 2,3,7,8-tetrachlorodibenzo-*p*-dioxin (TCDD) receptor ligands. *Carcinogenesis* 1991;11:2007–15.

144. Mekenyan OG, Veith GD, Call DJ, Ankley GT. A QSAR evaluation of Ah receptor binding of halogenated aromatic xenobiotics. *Environ Health Perspect* 1996;104:1302–10.

145. So SS, Karplus M. 3D-QSAR from molecular similarity matrices and genetic neural networks. *J Med Chem* 1997;40:4360–71.

146. Waller CL, McKinney JD. Comparative molecular field analysis of polyhalogenated dibenzo-*p*-dioxins, dibenzofurans, and biphenyls. *J Med Chem* 1992;35:3660–6.

147. Wagener M, Sadowski J, Gasteiger J. Autocorrelation of molecular surface properties for modeling corticosteroid binding globulin and cytosolic Ah receptor activity by neural networks. *J Am Chem Soc* 1995;117:7769–75.

148. Mhin BJ, Lee JE, Choi W. Understanding the congener-specific toxicity in polychlorinated dibenzo-*p*-dioxins: Chlorination pattern and molecular quadrupole moment. *J Am Chem Soc* 2002;124:144–8.

149. Lewis DFV, Jacobs MN, Dickins M, Lake BG. Molecular modelling of the peroxisome proliferator-activated receptor a ($PPAR_{\alpha}$) from human, rat and mouse, based on homology with the human $PPAR_{\gamma}$ crystal structure. *Toxicology* 2002;176:51–7.

150. Mankowski DC, Ekins S. Prediction of human drug metabolizing enzyme induction. *Curr Drug Metab* 2003;4:381–91.

151. Procopio M, Lahm A, Tramontano A, Pitea D. A model for recognition of polychlorinated dibenzo-*p*-dioxins by the aryl hydrocarbon receptor. *Eur J Biochem* 2002;269:13–8.

152. Vedani A, Dobler M, Lill MA. Virtual test kits for predicting harmful effects triggered by drugs and chemicals mediated by specific proteins. *ALTEX* 2005;22:123–34.

153. Vedani A, Dobler M, Lill MA. The challenge of predicting drug toxicity in silico. *Pharmacol Toxicol* 2006;99:195–208.

154. Lukacova V, Balaz S. Multimode ligand binding in receptor-site modeling: Implementation CoMFA. *J Chem Inf Comput Sci* 2003;43:2093–105.

155. Arulmozhiraja A, Morita M. Structure-activity relationships for the toxicity of polychlorinated dibenzofurans: Approach through density functional theory-based descriptors. *Chem Res Toxicol* 2004;17:348–56.

156. Hirokawa S, Imasaka T, Imasaka T. Chlorine substitution pattern, molecular electronic properties, and the nature of the ligand-receptor interaction: Quantitative property-activity relationships of polychlorinated dibenzofurans. *Chem Res Toxicol* 2005;18:232–8.

157. Wang Y, Liu H, Zhao C, Liu H, Cai Z, Jiang G. Quantitative structure-activity relationships models for the prediction of the toxicity of polybrominated diphenyl ether congeners. *Environ Sci Technol* 2005;39:4961–6.

158. Zheng G, Xiao M, Lu XH. QSAR study on the Ah receptor-binding affinities of polyhalogenated dibenzo-*p*-dioxins using net atomic-charge descriptors and a radial basis neural network. *Anal Bioanal Chem* 2005;383:810–6.

159. Sovadinova I, Blaha L, Janosek J, Hilscherova K, Giesy JP, Jones PD, et al. Cytotoxicity and aryl hydrocarbon receptor-mediated activity of *n*-heterocyclic polycyclic aromatic hydrocarbons: Structure-activity relationships. *Environ Toxicol Chem* 2006;25:1291–7.

160. Benfenati E. (2006, personal communication).

161. Bertilsson G, Heidrich J, Svensson K, Asman M, Jendeberg L. Sydow-Backman M, et al. Identification of a human nuclear receptor defines a new signaling pathway for CYP3A induction. *Proc Natl Acad Sci USA* 1998;95:12208–13.

162. Blumberg B, Sabbagh W Jr, Juguilon H, Bolado J Jr, van Meter CM, Ong ES, et al. SXR, a novel steroid and xenobiotic-sensing nuclear receptor. *Genes Dev* 1998;12:3195–205.

163. Kliewer SA, Moore JT, Wade L, Staudinger JL, Watson MA, Jones SA, et al. An orphan nuclear receptor activated by pregnanes defines a novel steroid signaling pathway. *Cell* 1998;92:73–82.

164. Staudinger J, Liu Y, Madan A, Habeebu S, Klaassen CD. Coordinate regulation of xenobiotic and bile acid homeostasis by pregnane X receptor. *Drug Metab Dispos* 2001;29:1467–72.

165. El-Sankary W, Gibson GG, Ayrton A, Plant N. Use of a reporter gene assay to predict and rank the potency and efficacy of CYP3A4 inducers. *Drug Metab Dispos* 2001;29:1499–504.
166. Dussault I, Lin M, Hollister K, Wang EH, Synold TW, Forman BM. A structural model of the constitutive androstane receptor defines novel interactions that mediate ligand-independent activity. *J Biol Chem* 2001;276:33309–12.
167. Drocourt L, Pascussi JM, Assenat E, Fabre JM, Maurel P, Vilarem MJ. Calcium channel modulators of the dihydropyridine family are human pregnane X receptor activators and inducers of CYP3A, CYP2B, and CYP2C in human hepatocytes. *Drug Metab Dispos* 2001;29:1325–31.
168. Moore JT, Kliewer SA. Use of the nuclear receptor PXR to predict drug interactions. *Toxicology* 2000;153:1–10.
169. Takeshita A, Koibuchi N, Oka J, Taguchi M, Shishiba Y, Ozawa Y. Bisphenol-A, an environmental estrogen, activates the human orphan nuclear receptor, steroid and xenobiotic receptor-mediated transcription. *Eur J Endocrinol* 2001;145: 513–7.
170. Coumol X, Diry M, Barouki R. PXR-dependent induction of human CYP3A4 gene expression by organochlorine pesticides. *Biochem Pharmacol* 2002;64: 1513–9.
171. Goodwin B, Moore LB, Stoltz CM, McKee DD, Kliewer SA. Regulation of the human CYP2B6 gene by the nuclear pregnane X receptor. *Mol Pharmacol* 2001;60:427–31.
172. Synold TW, Dussault I, Forman BM. The orphan nuclear receptor SXR coordinately regulates drug metabolism and efflux. *Nat Med* 2001;7:584–90.
173. Stoilov I. Cytochrome P450s: coupling development and environment. *Trends Genet* 2001;17:629–32.
174. Watkins RE, Wisely GB, Moore LB, Collins JL, Lambert MH, Williams SP, et al. The human nuclear xenobiotic receptor PXR: Structural determinants of directed promiscuity. *Science* 2001;292:2329–33.
175. Watkins RE, Maglich JM, Moore LB, Wisely GB, Noble SM, Davis-Searles PR, et al. 2.1 A crystal structure of human PXR in complex with the St. John's wort compound hyperforin. *Biochemistry* 2003;42:1430–8.
176. Watkins RE, Davis-Searles PR, Lambert MH, Redinbo MR. Coactivator binding promotes the specific interaction between ligand and the pregnane X receptor. *J Mol Biol* 2003;331:815–28.
177. Chrencik JE, Orans JO, Moore LB, Xue Y, Peng L, Collins JL, et al. Structural disorder in the complex of human pregnane X receptor and the macrolide antibiotic rifampicin. *Mol Endocrinol* 2005;19:1125–34.
178. Ekins S, Erickson JA. A pharmacophore for human pregnane X receptor ligands. *Drug Metab Dispos* 2002;30:96–9.
179. Schuster D, Langer T. The identification of ligand features essential for PXR activation by pharmacophore modeling. *J Chem Inf Model* 2005;45:431–9.
180. M.N. Jacobs. In silico tools to aid risk assessment of endocrine disrupting chemicals. *Toxicology* 2004;205:43–53.
181. Ingraham HA, Redinbo MR. Orphan nuclear receptors adopted by crystallography. *Curr Opin Struct Biol* 2005;15:708–15.

182. Watkins RE, Wisely GB, Moore LB, Collins JL, Lambert MH, Williams SP, et al. The human nuclear xenobiotic receptor PXR: Structural determinants of directed promiscuity. *Science* 2001;292:2329–33.

183. Huang W, Zhang J, Wei P, Schrader WT, Moore DD. Meclizine is an agonist ligand for mouse constitutive androstane receptor (CAR) and an inverse agonist for human CAR. *Mol Endocrinol* 2004;18:2402–8.

184. Dussault I, Yoo HD, Lin M, Wang E, Fan M, Batta AK, et al. Identification of an endogenous ligand that activates pregnane X receptor-mediated sterol clearance. *Proc Natl Acad Sci USA* 2003;100:833–8.

185. Suino K, Peng L, Reynolds R, Li Y, Cha JY, Repa JJ, Kliewer SA, et al. The nuclear xenobiotic receptor CAR: Structural determinants of constitutive activation and heterodimerization. *Mol Cell* 2004;16:893–905.

186. Shan L, Vincent J, Brunzelle JS, Dussault I, Lin M, Ianculescu I, et al. Structure of the murine constitutive androstane receptor complexed to androstenol: A molecular basis for inverse agonism. *Mol Cell* 2004;16:907–17.

187. Xu RX, Lambert MH, Wisely BB, Warren EN, Weinert EE, Waitt GM, et al. A structural basis for constitutive activity in the human CAR/RXR alpha heterodimer. *Mol Cell* 2004;16:919–28.

188. Ekins S, Mirny L, Schuetz EG. A ligand-based approach to understanding selectivity of nuclear hormone receptors PXR, CAR, FXR, LXR_{α}, and LXR_{β}. *Pharm Res* 2002;19:1788–800.

189. Forman BM, Tzameli I, Choi H-S, Chen J, Simha D, Seol W, et al. Androstane metabolites bind to and deactivate the nuclear receptor CAR-beta. *Nature* 1998;395:612–5.

190. Usui T, Saitoh Y, Komada F. Induction of CYP3As in HepG2 cells by several drugs. Association between induction of CYP3A4 and expression of glucocorticoid receptor. *Biol Pharm Bull* 2003;26:510–7.

191. Pascussi JM, Drocourt L, Fabre JM, Maurel P, Vilarem, MJ. Dexamethasone induces pregnane X receptor and retinoid X receptor-alpha expression in human hepatocytes: Synergistic increase of CYP3A4 induction by pregnane X receptor activators. *Mol Pharmacol* 2000;58:361–72.

192. Pascussi JM, Drocourt L, Gerbal-Chaloin S, Fabre JM, Maurel P, Vilarem MJ. Dual effect of dexamethasone on CYP3A4 gene expression in human hepatocytes: Sequential role of glucocorticoid receptor and pregnane X receptor. *Eur J Biochem* 2001;268:6346–58.

193. Pascussi JM, Gerbal-Chaloin S, Fabre JM, Maurel P, Vilarem MJ. Dexamethasone enhances constitutive androstane receptor expression in human hepatocytes: Consequences on cytochrome P450 gene regulation. *Mol Pharmacol* 2000;58: 1441–50.

194. Bledsoe RK, Montana VG, Stanley TB, Delves CJ, Apolito CJ, McKee DD, et al. Crystal structure of the glucocorticoid receptor ligand binding domain reveals a novel mode of receptor dimerization and coactivator recognition. *Cell* 2002;110:93–105.

195. Kauppi B, Jakob C, Farnegardh M, Yang J, Ahola H, Alarcon M, et al. The three-dimensional structures of antagonistic and agonistic forms of the glucocorticoid receptor ligand-binding domain: RU-486 induces a transconformation that leads to active antagonism. *J Biol Chem* 2003;278:22748–54.

196. Lewis DFV, Ioannides C, Parke DV, Schulte-Hermann R. Quantitative structure-activity relationships in a series of endogenous and synthetic steroids exhibiting induction of CYP3A activity and hepatomegaly associated with increased DNA synthesis. *J Steroid Biochem Mol Biol* 2000;74:179–85.

197. Lewis DFV, Ogg MS, Goldfarb PS, Gibson GG. Molecular modelling of the human glucocorticoid receptor (hGR) ligand-binding domain (LBD) by homology with the human estrogen receptor α (hER_{α}) LBD: Quantitative structure-activity relationships within a series of CYP3A4 inducers where induction is mediated via hGR involvement. *J Steroid Biochem Mol Biol* 2002;82:195–9.

13

APPLICATIONS OF QSAR METHODS TO ION CHANNELS

Alex M. Aronov, Konstantin V. Balakin, Alex Kiselyov, Shikha Varma-O'Brien, and Sean Ekins

Contents

Computational Toxicology: Risk Assessment for Pharmaceutical and Environmental Chemicals, Edited by Sean Ekins

13.1 INTRODUCTION TO ION CHANNELS

Voltage-gated ion channels for potassium, sodium, and calcium are allosteric proteins present in the outer membrane of many different cells such as those responsible for the electrical excitability and signaling in nerve and muscle cells [1]. They therefore represent validated therapeutic targets for anaesthesia, and several CNS and cardiovascular diseases [2]. Due to the ubiquitous expression of ion channels beyond the location of the therapeutic target, there is the requirement for selectivity; for example, drugs aimed at CNS ion channels should not interfere with cardiac ion channels [2]. This is perhaps difficult to predict, as multiple potassium, sodium, and calcium channels participate in the cardiac action potential. Because of this it is equally important to understand the recognition process of ligands for each of the ion channels that aid in selectivity with the potential to avoid life-threatening ion channel toxicity.

13.2 POTASSIUM CHANNELS

Numerous structural classes of potassium ion channels have been discovered. The six transmembrane domain proteins include voltage-gated channels (Kv), and Ca^{2+} activated K^+ channels (K_{ca}). However, there are also proteins with 2 transmembrane domains and 7 transmembrane domains [3]. The most widely studied potassium channel of relevance to toxicity and therefore the interests of the pharmaceutical industry to date appears to be K_v 11.1, the human ether-a-go-go-related gene (hERG), as is described in more detail below. Some computational modeling of other potassium channels is also addressed.

13.2.1 Predictive Modeling of hERG Blockers

A number of classes of drugs have been shown to prolong the QT interval, which reflects a slowing of repolarization of the ventricular myocardium [4,5]. Excessive QT interval prolongation can lead to the potentially life-threatening ventricular tachyarrhythmia, torsade de pointes. In cardiac tissue, inhibition of potassium channels are associated with QT interval prolongation [6,7]. The most common potassium channel linked to drug-induced QT interval prolongation is also responsible for the rapid component of the delayed rectifier potassium current (I_{Kr}). hERG is believed to encode the protein that underlies the delayed rectifier potassium current I_{Kr} [8,9], and many drugs associated with QT interval prolongation have been found to block hERG [10–12]. Drugs such as cisapride, terfenadine, astemizole, sertindole, and grepafloxacin have been withdrawn from the market in recent years in some degree due to cardiovascular toxicity associated with undesirable blockade of this channel. It is therefore important for drug discovery scientists to understand the structural requirements of molecules binding to this potassium channel to avoid potential toxicity.

To date the in vitro assessment of the drug-mediated interaction with these channels depends on various cell systems expressing the hERG channel and studies using methods such as patch clamping, radioligand binding, fluorescent probes, and rubidium flux [13,14]. These methods produce data of varying quality and reliability that can be potentially modeled computationally. Understanding the molecular features that confer hERG inhibition activity has therefore become a focus of considerable computational and statistical modeling efforts.

13.2.2 Homology Modeling of hERG

The only application of homology models to the a quantitative prediction of hERG blockade published thus far is that by Rajamani et al. [15]. The authors described a two-state homology model designed to represent the flexibility of the channel. Two homology models of the hERG channel were built, the partially open state (10° translation away from the KscA [16] reference state) and the fully open state (19° translation, closely corresponding to MthK [17,18]). IC_{50} values for 32 ligands assembled by Cavalli et al. [19] (hERG pIC_{50} ranging between 4 and 9) were used to derive the linear interaction energy (LIE) correlation. After the ligand set was docked to the homology models, the best poses (one for each ligand) were subjected to energy minimization, and used to derive the van der Waals and electrostatic contributions to the binding energy. Models corresponding to the two homology models resulted in poor fit (e.g., $r^2 = 0.24$ for the partially open state) when tested separately for their ability to predict hERG IC_{50} values. However, when the preferences of each compound for a particular state were compared in terms of the estimated interaction energy for each ligand in both states, separate better fits were obtained, with 21 ligands demonstrating a preference for the open state and 11 ligands preferring the partially closed state of the channel. The final combined model,

$$pIC_{50} = -0.163(\Delta vdw) + 0.0009(\Delta ele), \qquad (13.1)$$

resulted in $r^2 = 0.82$ (RMSD = 0.56) for 27 ligands, with five compounds identified as potential outliers. [15] The study demonstrated the difficulty of using hERG homology models for quantitative predictions of hERG blockade and the need for modeling techniques that capture the flexibility of the K^+ channel (see also Chapter 16).

13.2.3 Quantitative Structure-Activity Relationships

Ligand-based approaches have been extensively applied to understand the SAR of hERG channel blockers. One of the earliest structure-activity studies performed on a series of compounds that cause QT prolongation is by Morgan and Sullivan [20]. Although this work was carried out prior to the discovery of hERG as one of the leading drivers behind prolonged QT interval, it is now clear that the mechanism of action for most, if not all, of the compounds involves hERG blockade. As part of an overview of historical human and dog data on class III antiarrhythmic agents, the authors proposed a general structure of Q-phenyl-A-NR_1R_2. R_1 and R_2 are preferentially hydrophobes, R_1 as alkyl or phenylalkyl (R_1 = H is also acceptable for inclusion in the class), and R_2 as alkyl, arylalkyl, or heteroalkyl. Linker A is a 1-4 atom chain that may contain heteroatoms or be part of a ring. This linker chain was noted as one of the more variable regions in the molecule. Finally, *para*-substituent Q is an electron-withdrawing group, such as nitro, cyano, *N*-imidazole, or *N*-methylsulfonamide.

Another early example of hERG pharmacophore modeling appears to be the work by Matyus et al. [21] on elucidation of the pharmacophore for blockers of I_{Kr} current as class III antiarrhythmic agents. Eleven ligands were divided into two sets. The first set included the six most active agents, spanning a 1.5 log activity range, and the second set contained the less active compounds. Starting structures were obtained by energy minimization in the Tripos [22] force field. Sets of conformations were generated by the Multisearch option in DISCO [22] interface, up to a maximum of 100 conformations per ligand. As part of the conformation generation process, the various energy minima were sampled by random perturbation of torsions, followed by minimization and elimination of duplicates. Perturbed torsions were limited to those that have the potential to change the internal geometry of the atoms determining the DISCO features. A new feature definition was added to the hydrophobic class in DISCO to recognize hydrophobic aliphatic chains. DISCO was forced to consider five-point pharmacophores containing one donor atom, one receptor-associated acceptor site, and three hydrophobes, with dofetilide chosen as the reference. Fifty-three pharmacophores from DISCO were analyzed, and the best model was chosen. The basic nitrogen donor was within 5.19, 5.63, and 2.37 Å from the two aromatic features and the aliphatic feature, respectively. The less potent five-compound set was shown to satisfy only four points in the five-point pharmacophore, with one of the three hydrophobes missing.

Liu et al. [23] investigated activity data in the isolated guinea pig atrium assay for a series of 17 dofetilide analogues using CoMFA and CoMSiA. With dofetilide as a reference, its low-energy conformations were used as templates for molecule alignment. The optimized CoMFA model produced a correlation with $q^2 = 0.695$. From the CoMFA results the authors concluded that increasing hydrophobic bulk on the phenoxy moiety of dofetilide would lead to improved biological activity. A later study [24] of the 17 compounds involved three different techniques—CoMFA, CoMSiA, and hologram QSAR (HQSAR) [25]. In HQSAR, each molecule is divided into structural fragments that are counted in the bins of a fixed-length array to form a molecular hologram. These structural descriptors encoding compositional and topological molecular information are then used to derive a partial least squares

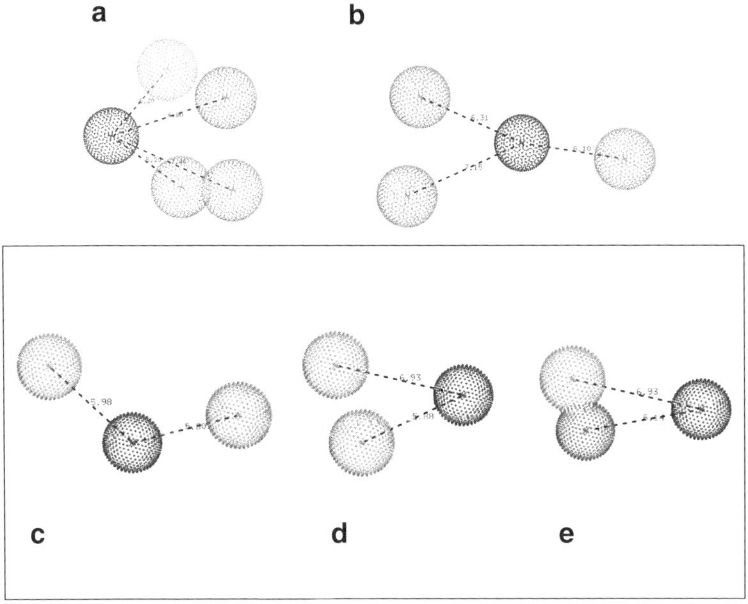

Figure 13.1 hERG potassium channel pharmacophores. (*a*) Ekins hERG pharmacophore [26]; blue sphere = positive ionizable, green spheres = hydrophobic. (*b*) Cavali hERG pharmacophore [19]. (*c*, *d*, *e*) Aronov hERG pharmacophores [55]; red feature = hydrogen bond acceptor. All pharmacophores recreated in PyMol [148] using distance geometry from the intrafeature distance (Å) tables supplied by the corresponding communications. See color plates.

regression model that correlates variation in structural composition with variation in experimental data. All three techniques were useful in predicting activity of 11 newly synthesized compounds ($r^2 = 0.943$, 0.891, and 0.809 for CoMFA, CoMSiA, and HQSAR, respectively), as well as six additional analogues.

The first hERG pharmacophore published since the renewed interest in predicting I_{Kr} blockade was described by Ekins et al. [26], using 15 molecules from the literature. The conformers were generated within Catalyst [27] (BEST mode) up to a maximum of 255 conformations per ligand. Hydrophobic, ring aromatic, donor, acceptor, and positive ionizable features were selected for possible inclusion. Ten Catalyst hypotheses were assessed, and the lowest energy cost hypothesis was deemed the best. It contained four hydrophobes that surrounded the central positive ionizable feature, and produced an r^2 value of 0.90. The proposed distances between the positive center and the hydrophobes were 5.2, 6.2, 6.8, and 7.5 Å (Figure 13.1*a*). The model was further applied to predict IC_{50} values for a test set of 22 mostly antipsychotic compounds known to inhibit hERG ($r^2 = 0.83$). A pharmacophore for antipsychotic drugs was also referred to and has now been published. This Catalyst model detailed three hydrophobic features and a ring aromatic feature. Interestingly the positive ionizable feature was absent, possibly due to a limitation of the method used [28]. Recently the initial training set from the original pharmacophore study was expanded to include 66 molecules, and a recursive partitioning method (ChemTree) was used with path length descriptors resulting in an observed-versus-predicted correlation of $r^2 = 0.86$. This model produced a correlation of $r^2 = 0.67$ when tested with a set of 25 additional molecules from the literature [29]. An updated version of this model was built with 99 literature molecules that has been used recently to rank the 23 sertindole analogues generated by Pearlstein et al. [30] (Spearman's rho = 0.74, $P < 0.0001$, $r^2 = 0.53$) [31]. The most recent testing of this tree model containing 99 molecules was with 35 diverse molecules [32]. Analysis of the 35 molecule test set resulted in a relatively low, though statistically significant, correlation (r^2 0.33, Spearman rho 0.55, $p = 0.0006$). From a Tanimoto similarity analysis of the test set molecules with Accord descriptors (Accelrys, San Diego, CA), it was observed that the log difference between observed and predicted log10 IC_{50} increased as the Tanimoto similarity declined. Therefore a Tanimoto similarity index larger than 0.77 was found to contain molecules with a log10 IC_{50} that were less than 1 log unit different from the observed data. These 18 remaining molecules alone produced dramatically improved testing correlation statistics (r^2 0.83, Spearman rho = 0.75, $p = 0.0003$) compared with using the whole test set [32].

Cavalli et al. [19] constructed a hERG pharmacophore based on a training set of 31 literature compounds previously implicated in QT prolongation. The conformational space was sampled by Monte Carlo analysis as implemented in the MacroModel software [33,34]. All of the dihedral angles of single linear bonds were allowed to move freely, and an unusually large 100 kJ/mol energy window was used to filter conformers. The procedure generated hundreds of conformers per ligand, which were then clustered using 1 Å root mean square displacement cutoff. A "constructionist" approach to pharmacophore generation was used with astemizole as the template onto which other ligands were sequentially superimposed by means of the commonality of geometric and spacial characteristics. The proposed pharmacophore contains three aromatic moieties connected by a nitrogen function that is a tertiary amine throughout the whole set of molecules. The nitrogen and the aromatic moieties are separated by distances of 5.2–9.1 Å, 5.7–7.3 Å, and 4.6–7.6 Å (Figure 13.1*b*). CoMFA analysis performed on the training set produced a correlation with $r^2 = 0.952$ ($q^2 = 0.767$), and its predictive ability was tested on a set of six additional compounds ($r^2_{pred} = 0.744$).

Pearlstein et al. [30] reported a CoMSiA model built using in-house patch clamp data for 28 compounds, 18 of them sertindole analogues ($q^2 = 0.571$). The conformational searching was performed with an MMFF94 force field. Ring centroids and the basic nitrogen were used as landmarks for superimposing the compounds onto sertindole by way of least squares fitting. The model indicated that decreasing the positive charge on the central nitrogen and increasing the steric bulk on the hydrophobic end of the molecule are two potential ways to reduce hERG blocking activity. The model was tested on a holdout set containing four sertindole analogues with widely varying potency for hERG. In addition, the authors constructed a homology model of the tetrameric pore region of hERG from the MthK [17] template and qualitatively docked the set of aligned inhibitor structures into the inner cavity of the channel. The proposed binding mode for the inhibitors within the intracellular region of the hERG pore resulted in a "drain-plug"-like occlusion as a possible mechanism for K^+ current disruption.

Keseru [35] used literature data on 55 compounds to train a QSAR model based on a number of calculated descriptors. Five descriptors were used: $c\log P$, calculated molar refractivity (CMR), partial negative surface area, and the VolSurf W2 (polarizability) and D3 (hydrophobicity) descriptors. A model of acceptable quality was obtained ($r^2 = 0.94$, SSE = 0.82) and tested on a 13 compound holdout set ($r^2 = 0.56$, SSE = 0.98). An HQSAR model was then created that made use of 2D fragment fingerprints (threshold hERG IC_{50} = 1 µM). The best HQSAR model was validated on a holdout set of 13 compounds ($r^2 = 0.81$, SSE = 0.67).

In a study of class III antiarrhythmics, Du et al. [36] selected a set of 34 compounds from the literature spanning a broad range of IC_{50} values. Using the features previously proposed to be important in characterizing hERG blockers [21], the authors turned to the HypoGen module in Catalyst. Models were allowed to contain at least one and at most two instances of every feature. The best pharmacophore hypothesis contained a positive ionizable feature, two aromatic rings, and a hydrophobic group. It was then applied to a test set of 21 compounds, which was split into three groups based on the hERG activity level (<1 µM, 1–100 µM, and >100 µM). The model predicted the activity of test ligands with $r^2 = 0.713$, with all highly active (<1 µM) compounds predicted correctly.

A simple two-component relationship for hERG blocking potency was proposed by Aptula and Cronin [37]. A set of 150 descriptors were calculated for 19 structurally diverse hERG blockers from the literature. The calculated variables included physicochemical parameters, topological indexes, and quantum chemical descriptors. Multiple linear regression was used to derive a relationship between hERG blocking potency and two descriptors, $-\log D$ and D_{max} (the maximum diameter of molecules):

$$\mathrm{pIC}_{50}(\mathrm{hERG}) = 0.58\log D + 0.30D_{max} - 0.36 \quad (13.2)$$

with a reasonable correlation ($r^2 = 0.87$, $q^2 = 0.81$). A further analysis of 81 chemicals from Redfern et al. [38] was described, but no results were shown. The authors suggest the maximum diameter cutoff of 18 Å for cases where hERG blocking activity is undesirable, and $D_{max} > 18$ Å for antiarrhythmics. The relationship described is rather intuitive, with hERG activity correlating with both hydrophobicity and ligand size [37]. More lipophilic ligands that have a large diameter are capable of binding in the hERG channel and engaging most of the residues implicated in hERG blockade. Although the overall findings with regard to lipophilicity are generally in agreement with earlier studies [39], the QSAR equation based on a study of only 19 ligands that contains no additional size constraints (e.g., a reasonable D_{max} limit beyond which the compound would be unable to enter the channel pore) appears to be a gross generalization.

Song and Clark [40] recently reported the development of a support vector regression model for hERG. The in-house descriptor set [41] included 261 counts of structurally diverse 2D fragments. The best model was built for 71 known hERG blockers ($q^2 = 0.636$), and testing on 19 additional compounds produced $r^2 = 0.849$ and RMSE = 0.597. Predictive power suffered when applied to 20 in-house ligands ($r^2 = 0.29$, RMSE = 1.26). Further analysis demonstrated that the best predictions were made for compounds with higher similarity to the training set, while results for a series chemically distinct from the training data were poor. This study clearly shows that the predictive scope of a global model is only as good as the training set, and development of local models for hERG can have an edge in some cases.

One of the larger hurdles for building QSAR models from literature data has been the large discrepancy observed for hERG IC_{50} values determined in different laboratories. Interlaboratory variability of greater than 10-fold is not uncommon, even in cases where inhibition was measured using the same cell line. Yoshida and Niwa recently generated 2D QSAR multiple linear regression models with 104 hERG ligands from different cell lines in the literature and interpretable descriptors such as ClogP, TPSA, diameter, summed surface area of atoms, and partial charges [42]. An indicator variable was also included to represent the different experimental conditions ($q^2 = 0.67$). Testing was performed using a leave-out group of 18 molecules repeated fivefold (average $r^2 = 0.66$, SD = 0.85) [43]. The same authors performed homology modeling and amino acid QSAR analysis for the Phe656 mutant hERG using IC_{50} data for three ligands from a published study [42]. The conclusion was that hydrophobic interactions between Phe656 and drugs are likely important [43]. Additional efforts in generating internally consistent hERG data sets—potentially including higher throughput validated measurements such as, but not limited to, planar patch that would be made available to the broad scientific community—are sorely needed to propel this field forward.

Local pharmacophore models have also been generated around narrow structural series such as the sertindole analogues [30]. For example, publications containing [^{3}H]-dofetilide binding data for the $5HT_{2A}$ class of molecules [44,45], and 3-aminopyrrolidinone farnesyltransferase inhibitors [46], have been used to produce individual pharmacophores that were ultimately combined to suggest common areas of positive ionizable features and hydrophobicity from aromatic rings [31,47].

13.2.4 Classification Methods

While QSAR methods aim to predict absolute compound activity, classification methods attempt to bin compounds by their potential for hERG inhibition. The first example of a hERG-based classification was reported by Roche et al. [48]. A total of 244 compounds representing the extremes of the data set (<1 μM and >10 μM for actives and inactives, respectively) were modeled with a variety of techniques such as substructure analysis, self-organizing maps, partial least squares, and supervised neural networks. The descriptors chosen included pK_a, Ghose-Crippen [49], TSAR [50], CATS [51], VolSurf [52], and Dragon [53] descriptors. The most accurate classification was based on an artificial neural network. In the validation set containing 95 compounds (57 in-house and 38 literature IC_{50} values) 93% of inactives and 71% of actives were predicted correctly.

Sun [54] reported a naïve Bayes classifier built around a training set of 1979 compounds with measured hERG activity from the Roche corporate collection. For the training set, 218 in-house atom-type descriptors were used to develop the model, and pIC_{50} = 4.52 was set as a threshold between hERG actives and inactives. Receiver operator curve (ROC) accuracy of 0.87 was achieved. The model was validated on an external set of 66 drugs, of which 58 were classified correctly (88% accuracy).

In a decision tree based approach to constructing a hERG model with calculated physicochemical descriptors, Buyck et al. [39] used three descriptors—Clog*P*, CMR, and the pK_a of the most basic nitrogen—to identify hERG blockers within an in-house data set. With IC_{50} = 130 nM as a cutoff, factors suggestive of hERG activity were determined to be $c\log P \geq 3.7$, $110 \leq$ CMR < 176, and pK_a max ≥ 7.3.

A combined 2D-3D procedure for identification of hERG blockers was proposed by Aronov and Goldman [55]. A 2D topological similarity screen utilizing atom pair [56] descriptors and an amalgamated similarity metric termed TOPO was combined with a 3D pharmacophore ensemble procedure in a "veto" format to provide a single binary hERG classification model. A molecule flagged by either component of the method was considered hERG active. Upon 50-fold cross validation of the model on a literature data set containing 85 actives (threshold HERG IC_{50} = 40 μM) and 329 inactives, 71% of hERG actives and 85% of hERG inactives were correctly identified. The model utilizing the TOPO metric was shown to be superior to a number of other 2D models using the ROC metric. Additionally five of eight (62.5%) hERG blockers were identified correctly in a 15 compound in-house validation set. Most of the statistically significant pharmacophores from the ensemble procedure were three-feature [aromatic]–[positive charge]–[hydrophobe] combinations (Figure 13.1*c*–*d*) similar to those reported by Cavalli et al. [19]. However, a novel three-point pharmacophore containing a hydrogen bond acceptor was also proposed (Figure 13.1*e*). The presence in hERG blockers of the acceptor functionality pointing toward the selectivity pore agrees with the previous observations of a potential for polar interactions with the side chains of Thr623 and Ser624 to stabilize the hERG-ligand complex [19,57,58].

Testai et al. [59] evaluated a set of 17 antipsychotic drugs, all of them associated with reports of torsadogenic cardiotoxicity. The search for a common

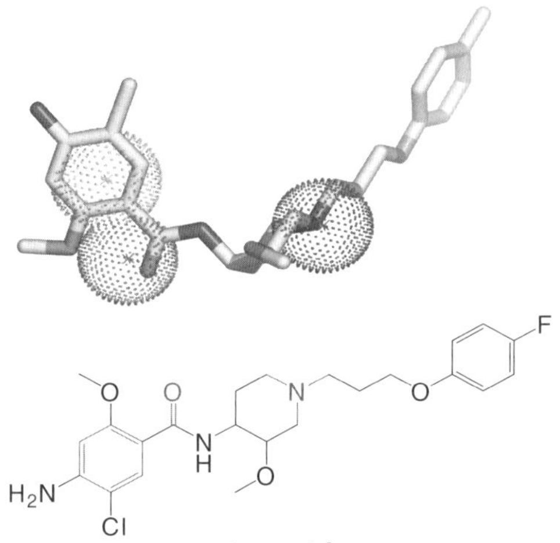

Figure 13.2 Cisapride mapped to the Aronov hERG pharmacophore described in Figure 13.1*e* [55] recreated in PyMol [148] using distance geometry from the intrafeature distance tables. See color plates.

molecular feature required for hERG blockade focused on measuring several different distances between atoms that could constitute a hERG-active template. The authors hypothesized that such a template for hERG-active ligands consists of a hydrocarbon chain, three or four atoms long, serving as a spacer between a basic sterically hindered nitrogen atom and a second, more variable moiety. Focusing on the distance between the basic nitrogen and this second moiety, Testai et al. observed that for all of the compounds in the data set the distance converged in the range between 4.32 and 5.50 Å (average = 4.87 Å). Starting from known small molecule X-ray structures, 3D structures of the antipsychotics were generated, followed by solvation and minimization. Interestingly the variable moiety present in 14 of the 17 ligands is a hydrogen bond acceptor, either a carbonyl oxygen or a heteroaromatic nitrogen, which is in agreement with the high information content seen for the acceptor-containing pharmacophore by Aronov and Goldman [55]. Indeed, the acceptor functionality can be found not only in antipsychotic agents, such as risperidone and droperidol [55], but also in other known hERG blockers, such as prokinetic agent cisapride (Figure 13.2). In the case of risperidone, the two groups pointed to different hydrogen bond acceptors that, incidentally, are located approximately the same distance from the basic nitrogen—the oxygen of the benzisoxazole [55] and the carbonyl of the pyrimidone [59].

Wang et al. [60] built a one-dimensional profile of a hERG-active compound from 10 known hERG blockers, and tested it for the ability to discriminate between hERG blockers and MDDR-derived decoys. One-dimensional profiling involves projecting a molecule from either 3D or 2D onto a single dimension using multidimensional scaling. Enrichment factors for a set of 92 hERG blockers ranged from 6 to 8 in the top 1–5%, which was quite similar to a 3D pharmacophore (based on that published by Ekins et al.) after the fraction of compounds screened increased above 15%.

Support vector machine (SVM) is a relatively recent data mining approach based on the structural risk minimization principle [61] from computational learning theory (see also Chapter 8). SVM constructs a hyperplane that separates two classes (this can be extended to multi-class problems). Separating the classes with a large margin minimizes a bound on the expected generalization error. Tobita et al. [62] selected 73 drugs with known hERG IC_{50} values for a training set, which they used to built an SVM classifier. Radial basis function was chosen as the SVM kernel. Fifty-seven 2D MOE [63] descriptors and 51 fragment count descriptors (subset of the 166-bit MACCS keys) were calculated. Two different separation boundaries were tried ($pIC_{50} = 4.4$, $pIC_{50} = 6$), with 95% and 90% accuracy for the classification, respectively. The model also predicted known cardiovascular side effects with an accuracy of approximately 70% when tested using an external set. Two topological patterns were proposed as contributing molecular fragments that correlate with hERG inhibition.

Bains et al. [64] applied genetic programming (GP) to build a predictive hERG model. GP, an evolutionary computing technique, "evolves" an algorithm in silico that matches input variables with the desired output and performs both the selection of relevant descriptors and algorithm building without human intervention. The data set totaled 124 compounds from the literature. Three types of descriptors were utilized: general molecular descriptors (e.g., molecular weight), topological fragments, and experimental parameter descriptors. The largest set, topological fragments, was generated automatically and exhaustively from the data set by application of a maximum common subgraph-based algorithm. Only fragments containing at least four heavy atoms and present in two or more ligands were kept. Experimental parameters were abstracted from experimental sections on the literature and added to the descriptor set. A total of 618 descriptors were used in the GP runs. GP algorithms trained on a training set were selected for their ability to make predictions on a generalization set and tested on an independent validation set. Model performance was judged using Akaike fitness criterion and the ROC. The best models achieved 85–90% accuracy in predicting on the validation set when $IC_{50} < 1\,\mu M$ was used as a threshold for hERG-active ligands, which is consistent with the results of previously published classification studies [48,55]. Analysis of correlations between topological descriptors and hERG IC_{50} revealed correlations between IC_{50} and the presence or absence of specific chemical fragments. While it is hardly possible to organize these observations into a stand-alone pharmacophore model, the authors pointed out that their results are broadly consistent with earlier publications [30,57,65,66].

Sammon maps and Kohonen maps have been used with a consistent training set (93 molecules) and test set (35 molecules) to compare the classification of high (log10 $IC_{50} < 0$) and low affinity (log10 $IC_{50} < 2$) compounds [32]. Sammon maps describe all relative distances between all pairs of compounds, and the distance of two points on the map directly reflects the similarity of the compounds. Sammon nonlinear maps are unique due to their conceptual simplicity and ability to reproduce the topology and structure of the data space in a faithful and unbiased manner [67]. Kohonen maps belong to a class of neural networks known as competitive learning or self-organizing networks. The Kohonen map consists of artificial neurons that are characterized by weight vectors with the same dimensionality as the descriptor set. The neurons are connected by a distance dependent function. In an unsupervised training algorithm the neurons self-organize until their pairwise neighborhoods represent the correct topology of the original data set. The average classification quality is high for both training and test selections: up to 86 % and 95 % of compounds were classified correctly in the corresponding data sets. At the same time, insufficient statistics prevent correct assignment of compounds belonging to the intermediate class between high and low affinity compounds. The Sammon mapping technique outperformed the Kohonen maps in classification of compounds from the external test set. In general, the Kohonen maps demonstrate a significant speed advantage compared to Sammon maps. They allow for the instant inclusion of new individual or multiple data points on the map without the need for re-computing the entire dataset, enabling visualization and analysis of larger databases or virtual libraries compared with Sammon maps. On the other hand, Sammon maps provide better distance and topology preservation compared with Kohonen maps, and the latter often contain gaps or undefined regions of chemical space.

13.2.5 Quantitative Structure-Activity Relationships of Other Potassium Channels

Du et al. [68] used a combination of homology modeling, docking, and QSAR to characterize the binding site of Kv7.1 channel, the functional part of the complex that generates the slowly activating cardiac K^+ current I_{Ks}. The homology model of Kv1.7 was built based on the structure of KcsA [16]. The training set consisted of 20 compounds with IC_{50} values ranging from 6 nM to 0.2 mM, and it was used to derive a pharmacophore model in Catalyst [27]. The model was allowed to select no more than four instances of each of the four features, which included hydrogen bond donor, hydrogen bond acceptor, aromatic ring, and a hydrophobe. The best hypothesis yielded $r = 0.953$ for the training set, and $r = 0.856$ for the additional test set consisting of 13 compounds from different activity classes. It contained a total of four features, one aromatic ring and three hydrophobes. Docking of three representative compounds from each of the major classes was performed with DOCK5 [69], and resulting orientations within the closed pore were broadly consistent with the proposed pharmacophore, with the nature of the residues lining the pore, as well as some of the mutagenesis data for Kv1.7. The notable exception was BMS-IKS, the oxadiazole-based Kv1.7 blocker, which only matched two of the three hydrophobic features.

Carosati et al. [70] performed a QSAR analysis of a series of 17 benzothiazine K_{ATP} channel openers. The data using three different biological readouts spanned over four orders of magnitude. 3D QSAR models built around GRIND [71] descriptors resulted in models of acceptable quality (q^2 ranged from 0.64 to 0.69, depending on the readout). The model allowed a number of conclusions relating to the role of various substituents around the benzothiazine core in the interaction with the channel. A similar study of benzopyrans as K_{ATP} channel openers was carried out by Uhrig et al. [72]. The primary focus of the study was the impact of C6-substitution on activity in the series. An equation was derived relating vasodilator activity to the direction of the dipole vector of the ligands ($r = 0.669$).

Researchers at Aventis described a pharmacophore for Kv1.5 channel blockers consisting of three hydrophobic centers in a triangular arrangement. First observed in an earlier active series [73], the pharmacophore was validated using a database of 423 Kv1.5 blockers [74]. The query was able to retrieve 58% of the known actives. A pharmacophore search of the corporate collection identified 27 clusters containing 1975 compounds. Screening the representatives of 18 clusters led to the discovery of an anthranilamide hit molecule (Kv1.5 $IC_{50} = 5.6\,\mu M$ in *Xenopus* oocytes) [74].

13.3 SODIUM CHANNELS

There are at least nine mammalian families of the sodium channel (Na_v) [3] that can be found in nerve axons, skeletal muscle fibers, cardiac myocytes, and neurons [75]. Whereas potassium channels only have one domain, sodium channels are large proteins consisting of four domains, each with six transmembrane domains. These proteins can switch between multiple states to enable selectivity such as resting, activation, and inactivation [76]. Sodium channels generate action potentials in axonal nerve fibers by transporting sodium and subsequently depolarize the membrane, which makes them important targets for treatment of epilepsy, neuroprotection, and analgesia. The sodium channel has between 6 and 9 distinct neurotoxin receptor sites and one for anaesthetics [3,77] that have been studied and used to understand the pore structure and selectivity filter [1]. However, due to their structural similarity with calcium channels there is a high degree of overlap between inhibitors of each of these channels [75,76].

13.3.1 Simple Pharmacophores for Ligand Interactions with Sodium Channels

There have been two reviews that have briefly mentioned the few modeling and pharmacophore studies that have been carried out for sodium channels

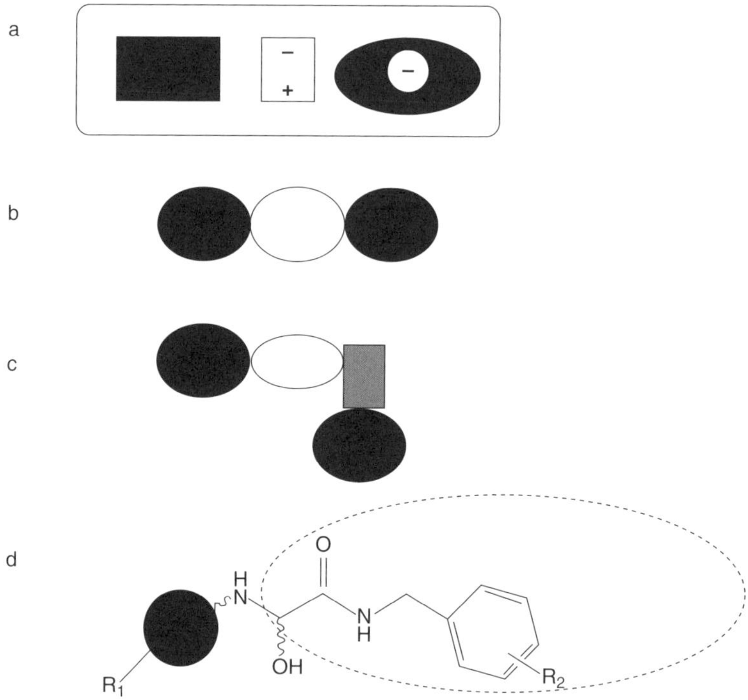

Figure 13.3 (*a*) Published pharmacophore for anesthetics binding at the sodium channel (dark objects = hydrophobic) [78]. (*b*) Schematic of anticonvulsant pharmacophore derived from ureylenes [85] (dark circles = hydrophobic, empty circle = hydrogen-bonding site). (*c*) Schematic anticonvulsant pharmacophore for semicarbazones (dark circles = hydrophobic, empty circle = hydrogen-bonding site, rectangle = electron donor). (*d*) Schematic anticonvulsant pharmacophore derived from γ-hydroxybutyric acid derivatives (dark circle = hydrophobic) [83].

[76,77]. These studies used the alignment of small number of molecules to create simple pharmacophores (Figure 13.3*a–d*). Possibly the earliest proposed pharmacophore for the voltage gated sodium channel was described by Khodorov et al. It consisted of a negatively charged site and two hydrophobic regions separated by a dipolar region [78] (Figure 13.3*a*). A conformational analysis using MM2 in MacroModel was carried out for two sodium channel modulators brevetoxin A and brevetoxin B which resulted in a ligand-receptor model for binding site 5 that is cigar shaped and 30 Å long, with hydrophobic and nonpolar binding interactions including hydrogen bonds [79]. Another more recent study by Unverferth et al. used five anticonvulsants, carbamazepine, phenytoin, lamotrigine, zonisamide, and rufinamide for which molecular mechanics and dynamics calculations were performed [80]. These molecules shared common pharmacophore features with deviations that were within 0.7 Å when superimposed: a hydrophobe, an electron donor group, and a hydrogen donor or acceptor unit. A further four molecules—a 3-aminopyr-

role, vinpocetine, dezinamide, and remacemide—also fulfilled the criteria of this pharmacophore. The lowest energy conformations calculated by MacroModel for phenytoin, carbamazepine, lamotrigine, and diclofenac resulted in a pharmacophore, which suggested that two phenyl and one tertiary amine were key determinants for binding [81]. Similarly a second group—including diphenhydramine, tripelennamine, imipramine, and benzotropine—have similar spatial orientations for these molecular features [81]. This study had generated sodium channel data with rat brain coronal slices. A pharmacophore was generated in Sybyl using 13 anticonvulsant molecules for which density functional calculations had been carried out with GAUSSIAN 98 [82]. The active analogue approach was used with carbamazepine and phenytoin as starting structures and also included valproic acid, felbamate, lamotrigine, remacemide, topiramate, zonisamide, ralitoline, ethosuximide, vinpocetine, rufinamide, and oxcarbazepine. All these molecules apart from ethosuximide bind to the sodium channel. The overlapping common area included a hydrophobic chain and a polar moiety. A further five molecules were shown to fit this pharmacophore, as well as nine valpromide derivatives for which sodium channel binding data was generated in rat forebrain membranes [82]. Four *N*-benzylamide anticonvulsant derivatives of γ-hydroxybutyric acid were superimposed using MOPAC in Alchemy 2000 [83]. This resulted in a basic pharmacophore (Figure 13.3*d*) with aromatic, hydrophilic, and benzylamide features and was tested with newly synthesized molecules that were evaluated with the mouse maximal electroshock seizure test [83]. Lowest energy conformations of three *N*-benzylated tocainide analogues were generated with SPARTAN PRO and indicated that the distance between the two aromatic rings was important along with the presence of a protonated amino group [84]. Pandeya et al. suggested an anticonvulsant pharmacophore (Figure 13.3*c*) based on isatin semicarbazone that consisted of two hydrophobes, a hydrogen bond donor, and an electron donor without using computational approaches [85].

13.3.2 Quantitative Structure-Activity Relationships

An early QSAR model used thirty 2-benzothiazolamines tested as inhibitors of sodium flux in rat cortical slices along with several physicochemical parameters such as molecular refractivity (MR), resonance, field, sigma, and pi [86]. The resultant regression equation indicated a trend of increasing potency with increased lipophilicity (pi), decreased size (MR), and increasing electron withdrawal (resonance). It should be noted that this model was not tested with external molecules, scrambled or used in any leave-one-out or leave-*n*-out paradigm [86]. Neuronal voltage sensitive sodium channel binding data for rat cerebral synaptosomes with 12 hydantoins yielded IC_{50} values that were correlated with clogP values ($r^2 = 0.64$) [87]. It was shown that different molecules with the same logP have different IC_{50}s, indicating that hydrophobicity alone does not determine binding activity. The same group followed up with a CoMFA model of neuronal voltage sensitive sodium channel using the same rat in vitro system with data for 14 molecules (non–cross-validated $r^2 = 0.988$), which was an improvement on $c\log P$ alone ($r^2 = 0.801$) [88]. Steric and electrostatic contributions were 0.75 and 0.25, respectively. A test set of eight structurally different molecules was predicted with the CoMFA model and resulted in a correlation $r^2 = 0.92$. This CoMFA model also lead to the design of a novel α-hydroxy-α-phenylamide. Recent research from Zha et al. expanded these two studies generating in vitro data for 50 hydantoins, hydroxyamides, oxazolidinediones, hydroxyl acids, and amino acids [89]. Pharmacophore alignments of selected compounds were undertaken in SYBYL and showed that the hydantoin ring was less important for molecules with long hydrophobic chains that occupy a hydrophobic region. It was also suggested that more potent inhibitors of the sodium channel may require a third hydrophobic interaction [89].

All the studies to date that described simple pharmacophores or QSAR models of the sodium channel have used data from rat or mouse in vitro or in vivo systems. This obviously represents a limitation in using these models to infer structural requirements for the human sodium channels. However, several recent patch clamping studies have published data for several compounds on the sodium channel in human atrial myocytes [28] or HEK293 cells expressing hNa1.3 [90]. In neither case was any computational modeling described, and it is likely that these datasets on their own will be too small for models of the cardiac and neuronal channels, although small sets of data (Figure 13.4*a*) may be useful for preliminary 3D-QSAR models in future. Previously a large number of antidepressants were identified as interacting with this channel using an in vitro method, an observation that was novel. It is possible that a computational model for this ion channel could aid in the discovery of other potential ligands that is more efficient than in vitro testing alone, as has been previously demonstrated for drug transporters [91]. For example, using Catalyst a HypoGen model was generated with good model statistics for the nine compounds ($r = 0.91$) and TTX, riluzole and terocaine mapped to the two hydrogen bond acceptors, and positive ionizable feature (Figure 13.4*b*). Although the training set is admittedly very small, this pharmacophore is in partial agreement with the earlier pharmacophores (Figure 13.3) indicating the requirement for hydrogen bonding sites. The preliminary model has also been used to map the known low μM inhibitors paroxetine (Figure 13.4*c*, predicted IC_{50} 0.29 μM) and sertraline (not shown, predicted IC_{50} 47 μM), which suggests it might have some predictive capability but likely needs further molecules in the training set. The hierachical use of a simple model to filter molecules that are then tested for sodium channel activity in vitro should certainly be considered.

13.3.3 Homology Models of the Sodium Channel

Homology models for the sodium channel pore have been created using the crystal structure of the KcsA bacterial potassium selective channel. This model

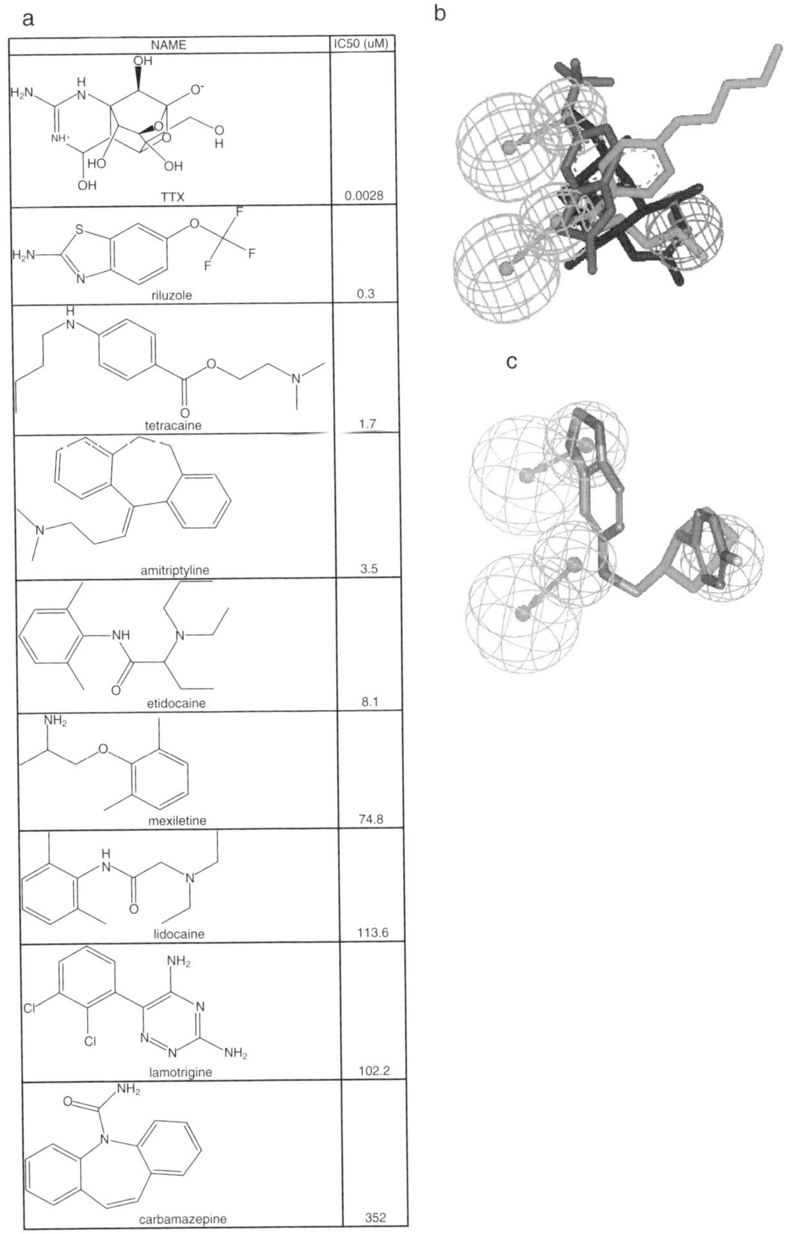

Figure 13.4 (*a*) Structures of molecules that bind the human sodium channel Na1.3 [90]. (*b*) Catalyst hypogen pharmacophore generated with nine molecules that bind the human sodium channel Na1.3. The molecule conformers were generated with the BEST algorithm with an energy level of 20kcal/mol. The model observed versus predicted correlation was $r = 0.91$, and the three most potent ligands TTX (blue), riluzole (red), and tetracaine (green) mapped to two hydrogen bond acceptor features (green) and one positive ionizable feature (blue). (*c*) Paroxetine aligned to the pharmacophore (predicted IC_{50} 0.29 μM). See color plates.

accommodates large toxin molecules such as tetrodotoxin and saxitoxin and was suggested as useful to provide ideas for amino acid residue changes for site directed mutagenesis experiments [92]. More recently the open and closed conformations of the $Na_v1.2$ and $Na_v1.8$ isoforms have been modeled using the KcsA and MthK potassium channels. Tetrodotoxin and teracaine were then docked in the local anaesthetic binding site 1 to suggest amino acid interactions [93]. To our knowledge, no combined pharmacophore and homology modeling efforts have been developed for the sodium channel, which should go some way toward further validating each approach.

13.4 CALCIUM CHANNELS

Calcium plays a pivotal role in cardiovascular function. The flow of calcium across cell membranes is necessary for cardiac automaticity, conduction and contraction, as well as maintenance of vascular tone. Thus calcium channel blockers (CCB) can interfere with calcium fluxes across cell membranes via direct blockage of calcium flow through L-type calcium channels (LCCs) found in the heart, vasculature, and pancreas. Although not considered further in this present discussion, N-type and T-type calcium channels located in neurons represent important targets for the treatment of pain, a topic that has recently been reviewed [94]. Interruption of calcium fluxes via LCCs leads to decreased intracellular calcium producing cardiovascular dysfunction that, in the most severe situations, results in cardiovascular collapse [95]. Some differences in toxic action exist between CCB drugs [96]; for example, diltiazem and especially verapamil tend to produce the most cases of hypotension, bradycardia, conduction disturbances, and deaths. By contrast, nifedipine and other dihydropyridines are generally less lethal and tend to produce sinus tachycardia instead of bradycardia with fewer conduction disturbances. CCB toxicity is often associated with significant hyperglycaemia and acidosis because of complex metabolic derangements related to these medications. A retrospective review by Cantrell et al. [97] demonstrated that the toxicity of CCBs following a therapeutic overdose can be highly variable and that the dose producing a toxic effect on the cardiovascular system may be within the maximum range of therapeutic doses. This may be the result of a number of factors, including the broad range of therapeutic doses as well as the preexisting conditions in patients taking these medications. Thus an unintentional overdose with a CCB may be lethal if the patient's cardiovascular ability to compensate for the toxic effects is compromised [98]. This variability makes caring for patients poisoned with these medications complicated [97].

QSAR studies of calcium channel agonists and antagonists can be viewed as useful tools for assessment of potential toxicity. Information on the primary structure of calcium channels has become increasingly accessible since their sequencing and 3D molecular models can be developed. Models based on the solved X-ray structure from the homologous voltage-gated potassium channel

[99] can provide detailed information about the binding site of the calcium channels. However, despite the availability of several homology models for calcium channels (e.g., [100–102]), there are no reported examples of their use for quantitative analysis of calcium channel agonists and antagonists. Therefore ligand-based approaches remain indispensable for understanding SAR dependencies within this large group.

13.4.1 Quantitative Structure-Activity Relationships

LCCs belong to a large family of voltage-gated calcium channels (VGCCs) that are located in the plasma membrane and form a highly selective conduit by which Ca^{2+} ions enter all excitable cells and some nonexcitable cells. Because of the pronounced therapeutic benefits associated with inhibition of LCCs [103], this subfamily has been extensively characterized by biochemical approaches. In particular, these studies revealed that the LCC complex is a heteropentamer consisting of $\alpha1$, β, $\alpha2/\delta$, and γ subunits [104,105]. For calcium channels to be effective, Ca^{2+} ions must enter selectively through the pore of the $\alpha1$ subunit, bypassing competition with other extracellular ions [104–106]. The structure of the $\alpha1$ subunit consists of four repeating motifs (MI–MIV), each motif comprising six hydrophobic segments (S1–S6). The $\alpha1$ subunit forms the conduction pore, the Ca^{2+} selective filter, the voltage sensor, and the known sites of channel regulation by drugs and toxins [107]. At least four distinct LCC $\alpha1$ subunits have been cloned and characterized [108]. Some CCBs (in particular, 1,4-dihydropyridines) have a different affinity for different tissues (cardiac and vascular) due to their ability to discriminate among the different $\alpha1$ subunits of LCC in various tissues [109].

4-Aryl-1,4-dihydropyridine (DHP) derivatives are the most widely studied class of calcium channel blockers that modulate Ca^{2+} permeation by stabilizing, respectively, the open and closed states of the LCCs. DHPs, exemplified by nifedipine, have been used in general medical practice worldwide for the treatment of hypertension and vasospastic angina for over two decades [110]. Radiolabeling experiments and site-directed mutagenesis reveal the $\alpha1$ subunit as the binding site of DHP derivatives [105]. Numerous recent examples of DHP agents that have progressed to clinical candidates have prompted extensive QSAR studies.

Classical Quantitative Structure-Activity Relationship Techniques The early QSAR models for calcium channel ligands were based on classical Hansch analysis and elucidated the structural requirements for the binding of molecules to their receptors [111–115]. It was found that various steric (B1, L), electronic (σ), and hydrophobic (π) parameters or their combination correlated well with the potency of various DHPs [111]. QSAR analysis of another set of DHPs revealed good correlations between electronic properties (*F*-constants) of the phenyl ring substituents and binding affinities or functional potency [112]; lipophilicity as well as ortho- and meta-substituents' inductivity and width were found to be the main factors affecting the receptor binding. A quadratic dependency of the Ca^{2+} antagonistic activity on the ClogP of terminal arylalkylamine moieties was suggested for a series of semotiadil congeners having a benzothiazine cyclic system [113]. Baxter et al. described compound FPL 64176 (methyl 2,5-dimethyl-4-[2-(phenylmethyl)benzoyl-1*H*-pyrrole-3-carboxylate) [114], the first example of a new class of calcium channel activators (CCAs) that does not act on any of the well-defined calcium channel modulator receptor sites, as typified by verapamil, diltiazem, and the DHPs. The potent activity of FPL 64176 was predicted using Hansch QSAR analysis on an initial set of eight less potent benzoylpyrroles [114]. Recently the Hansch analysis was applied to a wide series of nifedipine analogues containing nitroimidazolyl, phenylimidazolyl, and methylsulphonylimidazolyl groups at the C-4 position and different ester substituents at C-3 and C-5 positions of the dihydropyridine ring [116]; linear and quadratic relationships were obtained between the activity and hydrophobic, electronic, inductive, and resonance constants for the studied substituents.

The partial least squares (PLS) technique can be regarded as an advanced alternative to linear regression techniques commonly used in QSAR. In conjunction with special feature selection techniques, such as genetic algorithm (GA), this method can be used for generation of robust and easily interpretable models. Funatsu and coworkers have published a paper describing the application of the GA-PLS method in QSAR analysis of 35 DHP calcium channel antagonists [117]. In the studied compound set there were three variable positions (R^2, R^3, and R^4 substituents of the 4-phenyl ring), and the substitution pattern for each position was described by four descriptors: π, the hydrophobic substituent constant, σ_m, the Hammett σ constant, and B1 and L, and the sterimol parameters. The GA procedure enabled the removal of the redundant parameters, so only the most important properties entered the PLS analysis. Their results showed that the GA-PLS model based on 6 descriptors ($q^2 = 0.685$) was superior to the full 12 descriptor PLS model ($q^2 = 0.623$). In an external validation experiment employing the D-optimal criterion, a similar predictivity ($q^2 = 0.693$) was obtained for the 14 test compounds.

Nonlinear Quantitative Structure-Activity Relationship Modeling For the past 15 to 20 years, methods based on artificial neural networks have been shown to effectively deal with both linear and nonlinear dependencies that appear in real SAR problems. Hemmateenejad et al. applied a combination of GA, ANN, and principal component analysis (PCA) to a set of 124 DHPs bearing different ester substituents at the C-3 and C-5 positions of the DHP ring and nitroimidazolyl, phenylimidazolyl, and methylsulfonylimidazolyl groups at the C-4 position, with known Ca^{2+} channel binding affinities [118]. Ten different sets of descriptors (837 descriptors in total) were calculated for each molecule, and PCA was used to compress the descriptor groups into principal components (PCs). The GA was used for the selection of the best set of extracted PCs. A feed-forward backpropagated ANN was used to process the nonlinear relationship between the selected principal components and biological activity of the DHPs. A comparison between PC-GA-ANN and routine PC-ANN showed that the first model yielded improved predictions. The same group of researchers examined a set of quantum chemical descriptors for a QSAR study of 45 DHP calcium channel antagonists [119]. The best PC-GA-ANN models with five selected principal components were able to predict the activity of the molecules with prediction errors lower than ±5%. A database of 46 DHPs with known Ca^{2+} channel binding affinities was studied using the PC-ANN method [120]. A comparison of the developed approach with two other QSAR methods (multiple linear regression and a hologram QSAR model) showed that the PC-ANN approach can yield better predictions, once the right network configuration is identified. A group of DHP derivatives was used in order to compare the ANN modeling results with those obtained previously with PCA [121]. Calculated atomic and molecular descriptors using the semiempirical AM1 method were mainly used. It was shown that the predictive capability demonstrated by ANN were almost equivalent to that of PCA. Recently Si and coworkers described a novel machine learning algorithm, gene expression programming (GEP), and used it to develop QSAR model for a series of DHP calcium channel antagonists [122]. The heuristic method was used to search the descriptor space and select those responsible for activity. A nonlinear, six-descriptor model based on GEP with mean square errors 0.19 was set up with a predicted correlation coefficient $r^2 = 0.92$.

Three-Dimensional Quantitative Structure-Activity Relationship Approaches An early study by Seidel et al. showed that a certain 3D feature, such as a torsion angle, can be used to help explain calcium channel activity of a series of DHPs [123]. For a set of rigid analogues of nifedipine, their calcium channel affinities ($\log K_i$) were correlated with the deviation from 90° of the torsion angle between the two rings ($\Delta\alpha$):

$$\log K_i = 0.067(\pm 0.017)\Delta\alpha + 0.19(\pm 0.34), \tag{13.3}$$

where $n = 7$, $r = 0.88$, and $F = 16.5$. Since that time 3D-QSARs have been widely used for the prediction of in vitro or in vivo interactions between calcium channel ligands and their receptors [124–129]. Belvisi et al. described a 3D structure-activity relationship for eight CCBs [124]. Conformational analysis was carried out using a molecular mechanics method. A classification method was then used to select the most probable conformations linked to the biological activity and to build a model able to classify conformations according to their biological behavior. A comparative receptor surface analysis (CoRSA) was applied to study calcium channel antagonist activity of 35 DHP derivatives [125] in which the steric and electrostatic features of the most active compounds were used to generate a virtual receptor model, represented as points on a surface complementary to the van der Waals surface of the aligned compounds. The CoRSA structural descriptors, represented by the total interaction energies between each surface point of the virtual receptor and all atoms in a molecule, were used in a PLS data analysis to generate a 3D-QSAR model that gave an $r^2 = 0.928$ for calibration and $r^2_{cv} = 0.921$ for the leave-one-out cross validation.

CoMFA and CoMSIA were performed on a series of 24 isoxazolyl compounds as a potent T-type calcium channel blockers [126]. Four different conformations of the most active compound were used as template structures for the alignment. All CoMFA and CoMSIA models gave q^2 values greater than 0.5 and r^2 greater than 0.85. The predictive ability of the models was validated by an external test set of 10 compounds. The best predictions were obtained with a CoMFA model that gave predictive r^2 value of 0.87 for the test set. A series of DHPs possessing antitubercular activity were used for the derivation of CoMFA and CoMSIA 3D-QSAR models [127] with 33 compounds, of which 22 molecules were for training and 11 molecules were for a test set. Both methods produced comparable results with high internal as well as external predictivity. Steric and electrostatic fields of the inhibitors were found to be relevant descriptors. A series of 26 pyrrolo[2,1-*c*][1,4]benzothiazines with calcium antagonist activity were investigated using CoMFA [128]. The predictive ability of the CoMFA model was evaluated by a test set consisting of just three representative compounds. The best 3D-QSAR model found yielded a significant cross-validated, conventional, and predictive r^2 values equal to 0.703, 0.970, and 0.865, respectively, with the average absolute error of predictions being 0.26 log units. The robustness of this model was confirmed on a further test set of molecules consisting of diltiazem and nine pyrrolo[2,1-*d*][1,5]benzothiazepines with calcium antagonist activity. Davis et al. have described the use of the GRID method in the 3D-QSAR analysis of 36 4-benzoyl-2,5-dimethyl-1*H*-pyrrole-3-carboxylate calcium channel agonists [129]. PLS analysis of GRID maps showed the interaction energy between an alkyl hydroxyl probe and a series of agonists in 3D space and generated a predictive 3D-QSAR model with $r^2 = 0.86$ for the full 36 compound data set. The model significance was tested using the PRESS statistics.

One recent example serves as an illustration of 3D-QSAR modeling of a novel class of calcium channel blockers. Budriesi and coworkers synthesized a series of 8-aryl-8-hydroxy-8*H*-[1,4]thiazino[3,4-*c*][1,2,4]oxadiazol-3-ones (Figure 13.5) structurally related to diltiazem and pyrrolobenzothiazines, and assayed their calcium channel blocking activity in vitro [130]. To extract structural features related to activity, molecular descriptors based on the GRID force field were used, namely *gr*id-*in*dependent *d*escriptors (GRIND) [71]. These descriptors were introduced to overcome the main drawback of any CoMFA-like approach, which requires the alignment of compounds. The entire set of molecules was described by 870 descriptors, which composed the X data matrix used for QSAR analysis. Fractional factorial design was used to select the variables for the X-space, which finally amounted to 460 GRIND descriptors. Molecular descriptors were related to the compound potency by

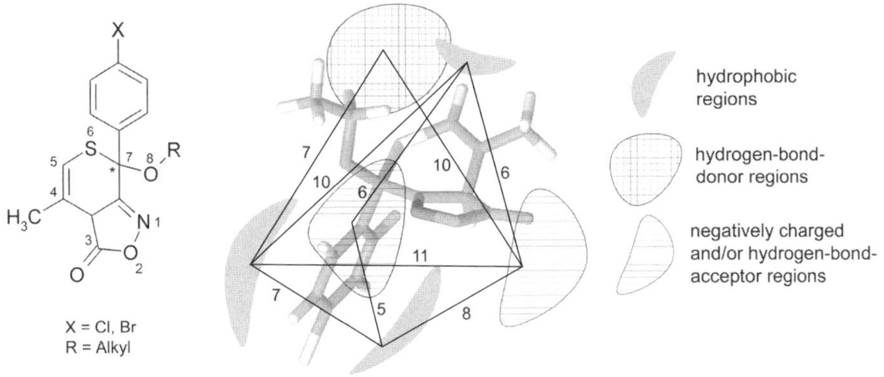

Figure 13.5 Structures of the analyzed thiazino[3,4-*c*][1,2,4]oxadiazol-3-ones and the virtual receptor scheme obtained from statistical analysis of GRIND descriptors (distances are expressed in Å). This illustration is a greatly modified version after a previously published figure [130].

means of PLS analysis. The optimal number of PLS components was chosen by monitoring the changes in the q^2 of the model, obtained by cross-validation procedures (leave one out and leave 20% out). The PLS model using two latent variables was optimal with r^2 (0.82), q^2_{loo} (0.66) and $q^2_{lo20\%}$ (0.63) values. The 3D-QSAR model was successfully validated by using an external set of 15 homologous and structurally similar compounds. The following pharmacophoric features were found to favorably affect the potency (Figure 13.5) [130]: (1) a small alkyl chain, such as ethyl or *i*-propyl, bound to the ether oxygen at position 8; (2) a methyl at position 5; (3) a weak hydrogen-bond acceptor at position 8 (e.g., the ether oxygen); (4) *para*-substituents on the phenyl ring at position 8; (5) a large hydrophilic region with hydrogen-bond accepting character in correspondence to the oxygen from the oxadiazolone moiety; and (6) a basic center, such as nitrogen, that is modeled in its protonated form.

Although the theoretically derived pseudoreceptor models are unlikely to represent the real binding site, they do allow interpretation of experimental data and reveal new hypotheses about channel activation on a molecular level. Thus an atomistic pseudoreceptor model was developed for a series of DHPs, which indicated a putative charge-transfer interaction for stabilizing the DHP/binding site complex [131]. Charge transfer (or electron–donor–acceptor) interactions indicate an electronic charge transfer from the HOMO of a donor molecule ($HOMO_D$) to the LUMO of an accepting neighbor molecule ($LUMO_A$). Small energy barriers between $HOMO_D$ and $LUMO_A$ increase the probability of charge transfer. For effective charge transfer, the corresponding molecular orbitals must be able to overlap and $HOMO_D$ and $LUMO_A$ must be energetically close. In the case of a charge–transfer interaction for the stabilization of the DHP/binding site complex, the electron-accepting LUMO should be located at the 4-phenyl ring of the DHPs, since the highest binding affinities are found for derivatives with electron withdrawing substituents at this position. This work demonstrated that charge–transfer interactions may play a major role in the receptor binding of DHPs; the experimentally derived free binding energies (ΔG) were correlated with the calculated LUMO* (the unoccupied molecular orbitals located on the 4-phenyl ring) energies and a highly significant correlation coefficient of 0.91 was obtained [131].

Chirality of DHP calcium channel blockers represents a difficult and significant problem for QSAR analysis [132]. A method for description of chirality has been developed as a continuous symmetry measure for chiral molecules [133]. The molecular structures of a series of DHPs were constructed and fully optimized with the Tripos force-field method [134]. The molecular structures corresponding to the relative minima on the conformational hypersurface were submitted to semiempirical MO calculations. For the most stable conformer of each molecule, pairwise similarity indices and partial atomic charges were calculated. The application of PCA to the complete $N \times N$ pairwise similarity matrices allowed definition of a chirality component and the computation of a chirality score in terms of the between-enantiomers difference in the component value. Of note, the molecular characteristics of small molecules that are essential for the activity of DHPs are also essential for the activity of their bioisosteric dihydropyrimidine analogues. A QSAR study on four different series of dihydropyrimidine analogues that mimic the DHP class [135] indicated the important characteristics are conformation of the molecule, the relative orientation of the aryl ring with respect to the pyrimidine ring, and some substituents capable of forming the hydrogen bonds with the receptor (but less bulky in nature) and high molar refractivity of the molecule.

13.4.2 Classification Methods

Several classification approaches have been described that can be useful for analysis of calcium channel antagonists. Thus, in the aforementioned work of Belvisi et al. [124], a classification of the 3D-QSAR approach was used to build a model able to classify conformations according to their calcium channel blocking activity. Cluster analysis on the active selected conformations subsequently allowed the identification of two different geometrical patterns for the active compounds. The least squares support vector machine (LSSVM) is a novel machine-learning algorithm that was used to develop a classification model for a novel series of DHP calcium channel antagonists [136]. Each compound was represented by calculated structural descriptors that encode constitutional, topological, geometrical, electrostatic, and quantum-chemical features. Good classification results were found using LSSVM: the percentage of correct predictions based on leave-one-out cross validation was 91.1%. The same method is also applicable for the generation of quantitative models. Thus a nonlinear, seven-descriptor model based on LSSVM resulted in a predicted correlation coefficient $r^2 = 0.87$, and a cross-validated correlation coefficient $r^2_{cr} = 0.82$.

Poroikov et al. developed a method for the analysis of molecular similarity based on a 2D description of molecules called multilevel neighborhoods of atoms [137]. Using this approach, they classified compounds according to their potential biological activity type. The prediction is based on the analysis of structure-activity relationships of the training set of more than 30,000 known biologically active compounds including 331 calcium channel blockers. In the leave-one-out cross-validation experiment they achieved a good level of accuracy with 94.1% of CCBs classified correctly [138]. Ivanenkov et al. applied Sammon nonlinear maps (NLMs) for classification and visualization of compounds according to their ion channel-specific activity [139]. The model was based on a preselected set of molecular descriptors encoding molecular lipophilicity, size, flexibility, surface charge distribution, and H-binding potential. The training set consisted of over 3000 known ligands for each ion channel (voltage-gated calcium, potassium, and sodium channels) and was used for generation of the NLM, followed by the study of distribution of different groups of VGIC ligands within the map. These ligand groups appeared to be located in different areas on the map (Figure 13.6), and this difference can be used for assessment of the ion channel-specific activity profile of test compounds. Like most of the other classification approaches described here, this method is primarily applicable for selection and synthesis of ion channel-focused libraries for bioscreening purposes.

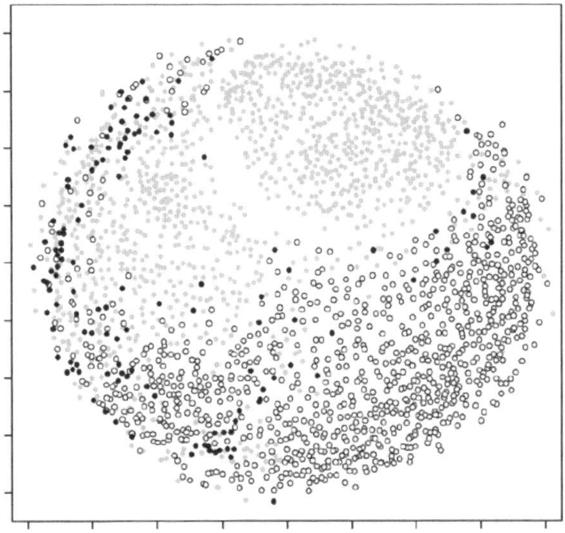

Figure 13.6 Distributions of VGIC-specific ligand groups within the Sammon nonlinear map: Calcuim channel antagonists (white circles, 1611 cmpds); potassium channel agonists and antagonists (gray circles, 1057 cmpds); sodium channel antagonists (black circles, 190 cmpds) [137].

13.4.3 ADME Related Issues and Calcium Channel Activity

It is well recognized that absorption, distribution, metabolism, and excretion-related properties of compounds can influence not only drug pharmacokinetics, toxicity, and selectivity but also, directly or indirectly, the pharmacological properties. For example, drug–membrane interactions resulting from incorporation of ion channel drugs into the phospholipid membrane can lead to a variety of biological responses. This may be related to several factors such as conformational changes during lateral diffusion of the drug to the receptor site within the bilayer; replacement of Ca^{2+} ions by intercalation of the drug with the membrane, provoking an indirect effect on embedded proteins; changes in the dynamics; and effects on the defect structures at the phase boundaries of the lateral phase-separated bilayer domains, often the binding locus of enzymes [140]. The effect of drug–membrane interactions on pharmacodynamics of CCBs can be illustrated by the following example. Retention times determined on immobilized artificial membrane (IAM) columns and expressed as chromatographic indexes $K_{\mathrm{w}}^{\mathrm{IAM}}$ have been used as descriptors of lipophilicity and polar interactions with phosphatidyl choline bilayer for a series of DHPs [141]. For the neutral CCBs, nifedipine, nitrendipine, nimodipine, and nisoldipine, receptor-binding values were determined on protein preparations from rat cortical brain and transformed into K_i values – $K_i = IC_{50}/(1 + LC/K_d)$, where LC is the ligand concentration, K_d the dissociation constant, and IC_{50} the concentration of the drug leading to a 50% inhibition of [^{3}H]-nimodipine-specific binding. Therefore the K_i value is a measure of the overall effect, including binding to the receptor as well as biomembrane permeation and interaction. The following equation was found:

$$\frac{\log 1}{K_i} = 1.581(\pm 0.225)\log K_{\mathrm{w}}^{\mathrm{IAM}} - 3.621(\pm 0.511), \qquad (13.4)$$

where $n = 4$, $r = 0.980$, and $s = 0.145$. Obviously the very small data set restricts generalizations. Nevertheless, these data suggest that membrane solubility of drugs may constitute the critical factor that allows the specific binding to the DHP-receptor.

QSAR analysis was applied to two series of 4-aryl and 4-heteroaryl DHPs with unsymmetrical ester substitutions on the DHP ring [142]. Regression analysis indicated the antihypertensive activity of an i.v. dose (determined in a spontaneously hypertensive rat model) correlated with the calculated octanol/water coefficent (clogP) but did not find a correlation between the in vitro potency and the clogP values. These data suggest the influence of ADME issues on the pharmacological activity of DHPs. In agreement with these observations are results obtained by Rojratanakiat and Hansch for a number of antischizophrenic drugs [143]. QSAR analysis of the binding of calcium antagonists to brain and heart tissue showed that relative binding to brain tissue increases with increasing octanol/water partition coefficients.

13.5 CONCLUSIONS

In the absence of crystal structures for the mammalian ion channels, QSAR methods such as pharmacophores, CoMFA, SVM, 2D-QSAR, GP, SOM, and recursive partitioning have been applied. However, to date L-type calcium channels and hERG appear to have been the most extensively studied. In contrast, there are far fewer examples for the sodium channel. With the focus on these three classes of ion channels as either therapeutic targets or as anti-targets, it is apparent that there is still some scope for developing and applying QSAR as more data becomes available, alongside other computational methods such as homology models. It is unclear whether the addition of more data to QSAR models will improve or degrade them as they are developed, and this will perhaps depend on whether they are local or global in nature.

The predictions generated with hERG QSAR models are approximately 70–85% accurate when used with external test sets, which are comparable with the different in vitro assays that have reasonably high false positive rates. A recent study has suggested that in silico methods for hERG do not as of yet have much of a role to play and cannot replace existing in vitro and in vivo testing [14]. However, contradictory evidence from the types of studies described above for hERG would appear to suggest that computational approaches such as QSAR are predictive and are at times at least as useful as the in vitro methods for identifying clinically relevant hERG inhibitors. Although it is unlikely that advocates of computational models for hERG would want to replace the source of new experimental data, we suggest that the case be made for careful use of models before in vitro and in vivo methods, in the same way that other computational models for drug-metabolizing enzymes and transporters, for example, can be used as a filter [144]. Clearly, there is still scope for educating the regulatory agencies in this regard. The published data on large numbers of diverse molecules tested with different hERG expressing systems, and analyzed using different methods in single studies [13,145], is expanding and represents a valuable resource for future QSAR modeling efforts. For example, the PubChem database (http://pubchem.ncbi.nlm.nih.gov/) now has a large hERG dataset for over 1000 molecules from a single laboratory. Some consideration should also be given to molecules that may inhibit hERG trafficking and may not be direct blockers of hERG [145], since these may require separate and unique pharmacophores.

The blockade of repolarizing currents, such as hERG, which can lead to delayed repolarization QT interval prolongation, can be offset if the drug also blocks depolarizing currents mediated by the sodium channel I_{Na} and/or calcium channel I_{Ca}. This shortens the cardiac action potential. Compensatory blockade of I_{Na} and I_{Ca} is applicable to metabolites as well as parent drugs. An example of this compensatory phenomenon is seen with verapamil [146] as well as other drugs [147] that block hERG and I_{Ca} at pharmacologically relevant concentrations. These drugs are not generally associated with high frequencies of occurrence of torsade de pointes. The different QSAR models for each ion channel can be used to make simultaneous predictions of interactions and result in improved predictions of torsade de pointes.

Knowledge of SAR dependencies is particularly advantageous for the combinatorial synthesis planning and in the lead optimization stages, and can be used for assessment of potential toxicity. The majority of the reported calcium channel models deal with derivatives of DHP and other congeneric data sets, and these models perform quite well. The development of a general calcium channel model that correctly predicts this activity of a diverse set of compounds looks unlikely. Conformational, steric, and electronic parameters of molecules are among the main factors affecting the calcium channel activity, although ADME-related properties may also be important.

Classical QSAR techniques, such as Hansch analysis and PLS, remain very useful for generation of QSAR dependencies for small congeneric data sets. Advanced data mining techniques, such as ANN and SVM, more effectively deal with nonlinear dependencies that appear in SAR analysis of more diverse data sets. Although the theoretically derived pseudoreceptor and pharmacophore models are unlikely to represent the real binding site, they do allow analysis and interpretation of experimental data and generation of the most reliable QSAR dependencies. More detailed information is expected from further refined models. The classification QSAR approaches may also prove particularly useful for rapid assessment of the potential for ion channel activity when dealing with scoring large chemical libraries. To our knowledge, there has been no single study using a common representation, such as a pharmacophore from one type of software, across all channels. Additionally individual studies use different tools that do not facilitate comparisons or amply describe the pharmacophore. Therefore it is problematic to summarize the differences and similarities of pharmacophores obtained as representatives of each channel. This needs to be remedied in future and will be important if we are to use such computational models of ion channels as tools for computational toxicology.

REFERENCES

1. Terlau H, Stuhmer W. Structure and function of voltage-gated ion channels. *Naturwissenschaften* 1998;85:437–44.
2. Triggle DJ, Gopalakrishnan M, Rampe D, Zheng W, editors. *Voltage-gated ion channels as drug targets*. Weinheim: Wiley-VCHV, 2006.
3. Catterall WA, Chandy KG, Gutman GA, editors. *The IUPHAR compendium of voltage-gated ion channels.* Leeds: IUPHAR Media, 2002.
4. Tan HL, Hou CJ, Lauer MR, Sung RJ. Electrophysiologic mechanisms of the long QT interval syndromes and torsade de pointes. *An Intern Med* 1995; 122:701–14.
5. Thomas SH. Drugs, QT interval abnormalities and ventricular arrhythmias. *Adverse Drug React Toxicol Rev* 1994;13:77–102.

6. Barry DM, Xu H, Schuessler RB, Nerbonne JM. Functional knockout of the transient outward current, long-QT syndrome, and cardiac remodeling in mice expressing a dominant-negative Kv4 alpha subunit. *Circ Res* 1998;83:560–7.
7. Jeron A, Mitchell GF, Zhou J, Murata M, London B, Buckett P, et al. Inducible polymorphic ventricular tachyarrhythmias in a transgenic mouse model with a long Q-T phenotype. *Am J Physiol Heart Circ Physiol* 2000;278:H1891–8.
8. Trudeau MC, Warmke JW, Ganetzky B, Robertson GA. HERG, a human inward rectifier in the voltage-gated potassium channel family. *Science* 1995;269:92–5.
9. Warmke JW, Ganetzky B. A family of potassium channel genes related to eag in *Drosophila* and mammals. *Proc Natl Acad Sci USA* 1994;91:3438–42.
10. Curran ME, Splawski I, Timothy KW, Vincent GM, Green ED, Keating MT. A molecular basis for cardiac arrhythmia: HERG mutations cause long QT syndrome. *Cell* 1995;80:795–803.
11. Rampe D, Murawsky MK, Grau J, Lewis EW. The antipsychotic agent sertindole is a high affinity antagonist of the human cardiac potassium channel HERG. *J Pharmacol Exp Ther* 1998;286:788–93.
12. Suessbrich H, Schonherr R, Heinemann SH, Attali B, Lang F, Busch AE. The inhibitory effect of the antipsychotic drug haloperidol on HERG potassium channels expressed in *Xenopus* oocytes. *Br J Pharmacol* 1997;120:968–74.
13. Diaz GJ, Daniell K, Leitza ST, Martin RL, Su Z, McDermott JS, et al. The [3H]dofetilide binding assay is a predictive screening tool for hERG blockade and proarrhythmia: Comparison of intact cell and membrane preparations and effects of altering [K+]o. *J Pharmacol Toxicol Meth* 2004;50:187–99.
14. Hoffmann P, Warner B. Are hERG channel inhibition and QT interval prolongation all there is in drug-induced torsadogenesis? A review of emerging trends. *J Pharmacol Toxicol Meth* 2006;53:87–105.
15. Rajamani R, Tounge BA, Li J, Reynolds CH. A two-state homology model of the hERG K(+) channel: application to ligand binding. *Bioorg Med Chem Lett* 2005;15:1737–41.
16. Zhou Y, Morais-Cabral JH, Kaufman A, MacKinnon R. Chemistry of ion coordination and hydration revealed by a K+ channel-Fab complex at 2.0 Å resolution. *Nature* 2001;414:43–8.
17. Jiang Y, Lee A, Chen J, Cadene M, Chait BT, MacKinnon R. Crystal structure and mechanism of a calcium-gated potassium channel. *Nature* 2002;417:515–22.
18. Jiang Y, Lee A, Chen J, Ruta V, Cadene M, Chait BT, et al. X-ray structure of a voltage-dependent K+ channel. *Nature* 2003;423:33–41.
19. Cavalli A, Poluzzi E, De Ponti F, Recanatini M. Toward a pharmacophore for drugs inducing the long QT syndrome: Insights from a CoMFA study of HERG K(+) channel blockers. *J Med Chem* 2002;45:3844–53.
20. Morgan TK, Jr., Sullivan ME. An overview of class III electrophysiological agents: A new generation of antiarrhythmic therapy. *Prog Med Chem* 1992;29:65–108.
21. Matyus P, Borosy AP, Varro A, Papp JG, Barlocco D, Cignarella G. Development of pharmacophores for inhibitors of the rapid component of the cardiac delayed rectifier potassium current. *Int J Quant Chem* 1998;69:21–30.
22. SYBYL. St. Louis, MO: Tripos Inc., 1997.

23. Liu H, Ji M, Jiang H, Liu L, Hua W, Chen K, et al. Computer-aided design, synthesis and biological assay of *p*-methylsulfonamido phenylethylamine analogues. *Bioorg Med Chem Lett* 2000;10:2153–7.
24. Liu H, Ji M, Luo X, Shen J, Huang X, Hua W, et al. New *p*-methylsulfonamido phenylethylamine analogues as class III antiarrhythmic agents: Design, synthesis, biological assay, and 3D-QSAR analysis. *J Med Chem* 2002;45:2953–69.
25. Tong W, Lowis DR, Perkins R, Chen Y, Welsh WJ, Goddette DW, et al. Evaluation of quantitative structure-activity relationship methods for large-scale prediction of chemicals binding to the estrogen receptor. *J Chem Inf Comput Sci* 1998;38:669–77.
26. Ekins S, Crumb WJ, Sarazan RD, Wikel JH, Wrighton SA. Three-dimensional quantitative structure-activity relationship for inhibition of human ether-a-go-go-related gene potassium channel. *J Pharmacol Exp Ther* 2002;301:427–34.
27. Catalyst. San Diego, CA: Accelrys, 2001.
28. Crumb Jr WJ, Ekins S, Sarazan D, Wikel JH, Wrighton SA, Carlson C, et al. Effects of antipsychotic drugs on I_{to}, I_{Na}, I_{sus}, I_{K1}, and hERG: QT prolongation, structure activity relationship, and network analysis. *Pharm Res* 2006;23:1133–43.
29. Ekins S. In silico approaches to predicting drug metabolism, toxicology and beyond. *Biochem Soc Trans* 2003;31:611–4.
30. Pearlstein RA, Vaz RJ, Kang J, Chen XL, Preobrazhenskaya M, Shchekotikhin AE, et al. Characterization of HERG potassium channel inhibition using CoMSiA 3D QSAR and homology modeling approaches. *Bioorg Med Chem Lett* 2003;13:1829–35.
31. Ekins S. Predicting undesirable drug interactions with promiscuous proteins in silico. *Drug Discov Today* 2004;9:276–85.
32. Ekins S, Balakin KV, Savchuk N, Ivanenkov Y. Insights for human ether-a-go-go-related gene potassium channel inhibition using recursive partitioning, Kohonen and Sammon mapping techniques. *J Med Chem* 2006;49:5059–71.
33. Mohamadi F, Richards NGJ, Guida WC, Liskamp R, Lipton M, Caufield C, et al. MacroModel—An integrated software system for modeling organic and bioorganic molecules using molecular mechanics. *J Comput Chem* 1990;1:440–67.
34. Chang G, Guida WC, Still WC. An internal coordinate Monte Carlo method for searching conformational space. *J Am Chem Soc* 1989;111:4379–86.
35. Keseru GM. Prediction of hERG potassium channel affinity by traditional and hologram qSAR methods. *Bioorg Med Chem Lett* 2003;13:2773–5.
36. Du LP, Tsai KC, Li MY, You QD, Xia L. The pharmacophore hypotheses of *I*(Kr) potassium channel blockers: Novel class III antiarrhythmic agents. *Bioorg Med Chem Lett* 2004;14:4771–7.
37. Aptula AO, Cronin MT. Prediction of hERG K+ blocking potency: Application of structural knowledge. *SAR QSAR Environ Res* 2004;15:399–411.
38. Redfern WS, Carlsson L, Davis AS, Lynch WG, MacKenzie I, Palethorpe S, et al. Relationships between preclinical cardiac electrophysiology, clinical QT interval prolongation and torsade de pointes for a broad range of drugs: Evidence for a provisional safety margin in drug development. *Cardiovasc Res* 2003;58:32–45.
39. Buyck C. An in silico model for detecting potential HERG blocking. In: Ford M, Livingstone D, Dearden J, Van de Waterbeemd H, editors, *EuroQSAR 2002*

Designing drugs and crop protectants: Processes, problems, and solutions. Oxford: Blackwell Publishing, 2003. p. 86–9.

40. Song M, Clark M. Development and evaluation of an in silico model for HERG binding. *J Chem Inf Model* 2006;46:392–400.
41. Clark M. Generalized fragment-substructure based property prediction method. *J Chem Inf Model* 2005;45:30–8.
42. Fernandez D, Ghanta A, Kauffman GW, Sanguinetti MC. Physicochemical features of the HERG channel drug binding site. *J Biol Chem* 2004;279:10120–7. Epub 2003 Dec 29.
43. Yoshida K, Niwa T. Quantitative structure-activity relationship studies on inhibition of HERG potassium channels. *J Chem Inf Model* 2006;46:1371–8.
44. Fletcher SR, Burkamp F, Blurton P, Cheng SK, Clarkson R, O'Connor D, et al. 4-(Phenylsulfonyl)piperidines: Novel, selective, and bioavailable 5-HT(2A) receptor antagonists. *J Med Chem* 2002;45:492–503.
45. Rowley M, Hallett DJ, Goodacre S, Moyes C, Crawforth J, Sparey TJ, et al. 3-(4-Fluoropiperidin-3-yl)-2-phenylindoles as high affinity, selective, and orally bioavailable h5-HT(2A) receptor antagonists. *J Med Chem* 2001;44:1603–14.
46. Bell IM, Gallicchio SN, Abrams M, Beese LS, Beshore DC, Bhimnathwala H, et al. 3-Aminopyrrolidinone farnesyltransferase inhibitors: Design of macrocyclic compounds with improved pharmacokinetics and excellent cell potency. *J Med Chem* 2002;45:2388–409.
47. Ekins S, Swaan PW. Development of computational models for enzymes, transporters, channels and receptors relevant to ADME/TOX. *Rev Comp Chem* 2004;20:333–415.
48. Roche O, Trube G, Zuegge J, Pflimlin P, Alanine A, Schneider G. A virtual screening method for prediction of the HERG potassium channel liability of compound libraries. *ChemBioChem* 2002;3:455–9.
49. Ghose AK, Crippen GM. Atomic physicochemical parameters for three-dimensional structure-directed quantitative structure-activity relationships: I. Partition coefficients as a measure of hydrophobicity. *J Comput Chem* 1986;7:565–77.
50. TSAR. Oxford, UK: Oxford Molecular Ltd.
51. Schneider G, Neidhart W, Giller T, Schmid G. "Scaffold-hopping" by topological pharmacophore search: A contribution to virtual screening. *Angew Chem Int Ed Engl* 1999;38:2894–6.
52. VolSurf. Perugia, Italy: Multivariate Infometric Analysis S.r.l.
53. DRAGON. Milano, Italy: Milano Chemometrics and QSAR group.
54. Sun H. An accurate and interpretable Bayesian classification model for prediction of hERG liability. *ChemMedChem* 2006;1:315–22.
55. Aronov AM, Goldman BB. A model for identifying HERG K+ channel blockers. *Bioorg Med Chem* 2004;12:2307–15.
56. Carhart RE, Smith DH, Venkataraghavan R. Atom pairs as molecular features in structure-activity studies: Definition and application. *J Chem Inf Comput Sci* 1985;25:64–73.
57. Mitcheson JS, Perry MD. Molecular determinants of high-affinity drug binding to HERG channels. *Curr Opin Drug Discov Devel* 2003;6:667–74.

58. Perry M, de Groot MJ, Helliwell R, Leishman D, Tristani-Firouzi M, Sanguinetti MC, et al. Structural determinants of HERG channel block by clofilium and ibutilide. *Mol Pharmacol* 2004;66:240–9.
59. Testai L, Bianucci AM, Massarelli I, Breschi MC, Martinotti E, Calderone V. Torsadogenic cardiotoxicity of antipsychotic drugs: A structural feature, potentially involved in the interaction with cardiac HERG potassium channels. *Curr Med Chem* 2004;11:2691–706.
60. Wang N, DeLisle RK, Diller DJ. Fast small molecule similarity searching with multiple alignment profiles of molecules represented in one-dimension. *J Med Chem* 2005;48:6980–90.
61. Vapnik V. *Statistical learning theory*. New York: Wiley, 1998.
62. Tobita M, Nishikawa T, Nagashima R. A discriminant model constructed by the support vector machine method for HERG potassium channel inhibitors. *Bioorg Med Chem Lett* 2005;15:2886–90.
63. Molecular Operating Environment (MOE). Montreal, Canada: Chemical Computing Group Inc., 2002.
64. Bains W, Basman A, White C. HERG binding specificity and binding site structure: Evidence from a fragment-based evolutionary computing SAR study. *Prog Biophys Mol Biol* 2004;86:205–33.
65. Pearlstein R, Vaz R, Rampe D. Understanding the structure-activity relationship of the human ether-a-go-go-related gene cardiac K+ channel: A model for bad behavior. *J Med Chem* 2003;46:2017–22.
66. Mitcheson JS, Chen J, Lin M, Culberson C, Sanguinetti MC. A structural basis for drug-induced long QT syndrome. *Proc Natl Acad Sci USA* 2000;97:12329–33.
67. Balakin KV, Ivanenkov YA, Savchuk NP, Ivaschenko AA, Ekins S. Comprehensive computational assessment of ADME properties using mapping techniques. *Curr Drug Disc Tech* 2005;2:99–113.
68. Du LP, Li MY, Tsai KC, You QD, Xia L. Characterization of binding site of closed-state KCNQ1 potassium channel by homology modeling, molecular docking, and pharmacophore identification. *Biochem Biophys Res Commun* 2005;332:677–87.
69. DOCK5. San Francisco: University of California, San Francisco.
70. Carosati E, Lemoine H, Spogli R, Grittner D, Mannhold R, Tabarrini O, et al. Binding studies and GRIND/ALMOND-based 3D QSAR analysis of benzothiazine type K(ATP)-channel openers. *Bioorg Med Chem* 2005;13:5581–91.
71. Pastor M, Cruciani G, McLay I, Pickett S, Clementi S. GRid-INdependent descriptors (GRIND): A novel class of alignment-independent three-dimensional Molecular descriptors. *J Med Chem* 2000;43:3233–43.
72. Uhrig U, Holtje HD, Mannhold R, Weber H, Lemoine H. Molecular modeling and QSAR studies on K(ATP) channel openers of the benzopyran type. *J Mol Graph Model* 2002;21:37–45.
73. Peukert S, Brendel J, Pirard B, Bruggemann A, Below P, Kleemann HW, et al. Identification, synthesis, and activity of novel blockers of the voltage-gated potassium channel Kv1.5. *J Med Chem* 2003;46:486–98.
74. Peukert S, Brendel J, Pirard B, Strubing C, Kleemann HW, Bohme T, et al. Pharmacophore-based search, synthesis, and biological evaluation of anthranilic

amides as novel blockers of the Kv1.5 channel. *Bioorg Med Chem Lett* 2004;14:2823–7.

75. Favre I, Moczydlowski E, Schild L. On the structural basis for ionic selectivity among Na+, K+, and Ca2+ in the voltage-gated sodium channel. *Biophys J* 1996;71:3110–25.
76. Anger T, Madge DJ, Mulla M, Riddall D. Medicinal chemistry of neuronal voltage-gated sodium channel blockers. *J Med Chem* 2001;44:115–37.
77. Li Y, Harte WE. A review of molecular modeling approaches to pharmacophore models and structure-activity relationships of ion channel modulators in CNS. *Curr Pharm Des* 2002;8:99–110.
78. Khodorov BI. Sodium inactivation and drug-induced immobilization of the gating charge in nerve membrane. *Prog Biophys Mol Biol* 1981;37:49–89.
79. Rein KS, Baden DG, Gawley RE. Conformational analysis of the sodium channel modulator, Brevetoxin A, comparison with brevetoxin B conformations, and a hypothesis about the common pharmacophore of the "site 5" toxins. *J Org Chem* 1994;59:2101–6.
80. Unverferth K, Engel J, Hofgen N, Rostock A, Gunther R, Lankau HJ, et al. Synthesis, anticonvulsant activity, and structure-activity relationships of sodium channel blocking 3-aminopyrroles. *J Med Chem* 1998;41:63–73.
81. Kuo CC, Huang RC, Lou BS. Inhibition of Na(+) current by diphenhydramine and other diphenyl compounds: Molecular determinants of selective binding to the inactivated channels. *Mol Pharmacol* 2000;57:135–43.
82. Tasso SM, Moon S, Bruno-Blanch LE, Estiu GL. Characterization of the anticonvulsant profile of valpromide derivatives. *Bioorg Med Chem* 2004;12:3857–69.
83. Malawska B, Kulig K, Spiewak A, Stables JP. Investigation into new anticonvulsant derivatives of alpha-substituted *N*-benzylamides of gamma-hydroxy- and gamma-acetoxybutyric acid: Part 5. Search for new anticonvulsant compounds. *Bioorg Med Chem* 2004;12:625–32.
84. De Luca A, Talon S, De Bellis M, Desaphy JF, Lentini G, Corbo F, et al. Optimal requirements for high affinity and use-dependent block of skeletal muscle sodium channel by *N*-benzyl analogs of tocainide-like compounds. *Mol Pharmacol* 2003;64:932–45.
85. Pandeya SN, Raja AS, Stables JP. Synthesis of isatin semicarbazones as novel anticonvulsants–Role of hydrogen bonding. *J Pharm Pharm Sci* 2002;5:266–71.
86. Hays SJ, Rice MJ, Ortwine DF, Johnson G, Schwarz RD, Boyd DK, et al. Substituted 2-benzothiazolamines as sodium flux inhibitors: Quantitative structure-activity relationships and anticonvulsant activity. *J Pharm Sci* 1994;83:1425–32.
87. Brown ML, Brown GB, Brouillette WJ. Effects of $\log P$ and phenyl ring conformation on the binding of 5-phenylhydantoins to the voltage-dependent sodium channel. *J Med Chem* 1997;40:602–7.
88. Brown ML, Zha CC, Van Dyke CC, Brown GB, Brouillette WJ. Comparative molecular field analysis of hydantoin binding to the neuronal voltage-dependent sodium channel. *J Med Chem* 1999;42:1537–45.
89. Zha C, Brown GB, Brouillette WJ. Synthesis and structure-activity relationship studies for hydantoins and analogues as voltage-gated sodium channel ligands. *J Med Chem* 2004;47:6519–28.

90. Huang CJ, Harootunian A, Maher MP, Quan C, Raj CD, McCormack K, et al. Characterization of voltage-gated sodium-channel blockers by electrical stimulation and fluorescence detection of membrane potential. *Nat Biotechnol* 2006;24:439–46.

91. Ekins S, Johnston JS, Bahadduri P, D'Souzza VM, Ray A, Chang C, et al. In vitro and pharmacophore based discovery of novel hPEPT1 inhibitors. *Pharm Res* 2005;22:512–7.

92. Lipkind GM, Fozzard HA. KcsA crystal structure as framework for a molecular model of the Na+ channel pore. *Biochemistry* 2000;39:8161–70.

93. Scheib H, McLay I, Guex N, Clare JJ, Blaney FE, Dale TJ, et al. Modeling the pore structure of voltage-gated sodium channelsin closed, open, fast -inactivated conformation reveals details of site 1 toxin and local anaesthetic binding. *J Mol Model* 2005;12:813–22.

94. McGivern JG. Targeting N-type and T-type calcium channels for the treatment of pain. *Drug Disc Today* 2006;11:245–53.

95. DeWitt CR, Waksman JC. Pharmacology, pathophysiology and management of calcium channel blocker and beta-blocker toxicity. *Toxicol Rev* 2004;23:223–38.

96. Kubota K, Pearce GL, Inman WH. Vasodilation-related adverse events in diltiazem and dihydropyridine calcium antagonists studied by prescription-event monitoring. *Eur J Clin Pharmacol* 1995;48:1–7.

97. Cantrell FL, Clark RF, Manoguerra AS. Determining triage guidelines for unintentional overdoses with calcium channel antagonists. *Clin Toxicol (Phila)* 2005;43:849–53.

98. Cantrell FL, Williams SR. Fatal unintentional overdose of diltiazem with antemortem and postmortem values. *Clin Toxicol (Phila)* 2005;43:587–8.

99. Doyle DA, Cabral JM, Pfuetzner RA, Kuo A, Gulbis JM, Cohen SL, et al. The structure of the potassium channel: Molecular basis of K+ conduction and selectivity. *Science* 1998;280:69–77.

100. Zhorov BS, Folkman EV, Ananthanarayanan VS. Homology model of dihydropyridine receptor: implications for L-type Ca2+ channel modulation by agonists and antagonists. *Arch Biochem Biophys* 2001;393:22–41.

101. Zhorov BS, Ananthanarayanan VS. Docking of verapamil in a synthetic Ca2+ channel: Formation of a ternary complex involving Ca2+ ions. *Arch Biochem Biophys* 1997;341:238–44.

102. Lipkind GM, Fozzard HA. Molecular modeling of interactions of dihydropyridines and phenylalkylamines with the inner pore of the L-type Ca2+ channel. *Mol Pharmacol* 2003;63:499–511.

103. Triggle DJ. L-type calcium channels. *Curr Pharm Des* 2006;12:443–57.

104. Catterall WA. Structure and function of voltage-sensitive ion channels. *Science* 1988;242:50–61.

105. Varadi G, Strobeck M, Koch S, Caglioti L, Zucchi C, Palyi G. Molecular elements of ion permeation and selectivity within calcium channels. *Crit Rev Biochem Mol Biol* 1999;34:181–214.

106. Randall A, Benham CD. Recent advances in the molecular understanding of voltage-gated Ca2+ channels. *Mol Cell Neurosci* 1999;14:255–72.

107. Striessing J, Grabner M, Mitterdorfer J, Hering S, Sinnegger MJ, Glossmann H. Structural bases of drug binding to L-Ca2+ channels. *Trends Pharm Sci* 1998;19:108–15.

108. Catterall WA. Structure and regulation of voltage-gated Ca2+ channels. *An Rev Cell Dev Biol* 2000;16:521–55.

109. Hu H, Marban E. Isoform-specific inhibition of L-type calcium channels by dihydropyridines is independent of isoform-specific gating properties. *Mol Pharmacol* 1998;53:902–7.

110. Striessing J. Pharmacology, structure and function of cardiac L-type calcium channels. *Cell Physiol Biochem* 1999;9:242–69.

111. Mahmoudian M, Richards WG. QSAR of binding of dihydropyridine-type calcium antagonists to their receptor on ileal smooth muscle preparations. *J Pharm Pharmacol* 1986;38:272–6.

112. Goll A, Glossmann H, Mannhold R. Correlation between the negative inotropic potency and binding parameters of 1,4-dihydropyridine and phenylalkylamine calcium channel blockers in cat heart. *Naunyn Schmiedebergs Arch Pharmacol* 1986;334:303–12.

113. Fujimura K, Ota A, Kawashima Y. Quantitative structure-activity relationships of Ca(2+)-antagonistic semotiadil congeners. *Chem Pharm Bull (Tokyo)* 1996; 44:542–6.

114. Baxter AJ, Dixon J, Ince F, Manners CN, Teague SJ. Discovery and synthesis of methyl 2,5-dimethyl-4-[2-(phenylmethyl)benzoyl]-1H-pyrrole-3-carboxylate (FPL 64176) and analogues: The first examples of a new class of calcium channel activator. *J Med Chem* 1993;36:2739–44.

115. Mager PP, Coburn RA, Solo AJ, Triggle DJ, Rothe H. QSAR, diagnostic statistics and molecular modelling of 1,4-dihydropyridine calcium antagonists: A difficult road ahead. *Drug Des Discov* 1992;8:273–89.

116. Hemmateenejad B, Akhond M, Miri R, Shamsipur M. Quantitative structure-activity relationship study of recently synthesized 1,4-dihydropyridine calcium channel antagonists. In: *Application of the Hansch analysis method*. Weinheim: Arch Pharm, 2002. p. 472–80.

117. Hasegawa K, Miyashita Y, Funatsu K. GA strategy for variable selection in QSAR studies: GA-based PLS analysis of calcium channel antagonists. *J Chem Inf Comput Sci* 1997;37:306–10.

118. Hemmateenejad B, Akhond M, Miri R, Shamsipur M. Genetic algorithm applied to the selection of factors in principal component-artificial neural networks: application to QSAR study of calcium channel antagonist activity of 1,4-dihydropyridines (nifedipine analogs). *J Chem Inf Comput Sci* 2003;43:1328–34.

119. Hemmateenejad B, Safarpour MA, Miri R, Taghavi F. Application of ab initio theory to QSAR study of 1,4-dihydropyridine-based calcium channel blockers using GA-MLR and PC-GA-ANN procedures. *J Comput Chem* 2004;25:1495–503.

120. Viswanadhan VN, Mueller GA, Basak SC, Weinstein JN. Comparison of a neural net-based QSAR algorithm (PCANN) with Hologram- and multiple linear regression-based QSAR approaches: application to 1,4-dihydropyridine-based calcium channel antagonists. *J Chem Inf Comput Sci* 2001;41:505–11.

121. Takahata Y, Costa MC, Gaudio AC. Comparison between neural networks (NN) and principal component analysis (PCA): Structure activity relationships of 1,4-dihydropyridine calcium channel antagonists (nifedipine analogues). *J Chem Inf Comput Sci* 2003;43:540–4.
122. Si HZ, Wang T, Zhang KJ, Hu ZD, Fan BT. QSAR study of 1,4-dihydropyridine calcium channel antagonists based on gene expression programming. *Bioorg Med Chem* 2006;Mar 29: published on the Web.
123. Seidel W, Meyer H, Born L, Kazda K, Dompert W. Rigid calcium antagonists of the nifedipine type: Geometrical requirements for the dihydropyridine receptor. In: Seydel J, editor, *QSAR and strategies in the design of bioactive compounds*. Weinheim: VCH, 1985. p. 366–9.
124. Belvisi L, Brossa S, Salimbeni A, Scolastico C, Todeschini R. Structure-activity relationship of Ca2+ channel blockers: A study using conformational analysis and chemometric methods. *J Comput Aided Mol Des* 1991;5:571–84.
125. Ivanciuc O, Ivanciuc T, Cabrol-Bass D. Comparative receptor surface analysis (CoRSA) model for calcium channel antagonists. *SAR QSAR Environ Res* 2001;12:93–111.
126. Doddareddy MR, Jung HK, Cha JH, Cho YS, Koh HY, Chang MH, et al. 3D QSAR studies on T-type calcium channel blockers using CoMFA and CoMSIA. *Bioorg Med Chem* 2004;12:1613–21.
127. Kharkar PS, Desai B, Gaveria H, Varu B, Loriya R, Naliapara Y, et al. Three-dimensional quantitative structure-activity relationship of 1,4-dihydropyridines as antitubercular agents. *J Med Chem* 2002;45:4858–67.
128. Corelli F, Manetti F, Tafi A, Campiani G, Nacci V, Botta M. Diltiazem-like calcium entry blockers: A hypothesis of the receptor-binding site based on a comparative molecular field analysis model. *J Med Chem* 1997;40:125–31.
129. Davis AM, Gensmantel NP, Johansson E, Marriott DP. The use of the GRID program in the 3-D QSAR analysis of a series of calcium-channel agonists. *J Med Chem* 1994;37:963–72.
130. Budriesi R, Carosati E, Chiarini A, Cosimelli B, Cruciani G, Ioan P, et al. A new class of selective myocardial calcium channel modulators: 2. Role of the acetal chain in oxadiazol-3-one derivatives. *J Med Chem* 2005;48:2445–56.
131. Schleifer K-J. Stereoselective characterization of the 1,4-dihydropyridine binding site at L-type calcium channels in the resting state and the opened/inactivated state. *J Med Chem* 1999;42:2204–11.
132. Goldmann S, Stoltefuss J. 1,4-Dihydropyridines: influence of chirality and conformation to the calcium antagonistic and agonistic effect. *Angew Chem* 1991;103:1587–605.
133. Benigni R, Cotta-Ramusino M, Gallo G, Giorgi F, Giuliani A, Vari MR. Deriving a quantitative chirality measure from molecular similarity indices. *J Med Chem* 2000;43:3699–703.
134. Clark M, Cramer III RD, Van Opdenbosch N. Validation of the general purpose tripos 5.2 force field. *J Comput Chem* 1989;10:982–1012.
135. Gupta SP, Veerman A, Bagaria P. Quantitative structure-activity relationship studies on some series of calcium channel blockers. *Mol Divers* 2004;8:357–63.

136. Yao X, Liu H, Zhang R, Liu M, Hu Z, Panaye A, et al. QSAR and classification study of 1,4-dihydropyridine calcium channel antagonists based on least squares support vector machines. *Mol Pharm* 2005;2:348–56.

137. Filimonov D, Poroikov V, Borodina Y, Gloriozova T. Chemical similarity assessment through multilevel neighborhoods of atoms: Definition and comparison with the other descriptors. *J Chem Inf Comput Sci* 1999;39:666–70.

138. Poroikov VV, Filimonov DA, Borodina YV, Lagunin AA, Kos A. Robustness of biological activity spectra predicting by computer program PASS for noncongeneric sets of chemical compounds. *J Chem Inf Comput Sci* 2000;40:1349–55.

139. Ivanenkov YA, Balakin KV, Savchuk NP, Nikolsky Y. Design of ion channel-targeted libraries using unsupervised machine learning approach. In: *Intelligent drug discovery and development*. Philadelphia, PA, 2003. p. 27–30.

140. Seidel JK. Drug-membrane interactions and pharmacodynamics. In Methods and Principles in medicinal Chemistry. In: Mannhold R, Kubinyi H, Folkers G, editors, *Methods and priniciples in medicinal chemistry*. Weinheim: Wiley-VCH, 2002. p. 217–89.

141. Barbato F, LaRotonda IM, Quaglia F. Chromatographic indexes on immobilized artificial membranes for local anaesthetics: relationships with activity data on closed sodium channels. *Pharm Res* 1997;14:1699–705.

142. Wikel JH, Bemis KG, Kurz K, Denney ML, Main BW, Moore RA, et al. Comparative QSAR studies of two series of 1,4-dihydropyridines as slow calcium channel blockers. *Drug Des Discov* 1994;11:1–14.

143. Rojratanakiat W, C. H. The relative dependence of calcium antagonists and neuroleptics binding to brain and heart receptors on drug lipophilicity. *J Pharm Pharmacol* 1990;42:599–600.

144. Ekins S, Waller CL, Swaan PW, Cruciani G, Wrighton SA, Wikel JH. Progress in predicting human ADME parameters in silico. *J Pharmacol Toxicol Meth* 2000;44:251–72.

145. Wible BA, Hawryluk P, Ficker E, Kuryshev YA, Kirsch G, Brown AM. HERG-Lite: A novel comprehensive high-throughput screen for drug-induced hERG risk. *J Pharmacol Toxicol Meth* 2005;52:136–45.

146. Zhang S, Zhou Z, Gong Q, Makielski JC, January CT. Mechanism of block and identification of the verapamil binding domain to HERG potassium channels. *Circ Res* 1999;84:989–98.

147. Kang J, Chen XL, Wang H, Ji J, Reynolds W, Lim S, et al. Cardiac ion channel effects of tolterodine. *J Pharmacol Exp Ther* 2004;308:935–40.

148. Delano WL. *The PyMol user's manual.* San Carlos, CA.: DeLano Scientific, 2002.

14

PREDICTIVE MUTAGENICITY COMPUTER MODELS

LAURA L. CUSTER, CONSTANTINE KREATSOULAS, AND STEPHEN K. DURHAM

Contents

14.1 INTRODUCTION

Numerous factors are responsible for the recent paradigm shift in the drug discovery and development process. First, and foremost, is the cost factor. The pharmaceutical industry can no longer sustain their former business model in the context of the exponential growth in costs associated with drug development as widely described by the investigators at Tufts University [1]. With this background we add the remarkable productivity of combinatorial chemistry

Computational Toxicology: Risk Assessment for Pharmaceutical and Environmental Chemicals, Edited by Sean Ekins

and high-throughput screening that sets the stage for the need for additional assessment on the development potential of an ever-growing list of drug candidates. In addition the pharmaceutical industry is now in search of drug targets that are less well defined; the so-called low-hanging fruit of the recent past no longer exists.

The pharmaceutical industry is also in constant pursuit of the ideal drug candidate that has all of the desired adsorption, distribution, metabolism, excretion, and toxicity (ADMET) properties. However, we rarely proceed into the clinical environment with a flawless drug candidate. In the realm of toxicology, one of the most important properties for early assessment is genotoxic potential, in particular, mutagenicity, which is the ability to cause permanent alteration in DNA sequence. This conclusion is based on the fact that very few mutagenic drug candidates will proceed into clinical trials and result in registration approval by regulatory agencies. The only routine exception to this "rule" is oncologic cytotoxic drugs, which may proceed based on the unmet serious medical need of the intended cancer patient population.

14.2 COMPUTATIONAL APPROACHES

There are two basic types of modeling approaches for predictive computational toxicity, quantitative structural-activity relationships (QSAR) and expert rule-based systems, or combinational hybrids of the two. Each of these approaches has had varying degrees of success in modeling toxicological endpoints. QSAR systems excel at modeling specific mechanisms, whereas expert rule-based systems are better for global types of modeling, where unknown or multiple mechanisms of action exist. Hybrid systems have been developed in an attempt to minimize the weaknesses inherent in each separate approach.

QSAR models (see Chapters 8–13) are based on analysis of chemical descriptors or characteristics derived from the molecular structure of compounds. These chemical descriptors represent calculations of molecular indices (e.g., counting the number of rotatable bonds, halogens, or heavy atoms) and properties (e.g., LUMO: lowest unoccupied molecular orbital; ease of electron cloud polarization, or hydrophobicity: logP). By using a multivariate statistical method approach, chemical descriptors are identified and ranked according to the best correlation with an observed endpoint, such as mutagenicity. Only those descriptors with the strongest correlation with the observed endpoint are incorporated into the model and used to predict mutagenicity of unknown compounds. This class of model provides a quantitative estimate of the likelihood of an unknown compound having mutagenic activity.

It is important to be familiar with how and to what purpose the QSAR model was developed in order to determine if the model is appropriate. Compounds used to build QSAR models are considered training-set compounds and should represent the range of structural diversity expected for compounds that the model will be predicting. For example, it would be inappropriate to use a model built entirely from a training-set of aromatic amines to predict the mutagenicity of furans. The two variables that most influence the success of QSAR models are (1) a large pool of high-quality data and (2) a well-defined limited number of mechanisms of action. As the number of compounds within the training set increases, the predictive power of the correlated descriptors increases. Conversely, if multiple mechanisms of action are responsible for the endpoint, then the likelihood of finding simple descriptors with a strong correlation to the endpoint is low, since the signal from any single mechanism is diluted in proportion to the total signal. In addition, when multiple mechanisms of action are incorporated into a single QSAR model, the same chemical substructure could enhance the toxicity of one mechanism while reducing the effects of another mechanism.

Expert system models are based on information from the collective experiences of experts working within the field. Toxicologists create rules based on a list of chemical structures that have been linked by cause-and-effect relationships. The most well known of these structure activity rules are the Ashby-Tennant rules [2–5]. Many different mechanisms of action and chemical substructures have been associated with mutagenicity; for example, an electrophilic fragment is located somewhere within the compound that can react with nucleophilic portions of DNA bases. Rules can be as broad as "rule-of-thumb" relationships (most molecules containing nitro groups are mutagenic), or as detailed quantitative data collected from compounds tested using a single consistent protocol. An example of an expert rule might be: IF a compound contains an aniline AND metabolic activation is present, THEN the compound is genotoxic, or ELSE nongenotoxic. A significant deficiency of the expert systems is that these models only provide a binary prediction: either the molecule has a toxic fragment or it does not have a toxic fragment. However, expert systems have an advantage over the QSAR method for the investigation and elimination of mutagenic activity because the predicted mutagenicity of an expert system is related to a specific mechanism and/or substructure within the compound. Identification of the problem mechanism or substructure enables the user to propose structural modifications to reduce or eliminate mutagenic activity from the compound. Although similar information is obtainable via QSAR methods, mutagenicity is usually derived from a mathematical function of multiple chemical descriptors. In the prediction of mutagenicity, some of the chemical descriptors are fragment-specific, but it is usually unclear which specific structural changes are required to reduce or eliminate the mutagenic potential. Expert systems also have the advantage that each rule is specific for a specific mechanism and structural class. In contrast, typical QSAR models are based on a single nonlinear model or several linear submodels [6–8].

Expert systems and QSAR methods can sometimes be combined to produce hybrid systems with greater predictive capability than either model alone. A QSAR model can be initially developed using chemical-substructure, fragment-based descriptors (e.g., partitioning a molecule at rotatable bonds) and then subsequently applying multivariate statistical methods to rank-order the fragments with the strongest correlation to the endpoint of interest. A hybrid model built from the most statistically significant fragments could then be used as the expert system (e.g., MCASE). These hybrid models have gained widespread acceptance by both industry and regulatory agencies with their routine incorporation into the safety assessment process.

14.3 SOURCES OF MUTAGENICITY DATA

The single most important element influencing the predicative performance of a model is the quantity and quality of the data set used to build the model. The *Salmonella* reverse-mutation assay, also known as the Ames test, is the most widely accepted mutagenicity assay [9–11]. This assay uses a set of genetically modified *Salmonella* strains that revert to a wild-type phenotype after exposure to a mutagenic agent. Since the Ames test is considered the gold standard against which all other mutagenicity assay results are evaluated, this is the ideal data for building predicative mutagenicity models. There are several publicly available databases containing Ames data [12–15] that may be suitable for modeling (Web links can be found in the references). To overcome resource limitations that often curtail the amount of data available to build robust models, individuals have turned to results from higher throughput mutagenicity screening assays to augment or serve as a surrogate for Ames data [16–18]. These screening assays were designed for a high correlation to the Ames test, but in reality they measure a different mechanism of action, which is the induction of reporter gene activity instead of a change in DNA sequence [19–22]. The model developer must therefore weigh the risk of modeling a surrogate mutagenic endpoint with a different mechanism of action as compared to using a limited but specific Ames data set.

14.4 COMMERCIALLY AVAILABLE SOFTWARE

There are several commercially available mutagenicity prediction programs representing the three major model systems (QSAR, expert, and hybrid systems), as well as newly evolving methods. This chapter will focus on the three most widely used predictive mutagenicity programs: TOPKAT (Toxicity Prediction by Komputer Assisted Technology), DEREK (Deductive Estimation of Risk from Existing Knowledge), and MCASE (Multiple Computer Automated Structure Evaluation). TOPKAT was created by Health Systems Inc., but is now developed and marketed by Accelrys (*http://www.accelrys.com/products/topkat/*). DEREK was originally created by Schering Agrochemical Company and then donated to Lhasa Limited, a nonprofit organization at the School of Chemistry, University of Leeds, UK (*http://www.lhasalimited.org/index.php*). Development of this program has been driven by a strong collaboration of representatives from academia, industry, and government. This collaboration has facilitated the sharing of information to enhance the predictive toxicity models, while still preserving confidentiality of each organization's proprietary data. MCASE/MC4PC (MCASE for Personal Computer) was developed by Gilles Klopman and his associates in the Department of Chemistry, Case Western University, OH. This program is now developed and marketed by MultiCASE Inc., a private company directed by Klopman and Case Western. All the programs listed above are considered global models because they were developed using compounds of diverse structure (noncongeneric) representing more than one biological mechanism. It is worthwhile to note that these models were designed using data training sets composed largely of literature data from non-pharmaceutical chemicals. The implication of this is that the user must consider whether a negative prediction is due to a limitation of the chemical space encompassed (did the training set contain sufficient compounds of similar structure to the one being analyzed). Additionally, these models cannot generate predictions for chemical mixtures or distinguish between chiral compounds because of their two-dimensional chemical descriptors basis.

14.4.1 TOPKAT

TOPKAT is a QSAR-based prediction system that uses 2D Kier and Hall chemical descriptors to predict mutagenic activity [23]. This system models the electrotopological state (*E*-values) and shape of the molecule to render the compound as a set of numerical descriptors, and generates endpoint predictions based on probability of mutagenicity ranging from 0 to 1 (100%). No mutagenic predictions can be generated for compounds containing inorganic atoms. TOKAT allows for some estimation of prediction reliability through a process called optimum prediction space (OPS) and through similarity searching. OPS analysis determines if the unknown compound is within the chemical space of the training set used to build the model. If the unknown compound is too dissimilar from compounds within the training set, the program is unable to generate a mutagenicity prediction. Because TOPKAT is a closed system, the user cannot augment the training set with structurally similar proprietary compounds of known mutagenicity. For added interpretation, a similarity search will show the user which compounds within the training set are most similar to the unknown compound. Unfortunately, because TOPKAT is a QSAR system based on electrotopological descriptors, a similarity search may return very structurally diverse compounds although they may be similar electrotopologically. In this situation it is very difficult to determine which portion of the molecule should be changed to reduce or eliminate mutagenic activity.

14.4.2 DEREK

DEREK (also see Chapter 18) is an expert or rule-based system with a list of chemical substructures correlated with mutagenic activity [24]. The model was built using published mutagenic mechanism-of-action data and a list of known mutagenic structural alerts derived from Ashby's list of DNA reactive electrophilic chemical fragments [2,24,25]. Because this is a rule-based system, the program generates a binary prediction: mutagenic or not mutagenic, along with information surrounding the origin of the rule (mode-of-action details, published references of mutagenic compounds containing the alerting fragment). The program only identifies "activating" structures not "deactivating," so negative predictions are based solely upon a lack of alerting substructures within the model. For these reasons the program's ability to accurately identify mutagenic activity (sensitivity) in unknown compounds depends on the inclusion of that particular mutagenic fragment or mechanism of action within the model's training-set. In contrast to TOPKAT, a closed system, DEREK is an open system that allows the user to add to the rules in the training set. Unlike TOPKAT, DEREK also generates a prediction for most unknown compounds, regardless of how dissimilar the unknown compound may be compared to the training set of molecules. For this reason care must be taken to determine if a negative prediction is due to a gap in the training set of similar compounds or is due to the fact that there are no known "activating" fragments or any known possible mutagenic mechanisms of action.

14.4.3 MCASE/MC4PC

MCASE (also see Chapter 18) is a hybrid model based on initially creating a list of chemical substructures (biophores) associated with mutagenicity (expert rule-base portion). The chemical descriptors (QSAR portion) of the identified substructures are then statistically analyzed to generate a quantitative prediction of mutagenic activity [26]. The chemical substructures represent contiguous 2 to 10 atom fragments that are present in active and inactive compounds. This hybrid strategy provides an advantage over straight rule-based systems in that both "activating" and "deactivating" biophores are incorporated into the quantitative prediction. MCASE provides predictions in the form of numerical CASE units. These CASE units represent the scaled mutagenic activity over the entire range of the training set. Because this is an open system that allows addition of compounds to the training set, users may wish to use interpretation values other than the suggested default values supplied with the software. Generically, activity values can be placed into three-tiers, <30 is considered inactive, 30–50 marginally active, and >50 active.

Under a Cooperative Research and Development Agreement (CRADA) between FDA and MultiCASE, new database modules have been developed for MCASE that contain an expanded training-set of compounds representing proprietary compounds from chemicals submitted to the agency for review and some mutagenic compounds that failed during development at pharmaceutical companies [27,28].

14.5 SENSITIVITY MODEL COMPARISONS

A hurdle to prospectively evaluating performance of predictive software packages lies in identifying sufficient data to test the system that were not used in

TABLE 14.1 Performance of Individual Programs to Predict Bacterial Mutagenicity of Pharmaceutical Development Compounds

	Test-Set	Sensitivity	Specificity	Concordance	Reference
TOPKAT		43% (10/23)	85% (257/316)	82% (277/339)	[24]
DEREK	375	52% (14/27)	75% (260/346)	74% (274/372)	
MCASE		48% (13/27)	93% (370/330)	90% (320/357)	
TOPKAT		63% (32/51)	76% (204/268)	74% (236/319)	[25]
DEREK	520	28% (21/76)	80% (353/444)	72% (374/520)	
CASETOX[a]		50% (16/32)	86% (156/181)	81% (172/213)	
DEREK	974	46% (41/90)	62% (547/882)	60% (588/972)	[23]
MCASE		30% (19/63)	84% (555/660)	79% (574/723)	
TOPKAT	416	40% (21/53)	66% (201/303)	73% (222/303)	[26]
DEREK		46% (38/82)	55% (226/409)	65% (264/409)	

[a] CASETOX is the close-ended version of MCASE.

construction of the model. Any test set containing a large proportion of the same compounds used to build the model would inherently result in the overestimation of prediction ability. Within the proprietary databases of drug companies lies a wealth of bacterial mutagenicity data not used to construct current commercially available software programs. In practice, only about 10% of pharmaceutical compounds tested are bacterial mutagens [29,30]. This low frequency represents a tough challenge for the predictive programs. Any system that arbitrarily assigns all compounds as negative would achieve around 90% overall concordance. Therefore the most conservative program would automatically appear to have the best overall concordance. Several researchers have evaluated TOPKAT, DEREK, and MCASE for mutagenic predictive ability using proprietary in-house bacterial mutagenicity data gathered during the drug development process (Table 14.1) [29–32].

All the predictive databases were found to have poor sensitivity for the pharmaceutical (low mutagen frequency) test sets. The sensitivities ranged from 40% to 63%, depending on the data set and software program with one exception: White et al. reported a low 28% sensitivity for DEREK. White hypothesized that the low sensitivity for this 520 compound data set was due to multiple occurrences, within the data set, of several closely related series of compounds that were repeatedly incorrectly predicted by DEREK. This demonstrates why caution must be exercised in how data from these programs is interpreted. DEREK generates a negative prediction whenever there is no rule indicating mutagenic activity, regardless of how dissimilar a compound may be from those in the training set.

There are several methods by which sensitivity can be increased. One way is to increase the number/diversity of compounds in the training-set and/or add modifications to some more generic DEREK alerts. MCASE designers recently published that dividing the general mutagenic alert model into 15 individual models, and increasing the number of training-set compounds, increased the percentage of mutagens identified to 94% but slightly decreased overall concordance [33]. At Bristol-Myers Squibb the aromatic amine alert was refined, and additional alerts were added to DEREK to supplement the program using in-house data to enrich the training-set. As observed with MCASE, sensitivity increased, but overall concordance decreased. Another way sensitivity can increase is when multiple programs agree on the prediction, although the number of indeterminate compounds increased as the number of programs used increased [29,34,35]. Some of the more commonly mis-predicted compounds are those that are reported to require metabolic activation for mutagenic activity. Although the training sets of these programs contain data from assays where metabolic activation was present, without using a metabolite-generating program (e.g., METEOR for DEREK), the predictive software does not evaluate potential metabolites formed from the parent compound.

It is unrealistic to expect greater accuracy from predictive programs than is found within the experimental data being modeled. When judging the predictive ability of a particular program or programs, consideration must be given to the reliability of the program and the inherent variability of the assay it is modeling. A strict positive-versus-negative analysis of inter- and intra-laboratory bacterial mutagenicity data indicated 85% reproducibility [36]. This reproducibility was determined when the same compounds were tested in different laboratories, or in the same laboratory on different dates without knowledge of the prior test. Compared to the 85% reliability of the biological assay being modeled, commercial software models produced between 60% and 90% overall concordance depending upon the dataset being evaluated (Table 14.1).

14.6 CONCLUSIONS

In drug development the goal is to identify failures early so that resources can be redirected to a viable drug candidate. As part of this algorithm to identify such failures, computational software has been widely incorporated by pharmaceutical companies to help manage resource allocation through more effective setting of testing priorities. All of the mutagenic predictive performance data presented in this chapter used "out-of-the-box" programs and had poor sensitivity (40%–63%) for the pharmaceutical (low mutagen frequency ~10%) test data sets. These programs were more robust when predicting compounds with limited structural diversity. Sensitivity can be enhanced by increasing the number and diversity of training-set compounds, or by using the programs in combination (consensus) and only accepting predictions when two or more systems were in agreement. A pitfall in using this type of consensus modeling is that it dramatically increases the number of indeterminate compounds.

The lack of sensitivity of these predictive programs should only be a problem if this information alone is relied upon to advance a candidate compound or to make human safety decisions. During the continuum of the drug development process, as interest in a compound increases, so does the amount of testing information. Therefore no molecule of interest would ever progress based solely on computational outcomes, but rather compounds of interest that contain mutagenic or chemical descriptor alerts would receive priority in vitro testing earlier in the development process. Therefore predictive mutagenicity paradigms can serve a useful purpose for rationally prioritizing work in an era of limited resources available to evaluate numerous drug candidates.

REFERENCES

1. DiMasi JA, Hansen RW, Grabowski HG. The price of innovation: New estimates of drug development costs. *J Health Econ* 2003;22:151–85.
2. Ashby J. Fundamental structural alerts to potential carcinogenicity or noncarcinogenicity. *Environ Mutagen* 1985;7:919–21.
3. Ashby J, Tennant RW. Chemical structure, *Salmonella* mutagenicity and extent of carcinogenicity as indicators of genotoxic carcinogenesis among 222 chemicals tested in rodents by the U.S. NCI/NTP. *Mutat Res* 1988;204:17–115.
4. Ashby J, Tennant RW. Definitive relationships among chemical structure, carcinogenicity and mutagenicity for 301 chemicals tested by the U.S. NTP. *Mutat Res* 1991;257:229–306.
5. Tennant RW, Ashby J. Classification according to chemical structure, mutagenicity to *Salmonella* and level of carcinogenicity of a further 39 chemicals tested for carcinogenicity by the U.S. National Toxicology Program. *Mutat Res* 1991;257:209–27.
6. Votano JR, Parham M, Hall LH, Kier LB, Oloff S, Tropsha A, Xie Q, Tong W. Three new consensus QSAR models for the prediction of Ames genotoxicity. *Mutagenesis* 2004;19:365–77.
7. Patlewicz G, Rodford R, Walker JD. Quantitative structure-activity relationships for predicting mutagenicity and carcinogenicity. *Environ Toxicol Chem* 2003; 22:1885–93.
8. Benigni R. Structure-activity relationship studies of chemical mutagens and carcinogens: Mechanistic investigations and prediction approaches. *Chem Rev* 2005;105:1767–800.
9. Mortelmans K, Zeiger E. The Ames *Salmonella*/microsome mutagenicity assay. *Mutat Res* 2000;455:29–60.
10. Krewski D, Leroux BG, Bleuer SR, Broekhoven LH. Modeling the Ames *Salmonella*/microsome assay. *Biometrics* 1993;49:499–510.
11. Kier LD. Use of the Ames test in toxicology. *Regul Toxicol Pharmacol* 1985;5:59–64.
12. Matthews E, Kruhlak N, Cimino M, Benz R, Contrera J. An analysis of genetic toxicology, reproductive and developmental toxicity, and carcinogenicity data: I.

identification of carcinogens using surrogate endpoints. <http://www.fda.gov/cder/Offices/OPS_IO/genrepcar.htm>. *Regul Toxicol Pharm* 2006;44:83–96.

13. National Library of Medicine US EPA GENE-TOX database <http://toxnet.nlm.nih.gov/cgi-bin/sis/htmlgen?GENETOX>, EPA (Environmental Protection Agency), 2006. <http://toxnet.nlm.nih.gov/cgi-bin/sis/htmlgen?GENETOX>.
14. *2006 Physicians' Desk Reference (PDR)*. Thomson PDR, 2006.
15. Gold L, Zeiger E, Editors. *Handbook of carcinogenic potency and genotoxicity databases*, Boca Raton, FL: CRC Press, 1997.
16. Mattioni BE, Kauffman GW, Jurs PC, Custer LL, Durham SK, Pearl GM. Predicting the genotoxicity of secondary and aromatic amines using data subsetting to generate a model ensemble. *J Chem Inf Comput Sci* 2003;43:949–63.
17. Mosier PD, Jurs PC, Custer LL, Durham SK, Pearl GM. Predicting the genotoxicity of thiophene derivatives from molecular structure. *Chem Res Toxicol* 2003; 16:721–32.
18. Mersch-Sundermann V, Klopman G, Rosenkranz HS. Chemical structure and genotoxicity: Studies of the SOS chromotest. *Mutat Res* 1996;340:81–91.
19. Verschaeve L, Van Gompel J, Thilemans L, Regniers L, Vanparys P, van der Lelie D. VITOTOX bacterial genotoxicity and toxicity test for the rapid screening of chemicals. *Environ Mol Mutagen* 1999;33:240–8.
20. Quillardet P, Hofnung M. The SOS chromotest: A review. *Mutat Res* 1993; 297:235–79.
21. Fluckiger-Isler S, Baumeister M, Braun K, Gervais V, Hasler-Nguyen N, Reimann R, Van Gompel J, Wunderlich HG, Engelhardt G. Assessment of the performance of the Ames II assay: A collaborative study with 19 coded compounds. *Mutat Res* 2004;558:181–97.
22. Reifferscheid G, Heil J. Validation of the SOS/umu test using test results of 486 chemicals and comparison with the Ames test and carcinogenicity data. *Mutat Res* 1996;369:129–45.
23. Kier LB, Hall LH. An electrotopological-state index for atoms in molecules. *Pharm Res* 1990;7:801–7.
24. Sanderson DM, Earnshaw CG. Computer prediction of possible toxic action from chemical structure: The DEREK system. *Hum Exp Toxicol* 1991;10:261–73.
25. Ridings JE, Barratt MD, Cary R, Earnshaw CG, Eggington CE, Ellis MK, Judson PN, Langowski JJ, Marchant CA, Payne MP, Watson WP, Yih TD. Computer prediction of possible toxic action from chemical structure: an update on the DEREK system. *Toxicology* 1996;106:267–79.
26. Klopman G, Frierson MR, Rosenkranz HS. Computer analysis of toxicological data bases: mutagenicity of aromatic amines in *Salmonella* tester strains. *Environ Mutagen* 1985;7:625–44.
27. Matthews EJ, Contrera JF. A new highly specific method for predicting the carcinogenic potential of pharmaceuticals in rodents using enhanced MCASE QSAR-ES software. *Regul Toxicol Pharmacol* 1998;28:242–64.
28. Matthews EJ, Kruhlak NL, Cimino MC, Benz RD, Contrera JF. An analysis of genetic toxicity, reproductive and developmental toxicity, and carcinogenicity data: II. Identification of genotoxicants, reprotoxicants, and carcinogens using in silico methods. *Regul Toxicol Pharmacol* 2006;44:97–110. Epub 2005 Dec 2013.

29. White AC, Mueller RA, Gallavan RH, Aaron S, Wilson AGE. A multiple in silico program approach for the prediction of mutagenicity from chemical structure. *Mutat Res/Gen Toxicol Environ Mutagen* 2003;539:77–89.

30. Greene N. Computer systems for the prediction of toxicity: An update. *Adv Drug Deliv Rev.* 2002;54:417–31.

31. Snyder RD, Pearl GS, Mandakas G, Choy WN, Goodsaid F, Rosenblum IY. Assessment of the sensitivity of the computational programs DEREK, TOPKAT, and MCASE in the prediction of the genotoxicity of pharmaceutical molecules. *Environ Mol Mutagen* 2004;43:143–58.

32. Cariello NF, Wilson JD, Britt BH, Wedd DJ, Burlinson B, Gombar V. Comparison of the computer programs DEREK and TOPKAT to predict bacterial mutagenicity: Deductive estimate of risk from existing knowledge. Toxicity prediction by computer assisted technology. *Mutagenesis* 2002;17:321–9.

33. Klopman G, Zhu H, Fuller MA, Saiakhov RD. Searching for an enhanced predictive tool for mutagenicity. *SAR QSAR Environ Res* 2004;15:251–63.

34. Durham SK, Pearl GM. Computational methods to predict drug safety liabilities. *Curr Opin Drug Discov Devel* 2001;4:110–5.

35. Dearden JC. In silico prediction of drug toxicity. *J Comput Aided Mol Des* 2003;17:119–27.

36. Zeiger E, Ashby J, Bakale G, Enslein K, Klopman G, Rosenkranz HS. Prediction of *Salmonella* mutagenicity. *Mutagenesis* 1996;11:471–84.

15

NOVEL APPLICATIONS OF KERNEL–PARTIAL LEAST SQUARES TO MODELING A COMPREHENSIVE ARRAY OF PROPERTIES FOR DRUG DISCOVERY

Sean Ekins, Mark J. Embrechts, Curt M. Breneman, Kam Jim, and Jean-Pierre Wery

Contents

Computational Toxicology: Risk Assessment for Pharmaceutical and Environmental Chemicals,
Edited by Sean Ekins

15.1 INTRODUCTION

We are currently witnessing a movement toward the wider utilization of various computational technologies at all stages in the process of drug discovery. For instance, various computational scoring methods for predicting ligand-protein interactions have been applied to docking or de novo growth in the binding site of therapeutic proteins [1–15]. Drug discovery requires more than binding to the target protein, namely molecules must be readily synthesizable with favorable molecular properties or what has been termed "drug likeness" [16–29]. Drug likeness studies are a clear attempt to understand the chemical properties that make molecules either successful or expensive clinical failures. Similarly the contribution of molecular properties which influence absorption, distribution, metabolism, excretion, and toxicity (ADME/TOX) are recognized alongside therapeutic potency as key determinants of whether a molecule can be successfully developed as a drug [20,30–39]. Many of these properties are used later as lead selection criteria that can be predicted before molecules are synthesized, purchased, or even tested in order to improve overall lead quality. Although some groups have used relatively simple filters like the rule of 5 (incorporating molecular weight, hydrogen bond donors, hydrogen bond acceptors, and logP) [40] to limit the types of molecules evaluated with other computational methods, or to focus libraries for high throughput screening (HTS), the proactive use of further computational models for ADME/TOX properties have been less widely described in conjunction with lead discovery [37,41].

Considerable research has focused on novel machine learning or computational algorithms that can be used for drug discovery for predicting these molecular properties. Such calculations can be performed with very large numbers of molecules and act as a molecule selection filter. Comparative molecular fields analysis (CoMFA) and pharmacophore approaches [42–53] have been used to model cytochrome P450 (CYP) enzymes involved in drug metabolism as well as transporters such as P-glycoprotein [54–56] and other transporters [57–59], nuclear hormone receptors [60,61], and ion channels [62–64] important for drug–drug interactions (see Chapters 8–13). These approaches have been rarely used to model more complex processes such as absorption, bioavailability, and clearance processes [65,66], which have required other more robust algorithms capable of dealing with larger data matrices [67–79] and hundreds to thousands of descriptors [80,81]. Recursive partitioning methods have been used extensively with these types of large sets of molecules and either continuous [82,83] or binary data, for therapeutic target endpoints as well as CYP inhibition [84] and toxicity properties such as AMES mutagenicity status [85]. Neural networks have also been used to predict CYP3A4 inhibition [86], human intestinal absorption [87], and human pharmacokinetic data [79,88]. Genetic algorithms have been used to predict metabolism in general to some extent [89]. Smaller quantitative structure activity relationship (QSAR) data sets have been used with inductive logic programming as a pharmacophore approach [90] but do not appear to have been used with ADME/TOX data sets to date. Kohonen self-organizing maps have been recently applied to model cytochrome P450-mediated drug metabolism [91], while *k*-nearest neighbors has been used to predict metabolic stability [92]. Genetic programming has also been applied to QSAR studies in which model complexity is controlled by a penalty function [93], although there are few published applications for ADME/TOX modeling to our knowledge.

At present, the computational method garnering the most interest for potential wider drug discovery applications is support vector machines (SVM) that have been adopted in the biosciences for pattern classification of cancer tissues [94], microarray gene expression [95], protein structure prediction [96], chromatography retention times [97], and structure activity relationships [98–102]. In most cases SVM have outperformed recursive partitioning methods [101] and provided similar results to artificial neural networks [103]. SVM has also been used successfully for active learning with bioactivity data for thrombin and CDK2 therapeutic targets [100,104]. The algorithm and general application of this technology has, however, also been patented by Vapnik and others [105,106], and this may ultimately represent a limitation of its commercial applications. To date there are limited studies where SVM have been applied to ADME/TOX properties like aqueous solubility and drug likeness [102,107] (see also Chapter 8).

In contrast to SVM, partial least squares (PLS) has been widely applied for QSAR studies in which there are relatively few data points but many descriptors [73,108–113]. Rosipal and Trejo have applied PLS with nonlinear regression and a kernel function [114] to produce K-PLS. Recently a number of machine learning approaches including PLS, SVM, K-PLS, and Kernel Ridge Regression have all been implemented in a single piece of software and used with several benchmark data sets [99]. One of the data sets used was a protein binding regression model [115] in which 94 molecules had 511 MOE and wavelet descriptors calculated. The K-PLS based model produced similar q^2 statistics and had faster execution times than the SVM models used for comparison [99]. These results with K-PLS indicate that it could be favorably applied to other data sets to enable QSAR model construction and aid drug discovery research. K-PLS is also without the potential impediment of patents to hinder exploitation and represents a viable alternative to SVM.

We have used transferable atom equivalent (TAE) descriptors [116,117] that encode the distributions of electron density based molecular properties, such as kinetic energy densities, local average ionization potentials, Fukui functions, electron density gradients, and second derivatives as well as the density itself. In addition autocorrelation descriptors (RAD) were used and represent the molecular geometry characteristics of the molecules, while they are also canonical and independent of 3D coordinates. The 2D descriptors alone or in combination with the latter 3D descriptors were calculated for 26 data sets collated by us from numerous publications. These data sets encompass various ADME/TOX-related enzymes, transporters, and ion channels as well as more complex processes and data previously used for modeling protein–ligand interactions. This comprehensive series of models ultimately provides a means to evaluate the wider utility of nonlinear K-PLS compared with the linear PLS method.

15.2 METHODS

15.2.1 Literature Data

Two data sets of percent inhibition data for recombinant CYP2D6 (1759 molecules) and recombinant CYP3A4 (1756 molecules) were purchased from Cerep (Redmond, WA) [84]. A set of 28 IC_{50} values for inhibition of recombinant CYP3A4 [118] and a set of 76 K_i values for inhibition of human CYP2C9 [110], data for the ester hydrolysis half-life in human blood of 71 molecules were all obtained from the literature [119]. A set of 484 IC_{50} values for inhibition of human and murine soluble epoxide hydrolase were kindly supplied by Professor Bruce Hammock (UC Davis, CA), and some of these data have been previously published [120]. A set 47 K_m values for human purified dopamine sulfotransferase (SULT 1A3) was obtained from the literature [121]. A P-gp data set for 192 substrates and nonsubstrates [122], 35 P-gp inhibitors (IC_{50}) of digoxin transport in Caco-2 cells [56], 100 ligands (K_d) for the noradrenaline transporter [123,124], 101 ligands (K_d) for the serotonin transporter [123,124], and 30 inhibitors (IC_{50}) of human organic cation transporter (hOCT1) [54] were all collated from the literature.

A set of 95 molecules with human serum binding data [115], 265 molecules with percent protein binding [125], 106 molecules with blood–brain barrier data [126], 183 molecules with percent fraction absorbed for humans [127–131], and 80 molecules with percent fraction absorbed for rat [127,128] were selected from the literature. The PHYSPROP database (Sept 2002 version, Syracuse Research Corporation) was used to extract 2868 molecules with experimentally determined aqueous solubility. Data for 75 drugs with human intrinsic clearance unbound [132], 116 molecules with K_d values [2] extracted from the Protein Data Bank, 110 drugs which are either aggregators or nonaggregators in vitro [133–135], 270 drugs with volume of distribution data in humans [125], 101 molecules with $\log D$ data [136,137], 2017 molecules with Ames mutagenicity data [85], and 99 molecules with human ether-a-go-go-related gene (hERG) IC_{50} patch clamping data [62,138,139] were all extracted from various literature sources. All activity data units were converted to log data when appropriate unless otherwise described. The 2D molecular structures were then generated in ChemDraw (CambridgeSoft, Cambridge, MA).

15.2.2 RECON Descriptors

Transferable atom equivalent (TAE) descriptors encode the distributions of electron density based molecular properties, such as kinetic energy densities, local average ionization potentials, Fukui functions, electron density gradients, and second derivatives as well as the density itself. Autocorrelation descriptors (RAD) represent the molecular geometry characteristics of the molecules, are canonical and independent of 3D coordinates. These descriptors and their calculation have been previously described in detail [116,117]. The 248 TAE and/or 135 RAD descriptors were routinely generated for molecules in training and testing sets after input of structures and activity data in an SDF file incorporating explicit hydrogens using RECON 5.5b and RECON 5.8, respectively. TAE descriptors were generated alone as well as with a combination of descriptor types.

15.2.3 Data Preprocessing

The matrix of molecular descriptors and biological activity data was scaled and variables with unchanging values were removed using feature selection with the StripMiner/Analyze software previously described [140,141]. From the descriptors with more that 95% correlation between each other (i.e., "cousin descriptors"), only the descriptor most correlated with the response was retained. In addition 4 sigma outliers were brought within 2.5 sigma.

15.2.4 K-PLS Modeling Method

The Analyze software uses the Kernel PLS method [114] with two key parameters, the number of latent variables and sigma. In this study these values were fixed at 5 and 10, respectively. K-PLS uses kernels and can therefore be seen as a nonlinear extension of the PLS method. The commonly used radial basis function kernel or Gaussian kernel was applied, where the kernel is expressed as [142]

$$K(\vec{x}, \vec{x}_i) = e^{-\|\vec{x}-\vec{x}_i\|^2/2\sigma^2}.$$

The K-PLS method can be reformulated to resemble support vector machines, but it can also be interpreted as a kernel and centering transformation of the descriptor data followed by a regular PLS method [99]. K-PLS was first introduced by Lindgren, Geladi, and Wold [143] in the context of working with linear kernels on data sets with more descriptor fields than data, in order to make the PLS modeling more efficient. Early applications of K-PLS were done mainly in this context [144–146]. The Parzen window, σ, in the formula above is a free parameter that is determined by hyper-tuning on a validation set. For each dataset σ is then held constant, independent of the various bootstrap splits.

15.2.5 Cross Validation

Analyze software enabled bootstrapping leave-out validation representing 10% or 50% of the training set with 10 or 100 random holdout groups. The standard q^2 and Q^2 statistics were calculated for the correlation of predicted and observed log IC_{50} data, where q^2 is defined as the correlation coefficient squared for the test set and Q^2 is defined according to

$$Q^2 = 1 - \frac{\sum_{i=1}^{n_{train}} (y_i - \hat{y}_i)^2}{\sum_{i=1}^{n_{train}} (y_i - \bar{y})^2},$$

where the y_i values are observations for an activity, and the corresponding $\hat{y}_i$ values are the predictions for an external test set.

A larger q^2 value is preferable and indicative of a good internally validated model. Data were also scrambled, whereby the descriptors and responses no longer corresponded. Analyze was used to then leave out 50% of the database 10 times prior to calculating the q^2 value.

15.2.6 Molecule Predictions

External test sets for some of the models were also selected from various sources. The CYP2D6 and CYP3A4 percent inhibition models were tested with previously published data for 100 molecules [147]. The blood–brain barrier data were tested with in vivo mouse brain penetration data for 110 molecules [148], and the human fraction absorbed model was tested with literature data not in the training set [149].

15.2.7 Confidence Estimation

We estimated the confidence in the K-PLS predictions using approaches called novelty detection and margin detection, which are two complementary methods. Reliable confidence estimates enable the ability to screen out uncertain predictions and to focus on high-confidence predictions. The goal of novelty detection is to identify compounds that are significantly different from those seen previously in the training set. The intuition is that it is harder to make predictions on "novel" compounds that were not represented in the training set during learning. Novelty detection can be implemented using any algorithm that can learn the distribution of the training set—a compound is considered novel if it falls outside of that distribution (Figure 15.1*a*). Our implementation uses the One-Class Support Vector Machine (SVM) algorithm [150,151] to learn a boundary around the training set distribution. The tightness of the boundary is controlled by a user parameter μ, which specifies the percentage of the training set to leave outside of the boundary. This parameter can also be used to measure the degree of novelty of a given compound: simply find the highest value of μ such that the given compound would not be considered novel if a One-Class SVM was trained on the training set. In this scenario a lower value of μ implies a more novel compound. In contrast,

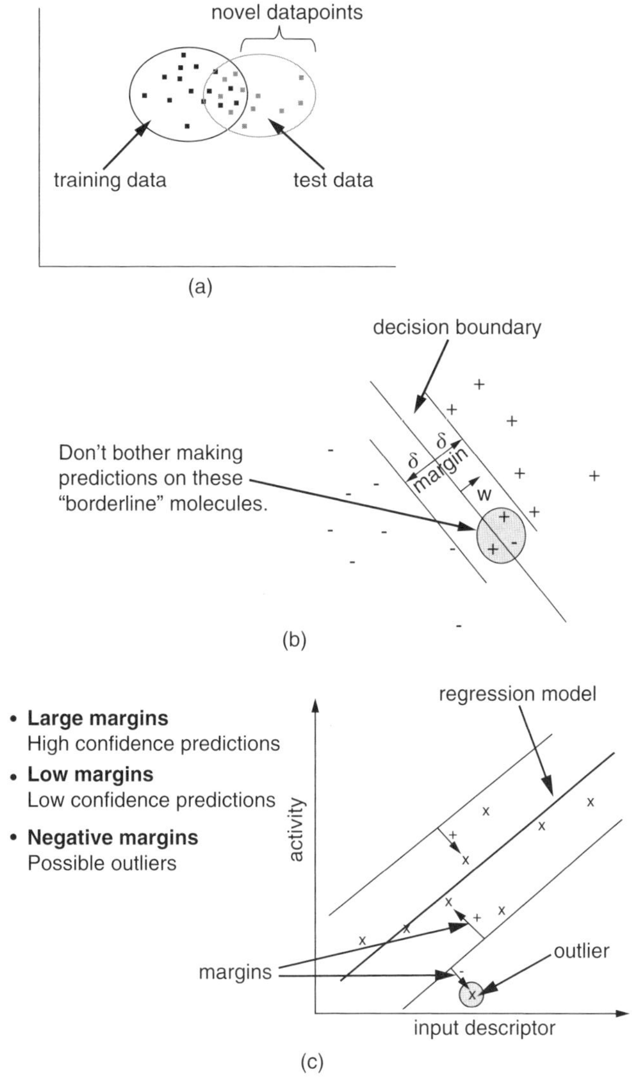

Figure 15.1 Figures to describe the basis of novelty and margin detection to provide a confidence measure for K-PLS predictions. (*a*) Novelty detection, (*b*) margin detection for classification, and (*c*) margin detection for regression.

margin detection does not require the assumption that predictions on novel compounds are less reliable, but instead uses the margin to directly measure the confidence of a prediction. In a classification task the margin is the distance of the prediction from the decision boundary—a larger margin implies higher confidence (Figure 15.1*b*). Determining the margin on a regression problem (i.e., learning a real-valued function) is significantly more difficult because it is necessary to estimate it from the margins of similar training points (Figure 15.1*c*). Instead, we can treat the problem like a classification task by assigning a classification cutoff, such as 40% inhibition, and by using the classification margin to determine the confidence of the predictions in terms of their ability to determine if a compound is below 40% inhibition or not.

15.3 RESULTS

Twenty-six ADME/TOX and bioactivity data sets were used with PLS and K-PLS and with 2D descriptors or a combination of 2D and 3D descriptors. The regression models were evaluated by measuring the q^2 value following leaving out either 10% or 50% of the molecules 100 times at random. The classification models were evaluated by measuring the area under the receiver operator curve (ROC) and then leaving out 10% of the molecules 100 times at random. The K-PLS models were additionally tested by scrambling the observed values and the descriptors 10 times at random, such that the q^2 becomes approximately zero, to indicate significance of the original model.

We found K-PLS to be superior to PLS with the large data sets for single point inhibition of CYP2D6 and CYP3A4 (Table 15.1, panels A, B) and there are only small changes in the q^2 statistic upon the addition of 3D descriptors. These K-PLS models were evaluated with a test set of 98 molecules not in the training sets, and Spearman's rho rank coefficients were calculated were the optimal value is 1. The CYP2D6 K-PLS model generated with 2D descriptors had a Spearman's rho of 0.37 ($p = 0.0001$), and the CYP2D6 model generated with 2D and 3D descriptors had a Spearman's rho of 0.44 ($p < 0.0001$). Similarly the CYP3A4 K-PLS model generated with 2D descriptors had a Spearman's rho of 0.59 ($p < 0.0001$), and the CYP3A4 model generated with 2D and 3D descriptors had a Spearman's rho of 0.62 ($p < 0.0001$). These results are the inverse of what we have demonstrated previously with models generated with recursive partitioning in which CYP2D6 gave the best results with the test set [84]. Using a cutoff of 40% inhibition, we estimated the confidence in the K-PLS predictions using novelty detection (Figure 15.1*a*) and margin detection (Figure 15.1*b*, *c*) to screen out uncertain predictions and to focus on high-confidence predictions. The positive effect of this process on the CYP2D6 and CYP3A4 models can be shown as a receiver operator curve (ROC) plots (Figure 15.2*a*–*d*).

The K-PLS model for CYP3A4 generated with a small set of IC_{50} data suggested that it was inferior to the PLS models and the addition of further 3D descriptors had no effect (Table 15.1, panel C). Conversely, the K-PLS model for CYP2C9 K_i data was preferable to the respective PLS model and once again the addition of 3D descriptors had a negligible effect (Table 15.1, panel D). The K-PLS model for human blood ester hydrolysis demonstrated an improvement upon the addition of the 3D descriptors and had better statistics than the PLS models (Table 15.1, panel E). The large soluble epoxide hydrolase IC_{50} data sets for both human and mouse also showed the superiority of

TABLE 15.1 Enzymes

	r^2	q^2 for 10 × 100 (RMSE)	q^2 for 50 × 100	q^2 Scrambling
A. Human recombinant expressed CYP2D6 percent inhibition for 1759 molecules				
2D–K-PLS	0.59	0.50 (21.72)	0.48 (22.22)	0.00005
2D PLS	0.37	0.34 (24.92)	0.34 (24.95)	
2D & 3D–K-PLS	0.64	0.52 (21.39)	0.50 (21.85)	0.00007
2D & 3D–PLS	0.39	0.35 (24.76)	0.35 (24.79)	
B. Human recombinant expressed CYP3A4 percent inhibition for 1756 molecules				
2D–K-PLS	0.49	0.37 (19.84)	0.34 (20.36)	0.00005
2D PLS	0.32	0.29 (20.97)	0.27 (21.17)	
2D & 3D–K-PLS	0.57	0.38 (19.72)	0.34 (20.45)	0
2D & 3D-PLS	0.36	0.29 (20.97)	0.27 (21.17)	
C. Human recombinant expressed CYP3A4 log IC_{50} data for 28 molecules [118].				
2D–K-PLS	1.00	0.27 (1.10)	0.21 (1.15)	0.056
2D PLS	0.93	0.41 (1.05)	0.29 (1.18)	
2D & 3D–K-PLS	1.00	0.28 (1.09)	0.21 (1.14)	0.04
2D & 3D–PLS	0.98	0.41 (1.03)	0.29 (1.14)	
D. Human recombinant expressed CYP2C9 log K_i data for 76 molecules [110]				
2D–K-PLS	0.98	0.65 (0.55)	0.54 (0.65)	0.0019
2D PLS	0.76	0.35 (0.80)	0.39 (0.81)	
2D & 3D–K-PLS	0.99	0.64 (0.55)	0.52 (0.67)	0.002
2D & 3D–PLS	0.80	0.33 (0.83)	0.39 (0.80)	
E. Human blood ester hydrolysis log half-life for 71 molecules [119]				
2D–K-PLS	0.85	0.56 (0.55)	0.32 (0.68)	0.0007
2D PLS	0.70	0.44 (0.63)	0.20 (0.78)	
2D & 3D–K-PLS	0.94	0.61 (0.52)	0.35 (0.66)	0.00083
2D & 3D–PLS	0.76	0.45 (0.63)	0.22 (0.77)	
F. Mouse soluble epoxide hydrolase log IC_{50} data for 484 molecules				
2D–K-PLS	0.75	0.60 (0.97)	0.58 (1.01)	0.00006
2D PLS	0.53	0.47 (1.13)	0.46 (1.13)	
2D & 3D–K-PLS	0.81	0.64 (0.93)	0.61 (0.96)	0.00067
2D & 3D–PLS	0.58	0.50 (1.09)	0.50 (1.1)	
G. Human soluble epoxide hydrolase log IC_{50} data for 484 molecules				
2D–K-PLS	0.73	0.57 (0.91)	0.55 (0.93)	0.00002
2D PLS	0.51	0.44 (1.03)	0.44 (1.03)	
2D & 3D–K-PLS	0.80	0.61 (0.86)	0.58 (0.89)	0.00005
2D & 3D–PLS	0.57	0.48 (0.99)	0.48 (1.00)	
H. Human dopamine sulfotransferase 1A3 log K_m data for 47 molecules [121]				
2D–K-PLS	0.97	0.58 (0.59)	0.45 (0.69)	0.0039
2D PLS	0.83	0.48 (0.69)	0.33 (0.83)	
2D & 3D–K-PLS	1.00	0.62 (0.58)	0.44 (0.70)	0.0062
2D & 3D–PLS	0.92	0.55 (0.63)	0.38 (0.76)	

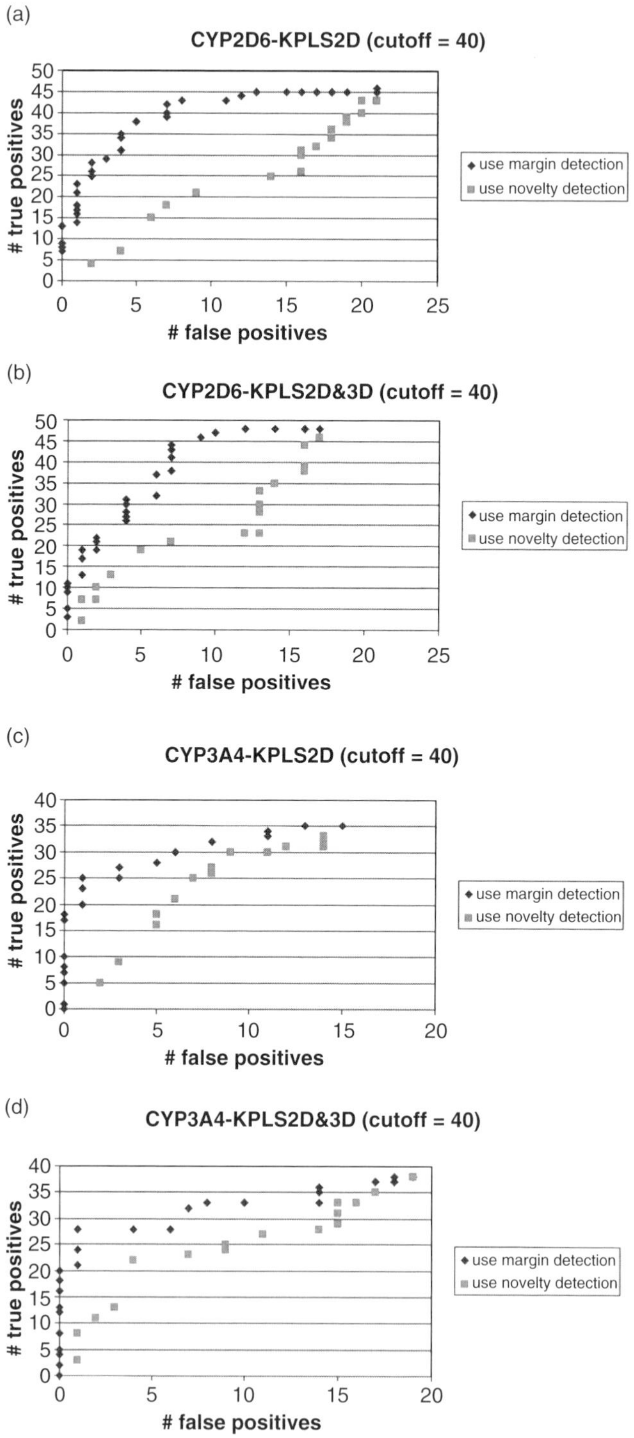

Figure 15.2 Novelty and margin detection applied to the 98 molecule test set for percent inhibition of CYP2D6 and CYP3A4. (*a*) CYP2D6 K-PLS model with 2D descriptors only. (*b*) CYP2D6 K-PLS model with 2D and 3D descriptors. (*c*) CYP3A4 K-PLS model with 2D descriptors only. (*d*) CYP3A4 K-PLS model with 2D and 3D descriptors.

2D and 3D descriptors with K-PLS and also illustrated that the more liberal leave out 50% 100 times behaved similarly (Table 15.1, panels F, G). The dopamine sulfotransferase K_m model showed a similar pattern to the epoxide hydrolase models (Table 15.1, panel H).

Various transporter proteins were assessed including P-glycoprotein (P-gp). The K-PLS classification models for P-gp substrates (Table 15.2, panel A) and K-PLS regression models for P-gp inhibitors (Table 15.2, panel B) were found to represent a slight improvement over PLS. The difference in favor of K-PLS is more pronounced for the noradrenaline transporter and serotonin transporter dissociation constant models (Table 15.2, panels C, D).

TABLE 15.2 Transporters

	10 × 100 ROC AUC	r^2	q^2 for 10 × 100 (RMSE)	q^2 for 50 × 100	q^2 Scrambling
A. P-gp data set for 192 substrates and nonsubstrates [122]					
2D–K-PLS	0.85				
2D PLS	0.82				
2D & 3D–K-PLS	0.85				
2D & 3D–PLS	0.80				
B. P-gp inhibition (log IC_{50}) of digoxin transport in Caco-2 cells for 35 molecules [56]					
2D–K-PLS		0.98	0.44 (0.80)	0.33 (0.90)	0.001
2D PLS		0.89	0.51 (0.77)	0.34 (0.98)	
2D & 3D–K-PLS		0.99	0.57 (0.69)	0.39 (0.86)	0.0032
2D & 3D–PLS		0.92	0.47 (0.82)	0.37 (0.96)	
C. Noradrenaline transporter log dissociation constants for 100 molecules [123,124]					
2D–K-PLS		0.95	0.42 (1.09)	0.39 (1.09)	0.00002
2D PLS		0.67	0.31 (1.23)	0.28 (1.26)	
2D & 3D–K-PLS		0.98	0.45 (1.04)	0.40 (1.06)	0.00007
2D & 3D–PLS		0.75	0.31 (1.25)	0.29 (1.25)	
D. Serotonin transporter log dissociation constants for 101 molecules [123,124]					
2D–K-PLS		0.93	0.42 (1.30)	0.33 (1.35)	0.0043
2D PLS		0.71	0.28 (1.50)	0.26 (1.53)	
2D & 3D–K-PLS		0.97	0.48 (1.23)	0.37 (1.30)	0.0041
2D & 3D–PLS		0.80	0.30 (1.50)	0.28 (1.52)	
E. hOCT1 log IC_{50} values for 30 molecules [54]					
2D–K-PLS		0.98	0.48 (0.65)	0.33 (0.74)	0.003
2D PLS		0.94	0.58 (0.58)	0.47 (0.67)	
2D & 3D–K-PLS		0.99	0.51 (0.64)	0.36 (0.73)	0.002
2D & 3D–PLS		0.98	0.68 (0.51)	0.57 (0.59)	

In marked contrast to the previous transporters, PLS and in particular the combination of 2D and 3D descriptors are preferable for the hOCT1 inhibitors (Table 15.2, panel E).

The PLS and K-PLS models for the small human serum binding data set were generally comparable (Table 15.3, panel A), whereas the trend for a

TABLE 15.3 Complex Processes

	10 × 100 ROC AUC	r^2	q^2 for 10 × 100 (RMSE)	q^2 for 50 × 100	q^2 Scrambling
A. Human serum binding log data for 95 molecules [115]					
2D–K-PLS		0.98	0.59 (0.38)	0.52 (0.44)	0.00064
2D PLS		0.79	0.56 (0.41)	0.52 (0.45)	
2D & 3D–K-PLS		0.99	0.56 (0.41)	0.47 (0.47)	0.00046
2D & 3D–PLS		0.84	0.58 (0.40)	0.55 (0.43)	
B. Percent protein binding data for 265 molecules [125]					
2D–K-PLS		0.83	0.48 (24.54)	0.44 (25.4)	0.00001
2D PLS		0.59	0.43 (25.52)	0.40 (26.39)	
2D & 3D–K-PLS		0.91	0.49 (24.05)	0.45 (24.92)	0.0002
2D & 3D–PLS		0.64	0.45 (25.09)	0.42 (25.99)	
C. Blood–brain barrier log data for 106 molecules [126]					
2D–K-PLS		0.93	0.57 (0.51)	0.51 (0.56)	0.00089
2D PLS		0.77	0.52 (0.55)	0.44 (0.62)	
2D & 3D–K-PLS		0.94	0.56 (0.52)	0.51 (0.55)	0.00086
2D & 3D–PLS		0.77	0.53 (0.54)	0.46 (0.60)	
D. Human percent fraction absorbed data for 183 molecules [127–131]					
2D–K-PLS		0.89	0.43 (24.46)	0.35 (26.79)	0.00004
2D PLS		0.61	0.42 (24.89)	0.35 (27.31)	
2D & 3D–K-PLS		0.95	0.55 (21.60)	0.44 (24.55)	0.00027
2D & 3D–PLS		0.65	0.43 (24.68)	0.35 (27.31)	
E. Rat percent fraction absorbed data for 80 molecules [127,128]					
2D–K-PLS		0.95	0.14 (29.77)	0.13 (29.45)	0.0017
2D PLS		0.76	0.33 (26.16)	0.22 (29.17)	
2D & 3D–K-PLS		0.99	0.28 (26.00)	0.20 (26.98)	0.0013
2D & 3D–PLS		0.82	0.35 (25.51)	0.23 (28.38)	
F. Experimentally determined aqueous solubility log data for 2868 molecules at 25°C[a]					
2D–K-PLS		0.77	0.73 (1.16)	0.73 (1.17)	0.00008
2D PLS		0.66	0.65 (1.32)	0.65 (1.34)	
2D & 3D–K-PLS		0.79	0.75 (1.12)	0.74 (1.14)	0.0003
2D & 3D–PLS		0.67	0.66 (1.32)	0.65 (1.33)	

TABLE 15.3 (*Continued*)

	10 × 100 ROC AUC	r^2	q^2 for 10 × 100 (RMSE)	q^2 for 50 × 100	q^2 Scrambling
G. 75 drugs with log human intrinsic clearance unbound drugs [132]					
2D–K-PLS		0.96	0.37 (0.63)	0.25 (0.69)	0.0026
2D PLS		0.66	0.32 (0.69)	0.21 (0.78)	
2D & 3D–K-PLS		0.99	0.36 (0.63)	0.24 (0.69)	0.0029
2D & 3D–PLS		0.73	0.29 (0.71)	0.21 (0.78)	
H. 116 molecules with log K_d values extracted from the Protein Data Bank [2]					
2D–K-PLS		0.83	0.46 (1.79)	0.44 (1.77)	0.002
2D PLS		0.61	0.31 (2.06)	0.29 (2.10)	
2D & 3D–K-PLS		0.91	0.52 (1.65)	0.49 (1.67)	0.002
2D & 3D–PLS		0.68	0.34 (2.02)	0.33 (2.03)	
I. 110 drugs which are either aggregators or non-aggregators in vitro [133–135]					
2D–K-PLS	0.89				
2D PLS	0.87				
2D & 3D–K-PLS	0.92				
2D & 3D–PLS	0.89				
J. 270 drugs with log volume of distribution data in humans [125]					
2D–K-PLS		0.72	0.30 (0.60)	0.24 (0.64)	0.00009
2D PLS		0.43	0.22 (0.62)	0.19 (0.66)	
2D & 3D–K-PLS		0.84	0.27 (0.61)	0.22 (0.64)	0.00011
2D & 3D–PLS		0.47	0.22 (0.63)	0.18 (0.67)	
K. 101 molecules with log D data [136, 137]					
2D- K-PLS		0.95	0.44 (1.32)	0.35 (1.42)	0.0009
2D PLS		0.71	0.49 (1.28)	0.40 (1.41)	
2D & 3D–K-PLS		0.97	0.48 (1.29)	0.38 (1.39)	0.001
2D & 3D–PLS		0.81	0.55 (1.19)	0.47 (1.30)	

[a] PHYSPROP database, Sept. 2002 version, Syracuse Research Corporation.

larger human protein binding data set favored K-PLS (Table 15.3, panel B). Similarly K-PLS models for predicting the blood–brain barrier permeation (Table 15.3, panel C) and human fraction absorbed (Table 15.3, panel D) showed K-PLS to slightly outperform PLS. The blood–brain barrier permeation K-PLS models were also tested with an external test set of 110 molecules with binary data, and novelty and margin detection were used after selecting a cutoff of zero (Figure 15.3*a*,*b*). The data show a positive shift to the left with novelty detection. The human fraction absorbed K-PLS models (Table 15.3, panel D) were also evaluated with a test set of 35 molecules not in the training set, and Spearman's rho rank coefficients were calculated. The Spearman's

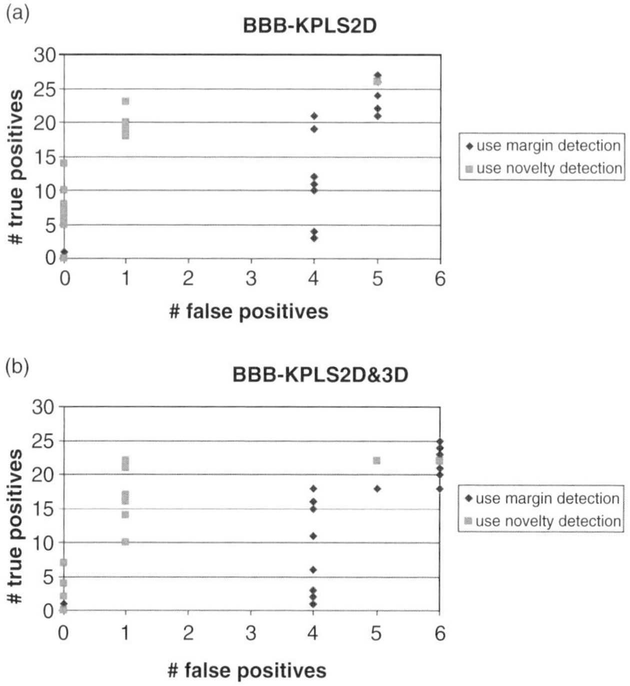

Figure 15.3 Novelty and margin detection applied to the 110 molecule test set for blood–brain barrier permeation (*a*) using 2D descriptors alone or (*b*) 2D and 3D descriptors.

rho for the predictions with 2D descriptors (0.33, $p = 0.047$) were slightly better than for the less significant combined 2D and 3D descriptor models (0.31, $p = 0.07$). Novelty and margin detection were used with this test set after selecting a cutoff of 40%, which appears to show novelty detection to be valuable when using K-PLS with the 3D descriptors (Figure 15.4*a*, *b*). A smaller fraction absorbed data set for rat gave noticeably poorer statistics overall, although PLS resulted in reasonable models (Table 15.3, panel E). A very large set of molecules with aqueous solubility values enabled very good q^2 statistics and RMSE values around 1 log unit for all models generated and demonstrated a marked improvement of K-PLS over PLS (Table 15.3, panel F). A small set of 75 drugs with human intrinsic clearance unbound data also showed K-PLS to be preferable (Table 15.3, panel G), while a set of molecules derived from PDB crystal structures and their in vitro binding affinities illustrated the importance of combining 2D and 3D descriptors with K-PLS. Despite the RMSE values being high, they were comparable to what has been attained in the literature with other methods (Table 15.3, panel H). The classification data set derived to predict the recently described phenomenon of molecules aggregating and causing false positives in high-throughput screens, also indicated that combining 2D and 3D descriptors with K-PLS gave slightly better results

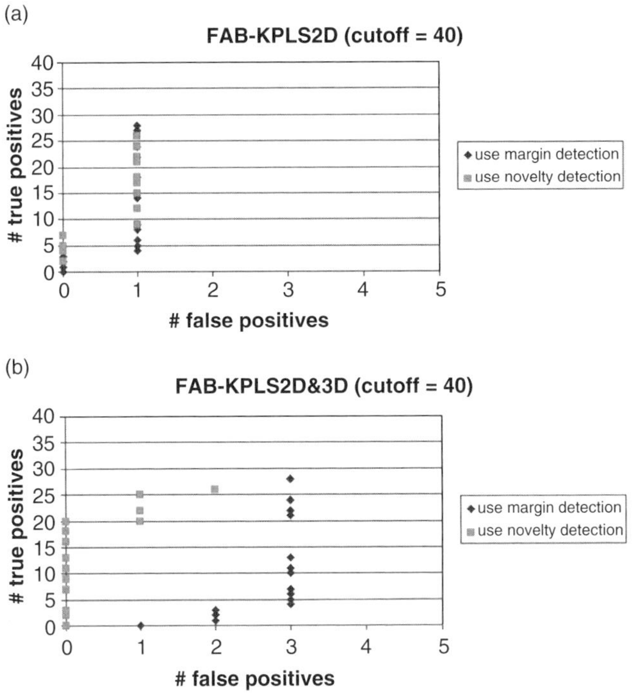

Figure 15.4 Novelty and margin detection applied to the 35 molecule test set for human fraction absorbed. (*a*) K-PLS model with 2D descriptors only; (*b*) K-PLS model with 2D and 3D descriptors.

TABLE 15.4 Toxicity: 2017 Molecules with Ames Mutagenicity Data [85]

	10 × 100 ROC AUC
2D–K-PLS	0.86
2D PLS	0.82
2D & 3D–K-PLS	0.86
2D & 3D–PLS	0.82

(Table 15.3, panel I). A large set of human volume of distribution data generated by many different groups represents a very noisy data set, and in this case suggested K-PLS gave improved models (Table 15.3, panel J). A set of molecules with a log D resulted in better q^2 statistics for PLS models (Table 15.3, panel K). Finally, toxicity-related data for Ames mutagenicity (Table 15.4) and the hERG ion channel again indicated a large improvement in using K-PLS, in contrast to PLS (Table 15.5). When scrambling the observed data with the structure descriptors was used with all K-PLS models, the q^2 values were approximately zero in all cases, indicating a further test of model significance.

TABLE 15.5 Ion Channels: 99 Molecules with Human Ether-a-go-go-related Gene (hERG) log IC_{50} Patch-clamping Data from Various Literature Sources [62,138]

	r^2	q^2 for 10 × 100 (RMSE)	q^2 for 50 × 100	q^2 Scrambling
2D–K-PLS	0.94	0.48 (1.01)	0.40 (1.12)	0.0063
2D PLS	0.70	0.27 (1.27)	0.24 (1.35)	
2D & 3D–K-PLS	0.97	0.49 (1.00)	0.41 (1.01)	0.0046
2D & 3D–PLS	0.79	0.30 (1.24)	0.27 (1.31)	

15.4 DISCUSSION

Computational approaches can help to identify novel molecules with activity for a particular therapeutic target or assist in avoiding potential toxic side effects caused by binding to an enzyme [11,152–154], receptor [61,155–159], transporter [55,56,122,160–164], or channel [62–64,165–167]. There are also other biological processes, some of which are more complex than just interacting with a single binding site and can be modeled computationally. These processes include absorption, blood–brain barrier permeation, toxicity, and pharmacokinetic properties that have been widely reviewed [30–32,35–38,77,126,168]. Many different computational methods have been published or are commercially available, and all have their relative strengths and weaknesses that will not be described here (see previous chapters).

In most previously published examples of these types of computational models a single data set was evaluated with one algorithmic approach, or alternatively, multiple algorithms were compared for a single data set. In the majority of cases these models were either evaluated by the leave-one-out approach or bootstrapping, with larger groups left out of the final model. In a very few examples a test set of molecules not included in the model was used for an external evaluation. The approach we have taken is to calculate a relatively narrow selection of 2D and 3D molecular descriptors [116,117,141] and use either the PLS [73,108,109,111,112,169–171] or K-PLS [99,114] algorithms to correlate the property with descriptors after removing any redundancies [99,116,140,141]. Out of the many published studies relating to ADME/TOX models, few of the training sets are readily available. We have opted to select as wide a range of properties as possible and included numerous enzymes such as CYPs, which are important drug-metabolizing enzymes that are frequently involved in drug–drug interactions [45,51,172]. Other enzymes involved in metabolism such as human blood esterases [119], human and mouse soluble epoxide hydrolases [120,173,174], and human dopamine sulfotransferase 1A3 [121] were also selected based on the availability of more recently published data.

Transporters, such as P-gp, play a key role in modulating the availability of drugs following absorption in the intestine and can also expel them from the brain [175–177]. Other transporters of interest were the monoamine recognizing serotonin and noradrenaline transporters [123,124] and hOCT1

[54,178]. More complex data sets were also selected including human serum binding [115,179], blood–brain barrier permeation [70,108,126,148,168,169, 180,181], absorption [34,69,72–78,131,149,182,183], and aqueous solubility [40,107], which have all been widely modeled. Human intrinsic clearance has been rarely modeled [32,65], while there are many scoring functions that relate diverse K_d values for binding different proteins [2,13,14,184–186]. The recently described aggregation of numerous ligands and some drugs as the mechanism of inhibition of various biological assays is of interest to anyone undertaking screening of commercially available compounds [133–135,187]. Properties such as volume of distribution [125,136] and $\log D$ [136,137] are available for some known drugs and were accessible. Very few toxicity related data sets are widely available with the exception of Ames mutagenicity [85] and several recently published hERG QSAR models [62–64,167]. Subsequently modeling of toxicity endpoints is currently limited to relatively small data sets or to methods using structural alerts [188–192] (see earlier chapters).

Following some preliminary studies comparing different machine-learning methods using the large CYP data sets and the protein-binding data, it was concluded that there were no significant differences between the cross-validated models generated with K-PLS and SVM (Dr. Kristin Bennett, personal communication). With the majority of these data sets we were able to generate acceptable computational models after internal validation with both K-PLS and PLS. In 21 out of the 26 data sets used, K-PLS resulted in generally improved q^2 and RMSE statistics over PLS. The effect of additional 3D descriptors was subtle and data dependent to result in slightly improved q^2 and RMSE statistics over just using 2D descriptors alone. There did not seem to be any apparent trends relating to the size of the training sets and the results with either PLS or K-PLS. All the models generated for each data set were statistically significant when compared to the scrambled models.

We have also taken steps to estimate the confidence in the K-PLS predictions using approaches called novelty detection and margin detection, which are two complementary methods (Figure 15.1*a–c*) that aid in identifying compounds that are significantly different from those seen previously in the training set. This approach was taken for the larger data sets for which test sets were available for CYP3A4 and CYP2D6 percent inhibition [84], blood–brain barrier permeation [148], and human fraction absorbed [149]. The CYP model predictions followed by the use of margin detection resulted in an improvement in the classification of true positives compared with novelty detection alone (Figure 15.2*a–d*). The mouse in vivo derived binary test set for blood–brain barrier permeation shows an improvement in the number of false positives after using novelty detection (Figure 15.3*a*, *b*). The smaller test set for fraction absorbed (Figure 15.4*a*, *b*) with K-PLS using 2D and 3D descriptors resulted in very few false positives, and novelty detection was superior to margin detection. Overall the application of these two approaches following predictions with the K-PLS models improves the classifications. It would be useful to combine them both into one method in future. It is interesting to note that although the internal validation q^2 values appear quite low, when tested with external molecules they are able to perform significantly better than random (Figures 15.2–15.4).

In conclusion, K-PLS has been applied broadly to different sized data sets for producing computational models for ADME/TOX and physicochemical properties. This and other studies indicate that K-PLS represents an acceptable alternative to SVM for use in drug discovery applications. The novel combination of K-PLS and SVM based novelty and margin detection methods undertaken may, however, enhance the prediction capabilities of K-PLS alone. This is analogous to the use of consensus scoring methods for ligand protein scoring and QSAR [14,184,193] and the addition of Tanimoto similarity scores for test molecules when making predictions with QSAR models [139,194]. The combination of a large number of diverse computational models for ADME/TOX and complex processes such as those described will be of value in the scoring and multiple optimization of virtual ligands for therapeutic targets [4].

ACKNOWLEDGMENTS

Prof. Bruce D. Hammock (UC Davis) and his laboratory are gratefully acknowledged for providing the published and unpublished soluble epoxide hydrolase data sets. Dr. Alexy Ishchenko, Dr. Maggie A.Z. Hupcey, and Ms. Katya Berezin are kindly acknowledged for generating most of the molecular structures used in these data sets. Ms. Jennifer Berbaum and Dr. Richard Harrison are acknowledged for providing the in vitro data for the CYP inhibition test set. Dr. Kristin Bennett, Dr. Michinari Momma, Dr. N. Sukumar (Rensselaer Polytechnic Institute), Dr. Nello Cristianini (UC Davis), and Dr. John Shawe-Taylor (Intelligent Stuff Ltd) are acknowledged for initial discussions and consultation on machine learning and novelty detection.

SE gratefully acknowledges Dr. Jun Shimada and Dr. Peter Lindblom for frequent stimulating discussions and Dr. Jack Baldwin and colleagues for their support of this work at Concurrent Pharmaceuticals Inc. (now Vitae Pharmaceuticals Inc.).

REFERENCES

1. Grzybowski BA, Ishchenko AV, Kim C-K, Topalov G, Chapman R, Christianson DW, et al. Combinatorial computational method gives new picomolar ligands for a known enzyme. *Proc Natl Acad Sci USA* 2002;99:1270–3.
2. Ishchenko AV, Shakhnovich EI. SMall Molecule Growth (SMoG2001): An improved knowledge-based scoring function for protein–ligand interactions. *J Med Chem* 2002;45:2770–80.
3. Shimada J, Ishchenko AV, Shakhnovich EI. Analysis of knowledge-based protein-ligand potentials using a self-consistent method. *Protein Sci* 2000;9:765–75.
4. Shimada J, Ekins S, Elkin C, Shaknovich EI, Wery J-P. Integrating computer-based de novo drug design and multidimensional filtering for desirable drugs. *Targets* 2002;1:196–205.

5. Caflisch A, Karplus M. Computational combinatorial chemistry for de novo ligand design: Review and assessment. *Perspect Drug Discov Des* 1995;3:51–84.
6. Stahl M, Todorov NP, James T, Mauser H, Boehm H-J, Dean PM. A validation study on the practical use of in silico de novo design. *J Comput Aided Mol Des* 2002;16:459–78.
7. Schmidt JM, Mercure J, Tremblay GB, Page M, Kalbakji A, Feher M, et al. De novo design, synthesis and evaluation of novel nonsteroidal phenanthrene ligands for the estrogen receptor. *J Med Chem* 2003;46:1408–18.
8. Pegg SC-H, Haresco JJ, Kuntz ID. A genetic algorithm for structure-based de novo design. *J Comput Aided Mol Des* 2001;15:911–33.
9. Sotriffer CA, Gohlke H, Klebe G. Docking into knowledge-based potential fields: a comparative evaluation of drugscore. *J Med Chem* 2002;45:1967–70.
10. Schneider G, Bohm H-J. Virtual screening and fast automated docking methods. *Drug Discov Today* 2002;7:64–70.
11. Rastelli G, Ferrari AM, Constantino l, Gamberini C. Discovery of new inhibitors of aldose reductase from molecular docking and database screening. *Bioorg Med Chem* 2002;10:1437–50.
12. Doman TN, McGovern SL, Witherbee BJ, Kasten TP, Kurumbail R, Stallings WC, et al. Molecular docking and high throughput screening for novel inhibitors of protein tyrosine phosphatase-1B. *J Med Chem* 2002;45:2213–21.
13. Wang R, Lu Y, Wang S. Comparative evaluation of 11 scoring functions for molecular docking. *J Med Chem* 2003;46:2287–303.
14. Bissantz C, Folkers G, Rognan D. Protein-based virtual screening of chemical databases: 1. evaluation of different docking/scoring combinations. *J Med Chem* 2000;43:4759–67.
15. Welch W, Ruppert J, Jain AN. Hammerhead: Fast, fully automated docking of flexible ligands to protein binding sites. *Chem Biol* 1996;3:449–62.
16. Clark DE, Pickett SD. Computational methods for the prediction of "drug-likeness." *Drug Discov Today* 2000;5:49–58.
17. Walters WP, Murcko MA. Prediction of "drug-likeness." *Adv Drug Deliv Rev* 2002;54:255–71.
18. Bemis GW, Murcko MA. The properties of known drugs: 1. Molcular frameworks. *J Med Chem* 1996;39:2887–93.
19. Bemis GW, Murcko MA. Properties of known drugs 2: Side chains. *J Med Chem* 1999;42:5095–9.
20. Veber DF, Johnson SR, Cheng H-Y, Smith BR, Ward KW, Kopple KD. Molecular properties that influence the oral bioavailability of drug candidates. *J Med Chem* 2002;45:2615–23.
21. Wenlock MC, Austin RP, Barton P, Davis AM, Leeson PD. A comparison of physicochemical property profiles of development and marketed oral drugs. *J Med Chem* 2003;46:1250–6.
22. Lipinski CA. Drug-like properties and the causes of poor solubility and poor permeability. *J Pharm Toxicol Meth* 2000;44:235–49.
23. Muegge I, Heald SL, Brittelli D. Simple selection criteria for drug-like chemical matter. *J Med Chem* 2001;44:1–6.

24. Ajay., Walters WP, Murcko MA. Can we learn to distinguish between "drug-like" and "nondrug-like" molecules? *J Med Chem* 1998;41:3314–24.
25. Oprea TI, Davis AM, Teague SJ, Leeson PD. Is there a difference between leads and drugs? A historical perspective. *J Chem Inf Comput Sci* 2001;41:1308–15.
26. Oprea TI. Current trends in lead discovery: Are we looking for the appropriate properties? *J Comput Aided Mol Des* 2002;16:325–34.
27. Blake JF. Examination of the computed molecular properties of compounds selected for clinical development. *Biotechniques* 2003:16–20.
28. Navia MA, Chaturvedi PR. Design principles for orally bioavailable drugs. *Drug Discov Today* 1996;1:179–89.
29. Brustle M, Beck B, Schindler T, W. K, Mitchell T, Clark T. Descriptors, physical properties, and drug likeness. *J Med Chem* 2002;45:3345–55.
30. Ekins S, Rose JP. In Silico ADME/TOX: The state of the art. *J Mol Graph* 2002;20:305–9.
31. Ekins S, Boulanger B, Swaan PW, Hupcey MAZ. Towards a new age of virtual ADME/TOX and multidimensional drug discovery. *J Comput Aided Mol Des* 2002;16:381–401.
32. Ekins S, Waller CL, Swaan PW, Cruciani G, Wrighton SA, Wikel JH. Progress in predicting human ADME parameters in silico. *J Pharmacol Toxicol Meth* 2000;44:251–72.
33. Johnson DE, Wolfgang GHI. Predicting human safety: Screening and computational approaches. *DDT* 2000;5:445–54.
34. Egan WJ, Merz KMJ, Baldwin JJ. Prediction of drug absorption using multivariate statistics. *J Med Chem* 2000;43:3867–77.
35. van de Waterbeemd H, Gifford E. ADMET in silico modelling: Towards prediction paradise? *Nat Rev Drug Discov* 2003;2:192–204.
36. Wessel MD, Mente S. ADME by computer. *An Rep Med Chem* 2001;36:257–66.
37. Darvas F, Dorman G, Papp A. Diversity measures for enhancing ADME admissibility of combinatorial libraries. *J Chem Inf Comput Sci* 2000;40:314–22.
38. Butina D, Segall MD, Frankcombe K. Predicting ADME properties in silico: Methods and models. *Drug Discov Today* 2002;7:S83–S8.
39. Kesuru GM, Molnar L. METAPRINT: A metabolic fingerprint. Application to cassette design for high-throughput ADME screening. *J Chem Inf Compu Sci* 2002;42:437–44.
40. Lipinski CA, Lombardo F, Dominy BW, Feeney PJ. Experimental and computational approaches to estimate solubility and permeability in drug discovery and development settings. *Adv Drug Deliv Rev* 1997;23:3–25.
41. Cheng A, Diller DJ, Dixon SL, Egan WJ, Lauri G, Merz Jr KMJ. Computation of the physico-chemical properties and data mining of large molecular collections. *J Comput Chem* 2002;23:172–83.
42. de Groot MJ, Ackland MJ, Horne VA, Alex AA, Jones BC. Novel approach to predicting P450-mediated drug metabolism: Development of a combined protein and pharmacophore model for CYP2D6. *J Med Chem* 1999;42:1515–24.
43. de Groot MJ, Ackland MJ, Horne VA, Alex AA, Jones BC. A novel approach to predicting P450 mediated drug metabolism: CYP2D6 catalyzed *N*-dealkylation

reactions and qualitative metabolite predictions using a combined protein and pharmacophore model for CYP2D6. *J Med Chem* 1999;42:4062–70.

44. de Groot MJ, Alex AA, Jones BC. Development of a combined protein and pharmacophore model for cytochrome P450 2C9. *J Med Chem* 2002;45:1983–93.
45. de Groot MJ, Ekins S. Pharmacophore modeling of cytochromes P450. *Adv Drug Deliv Rev* 2002;54:367–83.
46. Ekins S, Bravi G, Binkley S, Gillespie JS, Ring BJ, Wikel JH, et al. Three and four dimensional–quantitative structure activity relationship (3D/4D-QSAR) analyses of CYP2D6 inhibitors. *Pharmacogenetics* 1999;9:477–89.
47. Ekins S, Bravi G, Binkley S, Gillespie JS, Ring BJ, Wikel JH, et al. Three and four dimensional–quantitative structure activity relationship analyses of CYP3A4 inhibitors. *J Pharm Exp Ther* 1999;290:429–38.
48. Ekins S, Bravi G, Binkley S, Gillespie JS, Ring BJ, Wikel JH, et al. Three and four dimensional–quantitative structure activity relationship (3D/4D-QSAR) analyses of CYP2C9 inhibitors. *Drug Metab Dispos* 2000;28:994–1002.
49. Ekins S, Bravi G, Ring BJ, Gillespie TA, Gillespie JS, VandenBranden M, et al. Three dimensional–quantitative structure activity relationship (3D-QSAR) analyses of substrates for CYP2B6. *J Pharm Exp Ther* 1999;288:21–9.
50. Ekins S, Bravi G, Wikel JH, Wrighton SA. Three dimensional quantitative structure activty relationship (3D-QSAR) analysis of CYP3A4 substrates. *J Pharmacol Exp Ther* 1999;291:424–33.
51. Ekins S, de Groot M, Jones JP. Pharmacophore and three dimensional quantitative structure activity relationship methods for modeling cytochrome P450 active sites. *Drug Metab Dispos* 2001;29:936–44.
52. Jones JP, He M, Trager WF, Rettie AE. Three-dimensional quantitative structure-activity relationship for inhibitors of cytochrome P4502C9. *Drug Metab Dispos* 1996;24:1–6.
53. Rao S, Aoyama R, Schrag M, Trager WF, Rettie A, Jones JP. A refined 3-dimensional QSAR of P4502C9. *J Med Chem* 2000;43:2789–96.
54. Bednarczyk D, Ekins S, Wikel JH, Wright SH. Influence of molecular structure of substrate binding to the human organic cation transporter, hOCT1. *Mol Pharmacol* 2003;63:489–98.
55. Ekins S, Kim RB, Leake BF, Dantzig AH, Schuetz E, Lan L-b, et al. Application of three dimensional quantitative structure-activity relationships of *P*-glycoprotein inhibitors and substrates. *Mol Pharmacol* 2002;61:974–81.
56. Ekins S, Kim RB, Leake BF, Dantzig AH, Schuetz E, Lan L-b, et al. Three dimensional quantitative structure-activity relationships of inhibitors of *P*-glycoprotein. *Mol Pharmacol* 2002;61:964–73.
57. Swaan PW, Koops BC, Moret EE, Tukker JJ. Mapping the binding site of the small intestinal peptide carrier (PepT1) using comparative molecular field analysis. *Recept Channels* 1998;6:189–200.
58. Swaan PW, Szoka FC, Jr., Oie S. Molecular modeling of the intestinal bile acid carrier: A comparative molecular field analysis study. *J Comput Aided Mol Des* 1997;11:581–8.
59. Baringhaus KH, Matter H, Stengelin S, Kramer W. Substrate specificity of the ileal and the hepatic Na(+)/bile acid cotransporters of the rabbit: II. A reliable

3D QSAR pharmacophore model for the ileal Na(+)/bile acid cotransporter. *J Lipid Res* 1999;40:2158–68.

60. Ekins S, Mirny L, Schuetz EG. A ligand-based approach to understanding selectivity of nuclear hormone receptors PXR, CAR, FXR, LXRa and LXRb. *Pharm Res* 2002;19:1788–800.
61. Ekins S, Erickson JA. A pharmacophore for human pregnane-X-receptor ligands. *Drug Metab Dispos* 2002;30:96–9.
62. Ekins S, Crumb WJ, Sarazan RD, Wikel JH, Wrighton SA. Three dimensional quantitative structure activity relationship for the inhibition of the hERG (human ether-a-go-go-related gene) potassium channel. *J Pharmacol Exp Ther* 2002;301: 427–34.
63. Pearlstein RA, Vaz RJ, Kang J, Chen X-L, Preobrazhenskaya M, Shchekotikhin AE, et al. Characterization of HERG Potassium channel inhibition using CoMSiA 3D QSAR and homology modeling approaches. *Bioorg Med Chem* 2003;13:1829–35.
64. Cavalli A, Poluzzi E, De Ponti F, Recanatini M. Toward a pharmacophore for drugs inducing the long QT syndrome: Insights from a CoMFA study of HERG K+ channel blockers. *J Med Chem* 2002;45:3844–53.
65. Ekins S, Obach RS. Three dimensional-quantitative structure activity relationship computational approaches of prediction of human in vitro intrinsic clearance. *J Pharmacol Exp Ther* 2000;295:463–73.
66. Ekins S, Durst GL, Stratford RE, Thorner DA, Lewis R, Loncharich RJ, et al. Three dimensional quantitative structure permeability relationship analysis for a series of inhibitors of rhinovirus replication. *J Chem Inf Comput Sci* 2001; 41:1578–86.
67. Andrews CW, Bennett L, Yu LX. Predicting human oral bioavailability of a compound: Development of a novel quantitative structure-bioavailability relationship. *Pharm Res* 2000;17:639–44.
68. Yoshida F, Topliss JG. QSAR model for drug human oral bioavailability. *J Med Chem* 2000;43:2575–85.
69. Clark DE. Rapid calculation of polar molecular surface area and its application to the prediction of transport phenomena: 1. Prediction of intestinal absorption. *J Pharm Sci* 1999;88:807–14.
70. Kelder J, Grootenhuis PDJ, Bayada DM, Delbressine LPC, Ploeman J-P. Polar molecular surface as a dominating determinant for oral absorption and brain penetration of drugs. *Pharm Res* 1999;16:1514–9.
71. Krarup LH, Christensen IT, Hovgaard L, Frokjaer S. Predicting drug absorption from molecular surface properties based on molecular dynamics simulations. *Pharm Res* 1998;15:972–8.
72. Oprea TI, Gottfries J. Toward a minimalistic modeling of oral drug absorption. *J Mol Graph Model* 1999;17:261–74.
73. Norinder U, Osterberg T, Artursson P. Theoretical calculation and prediction of intestinal absorption of drugs in humans using MolSurf parameterization and PLS statistics. *Eur J Pharm Sci* 1999;8:49–56.
74. Palm K, Luthman K, Ungell A-L, Strandlund G, Beigi F, Lundahl P, et al. Evaluation of dynamic polar molecular surface area as a predictor of drug absorption:

Comparison with other computational and experimental predictors. *J Med Chem* 1998;41:5382–92.

75. Palm K, Stenberg P, Luthman K, Artursson P. Polar molecular surface properties predict the intestinal absorption of drugs in humans. *Pharm Res* 1997;14:568–71.
76. Stenberg P, Luthman K, Ellens H, Lee CP, Smith PL, Lago A, et al. Prediction of the intestinal absorption of endothelin receptor antagonists using three theoretical methods of increasing complexity. *Pharm Res* 1999;16:1520–6.
77. Stenberg P, Norinder U, Luthman K, Artursson P. Experimental and computational screening models for the prediction of intestinal drug absorption. *J Med Chem* 2001;44:1927–37.
78. Wessel MD, Jurs PC, Tolan JW, Muskal SM. Prediction of human intestinal absorption of drug compounds from molecular structure. *J Chem Inf Comput Sci* 1998;38:726–35.
79. Schneider G, Coassolo P, Lave T. Combining in vitro and in vivo pharmacokinetic data for prediction of hepatic drug clearance in humans by artificial neural networks and multivariate statistical techniques. *J Med Chem* 1999;42:5072–6.
80. Karelson M. *Molecular descriptors in QSAR/QSPR.* New York: Wiley-Interscience, 2000.
81. Todeschini R, Consonni V. *Handbook of molecular descriptors.* Weinheim: Wiley-VCH, 2000.
82. Chen X, Rusinko III A, Young SS. Recursive partitioning analysis of a large structure-activity data set using three-dimensional descriptors. *J Chem Inf Comput Sci* 1998;38:1054–62.
83. Chen X, Rusinko I, A., Tropsha A, Young SS. Automated pharmacophore identification for large chemical data sets. *J Chem Inf Comput Sci* 1999;39:887–96.
84. Ekins S, Berbaum J, Harrison RK. Generation and validation of rapid computational filters for CYP2D6 and CYP3A4. *Drug Metab Dispos* 2003;31:1077–80.
85. Young SS, Gombar VK, Emptage MR, Cariello NF, Lambert C. Mixture deconvolution and analysis of Ames mutagenicity data. *Chemom Intell Lab Sys* 2002;60:5–11.
86. Molnar L, Kesuru GM. A neural network based virtual screening of cytochrome P450 3A4 inhibitors. *Bioorg Med Chem Lett* 2002;12:419–21.
87. Niwa T. Using general regression and probalistic neural networks to predict human intestinal absorption with topological descriptors derived from two-dimensional chemical structures. *J Chem Inf Comput Sci* 2003;43:113–9.
88. Ritschel WA, Akileswaran R, Hussain AS. Application of neural networks for the prediction of human pharmacokinetic parameters. *Meth Find Exp Clin Pharmacol* 1995;17:629–43.
89. Klopman G, Tu M, Talafous J. META. 3. A genetic algorithm for metabolic transform priorities optimization. *J Chem Inf Comput Sci* 1997;37:329–34.
90. Marchand-Geneste N, Watson KA, Alsberg BK, King RD. New approach to pharmacophore mapping and qsar analysis using inductive logic programming: Application to thermolysin and glycogen phosphorylase b inhibitors. *J Med Chem* 2002;45:399–409.

91. Korolev D, Balakin KV, Nikolsky Y, Kirillov E, Ivanenkov YA, Savchuk NP, et al. Modeling of human cytochrome P450-mediated drug metabolism using unsupervised machine learning approach. *J Med Chem* 2003;46:3631–43.
92. Shen M, Xiao Y, Golbraikh A, Gombar VK, Tropsha A. Development and validation of *k*-nearest neighbour QSPR models of metabolic stability of drug candidates. *J Med Chem* 2003;46:3013–20.
93. Nicolotti O, Gillet VJ, Fleming PJ, Green DVS. Multiobjective optimization in quantitative structure-activity relationships: Deriving accurate and interpretable QSARs. *J Med Chem* 2002;45:5069–80.
94. Furey TS, Christianini N, Duffy N, Bednarski DW, Schummer M, Haussler D. Support vector machine classification and validation of cancer tissue samples using microarray expression data. *Bioinformatics* 2000;16:906–14.
95. Brown MPS, Grundy WN, Lin D, Christianini N, Sugnet CW, Furey TS, et al. Knowledge-based analysis of microarray gene expression data by using support vector machines. *Proc Natl Acad Sci USA* 2000;97:262–7.
96. Cai Y-D, Liu X-J, Xu X-B, Chou K-C. Support vector machines for the classification and prediction of β-turn types. *J Peptide Sci* 2002;8:297–301.
97. Song M, Breneman C, Bi J, Sukumar N, Bennett K, Cramer S, et al. Prediction of protein retention times in anion-exchange chromatography systems using support vector regression. *J Chem Inf Comput Sci* 2002;42:1347–57.
98. Bennet KP, Campbell C. Support vector machines: Hype or hallelujah? *SIGKDD explor* 2000;2:1–13.
99. Bennett KP, Embrechts MJ. An optimization perspective on kernel partial least squares regression. In: Suykens JAK, Horvath G, Basu S, Micchelli J, Vandewalle J, editors, *Advances in learning theory: Methods, models and applications*. Amsterdam: IOS Press, 2003. p. 227–50.
100. Warmuth MK, Liao J, Ratsch G, Mathieson M, Putta S, Lemmen C. Active learning with support vector machines in the drug discovery process. *J Chem Inf Comput Sci* 2003;43:667–73.
101. Burbidge R, Trotter M, Buxton B, Holden S. drug design by machine learning: Support vector machines for pharmaceutical analysis. *Comput Chem* 2001; 26:5–14.
102. Zernov VV, Balakin KV, Ivashchenko AA, Savchuk NP, Pletnev IV. Drug Discovery using support vector machines. The case studies of drug-likeness, agrochemical-likeness, and enzyme inhibition predictions. *J Chem Inf Comput Sci* 2003;43:2048–56.
103. Czerminski R, Yasri A, Hartsough D. Use of support vector machine in pattern classification: Application to QSAR studies. *Quant Struct-Act Rel* 2001;20:227–40.
104. Warmuth MK, Rasch G, Mathieson M, Liao J, Lemmen C. Active learning in the drug discovery process. In Dietterich TG, Becher S and Ghahramani Z, editors. *Advances in Neural Information Processing Systems* 2002;14:1449–56.
105. Boser B, Guyon I, Vapnik V. Pattern recognition system using support vectors. USA, 1997. US Patent 5,649,068.
106. Cortes C, Vapnik V. Soft margin classifier. USA, 1997. US Patent 5,640,492.
107. Lind P, Maltseva T. Support vector machines for the estimation of aqueous solubility. *J Chem Inf Comput Sci* 2003;43:1855–9.

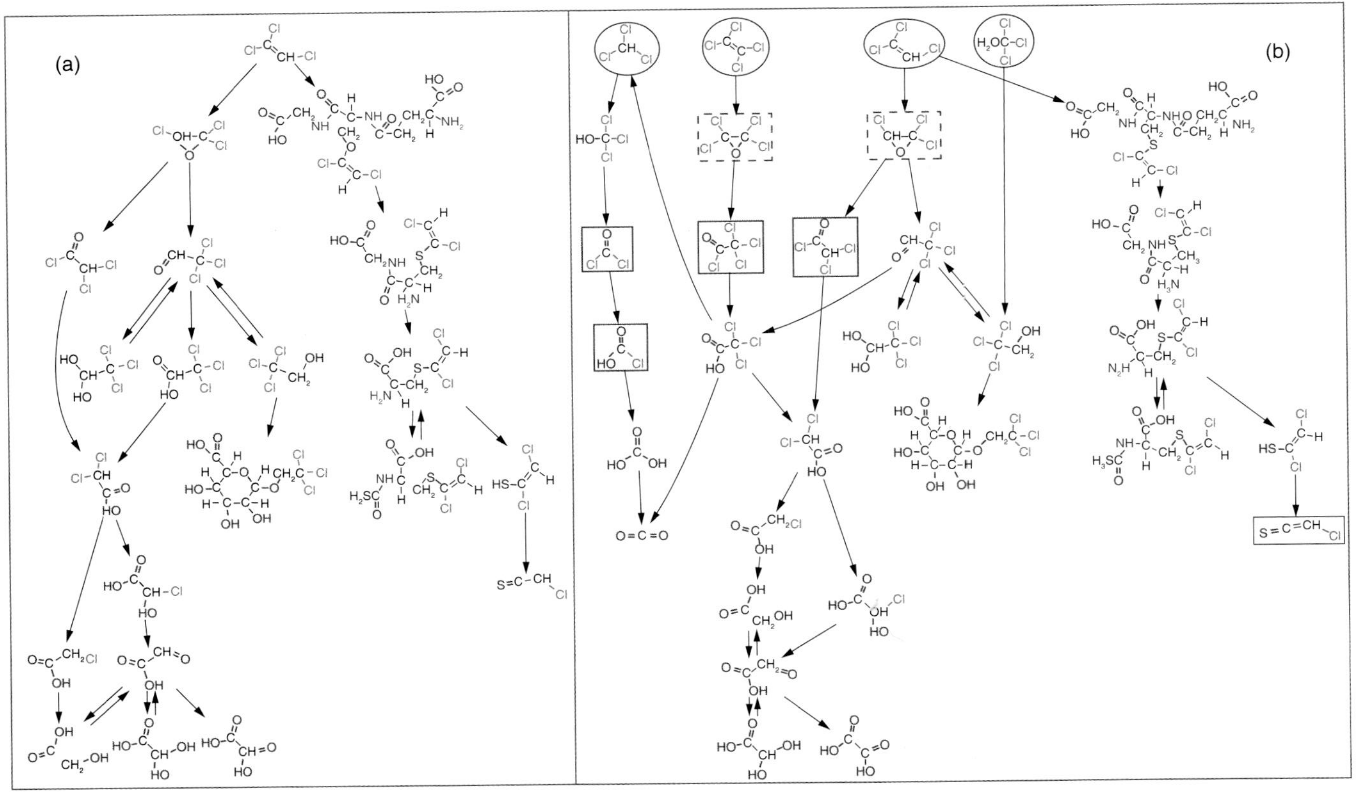

Figure 3.8 BioTRaNS-generated biotransformation pathways for (*a*) TCE and (*b*) a mixture of TCE, Perc, MC, and $CHCl_3$. For clarity, only selected agents (vEnzs and vAgnts) were used to generate these graphs. For (*b*), reactive metabolites are highlighted as follow: epoxides (brown, box, dashed); acid chlorides (orange, box, solid), thioketene (turquoise, box, solid); and starting chemicals (blue, ellipse, solid). (Reprinted from [111] with permission from ACS.)

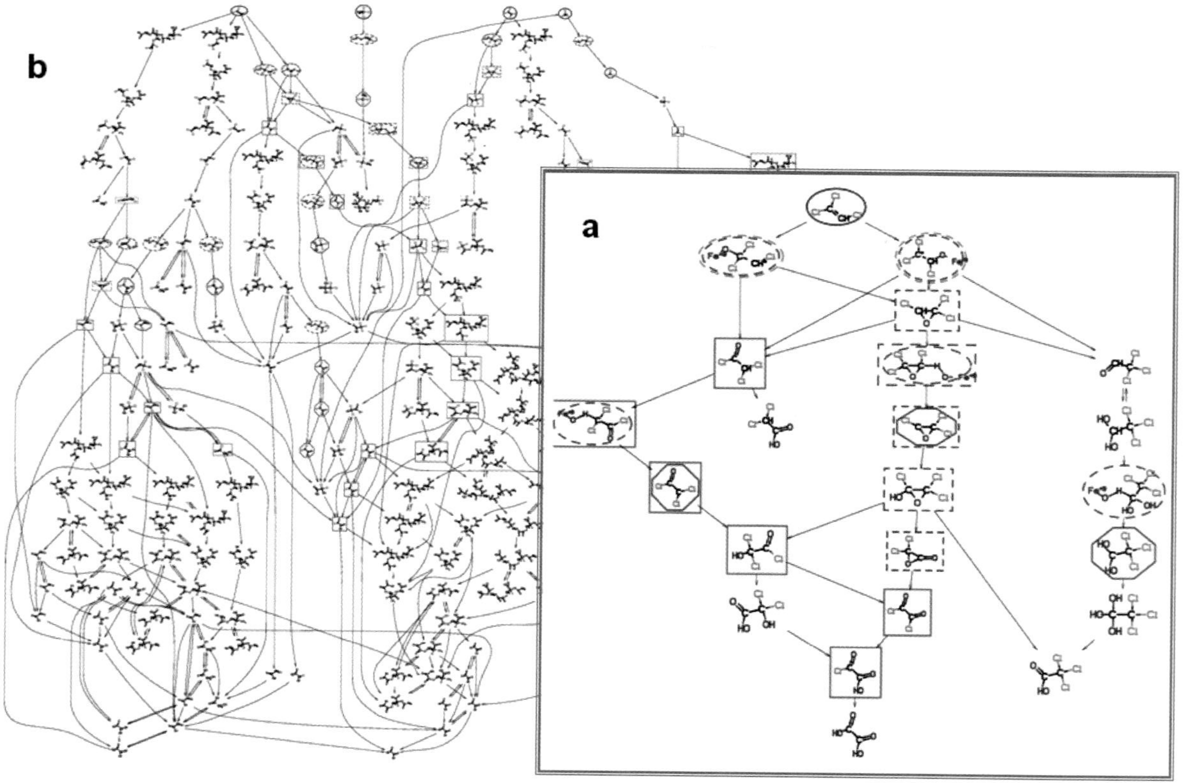

Figure 3.9 BioTRaNS-generated mechanism-based pathways for (*a*) the oxidation of TCE by CYP and (*b*) more comprehensive biotransformation pathways of TCE, Perc, MC, and $CHCl_3$. Chemicals are highlighted as follows: starting chemicals (blue, ellipse, solid); epoxides (brown, box, dashed); acid chlorides (orange, box, solid); radicals (red, octagon, solid); Fe-O-substrate complexes (magenta, ellipse, dashed), carbocations (salmon, ellipse, dashed). (Reprinted from [111] with permission from ACS.)

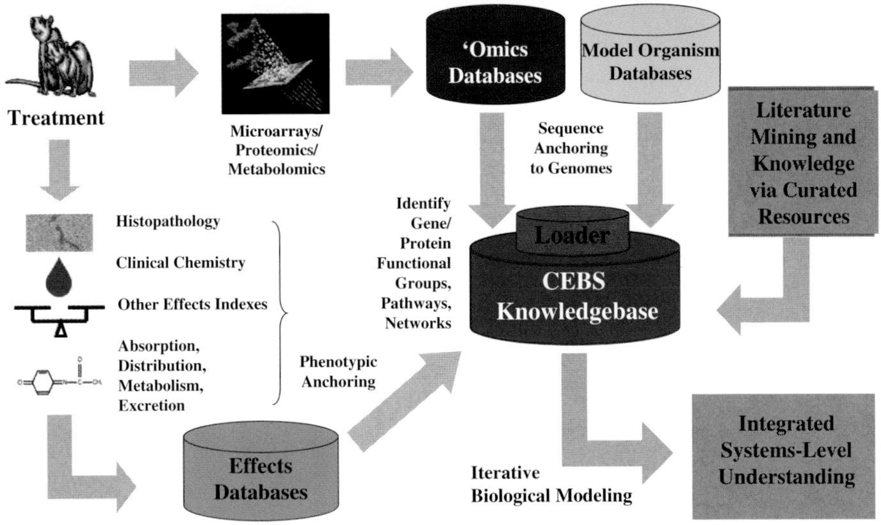

Figure 5.2 Framework for systems toxicology. The figure indicates the paths from the initial observation (rat in upper left) to an integrated toxicogenomics knowledgebase and thence to systems toxicology. The -omics data stream is shown by the clockwise path from rat to knowledgebase, and the traditional toxicology approach is shown in the counterclockwise path. The knowledgebase will integrate both data streams, along with literature knowledge, and, by virtue of iterative modeling, lead to a systems toxicology understanding. The framework involves "phenotypic anchoring" (to toxicological end-points and study design information) and "sequence anchoring" (to genomes) of multi-domain molecular expression datasets in the context of conventional indices of toxicology, and the iterative biological modeling of resulting data.

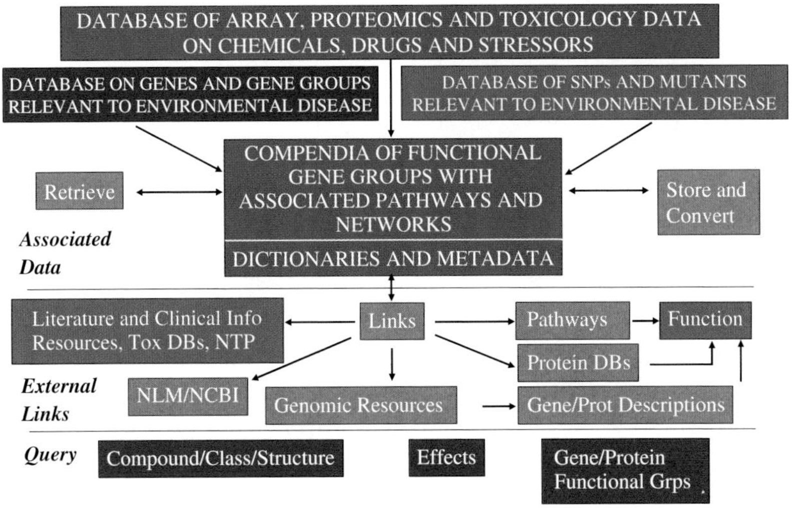

Figure 5.5 Conceptual framework for the development of the Chemical Effects in Biological Systems (CEBS) knowledgebase. CEBS knowledgebase is a cross-species reference toxicogenomics information system chemicals/stressors and their effects. The upper section indicates data associated in CEBS; the center section, external links from CEBS; and the lower section, sample query types that CEBS will support. The boxes in the upper section include primary data (blue), important genetic loci (red) and genetic markers such as SNPs (green). The tasks CEBS will carry out are shown with gray boxes. In the central section the links that are to databases are shown in gray, and the links to unstructured data are in green. Abbreviations: NTP, National Toxicology Program; NLM, National Library of Medicine; NCBI, National Center for Biotechnology Information.

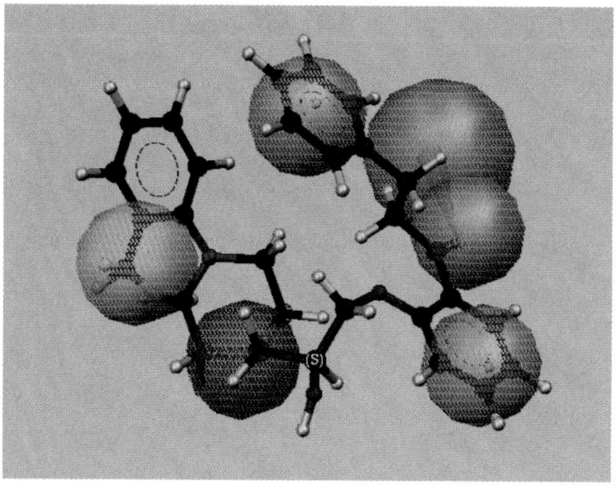

Figure 11.3 P-glycoprotein pharmacophore model derived from propafenones used for screening of the world drug index. Blue: hydrophobic/aromatic; green: hydrophobic; brown: H-bond acceptor; red: positive ionisable.

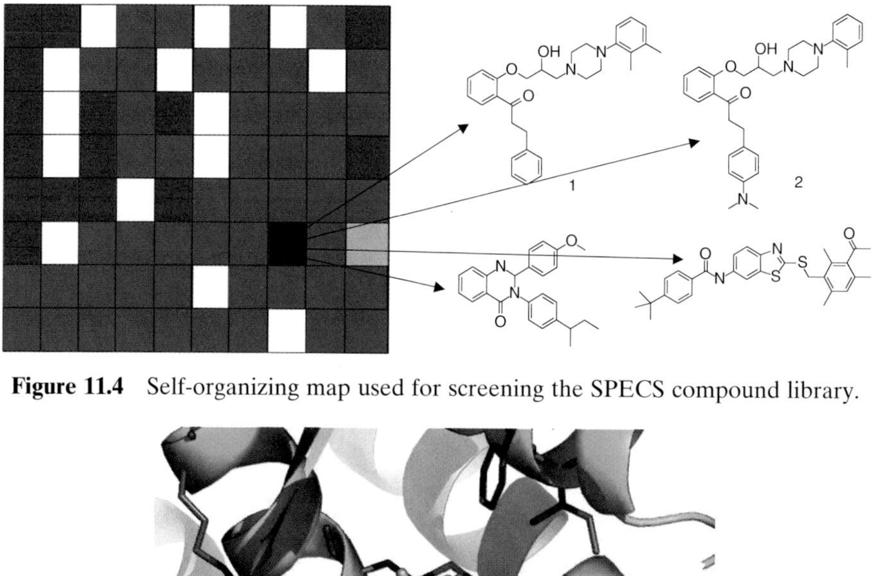

Figure 11.4 Self-organizing map used for screening the SPECS compound library.

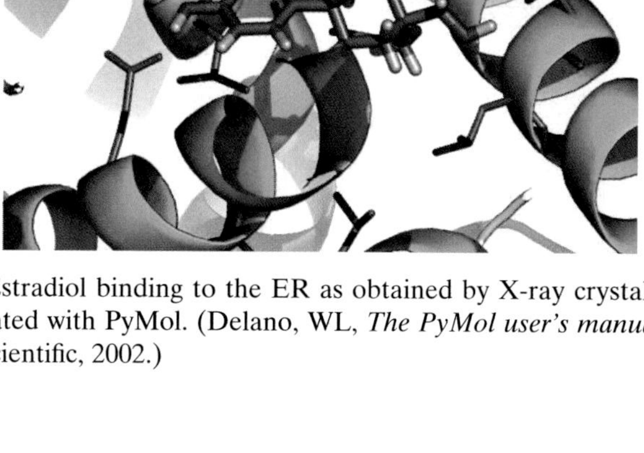

Figure 12.2 Estradiol binding to the ER as obtained by X-ray crystallography. The image was created with PyMol. (Delano, WL, *The PyMol user's manual*. San Cartos, CA: Delano Scientific, 2002.)

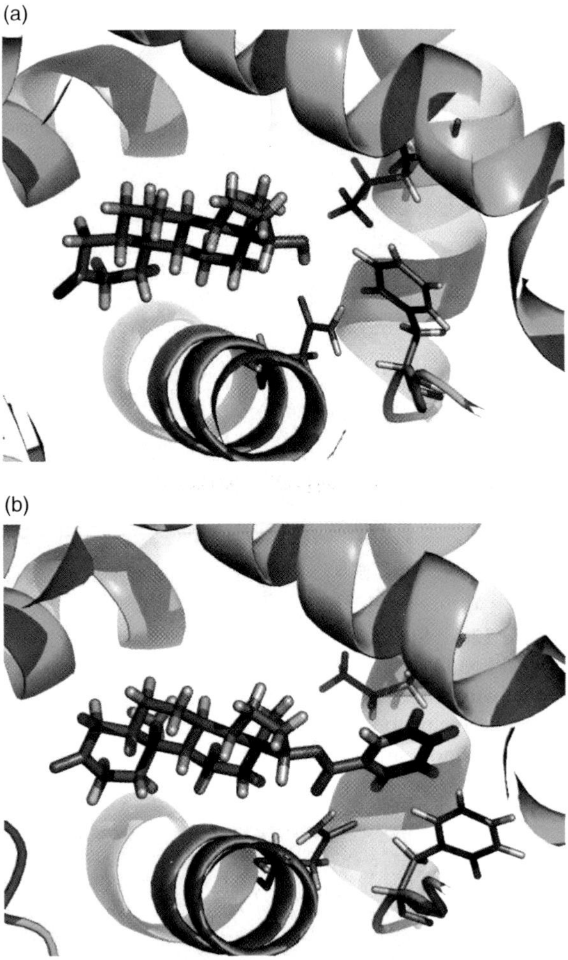

Figure 12.4 (*a*) Dihydrotestosterone (DHT) and DHT benzoate binding to the androgen receptor. Local induced fit is necessary to accommodate the additional volume of the benzoate group. The image was created with PyMol. See color plates.

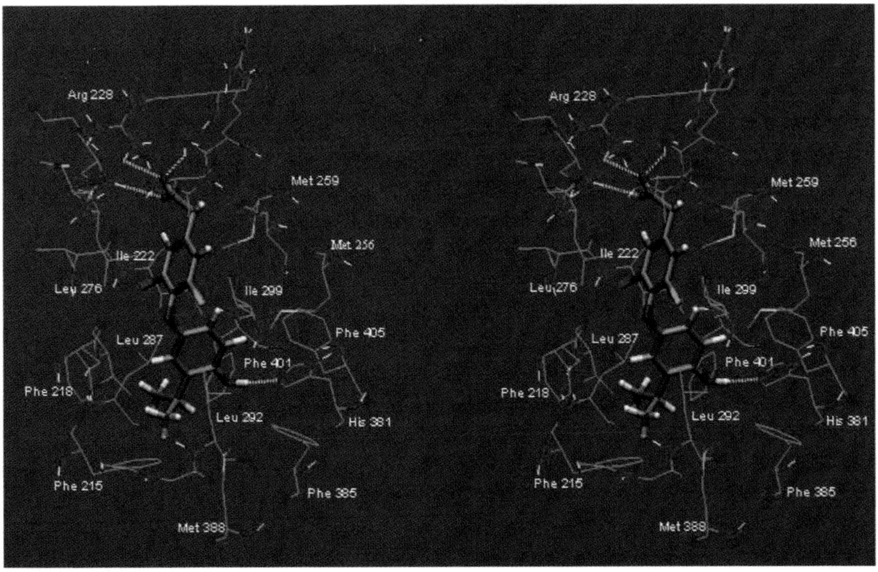

Figure 12.5 Stereo view of 3,5-dichloro-3′-isopropyl-thyronine bound to the thyroid hormone receptor α. Details of the binding pocket; dashed lines indicate stabilizing ligand-protein hydrogen bonds.

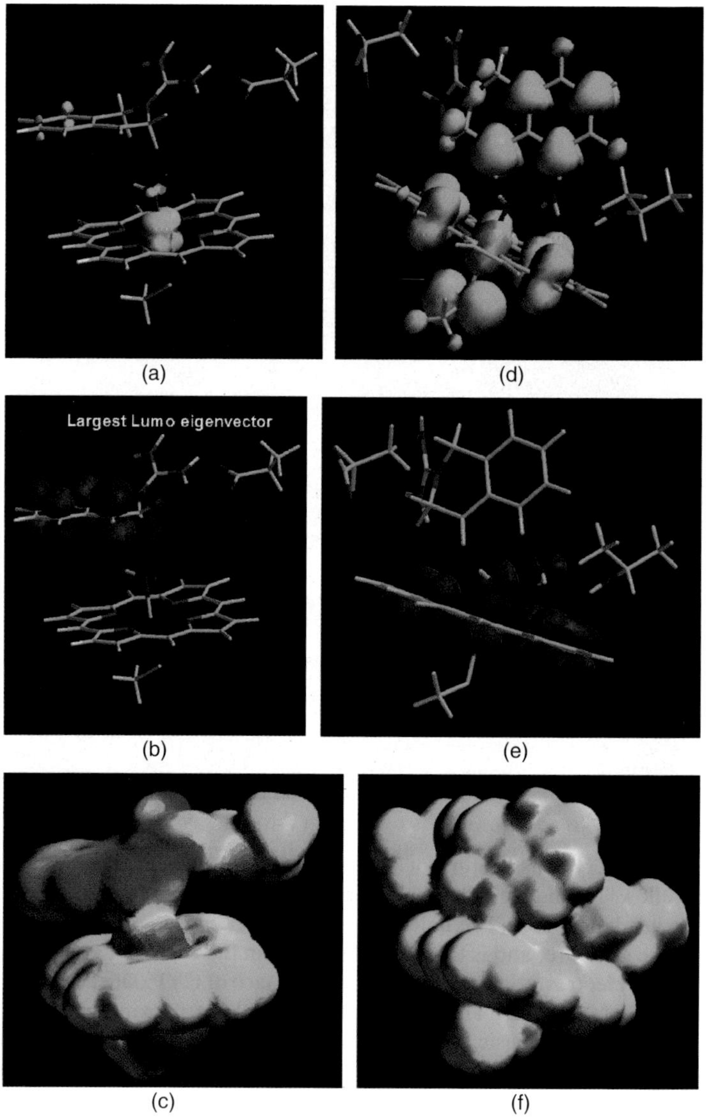

Figure 17.15 Ab initio calculations carried out on debrisoquine in a model reaction system derived on the left from a docking in our homology model and on the right, from the crystal structure of CYP2D6. (*a*) The spin density is centered on the iron. (*b*) The largest LUMO eigenvector is on the C4 position of debrisoquine, and (*c*) The largest positive electrostatic potential is again on C4. These features indicate a classic nucleophilic reaction. With the orthogonal approach likely from the crystal structure (*d*) debrisoquine is in a radical state, (*e*) the LUMO is centered on the heme, and (*f*) the charge is distributed around the complex.

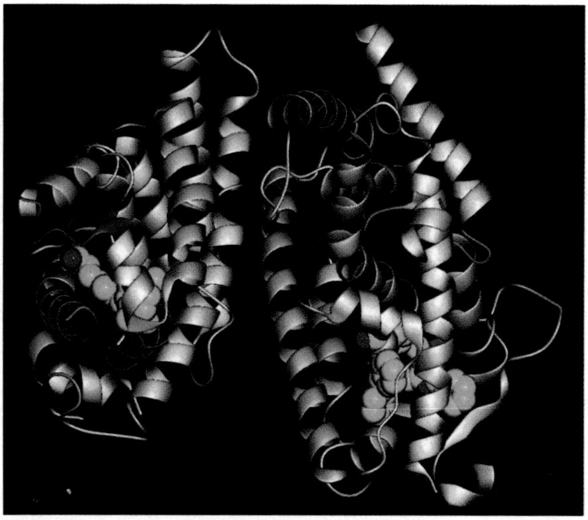

Figure 17.16 PPARγ/RXRα heterodimer crystal structure with retinoic acid bound to RXR (*left*) and rosiglitazone in PPARγ (*right*). Helix 12 can be seen in the front of the PPARγ structure crossing horizontally and blocking the active site pocket.

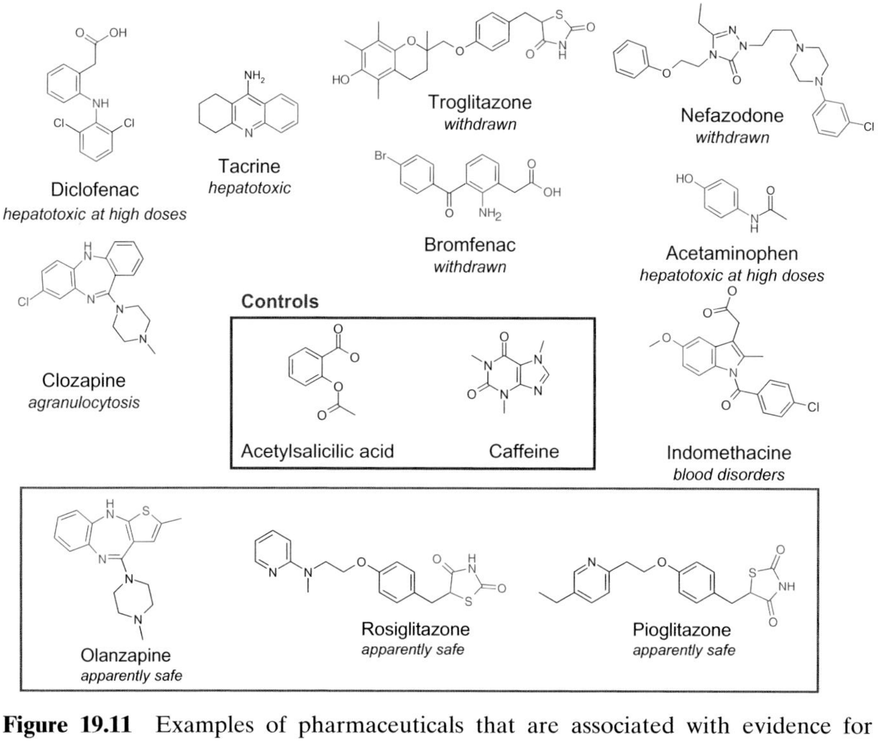

Figure 19.11 Examples of pharmaceuticals that are associated with evidence for human hepatotoxicity of varying levels of severity (caffeine and acetylsalicylic acid serve as negatives examples). Functional groups that are believed to be associated with metabolism pathways toward a reactive intermediate are highlighted in red. Note that despite structural alerts and proven bioactivation liabilities, some drugs given at comparably low dose are considered as safe (olanzapine, rosiglitazone, and pioglitazone). Their structural analogues given at relatively high doses (clozapine, troglitazone) have been associated with severe clinical side effects.

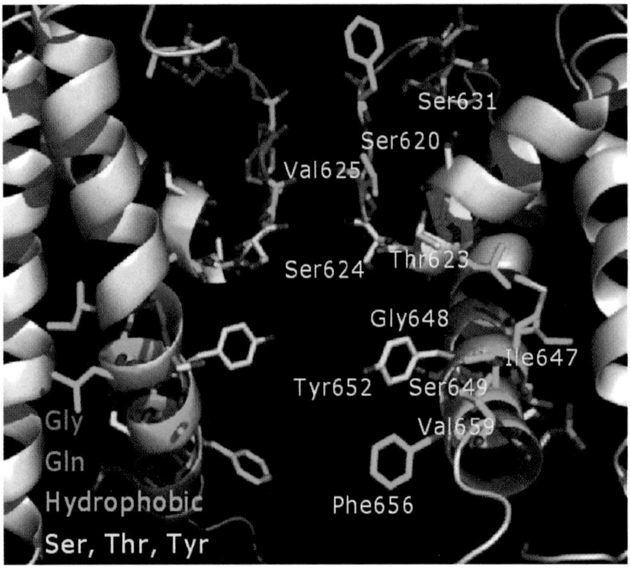

Figure 20.3 Modeled structure of a hERG channel. Amino acids are colored according to their physicochemical properties. Amino acids reported to participate in the binding of drugs [22] are labeled. For simplicity, only two of the four subunits that form the channel are shown.

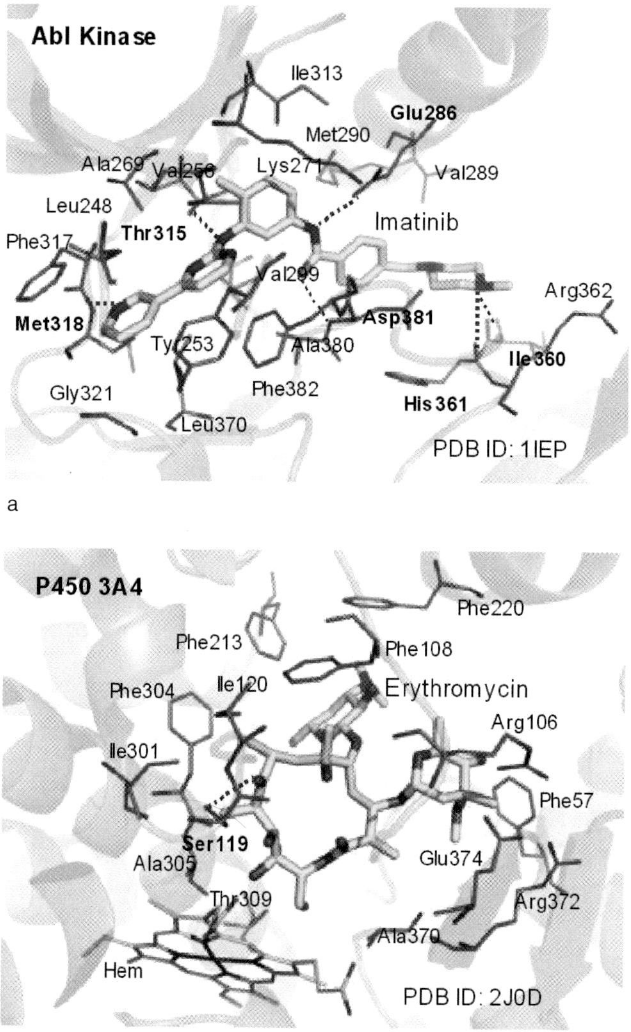

Figure 20.4 X-ray structures of the binding sites of the Abl kinase/imatinib and cytochrome P450 3A4 /erythromycin complexes (PDB: Protein Data Bank). The amino acids located within 4 Å of ligands are depicted. The hydrogen bonding interactions are shown by dashed lines, and the amino acids participating in hydrogen-bonding interactions are labeled in bold. The images were produced with Pymol (DeLano Scientific LLC).

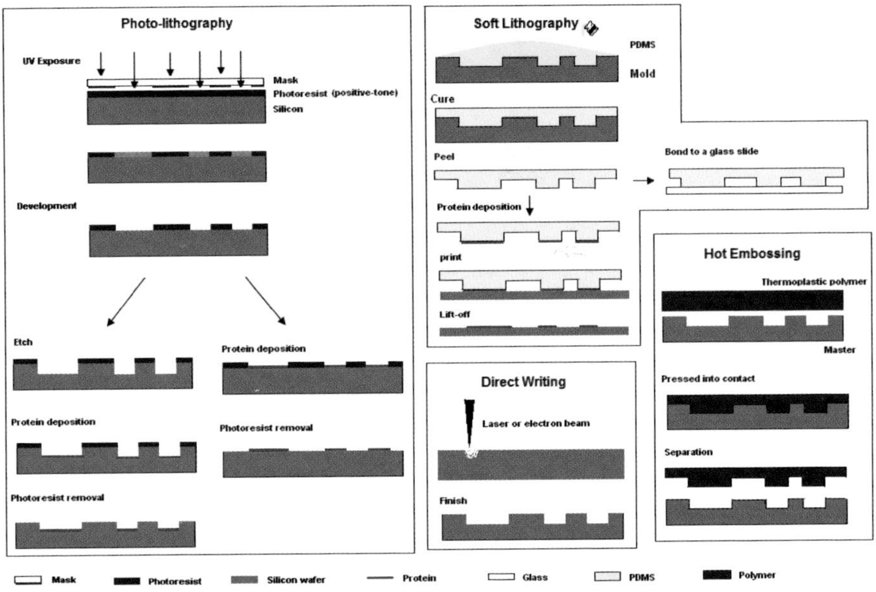

Figure 25.1 Schematics of typical nano and micro fabrication techniques: Photolithography, soft lithography, hot embossing, and direct writing.

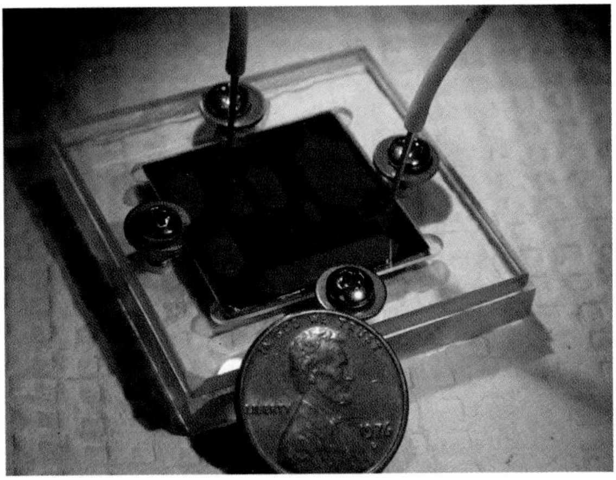

Figure 25.3 A typical μCCA device filled with red dye for visualization of fluidics pathways.

108. Luco JM. Prediction of brain–blood distribution of a large set of drugs from structurally derived descriptors using partial least squares (PLS) modeling. *J Chem Inf Comput Sci* 1999;39:396–404.

109. Wold S, Johansson E, Cocchi M. PLS-partial least squares projections to latent structures. In: Kubinyi H, editor. *3D-QSAR in drug design: Theory, methods and applications*. Leiden: ESCOM, 1993. p. 523–50.

110. Afzelius L, Masimirembwa CM, Karlen A, Andersson TB, Zamora I. Discriminant and quantitative PLS analysis of competitive CYP2C9 inhibitors versus non-inhibitors using alignment independent GRIND descriptors. *J Comput Aided Mol Des* 2002;16:443–58.

111. Klebe G, Abraham U, Mietzner T. Molecular similarity indices in a comparative analysis (CoMSIA) of drug molecules to correlate and predict their biological activity. *J Med Chem* 1994;37:4130–46.

112. Baroni M, Costantino G, Cruciani G, Riganelli D, Valigi R, Clementi S. generating optimal linear PLS estimations (GOLPE): An advanced chemometric tool for handling 3D-QSAR problems. *Quant Struct-Act Rel* 1993;12:9–20.

113. Kratochwil NA, Huber W, Muller F, Kansy M, Gerber PR. Predicting plasma protein binding of drugs: A new approach. *Biochem Pharmacol* 2002;64:1355–74.

114. Rosipal R, Trejo LJ. Kernel Partial Least Squares regression in reproducing Kernel Hilbert Space. *J Mach Learn Res* 2001;2:97–123.

115. Colmenarejo G, Alvarez-Pedraglio A, Lavandera J-L. Cheminformatic models to predict binding affinities to human serum albumin. *J Med Chem* 2001;44: 4370–8.

116. Breneman C, Sundling C, Sukumar N, Shen LQ, Katt WP, Embrechts M. New developments in PEST shape/property hybrid descriptors. *J Comput Aided Mol Des* 2003;17:231–40.

117. Whitehead C, Breneman C, Sukumar N, Ryan M. Transferable atom equivalent multicentered multipole expansion method. *J Comput Chem* 2003;24:512–29.

118. Riley RJ, Parker AJ, Trigg S, Manners CN. development of a generalized, quantitative physicochemical model of CYP3A4 inhibition for use in early drug discovery. *Pharm Res* 2001;18:652–5.

119. Buchwald P, Bodor N. Quantitative structure-metabolism relationships: Steric and non steric effects in the enzymatic hydrolysis of noncongener carboxylic esters. *J Med Chem* 1999;42:5160–8.

120. McElroy NR, Jurs PC, Morisseau C, Hammock BD. QSAR and classification of murine and human epoxide hydrolase inhibition by urea-like compounds. *J Med Chem* 2003;46:1066–80.

121. Dajani R, Cleasby A, Neu M, Wonacott AJ, Jhoti H, Hood AM, et al. X-ray crystal structure of human dopamine sulfotransferase, SULT1A3. *J Biol Chem* 1999;53:37862–8.

122. Penzotti JE, Lamb ML, Evenson E, Grootenhuis PDJ. A computational ensemble pharmacophore model for identifying substrates of *P*-glycoprotein. *J Med Chem* 2002;45:1737–40.

123. Tatsumi M, Groshan K, Blakely RD, Richelson E. Pharmacological profile of antidepressants and related compounds at human monoamine transporters. *Eur J Pharmacol* 1997;340:249–58.

124. Tatsumi M, Jansen K, Blakely RD, Richelson E. Pharmacological profile of neuroleptics at human monomaine transporters. *Eur J Pharmacol* 1999;368:277–83.

125. Thummel K, Shen DD. Design and optimization of dosage regimens: Pharmacokinetic data. In: Hardman JG, Limbird LE, Gilman AG, editors, *Goodman & Gilman's the pharmaceutical basis of therapeutics*. New York: McGraw-Hill, 2001. p. 1924–2023.

126. Norinder U, Haeberlein M. Computational approaches to the prediction of the blood-brain distribution. *Adv Drug Deliv Rev* 2002;54:291–313.

127. Chiou WL, Barve A. Linear correlation of the fraction of oral dose absorbed of 64 drugs between humans and rats. *Pharm Res* 1998;15:1792–5.

128. Chiou WL, Jeong HY, Chung SM, Wu TC. Evaluation of using dog as an animal model to study the fraction oral dose absorbed of 43 drugs in humans. *Pharm Res* 2000;17:135–40.

129. Chiou WL, Buehler PW. Comparison of oral absorption and bioavailability of drug between monkey and human. *Pharm Res* 2002;19:868–74.

130. Yazdanian M, Glynn SL, Wright JL, Hawi A. Correlating partitioning and Caco-2 cell permeability of structurally diverse small molecular weight compounds. *Pharm Res* 1998;15:1490–4.

131. Raevsky OA, Schaper K-J, Artursson P, McFarland JW. A novel approach for prediction of intestinal absorption of drugs in humans based on hydrogen bond desccriptors and structural similarity. *Quant Struct-Act Rel* 2001;20:402–13.

132. Harrison A, Gardner I, Morgan P. From rat to man: Pharmacokinetic predictions for compounds with multiple clearance mechanisms. *Drug Metab Rev* 2001;33:204.

133. McGovern SL, Caselli E, Grigorieff N, Shoichet BK. A common mechanism underlying promiscuous inhibitors from virtual and high-throughput screening. *J Med Chem* 2002;45:1712–22.

134. McGovern SL, Shoichet BK. Kinase inhibitors: Not just for kinases anymore. *J Med Chem* 2003;46:1478–83.

135. Seidler J, McGovern SL, Doman TN, Shoichet BK. Identification and prediction of promiscuous aggregating inhibitors among known drugs. *J Med Chem* 2003;46:4477–86.

136. Lombardo F, Obach RS, Shalaeva MY, Gao F. Prediction of volume of distribution values in humans for neutral and basic drugs using physicochemical measurements and plasma protein binding. *J Med Chem* 2002;45:2867–76.

137. Lombardo F, Shalaeva MY, Tupper KA, Gao F. ElogDoct: A tool for lipophilicity determination in drug discovery. 2. Basic and neutral compounds. *J Med Chem* 2001;44:2490–7.

138. Ekins S. In silico approaches to predicting metabolism, toxicology and beyond. *Biochem Soc Trans* 2003;31:611–4.

139. Ekins S, Balakin KV, Savchuk N, Ivanenkov Y. Insights for human ether-a-go-go-related gene potassium channel inhibition using recursive partitioning, Kohonen and Sammon mapping techniques. *J Med Chem* 2006;49:5059–71.

140. Embrechts M, Arciniegas F, Ozdemir M, Momma M. Scientific data mining with StripMiner. Proceedings of the IEEE Mountain workshop on soft computing in industrial applications. Virginia Tech, June 25–27, Blacksburg, Virginia, 2001. pp. 13–18, SMCiaOl.

141. Ozedmir M, Embrechts M, Arciniegas F, Breneman C, Lockwood L, Bennet K. Feature selection for in-silico drug design using genetic algorithms and neural networks. IEEE Mountain workshop on soft computing in industrial applications. Virginia Tech, Blackburg, Virginia, 2001.

142. Christianini N, Shawe-Taylor J. *Support vector machines and other kernel-based learning methods*. Cambridge: Cambridge University Press, 2000.

143. Lindgren F, Geladi P, Wold S. The kernel algorithm for PLS. *J Chemom* 1993;1993:45–59.

144. Ren S, Gao L. Simultaneous spectrophotometric deternmination of copper (II), lead (II) and cadmium (II). *J Automatic Chem* 1995;17:115–8.

145. Gao L, Ren S. Simultaneous spectrometric determination of manganese, zinc and cobalt by kernel partial least-squares method. *J Automatic Chem* 1998; 20:179–83.

146. Gao L, Ren S. Simultaneous spectrophotometric determination of four metals by the kernel partial least squares method. *Chemom Intell Lab Sys* 1999;45:87–93.

147. Ekins S, Berbaum J, Harrison RK. Generation and validation of rapid computational filters for CYP2D6 and CYP3A4. *Drug Metab Dispos* 2003;31:1077–80.

148. Crivori P, Cruciani G, Carrupt PA, Testa B. Predicting blood–brain barrier permeation from three-dimensional molecular structure. *J Med Chem* 2000;43:2204–16.

149. Zhao YH, Le J, Abraham MH, Hersey A, Eddershaw PJ, Luscombe CN, et al. Evaluation of human intestinal absorption data and subsequent derivation of a quantitative structure-activity relationship (QSAR) with the Abraham descriptors. *J Pharm Sci* 2001;90:749–84.

150. Scholkopf B, Platt JC, Shawe-Taylor J, Smola AJ, Williamson RC. Estimating the support of a high-dimensional distribution. Techical Report 99–87 *Microsoft Research* 1999.

151. Chang CC, Lin CJ. LIBSVM: A library for support vector machines. 2001.

152. Talele TT, Hariprasad V, Kulkarni VM. Docking analysis of a series of cytochrome P-450(14) alpha DM inhibiting azole antifungals. *Drug Des Discov* 1997;15:181–90.

153. Desiraju GR, Gopalakrishnan B, Jetti RKR, Nagaraju A, Raveendra D, Sarma JARP, et al. Computer-aided design of selective COX-2 inhibitors: comparative molecular field analysis, comparative molecular similarity indices analyses, and docking studies of some 1,2-diarylimidazole derivatives. *J Med Chem* 2002;45:4847–57.

154. Liu H, Huang X, Shen J, Luo X, Li M, Xiong B, et al. Inhibitory mode of 1,5-diarylpyrazole derivatives against cyclooxygenase-2 and cyclooxygenase-1: Molecular docking and 3D QSAR analyses. *J Med Chem* 2002;45:4816–27.

155. Yamamoto K, Masuno H, Choi M, Nakashima K, Taga T, Ooizumi H, et al. Three-dimensional modeling of and ligand docking to vitamin D receptor ligand binding domain. *Proc Natl Acad Sci USA* 2000;97:1467–72.

156. Lewis DFV, Ogg MS, Goldfarb PS, Gibson GG. Molecular modelling of the human glucocorticoid receptor (hGR) ligand-binding domain (LBD) by homology with the human estrogen receptor a (hERa) LBD: quantitative structure-activity relationships within a series of CYP3A4 inducers where induction is mediated via hGR involvement. *J Steroid Biochem Mol Biol* 2002;82:195–9.

157. Schapira M, Raaka BM, Samuels HH, Abagyan R. Rational discovery of novel nuclear hormone receptor antagonists. *Proc Natl Acad Sci USA* 2000;97: 1008–13.

158. Schapira M, Raaka BM, Samuels HH, Abagyan R. In silico discovery of novel retinoic acid receptor agonist structures. *BMC Structural Biol* 2001;1:1.

159. Flower DR. Modeling of G-protein coupled receptors for drug design. *Biochim Biophys Acta* 1999;1422:207–34.

160. Chang C, Swaan PW, Ngo LY, Lum PY, Patel SD, Unadkat JD. Molecular requirements of the human nucleoside transporters hCNT1, hCNT2 and hENT1. *Mol Pharmacol* 2004;65:558–70.

161. Kim KH. 3D-QSAR analysis of 2,4,5- and 2,3,4,5-substituted imidazoles as potent and nontoxic modulators of *P*-glycoprotein mediated MDR. *Bioorg Med Chem* 2001;9:1517–23.

162. Pajeva IK, Wiese M. Pharmacophore model of drugs involved in *P*-glycoprotein multidrug resistance: Explanation of structural variety (hypothesis). *J Med Chem* 2002;45:5671–86.

163. Chiba P, Ecker G, Schmid D, Drach J, Tell B, Goldenberg S, et al. Structural requirements for activty of propafenone-type modulators in *P*-glycoprotein-mediated multidrug resistance. *Mol Pharmacol* 1996;49:1122–30.

164. Stouch TR, Gudmundsson O. Progress in understanding the structure-activity relationships of *P*-glycoprotein. *Adv Drug Deliv Rev* 2002;54:315–28.

165. Hasegawa K, Miyashita Y, Funatsu K. GA strategy for variable selection in QSAR studies: GA-based PLS analysis of calcium channel antagonists. *J Chem Inf Comput Sci* 1997;37:306–10.

166. Kesuru GM. Prediction of hERG potassium channel affinity by traditional and hologram QSAR methods. *Bioorg Med Chem Lett* 2003;13:2773–5.

167. Roche O, Trube G, Zuegge J, Pflimlin P, Alanine A, Schneider G. A virtual screening method for the prediction of the hERG potassium channel liability of compound libraries. *ChemBioChem* 2002;3:455–9.

168. Clark DE. Rapid calculation of polar molecular surface area and its application to the prediction of transport phenomena: 2. Prediction of blood–brain barrier penetration. *J Pharm Sci* 1999;88:815–21.

169. Norinder U, Sjoberg P, Osterberg T. Theoretical calculation and prediction of brain–blood partitioning of organic solutes using molsurf parameterization and PLS statistics. *J Pharm Sci* 1998;87:952–9.

170. Luco JM, Ferretti FH. QSAR based on multiple linear regression and PLS methods for the anti-HIV activity of a large group of HEPT derivatives. *J Chem Inf Comput Sci* 1997;37:392–401.

171. Osterberg T, Norinder U. Theoretical calculation and prediction of *P*-glycoprotein-interacting drugs using MolSurf parameterization and PLS statistics. *Eur J Pharm Sci* 2000;10:295–303.

172. Ekins S, Ring BJ, Bravi G, Wikel JH, Wrighton SA. Predicting drug–drug interactions in silico using pharmacophores: A paradigm for the next millennium. In: Guner OF, editor, *Pharmacophore perception, development, and use in drug design*. San Diego: IUL, 2000. p. 269–99.

173. Morisseau C, Du G, Newman JW, Hammock BD. Mechanism of mammalian soluble epoxide hydrolase inhibition by chalcone oxide derivatives. *Arch Biochem Biophys* 1998;356:214–28.

174. Nakagawa Y, Wheelock CE, Morisseau C, Goodrow MH, Hammock BG, Hammock BD. 3D-QSAR analysis of inhibition of murine soluble epoxide hydrolase (MsEH) by benzoylureas, arylureas, and their analogues. *Bioorg Med Chem* 2000;8:2663–73.

175. Kim R, Fromm MF, Wandel C, Leake B, Wood AJ, Roden DM, et al. The drug transporter *P*-glycoprotein limits oral absorption and brain entry of HIV-1 protease inhibitors. *J Clin Invest* 1998;101:289–94.

176. Kim RB, Wandel C, Leake B, Cvetkovic M, Fromm MF, Dempsey PJ, et al. Interrelationship between substrates and inhibitiors of human CYP3A and *P*-glycoprotein. *Pharm Res* 1999;16:408–14.

177. Pauli-Magnus C, Von Richter O, Burk O, Ziegler A, Mettang T, Eichelbaum M, et al. Characterization of the major metabolites of verapamil as substrates and inhibitors of *P*-glycoprotein. *J Pharmacol Exp Ther* 2000;293:376–82.

178. Dresser MJ, Gray AT, Giacomini KM. Kinetic and selectivity differences between rodent, rabbit, and human organic catin transporters (OCT1). *J Pharmacol Exp Ther* 2000;292:1146–52.

179. Saiakhov R, Stefan LR, Klopman G. Multiple computer-automated structure evaluation model of the plasma protein binding affinity of diverse drugs. *Perspect Drug Discov Des* 2000;19:133–5.

180. Brewster ME, Pop E, Huang M-J, Bodor N. AM1-based model system for estimation of brain/blood concentration ratios. *Int J Quant Chem: Quant Biol Symp* 1996;23:1775–87.

181. Lombardo F, Blake JF, Curatolo WJ. Computation of brain–blood partitioning of organic solutes via free energy calculations. *J Med Chem* 1996;39:4750–5.

182. Palm K, Luthman K, Ungell A-L, Strandlund G, Artursson P. Correlation of drug absorption with molecular surface properties. *J Pharm Sci* 1996;85:32–9.

183. Sugawara M, Takekuma Y, Yamada H, Kobayashi M, Iseki K, Miyazaki K. A general approach for the prediction of the intestinal absorption of drugs: Regresssion analysis using the physiochemical properties and drug-membrane electrostatic interaction. *J Pharm Sci* 1998;87:960–6.

184. Terp GE, Johansen BN, Christensen IT, Jorgensen FS. A new concept for multidimensional selection of ligand conformations (multiselect) and multidimensional scoring (multiscore) of protein-ligand binding affinities. *J Med Chem* 2001;44: 2333–43.

185. Muegge I, Martin YC, Hajduk PJ, Fesik SW. Evaluation of PMF scoring in docking weak ligands to the FK506 binding protein. *J Med Chem* 1999;42:2498–503.

186. Wang R, Lai L, Wang S. Further development and validation of empiracal scoring functions for structure-based binding affinity prediction. *J Comput Aided Mol Des* 2002;16:11–26.

187. Roche O, Schneider P, Zuegge J, Guba W, Kansy M, Alanine A, et al. Development of a virtual screening method for identification of "frequent hitters" in compound libraries. *J Med Chem* 2002;45:137–42.

188. Barratt MD, Rodford RA. The computational prediction of toxicity. *Curr Opin Chem Biol* 2001;5:383–8.
189. Cronin MT. Computer-aided prediction of drug toxicity in high throughput screening. *Pharm Pharmacol Commun* 1998;4:157–63.
190. Cronin MTD, Schultz TW. Development of quantitative structure-activity relationships for the toxicity of aromatic compounds to tetrahymena pyriformis: Comparative assessment of the methodologies. *Chem Res Toxicol* 2001;14:1284–95.
191. Greene N. Computer systems for the prediction of toxicity: An update. *Adv Drug Deliv Rev* 2002;54:417–31.
192. Chen YZ, Ung CY. Prediction of potential toxicity and side effect protein targets of a small molecule by a ligand-protein inverse docking approach. *J Mol Graph Model* 2001;20:199–218.
193. So S-S, Karplus M. A comparitive study of ligand-receptor complex binding affinity prediction methods based on glycogen phosphorylase inhibitors. *J Comput Aided Mol Des* 1999;13:243–58.
194. Sheridan RP, Feuston BP, Maiorov VN, Kearsley SK. Similarity to molecules in the training set is a good discriminator for prediction accuracy in QSAR. *J Chem Inf Comput Sci* 2004;44:1912–28.

16

HOMOLOGY MODELS APPLIED TO TOXICOLOGY

Stewart B. Kirton, Phillip J. Stansfeld, John S. Mitcheson, and Michael J. Sutcliffe

Contents

16.1 INTRODUCTION

Experimental techniques—particularly X-ray crystallography, nuclear magnetic resonance (NMR) spectroscopy and electron microscopy—are able to determine the 3D structure of a protein. Co-crystallization of proteins with known inhibitors and substrates can give insight into protein–ligand interactions in the active site, enabling the inference of likely metabolites, how

Computational Toxicology: Risk Assessment for Pharmaceutical and Environmental Chemicals, Edited by Sean Ekins

modifications to the ligand and/or protein structure may potentially affect protein–ligand binding, and if there is scope for adverse ligand–ligand interactions. All this information provides a direct empirical means of assessing and predicting the potential toxicity of compounds.

However, the vast majority of proteins are not immediately amenable to these experimental techniques. Membrane-bound proteins are difficult to crystallize in the aqueous environment required for X-ray crystallography. Others are insufficiently soluble or too large for NMR studies, or are not robust enough to withstand the extreme pretreatment of samples required to obtain an electron micrograph. These factors, coupled with the rapid expansion and success of the various genome projects, have resulted in a large deficit between the knowledge of a protein amino acid sequence and their 3D structures. To address this deficit, alternative methods have been developed to help us understand the functional role of a protein in terms of its 3D structure, which gives important insight into its function. One such technique is homology (or comparative) modeling; other techniques include quantitative structure-activity relationships (QSAR) and the use of pharmacophores, both of which are discussed in other chapters.

This chapter focuses on the application of homology modeling (Section 16.2) to two aspects of toxicity: drug metabolism (Section 16.3) and ion channel block (Section 16.4). For drug metabolism, we focus on the cytochromes P450 and present CYP2D6 as a case study; for ion channel block, we focus on the human ether-a-go-go-related gene (hERG) K^+ channel as a case study.

16.2 HOMOLOGY MODELING

The premise underpinning homology modeling arises from the observation that proteins with similar amino acid sequences have a tendency to adopt similar 3D structures [1]. Therefore it is possible to predict the 3D structure of a protein based solely on knowledge of its amino acid sequence and the 3D structures of proteins with similar sequences.

Homology modeling requires (1) the amino acid sequence of the target protein (the protein for which there is no 3D structure) and (2) proteins with similar amino acid sequence to the target and known 3D structures (templates). This information is used to generate an approximation of the true 3D structure for the target protein (Figure 16.1). Homology modeling is an iterative process with validation possible at almost every level.

The first stage when building a homology model is to generate an alignment between the amino acid sequences of the target proteins and the amino acid sequences of the templates. This is the single most important step of the process, as misalignment of target and template sequences will introduce errors into any subsequent models that can lead to unnecessary misinformation and confusion. Once the amino acid sequence alignment is produced, the next step is to derive a 3D model by mapping the target sequence loosely onto

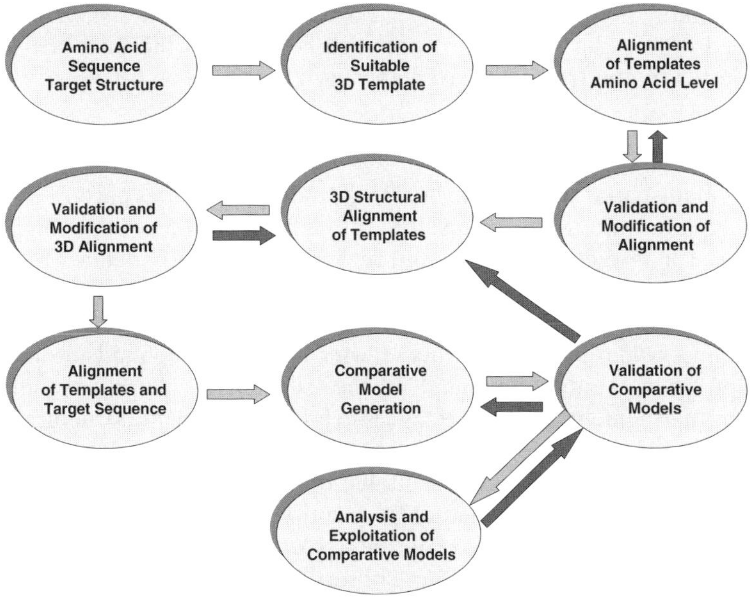

Figure 16.1 Overview of the comparative modeling process to demonstrate the iterative nature of the technique.

the known 3D structure of a homologous template. Selection of appropriate templates upon which to base the model is unique to each modeling study. If there is a single template with a significantly high sequence homology (or identity) to the target (typically greater than 80% identity), it is feasible and valid to base the models on that structure alone. Sometimes the target has relatively low sequence identity when compared to the templates. In this case increasing the number of templates allows a more comprehensive sampling of conformational space in the regions of low homology, and prevents overfitting of the amino acid sequence to the 3D structures. Consequently, using more than one template, if the sequence homology between templates and target is low, can generate a set of models that provide a better representation of the target protein than if a single low homology template is used. The CASP (critical assessment of structure prediction) experiments in particular [2–9] have shown that as a general rule if the sequence identity of the templates compared to the target is below 25%, the reliability and accuracy of any model is severely limited.

The quality of the models generated are critically dependent on (1) the choice of structural template(s) and (2) the sequence alignment used, emphasising the need for care in the initial stages. Homology modeling software tend to use one of two approaches—a fragment-based stepwise approach, (e.g., SWISSMODEL [10] and COMPOSER [11,12]), or a single-step probabilistic approach (e.g., Modeller [13]). Both approaches can produce high-quality models. However, the latter tends to be the method of choice as it enables experimentally derived restraints (e.g., from NMR-derived distances [14]) to be added during the modeling process, rather than in the post hoc fashion necessitated by the fragment-based approach.

16.2.1 Fragment-Based Homology Modeling

The fragment-based approach to homology modeling divides parts of the structural templates into three groups: the well-defined regions of the polypeptide backbone (or "structurally conserved regions"), the poorly defined regions of the polypeptide backbone (or "structurally variable regions"), and the amino acid sidechains. The first stage is to build the polypeptide backbone of the structurally conserved regions of the model, which are either defined automatically by the program or entered manually by the user. This results in a disjointed set of structural fragments. The next stage is to join these with the structurally variable regions of the polypeptide backbone (often "loops" joining secondary structural elements) or, where no suitable fragment exists in the templates, to scan a database of protein structures to identify a suitable fragment. The third stage is to alter the atoms in the amino acid sidechains, where necessary, so that they adopt geometries similar to those observed by crystal structure determination. This is followed by an energy minimisation of the model.

16.2.2 Single-Step Homology Modeling

The single-step approach represents the individual structural features in the model (e.g., main chain conformation and the position of hydrogen bonds) by probability distribution functions based partly on the structure of the template(s) and partly on known stereochemistry. These probabilities are used as restraints when the model is constructed. Since each feature is represented as a probability distribution function, a family of models consistent with this set of distributions, each with a slightly different conformation, is produced. If steric problems (particularly with bond lengths and, to a lesser extent, bond angles) occur consistently across the family of models, these are indicative of an error in the amino acid sequence alignment. The sequence alignment can be compared to the 3D alignment of the template structures with the models using interactive molecular graphics (e.g., PyMol [15] and Swiss-PDBViewer [16]), and the sequence alignment amended accordingly. The major advantage of this approach is that restraints derived from both experimental observations and analysis of previous models can be included in the modeling process in the form of distance restraints.

16.2.3 Validity Testing of Homology Models

Homology modeling is normally used when no experimentally derived structure is available, so it is impossible to compare the quality of a (theoretical) homology model with an existing crystal structure. Therefore it is necessary to subject the model to a number of complementary, yet independent, tests to obtain a measure of its validity. It is outside the scope of this review to consider our methods of choice in detail (e.g., see [17] for a detailed discussion). Although an ensemble of models, when considered as a whole, gives a more accurate insight into the system under investigation, it is usually more practical to consider only a single model at any one time. This is either the model deemed to have the lowest energy by the program used to produce it (e.g., Modeller [13]) or the model that, by some sort of clustering algorithm (e.g., NMRCLUST [18]), has been deemed most representative of the group. Single models selected in this way are validated by a number of tests to assess how well they represent the target protein.

Two types of checks are generally used for this: stereochemical quality—the less strained the structure, the more likely the model is to be correct—and side chain environment—the more thermodynamically favourable the configuration of atoms in the protein, the more likely the model is to be correct. A third check can also be used: the root mean square deviation (RMSD) between the main chain atoms of the model and of the most homologous template; this gives a very rough indication of how well the models produced have sampled conformational space. Examples of programs employed to check stereochemical quality of the models include PROCHECK [19], PROVE [20], and WhatIf? [21]. These are available through the Biotech Validation Suite [22]. Amino acid environment can be assessed using complementary programs such as Errat [23] and Verify 3D [24].

It is important to stress that results from all of these validation procedures must be contextualized by direct comparison with the results achieved by the templates used to generate the model. It is unreasonable to expect a model to perform better than any of the templates it was based upon, and as such comparable values between the templates and the models are indicative of a valid model. It has been shown that the RMSD between main chain atoms in the template with the highest homology to the target and the model is a good method of validation [25]. This builds on earlier work [1,26] that showed that for two proteins there was a relationship between the percentage sequence identity and the RMSD of the main chain atoms in homologous regions.

In addition to computational validation of the models, it is also important to validate the models against available experimental data. For example, site-directed mutagenesis data can give important insight into potential structural and functional roles of the respective amino acid(s)—these data can be used to support and/or refine the model. Experimental data from a range of ligands can also be used to support and/or refine the model. To facilitate validation, insight at the atomic level of how ligands bind to (models of) proteins—central to understanding toxicology at the molecular level—can be gained using molecular docking (e.g., for a review, see [27]).

16.3 CYTOCHROMES P450

The cytochromes P450 (CYPs) are a ubiquitous superfamily of enzymes. In mammals these enzymes are involved, among other things, in the metabolism of xenobiotic compounds including environmental toxins and therapeutic drugs. One of the most interesting characteristics of the CYPs is their promiscuity. Individual isoforms are capable of interacting with a wide range of chemically diverse substrates, and some CYPs have overlapping substrate specificities. This promiscuity is useful in terms of the defense of the organism against potentially harmful xenobiotics, but in some instances can lead to rapid drug clearance/inactivation, production of toxic compounds, and/or adverse drug–drug interactions. Therefore it is desirable to have an empirical or theoretical model for the structures of the drug-metabolizing CYPs.

In humans 90% of all of the drugs currently approved for clinical use are metabolized by one of seven CYP isoforms: CYP1A2, CYP2C9, CYP2C18, CYP2C19, CYP2D6, CYP2E1, and/or CYP3A4 [28–30]. Of these isoforms, CYP2D6 and CYP2C9 display polymorphisms that can result in the poor metabolism of drugs [31–37]. Having knowledge of the structural features of the active sites of these seven isoforms, could lead to a tool that is able to predict whether or not a drug candidate will interact with the CYPs and, if so, with which isoform the drug candidate may interact preferentially. This would improve the rational design of therapeutic drugs and target-specific inhibitors. It would also improve the risk assessment of xenobiotics and the avoidance of adverse drug–drug interactions, whereby one drug modulates the metabolism of another [30] by simple competition for the same active site, and/or binding in an allosteric region of the same enzyme. In addition active site knowledge for these enzymes could significantly reduce the failure rate in clinical trials by identifying any CYP liabilities in the early stages of drug development, and reduce the amount of time and money required to bring a new pharmaceutical to the market.

Over the years many homology models of the CYPs have appeared in the literature (Table 16.1). Up until the year 2000 all structural models for the human CYPs were based on the X-ray crystal structures of distantly related bacterial CYP isoforms. In 2000 a major breakthrough was achieved by Williams and coworkers [38] who were able to elucidate the structure of the rabbit enzyme CYP2C5. This structure was more closely related to the human isoforms than the bacterial isoforms, and by incorporating the new sequence into homology modeling experiments, the quality and accuracy of the homology models of the human CYPs was vastly improved (e.g., see [39]). More recently still, the determination of X-ray crystal structures for several human isoforms important in drug metabolism (CYP2A6 [40], CYP2C8 [41], CYP2C9 [42,43], CYP2D6 [44], and CYP3A4 [45,46]) has removed the need for homology models in some instances and has also, along with the availability of a second rabbit crystal structure (CYP2B4) [47,48], improved the quality of models for the other human isoforms.

By comparing the evolution of homology models of the CYPs and what we have learned from them with available experimental structures, we will be able to objectively assess the impact homology models have had on the understanding of protein–ligand interactions. In particular, as a case study we will focus

TABLE 16.1 Overview of Comparative Models of CYPs in the Scientific Literature

CYP Subfamily		Approach[a]	Authors & Year
CYP1A	1A1 & 1A2	Single (102A1)	Lewis et al., 1996 [134]
	1A1	Single (2C5)	Szklarz and Paulsen, 2002 [135]
			Liu et al., 2003 [136]
	1A2	Single (102A1)	Lozano et al., 1997 [137]
	1A2	Single (102A1)	Dai et al., 1998 [138, 139]
	1A2	Multiple (101A1, 102A1, 107A1, & 108A1)	de Rienzo et al., 2000 [140]
	1A2	Single (2C5)	Cho et al., 2003 [141]
	1A2	Single (2C5)	Lewis ct al., 2003 [142]
	1A2	Single (2C5)	Kim and Guengerich, 2004 [143,144]
	1A1 & 1A2	Multiple (102A1, 108A1, & 2C5)	Iori et al., 2005 [145]
	1A2 & 1A6	Single (102A1)	Lewis et al., 1999 [60]
CYP1B	1B1	Single (2C5)	Lewis et al., 2003 [146]
CYP2A	2A1, 2A4, 2A5, & 2A6	Single (102A1)	Lewis et al., 1995 [147]
	2A5	Single (2C5)	Stahl and Höltje, 2005 [148]
	2A6	Single (2C5)	Lewis et al., 2002 [149]
	2A6	Single (2C5)	Lewis et al., 2003 [150]
	2A6 & 2A13	Single (2C5)	He et al., 2004 [151]
CYP2B	2B1	Single (101A1)	Szklarz et al., 1994 [152]
	2B1	Multiple (101A1, 102A1, & 108A1)	Szklarz et al., 1995 [153]
	2B1	Multiple (102A1 & core regions of 101A1 & 108A1)	Dai et al., 1998 [139,154]
	2B1	Single (2C5)	Kumar et al., 2003 [155]
	2B1	Single (2B4)	Honma et al., 2005 [156] & Li et al., 2005 [157]
	2B1, 2B4, & 2B6	Single (102A1)	Lewis et al., 1997 [158]
	2B1, 2B4, & 2B5	Single (2C5)	Spatzenegger et al., 2001 [159]
	2B4	Multiple (101A1, 102A1, 107A1, & 108A1)	Chang et al., 1997 [160]
	2B4	Single (2C5)	Sechenykh et al., 2002 [161]
	2B4	Single (2C5)	Harris et al., 2004 [162]
	2B4	Multiple	Hodek et al., 2004 [163]

TABLE 16.1 (*Continued*)

CYP Subfamily		Approach[a]	Authors & Year
	2B6	Based on CYP2B1 homology model [153]	Domanski et al., 1999 [164]
	2B6	Single (102A1)	Lewis et al., 1999 [60]
	2B6	Single (102A1)	Lewis et al., 1999 [60, 165]
	2B6	Single (2C5)	Bathelt et al., 2002 [166]
	2B6	Single (2C5)	Lewis et al., 2002/2003 [149,167]
	2B6	Single (2C5)	Wang and Halpert, 2002 [168]
CYP2C	2C9	Multiple (101A1, 102A1, 107A1, & 108A1)	Payne et al., 1999 [169]
	2C9	Multiple (2C5 & part of 102A1)	de Groot et al., 2002 [170]
	2C9	Single (2C5)	Afzelius et al., 2001 [171]
	2C9 & 2C19	Single (102A1)	Lewis et al., 1998 [60, 172]
	2C9 & 2C19	Single (2C5)	Oda et al., 2004 [173]
	2C8, 2C9, & 2C19	Single (2C5)	Lewis et al., 2002/2003 [149, 167]
	2C8, 2C9, & 2C19	Single (2C5)	Tanaka et al., 2004 [174]
	2C18 & 2C19	Multiple (101A1, 102A1, 107A1, & 108A1)	Payne et al., 1999 [175]
	2C8, 2C9, 2C18, & 2C19	Single (2C5)	Ridderstrom et al., 2001 [176]
CYP2D	2D6	Single (101A1)	Koymans et al., 1993 [177]
	2D6	Multiple (101A1, 102A1, 108A1, & NMR restraints)	Modi et al., 1996 [14]
	2D6	Multiple (101A1, 102A1, & 108A1)	de Groot et al., 1996 [178]
	2D6	Single (102A1)	Lewis et al., 1997 [179]
	2D6	Multiple (101A1, 102A1, 107A1, 108A1, & NMR restraints)	Modi et al., 1997 [58]
	2D6	Multiple (101A1, 102A1, & 108A1)	de Groot et al., 1999 [180,181]
	2D6	Single (102A1)	Lewis et al., 1999 [60]
	2D6	Multiple (101A1, 102A1, 107A1, & 108A1)	de Rienzo et al., 2000 [140]
	2D6	Single (2C5)	Bapiro et al., 2002 [182]
	2D6	Multiple (101A1, 102A1, 107A1, 108A1, & 2C5)	Kirton et al., 2002 [39] & Kemp et al., 2004 [64]

TABLE 16.1 (*Continued*)

CYP Subfamily		Approach[a]	Authors & Year
	2D6	Single (2C5)	Lewis et al., 2002/2003 [149,167,183]
	2D6	Single (2C5)	Snyder et al., 2002 [184]
	2D6	Single (2C5)	Venhorst et al., 2003 [75]
	2D6	Single (2C5)	Yao et al., 2004 [185]
	2D6	Single (2C5)	Allorge et al., 2005 [186]
CYP2E	2E1	Single (102A1)	Lewis et al., 1997 [187]
	2E1	Single (102A1)	Tan et al., 1997 [188]
	2E1	Single (102A1)	Lewis et al., 1999 [60]
	2E1	Single (2C5)	Lewis et al., 2002 [149]
	2E1	Single (2C5)	Lewis et al., 2003 [189]
	2E1	Multiple (101A1, 102A1, 107A1, 108A1, 55A1, & 2C5)	Park and Harris, 2003 [190]
CYP3A	3A4	Single (101A1)	Ferenczy and Morris, 1989 [191]
	3A4	Single (102A1)	Lewis et al., 1996 [192]
	3A4	Multiple (101A1, 102A1, 107A1, & 108A1)	Szklarz et al., 1997 [193]
	3A4	Single (102A1)	Lewis et al., 1999 [60]
	3A4	Multiple (101A1, 102A1, 107A1, & 108A1)	de Rienzo et al., 2000 [140]
	3A4	Single (2C5)	Tanaka et al., 2004 [194]
CYP4A	4A11	Multiple (101A1, 102A1, 107A1, & 108A1)	Chang and Loew, 1999 [195]
	4A11	Single (102A1)	Lewis et al., 1999 [196]
CYP4F	4F3A & 4F11	Multiple (101A1, 102A1, 107A1, 108A1, & 2C5)	Kalsotra et al., 2004 [197]
CYP17	17 (17α)	Single (101A1)	Laughton et al., 1993 [198]
	17 (17α)	Multiple	Schappach and Höltje, 2001 [199]
CYP19	19 (aromatase)	Single (102A1)	Auvray et al., 2002 [200]
	19 (aromatase)	Single (2C5)	Chen et al., 2003 [201]
	19 (aromatase)	Single (2C9)	Favia et al. (2006) [202]
CYP27	27A1	Single (102A1)	Murtazina et al., 2002 [203]
	27B1	Single (2C5)	Yamamoto et al., 2004 [204]

[a] *Approach* refers to how the amino acid sequence alignment, on which the modeling is based, was produced: *single* indicates that the alignment is based on one 3D structure, whereas *multiple* indicates the alignment is based on more than one structure.

on the development of a model for CYP2D6, and comparison with the subsequently determined crystal structure.

16.3.1 CYP2D6: A Case Study

The CYP2D6 isoform is central to drug metabolism in humans. It is responsible for the clearance of at least 20% of the compounds in current clinical use, including antiarryhthmics, antidepressants, antipsychotics, beta-blockers, and analgesics [32]. This isoform is of particular interest to the pharmaceutical industry because it displays a genetic polymorphism, the consequence of which is large inter-individual and ethnic differences in the drug metabolism mediated by CYP2D6. This is highlighted by the defect in human called the debrisoquine/sparteine polymorphism [31,49]. This polymorphism arises from the culmination of various genetic mutations and affects a significant proportion of the Caucasian population [50]. It results in the defective metabolism of a number of clinical drugs, and inheritance of the "poor-metabolizer" phenotype has been linked with an increased susceptibility to Parkinson's disease and certain types of cancer [49,51]. As a result many pharmaceutical companies are interested in designing novel compounds that do not have a CYP2D6 liability. If a compound was preferentially metabolized by this isoform, it would increase the chances of failure as it progressed through clinical trials. Understanding the structure of CYP2D6, and potential protein–ligand interactions would aid in this rational metabolic design of potential drug candidates. Prior to the elucidation of the X-ray crystal structure, this was assisted by homology models.

Homology Models of CYP2D6 The first published homology model of CYP2D6 was created by Koymans et al. and appeared in the literature in 1993 [52]. It was based on the alignment of the CYP2D6 sequence against the bacterial enzyme CYP101A, isolated from *pseudomonas putida* (also commonly called $P450_{cam}$). The crystal structure was only distantly related to the human isoform (<25% sequence identity with CYP2D6) but was, at the time, the only CYP that had been studied in sufficient detail so that all of the residues implicated in protein–ligand binding had been identified. Combining the docking of the known CYP2D6 substrates dextromethorphan and debrisoquine into the homology model, the Dutch group was able to rationalize a previous in silico model for CYP2D6 inhibition [53] and postulate those residues that comprised the active site and those that were important in ligand binding. Even at this early stage, and despite the lack of sequence identity between the model and the bacterial template, the critical role of Asp 301 in forming a salt-bridge between the protein and the basic nitrogen atom of the substrate was postulated. The power of homology modeling became apparent at this point as observations from the model were used to direct "wet laboratory" investigations. Evidence supporting the proposed role of Asp 301 in ligand binding was gathered from further modeling experiments [14,54] and empirical mutagenesis data [55].

The publication and analysis of the X-ray structure of $P450_{cam}$ paved the way for the elucidation of other bacterial P450s and the X-ray crystal structures for CYP102A1 ($P450_{BM-3}$) isolated from *Bacillus megaterium* and CYP108A1 ($P450_{terp}$) from *Pseudomonas spp* quickly became publicly available. CYP102A1, in particular, caused a stir in the modeling community because it was the first (and as of yet the only) example of a class II bacterial P450, namely a P450 redox system akin to that of CYP isoforms in humans—comprising a two-component system made up of (1) a FAD-containing, flavin mononucleotide (FMN)-containing NADPH-dependent cytochrome P450 reductase and (2) a P450. Although the overall sequence identity between CYP102A1 and human CYP2D6 was not significantly improved compared to similarity with CYP101A, it was hoped that the close mechanistic behavior of the bacterial enzyme and the human isoform might give further insight into the function of the human CYP isoforms. As a result a spate of CYP2D6 homology models appeared in the literature between 1996 and 1999 incorporating the new bacterial crystal structures [14,56–60]. These models served to strengthen the resolve that Asp 301 was important in substrate binding, and were able to rationalize the metabolism of a number of known CYP2D6 substrates by using the homology models in conjunction with various docking algorithms. These experiments generated interest in other hydrophobic and hydrophilic residues that were believed to play a role in substrate recognition, selectivity and binding. For example, Modi et al. [58] used an homology model of CYP2D6 to rationally design a testosterone hydroxylase, the F483I mutant—based on a comparison with the sequence of CYP2D9—and then demonstrated that this CYP2D6 mutant did indeed possess the unprecedented ability to metabolize testosterone. It was apparent, even at this early stage, that despite the misgivings due to low sequence homology between the human CYP2D6 isoform and the bacterial templates, some interesting and useful insights were being garnered from the modeling studies.

The ability to accurately model the human CYP isoforms improved significantly with the publication of the first mammalian CYP crystal structure (CYP2C5 from rabbit) [38,61,62]. This sequence shared 40% identity with the human CYP2D6 isoform, and the second generation models that incorporated the crystal structure of CYP2C5 in the modeling process served to consolidate theories surrounding the importance of residues such as Glu216 in substrate recognition and binding [39], and the role of Phe120 as a steric influence in substrate binding [63]. The models were used to drive a raft of experimental work to test the hypotheses that they had generated. Gradually a great understanding of the factors influencing substrate selectivity and metabolism in CYP2D6 was achieved from a combination of homology models and experimental work based on site-directed mutagenesis.

Because of the increased confidence in the accuracy of previously ambiguous areas of the CYP2D6 homology model, the second-generation of homology models also provided the opportunity to derive quantitatively predictive models. In one publication [64] molecular docking was used in conjunction with the improved homology models of CYP2D6 to validate the accurate prediction of binding modes of those substrates that do not contain basic nitrogen atoms (e.g., [65]) and also to predict the relative strength of inhibition for a series of compounds. This is particularly noteworthy because, first, the docking of molecules were performed into a homology model, rather than a crystal structure. Hence the prediction was prone to inherent errors resulting from the differences in structure between the model and any experimental structure. Second, binding affinities are notoriously difficult to predict. With previous docking studies carried out on 11 different $P450_{cam}$ ligands, no clear correlation between predicted and experimentally determined binding affinities was found [66].

Docking Ligands into Models of CYP2D6 The importance of the use of docking with high-quality homology models in order to gain insight into metabolism, toxicology, and enzyme mechanism has been discussed. The relevance of docking to CYP2D6 (and the CYPs in general) is that the majority of the mainstream scoring functions used to rank and order binding orientations generated by docking algorithms have not been parameterized for heme-containing proteins, and as such do a poor job in predicting the crystallographically observed binding modes in a number of heme–protein ligand complexes. A study carried out in 2005 [67] using the docking algorithm GOLD [68,69] showed that for 45 heme-containing complexes, two well-known scoring functions—GOLDSCORE [68,69] and CHEMSCORE [70,71]—predicted only 57% and 64%, respectively, of the binding orientations correctly (defined as the highest ranked docking being within 2 Å RMSD of the crystallographically observed binding mode). This compares to an overall success rate for the GOLD validation set of 79% and can be attributed to a lack of parameterization for these specialist enzymes when the scoring functions were initially derived. The empirical evidence was used to define more stringent parameters, from which target-specific versions of GOLDSCORE and CHEMSCORE—implemented specifically for heme-containing proteins—were derived. These significantly improved the success rate for predicting the experimentally observed binding mode to 65% and 73% for GOLDSCORE and CHEMSCORE, respectively. The improvement in predicting metabolic data for the CYPs will enhance our ability to accurately assess in silico the potential toxicological issues for novel compounds. For example, these revised parameters were used recently [72] to investigate drug–drug interactions in CYP2D6 with anticancer treatments—the same study also successfully predicted a novel metabolite of metoclopramide, and suggested a particular CYP2D6 genotype/phenotype for those experiencing adverse reactions with metoclopramide, such as the extra-pyramidal syndrome.

Comparing the Models of CYP2D6 to the Crystal Structure Recently the crystal structure of CYP2D6 was determined [44]. The authors of this study carried out an exhaustive comparison of this structure with the homology models in the literature. The general conclusion was that the models provided

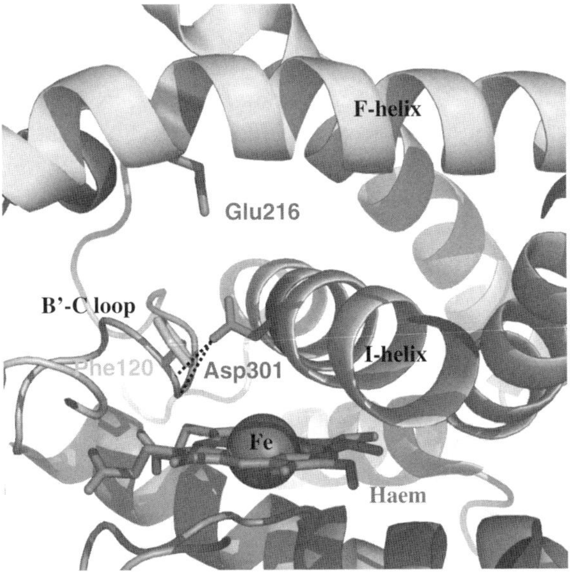

Figure 16.2 Schematic representation of the crystal structure of CYP2D6 [44] illustrating the position of amino acids identified as key by model building. Hydrogen bonds identified by modeling between Asp301 and the main chain amides of Val119 and Phe120 are denoted by dashed lines. (Image produced using Pymol [15].) See color plates.

a realistic picture of likely protein–ligand interactions. The hypothesis proposed independently by Kirton et al. [39] and Hanna et al. [73] that Asp301 is involved in hydrogen bonding to the B–C loop was verified when the crystal structure showed (Figure 16.2) that this residue did indeed form hydrogen bonds to the NH groups of Val119 and Phe120. The crystal structure (Figure 16.2) also supports the hypothesis [39] that both Asp301 and Glu216 are involved in substrate binding, although it is acknowledged that co-crystal data are required before this can be definitively established.

The role of active site aromatic residues—in particular, Phe483 and Phe481—are also explained by the crystal structure. The majority of homology models show Phe483 oriented toward the cavity and Phe481 remote from that position. This is consistent with the crystal structure (Figure 16.2) and does not explain why Phe481 has such a marked effect on the metabolism of a variety of substrates. However, a homology model produced by de Groot et al. [59] orients Phe481 into the active site cavity, while Phe483 is inaccessible. This model led to a constrained molecular dynamics study [44] that shows conformation in the active site can change to accommodate both Phe481 and Phe483. This allows Phe481 to be implicated in substrate recognition, while Phe483 is involved in reaction site binding. Without the model of de Groot et al.—and based solely on crystallographic evidence—this phenomenon may have gone unnoticed.

Phe120 is also interesting. The crystal structure places this residue in the active site cavity (Figure 16.2), as do the models based upon the bacterial $P450_{BM-3}$ structures. The more recent homology models, based on 2C5 invariably show Phe120 oriented away from the active site, and in retrospect this has been attributed to a misalignment of residues during the initial amino acid sequence alignments, emphasizing the importance of this stage of the homology modeling process. Alignment of the phenylalanines in both CYP2C5 and CYP2D6 B′–C loop regions gives the incorrect orientation of Phe120 in CYP2D6. An alternative alignment—which pairs Phe120 of CYP2D6 with an adjacent alanine in CYP2C5—does produce models with Phe120 in the active site cavity [63,74,75].

Moving away from the active site, the reductase binding region is conserved to a greater extent across the P450s than the *N*-terminal region. As such it is not surprising that the majority of homology models overlap particularly well in this region with the crystal structure.

Summary of CYP2D6 Modeling From CYP2D6, as an example, it is clear that homology modeling combined with molecular docking, active site characterization, bioinformatics analysis, and stringent well-paramaterized scoring functions can predict the sites of metabolism of a range of known substrates, and successfully identify key residues for substrate recognition and binding in the active site. It is also evident, by comparison with the recently available crystal structure of CYP2D6, that high-quality homology models can give a range of insights into the mechanism of an enzyme, and from this infer possible toxicological consequences for compounds of interest. However, it is important to remember that caution must be exercised in the initial stages of model building by ensuring that an accurate amino acid sequence alignment is obtained.

16.4 ION CHANNELS

Ion channels are membrane proteins that play essential roles in cell physiology and pharmacology (also see Chapter 13). Consequently they have been identified as potential targets for novel therapeutic compounds in a range of disease areas [76], such as cardiovascular disorders, central and peripheral nervous system disorders, and metabolic disorders. Unfortunately, they are also inextricably linked with the absorption, distribution, metabolism, and excretion/toxicology (ADME/Tox) profile of other therapeutic compounds—often through unforeseen drug–ion channel interactions [77].

Membrane proteins represent roughly 30% of the genome for most organisms [78], yet the structural information for the membrane domains of these proteins (e.g., see [79]) is limited to just over 200 (less than 1%) of the total number of entries within the protein data bank (PDB [80]). Nonetheless, there are 3D data for other non–membrane-bound domains, including some ligand binding sites. Recent developments have identified prokaryotic homologues of eukaryotic channels as potential targets for structural determination [81]. These prokaryotic channels are more simplistic in nature than their eukaryotic counterparts, and are therefore easier to obtain in quantities sufficient for crystallographic studies [81]. These structures can then be used as tools to aid in the elucidation of crystal structures of eukaryotic ion channels, such as the recently crystallized Kv1.2 channel [82].

A more accessible method used to exploit the prokaryotic structures as templates for the eukaryotic targets is comparative, or homology, modeling. This, in conjunction with ligand docking and molecular dynamics simulations, has an important role in novel drug design, in allowing visualization of the structure and drug interactions of both expected and serendipitous targets. As a case study we will now focus on the human ether-a-go-go-related gene (hERG) ion channel and its implications in toxicology.

16.4.1 hERG

On the whole, ion channels are seen as important targets for the pharmaceutical treatment of a number of disease areas. However, hERG in particular has been identified as an unintentional target for many drugs, resulting in adverse cardiotoxic side effects [83]. hERG is a member of the EAG family of voltage-gated potassium (K^+) channels [84]. It mediates the outward flow of K^+ ions identifiable as the rapid component of the delayed rectifier K^+ current (I_{Kr}) [85]. It participates in the complex interplay of ion channel currents that coordinate a normal cardiac action potential. Ion channel ligands principally act by blocking the conduction pore of these channels [86,87]. This prevents the efflux of K^+ ions, which slows the rate of cell repolarization and therefore causes a prolonged action potential [88]. The increase in action potential duration may be observed as a broadening of the interval between the Q and T waves on an electrocardiogram trace, and thus related disorders are known as Long QT syndromes (LQTS) [89]. In patients with LQTS there is a delay in the time it takes for the heart's electrical system to reset itself after each heartbeat. This may lead to arrhythmias such as Torsade de Pointes (TdP), which can degenerate further into a ventricular fibrillation and subsequent cardiac failure [90].

Recently a number of pharmaceutical drugs have been withdrawn from the market or not granted approval due to their pro-arrhythmic tendencies linked to inhibition of hERG. Two examples of this are terfenadine (*Seldane*) [91], a histamine H1-receptor antagonist, and cisapride (*Propulsid*) [92], a gastric pro-kinetic. For this reason the safety assessment profile of any novel compound must now consider the IC_{50} value of the compound for hERG.

Structure and Function of hERG By K^+ channel standards, hERG has a long amino acid sequence (1159 residues). It comprises two cytoplasmic

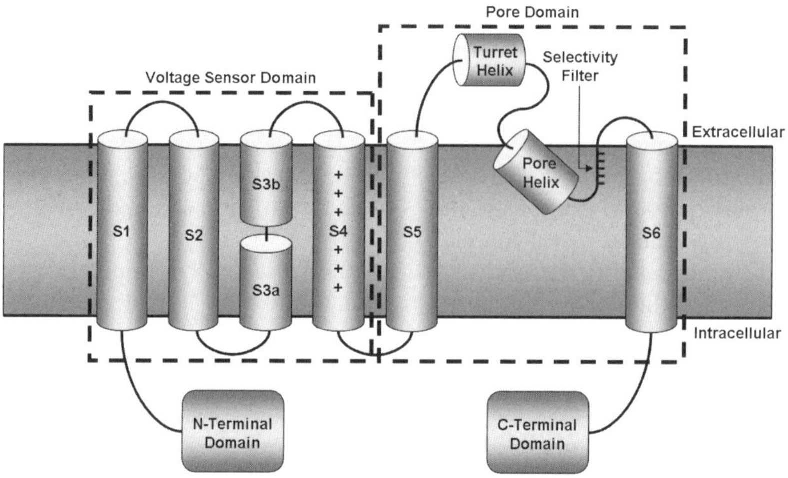

Figure 16.3 Schematic diagram of the transmembrane topology of a hERG subunit. The transmembrane portion of the hERG channel contains the voltage Sensor (comprising S1, S2, S3, and S4) and pore (comprising S5, turret helix, pore helix, selectivity filter, and S6) domains. A complete hERG channel consists of a tetramer of subunits, which fashions a membrane pore.

domains flanking a membrane domain that in turn consists of six transmembrane helices [84]. The membrane domain can be further subdivided into two functional units—the voltage-sensor (S1–S4) and pore (S5–S6) domains (Figure 16.3). The voltage-sensor domain responds to changes in membrane potential to open (activate) the channel [93]. This is largely driven by the S4 helix—which contains multiple basic (positively charged) amino acids that enable it to detect membrane potential and move towards the extracellular side of the membrane during action potentials. This movement is hypothesized to "pull" on the S4–S5 linker helix causing a swivel and a kink of the S6 helix that opens the channel [94,95].

A total of four subunits are required to form a functional channel, with the pore domains associating with one another to form a symmetrical tetrameric construct, encasing a conduction pore. The highly conserved pore helix and selectivity filter are critical to the selective and rapid throughput of K^+ ions [96]. These features appear as if mounted on the S5 and S6 transmembrane helices and constitute the extracellular portion of the membrane pore [97]. The intracellular portion of the pore is predominantly lined by the S6 helix to create an aqueous cavity below the selectivity filter [97]. It is to this site that compounds bind to prevent K^+ ion conduction [87,98].

Structural Templates for Modeling hERG Insight into the structure of the hERG drug binding site can be obtained from homology modeling. The majority of crystal structures available for use as templates are prokaryotic in origin and describe the pore domain of the channel in both the open and closed conformation (Table 16.2). Thus it is possible to model hERG in two different states. The conformational change that occurs during channel opening also impacts on the central cavity and the shape of the drug binding site is state dependent. As KcsA is the prototypical "closed" crystal structure [97], it has been used for the majority of hERG homology models of the closed state (Figure 16.4*a*) [87,99–105]. Conversely, the structures of MthK [103,106–108] and KvAP [109–111] have been used to obtain open state models of hERG (Figure 16.4*b*). A sequence alignment of hERG and these crystal structures is shown in Figure 16.5.

In addition to the pore domain, it is possible to create homology models of the voltage-sensor and *C*-terminal domains of hERG. There are also two structures of hERG itself, a crystal structure of the Per-Arnt-Sim (PAS) domain, found within the *N*-terminus [112], and an NMR solution structure of the S5 to pore helix linker (Turret) [113]. A full list of crystal structures that can be used to create homology models of hERG is shown in Table 16.2.

The Drug-Binding Site in hERG Compounds may only enter the central cavity from the intracellular side of the membrane, while the S6 helices are in the open conformation [114]. In the closed state access is prevented by a hydrophobic gate formed by the S6 transmembrane helices, which separates the drug binding site from the cytoplasm [115]. Since the selectivity filter is specifically designed for K^+ ion conduction, compounds are too large to traverse through the channel and thus they plug the conduction pore. In addition studies have shown that a compound may remain trapped within the central cavity even after the channel has returned to the closed state [115]. This suggests that the inner cavity is rather large and therefore may explain the promiscuity of drug block for hERG. Other Kv channels are unable to trap compounds as large as those that bind to hERG. This is suggested to be due to a Pro-X-Pro motif, within these other Kv channels, that generates a kink in the S6 helix, as observed in the Kv1.2 crystal structure (Figure 16.4*c*) [82]. The Pro-X-Pro motif alters the shape of the inner cavity. The structural changes—at least in the open state—however, are not as marked as those proposed by experimental data [116], so the influence of Pro-X-Pro on the shape of the cavity in the closed state remains unknown.

In 1998, KcsA became the first pore domain of a K^+ channel to be resolved by X-ray crystallography [97]. Based on homology with this channel it is possible to predict which residues face into the central cavity and thus are likely to interact with a bound drug. To test this prediction, four residues at the inner mouth of the selectivity filter and all the residues in the cytoplasmic half of the S6 helix were sequentially mutated to alanine [87]. This sequential (scanning) alanine mutagenesis has been repeated for a number of different compounds and identified Thr623, Ser624, and Val625—at the base of the selectivity filter—and Gly648, Tyr652, Phe656, and Val659—in S6—as important residues in drug binding [87,100,101,104,117–119].

TABLE 16.2 Regions of hERG for Which Homology Models Can Be Produced from Available Template Structures

Domain	K^+ Channel	Type of Organism	PDB ID[a]	Method	Resolution	Reference
S1–S6	Kv1.2	Eukaryotic	2A79	X-ray	2.9 Å	[82,132]
S1–S4	KvAP	Prokaryotic	1ORS	X-ray	1.9 Å	[109,131]
S5–S6	KvAP	Prokaryotic	1ORQ	X-ray	3.2 Å	[109,131]
S5–S6	KcsA	Prokaryotic	1K4C	X-ray	2.0 Å	[96,97]
S5–S6	MthK	Prokaryotic	1LNQ	X-ray	3.3 Å	[106,127]
S5–S6	KirBac1.1	Prokaryotic	1P7B	X-ray	3.6 Å	[205]
S5–S6	KirBac3.1	Prokaryotic	1XL4	X-ray	2.6 Å	[206]
S5–S6	NaK	Prokaryotic	2AHY	X-ray	2.4 Å	[206]
N-terminus (PAS)[b]	hERG	Eukaryotic	1BYW	X-ray	2.6 Å	[112]
Turret	hERG	Eukaryotic	1UJL	NMR	N/A	[113]
C-terminus (cNBD)	HCN2	Eukaryotic	1Q5O	X-ray	2.3 Å	[129]

[a] In cases such as KcsA, where there are many PDB entries for the same general structure, an example PDB ID is given.
[b] The Turret is not a domain per se, rather an extracellular loop between S5 and the pore helix of the pore domain.

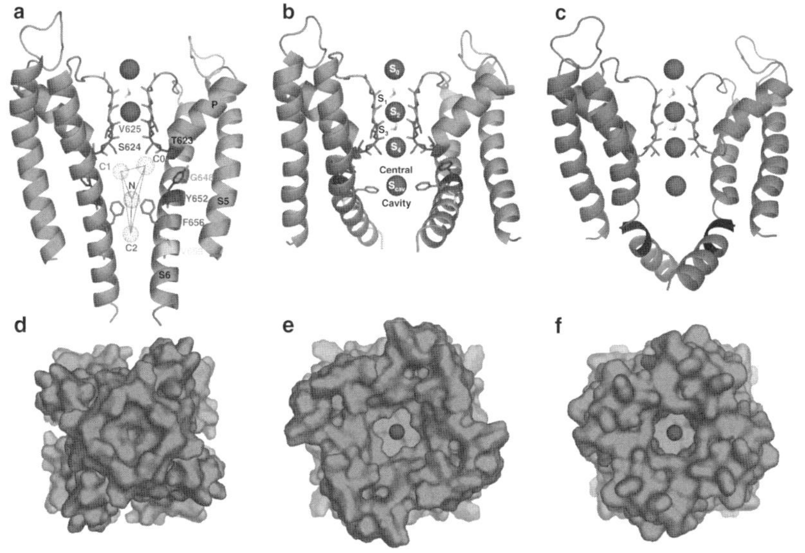

Figure 16.4 (*a*) A homology model of Pore domain of hERG (S5–S6) in the closed state [128], based on KcsA. Residues identified by mutagenesis to be important in drug binding are highlighted. The Cavalli pharmacophore (see text) is shown positioned within the central cavity. For clarity, only two of the four subunits of hERG are shown. (*b*) A homology model of the pore domain of hERG in the open state [128], based on KvAP. K^+ ion binding sites are shown as purple spheres interspersed by water molecules (these sites are denoted S_{ext}, S_1, S_1, S_3, S_4, and S_{cav}). (*c*) The crystal structure of Kv1.2 in the open state [82]. The Pro-X-Pro motif is highlighted in black to illustrate its influence on the central cavity. (*d–f*) Surface representations of the pore domains to illustrate the state of the modeled or crystallized structure, as viewed from the intracellular mouth. (Image produced using PyMol [15].) See color plates.

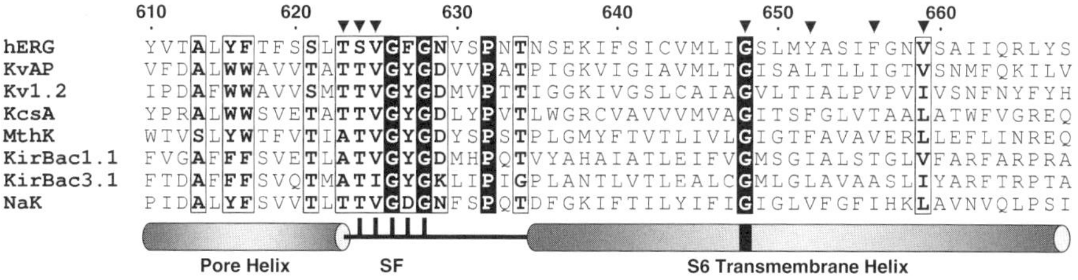

Figure 16.5 Sequence alignment between hERG and the sequences of the seven crystal structures that could be used to model the pore domain of hERG. The predicted secondary structure is shown below the alignment, with the cylinders representing helices and the horizontal line signifying a loop. The selectivity filter (SF) is marked by vertical lines, for each residue that contributes a carbonyl oxygen to K^+ coordination. Residues identified by mutagenesis are marked by filled triangles. (Image enhanced using ESPRIPT [133].)

Of these residues, the two aromatic residues—Tyr652 and Phe656—appear to be most important for the binding of compounds to the channel, so they have been subjected to a more detailed mutagenesis scan. In one study [118] the aromatic properties of Tyr652 were identified as important, but hydrophobicity was more important at the Phe656 position [118]. This is consistent with Tyr652 forming either cation-π or π-stacking interactions with the bound drug, whereas Phe656 is involved in hydrophobic contacts as well as potentially constituting part of the intracellular gate of hERG. Interestingly the two aromatic residues are only conserved at these positions within the ether-a-go-go (EAG) family of Kv channels [120]. Perhaps more intriguing, by homology, the EAG channel has an identical binding site to hERG—with all seven key residues retained—yet compounds bind to the EAG channel with a much lower affinity [121]. The primary difference between the two channels is that while ERG channels show a voltage-dependent inactivation process that closes the selectivity filter to K^+, EAG channels do not [121]. Residues important to inactivation have been identified in hERG. Consequently it is possible to eliminate inactivation by site-directed mutagenesis and, in doing so, reduce the affinity of a compound for hERG. Two examples of non-inactivating hERG mutants are S620T [121,122] and G648C:S631C [123]. All three of these mutated residues are distant from the central cavity and so are unlikely to directly interact with a bound drug. The converse is also possible for the EAG channels. The T432S:A443S bEAG double mutant induces inactivation and increases drug potency [124]. Interestingly another study suggests that hERG and (in this case) dEAG channels share a different alignment along the S6 helix—corresponding to a rotation of the helix—as can be induced by the process of inactivation [125]. Thus conformational changes in the inner cavity that occur with inactivation may be responsible for the differences in binding affinity of compounds for EAG and hERG channels.

While the two aromatic residues have been well characterized, the roles of the other residues implicated as crucial in drug binding are more difficult to define. In this instance bioinformatic techniques and comparative models provide valuable insight. Four of the five residues are highly conserved within the K^+ channel family and are believed to perform important roles in K^+ ion conduction (Thr623, Ser624, and Val625) or channel gating (Gly648) [126]. Modeling places the sidechain hydroxyls of Thr623 and Ser624 in suitable positions to form hydrogen bonds with a bound drug. In many cases, if the potency of a compound is influenced by S624A, it is also affected by T623A. Thus it is likely that both the Thr623 and Ser624 side chains point into the same area of the cavity when interacting with a bound ligand [87]. This is consistent with data for the T623A:S624A double mutant, which shows a greater reduction in drug affinity for the channel than in either of the single mutants, suggesting that the polar residues compensate for the absence of one another [104].

By homology, the carbonyl oxygen of Val625 is involved in the middle two K^+ coordination sites of the selectivity filter (S_2 and S_3). However, the side chain of Val625 is shielded from the central cavity, and thus it seems unlikely to directly interact with a bound drug [104]. In this instance it seems more likely that the alanine mutant induces structural changes that affect both Thr623 and Ser624. Interestingly this mutation also eliminates inactivation, which may explain the reduction in binding affinity [87].

Gly648 in hERG corresponds to the predicted K^+ channel glycine hinge, a point of flexibility that permits channel opening [126,127]. Although it is likely that a hinge point is retained in hERG, the opening of the G648A mutant is similar to wild-type despite the anticipated reduction in flexibility induced by the alanine (the side chain in alanine is a methyl group, so it hinders rotation of the polypeptide backbone compared with glycine in which there is no side chain). Since the channel may still open G648A is not predicted to reduce affinity by preventing drug access to the central cavity. In addition G648A does not influence the binding of a number of drugs such as terfenadine, cisapride, and propafenone, suggesting that drugs still reach the binding site [87,103]. Instead, it is likely that the additional methyl group of alanine inhibits the normal binding mode by either occluding an area of the cavity—Gly 648 is close to Thr623 and Ser624—or displacing other key residues [87].

hERG: Homology Modeling, Ligand Docking, and Molecular Dynamics To date, homology models of hERG have principally been produced to visually explain the effects of the site-directed mutagenesis, usually in conjunction with ligand docking studies. However, recently a number of molecular modeling studies of hERG have been published in their own right. These studies aim to provide a suitable method for screening hERG activity in novel compounds while also suggesting different binding modes for the docked compounds. The general approach involves docking the flexible compounds into rigid homology models of hERG, energy minimization of the ligand-receptor complex to permit protein relaxation and then calculation of binding scores or energies for the docked compound. Two such studies have created homology models from KvAP and then predicted binding energies for a set of sertindole analogues—in both cases a remarkably good correlation with the experimental IC_{50} values is recorded [110,111]. In another study [102] two models of hERG were created using KcsA as a template. In this case the S6 helices were manipulated to gradually open the channel, using the activated helices of MthK as a guide. This method created a collection of structures, of which two were selected, illustrating a fully open and partially open channel. A range of compounds were then docked into these models and by selectively choosing the optimum model for each compound to achieve a respectable data correlation. In our recent study [128] the structure of the hERG homology models was refined by rotating the S6 helices about the helical axis—as suggested by comparison of hERG with EAG [125]. This approach improves the agreement between ligands docked into the channel and the known mutagenesis data. In addition an improvement between the drug docking scores and the IC_{50} values was observed, when compared to the homology models without the imposed rotation. These studies suggest that such methods provide a useful in silico approach for predicting IC_{50} values for hERG compounds; increased levels of accuracy for predicting IC_{50} values for a large, diverse dataset of compounds could result from improved scoring functions (e.g., derived from ion channel-drug complexes as these experimental data become available) and/or improved homology models.

Homology Models and hERG Pharmacophores It is possible to compare the 2D pharmacophore and the 3D CoMFA QSAR model created by Cavalli et al. [128] with the mutagenesis data and the structures of the homology models. It is therefore feasible to position the scaffold of the pharmacophore within the drug binding site (Figure 16.4*a*). The Cavalli pharmacophore bears a high level of similarity to the other hERG pharmacophores (see Chapter 13 and figures therein) and consists of four features. The two principal features of the pharmacophore are a central positive charge (N) and an aromatic group (C0), which are present in most high-affinity blockers. The two other features (C1 and C2) are both either aromatic or hydrophobic in nature. In the central cavity, a K^+ ion may be coordinated by the negative dipole of the pore helices [97]. It therefore seems likely that the central nitrogen of the compound will locate close to the site of the cavity K^+ ion (S_{cav}) [96]. The aromatic nature of Tyr652 correlates with the C0 feature of the pharmacophore and suggests a π-stacking interaction. The C2 feature can then be positioned between the Phe656, to form hydrophobic interactions, this then places C1 in a position to form either hydrophobic or π-stacking interactions with Tyr652.

Homology Modeling Other hERG Domains Although the central cavity is the principal region of interest when considering the pharmacology of the hERG channel, the cyclic nucleotide binding domain (cNBD) on the *C*-terminus is also of potential interest when proposing the route a compound must take to gain access to the binding site. This domain hangs directly below the pore domain, on the intracellular side of the membrane, and forms a cytoplasmic pore, based on homology with the HCN2 cNBD crystal structure [129]. Initial assessment of the structure proposes that K^+ ions and drug molecules traverse directly through this cytoplasmic pore to the central cavity. However, a mutagenesis study of the HCN *C*-terminus [130] predicts that side entrances, which bypass the cytoplasmic pore, are used by ions instead. Such features—termed side-portals—are found in other Kv channels, this time in the *N*-terminus, which fulfills the role of the *C*-terminus in these channels [82]. It is likely that ions will take the same passageway in both hERG and HCN channels, although whether the drugs will also adopt this route has yet to be determined.

The voltage sensor domain is also significant for creating a more complete model of the hERG channel. However, this domain does not directly interact with the compounds. This domain may be modeled based on the structures of the isolated voltage sensor from KvAP [109,131] and the structure of the transmembrane domain of Kv1.2 [82,132].

Summary of hERG Modeling The unanticipated (and unwanted) interactions of compounds with membrane proteins have become a major issue when assessing the toxicology associated with novel drugs. In this section we have primarily discussed the cardiotoxic affects of hERG block. A wide range of unrelated compounds are known to block hERG with varying degrees of affinity. However, for the high-affinity block, compounds usually contain a basic group and are highly lipophilic. This correlates with the large hydrophobic volume of the central cavity designed for conducting cations. The residues in hERG that contribute to the binding site in the central cavity have been identified via mutagenesis studies, with two aromatic residues—Tyr652 and Phe656—proving to be influential for the majority of drugs. Due to the ever expanding wealth of structural information for K^+ channels it is possible to model with a high degree of accuracy the hERG K^+ channel and, in particular, its drug-binding site. Such molecular models are not only valuable from a visual perspective, but they also allow characterization of ligand-binding sites through ligand docking and molecular dynamics approaches. It remains to be established whether such methods are accurate enough to screen for compounds liable to bind to hERG. However, such tools have the potential to enable early identification of compounds predisposed to toxicity through hERG block.

16.5 CONCLUDING REMARKS AND FUTURE PERSPECTIVES

The use of homology modeling—in conjunction with molecular docking—has proved to be an appropriate and useful tool for those proteins that have yet to have their structures elucidated by experimental means. Not only has it facilitated rationalization of experimental results in a structural context, it has also been used to steer hypothesis-driven experimental studies. Its real strength is therefore in enriching the success of experimental studies by reducing the number of nonproductive "dead ends." Despite these successes there are still many areas where improvements could be made. Key challenges still exist in being able to accurately model protein flexibility and the role of water in protein–ligand binding. This, coupled with the ability to calculate quantitative binding affinities, would mark a big step forward in the use of homology modeling, in general, that can be applied to proteins involved in toxicity.

ACKNOWLEDGMENTS

P.J.S. is supported by a MRC/Novartis CASE Studentship. The authors would like to express their thanks to Gordon Roberts, Roland Wolf, Mark Paine, Jean-Didier Maréchal, Carol Kemp, Lesley McLaughlin, Matt Perry, and Peter Gedeck for stimulating discussions.

REFERENCES

1. Chothia C, Lesk AM. The relation between the divergence of sequence and structure in proteins. *EMBO J* 1986;5:823–6.
2. Moult J. A decade of CASP: Progress, bottlenecks and prognosis in protein structure prediction. *Curr Opin Struct Biol* 2005;15:285–9.
3. Moult J, Hubbard T, Bryant SH, Fidelis K, Pedersen JT. Critical assessment of methods of protein structure prediction (CASP): Round II. *Proteins* 1997;29(Suppl 1):2–6.
4. Moult J, Hubbard T, Fidelis K, Pedersen JT. Critical assessment of methods of protein structure prediction (CASP): Round III. *Proteins* 1999;37(Suppl 3):2–6.
5. Moult J, Fidelis K, Zemla A, Hubbard T. Critical assessment of methods of protein structure prediction (CASP): Round IV. *Proteins* 2001;45(Suppl 5):2–7.
6. Moult J, Fidelis K, Zemla A, Hubbard T. Critical assessment of methods of protein structure prediction (CASP)—Round V. *Proteins* 2003;53(Suppl 6): 334–9.
7. Moult J, Fidelis K, Rost B, Hubbard T, Tramontano A. Critical assessment of methods of protein structure prediction (CASP)—Round 6. *Proteins* 2005;61(Suppl 7):3–7.
8. Fischer D, Barret C, Bryson K, Elofsson A, Godzik A, Jones D, et al. CAFASP-1: Critical assessment of fully automated structure prediction methods. *Proteins* 1999;37(Suppl 3):209–17.
9. Fischer D, Rychlewski L, Dunbrack RL, Jr., Ortiz AR, Elofsson A. CAFASP3: The third critical assessment of fully automated structure prediction methods. *Proteins* 2003;53(Suppl 6):503–16.
10. Guex N, Peitsch MC. SWISS-MODEL and the Swiss-PdbViewer: An environment for comparative protein modeling. *Electrophoresis* 1997;18:2714–23.
11. Sutcliffe MJ, Haneef I, Carney D, Blundell TL. Knowledge based modelling of homologous proteins: Part I. Three-dimensional frameworks derived from the simultaneous superposition of multiple structures. *Protein Eng* 1987;1:377–84.
12. Sutcliffe MJ, Hayes FR, Blundell TL. Knowledge based modelling of homologous proteins: Part II. Rules for the conformations of substituted sidechains. *Protein Eng* 1987;1:385–92.
13. Sali A, Blundell TL. Comparative protein modelling by satisfaction of spatial restraints. *J Mol Biol* 1993;234:779–815.
14. Modi S, Paine MJ, Sutcliffe MJ, Lian LY, Primrose WU, Wolf CR, et al. A model for human cytochrome P450 2D6 based on homology modeling and NMR studies of substrate binding. *Biochemistry* 1996;35:4540–50.
15. DeLano WL. The PyMOL Molecular Graphics System. 2002 <http://www.pymol.org>.
16. Kaplan W, Littlejohn TG. Swiss-PDB Viewer (Deep View). *Brief Bioinform* 2001;2:195–7.
17. Kirton SB, Baxter CA, Sutcliffe MJ. Comparative modelling of cytochromes P450. *Adv Drug Deliv Rev* 2002;54:385–406.

18. Kelley LA, Gardner SP, Sutcliffe MJ. An automated approach for clustering an ensemble of NMR-derived protein structures into conformationally related subfamilies. *Protein Eng* 1996;9:1063–5.

19. Laskowski RA, Macarthur MW, Moss DS, Thornton JM. Procheck—A Program to check the stereochemical quality of protein structures. *J Appl Crystallogr* 1993;26:283–91.

20. Pontius J, Richelle J, Wodak SJ. Deviations from standard atomic volumes as a quality measure for protein crystal structures. *J Mol Biol* 1996;264:121–36.

21. Rodriguez R, Chinea G, Lopez N, Pons T, Vriend G. Homology modeling, model and software evaluation: three related resources. *Bioinformatics* 1998;14:523–8.

22. <http://biotech.ebi.ac.uk:8400>.

23. Colovos C, Yeates TO. Verification of protein structures: Patterns of nonbonded atomic interactions. *Protein Sci* 1993;2:1511–9.

24. Luthy R, Bowie JU, Eisenberg D. Assessment of protein models with three-dimensional profiles. *Nature* 1992;356:83–5.

25. Venclovas C, Zemla A, Fidelis K, Moult J. Criteria for evaluating protein structures derived from comparative modeling. *Proteins* 1997;29(Suppl. 1):7–13.

26. Martin AC, MacArthur MW, Thornton JM. Assessment of comparative modeling in CASP2. *Proteins* 1997;29(Suppl. 1):14–28.

27. Halperin I, Ma B, Wolfson H, Nussinov R. Principles of docking: An overview of search algorithms and a guide to scoring functions. *Proteins* 2002;47:409–43.

28. Nebert DW, Russell DW. Clinical importance of the cytochromes P450. *Lancet* 2002;360:1155–62.

29. Guengerich FP. Common and uncommon cytochrome P450 reactions related to metabolism and chemical toxicity. *Chem Res Toxicol* 2001;14:611–50.

30. Tanaka E. Clinically important pharmacokinetic drug–drug interactions: Role of cytochrome P450 enzymes. *J Clin Pharm Ther* 1998;23:403–16.

31. Mahgoub A, Idle JR, Dring LG, Lancaster R, Smith RL. Polymorphic hydroxylation of Debrisoquine in man. *Lancet* 1977;2:584–6.

32. Kroemer HK, Eichelbaum M. "It's the genes, stupid." Molecular bases and clinical consequences of genetic cytochrome P450 2D6 polymorphism. *Life Sci* 1995;56:2285–98.

33. Sullivan-Klose TH, Ghanayem BI, Bell DA, Zhang ZY, Kaminsky LS, Shenfield GM, et al. The role of the CYP2C9-Leu359 allelic variant in the tolbutamide polymorphism. *Pharmacogenetics* 1996;6:341–9.

34. Aithal GP, Day CP, Kesteven PJ, Daly AK. Association of polymorphisms in the cytochrome P450 CYP2C9 with warfarin dose requirement and risk of bleeding complications. *Lancet* 1999;353:717–9.

35. Kidd RS, Straughn AB, Meyer MC, Blaisdell J, Goldstein JA, Dalton JT. Pharmacokinetics of chlorpheniramine, phenytoin, glipizide and nifedipine in an individual homozygous for the CYP2C9*3 allele. *Pharmacogenetics* 1999;9:71–80.

36. Kidd RS, Curry TB, Gallagher S, Edeki T, Blaisdell J, Goldstein JA. Identification of a null allele of CYP2C9 in an African-American exhibiting toxicity to phenytoin. *Pharmacogenetics* 2001;11:803–8.

37. Takahashi H, Echizen H. Pharmacogenetics of warfarin elimination and its clinical implications. *Clin Pharmacokinet* 2001;40:587–603.
38. Williams PA, Cosme J, Sridhar V, Johnson EF, McRee DE. Mammalian microsomal cytochrome P450 monooxygenase: Structural adaptations for membrane binding and functional diversity. *Mol Cell* 2000;5:121–31.
39. Kirton SB, Kemp CA, Tomkinson NP, St-Gallay S, Sutcliffe MJ. Impact of incorporating the 2C5 crystal structure into comparative models of cytochrome P450 2D6. *Proteins* 2002;49:216–31.
40. Yano JK, Hsu MH, Griffin KJ, Stout CD, Johnson EF. Structures of human microsomal cytochrome P450 2A6 complexed with coumarin and methoxsalen. *Nat Struct Mol Biol* 2005;12:822–3.
41. Schoch GA, Yano JK, Wester MR, Griffin KJ, Stout CD, Johnson EF. Structure of human microsomal cytochrome P450 2C8: Evidence for a peripheral fatty acid binding site. *J Biol Chem* 2004;279:9497–503.
42. Williams PA, Cosme J, Ward A, Angove HC, Matak Vinkovic D, Jhoti H. Crystal structure of human cytochrome P450 2C9 with bound warfarin. *Nature* 2003;424:464–8.
43. Wester MR, Yano JK, Schoch GA, Yang C, Griffin KJ, Stout CD, et al. The structure of human cytochrome P450 2C9 complexed with flurbiprofen at 2.0-Å resolution. *J Biol Chem* 2004;279:35630–7.
44. Rowland P, Blaney FE, Smyth MG, Jones JJ, Leydon VR, Oxbrow AK, et al. Crystal structure of human cytochrome P450 2D6. *J Biol Chem* 2006;281: 7614–22.
45. Williams PA, Cosme J, Vinkovic DM, Ward A, Angove HC, Day PJ, et al. Crystal structures of human cytochrome P450 3A4 bound to metyrapone and progesterone. *Science* 2004;305:683–6.
46. Yano JK, Wester MR, Schoch GA, Griffin KJ, Stout CD, Johnson EF. The structure of human microsomal cytochrome P450 3A4 determined by X-ray crystallography to 2.05-Å resolution. *J Biol Chem* 2004;279:38091–104.
47. Scott EE, He YA, Wester MR, White MA, Chin CC, Halpert JR, et al. An open conformation of mammalian cytochrome P450 2B4 at 1.6-Å resolution. *Proc Natl Acad Sci USA* 2003;100:13196–201.
48. Scott EE, White MA, He YA, Johnson EF, Stout CD, Halpert JR. Structure of mammalian cytochrome P450 2B4 complexed with 4-(4-chlorophenyl)imidazole at 1.9-Å resolution: Insight into the range of P450 conformations and the coordination of redox partner binding. *J Biol Chem* 2004;279:27294–301.
49. Eichelbaum M, Spannbrucker N, Dengler HJ. Influence of the defective metabolism of sparteine on its pharmacokinetics. *Eur J Clin Pharmacol* 1979;16:189–94.
50. Daly AK, Leathart JB, London SJ, Idle JR. An inactive cytochrome P450 CYP2D6 allele containing a deletion and a base substitution. *Hum Genet* 1995;95:337–41.
51. Smith G, Stanley LA, Sim E, Strange RC, Wolf CR. Metabolic polymorphisms and cancer susceptibility. *Cancer Surv* 1995;25:27–65.
52. Koymans L, Donne-op den Kelder GM, Koppele Te JM, Vermeulen NP. Cytochromes P450: Their active-site structure and mechanism of oxidation. *Drug Metab Rev* 1993;25:325–87.

53. Koymans L, Vermeulen NP, van Acker SA, te Koppele JM, Heykants JJ, Lavrijsen K, et al. A predictive model for substrates of cytochrome P450-debrisoquine (2D6). *Chem Res Toxicol* 1992;5:211–9.

54. Islam SA, Wolf CR, Lennard MS, Sternberg MJ. A three-dimensional molecular template for substrates of human cytochrome P450 involved in debrisoquine 4-hydroxylation. *Carcinogenesis* 1991;12:2211–9.

55. Mackman R, Tschirret-Guth RA, Smith G, Hayhurst GP, Ellis SW, Lennard MS, et al. Active-site topologies of human CYP2D6 and its aspartate-301 → glutamate, asparagine, and glycine mutants. *Arch Biochem Biophys* 1996;331:134–40.

56. de Groot MJ, Vermeulen NP, Kramer JD, van Acker FA, Donne-Op den Kelder GM. A three-dimensional protein model for human cytochrome P450 2D6 based on the crystal structures of P450 101, P450 102, and P450 108. *Chem Res Toxicol* 1996;9:1079–91.

57. Lewis DF, Eddershaw PJ, Goldfarb PS, Tarbit MH. Molecular modelling of cytochrome P4502D6 (CYP2D6) based on an alignment with CYP102: Structural studies on specific CYP2D6 substrate metabolism. *Xenobiotica* 1997;27:319–39.

58. Modi S, Gilham DE, Sutcliffe MJ, Lian LY, Primrose WU, Wolf CR, et al. 1-methyl-4-phenyl-1,2,3,6-tetrahydropyridine as a substrate of cytochrome P450 2D6: Allosteric effects of NADPH-cytochrome P450 reductase. *Biochemistry* 1997;36:4461–70.

59. de Groot MJ, Ackland MJ, Horne VA, Alex AA, Jones BC. A novel approach to predicting P450 mediated drug metabolism. CYP2D6 catalyzed *N*-dealkylation reactions and qualitative metabolite predictions using a combined protein and pharmacophore model for CYP2D6. *J Med Chem* 1999;42:4062–70.

60. Lewis DF. Homology modelling of human cytochromes P450 involved in xenobiotic metabolism and rationalization of substrate selectivity. *Exp Toxicol Pathol* 1999;51:369–74.

61. Wester MR, Johnson EF, Marques-Soares C, Dijols S, Dansette PM, Mansuy D, et al. Structure of mammalian cytochrome P450 2C5 complexed with diclofenac at 2.1 Å resolution: Evidence for an induced fit model of substrate binding. *Biochemistry* 2003;42:9335–45.

62. Wester MR, Johnson EF, Marques-Soares C, Dansette PM, Mansuy D, Stout CD. Structure of a substrate complex of mammalian cytochrome P450 2C5 at 2.3 Å resolution: Evidence for multiple substrate binding modes. *Biochemistry* 2003; 42:6370–9.

63. Flanagan JU, Marechal JD, Ward R, Kemp CA, McLaughlin LA, Sutcliffe MJ, et al. Phe120 contributes to the regiospecificity of cytochrome P450 2D6: Mutation leads to the formation of a novel dextromethorphan metabolite. *Biochem J* 2004;380:353–60.

64. Kemp CA, Flanagan JU, van Eldik AJ, Marechal JD, Wolf CR, Roberts GC, et al. Validation of model of cytochrome P450 2D6: An in silico tool for predicting metabolism and inhibition. *J Med Chem* 2004;47:5340–6.

65. Guengerich FP, Miller GP, Hanna IH, Martin MV, Leger S, Black C, et al. Diversity in the oxidation of substrates by cytochrome P450 2D6: Lack of an obligatory role of aspartate 301-substrate electrostatic bonding. *Biochemistry* 2002;41: 11025–34.

66. Keseru GM. A virtual high throughput screen for high affinity cytochrome P450cam substrates: Implications for in silico prediction of drug metabolism. *J Comput Aided Mol Des* 2001;15:649–57.
67. Kirton SB, Murray CW, Verdonk ML, Taylor RD. Prediction of binding modes for ligands in the cytochromes P450 and other heme-containing proteins. *Proteins* 2005;58:836–44.
68. Jones G, Willett P, Glen RC. Molecular recognition of receptor sites using a genetic algorithm with a description of desolvation. *J Mol Biol* 1995;245:43–53.
69. Jones G, Willett P, Glen RC, Leach AR, Taylor R. Development and validation of a genetic algorithm for flexible docking. *J Mol Biol* 1997;267:727–48.
70. Eldridge MD, Murray CW, Auton TR, Paolini GV, Mee RP. Empirical scoring functions: I. The development of a fast empirical scoring function to estimate the binding affinity of ligands in receptor complexes. *J Comput Aided Mol Des* 1997;11:425–45.
71. Baxter CA, Murray CW, Clark DE, Westhead DR, Eldridge MD. Flexible docking using Tabu search and an empirical estimate of binding affinity. *Proteins* 1998;33:367–82.
72. Yu J, Paine MJI, Marechal J-D, Kemp CA, Ward CJ, Brown S, et al. In silico prediction of drug binding to cytochrome P450 2D6: Identification of a new metabolite of metoclopramide. *Drug Metab Dispos* 2006;34:1386–92.
73. Hanna IH, Kim MS, Guengerich FP. Heterologous expression of cytochrome P450 2D6 mutants, electron transfer, and catalysis of bufuralol hydroxylation: the role of aspartate 301 in structural integrity. *Arch Biochem Biophys* 2001;393:255–61.
74. McLaughlin LA, Paine MJ, Kemp CA, Marechal JD, Flanagan JU, Ward CJ, et al. Why is quinidine an inhibitor of cytochrome P450 2D6? The role of key active-site residues in quinidine binding. *J Biol Chem* 2005;280:38617–24.
75. Venhorst J, ter Laak AM, Commandeur JN, Funae Y, Hiroi T, Vermeulen NP. Homology modeling of rat and human cytochrome P450 2D (CYP2D) isoforms and computational rationalization of experimental ligand-binding specificities. *J Med Chem* 2003;46:74–86.
76. Hopkins AL, Groom CR. The druggable genome. *Nat Rev Drug Discov* 2002;1:727–30.
77. Vandenberg JI, Walker BD, Campbell TJ. HERG K+ channels: Friend and foe. *Trends Pharmacol Sci* 2001;22:240–6.
78. Wallin E, von Heijne G. Genome-wide analysis of integral membrane proteins from eubacterial, archaean, and eukaryotic organisms. *Protein Sci* 1998;7:1029–38.
79. <http://blanco.biomol.uci.edu/Membrane_Proteins_xtal.html>.
80. Berman HM, Westbrook J, Feng Z, Gilliland G, Bhat TN, Weissig H, et al. The Protein Data Bank. *Nucleic Acids Res* 2000;28:235–42.
81. MacKinnon R, Doyle DA. Prokaryotes offer hope for potassium channel structural studies. *Nat Struct Biol* 1997;4:877–9.
82. Long SB, Campbell EB, Mackinnon R. Crystal structure of a mammalian voltage-dependent Shaker family K+ channel. *Science* 2005;309:897–903.

83. Sanguinetti MC, Tristani-Firouzi M. hERG potassium channels and cardiac arrhythmia. *Nature* 2006;440:463–9.
84. Trudeau MC, Warmke JW, Ganetzky B, Robertson GA. HERG, a human inward rectifier in the voltage-gated potassium channel family. *Science* 1995;269:92–5.
85. Sanguinetti MC, Jiang C, Curran ME, Keating MT. A mechanistic link between an inherited and an acquired cardiac arrhythmia: HERG encodes the IKr potassium channel. *Cell* 1995;81:299–307.
86. Carmeliet E. Voltage-dependent and time-dependent block of the delayed K+ current in cardiac myocytes by dofetilide. *J Pharmacol Exp Ther* 1992;262: 809–17.
87. Mitcheson JS, Chen J, Lin M, Culberson C, Sanguinetti MC. A structural basis for drug-induced long QT syndrome. *Proc Natl Acad Sci USA* 2000;97: 12329–33.
88. Keating MT, Sanguinetti MC. Molecular and cellular mechanisms of cardiac arrhythmias. *Cell* 2001;104:569–80.
89. Curran ME, Splawski I, Timothy KW, Vincent GM, Green ED, Keating MT. A molecular basis for cardiac arrhythmia: HERG mutations cause long QT syndrome. *Cell* 1995;80:795–803.
90. Keating MT, Sanguinetti MC. Molecular genetic insights into cardiovascular disease. *Science* 1996;272:681–5.
91. Roy M, Dumaine R, Brown AM. HERG, a primary human ventricular target of the nonsedating antihistamine terfenadine. *Circulation* 1996;94:817–23.
92. Rampe D, Roy ML, Dennis A, Brown AM. A mechanism for the proarrhythmic effects of cisapride (Propulsid): High affinity blockade of the human cardiac potassium channel HERG. *FEBS Lett* 1997;417:28–32.
93. Smith PL, Yellen G. Fast and slow voltage sensor movements in HERG potassium channels. *J Gen Physiol* 2002;119:275–93.
94. Tristani-Firouzi M, Chen J, Sanguinetti MC. Interactions between S4-S5 linker and S6 transmembrane domain modulate gating of HERG K+ channels. *J Biol Chem* 2002;277:18994–9000.
95. Sanguinetti MC, Xu QP. Mutations of the S4-S5 linker alter activation properties of HERG potassium channels expressed in *Xenopus* oocytes. *J Physiol* 1999; 514:667–75.
96. Zhou Y, Morais-Cabral JH, Kaufman A, MacKinnon R. Chemistry of ion coordination and hydration revealed by a K+ channel-Fab complex at 2.0 Å resolution. *Nature* 2001;414:43–8.
97. Doyle DA, Morais Cabral J, Pfuetzner RA, Kuo A, Gulbis JM, Cohen SL, et al. The structure of the potassium channel: Molecular basis of K+ conduction and selectivity. *Science* 1998;280:69–77.
98. Lees-Miller JP, Duan Y, Teng GQ, Duff HJ. Molecular determinant of high-affinity dofetilide binding to HERG1 expressed in *Xenopus* oocytes: Involvement of S6 sites. *Mol Pharmacol* 2000;57:367–74.
99. Ishii K, Kondo K, Takahashi M, Kimura M, Endoh M. An amino acid residue whose change by mutation affects drug binding to the HERG channel. *FEBS Lett* 2001;506:191–5.

100. Sanchez-Chapula JA, Navarro-Polanco RA, Culberson C, Chen J, Sanguinetti MC. Molecular determinants of voltage-dependent human ether-a-go-go related gene (HERG) K+ channel block. *J Biol Chem* 2002;277:23587–95.

101. Perry M, De Groot MJ, Helliwell R, Leishman D, Tristani-Firouzi M, Sanguinetti MC, et al. Structural determinants of HERG channel block by clofilium and ibutilide. *Mol Pharmacol* 2004;66:240–9.

102. Rajamani R, Tounge BA, Li J, Reynolds CH. A two-state homology model of the hERG K+ channel: Application to ligand binding. *Bioorg Med Chem Lett* 2005;15:1737–41.

103. Witchel HJ, Dempsey CE, Sessions RB, Perry M, Milnes JT, Hancox JC, et al. The low-potency, voltage-dependent HERG blocker propafenone—Molecular determinants and drug trapping. *Mol Pharmacol* 2004;66:1201–12.

104. Perry M, Stansfeld PJ, Leaney J, Wood C, de Groot MJ, Leishman D, et al. Drug binding interactions in the inner cavity of HERG channels: Molecular insights from structure-activity relationships of clofilium and ibutilide analogs. *Mol Pharmacol* 2006;69:509–19.

105. Choe H, Nah KH, Lee SN, Lee HS, Lee HS, Jo SH, et al. A novel hypothesis for the binding mode of HERG channel blockers. *Biochem Bioph Res Commun* 2006;344:72–8.

106. Jiang Y, Lee A, Chen J, Cadene M, Chait BT, MacKinnon R. Crystal structure and mechanism of a calcium-gated potassium channel. *Nature* 2002;417:515–22.

107. Pearlstein RA, Vaz RJ, Kang J, Chen XL, Preobrazhenskaya M, Shchekotikhin AE, et al. Characterization of HERG potassium channel inhibition using CoMSiA 3D QSAR and homology modeling approaches. *Bioorg Med Chem Lett* 2003;13:1829–35.

108. Duncan RS, Ridley JM, Dempsey CE, Leishman DJ, Leaney JL, Hancox JC, et al. Erythromycin block of the HERG K+ channel: Accessibility to F656 and Y652. *Biochem Bioph Res Commun* 2006;341:500–6.

109. Jiang Y, Lee A, Chen J, Ruta V, Cadene M, Chait BT, et al. X-ray structure of a voltage-dependent K+ channel. *Nature* 2003;423:33–41.

110. Osterberg F, Aqvist J. Exploring blocker binding to a homology model of the open hERG K+ channel using docking and molecular dynamics methods. *FEBS Lett* 2005;579:2939–44.

111. Farid R, Day T, Friesner RA, Pearlstein RA. New insights about HERG blockade obtained from protein modeling, potential energy mapping, and docking studies. *Bioorg Med Chem* 2006;14:3160–73.

112. Morais Cabral JH, Lee A, Cohen SL, Chait BT, Li M, Mackinnon R. Crystal structure and functional analysis of the HERG potassium channel *N* terminus: A eukaryotic PAS domain. *Cell* 1998;95:649–55.

113. Torres AM, Bansal PS, Sunde M, Clarke CE, Bursill JA, Smith DJ, et al. Structure of the HERG K+ channel S5P extracellular linker: role of an amphipathic alpha-helix in C-type inactivation. *J Biol Chem* 2003;278:42136–48.

114. Spector PS, Curran ME, Keating MT, Sanguinetti MC. Class III antiarrhythmic drugs block HERG, a human cardiac delayed rectifier K+ channel: Open-channel block by methanesulfonanilides. *Circ Res* 1996;78:499–503.

115. Mitcheson JS, Chen J, Sanguinetti MC. Trapping of a methanesulfonanilide by closure of the HERG potassium channel activation gate. *J Gen Physiol* 2000;115:229–40.

116. del Camino D, Holmgren M, Liu Y, Yellen G. Blocker protection in the pore of a voltage-gated K+ channel and its structural implications. *Nature* 2000;403:321–5.

117. Kamiya K, Mitcheson JS, Yasui K, Kodama I, Sanguinetti MC. Open channel block of HERG K(+) channels by vesnarinone. *Mol Pharmacol* 2001;60:244–53.

118. Fernandez D, Ghanta A, Kauffman GW, Sanguinetti MC. Physicochemical features of the HERG channel drug binding site. *J Biol Chem* 2004;279:10120–7.

119. Sanchez-Chapula JA, Ferrer T, Navarro-Polanco RA, Sanguinetti MC. Voltage-dependent profile of human ether-a-go-go-related gene channel block is influenced by a single residue in the S6 transmembrane domain. *Mol Pharmacol* 2003;63:1051–8.

120. Warmke JW, Ganetzky B. A family of potassium channel genes related to eag in *Drosophila* and mammals. *Proc Nat Acad Sci USA* 1994;91:3438–42.

121. Ficker E, Jarolimek W, Kiehn J, Baumann A, Brown AM. Molecular determinants of dofetilide block of HERG K+ channels. *Circ Res* 1998;82:386–95.

122. Suessbrich H, Schonherr R, Heinemann SH, Lang F, Busch AE. Specific block of cloned Herg channels by clofilium and its tertiary analog LY97241. *FEBS Lett* 1997;414:435–8.

123. Smith PL, Baukrowitz T, Yellen G. The inward rectification mechanism of the HERG cardiac potassium channel. *Nature* 1996;379:833–6.

124. Ficker E, Jarolimek W, Brown AM. Molecular determinants of inactivation and dofetilide block in ether a-go-go (EAG) channels and EAG-related K(+) channels. *Mol Pharmacol* 2001;60:1343–8.

125. Chen J, Seebohm G, Sanguinetti MC. Position of aromatic residues in the S6 domain, not inactivation, dictates cisapride sensitivity of HERG and eag potassium channels. *Proc Natl Acad Sci USA* 2002;99:12461–6.

126. Shealy RT, Murphy AD, Ramarathnam R, Jakobsson E, Subramaniam S. Sequence-function analysis of the K+-selective family of ion channels using a comprehensive alignment and the KcsA channel structure. *Biophys J* 2003; 84:2929–42.

127. Jiang Y, Lee A, Chen J, Cadene M, Chait BT, MacKinnon R. The open pore conformation of potassium channels. *Nature* 2002;417:523–6.

128. Stansfeld PJ, Gedeck P, Gosling M, Cox B, Mitcheson JS, Sutcliffe MJ. Drug block of the hERG potassium channel: insight from modeling. *Proteins* 2007: in press.

129. Zagotta WN, Olivier NB, Black KD, Young EC, Olson R, Gouaux E. Structural basis for modulation and agonist specificity of HCN pacemaker channels. *Nature* 2003;425:200–5.

130. Johnson JP, Jr., Zagotta WN. The carboxyl-terminal region of cyclic nucleotide-modulated channels is a gating ring, not a permeation path. *Proc Natl Acad Sci USA* 2005;102:2742–7.

131. Jiang Y, Ruta V, Chen J, Lee A, MacKinnon R. The principle of gating charge movement in a voltage-dependent K+ channel. *Nature* 2003;423:42–8.

132. Long SB, Campbell EB, Mackinnon R. Voltage sensor of Kv1.2: Structural basis of electromechanical coupling. *Science* 2005;309:903–8.
133. Gouet P, Courcelle E, Stuart DI, Metoz F. ESPript: analysis of multiple sequence alignments in PostScript. *Bioinformatics* 1999;15:305–8.
134. Lewis DF, Lake BG. Molecular modelling of CYP1A subfamily members based on an alignment with CYP102: rationalization of CYP1A substrate specificity in terms of active site amino acid residues. *Xenobiotica* 1996;26:723–53.
135. Szklarz GD, Paulsen MD. Molecular modeling of cytochrome P450 1A1: Enzyme-substrate interactions and substrate binding affinities. *J Biomol Struct Dyn* 2002;20:155–62.
136. Liu J, Ericksen SS, Besspiata D, Fisher CW, Szklarz GD. Characterization of substrate binding to cytochrome P450 1A1 using molecular modeling and kinetic analyses: Case of residue 382. *Drug Metab Dispos* 2003;31:412–20.
137. Lozano JJ, Lopez-de-Brinas E, Centeno NB, Guigo R, Sanz F. Three-dimensional modelling of human cytochrome P450 1A2 and its interaction with caffeine and MeIQ. *J Comput Aided Mol Des* 1997;11:395–408.
138. Dai R, Zhai S, Wei X, Pincus MR, Vestal RE, Friedman FK. Inhibition of human cytochrome P450 1A2 by flavones: A molecular modeling study. *J Protein Chem* 1998;17:643–50.
139. Dai R, Pincus MR, Friedman FK. Molecular modeling of mammalian cytochrome P450s. *Cell Mol Life Sci* 2000;57:487–99.
140. de Rienzo F, Fanelli F, Menziani MC, De Benedetti PG. Theoretical investigation of substrate specificity for cytochromes P450 IA2, P450 IID6 and P450 IIIA4. *J Comput Aided Mol Des* 2000;14:93–116.
141. Cho US, Park EY, Dong MS, Park BS, Kim K, Kim KH. Tight-binding inhibition by alpha-naphthoflavone of human cytochrome P450 1A2. *Biochim Biophys Acta* 2003;1648:195–202.
142. Lewis DF, Lake BG, Dickins M, Ueng YF, Goldfarb PS. Homology modelling of human CYP1A2 based on the CYP2C5 crystallographic template structure. *Xenobiotica* 2003;33:239–54.
143. Kim D, Guengerich FP. Enhancement of 7-methoxyresorufin *O*-demethylation activity of human cytochrome P450 1A2 by molecular breeding. *Arch Biochem Biophys* 2004;432:102–8.
144. Kim D, Guengerich FP. Selection of human cytochrome P450 1A2 mutants with enhanced catalytic activity for heterocyclic amine *N*-hydroxylation. *Biochemistry* 2004;43:981–8.
145. Iori F, da Fonseca R, Ramos MJ, Menziani MC. Theoretical quantitative structure-activity relationships of flavone ligands interacting with cytochrome P450 1A1 and 1A2 isozymes. *Bioorg Med Chem* 2005;13:4366–74.
146. Lewis DF, Gillam EM, Everett SA, Shimada T. Molecular modelling of human CYP1B1 substrate interactions and investigation of allelic variant effects on metabolism. *Chem Biol Interact* 2003;145:281–95.
147. Lewis DF, Lake BG. Molecular modelling of members of the P4502A subfamily: Application to studies of enzyme specificity. *Xenobiotica* 1995;25:585–98.
148. Stahl GR, Holtje HD. Development of models for cytochrome P450 2A5 as well as two of its mutants. *Pharmazie* 2005;60:247–53.

149. Lewis DF. Homology modelling of human CYP2 family enzymes based on the CYP2C5 crystal structure. *Xenobiotica* 2002;32:305–23.

150. Lewis DF, Lake BG, Dickins M, Goldfarb PS. Homology modelling of CYP2A6 based on the CYP2C5 crystallographic template: Enzyme-substrate interactions and QSARs for binding affinity and inhibition. *Toxicol In Vitro* 2003;17:179–90.

151. He XY, Shen J, Hu WY, Ding X, Lu AY, Hong JY. Identification of Val117 and Arg372 as critical amino acid residues for the activity difference between human CYP2A6 and CYP2A13 in coumarin 7-hydroxylation. *Arch Biochem Biophys* 2004;427:143–53.

152. Szklarz GD, Ornstein RL, Halpert JP. Application of 3-dimensional homology modeling of cytochrome P450 2B1 for interpretation of site-directed mutagenesis results. *J Biomol Struct Dyn* 1994;12:61–78.

153. Szklarz GD, He YA, Halpert JR. Site-directed mutagenesis as a tool for molecular modeling of cytochrome P450 2B1. *Biochemistry* 1995;34:14312–22.

154. Dai R, Pincus MR, Friedman FK. Molecular modeling of cytochrome P450 2B1: Mode of membrane insertion and substrate specificity. *J Protein Chem* 1998;17: 121–9.

155. Kumar S, Scott EE, Liu H, Halpert JR. A rational approach to re-engineer cytochrome P450 2B1 regioselectivity based on the crystal structure of P450 2C5. *J Biol Chem* 2003;278:17178–84.

156. Honma W, Li W, Liu H, Scott EE, Halpert JR. Functional role of residues in the B′ region of cytochrome P450 2B1. *Arch Biochem Biophys* 2005;435:157–65.

157. Li W, Liu H, Scott EE, Gräter F, Halpert JR, Luo X, et al. Possible pathway(s) of testosterone egress from the active site of cytochrome 2B1: A steered molecular dynamics simulation. *Drug Metab Dispos* 2005;33:910–9.

158. Lewis DF, Lake BG. Molecular modelling of mammalian CYP2B isoforms and their interaction with substrates, inhibitors and redox partners. *Xenobiotica* 1997;27:443–78.

159. Spatzenegger M, Wang Q, He YQ, Wester MR, Johnson EF, Halpert JR. Amino acid residues critical for differential inhibition of CYP2B4, CYP2B5, and CYP2B1 by phenylimidazoles. *Mol Pharmacol* 2001;59:475–85.

160. Chang YT, Stiffelman OB, Vakser IA, Loew GH, Bridges A, Waskell L. Construction of a 3D model of cytochrome P450 2B4. *Protein Eng* 1997;10:119–29.

161. Sechenykh AA, Dubanov AV, Skvortsov VS, Ivanov AS, Archakov AI, Williams P, et al. Computer model of 3D structure of cytochrome P450 2B4. *Vopr Med Khim* 2002;48:526–38.

162. Harris DL, Park JY, Gruenke L, Waskell L. Theoretical study of the ligand-CYP2B4 complexes: effect of structure on binding free energies and heme spin state. *Proteins* 2004;55:895–914.

163. Hodek P, Sopko B, Antonovic L, Sulc M, Novak P, Strobel HW. Evaluation of comparative cytochrome P450 2B4 model by photoaffinity labeling. *Gen Physiol Biophys* 2004;23:467–88.

164. Domanski TL, Schultz KM, Roussel F, Stevens JC, Halpert JR. Structure-function analysis of human cytochrome P-450 2B6 using a novel substrate, site-directed mutagenesis, and molecular modeling. *J Pharmacol Exp Ther* 1999; 290:1141–7.

165. Lewis DFV, Lake BG, Dickins M, Edershaw PJ, Tarbit MH, Goldfarb PS. Molecular modelling of CYP2B6, the human CYP2B isoform, by homology with the substrate-bound CYP102 crystal structure: Evaluation of CYP2B6 substrate characteristics, the cytochrome b_5 binding site and comparisons with CYP2B1 and CYP2B4. *Xenobiotica* 1999;29:361–93.

166. Bathelt C, Schmid RD, Pleiss J. Regioselectivity of CYP2B6: Homology modeling, molecular dynamics simulation, docking. *J Mol Model* 2002;8:327–35.

167. Lewis DF. Essential requirements for substrate binding affinity and selectivity toward human CYP2 family enzymes. *Arch Biochem Biophys* 2003;409:32–44.

168. Wang Q, Halpert JR. Combined three-dimensional quantitative structure-activity relationship analysis of cytochrome P450 2B6 substrates and protein homology modeling. *Drug Metab Dispos* 2002;30:86–95.

169. Payne VA, Chang Y-T, Loew GH. Homology modeling and substrate binding study of human CYP2C9 enzyme. *Proteins* 1999;37:176–90.

170. de Groot MJ, Alex AA, Jones BC. Development of a combined protein and pharmacophore model for CYP2C9. *J Med Chem* 2002;45:1983–93.

171. Afzelius L, Zamora I, Ridderström M, Andersson TB, Karlén A, Masimirembwa CM. Competitive CYP2C9 inhibitors: Enzyme inhibition studies, protein homology modeling, and three-dimensional quantitative structure-activity relationship analysis. *Mol Pharmacol* 2001;59:909–19.

172. Lewis DFV, Dickins M, Weaver RJ, Eddershaw PJ, Goldfarb PS, Tarbit MH. Molecular modelling of human CYP2C subfamily enzymes CYP2C9 and CYP2C19: Rationalization of enzyme specificity and site-directed mutagenesis experiments in the CYP2C subfamily. *Xenobiotica* 1998;28:235–68.

173. Oda A, Yamaotsu N, Hirono S. Studies of binding modes of (S)-mephenytoin to the wild types and mutants of cytochrome P450 2C19 and 2C9 using homology modeling and computational docking. *Pharm Res* 2004;21:2270–8.

174. Tanaka T, Kamiguchi N, Okuda T, Yamamoto Y. Characterization of the CYP2C8 active site by homology modeling. *Chem Pharm Bull* 2004;52:836–41.

175. Payne VA, Chang Y-T, Loew GH. Homology modeling and substrate binding study of human CYP2C18 and CYP2C19 enzymes. *Proteins* 1999;37:204–17.

176. Ridderström M, Zamora I, Fjellström O, Andersson TB. Analysis of selective regions in the active sites of human cytochromes P450 2C8, 2C9, 2C18, and 2C19 homology models using GRID/CPCA. *J Med Chem* 2001;44:4072–81.

177. Koymans LMH, Vermeulen NPE, Baarslag A, Donné-Op den Kelder GM. A preliminary 3D model for cytochrome P450 2D6 constructed by homology model building. *J Comput Aided Mol Des* 1993;7:281–9.

178. de Groot MJ, Vermeulen NPE, Kramer JD, van Acker FAA, Donné-Op den Kelder GM. A three-dimensional protein model for human cytochrome P450 2D6 based on the crystal structures of P450 101, P450 102 and P450 108. *Chem Res Toxicol* 1996;9:1079–91.

179. Lewis DFV, Eddershaw PJ, Goldfarb PS, Tarbit MH. Molecular modelling of cytochrome P4502D6 (CYP2D6) based on an alignment with CYP102: Structural studies on specific CYP2D6 substrate metabolism. *Xenobiotica* 1997;27:319–40.

180. de Groot MJ, Ackland MJ, Horne VA, Alex AA, Jones BC. Novel approach to predicting P450 mediated drug metabolism: The development of a combined

protein and pharmacophore model for CYP2D6. *J Med Chem* 1999;42:1515–24.

181. de Groot MJ, Ackland MJ, Horne VA, Alex AA, Jones BC. A novel approach to predicting P450 mediated drug metabolism: CYP2D6 catalyzed *N*-dealkylation reactions and qualitative metabolite predictions using a combined protein and pharmacophore model for CYP2D6. *J Med Chem* 1999;42:4062–70.
182. Bapiro TE, Hasler JA, Ridderström M, Masimirembwa CM. The molecular and enzyme kinetic basis for the diminished activity of the cytochrome P450 2D6.17 (CYP2D7.17) variant: Potential implications for CYP2D6 phenotyping studies and the clinical use of CYP2D6 substrate drugs in some African populations. *Biochem Pharmacol* 2002;64:1387–98.
183. Lewis DF, Dickins M, Lake BG, Goldfarb PS. Investigations of enzyme selectivity in the human CYP2C subfamily: Homology modelling of CYP2C8, CYP2C9 and CYP2C19 from the CYP2C5 crystallographic template. *Drug Metab Drug Interact* 2003;19:189–210.
184. Snyder R, Sanger R, Wang J, Ekins S. Three-dimensional quantitative structure activity relationship for Cyp2d6 substrates. *Quant Struct-Act Rel* 2002;21:357–68.
185. Yao H, Costache AD, Sem DS. Chemical proteomic tool for ligand mapping of CYP antitargets: an NMR-compatible 3D QSARdescriptor in the *heme-based coordiante system*. *J Chem Inf Comput Sci* 2004;44:1456–65.
186. Allorge D, Bréant D, Harlow J, Chowdry J, Lo-Guidice J-M, Chevalier D, et al. Functional analysis of CYP2D6.32 variant: Homology modeling suggests possible disruption of redox partner interaction by Arg440His substitution. *Proteins* 2005;59:339–46.
187. Lewis DFV, Bird MG, Parke DV. Molecular modelling of CYP2E1 enzymes from rat, mouse and man: An explanation for species differences in butadiene metabolism and potential carcinogenicity, and rationalization of CYP2E substrate specificity. *Toxicology* 1997;118:93–113.
188. Tan Y, White SP, Paranawithana SR, Yang CS. A hypothetical model for the active site of human cytochrome P4502E1. *Xenobiotica* 1997;27:287–99.
189. Lewis DF, Sams C, Loizou GD. A quantitative structure-activity relationship analysis on a series of alkyl benzenes metabolized by human cytochrome P450 2E1. *J Biochem Mol Toxicol* 2003;17:47–52.
190. Park J-Y, Harris D. Construction and assessment of models of CYP2E1: Predictions of metabolism from docking, molecular dynamics, and density functional theoretical calculations. *J Med Chem* 2003;46:1645–60.
191. Ferenczy GG, Morris GM. The active site of cytochrome P-450 nifedipine oxidase: A model-building study. *J Mol Graphics* 1989;7:206–11.
192. Lewis DFV, Eddershaw PJ, Goldfarb PS, Tarbit MH. Molecular modelling of CYP3A4 from an alignment with CYP102: Identification of key interactions between putative active site residues and CYP3A-specific chemicals. *Xenobiotica* 1996;26:1067–86.
193. Szklarz GD, Halpert JR. Molecular modeling of cytochrome P450 3A4. *J Comput-Aided Mol Des* 1997;11:265–72.
194. Tanaka T, Okuda T, Yamamoto Y. Characterization of the CYP3A4 active site by homology modeling. *Chem Pharm Bull* 2004;52:830–5.

195. Chang YT, Loew GH. Homology modeling and substrate binding study of human CYP4A11 enzyme. *Proteins* 1999;34:403–15.
196. Lewis DFV, Lake BG. Molecular modelling of CYP4A subfamily members based on sequence homology with CYP102. *Xenobiotica* 1999;29:763–81.
197. Kalsotra A, Turman CM, Kikuta Y, Strobel HW. Expression and characterization of human cytochrome *P*450 4F11: Putative role in the metabolism of therapeutic drugs and eicosanoids. *Toxicol Appl Pharmacol* 2004;199:295–304.
198. Laughton CA, Zvelebil MJJM, Neidle S. A detailed molecular model for human aromatase. *J Steroid Biochem Mol Biol* 1993;44:399–407.
199. Schappach A, Höltje H-D. Investigations on inhibitors of human 17 alpha-hydroxylase-17,20-lyase and their interactions with the enzyme: Molecular modelling of 17 alpha-hydroxylase-17,20-lyase, Part II. *Pharmazie* 2001;56:835–42.
200. Auvray P, Nativelle C, Bureau R, Dallemagne P, Séralini G-E, Sourdaine P. Study of substrate specificity of human aromatase by site directed mutagenesis. *Eur J Biochem* 2002;269:1393–405.
201. Chen S, Zhang F, Sherman MA, Kijima I, Cho M, Yuan YC, et al. Structure-function studies of aromatase and its inhibitors: A progress report. *J Steroid Biochem Mol Biol* 2003;86:231–7.
202. Favia AD, Cavalli A, Masetti M, Carotti A, Recanatini M. Three-dimensional model of the human aromatase enzyme and density functional parameterization of the iron-containing protoporphyrin IX for a molecular dynamics study of heme-cysteinato cytochromes. *Proteins* 2006;62:1074–87.
203. Murtazina D, Puchkaev AV, Schein CH, Oezguen N, Braun W, Nanavati A, et al. Membrane-protein interactions contribute to efficient 27-hydroxylation of cholesterol by mitochondrial cytochrome P450 27A1. *J Biol Chem* 2002;227:37582–9.
204. Yamamoto K, Masuno H, Sawada N, Sakaki T, Inouye K, Ishiguro M, et al. Homology modeling of human 25-hydroxyvitamin D_3 1α-hydroxylase (CYP27B1) based on the crystal structure of rabbit CYP2C5. *J Steroid Biochem Mol Biol* 2004;89–90:167–71.
205. Kuo A, Gulbis JM, Antcliff JF, Rahman T, Lowe ED, Zimmer J, et al. Crystal structure of the potassium channel KirBac1.1 in the closed state. *Science* 2003;300:1922–6.
206. Shi N, Ye S, Alam A, Chen LP, Jiang YX. Atomic structure of a Na+- and K+-conducting channel. *Nature* 2006;440:570–4.

17

CRYSTAL STRUCTURES OF TOXICOLOGY TARGETS

Frank E. Blaney and Ben G. Tehan

Contents

Computational Toxicology: Risk Assessment for Pharmaceutical and Environmental Chemicals,
Edited by Sean Ekins

17.1 INTRODUCTION

In a recent review Guengerich described the classification of toxicity either by pathological effect (cell death, immunological hypersensitivity, cancer, etc.) or by mechanism [1]. In this chapter we consider toxicity in terms of the latter. According to Paracelsus, who is often called the father of toxicology, all chemicals, whether endogenous or xenobiotic, are toxic. This is a frightening concept without the knowledge that toxicity is very dependent on dose. However, there is sometimes a fine distinction between the dose level of a drug required for therapeutic effect and that for toxic response. Toxicity generally arises either through inhibition of enzymes or receptors or through the process of metabolism. Direct inhibition, of course, can be useful, as in the case of hydroxymethylglutaryl co-enzyme A (HMG-CoA) reductase, an enzyme involved in cholesterol synthesis. Statins act through inhibition of this enzyme in the liver, to lower cholesterol. Nevertheless, inhibition of the same enzyme elsewhere in the body gave rise to muscle toxicity, which was the reason that one statin was eventually withdrawn from the market. This is an example of "on-target" or mechanism-based toxicity. The metabolic enzymes are important in removing chemicals from the body, and if these are inhibited in any way, then dangerous levels of an otherwise useful drug can accumulate. A classic example of this is terfenadine which is metabolized by cytochrome P450 3A4 (CYP3A4). Inhibition of this enzyme can cause arrhythmias because of high levels of terfenadine which is a human cardiac ether-a-go-go (hERG) voltage-gated potassium channel inhibitor (see later and Chapters 13, 16, and 19).

Alternatively, the metabolite produced by the enzyme, may itself be more toxic than the parent compound. This is usually the case when the metabolic product is electrophilic or radical in nature. Electrophilic intermediates can often undergo irreversible covalent bond formation with proteins, e.g. 2,5-hexanedione, a cytochrome P450 (P450) mediated product of hexane metabolism, reacts with lysine residues to give pyrroles [2]. Covalent interactions in proteins can give rise to hypersensitivities and related immunological reactions. Radical products can interact with lipid and DNA molecules, and in the latter case can give rise to genotoxicity.

By far the most important group of drug-metabolizing enzymes is P450. In addition to toxicity arising from metabolism or inhibition, several human P450s, especially CYP2D6 and CYP2C9, exhibit polymorphism. These genetic variations can give rise to impaired metabolism of certain drugs. One variant of CYP2C9 shows a significant reduced ability to metabolize warfarin and phenytoin, leading eventually to toxic levels of these drugs [3]. A similar problem has been found with variants of CYP2D6 and the drugs debrisoquine and sparteine [4,5]. Since 2003, four of the main human P450s have been solved by X-ray crystallography, and much knowledge about their mechanism and substrate diversity has been gained (see also Chapter 10). A major part of this chapter will be devoted to this work.

Another type of mechanism-based toxicity, as defined by Guengerich, is "off-target" pharmacology [1]. The hERG potassium channel is a classic example of this. Inhibition of hERG by a wide variety of diverse drugs leads to prolonged QT syndrome, a condition that can result in death of the patient. A number of drugs have been withdrawn from the market because of their hERG liabilities. No crystal structure exists yet for hERG, although other potassium channels have been solved. A lot of modeling work has been published around homology modeling of hERG and its inhibitors, which has been covered elsewhere in this book (see Chapters 13, 16, 19 and 20).

The nuclear hormone receptors (NHRs or NRs) have been the targets of drug discovery for many years, as their endogenous ligands, such as the steroid hormones, thyroxin, vitamin D, and retinoic acid, are themselves small drug-like molecules of immense pharmacological importance. For example, one subfamily of NHRs, the peroxisome proliferator activated receptors (PPARs), are the target of the glitazone drugs that are important in the treatment of type 2 diabetes. It was discovered that several NHRs, including the pregnane X receptor (PXR) and constituitive androgen receptor (CAR), regulate the induction of important metabolizing enzymes, including some major P450s, and transporter proteins such as the multidrug-resistant proteins and P-glycoprotein. Quite often it has been found that a drug that is metabolized by a certain P450 can also act at the NHR that induces that enzyme (see Chapters 4 and 12). In its own right this is generally a useful mechanism of detoxification. However, if two drugs are co-administered, then it can give rise to adverse drug–drug interactions. This occurs because the induction of P450s or transporters by one drug can result in the abnormal metabolism or transport of the other.

A different form of toxicity occurs in the estrogen and androgen receptors when environmental chemicals have the ability to bind to them. Acting both as agonists or antagonists, these so-called endocrine disruptors are responsible for a number of hormone-related cancers, in addition to their roles in adversely affecting sexual development and reproductive fertility.

The ability to predict drug–drug interactions and likely endocrine disruptors remains a major goal of toxicity studies. In the last decade many of the NHRs have been solved by X-ray crystallography, and this has added greatly to our understanding of their modes of action (see also Chapter 12).

17.2 CYTOCHROMES P450

17.2.1 Background

One key aspect in the study of toxicity is the ability to predict the products of metabolism. In this respect P450s have been the major area of research for many years. The P450s have been shown to catalyze an almost bewildering number of reactions [6]. In addition to the common hydroxylation reactions at aliphatic and aromatic carbon centers, they can directly oxidize heteroatoms such as nitrogen, sulfur, phosphorous, and even iodine to yield heteroatom oxides. They are involved in heteroatom "release" via, for example, N- and O-dealkylation reactions, usually via the "formal" hydroxylation of the adjacent carbon atom, although direct electron abstraction followed by proton removal from the aminium radical may be a preferred possibility [7]. Epoxide formation is another commonly observed reaction, particularly important for toxicology in that epoxides are reactive electrophilic species capable of further covalent reaction with important proteins. P450s have been shown to catalyze the aromatization of alicyclic rings in steroids and to oxidize alcohols and aldehydes to ketones and carboxylic acids respectively.

Surprisingly P450s are also involved in a considerable number of reduction reactions including alkyl halides, azo compounds, N-oxides, hydroxylamines, and nitro groups. Further discussion of these and the many other, more unusual reactions attributed to P450s is beyond the scope of this chapter, and the reader is referred to the excellent review by Guengerich [6].

17.2.2 Structure of P450 Enzymes

The early bacterial structures showed that these P450s maintained a common fold of 12 α-helices, named A–L, and up to 5 β-sheets, although all of the latter are not always present. Extensions of some of the helices in the recent mammalian isoforms described below has led nowadays to a total of 16 helices. They include the important B′, F′, and G′ helices that are involved in ligand binding and/or recognition. These are illustrated with reference to the CYP2D6 crystal structure, in Figures 17.1 and 17.2. The heme iron is bound to a cysteine

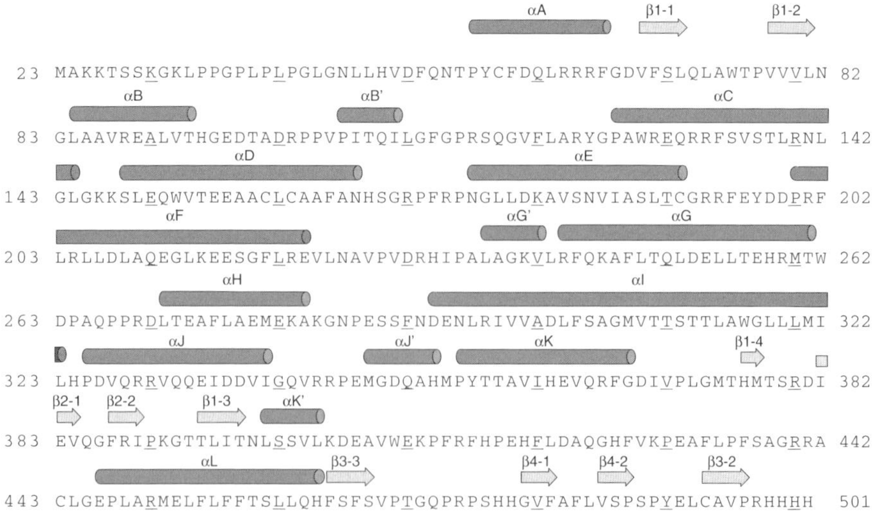

Figure 17.1 The sequence and secondary structure of CYP2D6 derived from the crystal structure.

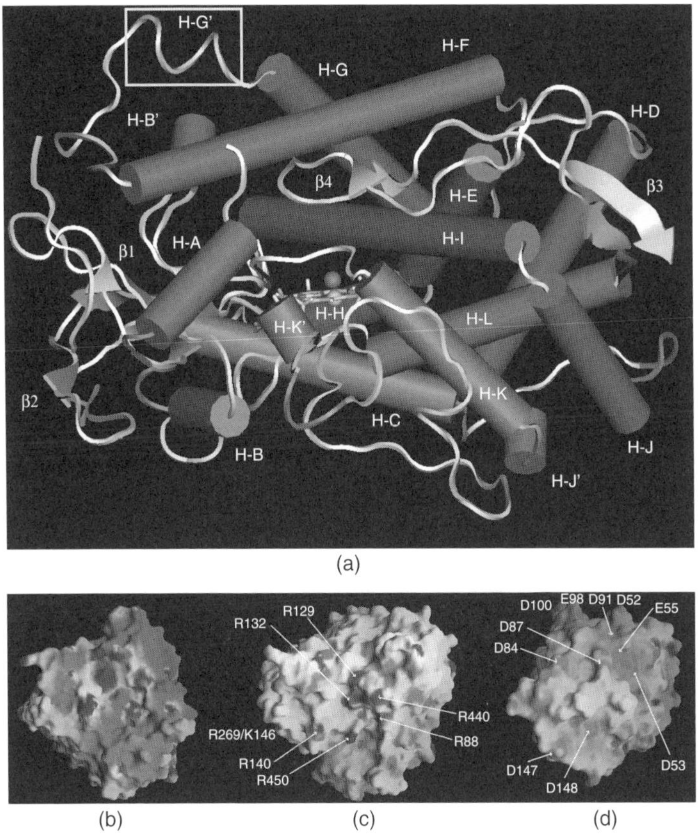

Figure 17.2 (*a*) Crystal structure of CYP2D6 with the helices in purple and the β-strands in yellow. The helices are labeled H–A through to H–L. The oxygenated heme can be seen just below H–I. The helical nature of H–G′ is evident although it was not detected in the software, and is therefore displayed in the cyan box at the top. (*b*) The largely negative electrostatic potential on the distal face of CYP2D6. Positively charged ligands are attracted to this. This is in contrast to (*c*) the proximal face of CYP2D6 where a large number of basic residues are found. This is a perfect compliment to the reductase partner in (*d*), which has clusters of acidic residues which form a tight salt-bridged interface with the cytochrome. See color plates.

residue in a highly conserved region. Just below this, the proximal face of the protein contains a considerable number of basic lysine and arginine residues, which form a perfect binding interface for the aspartate and glutamate clusters of the associated cytochrome reductase protein (Figure 17.2*c*, *d*). The substrate cavity varies considerably in size and in the nature of the residues lining it. On one side above the heme, the I helix contains a very highly conserved threonine residue that is believed by many to be a proton source during the dioxygen cleavage step. The other side is lined by the B′ helix and the B′–C loop, while the "roof" of the cavity is formed by the F and G helices and the

F–G loop. Many of the β-strands are also involved in substrate binding. In a classification derived originally from the bacterial structures, by Gotoh [8], six substrate recognition sites (SRS) are present and their positions have been conserved in the mammalian proteins. They are discussed in more detail later. Finally a varying number of channels are found in P450s that are necessary for substrate entry and product egress. The reader is referred to the recent article by Schleinkofer et al. for a full discussion of these channels [9].

17.2.3 Selectivity of P450 Substrates

It has long been recognized that some P450s have a preference for certain classes of substrates. CYP1A2, for example, prefers small planar aromatic molecules, CYP2D6 is said to prefer molecules with a basic center and an aromatic ring, and CYP2C9 is believed to favor acidic substrates (see Chapters 10 and 16). Lewis has carried out a survey of preferred substrates for a number of CYP450 isoforms, in particular, using measured and calculated physicochemical properties [10]. His conclusions are that a large number of substrates could have their preferences predicted by a decision tree that has only three properties: volume or size, pK_a, and surface area/depth, which is also a reflection on the planarity of the substrate. Volume alone can distinguish CYP3A4 (high volume) from CYP2E1 (low) and from the other CYP1 and CYP2 isoforms (medium). CYP2C8 was not considered in this analysis, which is unfortunate in that it has been shown to have an even larger substrate cavity than CYP3A4 [11]. For the unclassified (medium volume) substrates, the next decision is based on pK_a where basic substrates are classed as CYP2D6 and acidic substrates as CYP2C9 ligands. Neutral molecules are finally classified in terms of surface area/depth^2, which is considered as a measure of planarity. Highly planar molecules are classed as CYP1A2 substrates, low-value ones as CYP2B6, and medium as CYP2A6 substrates. Other properties have also been considered, but these three are the most successful in general. Of course, many exceptions are known for these rules, and certain substrates are metabolized by more than one isoform. However, when combined with docking studies, these rules are a useful starting point in predicting metabolism.

17.2.4 Bacterial Crystal Structures and Mechanism

The bacterial structures have proved to be invaluable in the early homology modeling of human CYP enzymes, although this has been largely superceded with the advent of mammalian crystal structures. In their model of human CYP2D6, Sutcliffe proposed that the B–C loop would be better built on the structure of the equivalent loop of $P450_{BM3}$, although the rest of the model could be based on the rabbit CYP2C5 [12]. Furthermore Lewis rebuilt all the human isoforms on the basis of the 2C5 structure but concluded that the sequence homology of 3A4 was still closer to $P450_{BM3}$ [13]. One bacterial structural study that continues to impact on the understanding of cytochrome

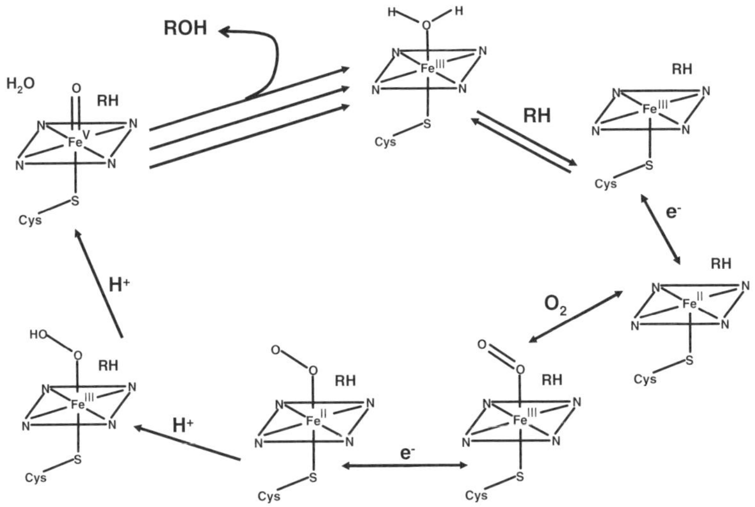

Figure 17.3 The P450 cycle showing the various electron and proton transfer stages, the oxygenation step and the final cleavage of the hydroperoxy intermediate to yield water and "compound 1". The substrate is shown as RH, which displaces the axial water ligand in the first step. ROH is the product. Several arrows in the final stage are shown to indicate that multiple mechanisms are possible.

P450 mechanism is the elegant elucidation of the oxygen intermediates in $P450_{CAM}$, by Schlichting et al. in 2000 [14].

The mechanism by which cytochromes bind and then activate oxygen involves two electron transfer and two protonation steps that together make up the "P450 cycle" (see Figure 17.3). An incoming substrate molecule initially displaces the iron-bound water to yield a pentacoordinate heme, which is readily reduced to Fe^{II} by the first electron transfer step. Schlichting showed that within the limits of resolution, the Fe–Cys sulfur bond remained at 2.2 Å with the iron sitting about 0.3 Å below the plane of the heme ring. In the next step the incoming oxygen molecule binds at an angle to the ring with the second atom oriented toward thr^{252}, on the I helix. The iron has now moved so as to lie slightly above the heme ring, which has flattened somewhat. Some displacement of the camphor substrate was observed, and in addition a new crystallographic water was seen that was hydrogen-bonded to the bound oxygen molecule and to the hydroxyl of thr^{252}. Surprisingly some large side chain shifts were also observed. The addition of the second electron and a proton produces a hydroperoxy Fe^{III} species. The experimental source of this second electron proved to be a fascinating challenge. In nature, it comes from a second protein, putidaredoxin, but this is not present in the crystallographic study. The first electron was provided chemically from the reducing agent, dithionite. Its use in the second electron transfer step was prohibited by the unstable nature of the dioxygen heme intermediate. The problem was solved by the use of X-ray radiolysis of water in the system. After a suitable radiation wavelength was chosen, large numbers of hydrated electrons could be produced that carried out the second reduction under conditions ideally suited for the crystallographic determination. The observed structure from this is consistent with the activated iron-oxygen species, commonly known as "compound 1" in the P450 cycle. The Fe–O bond distance was now 1.65 Å, compared to 1.80 Å in the Fe–O_2 intermediate. A new water molecule was observed in the crystal. This is hydrogen-bonded to thr^{252} and the backbone of gly^{248}, and it is almost certainly the result of the oxygen–oxygen cleavage following the second protonation step. The formation of this new water from the distal dioxygen atom allows the camphor to approach closer to compound 1 than before, because it is no longer sterically hindered by the oxygen molecule.

The results of this work by Schlichtlng et al. have had a major implication for docking studies carried out with P450 crystal structures or homology models. When looking at metabolic or toxicity problems arising from P450 inhibition, it is appropriate to use an unligated (pentacoordinate) heme system. However, if one is trying to use docking experiments as part of a metabolic prediction study, then it is essential to include the full compound 1 intermediate in the protein. The thr^{252} bound water molecule should also be included as this could have important steric or hydrogen-bonding effects in the approach of the substrate. Indeed it has been postulated that this water, or even the (hydro)peroxy intermediate, could provide alternative oxygen sources for some hydroxylation reactions (see later) [15].

17.2.5 CYP2C5—The First Mammalian Cytochrome P450 Structure

The successful solution of the many bacterial cytochrome crystal structures was largely due to the simple fact that they are soluble. Mammalian microsomal P450s, on the other hand, are membrane bound and hence insoluble. The *N*-terminal region of the microsomal isoforms is believed to form a single membrane-spanning α-helix that tethers it close to the other protein in the system, the NADP/H dependent cytochrome reductase. This reductase is the source of electrons for the catalytic mechanism described earlier.

By removing the first 22 residues of the protein and introducing a number of other "solubilizing" or aggregate inhibiting mutations, particularly in the F–G region, Eric Johnson's group were finally able to crystallize the first mammalian P450, namely rabbit CYP2C5, and solve its structure by X-ray crystallography [16]. The structure contained the same overall fold as the bacterial P450s, but some notable differences were observed. The greatest divergence is seen in the F–G loop region, close to the putative substrate entrance channel. This is actually closed in $P450_{CAM}$, but there is a clear pathway in CYP2C5. The *N*-terminal β-strands also show clear divergence from the observed bacterial protein structures. Of the six substrate recognition sites (SRS) defined by Gotoh [8], only SRS4, the region where the I helix crosses the heme, is structurally conserved. This forms one side of the base of the binding site, the other being defined by the pocket formed from helix K and β1–4 (SRS5). This is clearly different in CYP2C5 and will therefore affect the substrate selectivity and orientation during oxidation. The remaining four SRSs that define the overall binding pocket show RMS differences (between $P450_{BM3}$ and CYP2C5) ranging from 3.9 Å for SRS3 (the interior residues of helix G) to 6.4 Å for SRS2 (the interior residues of helix F). The latter two SRSs are closer to the roof of the substrate site and will determine aspects of substrate recognition and movement toward the final reactive orientation (see the discussion later on CYP2D6).

One particularly interesting observation by Williams et al. [17] was that after removal of the *N*-terminal membrane-spanning helix, the enzyme still exhibited membrane association. The short disordered region immediately before the first crystallographically observed residue, pro^{30}, contains a number of basic residues. These are believed to prevent translocation of the enzyme through the mitochondrial membrane, presumably by interaction with the lipid headgroups. This region is also close to the F–G and B–C loop regions. Together they make up the membrane interacting face of the cytochrome. This region is also presumed to form the substrate entrance channel and the implication is therefore that hydrophobic substrates such as progesterone are delivered, not from the cytoplasm but rather through the lipid bilayer.

Several other crystal structures of CYP2C5 with bound ligands have now been published [18–20]. In the first of these [18] the ligand, 4-methyl-*N*-methyl-*N*-(2-phenyl-2H-pyrazol-3yl)benzenesulphonamide (DMZ (1); Figure 17.4), was crystallized with CYP2C5 to 2.3 Å resolution. Comparison of this structure with that of the unliganded protein showed that considerable conformational changes had occurred during substrate binding. Helix I, for example, had now developed a pronounced bend, and helices F and G, had shifted to accommodate the substrate. One particularly interesting finding was that the density of the substrate could not be unambiguously assigned; instead, it was possible that an extended conformation of the ligand could be fitted in either of two antiparallel orientations. Together these two orientations accounted for all the density, although neither of them individually could do so. The more preferred orientation placed the 4-methyl group (the primary site of hydroxylation) within 4.4 Å of the heme iron (Figure 17.5*a*). A minor product of hydroxylation occurs on the *N*-phenyl ring and this can be explained by the alternate density-fitted orientation of the substrate. The phenyl ring now comes within 5.9 Å of the iron, and this greater distance could explain the much smaller amount of this metabolite observed by experiment.

Another conclusion was that the conformational changes observed in the substrate bound form arise as a result of direct induced interactions with the ligand. The B-factors in the flexible loop regions were generally better and previously poorly defined features, such as the existence of a B′ helix, were now observable.

The second structure of a CYP2C5-substrate complex was published a few months later, again from Johnson's group [19]. This was with the

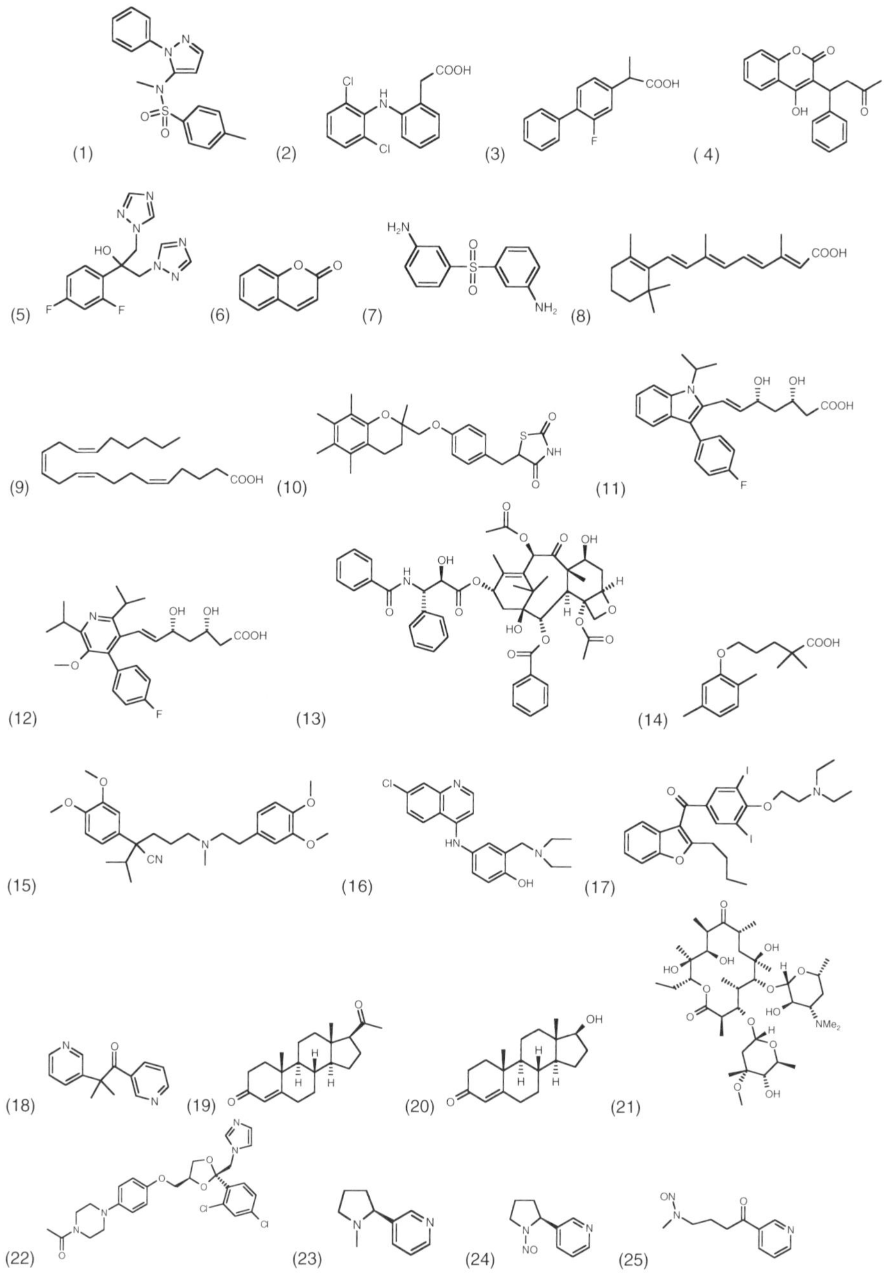

Figure 17.4 Compounds which are mentioned throughtout the chapter and numbered in the main text as follows: (1) DMZ, (2) diclofenac, (3) flurbiprofen, (4) warfarin, (5) fluconazole, (6) coumarin, (7) dapsone, (8) retinoic acid, (9) arachidonic acid, (10) troglitazone, (11) fluvastatin, (12) cerivastatin, (13) taxol, (14) gemfibrozil, (15) verapamil, (16) amodiaquine, (17) amiodarone, (18) metyrapone, (19) progesterone, (20) testosterone, (21) erythromycin, (22) ketoconazole, (23) nicotine, (24) *N*′-Nitrosonornicotine, (25) 4-(Methylnitrosamino)-1-(3-pyridyl)-1-butanone

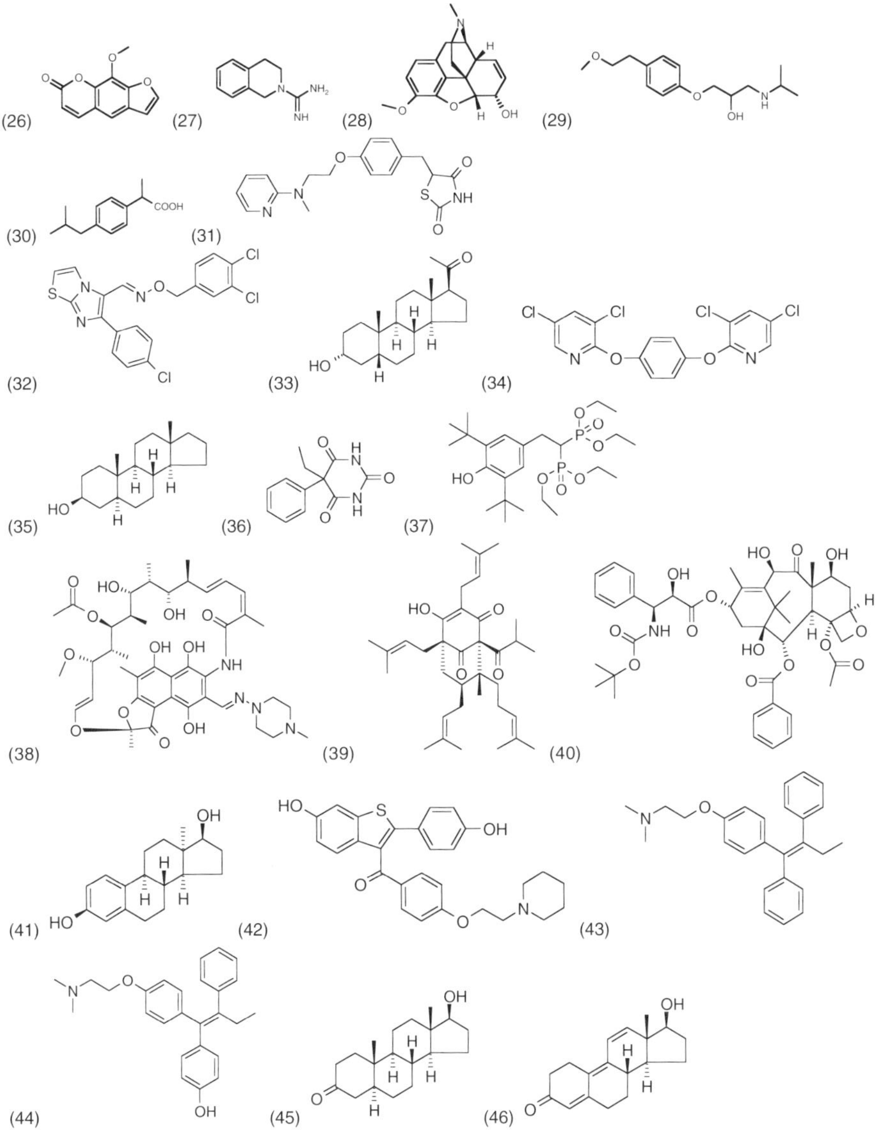

Figure 17.4 (*Continued*) (26) methoxsalen, (27) debrisoquine, (28) codeine, (29) metoprolol, (30) ibuprofen, (31) Rosiglitazone, (32) CITCO, (33) 5β-pregnanedione, (34) TCPOBOP, (35) 3α,5α-androstanol, (36) phenobarbital, (37) SR12813, (38) Rifampicin, (39) Hyperforin, (40) docetaxel, (41) 17β-estradiol, (42) raloxifene, (43) tamoxifen, (44) 4-hydroxytamoxifen, (45) dihydrotestosterone, (46) metribolone (R1881).

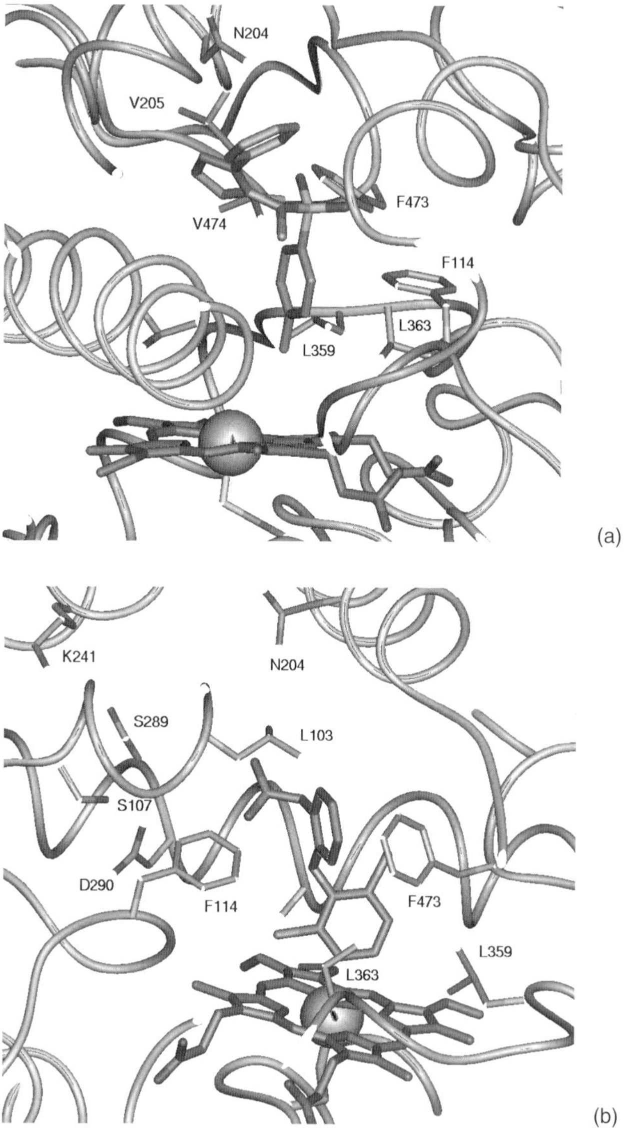

Figure 17.5 (*a*) DMZ bound in the CYP2C5 cavity; (*b*) diclofenac bound in the CYP2C5.

anti-inflammatory drug, diclofenac, at an improved resolution of 2.1 Å. Again evidence of an induced fit was observed and the main site of hydroxylation, the 4′-phenyl carbon came within 4.9 Å of the heme iron. Diclofenac (2) and other nonsteroidal anti-inflammatory drugs contain a carboxylic acid group, and this was expected to form a salt-bridge with a nearby arginine or lysine residue. It was a surprise therefore that no such interaction was observed in the crystal

prot												
cam	100											
bm3	17	100										
2c5	18	21	100									
2b4	19	18	51	100								
1a2	17	19	29	29	100							
2a6	18	20	51	52	28	100						
2c8	19	18	74	53	28	49	100					
2c9	20	18	77	50	29	49	78	100				
2c19	17	19	77	50	27	51	78	91	100			
2d6	19	17	41	43	30	35	41	40	40	100		
2e1	16	20	56	48	30	40	58	59	58	40	100	
3a4	20	27	23	26	22	20	23	23	23	20	23	100
prot	cam	bm3	2c5	2b4	1a2	2a6	2c8	2c9	2c19	2d6	2e1	3a4

Figure 17.6 Table of sequence identities between the various P450 isoforms giving some idea of the reliability in their use in homology modeling.

structure. Instead, the carboxylate was close to the aspartate residue, asp^{290} (Figure 17.5*b*). Rather than a direct interaction, it formed hydrogen bonds to a cluster of water molecules that were in the vicinity, not only of this aspartate but also of the residues asn^{204}, ser^{289}, and lys^{241}. It could only be concluded that the diclofenac carboxylate was not ionized in its bound form. Whether this is the case with its main human metabolizing enzyme, CYP2C9, remains to be established, since no crystal structure of this complex has been published. It is unlikely, however, as another NSAID, flurbiprofen (3), has been shown to interact with an arginine residue in CYP2C9 [21]. Some QM-MM calculations have also suggested that diclofenac is interacting with an arginine residue in its ionized form (A. Mulholland, personal communication).

One important use of the CYP2C5 structure has undoubtedly been in the construction of homology models of the functionally more important human isoforms, CYP2D6, CYP2C9, CYP2C19, and CYP3A4. Figure 17.6 shows the relative sequence identity of some bacterial isoforms with CYP2C5 and the main human cytochromes. Generally, it can be seen that sequence identities between CYP2C5 and the other human isoforms are better than with the bacterial enzymes. The exception as noted earlier is CYP3A4, where the identity is even lower than that with $P450_{BM3}$.

The role of homology modeling has been discussed elsewhere in this book (Chapter 16) and a review of the human P450 homology models published up to the end of 2004 was presented by de Graff et al. [22]. Many of these models were based on the CYP2C5 structure. The emergence of crystal structures of CYP2C8, CYP2C9, and CYP3A4 has obviously led to a decline in their use for these isoforms, although homology studies actively continued with CYP2D6 until its crystal structure was recently published [23].

17.2.6 CYP2C9—The First Human Isoform

In 2003 Williams et al. from Astex published the structure of CYP2C9, the first human isoform to be solved [24]. Two structures were actually reported, an apo form and a structure with the bound substrate, warfarin (4). It came as a surprise to many, however, that the substrate was in a pocket remote from the heme group, and this therefore probably represented an intermediate binding site prior to the substrate's accession to the heme. The enzyme showed some additional helical structure in the B–C and F–G loops with residues 101–106 forming a B′ helix, and residues 212–222 forming F′ and G′ helices. The familiar water molecule was hydrogen bonded to thr^{301} on helix I, where it is proposed to act as a proton source during the oxygen cleavage stage of the P450 cycle.

Previous reports using site-directed mutagenesis (SDM) have suggested that arg^{97} acts as a salt-bridge partner for the acidic groups found in typical CYP2C9 substrates [25]. However, in the crystal structure this formed hydrogen bonds to the propionate groups of the heme. The binding site of this structure is shown in Figure 17.7*a*. No other previously implicated basic residues such as arg^{105} and arg^{108} are oriented into the cavity, although there are numerous hydrophobic side chains such as phenylalanines 69, 100, and 476 and leucines 102, 208, 362, and 366 that can make interactions with the small lipophilic substrates.

Williams et al. have raised the intriguing idea that the unexpected warfarin binding site leaves enough unused volume for a second substrate or inhibitor to bind closer to the heme group [24]. Using docking studies, they were able to model either another warfarin or fluconazole (5) into this site. Support for this comes from experiments where S-warfarin was shown to increase the rate of metabolism of 7-methoxy-4-trifluoromethylcoumarin [26]. This type of cooperativity is certainly not new in the field of P450s. It has been demonstrated many times, especially with CYP3A4, where atypical kinetics are often found.

Some months after the Astex structure appeared, Wester et al. published a second structure of CYP2C9, complexed this time with the added substrate, flurbiprofen (3) [21]. This showed significant differences from the earlier CYP2C9 crystal structure of Williams et al. described above. The Scripps group have attributed these to the use of different constructs in the soluble protein generation. In particular, Williams et al. made use of a section of residues in the 206–224 range containing seven mutations. These were the same as those used in the generation of the soluble CYP2C5 construct, so they gave rise to a large change in the region encompassing part of the F helix and the F–G loop. The main difference was found in the conformation of the B–C loop—in particular, arg^{108} was now oriented into the binding cavity where it formed a salt-bridge with the carboxylate of flurbiprofen (Figure 17.7*b*). This was stabilized by further interactions between the arginine and asn^{289} and asp^{293} on helix I, placing the substrate within reaction distance to the heme iron. The binding cavity is shown in Figure 17.7*b*. Note that hydrophobic residues in five of Gotoh's substrate recognition sites (SRS1–SRS5) make contact with the ligand; only SRS6 does not appear to be involved.

As with the structure of Williams et al., there is still considerable available volume in the cavity. Experimental kinetic evidence is available to show that dapsone (7) can stimulate the oxidation of flurbiprofen. Use was made of the

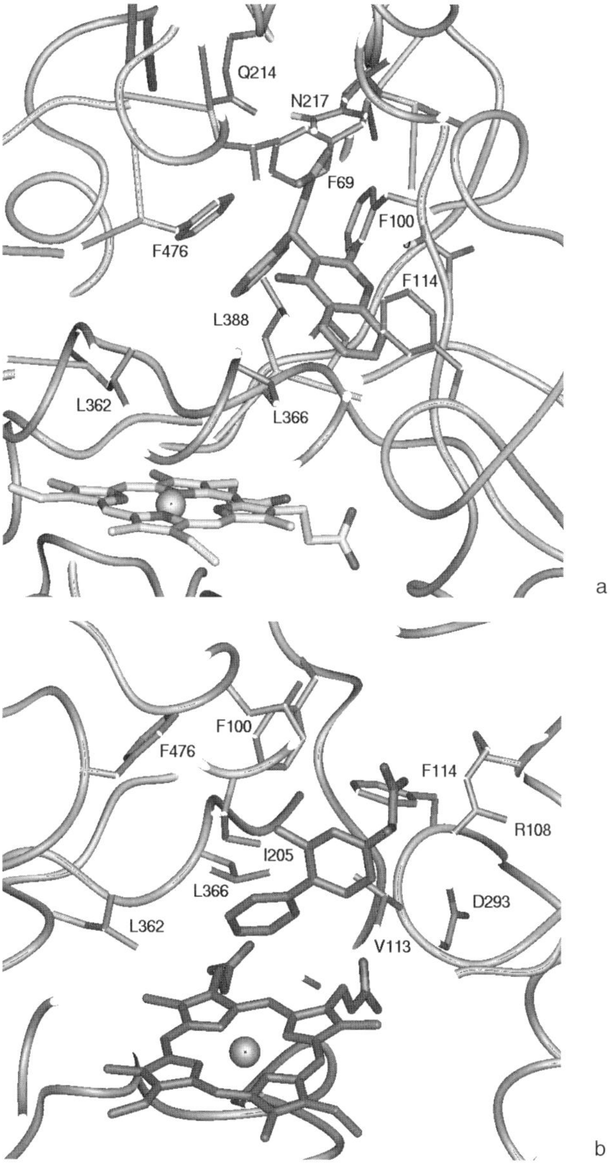

Figure 17.7 (*a*) Binding site of the Astex CYP2C9 structure; (*b*) Scripps CYP2C9 structure with flurbiprofen bound in the active site.

automated docking program, AUTODOCK [27], to simulate the binding of dapsone in the 2C9-substrate complex. It was postulated from the results that dapsone limited the movement of flurbiprofen in the binding pocket, keeping it close to the heme and preventing solvent from reacting with the oxy-iron compound 1 intermediate.

The Astex group has built an homology model of the CYP2C19 isoform based on their CYP2C9 crystal structures [28]. As can be seen in Figure 17.6, the sequences of these two enzymes are 91% identical, and in the active site only one residue, ile[99], has a nonconservative substitution to a histidine in CYP2C19. Despite this, CYP2C19 does not show any particular preference for a single class of substrate; in particular, CYP2C19 shows no affinity for small acidic ligands. Williams et al. have attributed this to the substitution of lys[77] in CYP2C9 to a glutamate in CYP2C19. This residue lies in the substrate access channel, and hence selectivity occurs in a recognition event rather than during the reaction itself. The presence of arg[108] in the active site of the Scripps's structure, and its conservation in CYP2C19, tend to support this suggestion.

17.2.7 CYP2C8

The next human CYP450 structure to appear, in 2004, was that of CYP2C8, again from the Scripps group [11]. This is an important enzyme with a wide and diverse substrate specificity. It is the main isoform responsible for the metabolism of unsaturated fatty acids, retinoic acid (8), and arachidonic acid (9), leading, in the latter case, to the synthesis of epoxyecosatrienoic acids that help in the control of blood pressure. It metabolizes the lipid-lowering drugs, cerivastatin (12) and fluvastatin (11). Thus inhibition of this enzyme by the potent blocker, gemfibrozil (14), has led to important toxic drug–drug interactions [29–31]. CYP2C8 is also responsible for metabolism of the antidiabetic glitazone drugs, such as troglitazone (10), the calcium channel blocker, verapamil (15), and miscellaneous other structures such as amodiaquine (16) (malaria) and amiodarone (17) (cardiac arrhythmia).

One of the first features to be mentioned by Schoch et al. [11] was the very large size of the active site cavity in CYP2C8. This was estimated to be about twice the size of the site in CYP2C5, which explains the enzyme's ability to accommodate and metabolize the large anticancer drug, taxol (13). The increased volume arises, in part, from inclusion of the F′ helix and β-sheet 1, as well as substitution by smaller polar residues in CYP2C8, compared to the larger hydrophobic ones found in the B′ helix and B′–C loop of CYP2C5. The active site of CYP2C8 is shown in Figure 17.8*a*.

Around the time that the CYP2C8 structure was published, Melet et al. described the generation of a CYP2C8 substrate pharmacophore and compared it with a 3D model based on the CYP2C5 structure and data obtained from site-directed mutagenesis (SDM) experiments [32]. The mutations were based on the docking of representative substrates into the homology model. The presence of the acidic groups in fluvastatin, retinoic acid, and troglitazone suggested an interaction with a basic residue in the active site, and thus arg[241], arg[97], and arg[105] were mutated. Other features of the pharmacophore included a large hydrophobic region and two hydrogen bonding sites. Phe[201] and phe[205] were chosen as the most likely residues to be involved in the hydrophobic interaction while ser[100], ser[103], ser[114], and asn[99] were thought of as possible sites

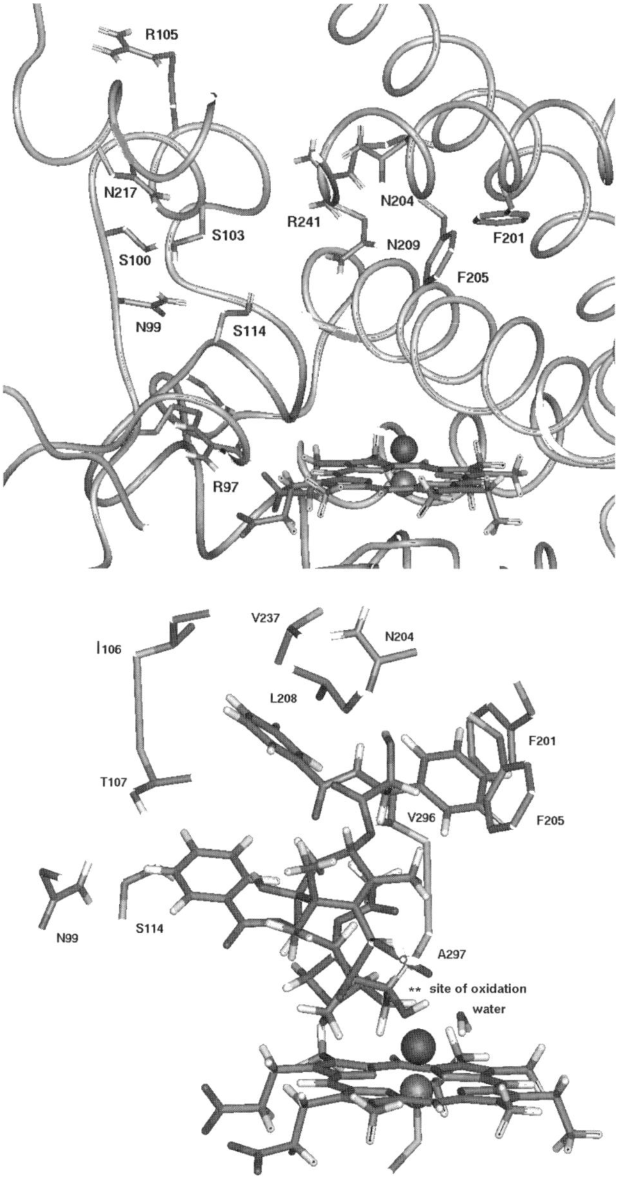

Figure 17.8 (*a*) Active site found in CYP2C8; (*b*) taxol manually docked in CYP2C8 showing key interaction residues.

of hydrogen bond formation (see Figure 17.8*a*). The arg[97] mutants had a dramatic effect on all substrates, but this was attributed to changes in heme binding. Arg[105] had little effect on oxidation, but arg[241] had a profound effect on the metabolism of anionic substrates. The effects of mutation of phe[201] and phe^{205} were very dependent on the substrate tested and could generally be described in terms of changes in the nature of hydrophobic interactions or steric clashes. Taxol, for example, showed an increase in K_m with F201L and F205I mutations that could be attributed to a loss of π–π interactions with the phenyl rings of the substrate. The mutation, F205A, on the other hand, showed a significant decrease in the K_m of taxol hydroxylation, which could arise as a result of an increase in volume in the active site, allowing a better fit of this very large substrate. Ser^{100} appeared to be involved in hydrogen bonding, but ser^{114} had a steric role. The S114F mutant, for example, was unable to catalyze hydroxylation of taxol. The N99L mutant had little effect on the substrates studied, with the exception again of taxol where a significant increase in K_m was observed.

In the same paper Melet et al. used the AUTODOCK program to study the various substrate binding positions in the published crystal structure of CYP2C8. AUTODOCK allows flexibility within the ligand but not the protein. Thus the large size of taxol proved a problem because it is likely that movements in the protein are necessary to accommodate it. The docking results of the other substrates, retinoic acid and fluvastatin, were reported to be in broad agreement with the SDM findings, although no direct role was found for arg^{241}, which was not in the vicinity of the substrate's carboxylate groups.

In contrast to the potential drawbacks of the automated docking approach, we have employed a combined QM-manual docking method that allows full conformational freedom of both ligand and protein (described later in association with CYP2D6). This was used to study the interaction of taxol with CYP2C8, and the results are shown in Figure 17.8*b*. One of the phenyl rings of taxol was sandwiched between F^{201} and F^{205}, explaining nicely the SDM results on these two residues. The amide formed a strong hydrogen bond with the sidechain of asn^{204}, with the adjacent phenyl ring occupying a hydrophobic pocket lined by ile^{106}, leu^{208}, and val^{237}. The third phenyl ring of taxol was in the vicinity of asn^{99} and ser^{114}. The observed increase in K_m with the two mutations of these residues described above is easily accounted for by steric clashes. The overall docking places the 3-hydrogen exactly within substrate reaction distance to the heme oxygen, thus explaining the observed regio- and enantio selectivity.

17.2.8 CYP3A4

In some ways CYP3A4 is the most important of the human isoforms. It is certainly responsible for the metabolism of more drugs and other xenobiotics than any other P450. The diversity of its substrates, which include large molecules such as steroids, cyclosporin, and macrolide antibiotics, and its continuing ability to exhibit complex non–Michaelis-Menten kinetics and cooperativity has led many to believe that the active site of this enzyme is very large with the ability to contain several substances at the same time. In 2004 the same two groups as before, at Scripps and Astex, almost simultaneously published

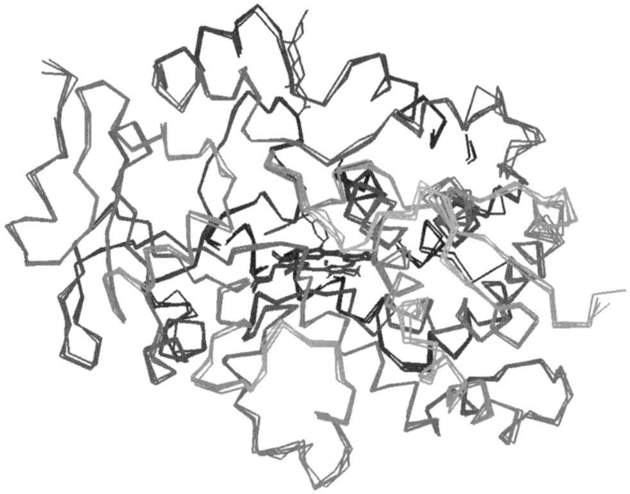

Figure 17.9 Overlay of the C-alpha chains of the four early CYP3A4 structures showing an almost perfect correspondence.

structures of CYP3A4. Yano et al. (Scripps) solved the apo enzyme to a resolution of 2.05 Å [33]. The apo form was also described by Williams et al. at a slightly lower resolution of 2.8 Å, but they also reported complexes of the enzyme with the inhibitor, metyrapone (18) and the substrate, progesterone (19) [34]. Unlike the previous case of CYP2C9, an overlay of these four structures, as depicted in Figure 17.9, showed that they were virtually identical.

The active site cavity of CYP3A4 is indeed large, being potentially about the same size as that in CYP2C8. It is accessible to the exterior by a number of channels, including one between the G′ helix and the B–C loop. This is the main pw2c egress channel as defined by Schleinkofer et al. for the 2C5 structure [9]. A second channel is formed by residues in the β-sheet 1 and F′ helix. A number of residues in this channel are involved in a hydrogen bonding network which includes asp^{61}, asp^{76}, glu^{374}, tyr^{53}, arg^{372}, and arg^{106}. One striking feature of the active site is the presence of a large cage of seven phenylalanines (residues 108, 213, 215, 219, 220, 241, and 304). This aromatic cage has the effect of making the active site seemingly smaller, but movement of the residues can easily result in a larger cavity. Some of these phenylalanines have been shown by SDM to be important for substrate metabolism and cooperativity [35–39]. In fact the active site is largely hydrophobic with the only major charged residue being arg^{212} (see Figure 17.10). Mutation of this residue to alanine, however, has no effect on the hydroxylation of testosterone (20) [38], and Yano et al. have suggested that the side chain can easily swing out of the way to accommodate the steroid substrate. In fact arg^{212} is in a different rotameric state in the CYP3A4 structures reported by Williams et al. [34].

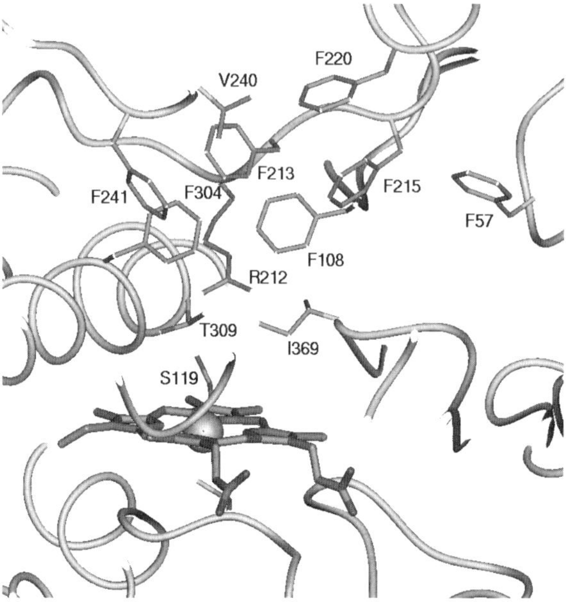

Figure 17.10 CYP3A4 active site showing the predominantly hydrophobic cavity with a cluster of phenylalanines at the top. Arg^{212} is oriented into the cavity in this structure.

Yano et al. performed their crystallization in the presence of erythromycin (21), but no density corresponding to this was found. They did use the AUTODOCK program, however, to confirm that this substrate can fit in an orientation that allows the expected metabolite to form. This docking pose was in agreement with SDM experiments [40–44]. It was proposed that the sugar of erythromycin formed hydrogen bonds with arg^{212}.

Two ligand-bound structures were reported by Williams et al., one with the classic inhibitor metyrapone and the other with the substrate, progesterone [34]. Metyrapone exhibits a classic type 2 UV spectrum suggesting direct coordination to the heme, and this is indeed confirmed in the crystal structure. The iron is coordinated by one of the pyridine rings with the rest of the ligand showing good shape complementarity with the enzyme (Figure 17.11*a*). The progesterone-bound structure is surprising in that the substrate does not reside in the active site. Instead, it sits on the outside of the protein, forming hydrophobic interactions with phe^{219} and phe^{220} and with the acetyl group hydrogen bonding to the backbone of asp^{214} (Figure 17.11*b*). The conclusion is that this acts as a substrate recognition site, although it was suggested that it may also be involved in modulating ligand cooperativity.

Very recently two new ligand bound CYP3A4 structures were reported, one with the inhibitor ketoconazole (22) and the second with the substrate erythromycin [45]. Unlike the metyrapone case, both of these structures dis-

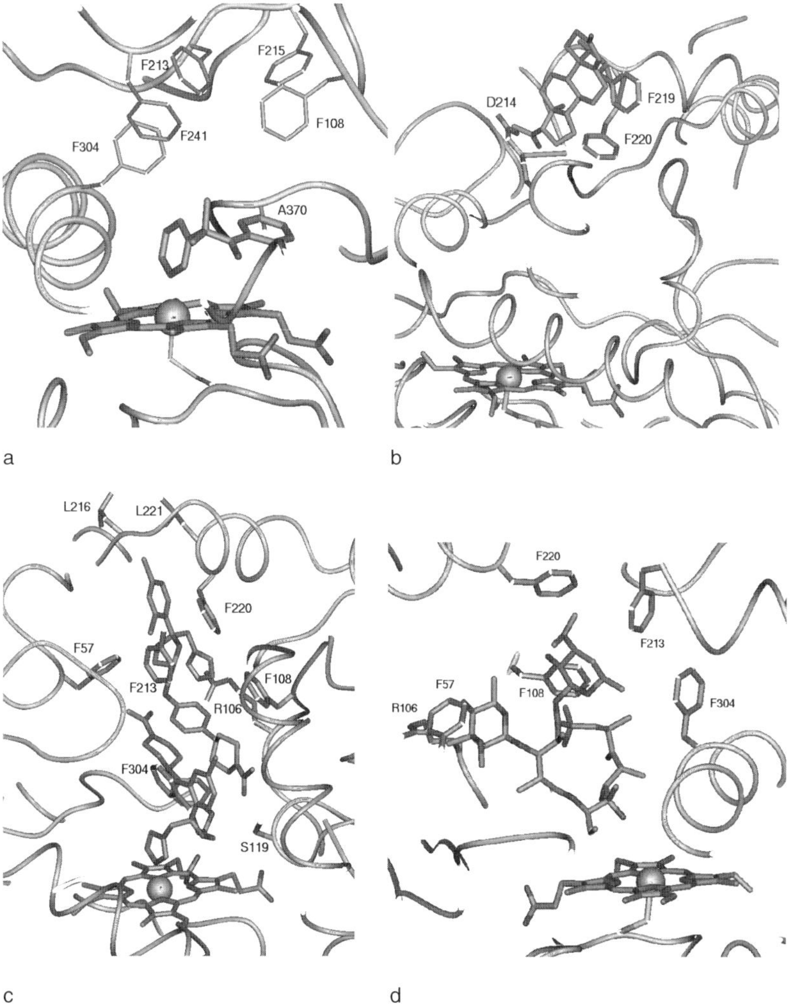

Figure 17.11 Ligands found in the various CYP3A4 crystal structures: (*a*) Metyrapone with the classic direct coordination to the heme iron, (*b*) progesterone sitting at the mouth of the entrance channel, (*c*) two molecules of ketoconazole sitting in a highly distorted active site, and (*d*) erythromycin sitting close to the heme but in the wrong orientation for metabolism.

played a large degree of protein flexibility. With the ketoconazole structure, large movements were seen in the F and G helices, and there was severe disruption of the aromatic cage mentioned above. The unstructured region (in the unliganded enzyme) of residues 210–213 in the F–F′ loop now formed an additional turn to the F helix that resulted in arg^{212} residing on the protein surface. A large movement was also observed in the I helix. One ketoconazole molecule was bound in the pocket as expected with the imidazole ring coordinated to the heme iron, in agreement with its type 2 UV spectrum. Other stabilizing polar interactions were observed between the acetyl carbonyl and arg^{106}, arg^{372}, and glu^{374}, and a π–π stacking was seen between phe^{304} and the dichlorophenyl group (Figure 17.11*c*). What was particularly surprising, however, was to find a second molecule of ketoconazole oriented above and antiparallel to the first (Figure 17.11*c*). Although it had been suspected for some time that CYP3A4 could accommodate multiple ligands that would account for its atypical kinetics, this was the first experimental evidence to confirm it.

With the erythromycin-bound structure, displacements from the apo conformation were not as pronounced, the main differences being in the F–G part. Compared to the ketoconazole structure, the F–F′ loop was disordered. The density was not good enough to unambiguously assign the position of erythromycin itself, so the substrate's crystal structure entry from CSD was rigidly fitted to the observed density with manual adjustment of the sugar rings. Experimentally, erythromycin is *N*-demethylated by CYP3A4, but the crystal structure places this dimethylamino group 17 Å from the heme iron (Figure 17.11*d*). However, there was additional density found in the map that could be explained by alternative orientations in the active site.

Although it will not be discussed further in this chapter, large conformational shifts upon ligand binding were also observed in the recently solved rabbit CYP2B4 structure [46–48]. These were discussed in terms of regions of plasticity and seemed to occur in parts of the protein closely associated with the lipid membrane to which it is attached [48]. One intriguing possibility proposed by Scott et al. was that conformational changes in helix C, which occurred during the binding of 4-(4-chlorophenyl)imidazole to CYP2B4, resulted in changes to the heme binding site and the binding interface of the P450 reductase partner. Thus ligand binding may also play a role in altering the catalytic activity by changing the efficiency through which electrons are passed from the reductase to the heme [47].

17.2.9 CYP2A6

The human cytochrome CYP2A6 has relatively few substrates. Those that are known are generally small, hydrophobic, and capable of adopting planar conformations. From a toxicity viewpoint, however, CYP2A6 is important in that it is one of the main enzymes responsible for nicotine (23) detoxification. In a negative sense, it is also responsible for the formation of mutagenic carcinogens arising from other compounds found in tobacco. These include *N*′-nitrosonornicotine (24) and 4-(methylnitrosamino)-1-(3-pyridyl)-1-butanone (25). Inhibition of this enzyme could therefore play an important role in smoking cessation and onset of smoking-related cancers.

In 2005 Yano et al. described the structures of two complexes of CYP2A6, one with coumarin (6) and the other with the inhibitor, methoxsalen (26) [49]. The active site cavity is much smaller than those of the other mammalian

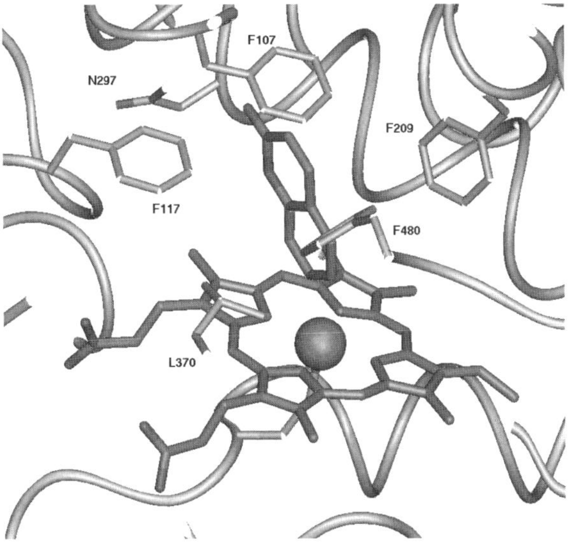

Figure 17.12 CYP2A6 with coumarin bound.

cytochromes, and is relatively hydrophobic, with only one important hydrogen bonding residue, asn[297], present. This asparagine forms interactions with the oxygens of both molecules studied. There are also a number of favorable hydrophobic interactions between, for example, phe[107] and the coumarin aromatic ring (Figure 17.12). Interestingly there appear to be no substrate or solvent channels in the structure.

17.2.10 CYP2D6

Many top-selling drugs used for CNS, cardiovascular, and other types of disorders act on the superfamily of receptors known as 7TM (transmembrane) receptors or GPCRs (G Protein-Coupled Receptors). Ligands for the main nonpeptide GPCR subfamily, the aminergic receptors, generally consist of a basic (protonated) nitrogen and one or more aromatic rings. These are the main recognition features for substrates of the human cytochrome, CYP2D6, and it is the importance of this enzyme's targets that has led it to be one of the most widely studied of the human P450s. Many groups have described homology models of CYP2D6 and used them in the rationalization or prediction of metabolism and inhibition, and in the design and interpretation of SDM experiments [12,50–54]. Unfortunately, CYP2D6 has proved to be one of the most elusive isoforms in terms of crystallography. It was not until 2006 that Rowland et al. published its structure at 3.0 Å resolution [23].

CYP2D6 maintains the normal fold of the mammalian P450s, with the main differences compared to CYP2C9, observed in the F helix, the F–G loop, the

B′ helix, and β sheets 1 and 4. CYP2D6 has been extensively studied by SDM, and all the main residues such as asp^{301}, glu^{216}, thr^{309}, phe^{483}, and phe^{120} are found in a well-defined active site cavity situated above the heme group. Perhaps more than any other isoform, SDM results have given rise to disagreements on the role of certain key residues, and the crystal structure has been able, at least in part, to resolve these issues. No single pharmacophore model could account for the observed range of products; for example, many compounds such as debrisoquine (27) and codeine (28) are metabolized at sites 5 to 7 Å distant from the protonated amine group [55], whereas a large group of others, such as metoprolol (29), are metabolized at sites 10 to 12 Å distant [56]. *N*-dealkylation reactions occur at distances much shorter than either of these pharmacophores can account for.

The 5 to 7 Å distant compounds are readily explained by an interaction of the basic center with asp^{301}, and its position is confirmed in the crystal structure [57]. However, several research groups have postulated that the primary interaction of amines is with the more distant glu^{216} [12,58], which explains the 10 to 12 Å distant compounds and the fact that asp^{301} plays a structural role in hydrogen bonding to the backbone of residues in the B′–C loop [12,59]. The crystal structure agrees with both of these facts but does not fully explain the SDM results. Mutation of either glu^{216} or asp^{301} to nonacidic residues results in loss of metabolism. However, both the mutants, E216D and D301E, are fully functional, albeit with shifts in the regioselective ratio of some products [60]. An asp^{216} mutant is much too distant to allow small substrates' access to the heme oxygen, and furthermore the longer chain of glu^{301} would only disrupt the stabilization of the B′–C loop observed in the X-ray structure. We have therefore argued that both the 5–7 Å and 10–12 Å compounds can be accommodated by binding to asp^{301} in either of its two common rotameric states, with the potential loss of the hydrogen bond to the B′–C backbone being more than compensated by hydrophobic stabilization of the ligands by phe^{120} (see below). The docking of debrisoquine and metoprolol in Figure 17.13 clearly shows this. Glu^{216} does not have a catalytic role; it instead acts as an essential substrate recognition site, which it is ideally situated to do, at the entrance of the substrate access channel. This is further confirmed by SDM data on the two phenylalanines, phe^{481} and phe^{483}. Mutation of either residue has an equally large detrimental effect on metabolism [61–63]. In both homology models and the crystal structure, phe^{483} is oriented into the active site where it regularly has been postulated to form favorable π–π interactions with the aromatic rings of substrates and inhibitors. Phe^{481}, however, is oriented in the opposite direction, away from the active site cavity, but during the recognition event it can form similar π–π interactions to those with phe^{483} when the ligand is bound to glu^{216} (Figure 17.13*c*).

The main difference between the active site of our CYP2D6 homology model and that of the crystal structure was in the conformation of the B–C region. The experimental structure places phe^{120} above the heme where it can play a pivotal role in orientating substrates correctly for subsequent reaction.

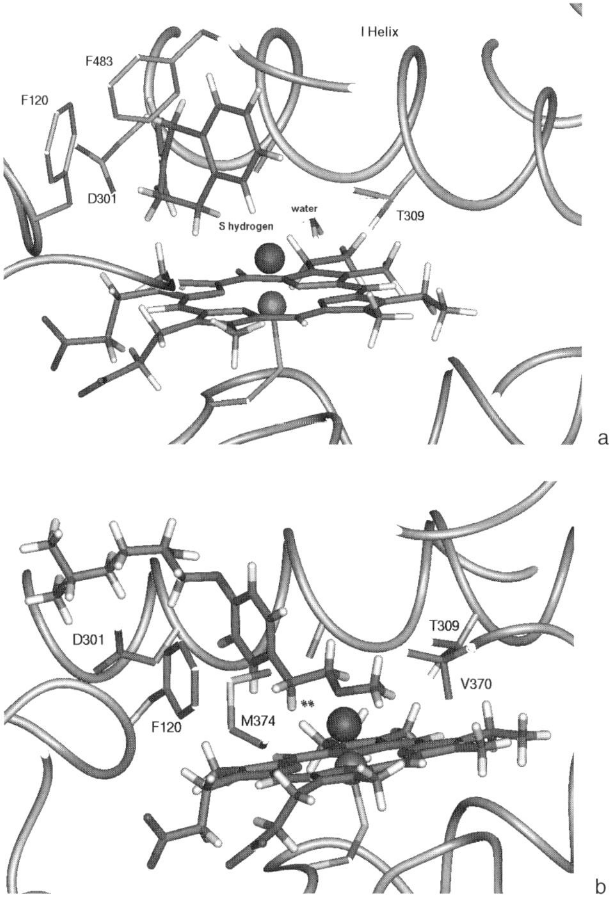

Figure 17.13 CYP2D6 with (*a*) debrisoquine and (*b*) metroprolol manually docked into the active site. Constraints were applied between the heme oxygen and the known metabolic sites. Both substrates can bind to asp^{301} and interact with phe^{120} and phe^{483}. (*c*) Debrisoquine docked into a substrate recognition site showing interactions with phe^{481}, phe^{483}, and glu^{216}.

The homology model had this residue outside the cavity, although admittedly the models of some other groups had it correctly placed [52,64].

17.2.11 Uses of Mammalian P450 Crystal Structures

Prediction of Metabolism and Inhibition In the absence of 3D structural information, many approaches have been used to predict metabolism and inhibition. Some models can be thought of as global, where statistical tools

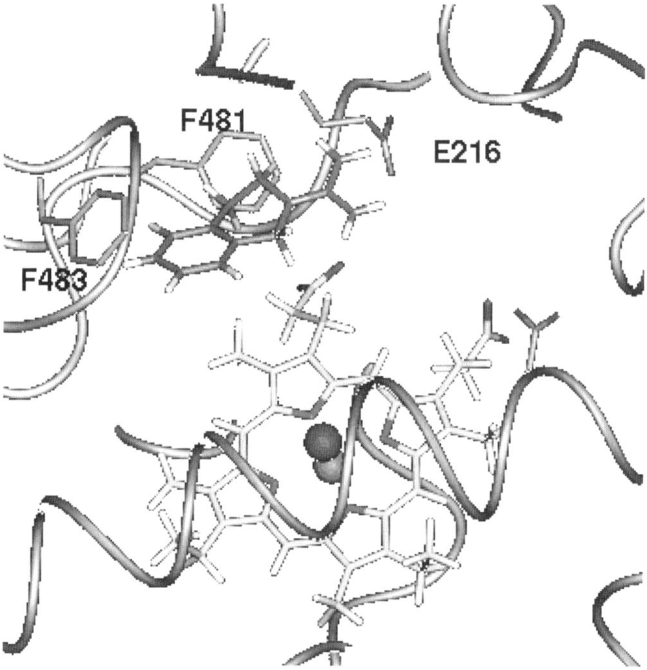

Figure 17.13 (*Continued*)

are used to combine a large number of physicochemical and/or 2D descriptors in some manner, so as to predict the likelihood of inhibition or even metabolism. They (hopefully!) do give better than random predictions but the basis of these predictions is often obscure; for instance, some models may do reasonably well in predicting P450 inhibition without the inclusion of any electronic descriptors. This is difficult for the chemist to rationalize, so their main use is in prediction of very large datasets, such as those arising from large virtual libraries.

Rule-based methods have been used with some success and several large commercial packages such as Meteor (LHASA Ltd., UK) and MetabolExpert (CompuDrug International Inc.) are available [65] (Chapter 18). Rather than using any 3D information, they contain large databases of rules and potential reactions and have a tendency to overpredict the number of metabolic paths and products. An alternative to empirical rules is the calculation of reactivity using quantum mechanical methods [66]. These can certainly produce correct results, but once again, they take no account of metabolic site accessibility, which is controlled by the enzyme. Pharmacophores have been widely used in all fields of ADMET prediction [67], but as was seen for the case of CYP2D6 above, multiple models may be necessary to explain regioselectivity. Furthermore they are based largely on geometric or, at best, simplistic chemical features and do not account for the substrate's varying reactivity. It is certainly difficult to explain the exquisite stereoselectivity so often observed with P450s via a pharmacophore approach alone.

With a crystal or homology structure, docking calculations can be performed that do better at least in predicting whether a particular site in a substrate can get within reaction distance of the heme. This alone is not enough, since the site of closest approach may not be reactive. The best approach is to combine the docking with some calculation, either empirical or quantum mechanical (QM) based, of a substrate's relative reactivity at its different potential sites.

In reaction with the active heme complex, compound 1, it is often assumed that the first stage is homolytic hydrogen abstraction, resulting in radical formation. This is certainly likely to occur with aliphatic hydroxylation. Ideally a calculation of the transition state should be carried out, but this is difficult in practice. Relative radical stabilities have therefore been used as an approximation. In a typical calculation one generates all possible radicals for a substrate, optimizes them, and determines their relative stabilities. They are then docked into the 3D protein structure using constraints between the heme and the sites predicted from the radical calculations. Early descriptions of this approach made use of homology models, but the same techniques can obviously be used with crystal structures.

In their papers on prediction of CYP2D6 [56,68] and CYP2C9 [69] metabolites, de Groot et al. used the semiempirical AM1 Hamiltonian to calculate relative radical strengths before docking. Their docking was aided by the use of a pharmacophore model. Semiempirical calculations tend to flatten the radical center, with associated errors in the energy, so we usually use *ab initio* HF with a 3-21G* basis set for the radical optimizations. Cramer has recently proposed an interesting alternative for the calculation of radical energies [70]. The formation of a radical species occurs via bond-stretching leading to eventual dissociation. This process is best described in terms of a Morse potential, which is an equation relating bond-stretching energy to interatomic distance, and which contains three variables from which the dissociation energy can be calculated. By calculating the energy semiempirically at a set of three or more interatomic distances, the variable values can be fitted, and hence the energy of radical formation can be calculated. This is orders of magnitude faster than using *ab initio* optimizations. Olsen et al. have recently compared the Morse function AM1 fitting method with much more detailed DFT calculations [71]. They found the results to be favorable.

A problem with calculating radical formation through hydrogen abstraction is that this is not the only mechanism by which CYP450s oxidize substrates. Hydrogen abstraction is unlikely to occur, for example, during the hydroxylation of aromatic rings. It is generally assumed that this proceeds through an electrophilic oxygen insertion directly into the ring. As such, simple calculations of HOMO orbital eigenvectors are often a good indicator of aromatic sites of hydroxylation [72] (Figure 17.14). It would be better, of course, to

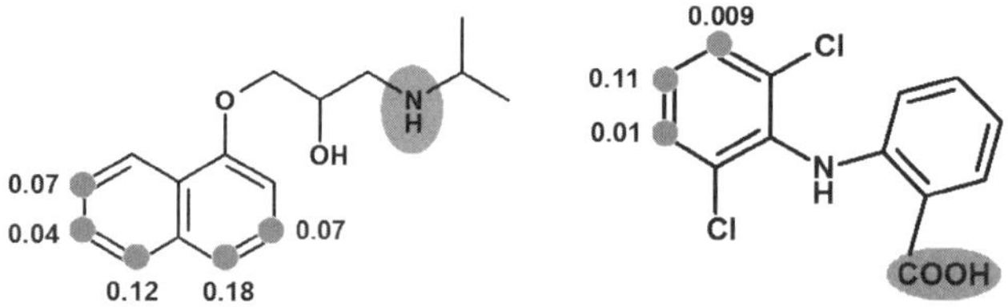

Figure 17.14 Semi-empirical HOMO orbital calculations on the aromatic substrates, propranolol (*left*) and diclofenac (*right*), used to predict the likely sites of hydroxylation.

calculate the transition state (TS) energy of the insertion reaction directly, but again, such calculations are difficult and very time-consuming. Bathelt et al., however, have carried out large transition state DFT calculations on a series of simple aromatics and showed that they follow a Hammett relationship [73,74]. They are however very orientation dependent, indicating that the enzyme environment can have a major influence. This is discussed in more detail later. Until computers and associated algorithms become much faster, accurate transition state calculations will be confined to single simple cases, and much work is necessary to develop approximate methods of estimating TS energies. Toward this goal Cruciani et al. have developed site reactivity indexes, based on detailed calculations of simple fragments, and used these in their Meta*S*ite program with good results [75].

Ideally the full prediction of metabolic fate by CYP450 enzymes should include modeling of the substrate/inhibitor in the access channel, its intermediate binding modes in the active site prior to the final "productive" interaction orientation with the metabolic site in some proximity to the heme group, and its egress from the site through one of the various exit channels. Ligand cooperativity has already been cited as a potential complication in this process, but essentially the whole modeling method relies on the docking algorithm(s) applied to the problem. Automated docking is by far the most common method used, but these authors feel that this is not the best way to approach the problem. The group at Scripps have repeatedly used the AUTODOCK program to predict binding modes in their crystal structures, whereas the Leicester group have used the genetic algorithm approach in GOLD [76] to suggest potential docking poses. Modifications have been made to better describe the interactions with the heme group, but essentially no major movement of the protein is allowed in these calculations, although ligands can be flexible [77]. We have attempted to use the FLO program [78], but although it allows some degree of protein flexibility, no means of including heme interactions was found. An interesting approach to docking was utilized by Cruciani et al. in the Metasite program [75]. Here the GRID flexible field method was used to probe the active sites with the classic probes for hydrophobic H-bonding and charge interactions. The original program made use of homology models, but the more recent versions have utilized the crystal structures of the human isoforms. GRID atom types corresponding to these probes were assigned to each atom of the substrate, and interatomic distances of these atoms were binned into a bitmap search that could be carried out in a fast and efficient manner, against the protein's grid.

One of the main uses of P450 crystal structures is in the prediction of metabolism and inhibition, mainly through docking-based approaches. The additional considerations of large conformational changes and multiple site occupancy add enormous challenges to an already difficult problem. Knowing that one molecule can act as a cooperative partner for another, as in the case of dapsone and flurbiprofen in CYP2C9, dual docking is certainly possible [21]. Afzelius et al., at AstraZeneca, have claimed successful predictions of drug metabolism by CYP3A4, using the ketoconazole-bound structure (Drug Metab. Rev., in press). However, at the time of writing, no further details were available.

All the more efficient metabolite prediction methods utilize some combination of substrate site reactivity and site accessibility to the heme. In our group we have made use of *ab initio* derived reactivity pointers (radical stabilities, HOMO eigenvectors, etc.), coupled to manual docking calculations. This is by necessity slow, but it does allow full control of ligand and protein flexibility. It also allows the user to make use of other information such as results of SDM experiments and UV spectra. Multiple starting orientations of the substrates are generated by hand, but these are largely driven by the expected site(s) of metabolism. The most important difference from other published methods in these calculations is that the full oxy-heme intermediate, and the associated I helix threonine-bound water are included in the calculations of substrate reactions. Appropriate parameters for the hemes in differing spin states have been calculated using *ab initio* HF calculations. Using the CHARMm program [79], distance constraints are set between the heme oxygen and the substrate's reaction site(s). Full relaxation of both ligand and protein is allowed, although constraints can also be applied to maintain alpha-helicity, preferred side chain rotamer values, or ligand conformations. Examples of this method have already been shown previously, with taxol in CYP2C8 and debrisoquine and metoprolol in CYP2D6. The results are in very good agreement with experimental observations.

In the case of P450 inhibitors, the heme species is not oxygenated and the threonine-bound water is not included. Functional groups such as pyridines or imidazoles are capable of coordinating to the heme, and the force-field parameters used will reflect a low-spin state. This will show up in the UV spectrum if available. Otherwise, a set of high-spin heme parameters are used. The docking procedure is the same as for substrates except for the lack of constraints to the heme. In choosing starting conformations, it should be remembered that inhibition can occur through binding anywhere in the active site, including the access and egress channels.

Detailed Studies of Reaction Mechanism From P450 crystal structures it has been possible to gain further information on the details of the catalytic mechanism(s). In the following paragraphs this is illustrated by some detailed studies we carried out on the mechanism of debrisoquine hydroxylation, before and after the availability of the CYP2D6 crystal structure.

Although it is commonly stated than the main product of CYP2D6-mediated debrisoquine metabolism is 4S-hydroxydebrisoquine, oxidation in fact occurs to a considerable extent at the 1, 3, 5, 6, 7, and 8 positions as well [80]. To study this, docking calculations were performed on the homology model, from which it was concluded that the substrate preferred to sit in an approximate coplanar orientation to the heme, with its guanidine group forming a salt bridge to asp^{301}. This meant that the pro-R 4-hydrogen would be abstracted to form the debrisoquine radical, and if the commonly accepted rebound mechanism has prevailed, the opposite enantiomer to that observed would be formed. Careful DFT calculations were therefore performed on a model system, consisting of a simple porphyrin with a single oxygen and a thiomethyl group as the axial ligands, a propionate anion (surrogate for asp^{301}), and debrisoquine in its docked orientation. Upon 4-hydrogen abstraction to form the radical, it was found that a 1-electron transfer occurred to the heme, yielding a complex with all spin density residing on the iron, and with the largest LUMO eigenvector and positive electrostatic potential on the 4-position of debrisoquine (see Figures 17.15*a–c*). The implication was that the substrate was set up for nucleophilic attack. If this were the case, the most likely source of oxygen would be water. Experiments were then carried out in labeled H_2O^{18}, but surprisingly no label was incorporated in the 4 position. The only explanation was that the oxygen source was from a tightly bound water or hydroxyl anion residing in the vicinity. This was obviously the water resulting from the oxygen cleavage, which was hydrogen bonded to thr^{309} and which was ideally situated to approach from above to form the 4-S-hydroxy enantiomer.

The interpretation, however, changed dramatically when the crystal structure became available. The presence of phe^{120} in the active site prevented the substrate from binding in a co-planar orientation. By forcing it to approach the heme orthogonally, the pro-S hydrogen was now the preferred site of abstraction. DFT calculations on this complex suggested that the intermediate was a radical with a large amount of spin density residing on the 4-position of the substrate (Figures 17.15*d–f*). It would appear therefore that the radical rebound mechanism is correct. The situation is more complex because in the labeling experiments mentioned above it was found that extensive ^{18}O incorporation occurred at the 3-position, suggesting that external water is the oxygen source here. This could be the water which is displaced from the heme during the first step in the P450 cycle.

Aromatic hydroxylation of debrisoquine should also occur via an orthogonal approach of the phenyl ring to the heme, guided again by phe^{120}. This is in agreement with work carried out by Bathelt at al, initially on model systems [73,74]. More recently in the same group Zurek et al. used the CYP2C9 crystal structure in a hybrid QM-MM approach to study the metabolism of ibuprofen (30). The transition states for multiple pathways were studied, and a clear

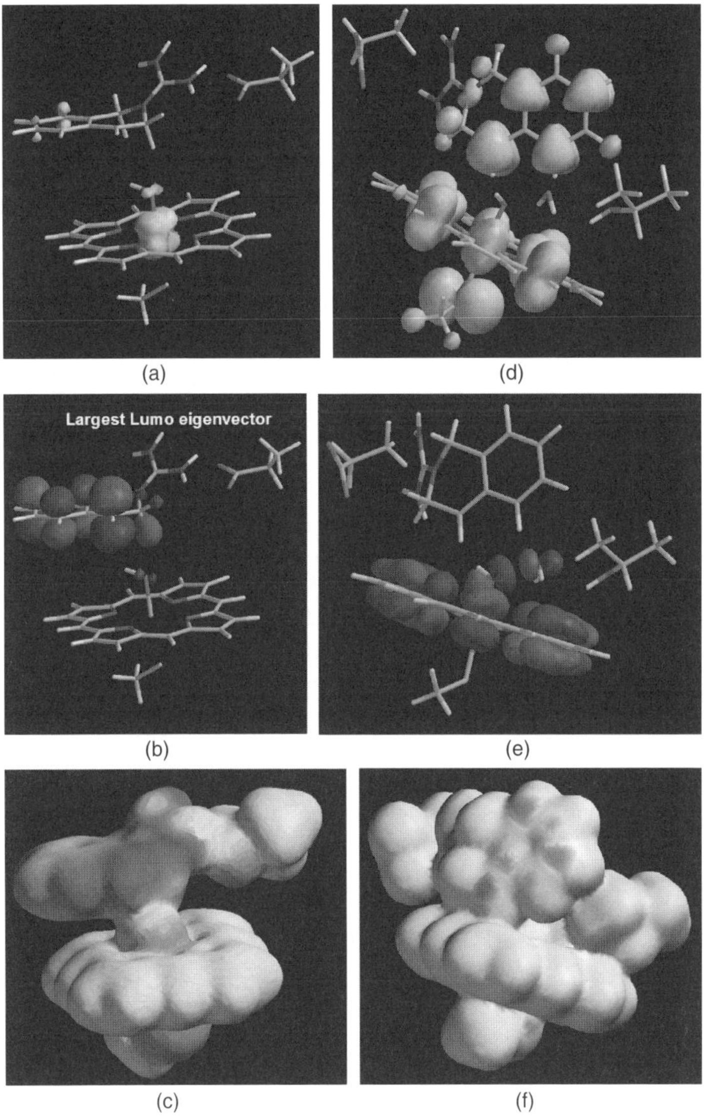

Figure 17.15 Ab initio calculations carried out on debrisoquine in a model reaction system derived on the left from a docking in our homology model and on the right, from the crystal structure of CYP2D6. (*a*) The spin density is centered on the iron. (*b*) The largest LUMO eigenvector is on the C4 position of debrisoquine, and (*c*) The largest positive electrostatic potential is again on C4. These features indicate a classic nucleophilic reaction. With the orthogonal approach likely from the crystal structure (*d*) debrisoquine is in a radical state, (*e*) the LUMO is centered on the heme, and (*f*) the charge is distributed around the complex. See color plates.

preference was found for oxygen insertion into an aromatic ring oriented orthogonal rather than coplanar to the heme group [81].

Prediction of Toxicity Arising from Allelic Variation One important cause of drug toxicity in P450s is impaired metabolism resulting from allelic variation in certain groups of individuals. Numerous polymorphisms have been described, for example for CYP2D6, some of which result in an inability to metabolize debrisoquine and sparteine. When this happens, toxic levels of drugs can then accumulate that gives rise to undesirable drug–drug interactions, with associated severe side effects. They often are linked to one or more single-point mutations. By using the crystal structures of P450s, it should now be possible to predict the effect of polymorphisms in selected subpopulations, and possibly even redesign drugs to get around the problems associated with them. For example, a common mutation found in several CYP2D6 polymorphs is R296C [82]. These mutants are invariably inactive. The crystal structure shows that this arginine sits at the mouth of the egress channel where it forms a salt-bridge with the adjacent glu^{293}. Mutation of this to a cysteine could significantly slow the rate of (basic) product egress through interaction with the glutamate. This arginine has in fact been implicated in inhibitor binding in a number of compounds (unpublished results). Recently we published a study on an inactive rare polymorph in which the key mutation was R440H. This residue was predicted from our homology model to be a key interaction between CYP2D6 and the reductase partner. Interaction energy calculations suggested that the histidine greatly reduced the strength of interaction between the two proteins, thus impairing the electron transfer process [50]. This was supported with the crystal structure, when solved, with an almost perfect agreement between the experimental and homology structures in the reductase binding region [23].

17.3 NUCLEAR HORMONE RECEPTORS

Nuclear hormone receptors (NRs) are ligand-activated transcription factors that regulate gene expression by interacting with hormone response elements on target genes (see also Chapter 12). This occurs via the formation of monomers, homodimers, or heterodimers generally with the retinoid X receptor (RXR) and interaction with the hormone response element [83,84].

All NRs have a common underlying structure of a variable *N*-terminus (A/B domain), a DNA-binding domain (DBD; C domain), a variable hinge region (D domain), a ligand-binding domain (LDB; E domain), and in some cases a variable *C*-terminus (F region). The DBD is a type-II zinc finger motif, consisting of two subdomains each containing a zinc ion coordinated to four cysteine, followed by an α-helix. The LDB is an antiparallel α-helical sandwich of 11 to 13 helices. The helices fold to form a hydrophobic cavity into which fatty acids, leukotrienes, prostaglandins, retinoic acids, steroid hormones, and thyroid hormones bind. This is generally believed to induce a conformational change in the receptor.

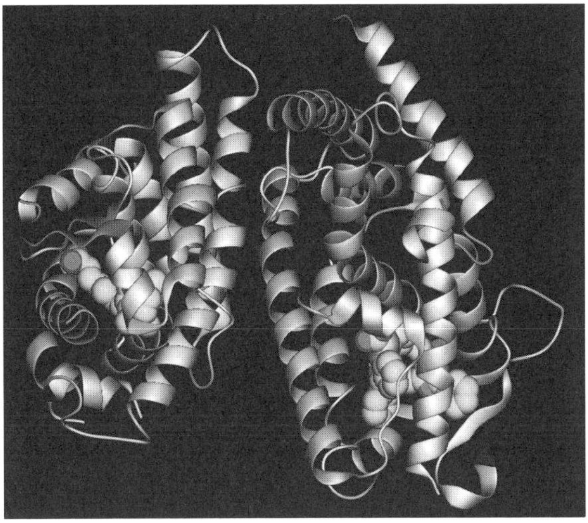

Figure 17.16 PPARγ/RXRα heterodimer crystal structure with retinoic acid bound to RXR (*left*) and rosiglitazone in PPARγ (*right*). Helix 12 can be seen in the front of the PPARγ structure crossing horizontally and blocking the active site pocket. See color plates.

The conformational change induced by the ligand has been described as a mouse trap mechanism [85]. This proposed mechanism of activation was originally based on the comparison between the apo-RXRα and *all trans*-retinoic acid (8) /RARγ crystal structures. It was later refined by the group of Grampe et al. [86] utilizing their structure of rosiglitazone (31) bound to PPARγ. In the apo-RXRα crystal structure an FFF motif toward the end of H10 stabilizes itself by inserting phe^{437} and phe^{438} into the center of the receptor, thus forming a break in the helical nature of H10. This forces the *C*-terminal activation function 2 (AF-2) or H12 helix away from the LBD of the receptor disrupting the binding site of the coactivator. In the case where retinoic acid is bound in the RXRα receptor, it effectively pushes the two phenylalanines, phe^{437} and phe^{438}, out of the center of the receptor. This exclusion by the ligand stops the FFF motif from causing a break in H10, which in turn enables the AF-2 helix to fold back onto the receptor LBD forming a hydrophobic cavity into which coactivator proteins can bind (Figure17.16).

It should be noted that not all NRs have the FFF motif present toward the end of H10, for PPARγ his^{449} fills the pocket occupied by phe^{437} and phe^{438} in RXRα [87,88]. Therefore although the mouse trap mechanism adequately describes the ligand-induced behaviour for RXRα, the conformational change of the AF-2 helix may not be quiet as significant for other NRs without the FFF motif.

17.3.1 Constitutive Androstane Receptor

CAR and PXR have recently become of interest to the pharmaceutical industry because of their ability to regulate the expression of detoxifying enzymes and transporters. In addition to these there are a number of other NRs implicated in similar roles, such as the vitamin D receptor (VDR), farnesoid X receptor (FXR), and the retinoid X receptor (RXR). The excellent review by Xie et al. [89] highlights evidence for this and gives associated references. Unfortunately, there is no literature on the use of VDR, FXR, and RXR for structure-based toxicity studies.

The crystal structure of the human constitutive androstane receptor was first released at the end of 2004 by Xu et al. [90]. In this paper two heterodimers of CAR/RXRα are presented with differing agonist ligands, CITCO (32) and 5β-pregnanedione (33), bound in the LBD of CAR. Comparison of these agonist-bound structures with the apo-CAR forms shows remarkable similarity, especially when compared to the murine CAR forms with the murine superagonist, TCPOBOP (34), and inverse agonist, androstanol (35), bound [91,92]. Xu et al. believe that this similarity between agonist and apo forms supports the proposal that the primary role of the ligand is to promote translocation as opposed to conformational change seen in other NRs. This is further supported by the mechanistic action of the inverse agonist androstanol. In the crystal structure [91], androstanol binding appears to be displacing tyr^{336} (murine) from the binding pocket. This in turn induces a kink between H10 and H11 and disrupts the salt-bridge, between K205 and the carboxylate terminus on ser^{358}, which was locking H12 in place. It provides an interesting contrast to the agonistic mechanism for other NRs in which a displacement of H12 prevents the binding of coactivators.

Surprisingly, little use has been made of these crystal structures for rational drug design. However, in the actual determination of the crystal of the CAR/RXRα complex with CITCO bound, the authors of the paper [90] do use in silico docking methods to determine the orientation of CITCO. As stated, the electron density of the CITCO binding region did not define the binding mode of CITCO unambiguously. Therefore the authors used the MVP docking program [93] to generate 118 low-energy conformations, which were placed with water into the binding site and then minimized. The two binding modes with the lowest energy were shown to fit the electron density. The only difference between these two modes was that their oxime linkers were mirror images of one another. Both modes are proposed to exist.

As mentioned earlier, with respect to CAR, the primary role of the ligand is to promote translocation, and one compound shown to do this through its competitive behavior with the CAR inverse agonist 3α,5α-androstanol is phenobarbital (36) [94]. Utilizing this knowledge, we manually docked phenobarbital into the agonist form of the CAR receptor [90], to see if any different interactions were formed by this series compared to the reference compound 5β-pregnanedione. As noted by Xu et al. [90], the binding within the CAR site is mediated primarily through hydrophobic contact. As with 5β-pregnanedione, the most significant hydrophobic contacts for phenobarbital within CAR were mediated by phe^{161}, ile^{164}, leu^{206}, phe^{217}, tyr^{224}, phe^{234}, and leu^{242}. However, in the case of phenobarbital binding, the aromatic residues phe^{161},

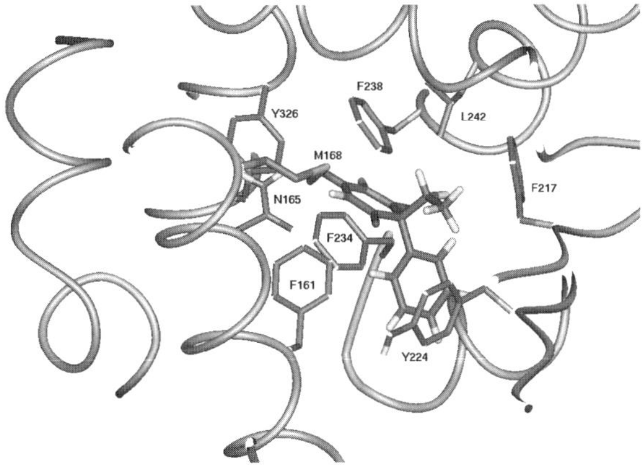

Figure 17.17 Phenobarbital manually docked into the CAR crystal structure.

phe[217], and tyr[224]are making additional π-stacking interactions with the 5-phenyl ring. Unlike 5β-pregnanedione binding, there is no hydrogen-bonding contact with his[203] for phenobarbital. Instead, the 1,3-diazinane-2,4,6-trione ring appears to be making two hydrogen-bonding contacts with the asn[165] head group (Figure 17.17). This docking illustrates how chemically unrelated CAR activating compounds can bind to the LBD.

17.3.2 Pregnane X Receptor

The first crystal structures of the human pregnane X receptor (hPXR) elucidated in 2001 by Watkins et al. [95] were the apo PXR and PXR/SR12813 complexes, determined at 2.5 and 2.75 Å, respectively. The PXR/SR12813 complex showed three distinct binding orientations of SR12813 (37) in a mainly hydrophobic ligand binding cavity of 19 residues. Alanine scanning suggested that in two of the three binding modes observed, the two polar residues ser[247] and his[407] were forming interactions, whereas four other polar residues ser[202], cys[284], gln[285], and arg[410] were only observed in one of the three other binding modes presented. A number of salt-bridges within the vicinity of the LBD were also examined, which interestingly revealed that due to the difference in the activation of the D205A mutant, alternate binding modes are likely to exist for SR12813 and rifampicin (38). This work provided the basis for structure-based drug design aimed at reducing off-target activity, and as the authors state "may be useful for in silico screening of drug candidates to predict and avoid dangerous side effects."

The first group to realize these words of Watkins et al. were Ekins and Erickson. In their paper they developed a Catalyst Hypogen model, consisting of four hydrophobic and one hydrogen-bonding features, from a set of 12 structurally diverse molecules [96]. The validity of the model was assessed by repositioning these inside the receptor using the crystal structure of SR12813 bound to PXR [97]. The model was consistent with at least one of the three experimentally determined orientations of SR12813 bound in PXR. The pharmacophore model was then used to predict the affinities of 28 known PXR ligands with reasonable success.

In the poster presented by Niederreiter and Langer [98] a receptor-based pharmacophore of PXR was developed from the published crystal structure [97]. In the derivation of this pharmacophore model the residues flanking the ligand present in the crystal structure, SR12813, were used to generate excluded volume spheres. Initially a very general model consisting of 3 hydrophobic regions, an acceptor region along with 15 excluded volume spheres was constructed and used to correctly identify 23 of 38 PXR ligands present in a screening set. Unfortunately, the selectivity of the model is poor as it also selected a number of inactive compounds present in the screening set. Building upon this initial model, hyperforin (39) was fitted and two additional hydrophobic regions were added to the pharmacophore. This revised model was only able to select highly active PXR ligands. The authors hypothesized that ligands selected using the general model could possibly activate PXR, whereas ligands selected using the revised model had a significantly greater chance of activating PXR because of a greater similarity to the known PXR ligands, SR12813 and hyperforin.

Following on from their poster presented in 2004, Schuster and Langer [99] have expanded upon their structure-based analysis of PXR to compare it to a number of different ligand-based design models. Utilizing their additional findings, they postulated that the only common feature shared by all PXR ligands is a direct or water-meditated hydrogen bond to gln^{285}, and additional activity may be gained via hydrogen-bonding interaction with his^{407} or a hydrophobic interaction with tyr^{306}. It was further proposed that this hydrophobic interaction with tyr^{306} might play a role in activation, as it serves to explain the different activities seen for paclitaxel (13) and docetaxel (40).

Probably one of the most thorough uses of multiple PXR crystal structures was performed by McLay and presented at the Catalyst user group meeting [100]. In this. Hypogen, HipHop, HipHopRefine, and structure-based models were constructed and compared to one another through the analysis of a screening set of 646 compounds, of which 82 are PXR active ligands. The structure-based pharmacophore model was generated from the crystal structures of SR12813 [97], hyperforin [101], and rifampicin bound to PXR. The crystal structures were overlaid on one another, and protein regions in common to all three were used to generate the excluded volumes. In addition the sites were explored by eye to identify five hydrophobic and three hydrogen bond acceptor regions. A number of experiments were carried out, varying the conformer search algorithm between fast and best [102] while also looking for complete matches or partial matches to the pharmacophore model (Table 17. 1).

The analyses showed that in a relatively large dataset, the retrieval of false positives was almost guaranteed if one was trying to identify a reasonable number of PXR active ligands. However, as mentioned in this presentation,

TABLE 17.1 Percentage of Actives Retrieved Using the PXR Structure-Based Pharmacophore

	Active (82)	Inactive (564)	Hit Rate
Structure based (fast)	4%	0%	100%
Structure based (best)	26%	1%	79%
Structure based partial match (fast)	49%	6%	54%
Structure based partial match (best)	72%	14%	42%
Three best partials with protein exclusions	46%	4%	62%

the hit rates observed for this method were as good as many hit rates presented for analysis of typical therapeutic targets. Notably the structure based pharmacophore hit rates were shown to be greater than any of the other methods presented in the talk.

17.3.3 Estrogen Receptor

The *Nature* paper of Brzozowski et al. in 1997 gave the first structural insights into the agonism and antagonism of the estrogen receptor [103]. The crystal structures reported were complexed with the endogenous agonist 17β-estradiol (41) and the antagonist raloxifene (42). They provided the framework in which structure-based design could be used to improve agonists and antagonists for the treatment of oestrogen-related diseases. Following this Shiau et al. [104] solved the crystal structure of the active metabolite of tamoxifen (43), 4-hydroxytamoxifen (OHT) (44), bound in ERα. This OHT/ ERα structure, when combined with the interesting tissue-specific behavior of tamoxifen shown by Fisher et al. [105] (the National Surgical Adjuvant Breast and Bowel Project—sponsored Breast Cancer Prevention Trial) in which there was a 49% reduction in breast cancer incidence, generated considerable interest in the use of structure-based design methods for estrogen receptor ligands.

In the same year that Shiau et al. [104] generated their structure, Gillner et al. [106] used the Brzozowski et al. crystallographic X-ray structure of the estrogen receptor to predict the binding affinities of environmental estrogens (PCBs, phenols, diethylstilbestrols, pesticides, androgen inhibitors, phytoestrogens, plant estrogens, and steroids), utilizing the CHARMm molecular mechanics protein–ligand interaction energy. The same authors, two years later, at the 2000 ACS National Meeting presented their work again and concluded molecular mechanics protein–ligand interaction energy was predictive with respect to ER ligand affinities, and might be used for identification of environmental estrogens [107].

Vedani et al. have utilized the Shiau structure as template for identifying potential binding modes of 106 investigated compounds [108,109] (see Chapter 12). The 80 training and 26 test compounds were made up of six different substance classes covering a range of seven orders of magnitude (IC_{50} 2.8 mM–0.2 nM). The group used an in-house developed technology, called *Quasar*, to analyze the compounds and develop a predictive model for endocrine disruptors. *Quasar* is a receptor-modeling concept based on 6D-QSAR that explicitly allows for simulation of induced fit [108,110,111]. The simulation reached a cross-validated r^2 of 0.895 and yielded a predictive r^2 (q^2) of 0.892. It forms part of the "virtual test kit" being developed by Biograf^{3R} [112] to identify adverse affects triggered by drugs and chemicals.

Also soon after the first crystal structure of ER, more traditional structure-based design work started to appear. The group of Stauffer et al. [113] used an acyclic amide structural template combined with the ER crystal structure to find additional hydrogen bonds and increase affinity. A number of separate groups, with just a few mentioned here [114–116], have utilized the crystal structure complexes to generate receptor-based alignment for a series of compounds, which were in turn used to generate predictive 3D-QSAR models.

In the last few years there have been a significant number of new papers utilizing both ERα and ERβ crystal structures. The Wyeth Research group have published a number of papers [117–119] and a patent [120] using ER crystal structures to design selective ERβ cpds. The papers give good examples of structure-based design (SBD) utilizing the two structures of ERα and ERβ to explain unexpected binding orientations from docking. These orientations are confirmed via crystallography, the differences are exploited and compounds with 100-fold selectivity are synthesized. The Wyeth patent also claims the use of crystallographic ERα to design ligands. Two interesting antirheumatic compounds are presented with unique pharmacology. A number of other very recent excellent reviews [121,122] and papers [123,124] also highlight successes with SBD methodologies for the estrogen receptor.

In addition to these SBD methods the ERα crystal structure has recently been used in evaluating linear interaction energy (LIE) methodology for lead optimization [125], and in calculating *ab initio* molecular interactions between ERα and 17β-estradiol [126].

Another region of the estrogen receptor crystal structure, not yet mentioned, is the coactivator binding region. It had previously been proposed from biochemical experiments that hydroxytamoxifen (HT) may occupy this region in ERα. Recently this has been solved crystallographically by Wang et al. [127] for ERβ. In this paper a crystal structure at 2.2 Å resolution is presented that shows two molecules of HT bound to ERβ, one in the LBD and the other bound to a site that overlaps the hydrophobic groove of the coactivator binding region. Wang et al. [127] used this information to design a small nonpeptide molecule that occupies the coactivator binding region. Previously another group from the University of Illinois had used the estrogen receptor crystal structure bound with a coactivator to successfully design coactivator binding inhibitors (CBI). Rodriguez et al. published in 2004 and 2002 on the design of CBI's [128,129]. Their 2004 paper [128] successfully used an outside-in design approach to make the first small molecule inhibitors of NR coactivator binding.

The ERα crystal structure has also been used to a reasonable extent in the construction of homology models of ERβ, PXR, AhR, and CAR. Jacobs et al. [130] used ERα to create models of ERβ, PXR, AhR, and CAR. The selective endogenous ligand and a number of different ligands known to fit closely within the ligand-binding site were interactively docked into the binding site, and the interactions with the receptor noted. A PPARα model generated from PPARγ was also presented in the paper. Two other groups [131,132] have generated ERβ models based on ERα. Hillish et al. [131] went on to use their model to develop ERα and ERβ selective ligands in which over 200-fold selectivity was achieved. Welsh et al. [132] used their ERβ model to accurately predict receptor binding capacity of various putative ligands. The molecular mechanics binding affinities were calculated for these and compared favourably to the experimentally observed relative binding affinities.

17.3.4 Androgen Receptor

Androgen receptors play an important role in male physiology as they bind the male sex steroids dihydrotestosterone (45) and testosterone (20), regulate male development, and are involved in disease states such as prostate cancer.

The first structure of the human androgen receptor (AR) complexed with the metribolone (R1881) (46) was solved in 2000 by Matias et al. [133]. The ligand-binding domain was analyzed in relation to mutations thought to give rise to the disease states androgen insensitivity syndrome (AIS) and prostate cancer. From analysis of the crystal structure complex it was proposed that the effects of most of the characterized mutants could be explained. One interesting mutation is the single-point mutation T877A associated with the prostate tumor cell line LNCaP. This causes a loss of the hydrogen bond to the 17β hydroxyl group in R1881, testosterone, or DHT. A number of other mutations involved in prostate cancer and complete, partial, and moderate AIS are also explained by the authors.

The "virtual test kit" being developed by Biograf^{3R} to identify adverse affects triggered by drugs and chemicals, mentioned earlier, also contains an androgen receptor based model. This model based on the crystal structure and 119 ligands, 88 in the training set and 26 in the test set, gave a cross-validated r^2 of 0.858 for the training set and a predictive r^2 (q^2) of 0.792 for the test set [134]. A further five compounds not belonging to any of the chemical classes used to train the model were then predicted with good success. This study is included in the "virtual test kit," and it was re-built using in-house technology called *Raptor*, this technology is similar to the afore-mentioned *Quasar*. The reader is referred to the Lill et al. paper for further details [135] (see also Chapter 12).

The crystal structure of AR has been used in a variety of ways, to aid in the development of QSAR models, to support hypotheses on agonism and antagonism, and to explore the design of potent AR ligands. The patent by Salvati et al. [136] describes the use of AR in the design of over 100 compounds, all exhibiting IC_{50}'s of less than 0.8 μM. Tamura et al. [137] suggested, through molecular modeling, that hydrogen-bond energies and the distance between two hydrogen-bonding sites influence the agonistic or antagonistic nature of the interaction with AR. Their hypothesis was compared to and supported by the results of recent crystallographic studies of agonist and antagonist bound AR complexes. More recently Soederholm et al. [138] used docking to identify preferred binding modes within the AR LBD. These binding modes were then used as the basis for a predictive 3D QSAR model.

17.4 CONCLUSION

While several aspects of ADMET-related properties, such as solubility and absorption, can be predicted by classical QSAR/QSPR-type approaches, the prediction of toxicity arising from enzyme metabolism or inhibition, or the interaction of various receptors with potentially toxic ligands, greatly benefits from 3D structural information of the targets. For many of these, such as the P450 cytochromes the major efforts have made use of homology models. In the last five to six years, however, an increasing number of these targets were solved by X-ray crystallography, often with important key ligands bound. These have been described in detail in this chapter. Work continues in the field of crystallography, and it is expected that other P450 cytochromes, with or without bound ligands, will continue to be elucidated in the near future. By a combination of site reactivity calculations and docking studies, it is becoming increasingly possible to accurately predict the products of metabolism or modes of inhibition of the cytochromes and other metabolizing enzymes. Current *ab initio* calculations of enzyme reactions can be slow, but with improvements in the methodology of coupled QM-MM programs and in computing power, these will undoubtedly become more routine. Other semiempirical methods described in this chapter are showing promise. At the same time improvements in docking methodology, in particular, the consideration of full protein and ligand flexibility and the inclusion of solvent, will impact greatly on the ability to predict metabolism. In contrast to metabolism, many of the liability target proteins, including the cytochromes and various transporter proteins, can be induced by interaction of ligands with transcription factors belonging to the nuclear receptor family. Again, many X-ray crystal structures have been solved, and insights into their interactions with potentially toxic compounds have emerged. Prediction of ligand interactions with the important inducing receptors, PXR and CAR, has been discussed. Endocrine disruptors, or environmental estrogens as they are often known, can activate several steroid nuclear receptors, including the estrogen and androgen receptors, and this has led to some types of cancer. The ability to predict binding and activation by these receptors is a continuing field of research.

Although not specifically described here, X-ray crystallography is impacting on other toxicity targets. The structures of many phase 2 metabolic enzymes, from the sulphotransferase family, have been solved [139,140]. These are important in the sulphation of numerous xenobiotics, steroid hormones and neurotransmitters. Another very exciting recent development was the publication of several multidrug transporter protein structures [141–143]. The impact of these on our ability to deal with drug resistance remains to be seen, but X-ray crystallography is certainly an area where increased activity is expected in the future.

REFERENCES

1. Liebler DC, Guengerich FP. Elucidating mechanisms of drug-induced toxicity. *Nat Rev Drug Discov* 2005;4:410–20.
2. Genter St Clair MB, Amarnath V, Moody MA, Anthony DC, Anderson CW, Graham DG. Pyrrole oxidation and protein cross-linking as necessary steps in the development of gamma-diketone neuropathy. *Chem Res Toxicol* 1988;1: 179–85.
3. Lee CR, Goldstein JA, Pieper JA. Cytochrome P450 2C9 polymorphisms: A comprehensive review of the in-vitro and human data. *Pharmacogenetics* 2002;12:251–63.
4. Lennard MS. Genetic polymorphism of sparteine/debrisoquine oxidation: A reappraisal. *Pharmacol Toxicol* 1990;67:273–83.
5. Meyer UA, Skoda RC, Zanger UM. The genetic polymorphism of debrisoquine/sparteine metabolism-molecular mechanisms. *Pharmacol Ther* 1990;46:297–308.
6. Guengerich FP. Common and uncommon cytochrome P450 reactions related to metabolism and chemical toxicity. *Chem Res Toxicol* 2001;14:611–50.
7. Okazaki O, Guengerich FP. Evidence for specific base catalysis in *N*-dealkylation reactions catalyzed by cytochrome P450 and chloroperoxidase. Differences in rates of deprotonation of aminium radicals as an explanation for high kinetic hydrogen isotope effects observed with peroxidases. *J Biol Chem* 1993;268: 1546–52.
8. Gotoh O. Substrate recognition sites in cytochrome P450 family 2 (CYP2) proteins inferred from comparative analyses of amino acid and coding nucleotide sequences. *J Biol Chem* 1992;267:83–90.
9. Schleinkofer K, Sudarko, Winn PJ, Ludemann SK, Wade RC. Do mammalian cytochrome P450s show multiple ligand access pathways and ligand channelling? *EMBO Rep* 2005;6:584–9.
10. Lewis DF. On the recognition of mammalian microsomal cytochrome P450 substrates and their characteristics: Towards the prediction of human p450 substrate specificity and metabolism. *Biochem Pharmacol* 2000;60:293–306.
11. Schoch GA, Yano JK, Wester MR, Griffin KJ, Stout CD, Johnson EF. Structure of human microsomal cytochrome P450 2C8: Evidence for a peripheral fatty acid binding site. *J Biol Chem* 2004;279:9497–503.
12. Kirton SB, Kemp CA, Tomkinson NP, St Gallay S, Sutcliffe MJ. Impact of incorporating the 2C5 crystal structure into comparative models of cytochrome P450 2D6. *Proteins* 2002;49:216–31.

13. Lewis DF. Modelling human cytochromes P450 involved in drug metabolism from the CYP2C5 crystallographic template. *J Inorg Biochem* 2002;91:502–14.
14. Schlichting I, Berendzen J, Chu K, Stock AM, Maves SA, Benson DE, Sweet RM, Ringe D, Petsko GA, Sligar SG. The catalytic pathway of cytochrome p450cam at atomic resolution. *Science* 2000;287:1615–22.
15. Zhang Z, Li Y, Stearns RA, Ortiz De Montellano PR, Baillie TA, Tang W. Cytochrome P450 3A4-mediated oxidative conversion of a cyano to an amide group in the metabolism of pinacidil. *Biochemistry* 2002;41:2712–8.
16. Williams PA, Cosme J, Sridhar V, Johnson EF, McRee DE. Mammalian microsomal cytochrome P450 monooxygenase: Structural adaptations for membrane binding and functional diversity. *Mol Cell* 2000;5:121–31.
17. Williams PA, Cosme J, Sridhar V, Johnson EF, McRee DE. Microsomal cytochrome P450 2C5: Comparison to microbial P450s and unique features. *J Inorg Biochem* 2000;81:183–90.
18. Marques-Soares C, Dijols S, Macherey AC, Wester MR, Johnson EF, Dansette PM, Mansuy D. Sulfaphenazole derivatives as tools for comparing cytochrome P450 2C5 and human cytochromes P450 2Cs: Identification of a new high affinity substrate common to those enzymes. *Biochemistry* 2003;42:6363–9.
19. Wester MR, Johnson EF, Marques-Soares C, Dijols S, Dansette PM, Mansuy D, Stout CD. Structure of mammalian cytochrome P450 2C5 complexed with diclofenac at 2.1 Å resolution: Evidence for an induced fit model of substrate binding. *Biochemistry* 2003;42:9335–45.
20. Wester MR, Johnson EF, Marques-Soares C, Dansette PM, Mansuy D, Stout CD. Structure of a substrate complex of mammalian cytochrome P450 2C5 at 2.3 Å resolution: Evidence for multiple substrate binding modes. *Biochemistry* 2003;42:6370–9.
21. Wester MR, Yano JK, Schoch GA, Yang C, Griffin KJ, Stout CD, Johnson EF. The structure of human cytochrome P450 2C9 complexed with flurbiprofen at 2.0-Å resolution. *J Biol Chem* 2004;279:35630–7.
22. de Graaf C, Vermeulen NP, Feenstra KA. Cytochrome p450 in silico: An integrative modeling approach. *J Med Chem* 2005;48:2725–55.
23. Rowland P, Blaney FE, Smyth MG, Jones JJ, Leydon VR, Oxbrow AK, Lewis CJ, Tennant MG, Modi S, Eggleston DS, Chenery RJ, Bridges AM. Crystal structure of human cytochrome P450 2D6. *J Biol Chem* 2006;281:7614–22.
24. Williams PA, Cosme J, Ward A, Angove HC, Matak VD, Jhoti H. Crystal structure of human cytochrome P450 2C9 with bound warfarin. *Nature* 2003;424: 464–8.
25. Ridderstrom M, Masimirembwa C, Trump-Kallmeyer S, Ahlefelt M, Otter C, Andersson TB. Arginines 97 and 108 in CYP2C9 are important determinants of the catalytic function. *Biochem Biophys Res Commun* 2000;270:983–7.
26. Stresser DM, Ackerman JM, Miller VP, Crespi CL. Flourometric cytochromes P450 2C8, 2C9 and 2C19 inhibition assays. <www.gentest.com/products/pdf/post_015.pdf>. 2006.
27. Morris GM, Goodsell DS, Halliday RS, Huey R, Hart WE, Belew RK, Olson AJ. Automated docking using a Lamarckian genetic algorithm and an empirical binding free energy function. *J Computat Chem* 1998;19:1639–62.

28. Williams PA, Cosme JM, Matak-Vinkovic D, Williams MG, Jhoti H. Crystals and three-dimensional structures of human cytochrome P 450 2C9 and their use in ligand design and homology modeling. Ed. (Astex Technology Ltd. U. 2003-426058[2004053383], 806. 20040318. US. 30-4-2003.

29. Prueksaritanont T, Tang C, Qiu Y, Mu L, Subramanian R, Lin JH. Effects of fibrates on metabolism of statins in human hepatocytes. *Drug Metab Dispos* 2002;30:1280–7.

30. Wang JS, Neuvonen M, Wen X, Backman JT, Neuvonen PJ. Gemfibrozil inhibits CYP2C8-mediated cerivastatin metabolism in human liver microsomes. *Drug Metab Dispos* 2002;30:1352–6.

31. Williams D, Feely J. Pharmacokinetic-pharmacodynamic drug interactions with HMG-CoA reductase inhibitors. *Clin Pharmacokinet* 2002;41:343–70.

32. Melet A, Marques-Soares C, Schoch GA, Macherey AC, Jaouen M, Dansette PM, Sari MA, Johnson EF, Mansuy D. Analysis of human cytochrome P450 2C8 substrate specificity using a substrate pharmacophore and site-directed mutants. *Biochemistry* 2004;43:15379–92.

33. Yano JK, Wester MR, Schoch GA, Griffin KJ, Stout CD, Johnson EF. The structure of human microsomal cytochrome P450 3A4 determined by X-ray crystallography to 2.05-A resolution. *J Biol Chem* 2004;279:38091–4.

34. Williams PA, Cosme J, Vinkovic DM, Ward A, Angove HC, Day PJ, Vonrhein C, Tickle IJ, Jhoti H. Crystal structures of human cytochrome P450 3A4 bound to metyrapone and progesterone. *Science* 2004;305:683–6.

35. Khan KK, He YQ, Domanski TL, Halpert JR. Midazolam oxidation by cytochrome P450 3A4 and active-site mutants: An evaluation of multiple binding sites and of the metabolic pathway that leads to enzyme inactivation. *Mol Pharmacol* 2002;61:495–506.

36. Domanski TL, Liu J, Harlow GR, Halpert JR. Analysis of four residues within substrate recognition site 4 of human cytochrome P450 3A4: Role in steroid hydroxylase activity and alpha-naphthoflavone stimulation. *Arch Biochem Biophys* 1998;350:223–32.

37. Stevens JC, Domanski TL, Harlow GR, White RB, Orton E, Halpert JR. Use of the steroid derivative RPR 106541 in combination with site-directed mutagenesis for enhanced cytochrome P-450 3A4 structure/function analysis. *J Pharmacol Exp Ther* 1999;290:594–602.

38. Harlow GR, Halpert JR. Alanine-scanning mutagenesis of a putative substrate recognition site in human cytochrome P450 3A4: Role of residues 210 and 211 in flavonoid activation and substrate specificity. *J Biol Chem* 1997;272: 5396–402.

39. Domanski TL, He YA, Khan KK, Roussel F, Wang Q, Halpert JR. Phenylalanine and tryptophan scanning mutagenesis of CYP3A4 substrate recognition site residues and effect on substrate oxidation and cooperativity. *Biochemistry* 2001;40:10150–60.

40. He YA, Roussel F, Halpert JR. Analysis of homotropic and heterotropic cooperativity of diazepam oxidation by CYP3A4 using site-directed mutagenesis and kinetic modeling. *Arch Biochem Biophys* 2003;409:92–101.

41. Domanski TL, He YA, Harlow GR, Halpert JR. Dual role of human cytochrome P450 3A4 residue Phe-304 in substrate specificity and cooperativity. *J Pharmacol Exp Ther* 2000;293:585–91.

42. Roussel F, Khan KK, Halpert JR. The importance of SRS-1 residues in catalytic specificity of human cytochrome P450 3A4. *Arch Biochem Biophys* 2000;374: 269–78.

43. Fowler SM, Riley RJ, Pritchard MP, Sutcliffe MJ, Friedberg T, Wolf CR. Amino acid 305 determines catalytic center accessibility in CYP3A4. *Biochemistry* 2000;39:4406–14.

44. Fowler SM, Taylor JM, Friedberg T, Wolf CR, Riley RJ. CYP3A4 active site volume modification by mutagenesis of leucine 211. *Drug Metab Dispos* 2002;30:452–6.

45. Ekroos M, Sjogren T. Structural basis for ligand promiscuity in cytochrome P450 3A4. *Proc Natl Acad Sci USA* 2006;103:13682–7.

46. Scott EE, He YA, Wester MR, White MA, Chin CC, Halpert JR, Johnson EF, Stout CD. An open conformation of mammalian cytochrome P450 2B4 at 1.6-Å resolution. *Proc Natl Acad Sci USA* 2003;100:13196–201.

47. Scott EE, White MA, He YA, Johnson EF, Stout CD, Halpert JR. Structure of mammalian cytochrome P450 2B4 complexed with 4-(4-chlorophenyl)imidazole at 1.9-Å resolution: Insight into the range of P450 conformations and the coordination of redox partner binding. *J Biol Chem* 2004;279:27294–301.

48. Zhao Y, White MA, Muralidhara BK, Sun L, Halpert JR, Stout CD. Structure of microsomal cytochrome P450 2B4 complexed with the antifungal drug bifonazole: Insight into P450 conformational plasticity and membrane interaction. *J Biol Chem* 2006;281:5973–81.

49. Yano JK, Hsu MH, Griffin KJ, Stout CD, Johnson EF. Structures of human microsomal cytochrome P450 2A6 complexed with coumarin and methoxsalen. *Nat Struct Mol Biol* 2005;12:822–3.

50. Allorge D, Breant D, Harlow J, Chowdry J, Lo-Guidice JM, Chevalier D, Cauffiez C, Lhermitte M, Blaney FE, Tucker GT, Broly F, Ellis SW. Functional analysis of CYP2D6.31 variant: Homology modeling suggests possible disruption of redox partner interaction by Arg440His substitution. *Proteins* 2005;59:339–46.

51. Snyder R, Sangar R, Wang J, Ekins S. Three-dimensional quantitative structure activity relationship for Cyp2d6 substrates. *Quant Struct-Act Rel* 2002;21: 357–68.

52. Venhorst J, ter Laak AM, Commandeur JN, Funae Y, Hiroi T, Vermeulen NP. Homology modeling of rat and human cytochrome P450 2D (CYP2D) isoforms and computational rationalization of experimental ligand-binding specificities. *J Med Chem* 2003;46:74–86.

53. Guengerich FP, Hanna IH, Martin MV, Gillam EM. Role of glutamic acid 216 in cytochrome P450 2D6 substrate binding and catalysis. *Biochemistry* 2003;42: 1245–53.

54. Lewis DF. Homology modelling of human CYP2 family enzymes based on the CYP2C5 crystal structure. *Xenobiotica* 2002;32:305–23.

55. Koymans L, Vermeulen NP, van Acker SA, te Koppele JM, Heykants JJ, Lavrijsen K, Meuldermans W, Donne-Op den Kelder GM. A predictive model

for substrates of cytochrome P450-debrisoquine (2D6). *Chem Res Toxicol* 1992;5:211–9.

56. de Groot MJ, Ackland MJ, Horne VA, Alex AA, Jones BC. A novel approach to predicting P450 mediated drug metabolism: CYP2D6 catalyzed *N*-dealkylation reactions and qualitative metabolite predictions using a combined protein and pharmacophore model for CYP2D6. *J Med Chem* 1999;42:4062–70.
57. Ellis SW, Hayhurst GP, Smith G, Lightfoot T, Wong MM, Simula AP, Ackland MJ, Sternberg MJ, Lennard MS, Tucker GT et al. Evidence that aspartic acid 301 is a critical substrate-contact residue in the active site of cytochrome P450 2D6. *J Biol Chem* 1995;270:29055–8.
58. Lewis DF, Eddershaw PJ, Goldfarb PS, Tarbit MH. Molecular modelling of cytochrome P4502D6 (CYP2D6) based on an alignment with CYP102: Structural studies on specific CYP2D6 substrate metabolism. *Xenobiotica* 1997;27:319–39.
59. Hanna IH, Kim MS, Guengerich FP. Heterologous expression of cytochrome P450 2D6 mutants, electron transfer, and catalysis of bufuralol hydroxylation: The role of aspartate 301 in structural integrity. *Arch Biochem Biophys* 2001;393:255–61.
60. Ellis SW, Anderson MC, Hartkoorn RC, Harlow JR, Chowdrry JE, Mather B, Blaney FE, Tucker GT. MGMS meeting on ADMET problems, Oxford, UK, Apr 14–16. 2003.
61. Hayhurst GP, Harlow J, Chowdry J, Gross E, Hilton E, Lennard MS, Tucker GT, Ellis SW. Influence of phenylalanine-481 substitutions on the catalytic activity of cytochrome P450 2D6. *Biochem J* 2001;355:373–9.
62. Chowdry J, Pucci MR, Harlow J, Tucker GT, Ellis SW. Evidence that both phenylalanine 481 and phenylalanine 483 are substrate-contact residues in the active site of cytochrome P450 206. *Br J Clin Pharmacol* 2002;53:443–4.
63. Lussenburg BM, Keizers PH, de Graaf C, Hidestrand M, Ingelman-Sundberg M, Vermeulen NP, Commandeur JN. The role of phenylalanine 483 in cytochrome P450 2D6 is strongly substrate dependent. *Biochem Pharmacol* 2005;70:1253–61.
64. Flanagan JU, Marechal JD, Ward R, Kemp CA, McLaughlin LA, Sutcliffe MJ, Roberts GC, Paine MJ, Wolf CR. Phe120 contributes to the regiospecificity of cytochrome P450 2D6: Mutation leads to the formation of a novel dextromethorphan metabolite. *Biochem J* 2004;380:353–60.
65. Langowski J, Long A. Computer systems for the prediction of xenobiotic metabolism. *Adv Drug Deliv Rev* 2002;54:407–15.
66. Korzekwa KR, Jones JP, Gillette JR. Theoretical studies on cytochrome P-450 mediated hydroxylation: A predictive model for hydrogen atom abstractions. *J Am Chem Soc* 1990;112:7042–6.
67. Ekins S. Predicting undesirable drug interactions with promiscuous proteins in silico. *Drug Discov Today* 2004;9:276–85.
68. de Groot MJ, Ackland MJ, Horne VA, Alex AA, Jones BC. Novel approach to predicting P450-mediated drug metabolism: Development of a combined protein and pharmacophore model for CYP2D6. *J Med Chem* 1999;42:1515–24.
69. de Groot MJ, Alex AA, Jones BC. Development of a combined protein and pharmacophore model for cytochrome P450 2C9. *J Med Chem* 2002;45:1983–93.

70. Lewin JL, Cramer CJ. Rapid quantum mechanical models for the computational estimation of C–H bond dissociation energies as a measure of metabolic stability. *Mol Pharm* 2004;1:128–35.

71. Olsen L, Rydberg P, Rod TH, Ryde U. Prediction of activation energies for hydrogen abstraction by cytochrome P450. *J Med Chem* 2006;49:6489–99.

72. Ackland MJ. Correlation between site specificity and electrophilic frontier values in the metabolic hydroxylation of biphenyl, di-aromatic and CYP2D6 substrates: A molecular modelling study. *Xenobiotica* 1993;23:1135–44.

73. Bathelt CM, Ridder L, Mulholland AJ, Harvey JN. Aromatic hydroxylation by cytochrome P450: model calculations of mechanism and substituent effects. *J Am Chem Soc* 2003;125:15004–5.

74. Bathelt CM, Ridder L, Mulholland AJ, Harvey JN. Mechanism and structure-reactivity relationships for aromatic hydroxylation by cytochrome P450. *Org Biomol Chem* 2004;2:2998–3005.

75. Cruciani G, Carosati E, De Boeck B, Ethirajulu K, Mackie C, Howe T, Vianello R. MetaSite: Understanding metabolism in human cytochromes from the perspective of the chemist. *J Med Chem* 2005;48:6970–9.

76. Jones G, Willett P, Glen RC, Leach AR, Taylor R. Development and validation of a genetic algorithm for flexible docking. *J Mol Biol* 1997;267:727–48.

77. Kirton SB, Murray CW, Verdonk ML, Taylor RD. Prediction of binding modes for ligands in the cytochromes P450 and other heme-containing proteins. *Proteins* 2005;58:836–44.

78. McMartin C, Bohacek RS. QXP: Powerful, rapid computer algorithms for structure-based drug design. *J Comput Aided Mol Des* 1997;11:333–44.

79. Brooks BR, Bruccoleri RE, Olafson BD, States DJ, Swaminathan S, Karplus M. CHARMM: A program for macromolecular energy, minimization, and dynamics calculations. *J Computat Chem* 1983;4:187–217.

80. Lightfoot T, Ellis SW, Mahling J, Ackland MJ, Blaney FE, Bijloo GJ, de Groot MJ, Vermeulen NP, Blackburn GM, Lennard MS, Tucker GT. Regioselective hydroxylation of debrisoquine by cytochrome P4502D6: Implications for active site modelling. *Xenobiotica* 2000;30:219–33.

81. Zurek J, Harvey JN, Mulholland AJ. Modelling drug metabolism in cytochrome P450 enzymes. 2006 ISQBP President's Meeting, Strasbourg. 2006.

82. Home Page of the Human Cytochrome P450 (CYP) Allele Nomenclature Committee. <www.cypalleles.ki.se>. 2006.

83. Gronemeyer H, Laudet V. Transcription factors 3: nuclear receptors. *Protein Profile* 1995;2:1173–308.

84. Mangelsdorf DJ, Thummel C, Beato M, Herrlich P, Schutz G, Umesono K, Blumberg B, Kastner P, Mark M, Chambon P, Evans RM. The nuclear receptor superfamily: The second decade. *Cell* 1995;83:835–9.

85. Renaud JP, Rochel N, Ruff M, Vivat V, Chambon P, Gronemeyer H, Moras D. Crystal structure of the RAR-gamma ligand-binding domain bound to all-trans retinoic acid. *Nature* 1995;378:681–9.

86. Gampe RT Jr, Montana VG, Lambert MH, Miller AB, Bledsoe RK, Milburn MV, Kliewer SA, Willson TM, Xu HE. Asymmetry in the PPARgamma/RXRalpha crystal structure reveals the molecular basis of heterodimerization among nuclear receptors. *Mol Cell* 2000;5:545–55.

87. Nolte RT, Wisely GB, Westin S, Cobb JE, Lambert MH, Kurokawa R, Rosenfeld MG, Willson TM, Glass CK, Milburn MV. Ligand binding and co-activator assembly of the peroxisome proliferator-activated receptor-gamma. *Nature* 1998;395:137–43.

88. Uppenberg J, Svensson C, Jaki M, Bertilsson G, Jendeberg L, Berkenstam A. Crystal structure of the ligand binding domain of the human nuclear receptor PPARgamma. *J Biol Chem* 1998;273:31108–12.

89. Xie W, Uppal H, Saini SP, Mu Y, Little JM, Radominska-Pandya A, Zemaitis MA. Orphan nuclear receptor-mediated xenobiotic regulation in drug metabolism. *Drug Discov Today* 2004;9:442–9.

90. Xu RX, Lambert MH, Wisely BB, Warren EN, Weinert EE, Waitt GM, Williams JD, Collins JL, Moore LB, Willson TM, Moore JT. A structural basis for constitutive activity in the human CAR/RXRalpha heterodimer. *Mol Cell* 2004;16: 919–28.

91. Shan L, Vincent J, Brunzelle JS, Dussault I, Lin M, Ianculescu I, Sherman MA, Forman BM, Fernandez EJ. Structure of the murine constitutive androstane receptor complexed to androstenol: A molecular basis for inverse agonism. *Mol Cell* 2004;16:907–17.

92. Suino K, Peng L, Reynolds R, Li Y, Cha JY, Repa JJ, Kliewer SA, Xu HE. The nuclear xenobiotic receptor CAR: Structural determinants of constitutive activation and heterodimerization. *Mol Cell* 2004;16:893–905.

93. Lambert MH. Docking conformationally flexible molecules into protein binding sites. In: *Practical application of computer-aided drug design*. New York: Dekker, 1997. p. 243–303.

94. Sueyoshi T, Kawamoto T, Zelko I, Honkakoski P, Negishi M. The repressed nuclear receptor CAR responds to phenobarbital in activating the human CYP2B6 gene. *J Biol Chem* 1999;274:6043–6.

95. Watkins RE, Wisely GB, Moore LB, Collins JL, Lambert MH, Williams SP, Willson TM, Kliewer SA, Redinbo MR. The human nuclear xenobiotic receptor PXR: Structural determinants of directed promiscuity. *Science* 2001;292:2329–33.

96. Ekins S, Erickson JA. A pharmacophore for human pregnane X receptor ligands. *Drug Metab Dispos* 2002;30:96–9.

97. Watkins RE, Davis-Searles PR, Lambert MH, Redinbo MR. Coactivator binding promotes the specific interaction between ligand and the pregnane X receptor. *J Mol Biol* 2003;331:815–28.

98. Niederreiter D, Langer T. Pharmacophore modeling for the prediction of compound activity on anti-targets. Case study: The pregnane X receptor (PXR). *QSAR and molecular modelling in rational design of bioactive molecules. Proceedings of the European symposium on structure-activity relationships (QSAR) and molecular modelling*, 15th, Istanbul, Turkey, Sep 5–10, 2004. p. 406–7, 2006.

99. Schuster D, Langer T. The identification of ligand features essential for PXR activation by pharmacophore modeling. *J Chem Inf Model* 2005;45:431–9.

100. McLay I. Pharmacophore models for developability prediction: A critical assessment. Catalyst user group meeting, Tokyo, Sep 7–9, 2005.

101. Watkins RE, Maglich JM, Moore LB, Wisely GB, Noble SM, Davis-Searles PR, Lambert MH, Kliewer SA, Redinbo MR. 2.1 A crystal structure of human PXR

in complex with the St. John's wort compound hyperforin. *Biochemistry* 2003;42:1430–8.

102. Accelrys. Catalyst. [Release 4.9]. 2003.
103. Brzozowski AM, Pike AC, Dauter Z, Hubbard RE, Bonn T, Engstrom O, Ohman L, Greene GL, Gustafsson JA, Carlquist M. Molecular basis of agonism and antagonism in the oestrogen receptor. *Nature* 1997;389:753–8.
104. Shiau AK, Barstad D, Loria PM, Cheng L, Kushner PJ, Agard DA, Greene GL. The structural basis of estrogen receptor/coactivator recognition and the antagonism of this interaction by tamoxifen. *Cell* 1998;95:927–37.
105. Fisher B, Costantino JP, Wickerham DL, Redmond CK, Kavanah M, Cronin WM, Vogel V, Robidoux A, Dimitrov N, Atkins J, Daly M, Wieand S, Tan-Chiu E, Ford L, Wolmark N. Tamoxifen for prevention of breast cancer: Report of the National Surgical Adjuvant Breast and Bowel Project P-1 Study. *J Natl Cancer Inst* 1998;90:1371–88.
106. Gillner M, Carlsson P, Greenidge P. Application of molecular mechanics to prediction of binding affinities of estrogen agonists to the human estrogen-a receptor using its x-ray crystallographic structure. *Organohalogen Compounds* 1998;37: 261–4.
107. Gillner M, Carlsson P, Greenidge P. Prediction of binding affinities of environmental estrogens for the human estrogen receptor-a using its X-ray structure and molecular mechanics. *Preprints of extended abstracts presented at the ACS National Meeting, American Chemical Society, Division of Environmental Chemistry* 2000;40:332–4.
108. Vedani A, Dobler M, Lill MA. Combining protein modeling and 6D-QSAR. Simulating the binding of structurally diverse ligands to the estrogen receptor. *J Med Chem* 2005;48:3700–3.
109. Vedani A, Dobler M, Lill MA. The challenge of predicting drug toxicity in silico. *Basic Clin Pharmacol Toxicol* 2006;99:195–208.
110. Vedani A, Zbinden P. Quasi-atomistic receptor modeling: A bridge between 3D QSAR and receptor fitting. *Pharmaceut acta Helv* 1998;73:11–8.
111. Vedani A, Dobler M. 5D-QSAR: The key for simulating induced fit? *J Med Chem* 2002;45:2139–49.
112. Biograf^{3R}. <http://www.biograf.ch/index.php>. 2006.
113. Stauffer SR, Sun J, Katzenellenbogen BS, Katzenellenbogen JA. Acyclic amides as estrogen receptor ligands: synthesis, binding, activity and receptor interaction. *Bioorg Med Chem* 2000;8:1293–316.
114. Yu SJ, Derington D, Welsh W. Computational studies of raloxifene (Evista) derivatives: 3D-QSAR/CoMFA models and binding energy calculation as a guide for selective estrogen receptor modulators (SERMs). Abstracts of Papers, 224th ACS National Meeting, Boston, Aug 18–22, 2002 COMP-160.
115. Sippl W. Receptor-based 3D QSAR analysis of estrogen receptor ligands—Merging the accuracy of receptor-based alignments with the computational efficiency of ligand-based methods. *J Comput Aided Mol Des* 2000;14:559–72.
116. Wolohan P, Reichert DE. CoMFA and docking study of novel estrogen receptor subtype selective ligands. *J Comput Aided Mol Des* 2003;17:313–28.

117. Manas ES, Unwalla RJ, Xu ZB, Malamas MS, Miller CP, Harris HA, Hsiao C, Akopian T, Hum WT, Malakian K, Wolfrom S, Bapat A, Bhat RA, Stahl ML, Somers WS, Alvarez JC. Structure-based design of estrogen receptor-beta selective ligands. *J Am Chem Soc* 2004;126:15106–19.

118. McDevitt RE, Malamas MS, Manas ES, Unwalla RJ, Xu ZB, Miller CP, Harris HA. Estrogen receptor ligands: Design and synthesis of new 2-arylindene-1-ones. *Bioorg Med Chem Lett* 2005;15:3137–42.

119. Mewshaw RE, Edsall RJ Jr, Yang C, Manas ES, Xu ZB, Henderson RA, Keith JC Jr, Harris HA. ERbeta ligands: 3. Exploiting two binding orientations of the 2-phenylnaphthalene scaffold to achieve ERbeta selectivity. *J Med Chem* 2005;48:3953–79.

120. Mosyak L, Xu ZB, Stahl M, Hum W-T, Somers WS, Manas ES. Crystal structure of estrogen receptor a bound to synthetic ligands in drug design. Ed. (USA). 2006-334982[2006160836], 127. 20060720. US. 18-1-2006.

121. Mukherjee S, Saha A. Molecular modeling studies of estrogen receptor modulators. *Curr Comput Aided Drug Des* 2006;2:229–53.

122. Manas ES, Mewshaw RE, Harris HA, Malamas MS. Isoform specificity: The design of estrogen receptor-b selective compounds. In Chapter 8: Editor, Hubbard, RE. *London Structure-based drug discovery: an overview*. RSC Publishing. 2006;219–56.

123. Renaud J, Bischoff SF, Buhl T, Floersheim P, Fournier B, Geiser M, Halleux C, Kallen J, Keller H, Ramage P. Selective estrogen receptor modulators with conformationally restricted side chains: Synthesis and structure-activity relationship of ERalpha-selective tetrahydroisoquinoline ligands. *J Med Chem* 2005; 48:364–79.

124. de Medina P, Boubekeur N, Balaguer P, Favre G, Silvente-Poirot S, Poirot M. The prototypical inhibitor of cholesterol esterification, Sah 58–035 [3-[decyldimethylsilyl]-n-[2-(4-methylphenyl)-1-phenylethyl]propanamide], is an agonist of estrogen receptors. *J Pharmacol Exp Ther* 2006;319:139–49.

125. Stjernschantz E, Marelius J, Medina C, Jacobsson M, Vermeulen NP, Oostenbrink C. Are automated molecular dynamics simulations and binding free energy calculations realistic tools in lead optimization? An evaluation of the linear interaction energy (LIE) method. *J Chem Inf Model* 2006;46:1972–83.

126. Fukuzawa K, Mochizuki Y, Tanaka S, Kitaura K, Nakano T. Molecular interactions between estrogen receptor and its ligand studied by the ab initio fragment molecular orbital method. *J Phys Chem B Condens Matter Mater Surf Interfaces Biophys* 2006;110:16102–10.

127. Wang Y, Chirgadze NY, Briggs SL, Khan S, Jensen EV, Burris TP. A second binding site for hydroxytamoxifen within the coactivator-binding groove of estrogen receptor beta. *Proc Natl Acad Sci USA* 2006;103:9908–11.

128. Rodriguez AL, Tamrazi A, Collins ML, Katzenellenbogen JA. Design, synthesis, and in vitro biological evaluation of small molecule inhibitors of estrogen receptor alpha coactivator binding. *J Med Chem* 2004;47:600–11.

129. Rodriguez AL, Tamrazi A, Katzenellenbogen JA. Substituted naphthalene scaffolds as estrogen receptor coactivator mimics. Abstracts of Papers, 224th ACS National Meeting, Boston, Aug 18–22, 2002 MEDI-360.

130. Jacobs MN, Dickins M, Lewis DF. Homology modelling of the nuclear receptors: Human oestrogen receptorbeta (hERbeta), the human pregnane-X-receptor

(PXR), the Ah receptor (AhR) and the constitutive androstane receptor (CAR) ligand binding domains from the human oestrogen receptor alpha (hERalpha) crystal structure, and the human peroxisome proliferator activated receptor alpha (PPARalpha) ligand binding domain from the human PPARgamma crystal structure. *J Steroid Biochem Mol Biol* 2003;84:117–32.

131. Hillisch A, Peters O, Kosemund D, Muller G, Walter A, Elger W, Fritzemeier KH. Protein structure-based design, synthesis strategy and in vitro pharmacological characterization of estrogen receptor alpha and beta selective compounds. *Ernst Schering Res Found Workshop*, 2004. p. 47–62.
132. Welsh WJ, DeLisle RK, Yu SJ, Nair A. Computational strategies for predicting the binding affinity of ligands to estrogen receptor (ER) subtypes a and b: Homology modeling of ERb. Book of Abstracts, 219th ACS National Meeting, San Francisco, Mar 26–30, 2000 ENVR-008.
133. Matias PM, Donner P, Coelho R, Thomaz M, Peixoto C, Macedo S, Otto N, Joschko S, Scholz P, Wegg A, Basler S, Schafer M, Egner U, Carrondo MA. Structural evidence for ligand specificity in the binding domain of the human androgen receptor: Implications for pathogenic gene mutations. *J Biol Chem* 2000;275:26164–71.
134. Lill MA, Winiger F, Vedani A, Ernst B. Impact of induced fit on ligand binding to the androgen receptor: A multidimensional QSAR study to predict endocrine-disrupting effects of environmental chemicals. *J Med Chem* 2005;48: 5666–74.
135. Lill MA, Vedani A, Dobler M. Raptor: Combining dual-shell representation, induced-fit simulation, and hydrophobicity scoring in receptor modeling. Application toward the simulation of structurally diverse ligand sets. *J Med Chem* 2004;47:6174–86.
136. Salvati ME. Synthesis of selective androgen receptor modulators and methods for their identification, design and use. Ed. (Bristol-Myers Squibb Co. U. 2001-US19665[2002000617], 140. 20020103. WO. 20-6-2001.
137. Tamura H, Yoshikawa H, Gaido KW, Ross SM, DeLisle RK, Welsh WJ, Richard AM. Interaction of organophosphate pesticides and related compounds with the androgen receptor. *Environ Health Perspect* 2003;111:545–52.
138. Soederholm AA, Lehtovuori PT, Nyroenen TH. Docking and three-dimensional quantitative structure-activity relationship (3D QSAR) analyses of nonsteroidal progesterone receptor ligands. *J Med Chem* 2006;49:4261–8.
139. Dajani R, Cleasby A, Neu M, Wonacott AJ, Jhoti H, Hood AM, Modi S, Hersey A, Taskinen J, Cooke RM, Manchee GR, Coughtrie MW. X-ray crystal structure of human dopamine sulfotransferase, SULT1A3: Molecular modeling and quantitative structure-activity relationship analysis demonstrate a molecular basis for sulfotransferase substrate specificity. *J Biol Chem* 1999;274:37862–8.
140. Gamage N, Barnett A, Hempel N, Duggleby RG, Windmill KF, Martin JL, McManus ME. Human sulfotransferases and their role in chemical metabolism. *Toxicol Sci* 2006;90:5–22.
141. Murakami S, Nakashima R, Yamashita E, Matsumoto T, Yamaguchi A. Crystal structures of a multidrug transporter reveal a functionally rotating mechanism. *Nature* 2006;443:173–9.

142. Dawson RJ, Locher KP. Structure of a bacterial multidrug ABC transporter. *Nature* 2006;443:180–5.

143. Seeger MA, Schiefner A, Eicher T, Verrey F, Diederichs K, Pos KM. Structural asymmetry of AcrB trimer suggests a peristaltic pump mechanism. *Science* 2006;313:1295–8.

18

EXPERT SYSTEMS

PHILIP N. JUDSON

Contents

Computational Toxicology: Risk Assessment for Pharmaceutical and Environmental Chemicals, Edited by Sean Ekins

18.1 INTRODUCTION

The terms "expert system" and "knowledge-based system" have meant different things at different times. It has always been the case that expert systems are those that seek to emulate a human expert making predictions. Some people have restricted the name to systems that base predictions on human knowledge. That being the case, the systems have alternatively been described as knowledge-based systems. Other systems base their predictions on generalized information that does not come from human experience but from the statistical analysis of data. The way in which the two kinds of systems use generalized information to solve a specific problem is analogous. So both tend to be called expert systems, and the source of information that they use is called the knowledge base by some people, whether or not it contains knowledge gleaned from humans. There is increasing use of terms like "knowledge base" and "knowledge management" for other, quite different kinds of database management systems with sophisticated search tools—particularly for searching textual data. For the purposes of this chapter, an "expert system" is one that uses generalized knowledge to make predictions in specific cases, and unless it is otherwise stated, a "knowledge base" contains generalized human knowledge.

In terms of practical application, expert systems overlap with systems for deriving and applying quantitative structure-activity relationship (QSAR) models or equations, and with systems using artificial neural networks (ANN) or genetic algorithms. The expert systems described in this chapter are characterized by their use of a generalized store of knowledge.

Expert system technology was one of the first branches of artificial intelligence [1,2]. It is by now well tried and mature. Nevertheless, expert systems are tools to support human thinking, not magic problem-solvers. While the name "expert system" is based on the notion that the systems behave *like* experts, it is generally considered that they should be used *by* experts or at least the well-informed.

18.2 3D VERSUS 2D

All structure-activity predictions, whether by human or computer, are based on the belief that the biological activity of a chemical is determined by its structure. At a general level, activity depends on properties such as fat–water partition coefficient, water solubility, acidity, and volatility, which are governed by the structure of a chemical and many of which can be calculated fairly well using algorithms based on structure-property relationships (see Chapter 9). At a more specific level, binding to proteins, including the active sites of enzymes, depends on features of chemical structure that may extend over most of the molecule or only over a relatively small part of it. Such features are commonly called pharmacophores or, in the case of toxins, toxicophores or sometimes toxophores.

Molecular modelers and others interested in how chemicals interact with biological sites point out that the world is three dimensional. It would seem then that to make successful predictions about biological activity one must use three-dimensional representations of chemical structures. But toxicologists have long been able to make meaningful predictions just by looking at structural formulas—apparently 2D representations. Researchers developing expert systems or choosing descriptors for QSAR models often report getting equally good or better results from using structural formulas than from using 3D structures (e.g., [3]). How can these things be?

Although a structural formula is a 2D image on a screen or a sheet of paper, it is not a 2D picture of the molecule: it is a graph. Many researchers prefer to call use of the chemical graph the "$2^1/_2$D" approach; others use the term "topological," which is more satisfying semantically provided that it is not confused with use of "topological index" to mean a specific type of numerical descriptor. The graph contains a lot of implicit information that a human expert, or a computer, can take into account: the presence of a letter "N"—a nitrogen atom—in the right environment hints at the potential presence of a positive charge at biological pH; if a nitrogen atom surrounded by small hydrocarbon fragments is connected through four bonds to a phosphorus atom in a phosphonate group, that suggests they are separated by a little over 5 Å—the right kinds of substituents at the right separation for an acetylcholine mimic and hence a potential neurotoxin.

Herein perhaps lies the reason why using the chemical graph can equal or outperform using the 3D structure. Chemical structures are dynamic; a single, frozen 3D representation is only one of, in some cases, tens of thousands of conformations of a structure. Three–dimensional analyses are limited to sets of the most likely conformations, which may be the wrong ones in unlucky cases. The distance between two atoms will differ between conformations, so ranges are used in analyses. The conformers are put into groups, commonly called "bins." One might, for example, list in overlapping bins molecules in which a nitrogen atom is separated from a phosphorus atom by 4.0–5.0 Å, 4.5–5.5 Å, 5.0–6.0 Å, and so on. An analysis might predict that molecules in the second bin are potential acetylcholinesterase inhibitors. Whether you say that the topological distance between two atoms is a certain number of bonds or the through-space distance falls within a certain range, you are making an equivalent approximation and you reach the same conclusion.

Both approaches have their hazards (e.g., the graph method as described here ignores constraints imposed by rings and double bonds). But, in practice, it seems that the graph approach can match the 3D approach, and it is much less computationally demanding. All of the systems to be described in this chapter depend wholly or primarily on chemical graphs rather than 3D representations. As an alternative to "toxicophores," chemical subgraphs associated with toxicity are widely called "alerts" or "structural alerts," and in the case of a particular set of sub graphs associated with genotoxicity, "Ashby alerts" [4].

18.3 TOXICITY PREDICTION BASED ON HUMAN KNOWLEDGE

A database contains specific information, such as that chloropropan-2-one (chloroacetone) vapor irritates the eyes. In response to the question "tell me about chloropropan-2-one," a database system can report that it irritates the eyes, or to the question "find me a chemical that irritates the eyes" can return the answer "chloropropan-2-one." If this one entry is the total content of the database, there are few other questions the system can answer. For example, it can return no information about bromopropan-2-one.

The knowledge bases of expert systems described in this chapter contain generalized information such as that "halocarbonyl compounds may irritate the eyes." The keying information, "halocarbonyl compounds" is stored as a representation of a chemical subgraph (see Figure 18.1). A chemical perception module in the system maps the stored subgraphs against the graph of a query compound. So, for example, if it is asked about bromopropan-2-one, it will find a match and report that "bromopropan-2-one may irritate the eyes." Note that if asked, it will similarly report that "chloropropan-2-one *may* irritate the eyes," whereas the database system was able to give the unqualified answer "chloropropan-2-one irritates the eyes"; from the contents of its knowledge base the expert system is able to make generalized predictions but only generalized ones. Note also that on the basis of this rule alone, the expert system will sometimes be wrong—for example, in this simple illustration the expert system would always predict potential eye irritation if, say, a haloester group were present, but a nonvolatile haloester might not irritate the eyes simply because it did not come into contact with them.

It is central to these systems that their chemical perception modules seek to perceive in a chemical graph the things perceived by a chemist—aromaticity and other aspects of electron distribution, potential for tautomerism, potential for protonation or deprotonation at biological pHs, implications of stereochemistry, and so on. This has caused research into and development of chemistry-related expert systems to become isolated to some extent from artificial intelligence work in other disciplines and may partly explain their distinct character. It may also have led to a bias in their uptake toward the chemistry-based disciplines and perhaps in their knowledge content, which in the early stages of their development sometimes lacked an appreciation of biological science. Both of these biases are now greatly diminished, perhaps

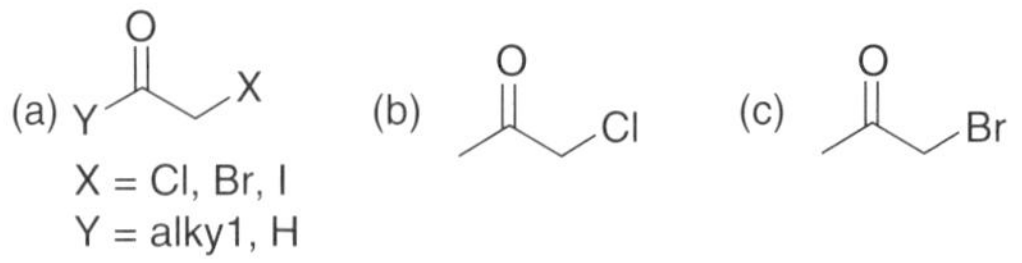

Figure 18.1 Representation of a structural alert and two chemicals containing it: (*a*) The alert; (*b*) chloropropan-2-one ($X = Cl$, $Y = CH_3$); (*c*) bromopropan-2-one ($X = Br$, $Y = CH_3$).

even eliminated, as researchers responsible for program and knowledge base development have brought in experts in toxicology and other areas of biology.

Practical expert systems contain meta-rules about the interactions between different pieces of knowledge, and some of them link to databases as well as knowledge bases so that they can be specific if something is known about a query. A system might contain a restriction rule that says that chemicals that are not put directly into contact with the eyes are unlikely to cause irritation unless they are volatile, and a means of estimating volatility. Bringing together this knowledge and the knowledge and data given above, the system can adjust its predictions and its confidence in them according to circumstance, reporting that chloropropan-2-one is known to irritate the eyes, bromopropan-2-one may irritate the eyes because it contains a relevant alert and it is expected to be volatile, and that some other, large, query molecule is unlikely to irritate the eyes in practice, even though it contains an alert for irritation because it is not expected to be volatile enough.

18.3.1 TOX-MATCH

The earliest-reported, expert system for predicting potential toxicity from human knowledge about toxicophores, or alerts, was TOX-MATCH [5]. A sister program, PHARM-MATCH, predicted useful pharmacological activity from a knowledge base of pharmacophores. Publications about the system continued for some years [6]. Joyce Kaufman gave a talk about it at a conference on reducing the use of whole animals in testing in 1991 but this idea was never commercialized. Predictions about toxicities had been developed in a language specific to a machine, and when they both became obsolete, the TOX-MATCH system was not re-implemented.

18.3.2 HazardExpert

HazardExpert [7] has a knowledge base of alerts that were originally identified or proposed by human experts working for the US Environmental Protection Agency [8] and supplemented by literature research. It uses a mathematical approach to assess the probability of activity against a toxicological endpoint. The probability that the presence of an alert will lead to expression of activity against an endpoint is calculated from the frequency of activity among compounds containing the alert that have been tested. So, if 72 out of 150 compounds are active and the rest are inactive, it is posited that the probability that a novel compound containing the alert will be active is 0.48 (the program actually expresses numerical probability as a percentage—48% in this case). It is further assumed that if a compound contains more than one alert, they each contribute to the probability of activity in accordance with conventional probabilistic arithmetic. So if the probability of activity against a given endpoint due to the presence of one alert is 48% and the probability due to a second alert is 25%, then the probability that a compound containing both alerts will be active is 61% [(0.48 + 0.25 − 0.48 × 0.25) × 100]. Some alerts are associated with more than one endpoint, and in those cases the probabilities for each endpoint can be different. The clean simplicity of this approach is appealing, but there may be pitfalls in practice: the laws of probability apply to chance events, but biological activity is not random.

In the current version of the program the potential for toxicity is finally expressed to the user as "nontoxic," "uncertain," "possible," "probable," or "highly probable," the classification being determined by preset percentage ranges (e.g., 60–100% is classified as "highly probable"). What is predicted is the probability that a compound will show activity, not the probability that the activity will have adverse consequences—for example the probability that a chemical will be a carcinogen, not the probability that an individual exposed to the chemical will develop cancer—and not the likely potency of the chemical (i.e., whether it is slightly toxic or very toxic).

HazardExpert estimates and takes into account the octanol/water partition coefficient (usually abbreviated to $\log P$ or $\log K_{ow}$) and pK_a of the query compound. The octanol–water partition coefficient is a surrogate for partition between water and fatty biological membranes, which has implications for transport of a chemical to its site of action and its capacity to bind at, or interact with, the site. The pK_a, a measure of acidity, indicates the readiness of the compound to ionize in solution, which can have a big influence on partition into membranes, since retention of the ionized form in the aqueous phase is likely to be favored. More recently modules that make use of predictions from ANN have been added to the software package available with HazardExpert.

18.3.3 Oncologic

Oncologic was developed by a team at the US Environmental Protection Agency (EPA) Office of Prevention of Pollution and Toxics in collaboration with LogiChem, Inc. [9,10]. It is available, still as a DOS system at the time of writing of this book, free of charge from the EPA. As its name implies, its predictions are confined to the carcinogenicity endpoint. They are driven by a set of decision trees written by human experts (as distinct from the binary decision trees created automatically by some data-mining computer systems). The user can build query molecules from a menu of chemical substructural features, each of which either directs decisions associated with nodes in the decision trees or can be used as a passive component to complete the framework of a structure. Structures containing fragments unknown to Oncologic cannot be drawn and entered. This feature places limitations on the usefulness of the system, although it does ensure that queries are within the system's prediction domain.

Oncologic makes predictions based on factors outside the scope normally covered by other expert systems—for example, it can predict the toxicity of asbestos-like materials on the basis of particle size and shape and surface charges. Modules can also make predictions for metals and polymers which are rarely covered by other systems.

18.3.4 DEREK

DEREK [11] was based on the technology developed for LHASA, a program to help with the planning of chemical synthesis routes that grew out of the earlier work at Harvard University [2], and it was developed in a collaboration between the LHASA group at Harvard and Lhasa Limited in Leeds, UK. It is a hazard prediction system based on the detection of structural alerts but takes into account in one or two cases factors that may influence potency, giving qualitative advice in those cases, such as "potentially strong." It provides limited supporting evidence in the form of comments from knowledge-base writers, who made use of the opportunity this gave them to include alerts having an indirect association with activity, since they could explain their thinking to the user. For example, if the user enters a structure containing an alpha-napthylamine, DEREK warns of potential carcinogenicity arising from contamination with the beta-isomer.

Although DEREK is still available from the LHASA group at Harvard University, it has not be updated in recent times, having been superseded by Derek for Windows from Lhasa Limited.

18.3.5 Derek for Windows

Derek for Windows [12] uses structural alerts to trigger warnings about potential toxicity, but its predictions are controlled by a reasoning system [13,14,15] based on the Logic of Argumentation. The system manipulates rules of the general form

If [grounds] is/are [threshold] then [proposition] is [force]

The items in square brackets can be assigned fixed values or can be variables determined by other rules of the same form. A simple example of a rule might be "if structural_alert_x_present is certain then skin_senzitisation is probable." A different rule might be "if log P < y then skin_sensitization is doubted." Rules involving the use of variables might be "if structural_alert_x_present is certain then skin_sensitization is partition_dependent_variable," "if log P < z then partition_dependent_variable is plausible," and "if log P > z then partition_dependent_variable is probable." The reasoning engine works out the interactions between rules in order to construct a report to the user. In the current system the potential for toxicity is expressed by the range of terms "impossible," "improbable," "doubted," "equivocal," "plausible," "probable," and "certain." These words have specific definitions in the system (e.g., "certain" means, in effect, "experimentally confirmed to be true," which is not as absolute as the meaning that it could have). "Open" means that there is no pertinent information on which to make a prediction. "Contradictory" represents the state in which there appears to be proof of both certainty and impossibility.

The system gives access to supporting evidence in the form of bibliographic references to papers describing the association of alerts with activity and mechanistic justification or interpretation, and papers reporting toxicological studies on compounds used as the basis for an alert.

18.3.6 The BfR Decision Support System

A decision support system to predict skin and eye irritation and corrosion was developed by staff at the former Bundesinstitut für gesundheitlichen Verbraucherschutz und Veterinärmedizin (German Institute for Consumer Health Protection and Veterinary Medicine) and is now with the Bundesinstitut für Risikobewertung (Federal Institute for Risk Assessment) [16]. The Bundesinstitut uses rules based on physicochemical properties, broad classifications of chemical groups, and structural alerts. For example, it can take into account $\log P$, surface tension, and melting point; predictions are grouped according to their applicability to chemicals containing only carbon, hydrogen, and oxygen, containing also nitrogen, and so on; prediction of activity on the basis of a structural alert can be restricted to compounds falling into one or more of the chemical groups. This approach has not been much exploited by other researchers. The program is also interesting because it predicts irritation and corrosion—endpoints that have not been given much attention by others but for which prediction is of increasing importance as legislation to limit or preclude some uses of animals in experiments comes into force in Europe. A difficulty with the program is that input of queries requires the use of a somewhat arcane, textual representation of chemical structures but this may change.

18.4 TOXICITY PREDICTION BASED ON COMPUTER-GENERATED MODELS

All the programs described in Section 18.3 use associations recognized by human experts between chemical substructures, of varying degree of detail, and toxicological endpoints. The experts form their ideas by looking at the structures of molecules that do or do not show activity and by picking out features that seem to be present only in the active ones. In some cases the experts will develop more complex models in which certain molecules that do contain a feature associated with activity are nevertheless inactive because of another, interfering feature. For example, there might be an electron-withdrawing group that radically changes the reactivity of the one associated with toxicity, or a bulky group that interferes with binding to a site of action.

Confidence in the validity of a toxicophore is much greater if there is a mechanistic rationale to explain its association with an endpoint, which hangs on expert human consideration. But, setting that aside, discovering toxicophores by looking for the coincidence of patterns within chemical structures that show common activity raises two questions. The first simply is, Can a computer do the same job? The second is, Given human limitations on scanning large data sets and remembering what is in them, can a computer do the same job better?

The programs described in this section mimic the human expert by seeking out patterns in data. They perhaps barely conform to the definition of an expert system as one that uses a generalized store of knowledge, but they are widely thought of as members of the class. The programs are included here partly for that reason and partly because they are important for their contributions to this area of research: the output from their analysis modules can support human experts working on the development of rules for the systems described above.

18.4.1 TOPKAT

TOPKAT [17,18] was one of the earliest systems for predicting toxicity to be described. It uses statistical analysis to identify substructural fragments associated with toxicity and derives Hansch QSAR equations (see Chapter 8) with which to predict toxic potencies against a range of endpoints, including carcinogenicity, mutagenicity, and acute lethal toxicity (LD_{50}). Physicochemical properties such as $\log P$ are taken into account (analyses would otherwise be skewed by the presence of molecules having different levels of activity on account of their physical properties rather than different interactions with biological sites). The fragments for which it searches are held in a library of several thousand, ranging from simple functional groups recognized by organic chemists, such as a nitro group and amide group, to complex ring systems. Groups in different environments are distinguished so that, for example, there are separate entries for a nitro group attached to an aromatic ring and one attached to an aliphatic chain.

The analysis and prediction processes are separated. The analysis phase generates sets of equations for groups found to be significant to activity in a training set of chemical structures. Conventional statistical measures of validity are computed. The application supplied to an end-user applies these models to query compounds and reports the predictions derived from the equations. The user can ask for information about the statistical quality of the models and about how well the query compounds fit into the prediction space of the models. Depending on whether models have been derived from public or confidential data, the user may be able to view the structures that were used in the training set.

18.4.2 MCASE and CASETOX

Gilles Klopman recognized that a weakness with models based on predefined structural fragments is potential bias in the fragment set. Chemists, for example,

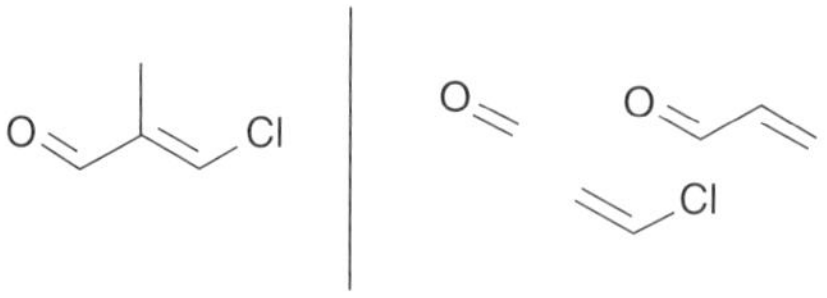

Figure 18.2 Types of fragments used by TOPKAT.

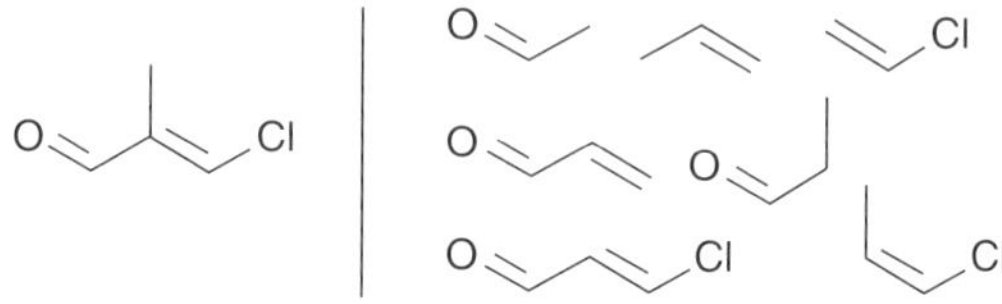

Figure 18.3 Types of fragments used by MCASE and CASETOX.

define functional groups that are useful when they are designing chemical syntheses, but such groups may not have biological significance. How does one decide a priori what might be biologically significant? MCASE [19] bases its predictions on fragments generated automatically from a training set of chemical structures. The fragment set comprises all linear atom and bond chains within a preset range of bond numbers (e.g., chains containing 3–12 bonds) that are found in the training set. The difference between this approach and the one used in TOPKAT is illustrated simplistically in Figures 18.2 and 18.3. The TOPKAT-like fragments in Figure 18.2 have been devised for the purposes of the illustration and may not actually be used in TOPKAT, and the list of MCASE-type fragments in Figure 18.3 is not complete. In MCASE, as in TOPKAT, the output from the analysis is a set of fragments of unknown interconnectivity associated with activity. This differs from the usual idea of a toxicophore in that a toxicophore is usually thought of as being a single fragment, albeit incorporating variable components. As systems of this kind are just as capable of finding predictors for useful biological activity as they are for toxicity, Gilles Klopman's group term the sets of fragments "biophores."

The program does not limit its analyses and predictions to positive associations between biophores and activity. It also seeks out sets of fragments that, while not conferring activity on molecules, may increase or decrease it, which Klopman's group call "modulators." Although the fragments are linear ones, branch-points are identified and included in the analyses to take some account of the effects of branching. Atom pairs may also be used (i.e., linear fragments in which only the terminal atoms are specified, the intervening atom and bond types being generic). As in the case of TOPKAT, properties such as $\log P$ are taken into account.

An assembly of a set of biophores, modulators, derived QSAR equations, statistical validation information, and supporting data is described as a knowledge base by Klopman's group. The supporting data may contain all or only some of the training set, depending on whether data are public or proprietary. CASETOX is an end-user application that depends on such knowledge bases to make predictions about novel compounds but does not have the analysis module that is part of MCASE.

18.4.3 LeadScope

The LeadScope program is a data mining and visualization package [20] that includes statistical analysis tools similar to those used in TOPKAT and MCASE. Users can classify chemicals in a variety of ways, including on the basis of chemical type selected from a predefined list of the kinds of features used as descriptors in TOPKAT. As well as being used as input to computational analyses, attributes of molecules such as physicochemical properties or biological activity can be visualized (e.g., as the variables in bar charts and/or by variation of color of the bars in a chart) to help the user to discover significant trends.

18.5 PREDICTING METABOLISM

Metabolism plays a big role in toxicological processes. A function of the liver and other organs in mammals is to convert potentially harmful chemicals into ones that can be easily excreted. That can lead to such efficient excretion that a chemical containing a toxicophore is harmless in practice, but the more usual consequences are either that the toxicophore itself becomes modified and thus deactivated or that some other, seemingly innocuous, part of the molecule becomes converted into a toxicophore. Indeed there is some ambiguity about what is meant by a toxicophore. A substructural feature that becomes associated in the minds of scientists with a toxicological effect could be directly responsible, or it could be converted through metabolism to a feature is responsible. In some cases such metabolic activation is known to be necessary because, for example, the result of an Ames test changes from negative to positive when it is conducted in the presence of liver S9 fractions that bring about metabolic reactions.

Making useful predictions about metabolic reactions presents considerable difficulties. As a generalization it is not unreasonable to propose that all metabolic conversions of a molecule for which appropriate enzymes exist in the study animal will occur. The questions are which metabolites will predominate and what will be the toxicological significance of the metabolites. To add to the complications, there may be discontinuities in the relationship between quantities of metabolites and dose level: if the enzyme responsible for the fastest metabolic conversion is present in fairly small amounts and becomes saturated, the product from a different enzyme may start to appear in large amounts. The need to be able to predict metabolism is nevertheless a

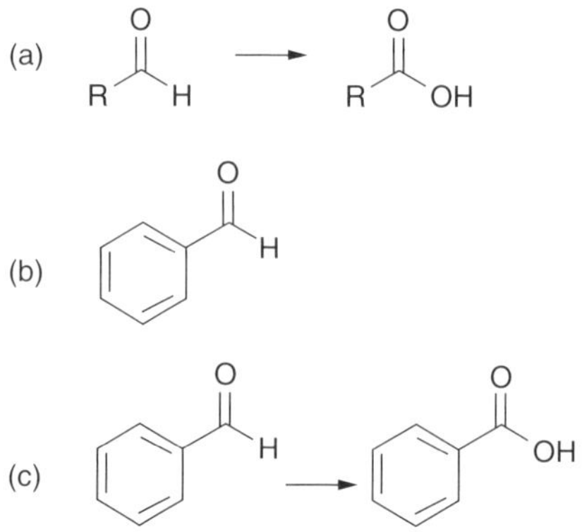

Figure 18.4 (*a*) A biotransformation—the keying fragment is the fragment to the left of the arrow; (*b*) a query molecule containing the keying fragment; (*c*) the reaction generated by applying the biotransformation to the query.

sufficiently important incentive for research groups to have invested effort into it, and modest success can now be claimed.

The three systems described here are knowledge-based expert systems. They all use the same technical approach. The knowledge base contains subgraph representations of metabolic reactions, or biotransformations. The program searches a query structure for the subgraphs of the precursors in the biotransformations (the keying fragments). If there is a match it uses the description of the biotransformation to generate the structure of the product (see Figure 18.4).

An open-ended search, in which all possible products are generated for a query structure and then all possible products of metabolism of those initial products are generated, and so on, can create a reaction tree that contains hundreds or even thousands of possible products. In some circumstances this may be useful—for example, to support an automated search for a structure that can explain an unexpected peak in the mass spectrum of a sample from a metabolism study—but more generally it is not. The programs provide automated and manual ways of limiting the output, the simplest and most obvious being to generate only the products from single-step reactions of a single structure and to let the user select which one(s) to resubmit for further processing. This is more helpful than it may appear to be because the program acts as an *aide mémoire*, reminding the user of all the possibilities, and someone with a good general understanding of metabolism will often be able to make judgments about which are the more likely in a given circumstance. However, the fact that a user can make such judgments is reason to expect the same of the software, since an expert system is supposed to capture and use human experience.

Metabolism experts distinguish between phase I and phase II reactions. As a generalization, phase II reactions involve the attachment of a fragment that confers water solubility on a compound to facilitate its rapid excretion—for example, glucuronidation, the attachment of a sugar residue. Phase I reactions introduce groups that make phase II reactions possible—for example, introduction of a hydroxyl group that can undergo glucuronidation. Because it moves out of membranes into the aqueous phase, and is transported away and excreted, a phase II product is much less likely to undergo further metabolism than a phase I product. Conversely, a phase I product is very likely to undergo further metabolism. There are exceptions to these rules but they hold as useful generalizations and all of the programs allow the user an option for reaction sequences to be terminated when a phase II product is generated.

Other ways of limiting output to the reaction sequences most likely to be seen are mentioned in the following sections about particular computer systems.

18.5.1 MetabolExpert

MetabolExpert [21] is a sister program to HazardExpert, and the two can be used in tandem. In addition to the restrictions mentioned above, the user can set an option to stop the program once a specified number of metabolites have been generated. The program compares the tree that it generates with trees for known, similar compounds, and depending on the degree of similarity, it assigns numerical probabilities to the products in the generated tree.

18.5.2 META

META [22,23] was developed by Gilles Klopman and coworkers. The mammalian model covers a wide range of reactions promoted by 26 types of enzymes. Metabolic products can be assessed automatically for potential carcinogenicity and the results reported to the user. Separate knowledge bases cover aerobic biodegradation, anaerobic biodegradation, and photodegradation.

The development team have described the use of a genetic algorithm to prioritize the predictions that META makes. [24]

18.5.3 Meteor

Meteor [25,26] is a sister program to Derek for Windows, and it uses reasoning based on the logic of argumentation to rank its predictions [15,27]. Structures are not automatically transferred between the two programs, but the user can manually select structures to transfer in either direction. Biotransformations are rated according to their likelihood were they to operate without competition from other biotransformations, using the same terms as those used in Derek for Windows, and rules taking account of properties such as $\log P$ modify the likelihood that a biotransformation will occur in a specific case in a similar way (see Section 18.3.5). Separately, where information is known, groups of biotransformations are ranked according to how likely they are to predominate over each other in competition. The predictions presented to the user take account of both assessments, and the user can set thresholds of likelihood and ranking position above which biotransformations are to be reported.

To support metabolic studies, there is an option to restrict the growth of reaction sequences to products containing a specified, radio labeled atom. There is also an option to construct a very large metabolic tree initially and then to prune it automatically to show only branches that lead to structures with specified molecular formulas coming from mass spectrometric studies.

18.6 THE INTERACTIONS BETWEEN QUALITATIVE AND QUANTITATIVE PREDICTION

Two schools of thought have long competed for ascendancy in science. One school holds that predictions—and indeed science—mean nothing unless they can be expressed in verifiable numbers. The other holds that numbers are more to do with a kind of logical accountancy and the philosophy is all.

Where numerical methods work and trustworthy data are available to drive them, they are practically useful and scientifically reassuring. But if, as can happen in fields such as toxicology, the only way to drive the models is to feed in dubious, or even near-fabricated data (e.g., qualitatively ranked observations re-expressed as exact numbers in order to be able to enter them into equations), they lose credibility. Qualitative methods can work well. In organic chemical synthesis, researchers successfully plan and implement series of reactions on novel compounds, or even introduce novel reactions, on the basis of qualitative reasoning about how chemicals usually behave—so-called "arrow-pushing" because of the representations the chemists use on paper. If you ask them about the likely outcome of a specific reaction using a novel starting material, they will usually be more right than wrong and they will be able to say how confident they are about the prediction. But, apart from giving some general guidance, they will not be able to predict how much heat must be dissipated from an exothermic reaction if there is not to be a dangerous incident in a scale-up plant. To do that requires quantitative measurements and calculations.

Hansch QSAR and related approaches belong to the world of numbers: conceptually, knowledge-based expert system approaches do not. Three areas of debate about expert systems have arisen from the distinction. Can you devise ways to generate qualified output from computer-based expert systems without hiding quantitative methods inside them? Assuming you can, how do you validate an expert system? How can you usefully combine output from different systems—some quantitative and some qualitative—to make predictions more reliable? The first of the questions is more a historical than a current one: some of the systems described earlier in this chapter demonstrate that it is possible to devise nonquantitative rules and reasoning methods that are of practical use. The other two questions are discussed further in Sections 18.6.1 and 18.6.2.

18.6.1 Evaluation and Validation of Expert Systems

Because of the complexities involved, predicting toxicity is an uncertain thing. A human expert making predictions within his or her sphere of work may be 80% correct. It takes a different expert to achieve that level of performance for a different endpoint or different kinds of chemicals. Some kinds of toxicity are easier to predict than others, and 80% accuracy is not often achieved. In unfavorable cases the performance of any individual may be no better than random. Individual computer programs similarly perform better in some areas than others, over the whole range from being surprisingly good to more or less useless.

Quantifying predictive performance is basic to statistics and has therefore naturally been a part of QSAR work. Accepted statistical measures of performance are routinely reported with QSAR models (see Chapter 8), and models for which the measures are not provided or for which they show poor performance are not trusted. Some of the mathematical methods that are used, both to develop QSAR models and to assess their performance, relate to predictions of potency. Others assess performance in terms of the correctness of simple binary (yes/no) prediction of whether a chemical is active or inactive, where chemicals are classed as active experimentally if the relevant measure exceeds a predetermined threshold.

Some of the systems described in this chapter, for example, TOPKAT and MCASE/CASETOX, use conventional, statistical methods to generate their models, and they make quantitative predictions. They can be validated in accordance with usual statistical practice. Knowledge-based expert systems have a different lineage. During their early development and use their purpose was hazard identification, rather than risk assessment, and that is still largely the case. Hazard is a nonquantitative concept, and human experts make practical predictions about it that are purely qualitative. Knowledge-based systems are designed to behave similarly. They do not contain the algorithms nor generate the numbers that support some of the validation methods used in QSAR. Even so, meaningful measures of performance are needed.

Some statistical methods may be appropriate for knowledge-based expert systems, but there are difficulties. One of the most basic requirements of a well-designed project to develop a statistical model is that there should be training sets and test sets. There are many views on how these sets should be chosen and on the interpretation of the ensuing validation. The test set may be chosen by random selection from a larger set, with the residue becoming the training set. "Leave-one-out" and "leave-out-many" approaches are variations in which all the members of the total set may be used for training or testing in parallel runs. In other studies the training set includes all the information available to the developer, with evaluation being conducted, often by someone else, against a test set that had not been available. There are great limitations on using these methods for knowledge-based expert systems. How can human experts ignore things selectively and without bias in order to build knowledge bases from subsets of the knowledge available to them? Is there any point in applying a test that will demonstrate shortcomings when the developer knows in advance not only what the shortcomings will be but that they are the consequence of deliberately ignoring key knowledge?

So evaluations of knowledge-based expert systems can only be one of two variants of the same kind: evaluation of performance against the full training set and (independent) evaluation against previously unpublished test sets. The developers of the systems described in this chapter use the first kind of evaluation as a routine part of their development work, report the results to their users or potential users, and may publish them. Independent evaluations of all of the systems described in this chapter have also been published—often in studies comparing two or more of the systems. Sometimes it is clear that the evaluations have been conducted against data that were not directly available to the knowledge base developers because the evaluations were based on in-house, proprietary data, but in other cases the position is less clear. Even when proprietary data are used, there is uncertainty because most of the developers will have included at least some proprietary data from other sources in their training sets and there is no way of knowing whether there is overlap in the data from the two sources, both being proprietary and confidential to their different owners.

If the sole or primary purpose of a system is to make predictions in truly novel circumstances, then statistical validation against a blind test set is the most convincing measure of performance. But all of the computer systems are limited to their prediction domain. QSAR models are expected to predict only chemicals that have features that were present in the training set(s). A knowledge-based system tells you what is known (or, more strictly, what has been recorded in its knowledge base), and it is predictive only in the sense that knowledge is generic and thus applicable to new, specific instances. Evaluation of performance of a knowledge-based system against the training set, or against publicly available data, which are likely at least to be a subset of the training set, may thus be more useful than it at first appears to be. It provides a test of how well the knowledge base team has captured current knowledge and how faithfully the computer program communicates the findings.

Common practice has been to assess knowledge-based systems and report their performance in terms of sensitivity, selectivity, and concordance. The statistics that generate these three measures all depend on classifying chemicals in the test set as active or inactive and recording whether or not the prediction made by the system in each case is in agreement. The problem of what to do about borderline chemicals is frequently discussed, and different solutions have been adopted in different circumstances: the problem of how to apply these measures if the computer system does not give straightforward binary (yes/no) predictions has received less attention.

With regard to the first problem, all borderline cases can be left out of a study, or they can arbitrarily all be classed as active or all as inactive, depending on whether overprediction or underprediction is the greater cause for concern in the intended area of use. Where a knowledge-based system gives some indication of its confidence in its predictions, this can also be taken into account. If the chemicals with borderline-measured toxicities are the ones about which the system is least confident, this is reassuring. Still care is needed, since the correspondence may be coincidental.

More generally, the question of what to do about programs that make predictions that are neither quantitative nor binary has yet to be answered satisfactory. The usual measures of performance require that a program either generate numerical estimates of potency or make a binary prediction—active or inactive. A program such as Derek for Windows may predict that activity is "plausible," "probable," "equivocal," "doubted," and so on. There is a case for treating "equivocal" as a prediction that activity will be borderline, but if activity is predicted to be "plausible," is that a "yes"? If it is, then what does "probable" mean?

It is possible to assess the performance of the program qualitatively by looking at the trends—where the program expresses stronger conviction, is it more usually correct in practice? But the formal methods for generating numerical measures of performance require unsatisfactory compromises that lose information. Arbitrary decisions are taken, for example, to class all predictions of "plausible" or stronger as "yes" and all others as "no." In some studies, researchers have ignored the computer assessment of how likely it is that chemicals will be active, and simply taken the detection of an alert as positive prediction of activity, but this overlooks the possibility that a substructural fragment might have been associated with *inactivity* and excludes important factors such as physicochemical properties that can modify the prediction that the computer ultimately makes.

The debate thus comes full circle, returning to the question of whether it is possible to evaluate predictions meaningfully without using rigorous numerical methods. In general, as a society, not only with regard to the prediction of toxicity, we have made little progress toward resolving the difficulties. There is a general assumption that science is about precision and that numbers must be provided and validated. The long-term success of knowledge-based expert systems may depend not on whether they work but on whether they satisfy criteria set from outside the artificial intelligence community. Legislators and users wanting to make sure that the systems they approve for critical applications meet the right standards may set statistical performance measures that are not valid for expert systems.

Yet we accept the reasoned judgments of doctors making diagnoses, law courts deciding on—in some jurisdictions—life or death issues, committees granting planning permission, and so on. In the opinions of many toxicologists, knowledge-based expert systems can match or even outperform numerically based programs, but they have no way of generating rigorous numerical statistics to back up their opinions. Finding and promoting alternative performance measures suitable for knowledge-based systems is important not only to defend their claim to respectability. Without valid performance measures, how are expert systems to be progressively improved? This is likely to be a key area for debate and new developments over the next few years.

18.6.2 Consensus Modeling and Related Approaches

Within the same area of expertise, different humans, and different computer programs, may disagree. Faced with a committee of humans, some arguing for toxicological activity and some against for the same chemical and endpoint, what is the most effective course of action? Whatever the answer, is it also an appropriate course to take when computer programs disagree? If the committee is unanimous in predicting one chemical to be toxic, but divided on another, does that mean that the first one is more likely to turn out to be active than the second one?

The potential consequences of being right or wrong influence the choice of solution when there is disagreement. In England, it used to be required in all cases that a jury reach a unanimous verdict before a defendant can be declared guilty, on the ground that an erroneous positive decision had consequences too serious for even one dissenting voice to be ignored. In the context of predicting the toxicity of chemicals in the pharmaceutical industry, whether it is better to under- or overpredict depends on the reason for making the prediction. A researcher deciding whether to leave chemicals unsynthesized on the ground of predicted toxicity does not want to leave out any chemicals that fit the specifications for activity against the pharmacological target of interest if they might turn out to be safe in practice. Someone responsible for ensuring the toxicological safety of a chemical progressing through development wants to be warned of every hazard, even if that means accepting a rather high percentage of false positive predictions, so that nothing will be overlooked during testing.

The unanimous agreement of a committee that a chemical will be unacceptably toxic would be a reasonable test to apply when deciding not even to synthesize the chemical. In practice, combinations of chemical and committee that would produce such a firm conclusion are rare. A unanimous prediction that a chemical is *likely to be* unacceptably toxic would be more typical but that could be accepted as sufficient reason to rule the chemical out of a research program. More usually there is disagreement within committees, whatever the subject of debate, and different cultures have different ways of dealing with it. The simple solution is a straightforward vote, with the majority view prevailing. But some individuals in a committee may command greater credence than others (rightly or wrongly), and more complicated voting schemes are designed to give their views greater weight.

There has been some research into combining the output from several computer systems that make predictions for the same toxicological endpoints on the basis of chemical structure and physicochemical properties. Until fairly recently researchers favoring numerical or nonnumerical methods have tended to work separately and to develop schemes better suited to one or other of the approaches.

Both groups share concerns over some basic questions that echo the ones with regard to the resolution of disagreements in a committee. If two out of three programs agree that a chemical will be positive against an endpoint, do you believe the two and ignore the third or do you conclude that the probability of activity is about 0.67? Should predictions be weighted to take account of differences in the performance of programs depending on, among other things, the type of query? There are no firm answers to these questions, and various approaches have been reported and shown to be useful in particular circumstances. As in the case of program evaluation, the solution depends to some extent on the intended use of the output. Researchers wanting to be alerted to all possible hazards may decide that if any one out of a group of programs gives a positive prediction, then that should be the overall output. Researchers wanting only to be alerted to cases where activity is, as near as possible, a certainty may decide that chemicals will be flagged as active only if all programs agree that they are.

A widely used term in connection with automated systems that combine output from multiple sources is "consensus modeling," but the term is not ideal. Consensus implies discussion within a committee leading all parties to a common view. None of the existing schemes for combining output from more than one source involve feedback and reassessment by the participants—in this case the prediction programs. All of them depend either on some kind of averaging or some kind of voting. Some of the problems are the same as those discussed in Section 18.6.1 in relation to validation techniques—the methods require input from the prediction programs in the form either of numerical potencies, or binary or ternary predictions of activity (yes/no/don't know), and so output from knowledge-based systems has to be interpreted in ways that can lead to loss of information and incorrect conclusions.

There is a danger to be aware of when making use of more than one prediction system. Many of the models used in QSAR and expert systems have been derived from published data or from proprietary data held by the same organizations. When programs agree with one another in their predictions, it is difficult to establish whether they agree simply because their models were developed from the same training set. If a third program disagrees, is that because it used a different training set or because it placed a different interpretation on the data in the same training set?

The use of reasoning may offer an alternative approach. Methods have been suggested for combining qualitative predictions with numerical probabilities [13], but little work has been reported in this area to date. In principle, a reasoning-based system could draw on knowledge, predicted values, and measured values to provide broadly based reports. The next generation of expert systems should bring together the strengths of different approaches and be capable of recognizing their weaknesses as well. They might be capable of issuing reports such as the following:

"Your chemical has already been tested and found to be active."

"Your chemical is expected to be a strong skin sensitizer because it contains an alert associated with activity by a mechanism that is widely accepted, and it is within the prediction space for QSAR model XXX, which predicts activity well above the threshold for labeling under EU regulations."

"Although your query chemical contains an alert associated with neurotoxicity and QSAR model YYY predicts activity, it is unlikely to be active in practice because it fits much better into the prediction space for QSAR model ZZZ, which was based on a different training set and predicts inactivity, and because its $\log P$ is so low that it is unlikely to penetrate biological membranes."

In the first case, the program would provide the data from the toxicity test and references or Web links to the relevant publications. In the second and third cases, although words have been used for the purposes of this illustration, the system would provide the numerical predictions generated by the QSAR models, together with validation measures such as R^2 and Q^2, and the measured or predicted $\log P$ value.

It is one thing to design a program with broad capabilities; it is another to gather together all the necessary data and knowledge and to make full use of those capabilities. It remains to be seen how far the next generation of software products will go toward the goals implied by the examples above, but progress in that direction is assured.

18.7 STRENGTHS AND WEAKNESSES OF EXPERT SYSTEMS

There are some differences in strengths and weakness between expert systems based on human knowledge and those based on computer-generated algorithms. The latter systems generate and use collections of unconnected molecular fragments associated with activity. These are less precise than the toxicophores to which they may relate, in which the fragments are connected in a specific way, so there is a risk of false positive predictions from matches against molecules containing all the fragments but not in the form of the toxicophore. It is not clear how far this is a problem in practice. Many of the toxicophores included in knowledge-based systems are fairly simple substructures unlikely to be subject to the problem.

A particular strength of knowledge-based expert systems over all others is that they can justify their predictions on the basis of mechanisms of action provided by the knowledge-base developers. This is not possible with systems where models are generated by computer, unless knowledge workers retrospectively rationalize the predictions and add information about their conclusions. But, in effect, by so doing, they are turning the system into a knowledge-based one. Indeed, some knowledge base content in existing systems derives from what researchers have learned or inferred by studying models generated by computer, augmented and re-expressed. It might be thought that if a system is shown to give statistically meaningful predictions, that is all that matters. However, scientists and the people they advise do not accept what they are told without question: predictions need to be supported by reasoned explanations, and formal statistical evidence alone may not convince nonstatisticians.

The performance of expert systems against some endpoints is better than against others, perhaps partly because the approaches they use suit some problems better but mainly because of the variable amount and quality of data and the different priorities attached to working on different endpoints. At the present time prediction of genotoxicity and skin sensitization are good enough to be in routine use in many organizations. The prediction of genotoxic carcinogenicity is considered useful to a degree, but the coverage of nongenotoxic carcinogenicity is patchy. In all cases users are much more confident if predictions are based on mechanistic understanding. For the same reason there is widespread doubt about the validity of predicting complex endpoints such as acute toxicity—especially quantitatively—since it arises from many mechanisms, each of which would be expected to obey different rules. Nongenotoxic carcinogenicity probably also falls into this category except where a mechanism has been proposed for a particular type of chemical and carcinoma.

Although there are claims to do better, it is doubtful that any of the expert systems achieves prediction success rates better than around 80–85% against any endpoint. But neither do most biological test methods, whether in vitro or in vivo, and so that may be an acceptable level of performance. From a different viewpoint, how can the performance of any system be shown to exceed the reliability of the measured data against which it is assessed? Prediction success rates are still well below 80% for some endpoints, and many endpoints are not even covered, but the success with endpoints on which work has concentrated suggests that weaknesses will be addressed as priorities change and toxicological knowledge becomes available. Advances in the biological sciences are leading to rapid expansion of knowledge, which augurs well for expert systems.

In summary, expert systems are now well enough developed to be in wide use and accepted as practical tools for the prediction of toxicity. Their current limitations are many, but so are the opportunities for their improvement, which continues apace.

REFERENCES

1. Shortliffe EH. *Computer-based medical consultations: MYCIN*. New York: Elsevier, 1976.
2. Corey EJ, Wipke WT. Computer-assisted design of complex organic syntheses. *Science* 1969;166:178–92.
3. Brown RD, Martin YC. The information content of 2D and 3D structural descriptors relevant to ligand-receptor binding. *J Chem Inf Comput Sci* 1997;37(1):1–9.
4. Ashby J, Tennant RW. Chemical structure, *Salmonella* mutagenicity and extent of carcinogenicity as indicators of genotoxic carcinogenesis among 222 chemicals tested in rodents by the US NCI/NTP. *Mutagenesis* 1988;204:17–115.
5. Kaufman JJ, Koski WS, Harihan P, Crawford J, Garmer DM, Chan-Lizardo L. Prediction of toxicology and pharmacology based on model toxicophores and pharmacophores using the new TOX-MATCH-PHARM-MATCH program. *Int J Quantum Chem, Quantum Biol Symp* 1983;10:375–416.
6. Koski WS, Kaufman JJ. TOX-MATCH/PHARM-MATCH prediction of toxicological and pharmacological features by using optimal substructure coding and retrieval systems. *Anal Chim Acta* 1988;210(1):203–7.
7. Smithing MP, Darvas F. HazardExpert—An expert system for predicting chemical toxicity. In: Finley JW, Robinson SF, Armstrong DJ, editors, *Food safety assessment*, ACS Symposium Series Vol. 484. Washington: American Chemical Society, 1992. p. 191–200.
8. Brink RH, Walker JD. EPA TSCA ITC interim report. Rockville: Dynamic Corporation, 1997.
9. Woo Y-T, Lai DY, Argos MF, Arcos JC. Development of structure activity relationship rules for predicting carcinogenic potential of chemicals. *Toxicol Lett* 1995;79:219–28.
10. Lai DY, Woo Y-T, Argos MF, Arcos JC. Cancer risk reduction through mechanism-based molecular design of chemicals. In: De Vito S, Garrett R, editors, *Designing safer chemicals*, ACS Symposium Series Vol. 640. Washington: American Chemical Society, 1996. p. 62–73.
11. Ridings JE, Barratt MD, Cary R, Earnshaw CG, Eggington E, Ellis MK, et al. Computer prediction of possible toxic action from chemical structure: An update on the DEREK system. *Toxicology* 1996;106:267–79.
12. Langton K, Marchant CA. Improvements to the Derek for Windows prediction of chromosome damage. *Toxicol Lett* 2005;158(1):S36–S37.
13. Judson PN, Vessey JD. A comprehensive approach to argumentation. *J Chem Inf Comput Sci* 2003;43(5):1356–63.
14. Judson PN, Marchant CA, Vessey JD. Using argumentation for absolute reasoning about the potential toxicity of chemicals. *J Chem Inf Comput Sci* 2003;43(5): 1364–70.
15. Judson PN. Using computer reasoning about qualitative and quantitative information to predict metabolism and toxicity. In: Testa B, Kramer SD, Wunderli-Allespach H, Volkers G, editors, *Pharmacokinetic profiling in drug research: Biological, physicochemical, and computational strategies*. Weinheim: Wiley-VCH. 2006.

16. Zinke S, Gerner I, Grätschel G, Schlede E. Local irritation/corrosion testing strategies: Development of a decision support system for the introduction of alternative methods. *ATLA* 2000;28:29–40.
17. Enslein K, Craig PN. A toxicity prediction system. *J Environ Toxicol* 1978;2: 115–21.
18. Enslein K, Gombar VK, Blake BW. Use of SAR in computer-assisted prediction of carcinogenicity and mutagenicity of chemicals by the TOPKAT program. *Mutation Res* 1994;305:47–61.
19. Klopman G, Chakravati SK, Zhu H, Ivanov JM, Saiakov RD. ESP: A method to predict toxicity and pharmacological properties of chemicals using multiple MCASE databases. *J Chem Inf Comput Sci* 2004;44(2):704–15.
20. Myatt GJ, Zhengming C, Yang C, Cross K, Blower P. Data mining and visualisation in drug discovery. *AAPS Newsletter*, Sep 2004, p. 16–19.
21. Darvas F, Marokhazi S, Kormos P, Kulkarni G, Kalasz H, Papp A. MetabolExpert: Its use in metabolism research and in combinatorial chemistry. In: Erhardt PW, editor, *Drug metabolism: Databases and high-throughput testing during drug design and development*. Oxford: Blackwell Science 1999. p. 237–71.
22. Klopman G, Dimayuga M, Talafous J. META I: A program for the evaluation of metabolic transformation of chemicals. *J Chem Inf Comput Sci* 1994;34:1320–25.
23. Talafous J, Sayre LM, Mieyal JJ, Klopman G. META 2: A dictionary model of mammalian xenobiotic metabolism. *J Chem Inf Comput Sci* 1994;34:1326–33.
24. Klopman G, Tu M, Talafous J. Meta3: A genetic algorithm for metabolic transform priorities optimization. *J Chem Inf Comput Sci* 1997;37:329–34.
25. Ali MA, Long A. Recent advances in Meteor: Metabolism prediction for heteroaromatic ring systems. *Drug Metab Rev* 2005;37(1):159.
26. Balmat A-L, Judson PN, Long A, Testa, B. Predicting drug metabolism—An evaluation of the expert system METEOR. *Chem Biodiver* 2005;2(7):872–85.
27. Button WG, Judson PN, Long A, Vessey JD. Using absolute and relative reasoning in the prediction of the potential metabolism of xenobiotics. *J Chem Inf Comput Sci* 2003;43(5):1371–77.

19

STRATEGIES FOR USING COMPUTATIONAL TOXICOLOGY METHODS IN PHARMACEUTICAL R&D

LUTZ MÜLLER, ALEXANDER BREIDENBACH, CHRISTOPH FUNK, WOLFGANG MUSTER, AND AXEL PÄHLER

Contents

Computational Toxicology: Risk Assessment for Pharmaceutical and Environmental Chemicals, Edited by Sean Ekins

19.1 ATTRITION CAUSES

Drug development is a process that normally takes 10 or more years from the early stages of target selection until the selection of a compound that is finally marketed and used by patients after a positive regulatory approval. Major causes for removing certain chemical structures from further clinical evaluation relate to (pre)clinical toxicity or unfavorable absorption, distribution, metabolism, and excretion (ADME) properties. According to various figures, the failure rates for safety (toxicity) are currently estimated at around 20% for the early stages of clinical evaluation (see other chapters in this book). This compares to other major causes of attrition such as efficacy at about 40% and economics at about 35%. A clear improvement of these figures, on the early selection of the right candidate with a high probability of good clinical safety and efficacy, is expected from various areas including structure-activity relationships or "in silico safety."

19.2 CHEMISTRY AND TOXICITY

Drug-induced adverse drug reactions (ADRs) are classified as pharmacology related (type A) or as direct acting type B, C, and D. In the case of pharmacology-related mechanisms of toxicity the effect may be related either to the primary pharmacology of the desired target (type A1) or to other secondary pharmacological effects (type A2, lack of selectivity due to activity vs. non-intended receptors). In most cases the pharmacological side effects are due to exaggerated response in local tissues because of overdose, unfavorable tissue kinetics, or the accumulation of active molecules (parent or metabolite). Usually these cases of toxicity involve the parent drug or metabolites with minor metabolic alterations or conjugation reactions during drug metabolism and are typically discovered significantly later than the pharmacological properties of the parent drug. Compounds falling into the categories of B, C, and D types of toxicity typically share a common feature, namely the bioactivation of the parent drug molecule to toxic reactive, typically electrophilic, metabolites capable of covalent binding to cellular macromolecules. Depending on the cellular target that is modified, the consequences may be acute (type C like in the case of acetaminophen overdose and covalent binding of its *N*-acetyl-*p*-benzo-quinone imine (NAPQI) metabolite to cellular proteins), delayed (type D, e.g., for genotoxic agents that covalently modify tissue DNA via reactive metabolites like the anti-androgen cyproterone acetate) or due to immunological response (type B, so-called idiosyncratic ADRs). Whereas type D toxicities like genotoxicity and teratogenicity may also be related to secondary pharmacological or endocrinological effects, type B and C toxicities involve reactive metabolite formation in most incidences. Chemical reactivity of electrophilic metabolites with cellular macromolecules follows well-understood mechanisms of chemistry and target different amino acids in proteins or DNA bases depending on the nature of the electrophile formed [1]. It is a widely accepted feature of certain functional groups to be susceptible to reactive metabolite formation, mostly during phase I drug metabolism reactions. These so-called structural alerts include aromatic amines (prone to N-oxidation), parahydroxy amines or amides (e.g., in acetaminophen), thiophenes (tienilic acid), or simple functionalities like carboxylic acids (conjugated to reactive acyl glucuronides as for bromfenac) prone to reactive metabolite formation (see later). Covalent alterations of cellular structures or redox cycling of reactive intermediates such as quinines have been implicated in a variety of drug-related toxicities. Direct acting agents such as acetaminophen follow mostly classical dose–response relationships in their disruption of critical cellular functions. Type B toxicities do not necessarily follow classical dose–response curves, although some evidence suggests at least a partial contribution of high doses to the risk associated with idiosyncratic toxicities [2,3]. Toxicities that are largely dependent on individual susceptibility are rare occurrences and hence currently are not amenable to prediction from animals or in vitro data. The involvement of reactive metabolites in drug-induced ADRs has been attributed to numerous cases of postmarketing attrition especially for those involving hepatic toxicity. The underlying molecular mechanisms of drug bioactivation leading to covalent modifications of cellular macromolecules are well documented [4,5,6]. Although the causal link between reactive metabolite formation and clinically manifested ADRs is mainly anecdotal in nature, an early assessment and possible avoidance of bioactivation reactions has become an integral part of the drug discovery and development process. Strategies for the preclinical characterization of bioactivation liabilities of NCEs have been developed in recent decades and are being pursued in many pharmaceutical companies.

19.3 TARGETS AND TOXICITY

Ideally a pharmaceutical compound is designed to interact specifically with one target (although there is some interest in molecules that possess activity towards multiple related targets such as kinases). The interaction with the target/s, whether it is as an antagonist or agonist, may also be associated with undesirable toxicities. A good example for target-organ toxicities that seem to be indistinguishable from the therapeutic target is the therapeutic intervention in Alzheimer's disease via inhibition of γ-secretase [7]. The inhibition of this target seems to be inevitably associated with inhibition of Notch signaling [8,9]. There appears to be little one can do with chemistry, and hence structure-activity relationships (SAR), to avoid this other than, if possible, to optimize ADME characteristics for tissue-selective pharmacological behavior. This approach would direct the compound more toward the target in a specific tissue than to other tissues, in which undesired toxicities play a major role. For therapeutic intervention at the γ-secretase target in the brain, this would mean the optimization of brain penetration after oral intake in a way that avoids potentially problematic interactions with the immune system and the gastrointestinal tract [10,11]. Some other well-known receptors that are known to be associated with multiple aspects of toxicity and yet are particularly important pharmacological targets are the glucocorticoid receptor, the peroxisome proliferator activated receptor (PPAR), and the aryl hydrocarbon receptor (AHR). Structural and computational molecular modeling of the interaction of small organic molecules with these receptors has helped to build good quantitative structure-activity relationship (QSAR) models to systematically rank compounds according to their suggested toxicities via a specific interaction with these receptors [12].

As a consequence the pharmaceutical industry has developed a more or less uniform way of profiling drug-like structures in ADME and toxicity assays to filter them prior to administration to human volunteers or patients. A generic view of the strategies as currently used in the pharmaceutical industry for toxicity testing and ADME screening are depicted in Figures 19.1 and 19.2.

19.4 TOXICOLOGY AND ADME ASSAYS AND THEIR EXPECTATIONS

19.4.1 Approaches for Metabolite Prediction Relevant to Drug Safety

In silico approaches to predict drug metabolism are of particular interest to the pharmaceutical industry as having the potential to impact the early drug discovery process as well as in the candidate selection phase. The identification of metabolic soft spots in a new chemical entity could help improve the hepatic

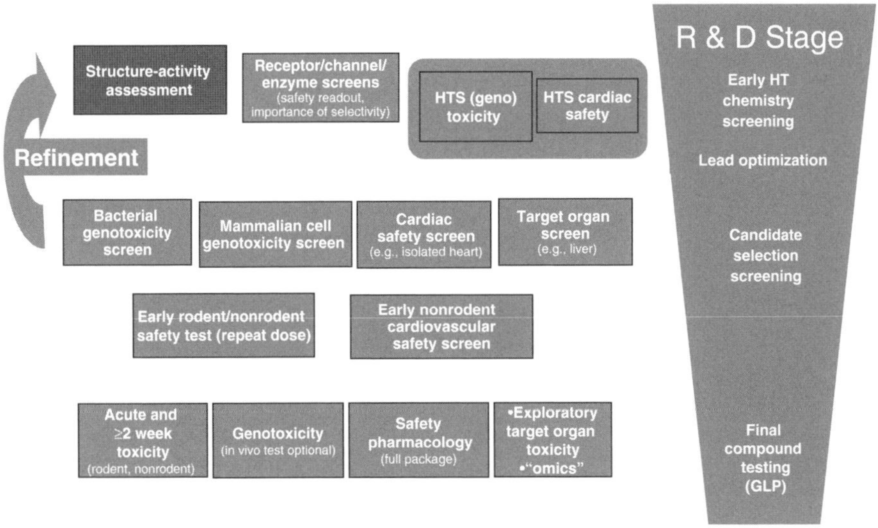

Figure 19.1 Schematic view of the toxicity profiling process in drug discovery and stages of preclinical evaluation prior to going into first human trials (note that the results of in vitro and in vivo toxicity assays are used to refine in silico models).

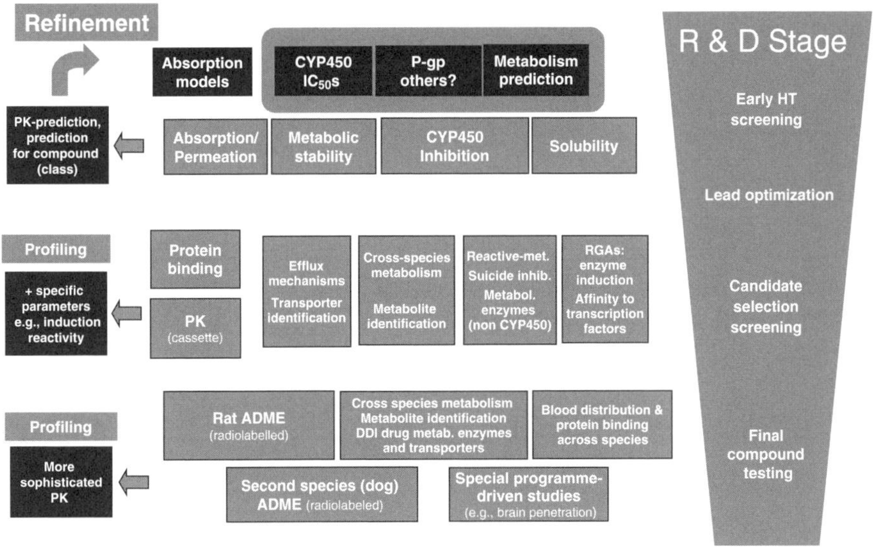

Figure 19.2 Schematic view of the ADME profiling process in drug discovery and stages of preclinical evaluation prior to going into first human trials (note that the results of in vitro and in vivo toxicity assays are used to refine in silico models).

metabolic clearance properties. In addition the identification of potentially toxic structural motifs ("toxicophores") are the two most intriguing applications of in silico tools related to ADME prediction. The apparent advantage of computerized systems to evaluate a large number of chemical entities before chemical synthesis makes these approaches appealing to the drug discovery process. Numerous commercially available tools for the prediction of metabolites exist, such as METEOR, MetabolExpert, and MetaSite (see Chapter 18). Most software packages correctly predict metabolites that are detected experimentally. However, a relatively high incidence of false positive and false negative predictions of metabolites is common to most computerized systems. Whereas the majority of applications add value to the identification of the most probable site of metabolism (metabolic soft spot), a major drawback of these expert systems is that the relative formation rates of metabolites or absolute abundance of metabolites cannot be predicted with these methods. The false positive prediction of metabolites might be attributed to the fact that most software packages cannot account for rate constants of metabolite formation and biological processes such as tissue distribution. Thus in the hand of the drug metabolism expert these software packages have a certain value in guiding the investigators to experimental approaches for the identification of drug metabolites. Still the false negative prediction of drug metabolites remains a major drawback. However, the generation of additional new local rules specific to a particular company may improve the predictive power of some of these applications such as METEOR [12,13].

Preclinical tools for the assessment of metabolism with regard to reactive intermediate formation are applied to assess liabilities associated with covalent binding. This process might also help establish SAR interrelationships for such liabilities at least for local systems and allows medicinal chemists to find compounds with improved reactive metabolite formation that can be supported by early metabolite identification efforts. Although isolated cases of successful predictions of reactive metabolite formation and adverse drug reactions exist [15,16], the most commonly used approach would include the recognition of structural alerts associated with reactive metabolites formation. This *in cerebro* approach builds on the expertise of medicinal chemists and drug metabolism specialists to prioritize compounds for testing in the appropriate in vitro tools for the characterization of metabolic liabilities.

Higher throughput in vitro methods such as the Cytochrome P450 time-dependent inactivation assay and the screening for drug-glutathione adducts have emerged over the last decades and made it possible to address bioactivation liabilities experimentally early in drug discovery [13,17,18,19,20,21]. Still the most promising approach for the development of safe and efficacious new drugs is the discovery of highly potent and selective molecules because many adverse drug reactions involving reactive metabolite formation have been attributed preferentially to those medicines that are used at higher doses [3,22].

19.4.2 Drug Metabolism Studies Supporting Drug Safety Evaluations

The main goal of drug metabolism for the body is to eliminate potentially harmful xenobiotics via urine and/or bile. This is typically realized in a stepwise process, the often lipophilic drug molecules are metabolized to mostly inactive, nontoxic and more hydrophilic products that can then be readily excreted into urine or bile [23]. In some cases, however, drug metabolites might be toxic or represent activated products (e.g., acyl-glucuronides) that can potentially lead to organ toxicity or idiosyncratic/immune-mediated toxicity. Therefore the extensive knowledge of drug metabolism cannot be underestimated for the overall understanding of the pharmacological and safety properties of new drug molecules [24]. Exposure to parent drug and metabolites during the course of an experiment is key to the interpretation of toxicological findings. Emphasis is generally placed on the qualitative similarity in metabolite profiles between humans and the toxicological species in order to ensure that both are exposed to the parent drug as well as the same metabolites, any of which may contribute to toxicity.

19.4.3 In vitro Drug Metabolism Studies Guiding Early Drug Safety Evaluations

During preclinical development the metabolite structures of potential new drug candidates are typically identified via in vitro tools such as subcellular liver fractions (microsomes, S9), expressed human enzymes, or cellular systems (e.g., hepatocytes, tissue slices) [25]. The metabolism observed in the different systems is compared across species, with the aim that all the metabolites formed in human in vitro systems can also be identified in the systems originating from the animal species used in major toxicity studies.

The metabolism might show considerable quantitative and qualitative species differences in the in vitro systems used, as exemplified in Figure 19.3.

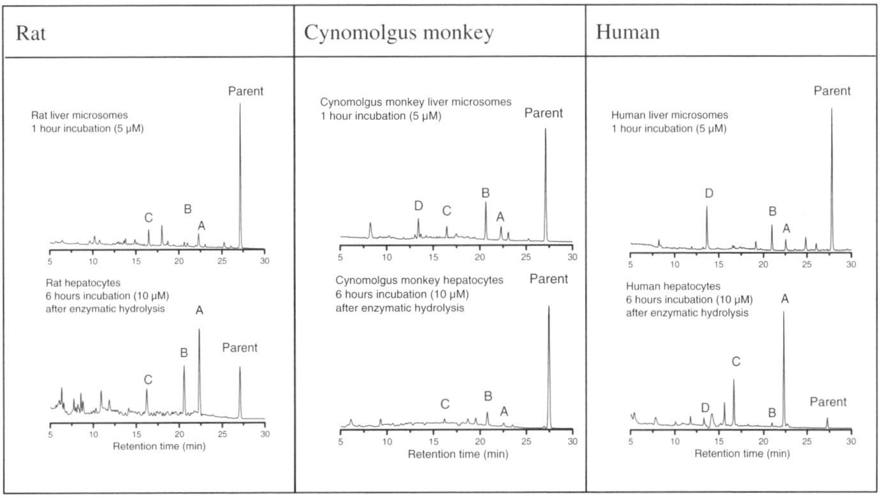

Figure 19.3 In vitro metabolite patterns obtained by incubation of liver microsomes and hepatocytes from rat, cynomolgus monkey, and human with a test drug. The identity of parent drug and metabolites A, B, C, D, were confirmed by LC-MS/MS analysis.

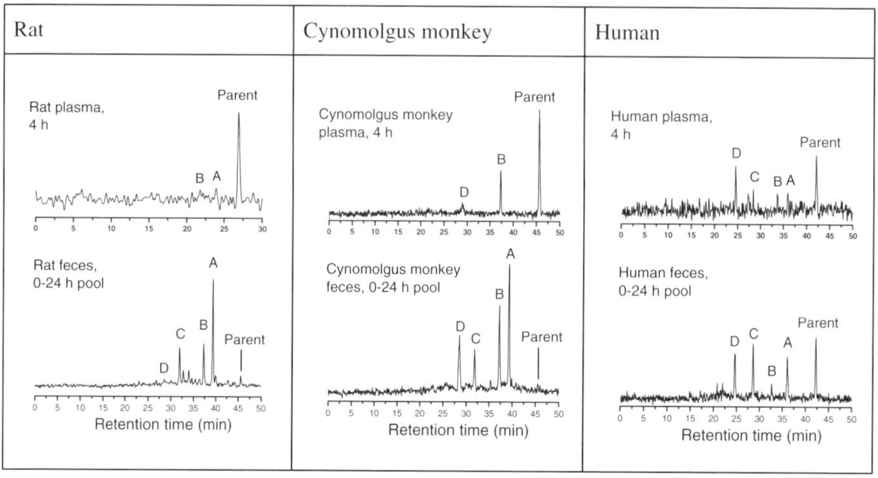

Figure 19.4 Metabolite patterns in plasma and feces collected from rat, cynomolgus monkey, and human excretion balance studies with a test drug (same as Figure 19.3). The identity of parent drug and metabolites A, B, C, D, were confirmed by LC-MS/MS analysis.

While primary hydroxy metabolites A and B of a test drug were formed in all species, the further oxidized metabolite D was mainly observed in cynomolgus monkey and human microsomal incubations but not in hepatocytes. Based on such in vitro data an extrapolation of the relevance of these metabolites to the in vivo situation (shown for this test compound in Figure 19.4) is difficult as is an in silico prediction of such differences.

19.4.4 In vivo Animal Models Used for ADME Studies in Support of Drug Safety Evaluation

The plasma exposure of a parent drug and major metabolites in the main toxicological species along with any related interspecies differences are of special interest for interpretation of toxicity studies. It is, however, only the data from the human ADME study with radiolabeled material that allows the final establishment of safety margins for all the relevant metabolites, as only then the systemic availability of all metabolites formed in human, is then known. Special attention has to be drawn to major human metabolites, which account for a considerable amount of the exposure (AUC) relative to the parent drug, and human-specific metabolites (Baillie [26]). For the human-specific metabolites and major metabolites that do not reach comparable systemic exposure in at least one animal species used in the different toxicity studies, separate toxicity studies should also be considered [26].

Major metabolites in plasma are not necessarily the major metabolites overall, as both the rates of formation and elimination as well as other kinetic parameters are important factors. An example of species-specific differences in the exposure of major metabolites is outlined in Figure 19.4. The major part of the radiolabeled dose of the test drug was equally eliminated into feces in both human and cynomolgus monkey. The oxidative metabolite D was formed in addition to other oxidative metabolites in both species to a similar extent (~15% of the dose) based on the analysis of pooled feces samples. However, this metabolite was only a minor peak in cynomolgus monkey plasma, while in human it reached systemic plasma exposures comparable to or exceeding that of the parent drug. Species-specific differences in the rates of elimination of this metabolite might be one reason for this difference, which was only apparent once human plasma samples were analyzed quantitatively for these metabolites.

19.5 EXPECTATIONS FROM "IN SILICO" MODELS ON TOXICITY PREDICTION

19.5.1 General Considerations on Feasibility

Effects that are buried in the chemical structure or are related to the metabolic conversion of it should be generally amenable to in silico prediction. This may be particularly true for drug-like structures that occupy the chemical space as defined by the so-called Lipinski rule of five [27]. Hence, if enough knowledge is generated on typical representatives of the chemical space, then the properties, including unwanted toxicities, of other, not yet tested, representatives of that space within this chemical space should be predictable. There are two general obstacles to this: (1) many new pharmacological targets require chemistry (e.g., high lipophilicity) that violates the Lipinski rules; (2) a measure of "coverage" is required from the SAR tool used.

19.5.2 Availability of Gold Standard Databases

Predictability levels are determined by the ability to compare unknown compounds against a set reference database of known toxicities, also referred to as a training set. The availability and, most important, the quality of large training sets (e.g., gold standard databases) are indispensable prerequisites for the development of predictive in silico tools. There are content vendors, service companies, individual pharmaceutical companies, as well as regulatory agencies working on the daunting task to compile and organize relevant datasets for all relevant toxicological endpoints. Investment in the future development of such databases is not likely to be by individual pharmaceutical companies, but by consortia. There are also public initiatives, mainly by the government and consortia, to build these content databases. These efforts seem to be focused so far on convincing pharmaceutical companies to deposit their data from past projects in a public database; however, this implies a number of patent and legal issues.

Genotoxicity/mutagenicity, carcinogenicity, primary irritation, and skin sensitization are the most well-covered toxicological endpoints as standardized assays for their assessment were established several years ago. There are various databases of such data that can be freely accessed via the internet and a selection of these is listed here (also see Chapters 5 and 6):

- TOXNET, a cluster of databases on toxicology, hazardous chemicals and related areas by the National Library of Medicine (CCRIS, GENETOX, DART, ITER, IRIS, and HSDB) (*http://toxnet.nlm.nih.gov/*).
- The US National Toxicology Program (NTP) has converted study reports into an electronic format which can be accessed freely from the Web site (*http://ntp-apps.niehs.nih.gov/ntp_tox/index.cfm*).
- Distributed Structure–Searchable Toxicity (DSSTox) public database network by the US EPA, as well as GeneTox and ECOTOX [28].
- HazDat database by the Agency for Toxic Substances and Disease Registry's (*http://www.atsdr.cdc.gov/hazdat.html*).
- Genotoxicity Database of Environmental Chemical Substances ("The Mutants" database sponsored by Dr. Motoi Ishidate) consisting of a summary of genotoxicity/mutagenicty data of about 2000 different chemical substances (*http://members.jcom.home.ne.jp/mo-ishidate/*).
- The Carcinogenic Potency and Genotoxicity Database (*http://potency.berkeley.edu/*).
- The Carcinogenicity and Genotoxicity eXperience database (CGX data base at *http://www.lhasalimited.org/*). [29]

Such databases have enabled computational scientists to develop global QSARs for these endpoints. As there is an increasing demand for fast and efficient ways to access such information commercial structure integrated databases and management systems have been developed by different vendors. VITIC is a database that stores toxicological information in a structure and substructure searchable format and enables analysis of data and extracting previously unknown relationships (see later) [30]. Leadscope Inc. has recently released the FDA Genetox Databases of the Center for Food Safety and Applied Nutrition (CFSAN) with around 500 chemical structures and of the Center for Drug Evaluation and Research (CDER) containing genetic toxicity information extracted from pharmacological reviews contained in New Drug Approvals (NDA) (*http://www.leadscope.com/fdadb_cat.php*). The Elsevier MDL toxicity database is a structure-searchable bioactivity database of toxic chemical consisting of general chemical information, six categories of toxicity data and references to the original publications and to relevant review articles (*http://www.mdl.com/products/predictive/toxicity/index.jsp*). RTECS (Registry of Toxic Effects of Chemical Substances) created by NIOSH is a compendium of data extracted from the open scientific literature and is maintained and marketed by MDL (*http://www.mdl.com/products/predictive/rtecs/index.jsp*).

In general, commercially compiled databases like VITIC, Leadscope and MDL toxicity databases, which underwent a thoroughly conducted quality check, can be used as a basis to establish new QSARs. The chance of success is even higher if in-house data are used that have been conducted over several years by the same or only a small number of different laboratories applying stringent and standardized test protocols.

It has to be emphasized that comprehensive, high-quality datasets for more complex toxicological endpoints like organ toxicities (e.g., liver, kidney), cardiac safety, and teratogenicity are still not really available. Extraction of all relevant data from different sources and structured storage to enable automated data mining and analysis must be the first step. This would be crucial for any further progress in the field of in silico toxicity predictions.

19.5.3 Global SAR/Rule-Based Models

Generally applicable structure-activity relationships (SARs), so-called global SARs/models, have been developed using a noncongeneric set of chemicals encompassing a number of different biological mechanisms. Several commercially available toxicity prediction systems like DEREK (Deductive Estimation of Risk from Existing Knowledge) and MCASE (Multiple Computer Automated Structure Evaluation) are available and regarded as global models (see also Chapters 8 and 18). They are typically developed from training sets based on data taken from public sources. They were not limited to predicting within a specific chemical structural class.

Global SAR systems have enabled the pharmaceutical industry to move all of their predictive toxicology tests as early in discovery as possible. Global SARs are validated with a large training set of diverse structures and can therefore be used before any wet laboratory results. Expert systems like DEREK and MCASE offer the capability of screening thousands of virtual compounds. This has shifted drug companies' efforts toward applying these new dry biology technologies in early lead optimization.

19.5.4 Local SAR Rule and Descriptor-Based Models

In contrast to generally applicable global SARs, "local" SARs are valid only within a limited, congeneric dataset, thus constraining predictive space and value, and they are generally used later in the development process. In general, if certain compound data are used to make a predictive model, that model only works in the same chemical space as those compounds. Therefore, when changing to a new chemical space, the model needs to be rebuilt. For a specific drug discovery project, a local model can be very valuable, since the predictive values within a homologous data set are often very high in order to assist the discovery chemists in selective improvement of synthesis planning.

There are a number of commercial QSAR modeling systems available that can help scientists establish reliable QSARs and structure-property relationships (QSPRs). Such tools analyze the relationship between structural descriptors and bioactivity. Clustering and similarity methods can then be used to identify outliers and select the appropriate models for the prediction of newly synthesized structures within a covered chemical space.

To facilitate easy and quick generation of local QSARs, an in-house QSAR tool has been developed within Hoffman-La Roche. This tool allows all project teams to build a local model. Local SARs have been developed for several projects using hERG assay data and the in vitro screen for chromosomal damage in mammalian cells. These local SARs can also be applied in a more general sense.

It is best to apply QSAR analysis as early as possible to measured wet laboratory results, although the generation of many data points is not as resource intensive when using high-throughput screening. Correlations of measured in vitro activity with molecular descriptors can normally be established in a narrow chemical space starting with 10 data points or fewer as demonstrated in many early Hansch-type QSARs. The continuous re-feeding of measured parameters in the model normally leads to a very effective optimization of the model with additional chemical structures.

19.6 SAFETY PHARMACOLOGY ASSAYS AND THEIR EXPECTATIONS

It has been estimated that about 75% of acute ADRs in humans can potentially be predicted by primary, secondary, and safety pharmacology studies [35]. To protect healthy volunteers and patients, pharmaceuticals intended for human use are required to be investigated for their "potential undesirable pharmacodynamic effects . . . on physiological functions in relation to exposure in the therapeutic range and above" [36]. This includes functional and affinity-based central nervous system and cardiovascular profiling, respiratory studies, and investigations on renal and gastrointestinal function as main aspects. Target selectivity of lead candidate molecules helps to rule out well-known mechanisms of ADRs. However, unwanted pharmacodynamic properties of unknown or mixed modes of action are currently difficult to predict in silico, and this requires continued in vitro and in vivo safety testing.

19.6.1 Cardiovascular System (Cardiac Channels)

In recent years proarrhythmic properties of some noncardiovascular drugs received particular regulatory and pharmaceutical industry attention. It has been recognized that the highest frequency of drug withdrawals from the market is attributable to a single adverse drug reaction: fatal ventricular tachyarrhythmias or torsade de pointes (TdP) type [37,38]. The FDA safety database for 1969 to 1998 lists 2194 cases of which 27.9% were life-threatening, 16.2% associated with serious condition, and 9.8% with fatal outcome [39].

Overall, the rate of observed TdP varies dramatically from 1 in 1000 to 1 in 100,000 patients [40]. By definition, the underlying cause for the development of TdP is a delayed cardiac repolarization, which can be determined as the prolongation of the QT-interval on the surface electrocardiogram (ECG) [41]. Meanwhile it is well established that the mechanism is the same as for the therapeutic effect of class III antiarrhythmic agents. In fact most, if not all, of the non–anti-arrhythmic agents associated with the liability to induce TdP prolong the QT-interval with the same mechanism, namely a block of the potassium current conducted by the channel encoded by the human ether-a-go-go-related gene (hERG) [42] that conducts the rapid component of the delayed rectifier potassium current (I_{kr}, see also Chapters 13 and 16). Although the relationship between hERG block, QT-interval prolongation, and the occurrence of TdP is weak [40], dedicated regulatory guidelines that demand pre-clinical and clinical investigations of the potential to delay ventricular repolarization were issued recently [43,44].

Because of the high degree of uncertainty about whether hERG blocking activity of a molecule translates into a proarrhythmic potential downstream in the development process, the challenge for the safety pharmacologist in concert with the medicinal chemist is to weed out as early as possible such molecules of reduced value that might fail in clinical development due to QT-interval prolongation. As long as even a small QT-interval prolongation is considered to be a warning signal, and any conclusion about a drug being safe regarding the real endpoint (i.e., TdP) cannot be achieved prior to marketing [40], molecules with this property bear an increased risk of failure during development and in the regulatory approval process. Clearly, the down side of this strategy is that potentially valuable compounds will be eliminated prior to proof that such block of hERG with or without prolonging of the QT-interval ultimately translates into a real proarrhythmic risk in patients.

In the past costly and time-consuming patch-clamp techniques constituted the gold standard to evaluate a drug's propensity to block hERG. The advent of semi-automated and automated patch-clamp tools has increased the throughput significantly [45]. However, for a strategy that attributes importance to the removal of potential hERG blockers from the development pipeline across all therapeutic areas there is great need for computational models to support medicinal chemists in their efforts to optimize lead molecules. In this effort across the industry the challenge remains to minimize the potential for hERG channel block but not at the expense of discarding therapeutically innovative drugs [42]. The latest example that contradicts the latter statement is the labeling of ranolazine even though blocking hERG and prolonging QT-interval proved negative in numerous preclinical studies on any proarrhythmic potential [46]. Also it has to be appreciated that there are QT-interval prolonging drugs on the market that do not induce TdP (e.g., ziprasidone). The resulting dilemma for the pharmaceutical industry and regulators is that almost all drug classes were associated with drugs that showed repolarization delay [47].

19.6.2 Expectations from "In silico" Models for Cardiac Channel Safety Pharmacology

The value of computational tools arises from their applicability early in development. An excellent correlation with wet laboratory data, an easy to use and interpretable model, high sensitivity, as well as high specificity are key requirements for a useful in silico model. As a nonexpert tool it should be available to the medicinal chemist via computer networks. Ideally such potentially powerful tools can be used to predict liabilities to induce an adverse drug reaction but also to guide the chemists to structurally modify the molecules via discovering the features that prevent the binding to the ion channel and immediately verify any successful chemical optimization step.

The ultimate goal nevertheless is to model effects of investigational drugs on complex tissues (e.g., the heart) or even on organisms. So the real endpoint is whole tissue and organ modeling to simulate TdP instead of hERG block. The first attempts at approaches for systems biology modeling in cardiac safety appear promising [48,49,50], and with the increasing computational capacity similarly complex functions might be modeled in the future.

19.6.3 General Considerations on Feasibility for Cardiac Channel SAR

To date, most of the models used to understand the structure-activity relationship (SAR) of hERG that have been reported were developed based on literature data [51,52,53]. However, variable hERG IC_{50} values obtained from the literature are a major drawback. The best approach to gaining reliable training sets for the generation of in silico models is to use in-house in vitro data that have been elaborated with the same cells and according to a standard test protocol. Hence in the last few years computational tools have supported the optimization of compounds from a specific chemical class with a small amount of experimental hERG data. These tools can have a significant impact on the avoidance of hERG inhibition, or at least they alert the chemists early if the SAR for the intended pharmacological target and hERG inhibition overlap. Models can already be generated with a small training set that ensures flexibility and speed in drug discovery.

19.6.4 Availability of Gold Standard Databases for Cardiac Channel SAR

As indicated previously, the current problem in early in vitro profiling is the interpretation of findings regarding ADR's. BioPrint®, a commercially available database combined with a pharmacoinformatics tool that contains ADR information for 2500 molecules and links in vitro binding results (including hERG binding data for a subset of molecules) with 940 different unwanted side effects [54]. The great value of this database lies in the inherent human experience for most of the drugs. Qualitative as well as quantitative information offers a huge source that can be used to understand and relate binding information to ADRs and link the potency to a probability that a certain ADR will become apparent in the clinical setting. This database is considered to be valuable for drug safety and for other applications throughout the development process. Some other well-established expert tools contain only limited safety pharmacology data. The VITIC database also contains hERG data (*http://www.lhasalimited.org/*) and substructure queries are possible to identify hERG liability and/or to obtain hERG IC_{50} values.

MCASE and DEREK are still in their infancy regarding hERG queries. For both systems their own rules can be developed to create modules that, upon entering substructures, will generate alerts [55]. Since most of the larger pharmaceutical companies possess huge internal databases, it is expected that internal data-sharing initiatives will evolve with more searchable tools to complement commercial databases.

19.6.5 Global SAR or Rule-Based Models for Safety Pharmacology for Cardiac Channels

Since a global model that fits all drug-like structures would be useful to predict any interaction with the hERG ion channel, various global approaches have been investigated [51,52,53,56,57]. All these models have contributed to an understanding of the complex interaction between ligand and hERG. However, in our opinion, none of these models have proved to be as globally applicable as would be necessary to become a versatile tool for early preclinical cardiac safety assessment. It also has become apparent that such models work satisfactorily if the training set of structures is more or less similar to the test set.

Another disadvantage is the size of the training set required to obtain a useful model. Reports range from 15 [58] to 1979 compounds [59] used build a hERG prediction model. For all the QSAR approaches that were established based on a limited dataset taken from the literature, the reliability needs to be tested on a big database of homogeneous biological data not generated in different laboratories with unique protocols. Obviously the complexity of mechanisms by which chemicals can interact with this ion channel has proved it very difficult to construct global models. Even with a training set as large as 1979 compounds to establish a model, a smaller set of reference compounds was correctly classified at 87.9%. However, a considerably and potentially inappropriately high cutoff IC_{50} value of 30 μM was chosen to define positive and negative hERG blockers by ignoring the fact that margins of safety might be such that drugs are safe with lower IC_{50} values and unsafe with higher IC_{50} values. By this approach, since these concentrations are not achieved under therapeutic conditions, many safe compounds would be eliminated, although they might not induce QT-interval prolongation. We favor a ratio between an efficacious concentration at the target and the activity at the hERG channel that should exceed 30-fold. Any chemical exceeding this ratio is considered to possess a good cardiac safety margin [35,37] provided that at that early development stage this conclusion bears some uncertainties. No details on whether the molecules of the training set were positive under clinical conditions have been made available.

Site-directed mutagenesis together with homology models based on crystal structures of related potassium ion channels have provided an important molecular insight into interactions between drugs and key binding site residues [60] (see Chapter 16). However, identification of these key binding sites has not yet provided a global knowledge for the prediction of potent binding to the hERG channel by all different molecular chemotypes. This limits the use in preclinical safety practice where huge numbers of structurally diverse molecules are investigated.

Roche et al. [57] applied a variety of techniques to model hERG activities of 472 drug-like compounds. Still 29% of the actives were not predicted correctly, which resulted in a termination of this approach due to a poor correlation. Even though some of the models didn't provide the predictive accuracy that is required to be used in lead optimization, some of them can still be of use if the size of large virtual libraries has to be reduced. This might increase the likelihood that a larger portion of molecules with improved hERG blocking properties will enter the lead identification phase.

Yap et al. [61] generated a predictive model for TdP. However, the mechanism behind the development of TdP is still poorly understood, and therefore any prediction needs to be considered with caution. The accuracy to identify TdP+ molecules was 97.4% and for TdP– molecules 84.6%, suggesting that this tool has value for the prediction of TdP. We have no data yet on its developmental use. Gepp and Hutter [62] also modeled TdP drugs with a training set of 264 molecules. The applicability is limited because 123 and 124 descriptors were applied, which makes interpretation very difficult. Interestingly, if a tertiary or secondary amine is used, followed by an aliphatic carbon that is separated by one, two, or three atoms (but not oxygen from an aromatic carbon), 71% of the molecules can be identified correctly.

The impact of small chemical modification on the potential of a compound to interact with amino acids in the hERG channel can be nicely depicted by the fundamental differences between terfenadine and its metabolite fexofenadine. In hERG expressing mammalian cells in vitro, terfenadine displays an IC_{50} of around 200nM while fexofenadine has an IC_{50} of around 13,000nM, showing a nearly 100-fold difference. Apparently the carboxyl function in fexofenadine effectively shields the positive charge on the otherwise freely accessible protonated N^+ in terfenadine. A visual representation of this mechanism is displayed in Figure 19.5 with an ab initio calculation of the likely molecular conformation of terfenadine versus fexofenadine under physiological conditions.

19.6.6 Local SAR Rule and Descriptor-Based Models in Cardiac Channel Safety Pharmacology

A more pragmatic approach that makes it easier to perform QSAR with a limited number of relevant descriptors is currently considered as the right

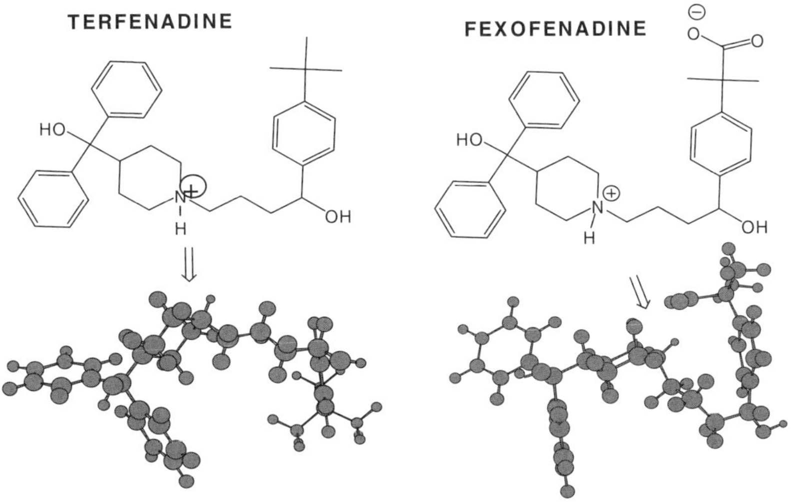

Figure 19.5 Chemical structures of terfenandine and its metabolite fexofenadine displayed traditionally and by ab initio calculation (note that the ab initio calculation demonstates the shielding of the positive charge on the N^+ by the carboxyl function in fexofenadine whereas the positive charge is freely accessible for hERG channel protein interaction on terfenadine).

balance between cost and benefit (see also Chapter 20). Overall, in our experience, physicochemical properties played an important role for 80% of the structures investigated while the decisive substructures were clearly identified in only 20%. Limiting the descriptors to experimental p*K*a, log *D*, *C*log *P*, hydrophobic surface and volume, PSA, amphiphilicity, molecular shape parameters, and specific structural moieties proved to be successful when QSARs were developed. The drawback is that each model is project specific, meaning a training set of 5 to 10 hERG values needs to be generated. However, once established such a tool can be made available globally for ease of use by the medicinal chemists. Spot-checking by measuring hERG effects experimentally is recommended to confirm that the QSAR model is still valid for a particular series and project. In a feedback process the accuracy of the model should be constantly re-appraised by the project team.

19.7 SUCCESS/FAILURE EXAMPLES WITH COMPUTATIONAL APPROACHES IN THE DRUG DEVELOPMENT PROCESS

19.7.1 Genotoxicity

Currently better predictions can be obtained for mutagenicity, carcinogenicity, skin sensitivity, and primary irritancy than for hepatoxicity, cardiotoxicity,

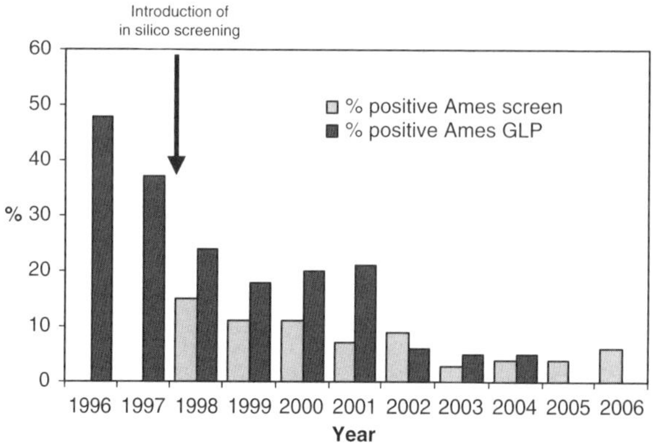

Figure 19.6 Decrease of percentage of positive results in the Ames test by applying an in silico pre-filter in Hoffmann-La Roche Ltd.

kidney toxicity, and neurotoxicity. This is because decisions in toxicology are subjective by nature and people understand better the former types of toxicities.

Prediction of genotoxicity is one of the success stories within the area of QSAR tools, mainly because there is a large amount of reliable data available (see also Chapter 14) and the mechanisms of direct-acting mutagens are closely linked to the reactivity of the molecule. In particular, the Ames assay, which is required for pharmaceutical registration worldwide, is widely used as a predictor of potential mutagenicity with high concordance values. The impact for drug development is obvious. A positive Ames result could be a major hurdle for drug development candidates, depending on the indication. Use of a rule-based expert system like DEREK for Windows, which can be adapted to in-house rules, is summarized in Figure 19.6. Over the last 10 years the percentage of positive Ames tests has been reduced significantly. Even the screening version of the Ames test, which is normally conducted in a very early project phases, is yielding currently about 5% positives per year. The prediction of *Salmonella* mutagenicity by an in silico approach is regarded as reliable enough as an initial mutagenicity assessment, as long as no "wet laboratory" data are generated.

Several structure-activity investigations during the development process have led to the development of new "rules" for the expert systems. During the development of a CNS drug it became apparent that all compounds with allyl groups adjacent to an oxadiazole ring with a particular arrangement of the heteroatoms were mutagenic in at least one *Salmonella* strain. Compounds containing allyl groups in the side chain and an ON–arrangement in the aromatic heterocycle are stronger positives than NO–arranged heteroaromatics (Figure 19.7) [64]. A new DEREK rule for the mutagenicity of 3-aminomethyl-1,2,4-oxadiazole has been integrated in the rule base.

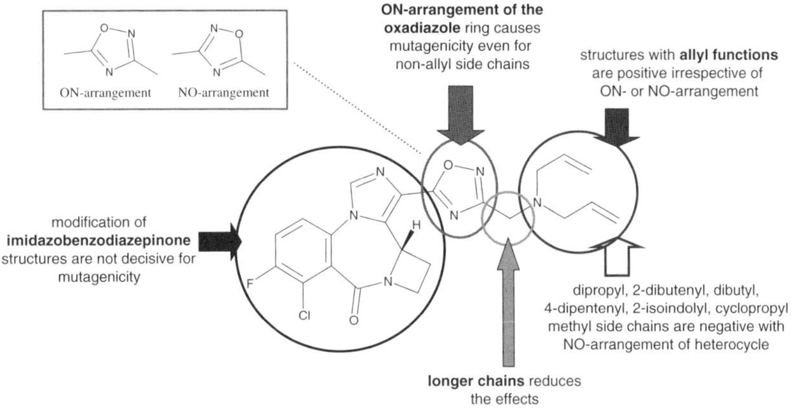

Figure 19.7 Structure-activity relationship (SAR) of 1,2,4-oxadiazol-CH2-*N*-allyl derivatives in the Ames test.

A structure-activity relationship investigation with a number of serotonin receptor ligands was undertaken to search for compounds without mutagenic liability in the Ames test. For several three-ringed heterocyclic structures, increases in the number of revertant colonies were observed in strain TA1537, and evidence was provided that the observed genotoxic effects are strongly influenced by the intercalating properties of the compounds. The highest mutagenic response was seen with a compound possessing a central aromatic ring. The mutagenic activity of the naphthaleno-derivatives appeared to be stronger when compared with the indeno-compounds, probably because of the less curved structure. Dimethyl-substitution of the indeno-substructure proved to reduce the intercalating ability of the compounds and lead to loss of mutagenic activity (Figure 19.8) [65]. Consequently a clear SAR was established that could be translated to an in silico rule. Such rules implemented in the prediction tools would assist discovery chemists in avoiding synthesis of new structures with known liabilities.

Concordance values for genotoxicity prediction of trained QSAR tools are normally between 85% and 95%. Depending on how the expert systems are trained, the main failures are structures incorrectly predicted to be positive (oversensitive) or mutagens that are not identified by the programs (low sensitivity). The latter can be easily related to the difficulty in translating clastogenic events into simple rules. Nearly all mutagenic compounds that are classified negative by the systems showed effects only in clastogenicity tests, such as the in vitro micronucleus and the chromosomal aberration tests. In contrast to direct-acting mutagens, there are various mechanisms known to lead to clastogenic readouts in the in vitro tests. Cell cycle disturbance, imbalance of nucleotide pools, and aneugenic activity are only a few such examples. Therefore attempts to improve these tools must take into account the various underlying mechanisms. Predictions that combine distinctive mechanisms of

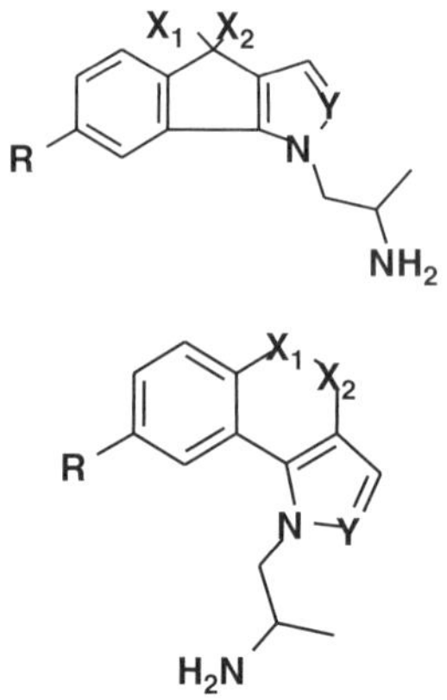

Figure 19.8 SAR of indeno-/naphthaleno-pyrroles/pyrazoles in the Ames test. The mutagenic activity of the naphthaleno derivatives are stronger compared to the indeno compounds probably because of a less curved structure. The mutagenicity of the indeno compounds is dependent on the absence of a dimethyl substitution at position 4. All compounds without dimethyl substitution are mutagenic, whereas those compounds with a dimethyl substitution are nonmutagenic. It can be assumed that the dimethyl substitution at position 4 strongly reduces the intercalating properties of the compounds. Pyrazole analogues of both, indeno and naphthaleno structures appear to proclude stronger mutagenic responses than the pyrrole derivatives.

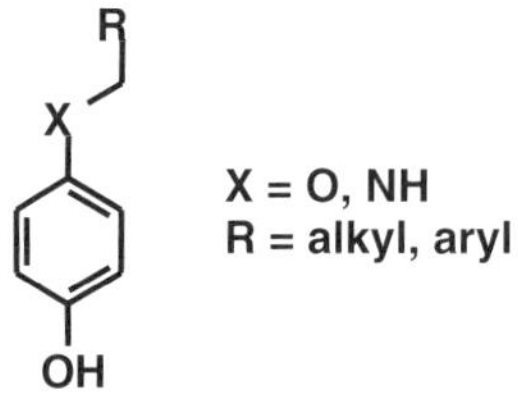

Figure 19.9 SAR of p-hydroxy phenylamino (phenoxy) ethyl derivatives in the mouse lymphoma *tk+/–* assay and the chromosomal aberration test. The para-hydroxy phenoxy(phenylamino) moiety of the molecules is responsible for the observed in vitro effects. Nitrogen enhances the mutagenic effect while substitution of oxygen or nitrogen in a para-position to the hydroxy group by a carbon or a sulfonyl-group abolishes mutagenicity. The mechanism may be based on radical formation via a semiquinone-type species and generation of reactive oxygen species. The latter mechanism being confirmed by the Comet assay used in combination with formamidopyrimidine glycosylase to specifically determine oxidative DNA base damage.

action could lead to major improvements in genotoxicity predictions. Already this has been observed from the clastogenic activity of para-hydroxy phenoxy (phenylamino) compounds in the chromosomal aberration in vitro as well as from the mouse lymphoma test (see Figure 19.9) [66].

Phenylamino-derivatives are even more potent clastogens then the phenoxy compounds. Substitution of the oxygen or nitrogen in a position para to the hydroxy group with a carbon or sulfonyl group completely inhibits the observed mutagenic response. The structure-activity relationship, together with the extreme and extended exposure conditions, did not lead to the same effect that was observed in vivo. This suggests a mechanistic hypothesis, that a radical mechanism via a semiquinone-like structure producing reactive oxygen species (ROS) is responsible for the mutagenic/clastogenic effects in vitro [66]. In summary, if the clastogenic activity could be based on a distinct mechanism, then a clear-cut SAR could be established. Regarding the performance of the computational programs DEREK, TOPKAT, and MCASE to predict genotoxicity, Snyder [67] suggests that "limitations are primarily a consequence of incomplete understanding of the fundamental genotoxic mechanisms of non-structurally alerting drug rather than inherent deficiencies in the computational programs."

One remaining problem in the knowledge base of DEREK is the discrepancy between mutagenicity and carcinogenicity predictions. That is, while mutagenicity rules have been continually refined over the last few years, carcinogenicity rules have remained too generic (e.g., polycyclic aromatic carbons, aromatic amines). As a consequence structures are alerted for carcinogenicity without a mutagenic potential. This leads us to the widest gaps within genotoxicity prediction failures in the structural class of aromatic amines/amides. There are a number of publications currently trying to explain the divergent results of aromatic amines/amides by the electrophilicity of aromatic compounds or the extent of the aromatic system and substituents with electron-donating and electron-withdrawing properties [68,69,70]. A consortium of pharmaceutical companies is also attempting to improve the prediction by compiling a comprehensive dataset of genotoxicity in vitro results following treatment with aromatic amines/amides (personal communication). For the time being, it seems that there is agreement over a basic set of structurally alerting groups (Figure 19.10) [71] that generate mutations in bacterial systems (Ames test), and many representatives of this group have been tested positive for tumor induction in a rodent model.

19.7.2 Phototoxicity

Phototoxicity or chemical phototoxicity is the term used for an acute reaction that can be induced by a single treatment with a chemical and UV or visible radiation. The basic mechanism of phototoxicity can be described as an increase in toxicity of a chemical induced by exposure to UV or visible radiation. Therefore the phototoxic potential of a chemical can be measured as an increase in cytotoxicity after exposure to UV or visible light. The 3T3 NRU PT test is a validated and robust in vitro phototoxicity test according to the criteria laid down by the ECVAM Workshop on practical aspects of the validation of toxicity test procedures and the conclusion of the ECVAM/COLIPA validation study [72]. Owing to the convincing performance of the 3T3 NRU PT test, the test is now established and in use in industry laboratories to screen

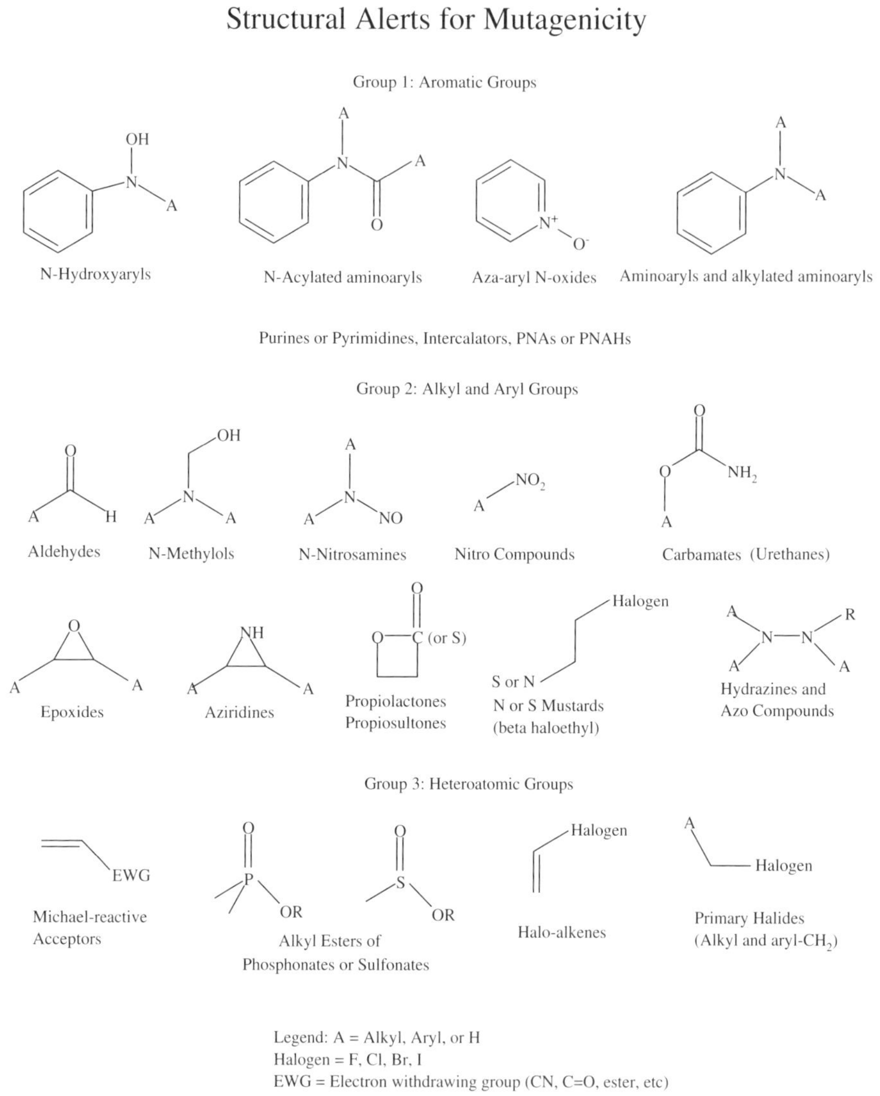

Figure 19.10 Some examples of structurally alerting functional groups that are known to be involved in reactions with DNA [71].

for phototoxic potential and for regulatory purposes. Therefore the development of an in silico prediction tool for the 3T3 NRU PT test would help avoid phototoxic liabilities at a very early project phase.

Since there is a causal link between the chemical structure and UV absorbance that is a prerequisite for photo-induced toxicity, the scientific rationale for a successful global model is obvious. An in silico tool, combined with UV spectra, has been developed with a high-quality database of approximately

700 in-house in vitro results. The concordance value of this combined system is significantly higher (95% vs. 85% compared with the in silico tool alone). The number of in vitro phototoxicity screens could be significantly reduced by implementing the combined system in the development process. Detailed analyses of the information from the QSAR model could have an impact on future phototoxicity safety guidelines regarding the threshold of UV absorption or relevant wave length for photo-induced toxicities. Since phototoxicity seems to be intrinsically linked to photogenotoxicity [73,74], prediction for phototoxic behavior can further help in early assessment of potential liabilities associated with photogenotoxic effects of small molecules.

19.7.3 Immunotoxicity/Covalent Binding, Hepatotoxicity

Covalent binding of drugs to critical cellular macromolecules has been implicated in the toxicity of a number of drugs, although the molecular mechanism underlying the toxicity has not been clearly identified. Sometimes the drug itself is protein reactive, but in most cases it is the reactive metabolite that binds to the cellular target (e.g., CYPs and other enzymes or proteins). The reactive metabolites are generated by a number of drug-metabolizing enzymes, including cytochromes P450, UDP glucuronosyl transferase, sulfotransferases, and peroxides. Once generated, electrophilic metabolites react with nucleophilic target sites and form covalent adducts. Not every covalent interaction between a drug or drug metabolite and cellular proteins necessarily leads to toxicity. It has been generally recognized that there is a correlation between the occurrences of covalent protein adduct formation and certain forms of hepatic toxicity. Covalent adduct formation of reactive metabolites with liver proteins could lead to a number of downstream events that may be implicated in liver injury. These include (1) direct toxicity by binding to or inactivating a specific protein critical for cell survival, (2) direct toxicity by binding to a large number of proteins resulting in disruption of cell homeostasis, (3) covalent modification of protein leading to the formation of a neo-antigen and possibly to an immune response, and (4) activation of a signaling cascade. However, in most cases covalent binding to cellular proteins appears to remain asymptomatic and "silent."

Many cases of drug-induced liver toxicity have been demonstrated in humans through covalent binding of protein by reactive metabolites. This is not only true for intrinsically hepatotoxic drugs given at high doses (e.g., acetaminophen, methapyrilene) but also for many drugs causing idiosyncratic hepatic toxicity (e.g., halothane, dihydralazine, tienilic acid, or troglitazone). These considerations indicate that covalent binding to proteins is an important biomarker for the presence of potentially harmful reactive metabolites (i.e., leading to the time-dependent inactivation of cytochrome P450, etc.). However, the interpretation of covalent binding data must be performed with care for the following reasons: First, there are numerous drugs that covalently bind to proteins that are not hepatotoxic. Second, many other mechanisms such as oxidative stress have been implicated in drug-induced liver injury, so the effects of covalent binding may only become effective in conjunction with these other factors. Third, the nature of the target protein is crucial. Most binding to proteins is selective. It is not necessarily the absolute amount of a protein adduct that dictates whether a toxic effect ensues but rather the fraction of the total protein that it modifies. Fourth, many protein adducts are short-lived and are rapidly degraded whereas others are more persistent over time. Accumulating drug adducts to a protein species with a long half-life are expected to be more toxic than modified proteins with rapid turnover rates. Finally, there exist many cellular detoxification systems and nucleophilic scavengers (i.e., glutathione) that inactivate electrophilic metabolites within a cell.

Hepatotoxicity is one of the major causes for drug withdrawal. At present, no assay can predict the ability of a new chemical entity to cause idiosyncratic reaction(s) or severe ADRs with respect to hepatic injury in the clinic. Circumstantial evidence, however, indicates that reactive metabolites and subsequent covalent binding to cellular targets (hapten formation) are involved in the vast majority of cases. However, additional common risk factors seem to be involved in drug-induced hepatotoxicity representing a so-called danger signal for the cell [75]. These include the potential of a new chemical entity to (1) induce hepatic oxidative stress, (2) disrupt the mitochondrial membrane potential, (3) interfere with hepatic active transport processes like the elimination of bile acids leading to intra-hepatic cholestasis, or (4) the induction of cellular necrosis or apoptosis leading to cell death. These risk factors might be attributed to an underlying disease or to a susceptible subpopulation that is at particular risk of hepatotoxicity. For almost all drugs associated with severe liver ADRs, bioactivation mechanisms have been proposed and/or reactive metabolites have been identified. It is important to appreciate that a compound that tests positive in the assays currently available for reactive metabolite assessment (covalent binding to microsomes, glutathione adduct formation, or time-dependent inhibition) may not necessarily induce hepatic or idiosyncratic toxicity in humans. Furthermore hepatic injury occurring in the clinic is often not seen in preclinical toxicity studies. Reactive metabolite formation and covalent binding alone rarely seem to be the cause of idiosyncratic drug reactions. Nevertheless, it is believed that the risk for adverse events due to covalent binding can be significantly reduced by avoiding chemical functionalities known to be susceptible to reactive metabolite formation. This approach has become the gold standard in pharmaceutical industry [20,76]. Depicted in Figure 19.11 are a number of structural features thought to be associated with reactive metabolite formation and covalent protein-binding intermediates.

Troglitazone as Example for Multi-factorial Hepatotoxicity A prominent case of postmarketing withdrawal of a drug due to hepatotoxicity is the thiazolidinedione insulin sensitizer drug troglitazone. Bioactivation pathways

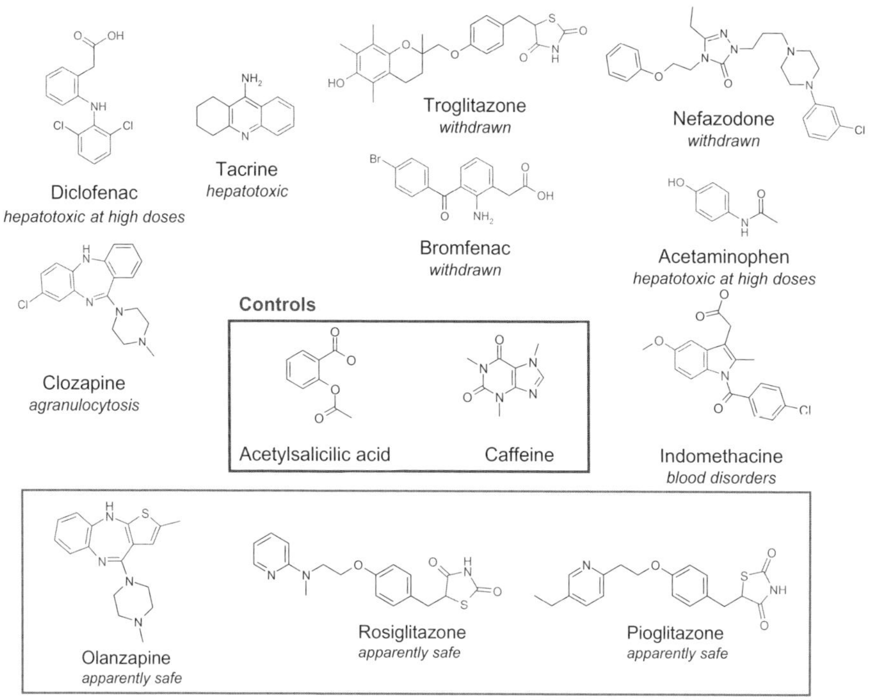

Figure 19.11 Examples of pharmaceuticals that are associated with evidence for human hepatotoxicity of varying levels of severity (caffeine and acetylsalicylic acid serve as negatives examples). Functional groups that are believed to be associated with metabolism pathways toward a reactive intermediate are highlighted in red. Note that despite structural alerts and proven bioactivation liabilities, some drugs given at comparably low dose are considered as safe (olanzapine, rosiglitazone, and pioglitazone). Their structural analogues given at relatively high doses (clozapine, troglitazone) have been associated with severe clinical side effects.

have been proposed for this drug, although the hepatotoxicity seems multifactorial [77]. Formation of quinone and quinone-methide-type reactive intermediates has been linked to the pro-oxidant activity of troglitazone in rat primary hepatocytes. This effect was significantly higher for troglitazone compared with other vitamin E analogues. Cytotoxicity in hepatic cells and oxidative stress-inducing properties for thiazolidinedione derivatives were dependent on the presence of the 6-hydroxychromane moiety, suggesting a link with this specific molecular structure of troglitazone. In addition rat hepatoma cells were more sensitive to troglitazone than to other derivatives lacking the 6-hydroxychromane moiety. Besides the potential direct cytotoxic effect of troglitazone-derived reactive metabolites on cellular structures, the drug was demonstrated to disrupt mitochondrial function in rat hepatocytes [78,79]. Funk and others have demonstrated that troglitazone and its major hepatic metabolite troglitazone-sulfate competitively inhibit the cannalicular bile salt export pump. This effect has been associated with the induction of intrahepatic cholestasis in rats [80]. Although these molecular properties of troglitazone do not represent a cause for hepatotoxic events per se, it seems likely that certain individuals are more susceptible to troglitazone-induced hepatotoxicity, especially in disease populations. Diabetic patients with a history of chloestasis are especially at risk from troglitazone [81]. The relatively higher clinical dose as well as other risk factors might contribute to the hepatotoxicity of this compound, in contrast to the lack of toxicity with the chemically related drugs rosiglitazone and pioglitazone.

The example of troglitazone illustrates the interplay between drug metabolism properties with the complex physiological processes that are in most cases, for drug-induced hepatotoxicity, multi-factorial. Although toxic properties are associated with certain chemical motifs, the mechanisms by which such problematic drugs elicit their toxicities carry strong host-associated components that are not predictable. This complexity of molecular and biochemical interactions are hardly reproducible in a test tube or a computer simulation. Thus mechanistic toxicological investigations and the in-depth biochemical understanding of preclinical safety findings in multiple species will in the future help provide more predictive insight to enable the development of effective and safe drugs for humans.

19.7.4 Phospholipidosis

Phospholipidosis, a phospholipid storage disorder, is characterized by an excessive accumulation of intracellular phospholipids. Compounds that induce phospholipidosis include a wide variety of pharmacological agents (antipsychotics, antidepressants, antiarrhythmics, and cholesterol-lowering agents). These compounds are of concern for the pharmaceutical industry because a candidate pharmaceutical agent can be rejected because of evidence of inducing phospholipidosis in a preclinical animal study. Phospholipidosis is widely reported in rats and is identified by the accumulation of phospholipids in the lysosomes of many cell types.

Each of the phospholipidosis-inducing compounds like amiodarone, imipramine, propranolol, chlorpromazine, and fluoxetine shares mainly two common physicochemical properties, namely a charged cationic amine group and a distinct amphiphilic nature. The calculated amphiphilic vector for a molecular structure gives a first hint that it might show phospholipidosis in vitro. Amphiphilicity of a compound is defined as the distance between the charged residue and the more remote hydrophobic residues. From the charged group (identified by a pK_a program) a vector is calculated to each atom within a molecule and weighted with respect to its hydrophobic/hydrophilic property on the basis of an atom contribution method. The sum of the calculated vectors is calibrated by means of measured amphiphilicities. To investigate the conformational effect, a number of conformers were generated (with Catalyst software) and compared with measured values for a set of drugs.

Amphiphilicity is expressed in terms of free energy (ΔG_{AM}). Compounds with calculated basic pK_a values smaller than seven and a free energy of amphiphilicity (ΔG_{AM} of smaller than –6 kJ/mol) show no potential hazard in the phospholipidosis assay [82]. CAFCA, a novel tool for the calculation of amphiphilic properties of charged drug molecules, has been developed mainly as an in vitro test for phospholipidosis. This semiquantitative evaluation microscopically detects lysosomal lipid inclusions of fibroblast cultures when exposed for 72 h to the test compound. A binary alert is then generated. Calculated data and molecular modeling have been shown to be in good accordance with experimentally derived published values. CAFCA can be used to estimate preferred conformations as well as orientations of molecules in biological membranes and to quantify amphiphilic properties of molecules [83]. For a complete analysis of the phospholipidosis potential, additional pharmacokinetic parameters like volume of distribution and metabolic stability, as well as envisioned length of therapy, have to be considered [83].

19.8 LOCAL VERSUS GLOBAL SAR MODELS FOR TOXICITY PREDICTION

As is evident, global rule-based models can serve the fundamental purpose of displaying associations of certain structural elements with toxicity. Openly available databases and commercially available systems such as DEREK can even facilitate a limited type of knowledge sharing. This mode of improvement of SAR for toxicities has been promoted and practiced by Lhasa in this company's development of rules for chemistry-associated toxicities and metabolism processes. Databases such as VITIC will enable confidential data to be shared within a user community with pre-agreed terms and conditions [30]. All stakeholders will benefit as a result, since use of the database avoids repetition of testing for the pharmaceutical industry when dealing with the issue of genotoxicity of intermediate-related impurities in the active pharmaceutical ingredient [84]. In the long run, such a database could be used to build better QSAR models that relate the structure modifications surrounding an alerting functional group. An example for a rule-based model developed in-house by a major pharmaceutical company was published by Muehlbacher et al. [85]. This in-house model was shown to link structurally alerting moieties with literature and in-house data on toxicity (and the absence thereof). Intrinsic to this approach is a ranking of toxicities by their supposed validity when structural modifications are made, combined with animal and human toxicity findings on certain associated alerting functional groups. By this approach, various problems associated with SARs could be addressed for toxicity: (1) data from internal failed compounds or series could be leveraged and linked, (2) positive and negative experience relative to an alerting moiety could be displayed for better decision making, (3) all scientists within the company could communicate via this platform, and (4) the software helped develop a more open exchange of information between medicinal chemists and toxicologists, a notorious problem in the industry.

19.9 SHORTCOMINGS OF CURRENT PROCESSES AND METHODS; IMPROVEMENT IN EDUCATION

Computational methods have only recently been applied in toxicity prediction, so quantitative analysis is less mature in toxicology compared with similar uses in target design and pharmacology. Part of the explanation is that toxicity is thought of as a multifactorial process compared to pharmacological effects, which are mostly due to the interaction of a drug with only a single target. An additional explanation is that in vivo subchronic or chronic toxicity testing is viewed as a low-throughput science, so the available databases and structure diversity information have been very limited. Nevertheless, chronic animal experiments are often required to identify target organ toxicities (e.g., to the liver and kidney), and these studies have become the basis for structure–toxicity relationship analysis. Moreover, it is much more difficult to predict toxicities from only structural information than from pharmacological properties. The amount of material needed for high-dose and chronic toxicity studies with a sufficient number of animals is usually a multiple of what is needed for pharmacology experiments. Hence the availability of gold standard data on which to build QSAR approaches is limited, and existing data frequently do not cover the desired chemical space. Openly accessible databases very often show preponderance toward data from simple industrial chemicals while the more complex chemistry space of drug-like molecules is often not very well represented [86].

Since chemistry-related toxicities are a major contributor to the failure of drug candidates progressing to clinical trials, it is desirable that compound selection be improved by a better filter process of compounds that do not show an acceptable exposure window between toxicity and efficacy. For this to occur, several critical improvements are needed:

1. Use of toxicological databases that link effects with commonly agreed-upon denominators of toxicity.
2. Sharing of positive and negative toxicity data in conjunction with exposure for inclusion in these databases.
3. Improvement of coverage of the chemical space within a drug-like space and away from the industrial chemicals space that is currently dominating the toxicity databases.

These critical issues will require open sharing of data across the industry. Ultimately, the goal is to move away from the descriptive characteristics and assessments of single compounds to a broader characterization of physicochemical denominators of toxicity in certain structural groups.

19.10 REGULATORY ASPECTS

So far in the regulatory context there appears to be very little impact of in silico toxicology on the decision-making processes. Product dossiers for pharmaceuticals, chemicals, pesticides, cosmetics and hair dyes, and so on, are normally evaluated at the end of a process of testing and evaluation, so they include a wealth of in vitro and in vivo data. Regulatory decisions normally do not refer to in silico methods for toxicity evaluation, and there has been little regulatory recognition of in silico methods as a means to eliminate certain animal experiments for toxicity or to encourage evaluations in cases to here toxicity studies would otherwise not be performed. Nevertheless, one active group of scientists on the regulatory side has actively constructed in silico models for toxicity prediction and has promoted their use for testing strategies and risk assessment. Most members of this group are active in the FDA's Office of Pharmaceutical Sciences in the Informatics and Computational Safety Analysis Staff (ICSAS). Their Web site (*http://www.fda.gov/cder/Offices/OPS_IO/default.htm*) mentions a number of projects related to computational toxicity prediction [87], some aspects of which were discussed in this chapter. The important activities of this group not only include computational predictions from animal and in vitro toxicity data but also predictions based on databases that specifically record human side effects. In cooperation via a Cooperative Research and Development Agreement (CRaDA) with MCASE (*http://www.multicase.com/*) and MDL (*http://www.mdli.com/*), modules have been released that facilitate a computational prediction of human liver toxicities and the maximum recommended human dose [88]. This work is extending in silico prediction into a new direction in several ways: (1) since human, not animal, data are used for prediction, there is more relevance for human safety, (2) since human data are taken for the training set, there are limited possibilities to test the predictions in animal studies, and (3) specific human safety studies are being triggered from predictions of unwanted toxicities based on animal data that could not have been otherwise identified. Apart from this effort, there is little evidence that regulatory agencies are considering structure-activity relationships and hence potentially in silico tools in their industry guidances responsibilities. Some hope for improvement comes from two recent examples that suggest recognition on the regulatory side of the importance of software models such as SAR by both the US FDA and the European Union. The FDA has communicated in a recent draft guideline on how to test pharmaceuticals for safety of (human) metabolites as follows [89]: the use of "With the availability of computational software designed to predict activity relative to a known structure, the mutagenic, carcinogenic, or teratogenic potential of a drug or a metabolite can be evaluated as soon as a structure is identified. Although structure activity relationship analyses are not considered a substitute for actual testing, we encourage submission of the results from these analyses." The example of the European regulatory guideline on limits of genotoxic impurities in pharmaceuticals is as follows [84]:

"Guided by existing genotoxicity data or the presence of structural alerts, potential genotoxic impurities should be identified. When a potential impurity contains structural alerts, additional genotoxicity testing of the impurity, typically in a bacterial reverse mutation assay, should be considered." While so far it is clear that these authorities have only limited knowledge about the use of computational toxicology that can regulate the pharmaceutical industry, these two regulatory guideline examples are a good indication that the regulatory world is ready to accept the use of in silico systems in toxicity testing strategies. Consequently there is a need for training and education of regulatory and industry scientists on the use and limitations of the models and systems so as to ensure a unified interpretation. In summary, in silico toxicology and in silico safety pharmacology, when based on good in vitro and in vivo training data sets are excellent tools for drug selection and development. However, in the near future it does not yet seem possible for the complex and multifactorial processes that are usually associated with adverse toxic effects to be completely modeled in silico and thus for animal experimentation to be entirely dispensed with.

REFERENCES

1. Dipple A. DNA adducts of chemical carcinogens. *Carcinogenesis* 1995;16:437–41.
2. Uetrecht J. Role of drug metabolism for breaking tolerance and the localization of drug hypersensitivity. *Toxicology* 2005;209:113–8.
3. Walgren JL, Mitchell MD, Thompson DC. Role of metabolism in drug-induced idiosyncratic hepatotoxicity. *Crit Rev Toxicol* 2005;35:325–61.
4. Nelson SD. Structure toxicity relationships—How useful are they in predicting toxicities of new drugs? *Adv Exp Med Biol* 2001;500:33–43.
5. Kalgutkar AS, Gardner I, Obach RS, Shaffer CL, Callegari E, Henne KR, Mutlib AE, Dalvie DK, Lee JS, Nakai Y, O'Donnell JP, Boer J, Harriman SP. A comprehensive listing of bioactivation pathways of organic functional groups. *Curr Drug Metab* 2005;6:161–225.
6. Zhou S, Chan E, Duan W, Huang M, Chen YZ. Drug bioactivation, covalent binding to target proteins and toxicity relevance. *Drug Metab Rev* 2005; 37:41–213.
7. Siemers E, Skinner M, Dean RA, Gonzales C, Satterwhite J, Farlow M, Ness D, May PC. Safety, tolerability, and changes in amyloid b concentrations after administration of a g-secretase inhibitor in volunteers. *Clin Neuropharmacol* 2005; 28:126–32.
8. Milano J, McKay J, Dagenais C, Foster-Brown L, Pognan F, Gadient R, Jacobs RT, Zacco A, Greenberg B, Ciaccio PJ. Modulation of Notch processing by g-secretase inhibitors causes intestinal goblet cell metaplasia and induction of genes known to specify gut secretory lineage differentiation. *Tox Sci* 2004;82:341–58.
9. De Smedt M, Hoebeke I, Reynvoet K, Leclercq G, Plum J. Different thresholds of Notch signaling bias human precursor cells toward B-, NK-, monocytic/dendritic-, or T-cell lineage in thymus microenvironment. *Blood* 2005;108:3498–506.

10. Wong GT, Manfra D, Poulet FM, Zhang Q, Josien H, Bara T, Engstrom L, Pinzon-Ortiz M, Fine JS, Lee HJJ, Zhang L, Higgins GA, Parker EM. Chronic treatment with the γ-secretase inhibitor LY-411,575 inhibits amyloid peptide production and alters lymphopoiesis and intestinal cell differentiation. *J Biol Chem* 2004;279: 12876–82.
11. Maillard I, Fang T, Pear WS. Regulation of lymphoid development, differentiation and function by the Notch pathway. *An Rev Immunol* 2005;23:945–74.
12. Dobler M, Lill MA, Vedani A. From crystal structures and their analysis to the in silico prediction of toxic phenomena. Helv Chim Acta 2003;86:1554–68.
13. Caldwell GW, Yan Z. Screening for reactive intermediates and toxicity assessment in drug discovery. *Curr Opin Drug Discov Devel* 2006;9:47–60.
14. Yamashita F, Hashida M. In silico approaches for predicting ADME properties of drugs. *Drug Metab Pharmacokinet* 2004;19:327–38.
15. Hatch FT, Colvin ME. Quantitative structure-activity (QSAR) relationships of mutagenic aromatic and heterocyclic amines. *Mutat Res* 1997;376:87–96.
16. Sabbioni G. Hemoglobin binding of aromatic amines: molecular dosimetry and quantitative structure-activity relationships for *N*-oxidation. *Environ Health Perspect* 1993;99:213–6.
17. Baillie TA, Davis MR. Mass spectrometry in the analysis of glutathione conjugates. *Biol Mass Spectrom* 1993;22:319–25.
18. Baillie TA, Kassahun K. Biological reactive intermediates in drug discovery and development: a perspective from the pharmaceutical industry. *Adv Exp Med Biol* 2001;500:45–51.
19. Day SH, Mao A, White R, Schulz-Utermoehl T, Miller R, Beconi MG. A semi-automated method for measuring the potential for protein covalent binding in drug discovery. *J Pharmacol Toxicol Meth* 2005;52:278–85.
20. Evans DC, Watt AP, Nicoll-Griffith DA, Baillie TA. Drug-protein adducts: An industry perspective on minimizing the potential for drug bioactivation in drug discovery and development. *Chem Res Toxicol* 2004;17:3–16.
21. Yan Z, Rafferty B, Caldwell GW, Masucci JA. Rapidly distinguishing reversible and irreversible CYP450 inhibitors by using fluorometric kinetic analyses. *Eur J Drug Metab Pharmacokinet* 2002;27:281–7.
22. Smith DA, Obach RS. Seeing through the mist: Abundance versus percentage. Commentary on metabolites in safety testing. *Drug Metab Dispos* 2005;33: 1409–17.
23. Lin J, Sahakian DC, de Morais SMF, Xu JJ, Polzer RJ, Winter SM. The role of absorption, distribution, metabolism, excretion and toxicity in drug discovery. *Curr Top Med Chem* 2003;3:1125–54.
24. Garattini S. Drug metabolism: from experiments to regulatory aspects. *Exp Toxic Pathol* 1996;48(Suppl II):142–51.
25. Eddershaw PJ, Beresford AP, Bayliss MK. ADME/PK as part of a rational approach to drug discovery. *Drug Discov Today* 2000;5(9):409–14.
26. Baillie TA, Cayen MN, Fouda H, Gerson RJ, Green JD, Grossman SJ, Klunk LJ, LeBlanc B, Perkins DG, Shipley LA. Drug metabolites in safety testing. *Toxicol Appl Pharmacol* 2002;182:188–96.

27. Lipinksi CM, Lombardo F, Dominy BW, Feeney PJ. Experimental and computational approaches to estimate solubility and permeability in drug discovery and development settings. *Adv Drug Deliv Rev* 2001;46:3–26.
28. Richard AM, Williams CR. Distributed structure-searchable toxicity (DSSTox) public database network: A proposal. *Mutat Res* 2002;499:27–52.
29. Kirkland D, Aardema M, Henderson L, Müller L. Evaluation of the ability of three in vitro genotoxicity tests to discriminate rodent carcinogens and non-carcinogens: I. Sensitivity, specificity and relative predictivity. *Mutat Res* 2005;584:1–257.
30. Judson PN, Cooke PA, Doerrer NG, Greene N, Hanzlik RP, Hardy C, Hartmann A, Hinchcliffe D, Holder J, Müller L, Steger-Hartmann T, Rothfuss A, Smith M, Thomas K, Vessey JD, Zeiger E. Towards the creation of an international toxicology information centre. *Toxicology* 2005;213:117–28.
31. Sanderson DM, Earnshaw CG. Computer prediction of possible toxic action from chemical structure. *Hum Exp Toxicol* 1991;10:261–73.
32. Greene N, Judson PN, Langowski JJ, Marchant CA. Knowledge-based expert systems for toxicity and metabolism prediction: DEREK, StAR, and METEOR. SAR & QSAR. *Environ Res* 1999;10:299–313.
33. Klopman G, Rosenkranz H. Approaches to SAR in carcinogenesis and mutagenesis —Prediction of carcinogenicity/mutagenicity using Multi-Case. *Mutat Res* 1994;305(1):33–46.
34. Matthews EJ, Contrera JF. A new highly specific method for predicting the carcinogenic potential of pharmaceuticals in rodents using enhanced MCASE QSAR-ES software. *Regul Toxicol Pharmacol* 1998;28(3):242–64.
35. Redfern WS, Wakefield ID, Prior H, Pollard CE, Hammond TG, Valentin, J-P. Safety pharmacology—A progressive approach. *Fundam Clin Pharmacol* 2002;16:161–73.
36. ICH S7A, Safety pharmacology studies for human pharmaceuticals. 2000. <http://www.ich.org>.
37. Crumb W, Cavero I. QT interval prolongation by non-cardiovascular drugs: Issues and solutions for novel drug development. *Pharm Sci Technol Today* 1999; 2:270–80.
38. Shah RR. Drugs, QT interval prolongation and ICH E14: The need to get it right. *Drug Saf* 2005;28:115–25.
39. Fung MC, Hsiao-hui Wu H, Kwong K, Hornbuckle K, Muniz E. Evaluation of the profile of patients with QTc prolongation in spontaneous adverse event reporting over the past three decades—1969–98. *Pharmacoepidemiol Drug Saf* 2000;9 (Suppl 1):S24.
40. Malik M, Camm AJ. Evaluation of drug-induced QT interval prolongation: Implications for drug approval and labelling. *Drug Saf* 2001;24:323–51.
41. Dessertenne F. La tachycardie ventriculaire à deux foyers opposés variable. *Arch Mal Coeur Vaiss* 1996;59:263–72.
42. Recanatini M, Poluzzi E, Masetti M, Cavalli A, De Ponti F. QT prolongation through hERG K+ channel blockade: Current knowledge and strategies for the early prediction during drug development. *Med Res Rev* 2005;25:133–66.
43. ICH E14. The clinical evaluation of QT/QTc interval prolongation and proarrhythmic potential for nonantiarrhythmic drugs. 2005. <http://www.ich.org>.

44. ICH S7B. The non-clinical evaluation of the potential for delayed ventricular repolarization (QT interval prolongation) by human pharmaceuticals. 2005. <http://www.ich.org>.

45. Bennett PB, Guthrie HR. Trends in ion channel drug discovery: Advances in screening technologies. *Trends Biotechnol* 2003;21:563–9.

46. Antzelevitch C, Belardinelli L, Zygmunt AC, Burashnikov A, Di Diego JM, Fish JM, Cordeiro JM, Thomas G. Electrophysiological effects of ranolazine, a novel antianginal agent with antiarrhythmic properties. *Circulation* 2004;110:904–10.

47. Shah RR. The significance of QT interval in drug development. *Br J Clin Pharmacol* 2002;54:188–202.

48. Muzikant AL, Penland RC. Models for profiling the potential QT prolongation risk of drugs. *Curr Opin Drug Discov Devel* 2002;5:127–35.

49. Bottino D, Penland RC, Stamps A, Traebert M, Dumotier B, Georgiva A, Helmlinger G, Lett GS. Preclinical cardiac safety assessment of pharmaceutical compounds using an integrated systems-based computer model of the heart. *Prog Biophys Mol Biol* 2006;90:414–43.

50. Noble D. Systems biology and the heart. *Biosystems* 2006;83:75–80.

51. Ekins S, Crumb WJ, Sarazan RD, Wikel JH, Wrighton SA. Three-dimensional quantitative structure-activity relationship for inhibition of human ether-a-go-go-related gene potassium channel. *J Pharmacol Exp Ther* 2002;301:427–34.

52. Cavalli A, Poluzzi E, De Ponti F, Recanatini M. Toward a pharmacophore for drugs inducing the long QT syndrome: Insights from a CoMFA study of HERG K(+) channel blockers. *J Med Chem* 2002;45:3844–53.

53. Keserü GM. Prediction of hERG potassium channel affinity by traditional and hologram qSAR methods. *Bioorg Med Chem Lett* 2003;13:2773–5.

54. Krejsa CM, Horvath D, Rogalski SL, Penzotti JE, Mao B, Barbosa F, Migeon JC. Predicting ADME properties and side effects: The BioPrint approach. *Curr Opin Drug Discov Devel* 2003;6:470–80.

55. Johnson DE, Rodgers AD. Computational toxicology: Heading toward more relevance in drug discovery and development. *Curr Opin Drug Discov Devel* 2006;9:29–37.

56. Pearlstein RA, Vaz RJ, Kang J, Chen XL, Preobrazhenskaya M, Shchekotikhin AE, Korolev AM, Lysenkova LN, Miroshnikova OV, Hendrix J, Rampe D. Characterization of HERG potassium channel inhibition using CoMSiA 3D QSAR and homology modeling approaches. *Bioorg Med Chem Lett* 2003;13:1829–35.

57. Roche O, Trube G, Zuegge J, Pflimlin P, Alanine A, Schneider G. A virtual screening method for prediction of the HERG potassium channel liability of compound libraries. *Chembiochem* 2002;3:455–9.

58. Ekins S. In silico approaches to predicting drug metabolism, toxicology and beyond. *Biochem Soc Trans* 2003;31:611–4.

59. Sun H. An accurate and interpretable Bayesian classification model for prediction of hERG liability. *Chem Med Chem* 2006;1:315–22.

60. Mitcheson JS, John S, Chen J, Lin M, Culberson C, Sanguinetti MC. A structural basis for drug-induced long QT syndrome. *Proc Natl Acad Sci USA* 2000; 97(22):12329–33.

61. Yap CW, Cai CZ, Xue Y, Chen YZ. Prediction of torsade-causing potential of drugs by support vector machine approach. *Toxicol Sci* 2004;79:170–7.
62. Gepp MM, Hutter MC. Determination of hERG channel blockers using a decision tree. *Bioorg Med Chem* 2006;14:5325–32.
63. Aronov AM. Predictive in silico modeling for hERG channel blockers. *Drug Discov Today* 2005;10(2):149–55.
64. Muster W, Albertini S, Gocke E. Structure-activity relationship of oxadiazoles and allylic structures in the Ames test: An industry screening approach. *Mutagenesis* 2003;18(4):321–9.
65. Albertini S, Bös M, Bos M, Gocke E, Kirchner S, Muster W, Wichmann J. Suppression of mutagenic activity of a series of 5HT2c receptor agonists by the incorporation of a gem-dimethyl group: SAR using the Ames test and a DNA unwinding assay. *Mutagenesis* 1998;13(4):397–403.
66. Muster W, Chételat AA, Kirchner S, Rothfuss A, Albertini S, Speit G, Gocke E. Mutagenicity testing and investigations on the mechanism of genotoxicity for a NMDA receptor antagonist. In preparation.
67. Snyder RD, Pearl GS, Mandakas G, Choy WN, Goodsaid F, Rosenblum IY. Assessment of the sensitivity of the computational programs DEREK, TOPKAT, and MCASE in the prediction of the genotoxicity of pharmaceutical molecules. *Environ Mol Mutagen* 2004;43:143–58.
68. Cronin MTD, Manga N, Seward JR, Sinks GD, Schultz TW. Parametrization of electrophilicity for the prediction of the toxicity of aromatic compounds. *Chem Res Toxicol* 2001;14(11):1498–505.
69. Franke R, Gruska A, Giuliani A, Benigni R. Prediction of rodent carcinogenicity of aromatic amines: A quantitative structure-activity relationships model. *Carcinogenesis* 2001;22(9):1561–71.
70. Benigni R, Passerini L. Carcinogenicity of the aromatic amines: from structure-activity relationships to mechanisms of action and risk assessment. *Mutat Res* 2002;511:191–206.
71. Müller L, Mauthe RJ, Riley CM, Andino MM, De Antonis D, Beels C, DeGeorge J, De Knaep AGM, Ellison D, Fagerland JA, Frank R, Fritschel B, Galloway S, Harpur E, Humfrey CDN, Jacks AS, Jagota N, Mackinnon J, Mohan G, Ness DK, O'Donovan MR, Smith MD, Vudathala G, Yotti L. A rationale for determination, testing and control of genotoxic impurities in pharmaceuticals. *Regul Toxicol Pharmacol* 2006;44:198–211.
72. Spielmann H, Balls M, Dupuis J, Pape WJ, Pechovitch G, De Silva O, Holzhutter HG, Clothier R, Desolle P, Gerberick F, Liebsch M, Lovell WW, Maurer T, Pfannenbecker U, Potthast JM, Csato M, Sladowski D, Steiling W, Brantom P. The international EU/COLIPA in vitro phototoxicity validation study: Results of phase II (blind trial): Part 1. The 3T3 NRU phototoxicity test. *Toxicol In Vitro* 1998;12(3):305–27.
73. Gocke E, Albertini S, Chetelat AA, Kirchner S, Muster W. The photomutagenicity of fluoroquinolones and other drugs. *Toxicol Lett* 1998;102–3:375–81.
74. Gocke E, Chetelat AA, Csato M, McGarvey DJ, Jakob-Roetne R, Kirchner S, Muster W, Potthast M, Widmer U. Phototoxicity and photogenotoxicity of nine pyridone derivatives. *Mutat Res* 2003;535(1):43–54.

75. Uetrecht JP. Is it possible to more accurately predict which drug candidates will cause idiosyncratic drug reactions? *Curr Drug Metab* 2000;1:133–41.
76. Ju C, Uetrecht JP. Mechanism of idiosyncratic drug reactions: Reactive metabolites formation, protein binding and the regulation of the immune system. *Curr Drug Metab* 2002;3:367–77.
77. Smith MT. Mechanisms of troglitazone hepatotoxicity. *Chem Res Toxicol* 2003;16:679–87.
78. Narayanan PK, Hart T, Elcock F, Zhang C, Hahn L, McFarland D, Schwartz L, Morgan DG, Bugelski P. Troglitazone-induced intracellular oxidative stress in rat hepatoma cells: A flow cytometric assessment. *Cytometry A* 2003;52:28–35.
79. Tirmenstein MA, Hu CX, Gales TL, Maleeff BE, Narayanan PK, Kurali E, Hart TK, Thomas HC, Schwartz LW. Effects of troglitazone on HepG2 viability and mitochondrial function. *Toxicol Sci* 2002;69:131–8.
80. Funk C, Ponelle C, Scheuermann G, Pantze M. Cholestatic potential of troglitazone as a possible factor contributing to troglitazone-induced hepatotoxicity: In vivo and in vitro interaction at the canalicular bile salt export pump (Bsep) in the rat. *Mol Pharmacol* 2001;59:627–35.
81. Menon KVN, Angulo P, Lindor KD. Severe cholestatic hepatitis from troglitazone in a patient with nonalcoholic steatohepatitis and diabetes mellitus. *Am J Gastroenterol* 2001;96:1631–4.
82. Fischer H., Kansy M, Potthast M, Csato M. Predition of in vitro phospholipidosis of drugs by means of their amphiphilic properties. In: Höltjc H-D and Sippl W, editors. Proceedings of the 13th European Symposium on QSAR: Application in chemometrics, Barcelona, Spain. 27 August–1 September 2000. Barcelona: Prous Science, 2001. p. 286–9.
83. Fischer H, Kansy M, Bur D. CAFCA: A novel tool for the calculation of amphiphilic properties of charged drug molecules. *Chimica* 2000;54(11):640–5.
84. CHMP (2006) Guideline on the limits of genotoxic impurities (CHMP/SWP/5199/02, June 2006).
85. Muehlbacher JP, Ertl P, Selzer P, Müller L, Glowienke S, Schuffenhauer A. Toxizitätsvorhersage im Intranet. *Nachrichten aus der Chemie* 2004;52:162–4.
86. Yang C, Richard AM, Cross KP. The art of data mining the minefields of toxicity databases to link chemistry to biology. *Curr Comput Aided Drug Des* 2006;2: 135–50.
87. Matthews EJ, Kruhlak NL, Cimino MC, Benz RD, Contrera JF. An analysis of genetic toxicity, reproductive and developmental toxicity, and carcinogenicity data: II. Identification of genotoxicants, reprotoxicants and carcinogens using in silico methods. *Regul Toxicol Pharmacol* 2006;44:97–110.
88. Contrera JF, Matthews, EJ, Kruhlak, NL, Benz RD. Estimating the safe starting dose in phase I clinical trials and no observed effect level based on QSAR modeling of the human maximum recommended daily dose. *Regul Toxicol Pharmacol* 2004;40:185–206.
89. FDA. *Draft Guideline for industry: Safety testing of drug metabolites.* US FDA, CDER June 2005.

20

APPLICATION OF INTERPRETABLE MODELS TO ADME/TOX PROBLEMS

Tomoko Niwa and Katsumi Yoshida

Contents

20.1 INTRODUCTION

Experimental screening of compounds for biological activity is usually time-consuming and expensive. An attractive alternative is in silico screening; that is, using computers to predict the effects of drugs. In silico screening allows easy and rapid handling of large numbers of compounds. Another big advantage of in silico screening is its ability to predict the activities of compounds not yet synthesized.

Computational Toxicology: Risk Assessment for Pharmaceutical and Environmental Chemicals, Edited by Sean Ekins

A variety of computational methods have recently been applied to the construction of predictive models in the field of ADME/Tox (Absorption, Distribution, Metabolism, Excretion, and Toxicology), and these models can be classified into two types:

- Complex, high-performance models suitable for screening
- Simple, interpretable models suitable for assisting our understanding of structural factors

The purpose of predictive models is to correctly evaluate the activities of compounds. The major concern when developing models is their performance as judged by the percentage of correctly predicted compounds or the quality of the correlation between experimental and predicted values. Much effort has been devoted to enhancing the predictability of in silico models. Modern multivariate methods, such as neural networks and support vector machines (SVMs), are especially powerful tools for building high-performance models. The resulting models are very useful for filtering and prioritizing a large number of compounds before biological screening begins. However, these models have the disadvantage that they work like black boxes, so their meaning is difficult to understand. The use of complex descriptors leads to the generation of predictive models that, though powerful, are difficult to interpret.

It often happens during chemical modification and optimization that compounds exhibit unexpected unfavorable effects, even when the lead compounds are free from such effects. In such cases the most practical way to reduce the unfavorable effects is further chemical modification. To perform effective and efficient chemical modification, it is important for medicinal chemists to understand the structural factors responsible for unfavorable effects. To practically assist medicinal chemists in understanding these structural factors and to guide their efforts to reduce unfavorable effects through chemical modification, predictive models should be simple and easy to understand.

On the face of it, building simple models should be easier than building complex ones. However, because the kinds of descriptors and computational methods that can be used in simple models are limited, more careful selection of the descriptors and computational methods has to be made. The quality of simple models can be greatly affected by the combination of descriptors and computational methods used. This chapter is focused on the application of interpretable models to ADME/Tox problems, and it deals with the kinds of models that are in wide use, ways to use these models effectively to accelerate drug development, and what can be learned from the models.

20.2 INTERPRETABLE METHODS AND DESCRIPTORS

Chris Lipinski [1] identified a set of structural features commonly found in orally active drugs. Because most of the features involve the number five, the rule was named Lipinski's "rule of five." By this rule, an orally active drug generally has the following characteristics:

- Not more than 5 hydrogen-bond donors (OH and NH groups)
- Not more than 10 hydrogen-bond acceptors (notably N and O)
- A molecular weight under 500 Da
- A partition coefficient $\log P$ under 5

This rule of five is popular among medicinal chemists. It is not only useful in chemical-modification studies but also easy to understand. The simplicity of the rule largely contributes to its widespread use. To have practical application in chemical-modification studies, a model should be simple and easy to understand. Furthermore, because medicinal chemists have much experience in reading graphical patterns, it is highly desirable to have predictive models that represent 2D or 3D structures. Candidate model types that exploit this experience and leverage medicinal chemists' efforts are:

- 2D chemical structural models
- 3D structural models
- Multivariate models based on simple descriptors

These types of models are useful not only to medicinal chemists but also to those responsible for evaluating the ADME/Tox characteristics of drugs.

Palm et al. [2] found excellent sigmoidal correlations between the fraction of drugs absorbed after oral administration to humans (FA) and the polar surface area (PSA) of the drugs. These authors defined PSA as the sum of the surface areas of the polar atoms, usually oxygens, nitrogens, and attached hydrogens, in a molecule [3]. PSA has been found to correlate with drug-transport properties of drugs, such as intestinal absorption and penetration of the blood–brain barrier [4,5]. PSA is a good example of a descriptor suitable for use in interpretable models because it is easy to understand, has a clear physicochemical meaning, and represents a fundamental property of drugs. Although certain descriptors are sometimes indispensable for building predictive models, the selection of descriptors with clear physicochemical meanings and that represent fundamental properties of drugs is important in the generation of useful interpretable models.

Palm et al. [3] took into account the flexibility of molecules by using molecular mechanics to calculate an averaged PSA according to a Boltzmann distribution. Later Clark [4,5] found that the use of a representative conformation was sufficient for the calculation of reliable PSA values. Ertl [6] developed a method to calculate PSA as the sum of fragment contributions and proposed a topological PSA (TPSA). The advantage of TPSA is that it can be directly calculated from the 2D chemical structure, which makes the calculation rapid and reproducible.

20.3 2D CHEMICAL STRUCTURAL PATTERNS

The most direct way to gain an understanding of the structural factors causing an unfavorable effect of a drug is to identify the 2D chemical structural patterns responsible for the effect. For example, Song and Clark [7] determined the molecular fragments relevant to the variations in the binding affinity of the hERG (human ether-a-go-go-related gene) cardiac potassium channel by developing quantitative structure-activity relationship (QSAR) models based on molecular fragments and the sparse ν-SVR (support vector regression) [7] method. SVR is derived from SVMs, which are a class of supervised learning algorithms originally developed to recognize patterns [8]. These authors collected in vitro hERG inhibition data (pIC_{50}) for 90 structurally diverse drugs from the literature. The descriptors used were 261 structurally diverse fragments that frequently appear in bioactive compounds [9]. The sparse ν-SVR method was used to vary the coefficients for each fragment and to remove fragments with low contributions to the model. Linear SVR predictive models based on 45 selected fragments (shown in Figure 20.1) were simultaneously constructed. SVR minimizes a regularized error that controls both the training error and the complexity of the model, so it is relatively robust to noisy data and less subject to overfitting.

The performance of the generated model was found to be good for both the training ($r^2 = 0.912$, RMSE = 0.440) and test ($r^2 = 0.849$, RMSE = 0.597) sets. The coefficient associated with a fragment indicates the role of that fragment in hERG binding affinity. Fragments with positive coefficients are those that are predicted to increase the hERG pIC_{50} values, and the magnitude of a coefficient is a measure of the contribution made by the fragment (Figure 20.1). Four general trends are found:

- Hydrophobic fragments such as fluorobenzene increase hERG binding affinity.
- Hydrophilic fragments such as amide and carboxylic acid decrease hERG binding affinity.
- Fluorine or methanesulfonamide fragments increase hERG binding affinity.
- The effects of amines are complex and can be modulated by other structural features such as charge and charge shielding.

Each fragment has different characteristic physicochemical properties. For example, representative properties of primary amines are hydrophilicity, the ability to form cations, and the ability to form hydrogen bonds. It is often difficult to know which property is the most important one. Nevertheless, representation by the 2D chemical structure is a good way to generate ideas for further chemical modification.

Another example of the use of structural patterns is the parsing of ring fragments prevalent in toxicological databases. Thus Kho et al. [10] used

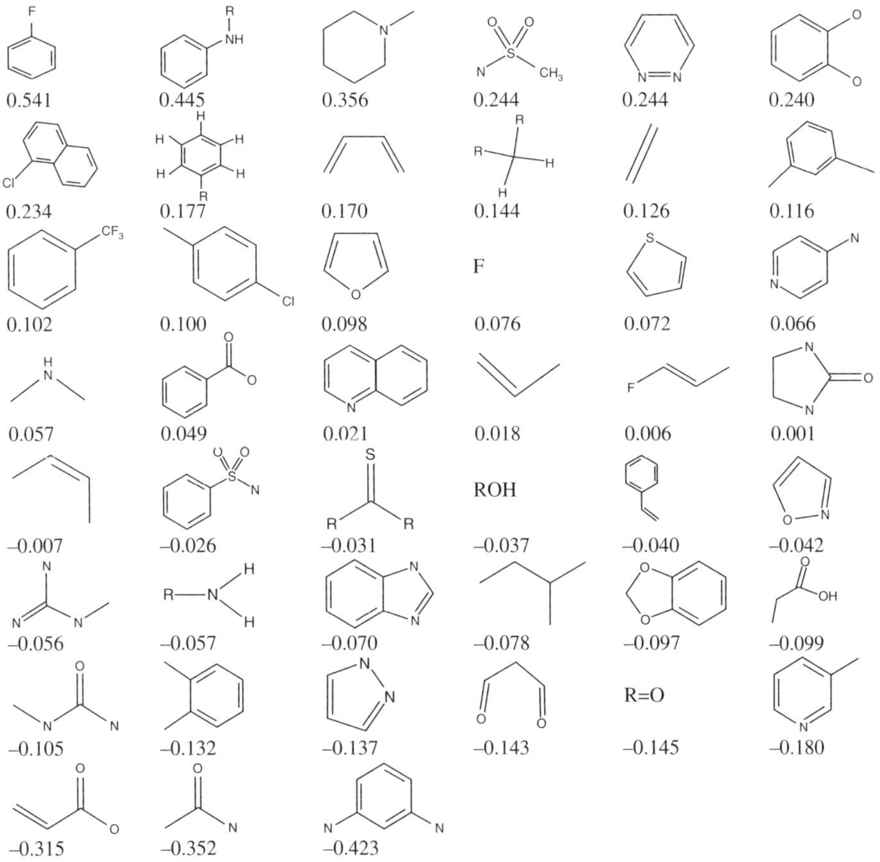

Figure 20.1 Molecular fragments identified as the most relevant descriptors for hERG binding affinity and the corresponding weights [7]. (Published with permission from the American Chemical Society.)

Ames-test data for 6039 compounds taken from the Chemical Carcinogenesis Research Information System/Toxicology Data Network (CCRIS/TOXnet) [11]. They were able to categorize the Ames data obtained in four strains of *Salmonella typhimurium* (TA98, TA100, TA1535, TA1537) into two classes, Ames negative and Ames positive. The data were analyzed as follows:

1. Identification of ring fragments prevalent in each class
2. Determination of the frequency and calculation of the odds ratio for the ring fragments
3. Identification of rules

The odds ratio (*OR*) is commonly used in epidemiological and clinical studies to express the relationship between two populations [12,13]. It is defined as the ratio of the odds of an event's occurring in one class to the odds of its occurring in another class. Kho et al. [10] used a program called SARvision Plus [14] to parse the fragments and organize them hierarchically. The highest levels of the tree structures generated provide a list of simple ring fragments. For these fragments, $\log OR$ was calculated for Ames-negative versus Ames-positive compounds. With this $\log OR$ value it is possible to identify the quantitative contribution of each fragment. Another application of this method is to compare the odds for two fragments so that medicinal chemists can easily evaluate which fragment contributes more to an unfavorable effect. The information can then be used to guide efforts to eliminate such effects by chemical modification. The use of OR and molecular fragments is a good way of gaining a visual understanding of the relevant structural factors when sufficient numbers of molecules are available for analysis.

20.4 MULTIVARIATE MODELS USING SIMPLE 2D DESCRIPTORS

2D-QSAR models based on multiple linear regression (MLR) are easy to understand, especially when interpretable descriptors are used. MLR is not as powerful as more complicated methods, such as neural networks, so careful formulation of descriptors is critical for the derivation of models that are both predictive and practically useful. Here is a set of guidelines that we have found to be useful in the formulation of descriptors for interpretable models:

- Descriptors with clear physicochemical meaning and in common use among medicinal chemists should be used.
- Ease of calculation of descriptors is desirable so that the models derived can be conveniently used by medicinal chemists. Descriptors that can be calculated from 2D chemical structures without the use of software are recommended.
- The use of sparse descriptors makes the models generated unstable, especially when the number of data points is small. Descriptors with nonzero values for most or all of the data points, such as molecular weight, are recommended.
- To build widely applicable models, descriptors very specific to particular data sets should be avoided.
- Descriptors should not be highly collinear with each other, since this causes problems in estimating regression coefficients.
- Descriptors for which high collinearity is often observed, such as molecular weight and molecular surface area, should not be used together even if such collinearity is not found in the particular data set used in the modeling.
- To avoid chance correlations, the use of at least five compounds per descriptor is required [15].

The preparation of the input data is also important. Experimental data taken from the literature need to be carefully checked so as to minimize differences in the experimental conditions. Special caution is required when the experimental data is derived from different kinds of cells.

Using carefully selected descriptors, Yoshida et al. [16] constructed simple and easy-to-interpret 2D-QSAR models for hERG inhibition:

$$\begin{aligned} \mathrm{pIC}_{50} = {} & 0.221(\pm 0.097)\mathrm{C\log P} - 0.017(\pm 0.005)\mathrm{TPSA} \\ & + 0.231(\pm 0.055)\,\mathrm{diameter} + 0.037(\pm 0.016) \\ & \mathrm{PEOE\_VSA\text{-}4} - 0.793(\pm 0.330)\mathrm{Cell} + 2.993(\pm 0.719) \\ n = 104,\ & r = 0.839,\ r^2 = 0.704,\ q^2 = 0.671,\ s = 0.763,\ F = 46.6 \end{aligned} \tag{20.1}$$

In the equation, n is the number of data points, r is the correlation coefficient, r^2 is the goodness of fit, q^2 is the leave-one-out cross-validated correlation coefficient expressing the goodness of prediction, s is the standard deviation, and F is the ratio of the variance of the calculated values to that of the observed values. The numbers in parentheses are the 95% confidence intervals. According to equation (20.1), the following structural factors affect the affinity of ligands for hERG:

- Hydrophobicity (ClogP) [17] increases hERG binding affinity.
- Increased topological polar surface area (TPSA) [6] decreases hERG binding affinity.
- Increased length of the drug (diameter) [18] increases hERG binding affinity.
- Atoms bearing charges [19] in a certain range (PEOE_VSA-4) [20] (Figure 20.2) increase hERG binding affinity.
- pIC_{50} values determined in CHO cells are about 0.8 log unit lower than those determined in HEK cells.

The model derived clearly reveals the structural factors responsible for hERG binding affinity. It also justifies the combined use of data measured with hERG channels expressed in HEK and CHO cells, and the statistical significance of the "cell" term reflects the differing permeabilities of the cell membranes. Because the descriptors were carefully chosen and the input data carefully checked, the model derived is also applicable to compounds not

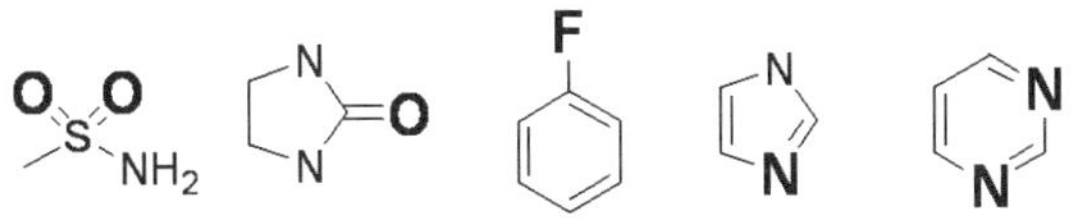

Figure 20.2 Examples of atoms with nonzero PEOE_VSA-4 values. The atoms shown in bold have atomic partial charges [19] in the range –0.25 to –0.20. (Published with permission from the American Chemical Society.)

included in the training set [16]. Of course, this type of modeling is not feasible for every target, but it is worth checking the possibility of building simple models before moving to more complicated ones.

20.5 DEEPER INSIGHT FROM COMBINED USE OF MULTIPLE INTERPRETABLE MODELS

Different kinds of interpretable models are available to analyze molecular interactions. So it is of interest to compare the models to get further insight into the modes of interaction. A good example is provided by studies of hERG binding affinity, where the molecular fragments reported by Song et al. [7] can be compared with the QSAR model of Yoshida et al. [16]. According to the QSAR model, hydrophobicity increases the affinity, whereas hydrophilicity decreases it. The molecular fragments shown in Figure 20.1 provide typical examples illustrating hydrophobicity (such as fluorobenzene) and hydrophilicity (such as amide fragments). The atoms shown in Figure 20.2 that were found to increase hERG affinity also appear in Figure 20.1. The QSAR model helps us understand the physicochemical significance of the molecular fragments in Figure 20.1. For example, the secondary and tertiary amines in Figure 20.1 increase hERG affinity, while primary amines decrease the affinity. Accordingly, the QSAR model suggests possible contributions from hydrophobic and hydrogen-bonding interactions.

It is instructive to study the correspondence between the molecular determinants derived from the QSAR model and the 3D structural characteristics of the binding site of the target protein. The 3D structure of the putative binding site in a homology-modeled hERG channel [16] based on the crystal structure of the *Methanobacterium thermoautotrophicum* potassium (MthK) channel in the open state [21], along with representative amino acids reported to participate in the binding of drugs [22], are shown in Figure 20.3. Comparison of the QSAR model and the 3D structure of the putative binding-site reveals the following characteristics:

- In the pore region, half of the amino acids are hydrophobic (shown in green in Figure 20.3), and there are no ionizable amino acids, so that the pore is hydrophobic. It is probable that hydrophobicity (as expressed by ClogP) increases hERG binding affinity and that hydrophilicity (expressed by TPSA) decreases it.
- Because the pore region is large, drugs with increased length (expressed by their diameter) appear to fill the large inner pore well by conforming their shape to the available space.
- The amino acids that are candidates for forming a hydrogen bond with the atoms shown in Figure 20.2 are Thr623, Ser624, Ser649, and Tyr652 (shown in yellow in Figure 20.3).

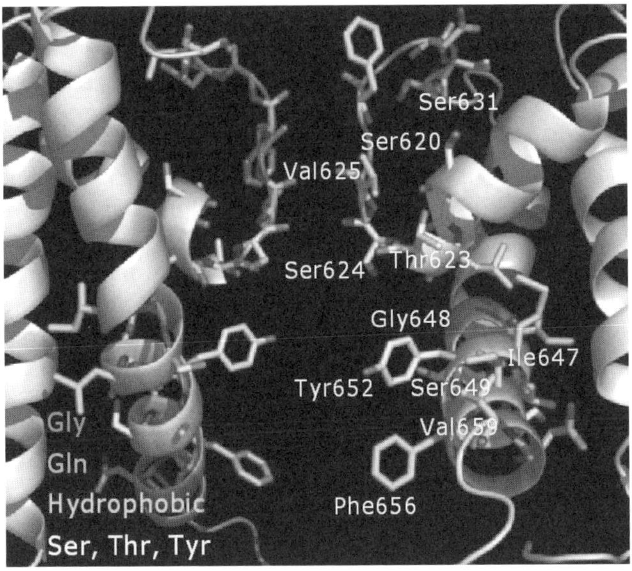

Figure 20.3 Modeled structure of a hERG channel. Amino acids are colored according to their physicochemical properties. Amino acids reported to participate in the binding of drugs [22] are labeled. For simplicity, only two of the four subunits that form the channel are shown. See color plates.

Substantial insight can be gained from the combined use of multiple models, especially if each model is easy to understand. A combination of such interpretable models permits deeper insight into the structural characteristics of binding interactions than any one model alone, as well as providing ideas for the further development of candidate drugs.

20.6 RECENT ADVANCES IN 3D STRUCTURAL MODELS

Three-dimensional X-ray structures and computationally generated models can provide much information on the modes of interaction between target proteins and their ligands at an atomic level (see also Chapter 17). Because 3D structural models are visually understandable, they are the models of choice for medicinal chemists. The most reliable method for obtaining 3D structures is X-ray crystallography. However, the purification and crystallization of proteins is often costly and time-consuming, and the crystallization of membrane proteins is still very difficult to achieve. An attractive alternative to X-ray crystallography is the computational prediction of binding interactions.

Computational protein–ligand docking is widely used nowadays to select compounds for biological screening [23]. In the field of ADME/Tox, where

X-ray crystallographic data are not abundant, docking techniques are especially attractive. However, there is a major problem with this approach. The structures of proteins are not rigid, but flexible. For example, the hERG channel changes its structure from an open state to a closed state during the capture of ions or drugs [24], and cytochromes P450 are reported to change their structure according to the shapes of the drugs that bind to them [25,26]. In part, because of their structural flexibility these proteins can accommodate the wide diversity of drugs that are observed to bind to them. However, one drug may bind to a target protein in several different ways [26]. Therefore the use of a single protein structure in a docking study is sometimes insufficient, and the use of multiple protein structures is recommended for a thorough analysis.

Rajamani et al. [27] built a two-state homology model of the hERG channel (see also Chapter 16). When inhibitors were docked with the automatic docking software Glide [28], good agreement was obtained between the predicted affinity and the experimentally determined affinity.

Farid et al. [29] performed more exhaustive modeling of hERG channels. First, the ligand was docked into the hERG channel with Glide to produce a number of poses of the ligand. At that time steric clashes were reduced by the scaling of Van der Waals interactions or mutating critical amino acids to alanine. The optimal conformation of the protein was constructed with Prime [28] for each pose of the ligand, and the ligand was then docked again. From the ligand–protein complexes generated, candidate structures were selected using an energy-based scoring function. Excellent correlations between the calculated and experimentally determined inhibitory activity were made for a variety of drugs.

Flexible-docking procedures are time-consuming. There remains room for improvement, but such procedures yield more reliable structures of complexes when the target protein is structurally flexible than do rigid docking procedures. Furthermore, obtaining more than one docked structure aids our understanding of the binding interaction, provides insight into to the structural nature of both ligand and binding site, and helps generate further ideas for drug development.

20.7 PHYSICOCHEMICAL PROPERTIES LEADING TO PROMISCUOUS BINDING

Among the various interactions between proteins and ligands [30], the most selective are hydrogen-bonding interactions, which occur when strict chemical and geometrical conditions (element types, distances, and angles) are fulfilled [31]. Ionic interactions are also selective, since positively and negatively charged amino acids must be located in appropriate relative positions. The shape of the binding site plays an important role in ligand recognition. If the shape of the ligand does not fit that of the binding site, the ligand will be unable to bind to the protein effectively.

Hydrophobic interactions are also important in biological processes, and octanol/water partition coefficients (log *P*) are widely used to represent the hydrophobicity of molecules. Various computational methods have been developed to estimate log *P* values. In these methods, log *P* is calculated as a sum of fragment-based or atom-based contributions [32] (also see Chapter 9). The computations clearly show that the overall shapes of molecules have little effect on their hydrophobicity, and hence that hydrophobic interactions are not as specific as hydrogen-bonding interactions. In addition desolvation always accompanies the binding of a drug to a protein so that higher hydrophobicity generally enhances binding interactions.

Many kinds of interactions are possible for aromatic amino acids. Aromatic $\cdots$ aromatic ($\pi \cdots \pi$) interactions and cation $\cdots \pi$ interactions [33] are well known. Weak hydrogen-bonding interactions such as CH $\cdots \pi$, NH $\cdots \pi$, OH $\cdots \pi$, and CH $\cdots$ O play important roles in biological systems [31]. In effect aromatic amino acids can recognize a variety of structures. Moreover, as mentioned above, the high hydrophobicity of aromatic amino acids enhances binding interactions.

Proteins such as cytochromes P450 and the hERG channel accommodate structurally diverse drugs to the extent that the promiscuous binding of drugs to these proteins is a major problem in drug development. Recently highly selective inhibitors have been developed [34,35].

To aid our understanding of promiscuous binding interactions, it is helpful to compare highly selective interactions and promiscuous interactions as revealed by the X-ray crystallographic structures of the complexes formed. An example of a highly selective interaction is the binding site of the complex formed between imatinib, which is used in the treatment of patients with chronic myeloid leukaemia [34], and Abl kinase [36] is shown in Figure 20.4*a*. The amino acids locating within 4 Å of imatinib (Figure 20.4*a*) or erythromycin (Figure 20.4*b*) are depicted. Note the three characteristic interactions of the imatinib/Abl kinase complex:

- Six hydrogen bonds (shown by dashed lines) tightly connect imatinib to Abl kinase.
- The shape of the binding site of Abl kinase is long and narrow, so that imatinib, which is long, fits well into the binding site.
- Imatinib (molecular weight: 493.6 Da) is surrounded by 21 amino acids.

Although a cluster of aromatic rings is formed by three aromatic amino acids and the terminal hetero rings of imatinib, mutagenesis experiments suggest that aromaticity is not a critical property of the amino acid at position 317 [37].

As an example of a promiscuous interaction, the binding site of the complex formed between erythromycin and cytochrome P450 3A4 [26] is shown in Figure 20.4*b*. Characteristic interactions of this complex are the following:

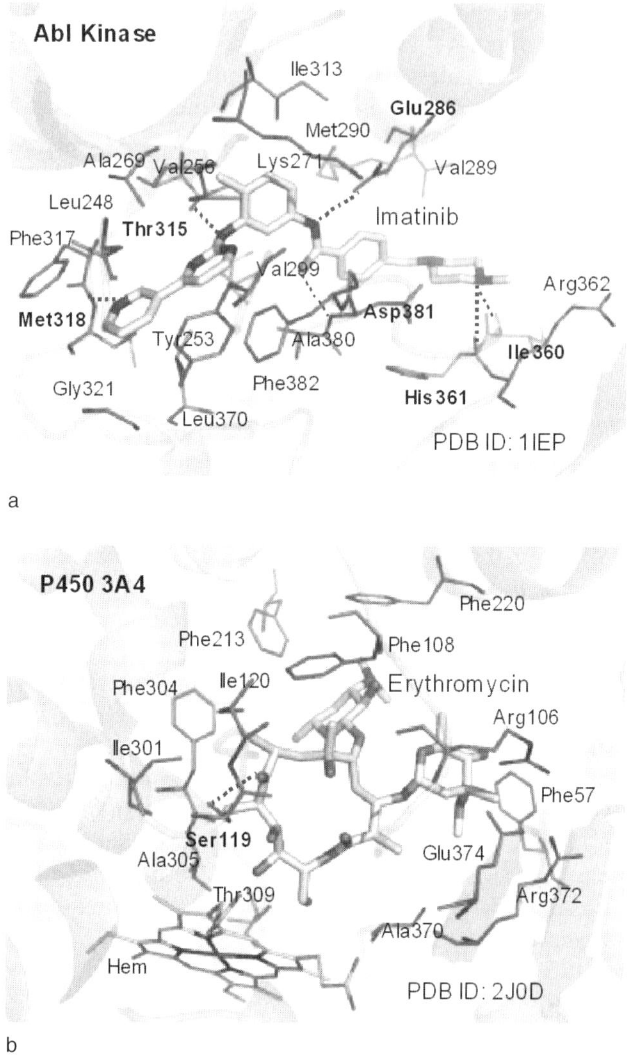

Figure 20.4 X-ray structures of the binding sites of the Abl kinase/imatinib and cytochrome P450 3A4 /erythromycin complexes (PDB: Protein Data Bank). The amino acids located within 4 Å of ligands are depicted. The hydrogen bonding interactions are shown by dashed lines, and the amino acids participating in hydrogen-bonding interactions are labeled in bold. The images were produced with Pymol (DeLano Scientific LLC). See color plates.

- Five aromatic amino acids form an aromatic cluster surrounding erythromycin.
- Hydrophobic and aromatic amino acids form a large, round binding site.
- Only one hydrogen bond is present.

- Erythromycin (molecular weight: 733.9 Da) is surrounded by 14 amino acids.

There are three ionic amino acids near erythromycin, but their ionic groups do not directly interact with erythromycin. Even though erythromycin is much larger than imatinib, fewer amino acids surround erythromycin in P450 3A4 than surround imatinib in Abl kinase.

These results show that Abl kinase strictly recognizes imatinib, and a complex is formed in which hydrogen-bonding interactions and shape complementarities are important. By contrast, P450 3A4 only loosely recognizes erythromycin, and a complex is formed in which hydrophobic and aromatic interactions are important. These examples suggest that promiscuous binding is largely governed by the physicochemical characteristics of the binding sites. The differences in binding interactions observed in these examples are largely attributable to differences in the nature and number of the amino acids at the binding sites. It is often found that the hydrophobicity and aromaticity of drugs are associated with unfavorable effects. For example, hydrophobic interactions are reported to increase the affinity of binding to hERG [16] and cytochrome P450 [38].

20.8 CHEMICAL MODIFICATION AND TARGET SELECTION

Ironically, while subtle chemical modification often results in the loss of pharmacological activity, even intensive chemical modification often fails to decrease unfavorable interactions with promiscuous proteins. Such a finding is not unexpected, since promiscuous proteins can recognize a variety of structures. For the recognition of diverse structures, unspecific interactions such as hydrophobic interactions are more suitable than specific interactions such as hydrogen-bonding interactions. Consequently one effective way to avoid undesirable interactions is to increase as much as possible the proportion of specific interactions among the binding forces.

Drugs with simple or common pharmacophores cannot discriminate small differences among binding sites. Pharmacophores for ADME/Tox proteins are generally not complex [39]. By contrast, drugs with complex or unusual pharmacophores have the potential to bind selectively to target proteins. Therefore chemical modification intended to increase the complexity of the pharmacophore is one way to avoid unfavorable effects arising from promiscuous interactions.

Although it is possible to modify the chemical structures of drugs, it is of course not possible to change the structure of a protein. The probability of encountering unfavorable effects is expected to be high when the binding site of the target protein is structurally very similar to that of an ADME/Tox protein. Accordingly, the shape of the binding site of the target protein can greatly influence the fate of a drug-development project.

The number of available X-ray structures of proteins is increasing, and computationally modeled structures of proteins have recently become suitable for practical use. These developments in exploring the structures of the binding sites of proteins can now aid in the selection of target proteins. Usually the best way to avoid undesirable interactions is to discard target proteins with problematic binding sites. However, there are many cases where the target protein is so attractive for drug development that it is hard to give it up as a target. In such cases it may be worth exploring the binding site to search for a way of avoiding the problem of undesirable interactions. If a part of the binding site of the target protein is structurally unique, and it is possible to redesign drugs to enable them to interact with the unique part so as to enhance their activity at the target in a sufficiently selective manner, the problem may be overcome. Furthermore the physicochemical properties of the protein surfaces near the binding site could influence the binding affinity to some extent, so consideration of the surface properties may be useful.

There are many kinds of binding sites in target proteins. Some are large and hydrophobic, and others are small and hydrophilic. When the binding site of a protein is composed of a large number of hydrophobic amino acids and a small number of hydrophilic or polar amino acids, the protein prefers to bind large, hydrophobic drugs, that is, drugs that can interact well with hydrophobic amino acids and fill the large binding space available. As mentioned above, hydrophobic interactions are a very common mechanism of molecular recognition, so proteins with hydrophobic binding sites tend to recognize drugs rather unselectively. In contrast, proteins with very hydrophilic binding sites recognize hydrophilic drugs well, yet drugs that are too hydrophilic tend to be poorly transported across membranes [40]. Proteins that can recognize drugs with complex pharmacophores are suitable targets for the development of specifically interacting drugs. Consideration of four structural characteristics of binding sites for drug development is recommended:

- No resemblance to binding sites of ADME/Tox proteins
- Not too large and not too small
- Not too hydrophobic and not too hydrophilic
- Sufficiently complex

The number of available structures of ADME/Tox proteins is increasing, so it is especially important at the beginning of a project to examine the binding sites of target proteins and carefully select target proteins while giving due consideration to ADME/Tox concerns.

20.9 CONCLUSIONS

The best way to overcome ADME/Tox problems is to understand the underlying causes of the problem, and for this purpose simple and clearly understandable models are likely to be the most useful. They are invaluable to medicinal chemists struggling to reduce unfavorable effects of drugs, as well as to those responsible for evaluating the ADME/Tox characteristics of drugs. In addition the combined use of various kinds of interpretable models greatly enhances their value.

The promiscuous recognition of structurally diverse drugs by certain proteins is a critical problem in the field of ADME/Tox. In the recognition of diverse structures, unspecific interactions such as hydrophobic interactions are more suitable than specific interactions such as hydrogen-bonding interactions. An effective way to avoid such unspecific interactions is to increase as much as possible the proportion of specific interactions among the binding forces. It is also important at the beginning of a project to closely examine the binding sites of target proteins to determine whether they structurally resemble the binding sites of ADME/Tox proteins. Even if such a resemblance is found, it is worth exploring the target binding site carefully, to look for a part of the site that can be exploited to enhance the activity of a drug toward the target protein. In the future such computational methods should become widespread among companies exploring drugs that target two or more proteins simultaneously for their clinical effect.

REFERENCES

1. Lipinski CA, Lombardo F, Dominy BW, Feeney PJ. Experimental and computational approaches to estimate solubility and permeability in drug discovery and development settings. *Adv Drug Deliv Rev* 2001;46:3–26.
2. Palm K, Stenberg P, Luthman K, Artursson P. Polar molecular surface properties predict the intestinal absorption of drugs in humans. *Pharm Res* 1997;14:568–71.
3. Palm K, Luthman K, Ungell AL, Strandlund G, Artursson P. Correlation of drug absorption with molecular surface properties. *J Pharm Sci* 1996;85:32–9.
4. Clark DE. Rapid calculation of polar molecular surface area and its application to the prediction of transport phenomena: 1. Prediction of intestinal absorption. *J Pharm Sci* 1999;88:807–14.
5. Clark DE. Rapid calculation of polar molecular surface area and its application to the prediction of transport phenomena: 2. Prediction of blood–brain barrier penetration. *J Pharm Sci* 1999;88:815–21.
6. Ertl P, Rohde B, Selzer P. Fast calculation of molecular polar surface area as a sum of fragment-based contributions and its application to the prediction of drug transport properties. *J Med Chem* 2000;43:3714–7.
7. Song M, Clark M. Development and evaluation of an in silico model for hERG binding. *J Chem Inf Model* 2006;46:392–400.
8. Cristianini N, Shawe-Taylor J. *An introduction to support vector machines and other kernel-based learning methods.* Cambridge: Cambridge University Press, 2000.
9. Clark M. Generalized fragment-substructure based property prediction method. *J Chem Inf Model* 2005;45:30–8.

10. Kho R, Hodges JA, Hansen MR, Villar HO. Ring systems in mutagenicity databases. *J Med Chem* 2005;48:6671–8.
11. <http://toxnet.nlm.nih.gov/cgi-bin/sis/htmlgen?CCRIS>.
12. Lilienfeld DE, Stolley PD. *Foundations of epidemiology*, 3rd edition. New York: Oxford University Press, 1994.
13. Wolter KM. *Introduction to variance estimation*. New York: Springer-Verlag, 1985.
14. <http://www.chemapps.com/products.html>.
15. Topliss JG, Costello RJ. Change correlations in structure-activity studies using multiple regression analysis. *J Med Chem* 1972;15:1066–8.
16. Yoshida K, Niwa T. Quantitative structure-activity relationship studies on inhibition of HERG potassium channels. *J Chem Inf Model* 2006;46:1371–8.
17. <http://www.biobyte.com/bb/prod/clogp40.html>.
18. Petitjean M. Applications of the radius-diameter diagram to the classification of topological and geometrical shapes of chemical compounds. *J Chem Inf Comput Sci* 1992;32:331–7.
19. Gasteiger J, Marsili M. Iterative partial equalization of orbital electronegativity—A rapid access to atomic charges. *Tetrahedron* 1980;36:3219–28.
20. <http://www.chemcomp.com/>.
21. Jiang Y, Lee A, Chen J, Cadene M, Chait BT, MacKinnon R. The open pore conformation of potassium channels. *Nature* 2002;417:523–6.
22. Recanatini M, Poluzzi E, Masetti M, Cavalli A, De Ponti F. QT prolongation through hERG K(+) channel blockade: Current knowledge and strategies for the early prediction during drug development. *Med Res Rev* 2005;25:133–66.
23. Stahl M. Structure-Based Library Design. In: Böhm HJ, Schneider G, editors, *Virtual screening for bioactive molecules.* Weinheim: Wiley-VCH, 2000. p. 229–64.
24. Sanguinetti MC, Tristani-Firouzi M. hERG potassium channels and cardiac arrhythmia. *Nature* 2006;440:463–9.
25. Scott EE, He YA, Wester MR, White MA, Chin CC, Halpert JR, et al. An open conformation of mammalian cytochrome P450 2B4 at 1.6-Å resolution. *Proc Natl Acad Sci USA* 2003;100:13196–201.
26. Ekroos M, Sjogren T. Structural basis for ligand promiscuity in cytochrome P450 3A4. *Proc Natl Acad Sci USA* 2006;103:13682–7.
27. Rajamani R, Tounge BA, Li J, Reynolds CH. A two-state homology model of the hERG K^+ channel: Application to ligand binding. *Bioorg Med Chem Lett* 2005;15:1737–41.
28. <http://www.schrodinger.com/Products.php?mID=6&sID=0&cID=0>.
29. Farid R, Day T, Friesner RA, Pearlstein RA. New insights about HERG blockade obtained from protein modeling, potential energy mapping, and docking studies. *Bioorg Med Chem* 2006;14:3160–73.
30. Lewis DF, Dickins M. Substrate SARs in human P450s. *Drug Discov Today* 2002;7:918–25.
31. Desiraju GR, Steiner T. The weak hydrogen bond. In: *Structural chemistry and biology*. New York: Oxford University Press, 1999.

32. Wildman SA, Crippen GM. Prediction of physicochemical parameters by atomic contributions. *J Chem Inf Comput Sci* 1999;39:868–73.

33. Gallivan JP, Dougherty DA. Cation-π interactions in structural biology. *Proc Natl Acad Sci USA* 1999;96:9459–64.

34. Druker BJ, Sawyers CL, Kantarjian H, Resta DJ, Reese SF, Ford JM, et al. Activity of a specific inhibitor of the BCR-ABL tyrosine kinase in the blast crisis of chronic myeloid leukemia and acute lymphoblastic leukemia with the Philadelphia chromosome. *N Engl J Med* 2001;344:1038–42.

35. Kimura S, Naito H, Segawa H, Kuroda J, Yuasa T, Sato K, et al. NS-187, a potent and selective dual Bcr-Abl/Lyn tyrosine kinase inhibitor, is a novel agent for imatinib-resistant leukemia. *Blood* 2005;106:3948–54.

36. Nagar B, Bornmann WG, Pellicena P, Schindler T, Veach DR, Miller WT, et al. Crystal structures of the kinase domain of c-Abl in complex with the small molecule inhibitors PD173955 and imatinib (STI-571). *Cancer Res* 2002;62:4236–43.

37. Burgess MR, Skaggs BJ, Shah NP, Lee FY, Sawyers CL. Comparative analysis of two clinically active BCR-ABL kinase inhibitors reveals the role of conformation-specific binding in resistance. *Proc Natl Acad Sci USA* 2005;102:3395–400.

38. Lewis DF, Jacobs MN, Dickins M. Compound lipophilicity for substrate binding to human P450s in drug metabolism. *Drug Discov Today* 2004;9:530–7.

39. de Groot MJ, Ekins S. Pharmacophore modeling of cytochromes P450. *Adv Drug Deliv Rev* 2002;54:367–83.

40. Hansch C, Leo A, Mekapati SB, Kurup A. QSAR and ADME. *Bioorg Med Chem* 2004;12:3391–400.

PART IV

APPLYING COMPUTERS TO TOXICOLOGY ASSESSMENT: ENVIRONMENTAL

21

THE TOXICITY AND RISK OF CHEMICAL MIXTURES

JOHN C. LIPSCOMB, JASON C. LAMBERT, AND MOIZ MUMTAZ

Contents

21.1 TOXICOLOGY

Environmental media such as water and air are subject to contamination with myriad chemical and biological pollutants. Thus it is inevitable that in a lifetime humans are exposed to a diverse array of chemical mixtures through occupational, recreational, and/or domestic exposures. These mixtures may be simple, consisting of two or more definable compounds, or may be more complex containing several hundred related congeners and/or unrelated

Computational Toxicology: Risk Assessment for Pharmaceutical and Environmental Chemicals, Edited by Sean Ekins

compounds. Due to both the complexity of mixtures (number of component chemicals, physicochemical properties, environmental fate, etc.) and chemical-to-chemical differences in toxicokinetic and toxicodynamic properties of component chemicals, toxicities associated with exposure to mixtures are often difficult to characterize. Indeed associations between mixture exposure and biological alteration are typically drawn from empirical observations of single chemical toxicities in animal studies. Many regulatory agencies, including the US EPA, FDA, OSHA, and ATSDR, are highly dependent on human epidemiological evidence when available and/or animal toxicity data to establish standards, or acceptable levels (AL) of chemical exposure, for protection of human health. Reference values such as an oral reference dose (RfD), threshold limit value (TLV), or minimal risk level (MRL) are derived based on identification of a reliable point of departure (POD), which is typically the dose or concentration of a chemical that is without effect in a human population or suitable animal species (no-observed-adverse-effect level, NOAEL).

Estimations of human risk from chemical exposure using animal NOAELs as PODs require use of adjustment factors to account for uncertainty in the extrapolation. A scientifically defensible POD and the composite uncertainty factor are used to derive an AL of exposure for chemicals, which may be several orders of magnitude lower in dose than the highest dose known to be without adverse effect in animals. It should be noted that NOAELs are often limited by dose selection in animal studies. In particular, NOAEL values do not inform the shape of the dose–response curve in the low-dose region. A statistical modeling approach, benchmark dose, is often employed to identify a dose–response level that approximates the NOAEL, more reflective of the entire data set, and more health protective. The lower statistical bound of this modeled benchmark dose (e.g., $BMDL_{10}$) is then used as the POD for derivation of a reference value. An additional complicating factor in mixtures risk assessment is how to integrate data not typically used in single chemical risk assessment activities.

For toxicity and risk assessment of chemical mixtures, several terms and definitions should be understood (see Appendix 1). Characterizing the relative contribution of mixture components is highly dependent on identification of the target organ or tissue dose, mode of action, and duration of effect. Consideration of data describing mode of action (MOA) or mechanism of action is the first step of hazard identification in risk assessment. For risk assessment of mixtures, chemicals are separated into groups according to their MOA. While MOA data are typically not available, they are the basis for dividing chemicals in a mixture into groups for risk analysis. When the data are not available, it is common practice to assume that the chemicals act via an independent (dissimilar) MOA. While there is a difference between MOA and mechanism of action, for the purpose of a chemical mixture risk assessment, the difference is moot—the grouping can be made on the basis of target tissue (as described later).

The US EPA guidance on mixtures [1] defines MOA as a series of "key events" and processes, starting with interaction of an agent with a cell and proceeding through operational and anatomical changes causing disease formation. A key event, as defined in the 2005 *US EPA Guidelines for Carcinogen Risk Assessment* [2], is "an *empirically observable precursor step that is itself a necessary element of the mode of action* or is *a biologically based marker* for such an element". By contrast, the mechanism of toxic action requires a more detailed understanding and description of events, often at the molecular and cellular level [3,4]. For well-defined modes of action, mechanistic data may provide merely more detail in support of identified key events. However, for less well-defined toxic modes of action, mechanistic data can lead to identification of previously unknown obligatory steps in the causal pathway to toxicity. Regardless of the level of detail (MOA vs. mechanism) available, the risk assessor must defend the parsing of mixture components into groups. A similarity in either MOA or mechanism of action may be used to combine chemicals into groups.

When mixture component chemicals are determined to act through independent toxic modes of action, response addition is recommended. When component chemicals act via a similar toxic mode of action, dose addition is recommended [1]. In response addition, the net response estimated for the mixture is equivalent to the sum of the individual component responses [1]. The assumption is that for a toxicity following exposure to a mixture, the type or degree of adverse effect caused by one mixture component has no direct impact on the type or degree of adverse effect caused by a second component. Response addition requires dose–response data for all components of a mixture sufficient to define the slope of the dose–response carve with a degree of certainty sufficient to support predictions of risk at uncharacterized, environmentally relevant doses.

For mixtures of components that are determined to act through a similar (common) mode of action, the likelihood of toxicity associated with a mixture is determined by adding the normalized doses of the components, where the concept of threshold is applied to the mixture rather than to the individual components as is done in response addition. The assumption for dose addition is that components are essentially toxicological "clones" of one another such that the relative proportions of each in a mixture are treated as dilutions of one another [1,4,5]. Toxicity data for the components must be sufficient to support their grouping according to a common mode or mechanism of toxicity.

These addition models are more applicable to mixtures of chemicals at low environmental doses or concentrations where the likelihood of interactions such as potentiation, synergy, and antagonism is low. However, at higher doses the potential exists for direct chemical and/or biological interactions among individual compounds within a mixture that may alter toxicokinetic or toxicodynamic properties, rendering additivity models inappropriate for estimation of adverse mixture effects [1,6].

21.2 RISK ASSESSMENT

21.2.1 The Completed Exposure Pathways

Mixtures Risk Assessment The mere occurrence of chemicals and contaminants in the environment does not increase the potential of risk to human or environment, but their exposure does. A five-step process is used to determine the extent, route, and duration of exposure and includes its environmental fate and transport. This process allows identification of likely site-specific exposure to chemicals and chemical mixtures, the extent of exposure, and the conditions under which the exposure occurred. This way contaminants of concern can be identified in a systematic manner by combining the chemical hazard and exposure data [7].

Step 1. The contaminant source or release point, such as landfills or storage drums, are identified.

Step 2. The environmental fate and means of transport are substantiated, including movement of contaminants through different environmental media and change in their characteristics, such as degradation as a function of environmental influences.

Step 3. A specific point or area of exposure such as a geographic location is identified.

Step 4. The route or means by which contaminants come in contact with human populations such as inhalation, ingestion, or dermal contact are identified.

Step 5. The potentially exposed population and characterization of the population that may come or have come in contact with the contaminant are identified.

This process helps set up a system of pathways from the source of contamination, to the mechanism of transport through environmental media, to routes of exposure, and to the exposed or potentially exposed population. The duration and frequency of exposure, chemical classes associated, and the significance of each pathway are documented as well.

Exposures to chemicals are not limited to unintentional exposure to environmental contaminants. Through eating, drinking, and eating, we are exposed to a variety of other chemicals including drugs, pharmacologically active compounds present in our foods, medications, and recreational use such as alcohol and tobacco. The recent National Health and Nutrition Examination Survey (NHANES) reports published by the Centers for Disease Control and Prevention (CDC) provide identification, detection, and quantitation of over 100 chemicals in human populations [8]. Various classes of exogenous and synthetic as well as naturally occurring chemicals are included, and additionally metals, pesticides, organochlorines, pyrethroids, organophosphates, herbicides, phthalates, phytoestrogens, polycyclic aromatic hydrocarbons, and tobacco smoke products. Regardless of how these chemicals enter the body, they are absorbed and distributed within common compartments, and hence are present as mixtures.

21.2.2 The Risk Assessment Methods

As can be expected, some chemicals are data-rich while others are data-poor. Often they are data-poor, especially for chemical mixtures where the data may be adequate, barely adequate or nonexistent for a specific mixture. Thus limited methods are available for the toxicity assessment of chemical mixtures [1,9]. Even though there are various uncertainties and assumptions embedded in these methods, the following three approaches have gained acceptance by the regulatory agencies and the regulated community for risk assessment of mixtures of industrial, occupational, and environmental chemicals. The method employed is on a case-by-case basis, it depends on the exposure scenario and the quality of available data on exposure and toxicity.

Whole Mixtures Approaches

The Mixture of Concern Approach This approach can be used only if adequate toxicity data on the specific chemical mixture of concern are available. Having the data means that sufficient and appropriate experimental testing has been done and that the results are on record about this mixture's toxicity. This is the most direct and simplest approach, with the fewest uncertainties to derive a criterion or acceptable level for stable mixtures such as fuel oils, jet fuels, mixtures of polychlorinated biphenyls (PCBs), and polybrominated biphenyls (PBBs) [10]. However, working from available data may leave uncertainties about the precise composition of a mixture undetermined. Very few mixtures other than those mentioned above have been studied adequately for toxicity assessment. Hence this data-mining approach is the least frequently used method.

The Similar Mixture Approach On a case by case basis a candidate mixture or groups of mixtures could act similarly [1,11]. Analogies are used when adequate information is not available for the mixture of concern, and this approach is often applied to complex mixtures that have been extensively investigated, such as coke oven emissions, diesel exhaust, and woodstove emissions. However, information should be available to ascertain that a mixture is sufficiently similar to the mixture of concern. There are no quantitative criteria to decide when a mixture is sufficiently similar, though it is recognized that some key components should be represented in similar proportions [1]. As in the previous approach, the mixture must be treated as an individual chemical because the whole mixture has to be experimentally tested to some extent.

Component-Based Approaches The design and implementation of a complex mixture toxicity testing regimen is a laborious undertaking [12]. Once completed for the mixture of concern or a defensible toxicologically similar mixture [1,11], the empirical data for the whole mixture can be used with fewer uncertainties in risk assessment. However, often it is the case that the absence of whole mixture data forces consideration of data from single chemical components of the mixture. These data can be combined with appropriate assumptions. Because most toxicity data are generated using a single chemical exposure approach, in the remainder of this chapter we will mostly discuss component-based approaches to chemical mixture risk assessment.

The initial process in the application of toxicity (dose–response) data in risk assessment is the extrapolation of findings to establish acceptable levels (AL) of human exposure. These levels may be reference values (inhalation reference concentrations, RfC; or oral reference doses, RfD), minimal risk levels (MRL) values, occupational exposure limits, and so on. When the toxicity data are derived from animals, the lowest dose representing the NOAEL (preferably) or the LOAEL defines the point of departure (POD). In setting human RfD, RfC, or MRL values, the POD requires several extrapolations (see [13] and revisions). Extrapolations are often made for interspecies differences, intraspecies variability, duration of exposure, and effect level. Each area is generally addressed by applying a respective uncertainty factor having a default value of 10; their multiplicative value is called the composite uncertainty factor (UF). The UF is mathematically combined with the dose at the POD to determine the reference value:

$$\text{Reference value} = \frac{\text{POD}}{\text{UF}}. \tag{21.1}$$

The development of AL values can aid the component-based approaches to risk assessment based on dose addition. AL values can be developed for the critical effect and for secondary effects. For chemicals with older AL values developed from point estimates of the POD (e.g., NOAEL values), when more recent and more thorough dose response data are available, the AL should be re-derived using more advanced (i.e., benchmark dose analysis) approaches to estimating the POD.

The US EPA has developed several methods to estimate the risks or the potential for risks from chemical mixture exposures. The hazard index (HI) provides an indicator of potential health risk and is based on data for the critical effect [14]. There are several variants of the HI approach. In each of the variants a hazard quotient (HQ) is developed by dividing the human exposure by the AL. Thus, in all cases, when the HQ or a sum of several HQs (an HI) is greater than 1, this indicates a potential for elevated human health risks. The screening HI approach is the most conservative—it combines all chemicals in the mixture into a single group and develops the HI by summing all the HQ values for the mixture components, regardless of their critical effect. The second approach is the HI approach in which the HQ values for each chemical are segregated by organ or system affected; HQ values are developed for each affected organ or system. The third is the target-organ toxicity dose (TTD) approach, and it differs in that it includes data for each organ or tissue affected, whether or not it is the critical target. Of these approaches, the TTD approach takes best advantage of the available toxicity data. To illustrate these points, we have here developed and analyzed toxicity data for a hypothetical eight-chemical mixture.

When data are available and sufficient to allow grouping of chemicals according to similarities in mechanism or mode of action, such should be accomplished. However, that level of detail is seldom available, and more pragmatic chemical groupings have to be developed—the HI approach groups chemicals at the level of the effected organ, tissue, or system [15]. Thus the HI approach (compared to a MOA-based grouping of components) is accompanied by a lower degree of confidence (higher degree of uncertainty), and decisions made from HI data should refrain from "bright-line" type approaches. Considerable judgment is needed for interpretation of results. Further, compared to whole mixture approaches, these approaches rely on human exposure levels (e.g., MRL values), which are often based on single-chemical toxicity tests in animals and extrapolated through the use of uncertainty factors. Chemical mixtures can contain components that are structurally similar, are metabolized by the same enzyme(s), and act through similar modes and or mechanisms. Chemical components may exacerbate or antagonize the effects caused by one another. These interactions can result in chemical interactions resulting in modifications of tissue doses or tissue responses. Frequently this level of detail is not known. Given recent subdivision of the uncertainty factors governing animal to human extrapolation (UFA) and human interindividual variability (UFH) into toxicokinetic (TK) and toxicodynamic (TD) components, the process of extrapolation of effect levels noted in single chemical (animal) exposures ignores potential chemical interactions that may occur in extrapolation steps addressed by applying uncertainty factors. In comparison, whole mixtures approaches rely on toxicity data developed in the test species following exposure to the mixture or a toxicologically similar mixture.

Table 21.1 shows a hypothetical data set for eight chemicals. Note that while some of the compounds affect tissues at very nearly equal doses, others are separated by one to two orders of magnitude. This example makes use of oral reference dose values, but in practice, values for acceptable levels could be TLVs, OELs, RfD, MRLs, or TTDs. In this example the lowest dose identifies the critical target organ and represents the oral RfD for each compound (denoted by *); secondary effects are also listed. Data on multiple effects for a single chemical will be used differently in the HI approaches presented, with differing results.

The HI Approach For historical reasons seemingly based on data availability, the HI approach is the most often used approach, and it is traditionally applied to noncarcinogens. The hazard index is found by integratings the exposure level and the related toxicity into a single value with potency-weighted dose additions. The HI approach is simple to implement but somewhat limited in its scope, however, since it can underpredict or overpredict risk estimates. Initially

TABLE 21.1 Organ/Tissue Response Data Extrapolated to Human Oral Reference Dose and Target Organ Tissue Dose Values (in mg/kg-day)

Chemical	Developmental	Kidney	Heart	Spleen
A	5E-2*	Not affected	8E-2	6E-2
B	4.5E-2	1E-2*	4E-2	Not affected
C	Not affected	8E-2	2E-2*	3E-2
D	8E-2	5E-2	8E-3*	1E-2
E	2E-2*	Not affected	Not affected	Not affected
F	1E-3	4E-3	6E-4*	8E-4
G	1E-1	4E-2*	8E-2	6E-2
H	Not affected	8E-3*	8E-2	4E-3

Note: *Reference dose value; other values are TTD values for secondary effects.

the potential health hazard from exposure to each chemical is estimated by calculating its individual hazard quotient (HQ). The HQ is derived by dividing a chemical's exposure level (E) by its acceptable/allowable exposure level (AL) such as a MRL, RfD, an occupational exposure limit (OEL), or a TLV [1], as follows:

$$HQ_i = \frac{E_i}{AL_i}, \quad (21.2)$$

where, for chemical *i*, E is the human exposure in mg/kg-day and AL is the acceptable level for exposure (e.g., the RfD value). The HI of the mixture is then calculated by summing the component hazard quotients.

In a manner analogous to the hazard index approach for noncarcinogens, hazard quotients for carcinogenic mixture components can be estimated by dividing chemical exposure levels by doses (DR) associated with a set level of cancer risk; the HI is the sum of the HQ values [9,16]:

$$HQ_i = \frac{E_i}{DR_i}. \quad (21.3)$$

The HI approach assumes that all components have similar joint action; that is, their uptake, pharmacokinetics, and dose–response curves are similar in shape.

Two other methods are based on the tenets of dose addition but differ from the HI approach. Those are the relative potency factor (RPF) and the toxicity equivalent factor (TEF) approaches.

The RPF Method The RPF method defines similarity at the level of the type of effect noted (e.g., cholinesterase inhibition) to group chemicals into RPF sets. Chemicals in a given RPF set should (1) share a common the sequence of key cellular and biochemical events (measurable parameters) that result in a toxic effect and (2) demonstrate similarly shaped dose–response curves—at least within the range of doses relevant to the exposures of interest [1,5]. For RPF values, response is measured as the toxicologic outcome rather than as response level to a specific measure of chemical interaction (i.e., a measure of some mechanistic event). An index chemical is identified based on the extent of its database and the extent to which its characteristics best represent the overall characteristics of the RPF set. Once a response level is chosen, the relative potency of individual components is expressed as a ratio of the levels of the effective dose of the individual component to the dose of the index chemical resulting in the same level of response. Finally the dose of the individual component is converted to an index chemical equivalent dose (ICED) by multiplying the RPF value for that component by the exposure dose of the component. ICED values are summed and the response estimated by interpreting the response from the dose–response curve describing the index chemical. The RPF method has been used for organophosphate insecticides [5] and polycyclic aromatic hydrocarbons (PAHs) [17].

The TEF Method The TEF method is a variant of the RPF method, but it relies on more data than the RPF method. To develop TEF values, the available data must support a grouping of chemicals into the same class, and must go so far as to describe and defend similarity at the level of the molecular basis for the toxicologic effect. Under the TEF approach, differences in potencies of components are defined on the basis of differences in response quantified at some mechanistic level of detail (e.g., receptor binding). In contrast, the RPF approach quantifies differences at the level of the effect (i.e., expression of an adverse health outcome). Another key difference between TEF and RPF approaches is that TEF values apply to all effects and exposure routes for the components, whereas RPF values are generally defined as route and effect-specific. To mathematically estimate a mixture risk is analogous to that for estimating risks via the TEF (or RPF) approach. An index chemical is identified from the TEF class. Then the concentration of each component of the mixture is scaled for relative potency to the index chemical by multiplying its TEF value by its exposure dose to generate toxic equivalents (TEQs). These values are summed to produce a value defined as the total TEQ for the mixture [1,18]. The TEF approach has been used to assess the health risks of chlorinated dibenzo-*p*-dioxins (CDDs).

There are two types of HI that are typically performed: the screening level HI [14] and the HI [16]. The screening level HI is often used for site-specific applications [14], and it differs from the HI approach in that it is more health conservative (it screens across all effects) and is accompanied by a higher degree of uncertainty. Under the screening HI approach all chemicals in the mixture are evaluated as if they exerted toxicity at the level of the organism. For each chemical, the AL used in the calculations is the lowest of

TABLE 21.2 Screening Hazard Index Calculation

Chemical	Exposure*	RfD*	Organ/Tissue	HQ
A	6E-3	5E-2	Developmental	0.12
B	1.5E-3	1E-2	Kidney	0.15
C	9E-3	2E-2	Heart	0.45
D	6E-4	8E-3	Heart	0.075
E	8E-3	2E-2	Developmental	0.4
F	4E-5	6E-4	Heart	0.067
G	6E-3	4E-2	Kidney	0.15
H	4E-3	8E-3	Kidney	0.5
Hazard index				**1.91**

Source: Risk Assessment Guidance for Superfund [14].

Note: *Both exposure and RfD are in units of mg/kg-day.

the AL values developed for the affected tissues; it will represent the AL for the critical effect. For example, exposure levels for chemicals affecting the liver, kidney, brain, gastrointestinal tract, and immune system are divided by the AL for the critical effect to calculate HQ values. The HQ values for each chemical, regardless of the system affected, are summed to produce the HI. This approach may be sufficient if the HI value is below 1.0 [14]; if the HI value is above 1.0, then a more refined approach to estimating HI should be undertaken. Based on a set of hypothetical values, table 21.2 demonstrates the screening HI approach, and the HQ is 1.91.

For this mixture of chemicals, differences in the outcomes between the screening HI approach and the HI approach become apparent. If a value of 1.0 for the HI is chosen as a decision point, then the screening HI evaluation with the HQ value of 1.91 indicates that an additional examination is warranted [14]. Table 21.3 demonstrates the application of the HI approach to this same set of data (toxicity data from Table 21.1, *exposure data* from Table 21.2). Here chemicals are grouped by affected organ or tissue, with chemicals A and E having their critical effect in developing tissues; chemicals B, G, and H having their critical effect manifest in the kidney; and chemicals C, D, and F demonstrating their critical effect in the heart. Developmental toxicity is not produced by chemicals C and H, kidney is not affected by chemicals A and E, heart is not affected by chemicals E, and the spleen is not affected by chemicals B and E. These differences are responsible for differences in the HI values estimated by the following methods. Note that none of the chemicals have a critical effect in the spleen. The highest (organ-specific) HQ value is 1.39, somewhat reduced from 1.91. The "standard" HI approach where hazard quotients are segregated by target organ effect and/or mode or mechanism of action may provide a more realistic estimate of risk, but could potentially *overestimate* the hazard to the intact organism [14]. This concern is addressed in the target organ toxicity dose (TTD) approach, below. In this hypothetical 8 component mixture, segregating the hazard quotients according to target

TABLE 21.3 Hazard Index Calculation

			Organ/Tissue HQ			
Chemical	Exposure*	RfD*	Developmental	Kidney	Heart	Spleen
A	6E-3	5E-3	0.12**	0.12	0.12	0.12
B	1.5E-3	1E-3	0.15	0.15**	0.15	Not affected
C	9E-3	2E-2	Not affected	0.45	0.45**	0.45
D	6E-4	8E-3	0.075	0.075	0.075**	0.075
E	8E-3	2E-2	0.4**	Not affected	Not affected	Not affected
F	4E-5	6E-4	0.067	0.067	0.067**	0.067
G	6E-3	4E-2	0.15	0.15**	0.15	0.15
H	4E-3	8E-3	Not affected	0.5**	Not affected	0.5
Hazard index			**0.96**	**1.39**	**1.01**	**1.36**

Source: [14].

Note: *Both exposure and RfD are in units of mg/kg-day; **critical effect.

organ (using the HI approach) leads to a decrease in the aggregate hazard indices compared to that developed under the screening HI approach. This decrease in the HI value is supportive of a decreased concern for the risk from exposure to this mixture under the exposures evaluated. However, neither of these approaches employed the full spectrum of actual toxicity (dose–response) data for effects other than the critical effect. Each focused only on the dose–response data for the critical effect, ignoring dose–response data for other tissues (secondary effects) that may be only minimally less sensitive.

The Target-Organ Toxicity Dose (TTD) Realistic risk estimation is vital for the acceptance and communication of risk. An overestimation of risk due to ignoring dose–response data for secondary effects is possible when only MRLs or RfDs are used that are based on dose–response data for the critical effect [19]. To circumvent this problem, the concept of TTD was developed and employed [1,20]. TTD is in essence an effect or organ-specific criterion that uses the same methodology and process as the MRLs or RfDs. For a given chemical there can be an MRL for hepatotoxicity and a series of TTDs for nephrotoxicity and reproductive toxicity resulting from the same mixture exposure [21,22]. Thus the TTD method is a simple modification of the HI approach and yields a series of HIs for various toxic effects, including response levels that represent the effect, regardless of whether it is the critical effect or a secondary effect.

The same data for this hypothetical chemical mixture were subjected to a TTD analysis (Table 21.4). Here a HI value for each affected organ, tissue, or system is determined and combined with the exposure value to determine a HI for each effect of each chemical. By this more robust approach, concern is raised if the HI is a value above 1.0. Under the TTD approach, kidney, heart, and spleen would each have HI values unlikely to lead to a level of concern.

The basic limitation of the equation for potency-weighted dose additivity used in the HI and TTD approaches is that it does not show the influence of interactions on the overall joint toxicity of mixture. As a result the mixtures risk estimates can be either under- or overpredicted [1,14,20,21]. The National Academy of Sciences has proposed the use of additional safety factors if synergistic interactions are of concern [23].

21.2.3 Chemical Interactions

A chemical interaction is the combined effect of two chemicals resulting in a greater (synergistic, potentiation, or supra-additive) or lesser (antagonistic, inhibitive, sub-additive, or infra-additive) effect than expected based on additivity (see [24]) Bioaccessibility, bioavailability, and bioaccumulation, three chemical-specific properties, play a key role in their potential interactions. The co-occurrence and exposure of various chemicals in the environment can lead

TABLE 21.4 Target Organ Toxicity Dose Calculation

	Developmental			Kidney			Heart			Spleen		
Chemical	TTD	Exposure	HQ	TTD	Exposure	HQ	TTD	Exposure	HQ	TTD	Exposure	HQ
A	5E-2*	6E-3	0.12	NA	6E-3	ND	8E-2	6E-3	0.075	6E-2	6E-3	0.1
B	4.5E-2	1.5E-3	0.033	1E-2*	15E-3	0.15	4E-2	1.5E-3	0.038	NA	1.5E-3	ND
C	NA	9E-3	ND	8E-2	9E-3	0.113	2E-2*	9E-3	0.45	3E-2	9E-3	0.3
D	8E-2	6E-4	0.0075	5E-2	6E-4	0.012	8E-3*	6E-4	0.075	1E-2	6E-4	0.06
E	2E-2*	8E-3	0.4	NA	8E-3	ND	NA	8E-3	ND	NA	8E-3	ND
F	1E-3	4E-5	0.04	4E-3	4E-5	0.01	6E-4*	4E-5	0.067	8E-4	4E-5	0.05
G	2.5E-1	6E-3	0.024	4E-2*	6E-3	0.15	8E-2	6E-3	0.075	6E-2	6E-3	0.1
H	NA	4E-3	ND	8E-3*	4E-3	0.5	NA	4E-3	ND	2E-2	4E-3	0.2
Hazard index			**0.62**			**0.94**			**0.78**			**0.81**

Source: [20].

Note: * RfD value; NA = not affected; ND = calculation of TTD for this chemical/endpoint not warranted, value is zero.

to interactions at various levels in the body throughout the processes of absorption, distribution, metabolism, and excretion (ADME), resulting in changes in their bioavailability and interactions with macromolecules such as receptors and proteins in the body. They may interact synergistically, additively, or antagonistically depending on their mechanism of action, concentrations, and ratios. Such interactions can have direct implications to the joint toxicity of the mixture and evaluation of the potential adverse effects. These interactions can be important to public health and environmental scientists, who consider both low as well as high dose exposures. Each chemical component of a mixture has the potential to influence the toxicity of the other component. However, the magnitude or the capacity of this potential to interact is not so easy to determine and frequently unknown because of the various complexities in the mechanisms of interactions. There is very little guidance on how these interactions should be evaluated or incorporated into the overall risk assessment.

For whole mixtures, toxicity data are rare. Therefore human health risk assessment of mixtures is often based on the toxicokinetic and toxicodynamic properties of component chemicals. A complex aspect of component-based mixtures risk assessment is incorporation and integration of a potential myriad biological processes and chemical characteristics from single chemical toxicity data. One must consider characteristics of chemicals in a mixture (number of component chemicals, physicochemical properties such as lipophilicity or hydrophilicity, environmental fate, etc.) as well as route(s) of exposure, temporal association between exposures, and biological longevity of constituents within physiological compartments. In addition toxicokinetic and toxicodynamic properties may be exceedingly diverse both among species and within human populations. Therefore, for mixtures risk assessment, it is especially important to characterize the ADME (toxicokinetics) of toxins, which may include both parent compound/s and metabolite/s, and the biological interaction of toxins within a target organ or tissue (toxicodynamics) that encompasses all molecular and cellular processes involved in a toxicity of concern (i.e., mechanism and MOA).

A Weight-of-Evidence (WOE) Method To provide further guidance on the evaluation of chemical interactions and their impact on risk values calculated from data not reflective of the mixture, a weight-of-evidence (WOE) method was developed [25,26]. The assessment of WOE for interactions enables assessors to judge if the interactions influence the overall toxicity of the mixture and if the anticipated joint toxicity will be greater than or less than expected based on the principle of additivity. The WOE method yields a composite representation of all the evidence on toxicologic interactions from human studies to animal bioassay data; relevance of route, duration, and sequence; and the significance of interactions. The method consists of a classification scheme used to provide a qualitative and, if needed, a quantitative estimation of the effect of interactions on the aggregate toxicity of a mixture. The first two components of the scheme are major ranking factors for the quality of the mechanistic information that supports the assessment and the toxicologic significance of the information. The last three components of the WOE are modifiers that express how well the available data correspond to the conditions of the specific risk assessment in terms of the duration, sequence, routes of exposure, and the animal models. This method evaluates data relevant to joint action for each possible pair of components of a mixture, and as such, it requires mechanistic information and direct observation of toxicologically significant interactions. Initially, for each pair of component chemicals of a mixture, two binary weight of evidence (BINWOE) determinations are made to estimate the effect of the first chemical on the second chemical's toxicity, and also to estimate the effect of the second chemical on the toxicity of the first chemical. A BINWOE determination is a qualitative judgment, based on empirical observations and mechanistic considerations, that categorizes the most plausible nature of any potential influence of one compound on the toxicity of another for a given exposure scenario [12,21,25]. Once all of the qualitative WOE determinations have been made for each pair of compounds in the mixture, they can be used for a qualitative assessment of the overall mixture.

21.3 TOXICOKINETICS

The considerations above focus on effects at the tissue level, but do not give explicit consideration to dosimetry. TK is the field of science concerned with the ADME of chemicals in the body. Thus TK interactions can mediate alterations in tissue dosimetry and can then change tissue responses. Among the TK processes the most likely source of interaction is in metabolism. While a given chemical can produce several metabolites, each metabolic step is catalyzed by a single enzyme. Because of the multifunctional nature of enzymes, each enzyme may be responsible for metabolizing a great many compounds. A good example of this is provided by cytochrome P450 2E1, a membrane-bound enzyme that is responsible for the oxidation of a great many low molecular weight halogenated organic compounds. Although suicide (mechanistic) inhibition by substrates can occur, resulting in irreversible loss of enzyme function, the more common type of inhibition is competitive inhibition (CI). In CI, the degree of inhibition is determined by the relationship between the Michaelis constant (K_m) and the substrate concentration. In the instance where two substrates have the same K_m value, the substrate with the higher concentration will be metabolized at a higher fraction of its maximal initial rate, which can be predicted by the Michaelis-Menten rate equation below:

$$MR = \frac{(V_{max} * [s])}{K_m + [s]}, \tag{21.4}$$

where MR is the metabolic rate, V_{max} is the theoretical maximal initial velocity of the reaction and $[s]$ is the substrate concentration. Competitive inhibition is common among mixtures of similar components (e.g., low molecular weight halogenated solvents). Competitive inhibitors compete for the same enzymes and binding sites, and the steady state equation for competitive inhibition is shown below:

$$V = \frac{(V_{max} * [s])}{K_m\left(1 + \frac{(I)}{K_i}\right) + [s]} \tag{21.5}$$

While rate equations can be solved and metabolic rates can be estimated and determined easily in vitro because of the control of variables (specifically, substrate concentration), the situation is much more difficult in vivo. Here the advent of physiologically based pharmacokinetic (PBPK) models has offered a powerful tool for examining metabolic interactions that occur following exposure to chemical mixtures (see Chapter 3).

PBPK models are constructed to represent the anatomy, physiology, and biochemistry of (usually) mammalian test species and humans. They contain ordinary differential equations, simultaneously solved by computer-based programs like Advanced Continuous Simulation Language (ACSL; Aegis Technologies, Huntsville, AL). Models can also be supported by other programs, including spreadsheet applications. The model contains compartments that may represent groups of tissues or tissues or organs that have importance to the chemical under evaluation (e.g., liver as an important site of metabolism and toxicity). Models contain values for physiological parameters that are generally applicable to the species (blood flows, organ volumes, respiratory rates) and biochemical (or chemical-specific) values that are unique to the chemical under investigation (partition coefficients, metabolic rate constants). Typically metabolism is modeled as if it only occurs in the liver, which represents a simplification. PBPK models may be used to develop metabolic rate constants from in vivo data (closed chamber gas uptake studies), or in vitro data may be extrapolated for use in developing metabolic rate constants for inclusion in PBPK models. These models offer an advantage over strictly extrapolated in vitro metabolic rate constants because they can also simulate the effects of extrahepatic effects (like sequestration and exhalation) on chemical (substrate) availability to the liver.

Several examples demonstrate the application of PBPK modeling to evaluate the tissue distribution of mixture components. Trichloroethylene (TCE) and 1,1-dichlorethylene (DCE) interactions were characterized by Andersen et al. (1987). In vivo data on serum enzyme levels (a marker for toxicity) were available from exposures to DCE alone and DCE in combination with TCE. A comparison of multiple possible interaction models demonstrated a convincing fit between the competitive inhibition model for hepatic metabolism and resulting toxicity data. This study offers an example of how PBPK modeling can be used to evaluate metabolic interactions for a binary mixture.

A PBPK model was developed to study the interactions between carbon tetrachloride (CCl4) and kepone in rats [28]. Previous experimental findings demonstrated that kepone co-exposure amplified the CCl4-induced lethality 67-fold. Additional laboratory studies with rats fed a control diet or one containing kepone, with CCl4 administration i.p. were conducted to characterize liver pathology resulting from CCl4 metabolites and to provide data on the exhalation of CCl4 used to indicate metabolism. A pharmacodynamic model was developed and incorporated that characterized the injury, repair, and death sequence for cells and animals. The results of this PBPK-PD model demonstrated an approximate 61- to 142-fold increase in CCl4 lethality with kepone treatment. However, when the results were adjusted for PK interactions at the level of CCl4 metabolism, the increase in lethality was reduced to approximately 4-fold. These results demonstrate that the kepone–CCl4 interaction is primarily based on a TK interaction but does also rely to some extent on a TD interaction.

Dobrev and colleagues [29] constructed an analysis to estimate interaction thresholds in the rat for three commonly found environmental contaminants, TCE, tetrachloroethylene (Perc), and 1,1,1-trichloroethane (MC). These solvents were chosen because of their ubiquitous finding as environmental contaminants, as well as their dependence on CYP-mediated oxidation as a primary metabolic process. PBPK models were constructed for each chemical. PBPK modeling of gas uptake data was applied, and interactions were characterized as concentrations of Perc and MC required to increase blood TCE concentrations by 10%. Interaction models were constructed for competitive, noncompetitive, and uncompetitive inhibition for metabolism. Although each model fit the data to some extent, the similarity between model-predicted (optimized) values for K_i and K_m values indicated that the competitive model best applied. This model was implemented to interpret gas uptake data for binary and ternary mixtures. At the TLV concentration for TCE (50 ppm), the calculated thresholds for interaction were 25 ppm for Perc and 135 ppm for MC. When exposed to TCE at the TLV concentration of 50 ppm and MC versus TCE, Perc (at its TLV of 25 ppm) and MC, the latter exposure reduced the interaction threshold from 175 to 130 ppm MC. These results demonstrate the value of PBPK modeling in refining the descriptions of toxicologically important (metabolic, toxicokinetic) interactions among components of environmentally-important mixtures.

Krishnan and colleagues developed an approach to the PBPK analysis of complex mixtures in which the toxicologic interactions of binary mixtures are first combined in the affected and modeled target tissue [30]. An example for a mixture of volatile organic chemicals (m-xylene, toluene, ethylbenzene, dichloromethane, and benzene) that interact via competitive inhibition was developed. The approach requires that PBPK models be available for each component; models are interconnected by K_i values and, naturally, by substrate concentrations—here, in the liver. It is necessary that all binary interactions be characterized. For a mixture of n components, the number of binary interaction experiments performed to determine K_i values will be $n(n - 1)/2$. While not indicated, it seems that some reduction of resources would be possible by determining K_i values in vitro and extrapolating them to the in vivo setting. In addition a simpler approach to evaluating metabolic interactions may be developed—for the risk analysis of mixtures, the maximal impact of metabolic interactions (inhibition) could be estimated by reducing the hepatic extraction ratio to zero. Alternately, the impact of enzyme induction could be simulated by raising the hepatic extraction ratio to 1.0. Even if a level of uncertainty in the resulting model predictions exist, this approach can be used as a minimum to guide the construction of in vivo complex mixtures exposures undertaken to develop TK or toxicity data on the mixture under evaluation.

Humans may be exposed to very complex mixtures, like the one posed by gasoline. The same principles of binary interactions apply to mixtures like gasoline, and Dennison and colleagues [31] have taken advantage of the principle of chemical lumping. In this analysis, five components of gasoline were of primary interest (benzene, toluene, ethylbenzene, xylene, and hexane). The remaining components were lumped into a single group for analysis—the analysis comprised six lumps, five of which contained a single chemical. A PBPK model was developed and parameter values, including K_i values, were incorporated. Whole animal gas uptake experiments were conducted using fractions of gasoline and decreases in chamber concentrations of components were used as the basis for model predictions of metabolic interactions. Given the molecular attributes of these compounds, the finding of competitive inhibition was not surprising. The results of this study indicate that metabolic inhibition of components occurred at 200 ppm, and for some, at concentrations as low as 100 ppm. The approach appears to be valid, but the specific application should be tailored to address toxicologically or toxicokinetically important chemicals individually (one chemical per lump).

While there are numerous PBPK and TK approaches to chemical mixtures, these representative examples demonstrate a spectrum of approaches, from simple binary mixtures to complex, often uncharacterized mixtures. The underlying commonality is the need for a validated TK model structure, the inclusion of defensible parameter values, and some external measure of metabolic interaction that can serve as the basis for starting values for model parameters. Within a mixture, PBPK modeling approaches are most valid for the chemical of concern when that chemical is specifically addressed by the model.

21.4 COMPUTATIONAL APPROACHES

Because of the relative dearth of toxicological testing data, hazard identification for environmental pollutants becomes a challenging task. Until such data become available, computational approaches and tools are being routinely used. The US EPA's MIXTOX database [14,32,33] is a collection of bibliographic summaries of chemical interaction studies, most of which are studies of binary mixtures. Veteran mixtures risk assessors may be familiar with the database—it was previously available in an MS-DOS operated version. Chemicals of interest were entered and the database search engine reported interactions data for identified chemicals. The US EPA undertook a similar approach addressing carcinogen interactions [34]. These databases have not been updated and are only scarcely available (or used) at the present. In addition to these, a few newer and presently available computer-assisted programs less directly related to chemical mixtures interactions are presented.

It is conceivable that quantitative structure-activity (QSAR) approaches (e.g., TOPKAT; see Chapter 7) could be applied to predict response levels for uncharacterized contaminants for use in the HI approach. Further, specific submodels existing (e.g., that for developmental toxicity) could be applied to estimate system-specific response levels for application in the TTD approach. To our knowledge, there are no computer-assisted programs available that can automate the prediction of toxicity for mixtures. Much of the reason may reside in the relative lack of empirical observations and characterizations of chemical interactions. Many QSAR approaches rely on "training set" approaches to the development of automated programs. Another impediment may be the many examples of the levels, types and biochemical bases for chemical interactions, the intricacies of which would benefit from an automated approach. This area is a useful area for exploration.

Many early PBPK modeling efforts were based on the Simusolv software, and support for this seems not readily available at the present time. More recently the ACSL and Berkeley Madonna (University of California, Berkley, CA) have become more widely used. In addition to these computer software packages, Haddad et al. [35] demonstrated the application of a spreadsheet program to support a PBPK model, and Trent University (Peterborough, Ontario, Canada, updated 2003) made available a spreadsheet program to run PBPK models. Further there are several computer-assisted applications, several as freeware, to perform pharmacokinetic analyses and interpret in vitro enzyme kinetic data (see Chapter 3).

21.5 CONCLUSIONS

The ultimate goals of a risk assessment depends on its scope whether it is environmental protection, worker safety, or public health [36]. Even though each one has a slightly different perspective, the overall goal is to perform an assessment appropriate for the exposure scenario that could be short term or chronic, continuous or episodic, single chemical or multiple chemical, environmental contaminants or occupational chemicals. Whatever the case may be, the risk posed by unintentional exposures must be integrated with the risk posed by intentional exposures such as the use of medicines/drugs, recreational/social use, and chemicals routinely found in foods. In summary, due consideration has to be given to multiple sources and routes of exposure, cumulative characteristics of chemicals in fat and bones, and the total risk posed by aggregate exposures to synthetic and naturally occurring chemicals.

A great many, if not most, environmental contaminants have poorly understood toxicities and tissue distribution patterns. Several methods exist by which computational approaches to predictions of hazard (toxicity) and distribution (pharmacokinetics) may be developed. These results could be integrated to aid in determining the likelihood of chemical interactions and toxicologic responses following the exposures to mixtures. To date, there seem to be no integrated approaches readily available, and this area represents a significant target for research.

APPENDIX (From [9])

Critical Effect Response occurring at the lowest point on the dose scale, as dose increases.

Secondary Effect Any effect occurring at a higher dose than the critical effect.

Acceptable/Allowable Human Exposure Limit (AL) Term used in the denominator of the hazard quotient equation. It may be any of several established or derived values.

Hazard Quotient (HQ) Ratio of human exposure to acceptable limits (E/AL). HQ can be developed for individual chemicals and effects.

Hazard Index (HI) Sum of all hazard quotient values developed for components of a chemical mixture.

Hazard Index Approach A chemical mixtures risk assessment method where hazard quotients for component chemicals are only developed using the critical effect. Hazard quotient values are grouped by critical effect and summed. Multiple hazard indexes are developed, one for each affected target organ or system.

Screening Level Hazard Index Approach A chemical mixtures risk assessment method where hazard quotients for component chemicals are only developed using the critical effect. All hazard quotient values are summed, regardless of the target organ.

Target Organ Toxicity Dose (TTD) A variant of the hazard index approach where effects for all components for all affected tissues. The target organ toxicity dose is summed regardless of whether they represent the critical effect or a secondary effect. Hazard index values are developed for each affected organ or system.

Components Chemicals that make up a mixture.

Simple Mixture A combination of a relatively small number of chemicals (no more than 10) that have been identified and quantified (e.g., the components of concern for a receptor population near a hazardous waste site may constitute a simple mixture).

Complex Mixture A combination of so many chemicals that the composition of the mixture is not fully characterized, either qualitatively or quantitatively, and may be variable (e.g., cigarette smoke, diesel exhaust, gasoline).

Similar Mixtures Mixtures having the same chemicals but in slightly different proportions or having most but not all chemicals in common and in highly similar proportions. Similar mixtures are expected to have similar fate, transport, and health effects (e.g., the jet fuel JP-5 from different sources).

Index Chemical Chemical selected as the basis for standardization of toxicity of components in a chemical class (e.g., 2,3,7,8-tetrachlorodibenzo-*p*-dioxin [TCDD] for the assessment of dioxin-like compounds; benzo[a]pyrene for the assessment of carcinogenic polycyclic aromatic hydrocarbons [PAHs]).

Interaction Effect of a mixture that is different from additivity based on the dose–response relationships of the individual components.

Additivity Effect of the mixture that can be estimated from the sum of the exposure levels (weighted for potency) or the effects of the individual components.

Synergism Effect of the mixture that is greater than that estimated for additivity on the basis of the toxicities of the components.

Potentiation A component that does not have a toxic effect on an organ system but increases the effect of a second chemical on that organ system.

Antagonism Effect of the mixture that is less than that estimated for additivity on the basis of the toxicities of the components.

Inhibition A component that does not have a toxic effect on a certain organ system but decreases the apparent effect of a second chemical on that organ system.

ACKNOWLEDGMENTS

The views expressed in this paper are those of the authors and do not necessarily reflect the views and policies of the US Environmental Protection Agency or the Agency for Toxic Substances and Disease Registry. Mention of trade names or commercial products does not constitute endorsement or recommendation for use. The authors are grateful to Rick Hertzberg and Linda Teuschler for thoughtful comments during manuscript development and to Bette Zwayer and Lana Wood for expert technical support during manuscript preparation.

REFERENCES

1. US EPA. *Supplementary guidance for conducting health risk assessment of chemical mixtures.* EPA/630/R-00/002. Washington: US Environmental Protection Agency, Office of Research and Development, 2000.

2. US EPA. *Guidelines for carcinogen risk assessment.* EPA/630/P-03/001B. Washington: US Environmental Protection Agency, Risk Assessment Forum, 2005. Available at <http://www.thecre.com/pdf/20050404_cancer.pdf>.

3. US EPA. *Guidance for identifying pesticide chemicals that have a common mechanism of toxicity.* Washington: US Environmental Protection Agency, 1999.

4. US EPA. *The feasibility of performing cumulative risk assessments for mixtures of disinfection by-products in drinking water.* EPA/600/R-03/051. Cincinnati, OH: US Environmental Protection Agency, National Center for Environmental Assessment, Office of Research and Development, 2003.

5. US EPA. *Developing relative potency factors for pesticide mixtures: biostatistical analyses of joint dose-response.* EPA/600/R-03/052. Cincinnati, OH: US Environmental Protection Agency, National Center for Environmental Assessment, Office of Research and Development, 2003.

6. Feron VJ, Groten JP, Jonker D, Cassee FR, van Bladeren PJ. Toxicology of chemical mixtures: Challenges for today and the future. *Toxicology* 1995;105:415–27.

7. ATSDR (Agency for Toxic Substances and Disease Registry). *ATSDR public health assessment guidance manual.* Atlanta: Agency for Toxic Substances and Disease Registry, Department of Health and Human Services; 2005. Available at <http://www.atsdr.cdc.gov/HAC/PHAManual/toc.html>.

8. CDC (Centers for Disease Control and Prevention). *Third national report on human exposure to environmental chemicals.* Atlanta: Centers for Disease Control and Prevention, Department of Health and Human Services; 2005. Available at <http://www.cdc.gov/exposurereport/3rd/pdf/thirdreport.pdf> (accessed Sep. 2006).

9. ATSDR (Agency for Toxic Substances and Disease Registry). *Guidance manual for the assessment of joint toxic action of chemical mixtures.* Atlanta: Agency for Toxic Substances and Disease Registry, US Department of Health and Human Services; 2004. Available at <http://www.atsdr.cdc.gov/interactionprofiles/ipga.html>.

10. Pohl H, Hansen H, Chou S. Public health guidance values for chemical mixtures: current practice and future directions. *Regul Toxicol Pharmacol* 1997;26:322–9.

11. Eide I, Neverdal G, Thorvaldsen B, Grung B, Kvalheim OM. Toxicological evaluation of complex mixtures by pattern recognition: Correlating chemical fingerprints to mutagenicity. *Environ Health Perspect* 2002;110:985–8.

12. Simmons JE, Richardson SD, Speth TF, Miltner RJ, Rice G, Schenck KM, Hunter ES III, Teuschler LK. Development of a research strategy for integrated technology-based toxicological and chemical evaluation of complex mixtures of drinking water disinfection byproducts. *Environ Health Perspect* 2002;110:1013–24.

13. US EPA. *Reference dose (RfD): Description and use in health risk assessments.* Washington: US Environmental Protection Agency, Office of Health and Environmental Assessment, Environmental Criteria and Assessment; 1993. Available at <http://www.epa.gov/iris/rfd.htm>.

14. US EPA. *Risk assessment guidance for superfund.* Vol. I (RAGS). *Human health evaluation manual: Part A. Interim final.* Dec. OSWER 9285.7-02B, EPA/540/1-89/002. Washington: US Environmental Protection Agency, Office of Emergency and Remedial Response, 1989.

15. US EPA. *Technical support document on health risk assessment of chemical mixtures.* EPA/600/8-90/064. Washington: US Environmental Protection Agency, 1990.
16. US EPA. *Guidelines for the health risk assessment of chemical mixtures.* EPA 630/R-98/002. Washington: US Environmental Protection Agency, 1986.
17. Schoeny RS, Margosches E. Evaluating comparative potencies: Developing approaches to risk assessment of chemical mixtures. *Toxicol Ind Health* 1989; 5(5):825–37.
18. Van den Berg M, Birnbaum LS, Denison M, De Vito M, Farland W, Feeley M, Fiedler H, Hakansson H, Hanberg A, Haws L, Rose M, Safe S, Schrenk D, Tohyama C, Tritscher A, Tuomisto J, Tysklind M, Walker N, Peterson RE. The 2005 World Health Organization reevaluation of human and mammalian toxic equivalency factors for dioxins and dioxin-like compounds. *Toxicol Sci.* 2006;93(2):223–41.
19. ATSDR (Agency for Toxic Substances and Disease Registry). *Minimal risk levels (MRLs) for hazardous substances.* Atlanta: Agency for Toxic Substances and Disease Registry, Department of Health and Human Services, 2006. Available at <http://www.atsdr.cdc.gov/mrls.html> (accessed November 2006).
20. Mumtaz MM, Poirier KA, Colman JT. Risk assessment for chemical mixtures: Fine-tuning the hazard index approach. *J Clean Technol Environ Toxicol Occup Med* 1997;6:189–204.
21. ATSDR (Agency for Toxic Substances and Disease Registry). Interaction profiles for persistent chemicals found in fish: Chlorinated dibenzo-*p*-dioxins (CDDs), hexachlorobenzene, dichlorodiphenyl dichloroethane (*p,p* = –DDE), methyl mercury, and polychlorinated biphenyls (PCBs) (draft for public comments). Atlanta: Agency for Toxic Substances and Disease Registry, US Department of Health and Human Services, 2001. Available at <http://www.atsdr.cdc.gov/interactionprofiles/ip01.html>.
22. ATSDR (Agency for Toxic Substances and Disease Registry). Interaction profiles for persistent chemicals found in breast milk: Chlorinated dibenzo-*p*-dioxins (CDDs), hexachlorobenzene, dichlorodiphenyl dichloroethane (*p,p* = –DDE), methyl mercury, and polychlorinated biphenyls (PCBs) (draft for public comments). Atlanta: Agency for Toxic Substances and Disease Registry, US Department of Health and Human Services, 2001. Available at <http://www.atsdr.cdc.gov/interactionprofiles/ip03.html>.
23. NAS (National Academy of Sciences). Risk assessment of mixtures of systemic toxicants in drinking water. In: *Drinking water and health*, Vol. 9. Washington: National Academy Press, National Research Council, Safe Drinking Water Committee, 1989. p. 121–32.
24. Hertzberg RC, MacDonell MM. Synergy and other ineffective mixture risk definitions. *Sci Total Environ* 2002;8:31–42.
25. Mumtaz M, Durkin PR. A weight of the evidence approach for assessing interactions in chemical mixtures. *Toxicol Ind Health* 1992;8(6):377–406.
26. Mumtaz MM, De Rosa CT, Durkin NP. Approaches and challenges in risk assessments of chemical mixtures. In: Yang RSH, editor, *Toxicology of chemical mixtures: From real life examples to mechanisms of toxicologic interactions.* Orlando, FL: Academic Press, 1994. p. 565–97.

27. Kedderis, GM. In vitro to in vivo extrapolation of metabolic rate constants for physiologically based pharmacokinetic models. In: JC Lipscomb and EV Ohanian, editors *Toxicokinetics and Risk Assessment*. Informa Healthcare Publishers, New York, 2007.

28. El-Masri HA, Thomas RS, Sabados GR, Phillips JK, Constan AA, Benjamin SA, Andersen ME, Mehendale HM, Yang RSH. Physiologically based pharmacokinetic/pharmacodynamic modeling of the toxicologic interaction between carbon tetrachloride and kepone. *Arch Toxicol* 1996;70:704–13.

29. Dobrev ID, Andersen MA, Yang RSH. Assessing interaction thresholds for trichloroethylene in combination with tetrachloroethylene and 1,1,1-trichloroethane using gas uptake studies and PBPK modeling. *Arch Toxicol* 2001;75:134–44.

30. Krishnan K, Haddad S, Beliveau M, Tardif R. Physiological modeling and extrapolation of pharmacokinetic interactions from binary to more complex chemical mixtures. *Environ Health Perspect* 2002;110(Suppl 6):989–94.

31. Dennison JE, Andersen ME, Clewell HJ, Yang RSH. Development of a physiologically based pharmacokinetic model for volatile fractions of gasoline using chemical lumping analysis. *Environ Sci Technol* 2004;38:5674–81.

32. Marnicio RJ, Hakkinen PJ, Lutkenhoff SD, Hertzberg RC, Moskowitz PD. Risk analysis software and databases: Review of Riskware '90 conference and exhibition. *Risk Anal* 1991;11:545–60.

33. Teuschler LK, Hertzberg RC. Current and future risk assessment guidelines, policy, and methods development for chemical mixtures. *Toxicology* 1995;105: 137–44.

34. Arcos JC, Woo YT, Lai DY. Database on binary combination effects of chemical carcinogens. *Environ Carcin Revs: (C J Environ Sci Health)* 1988;C6(1):1–164.

35. Haddad S, Pelekis ML, Krishnan K. A methodology for solving physiologically-based pharmacokinetic models without the use of simulation software. *Toxicol Lett* 1996;85:113–26.

36. Monosson E. Chemical mixtures: Considering the evolution of toxicology and chemical assessment. *Environ Health Perspect* 2005;113:383–90.

22

ENVIRONMENTAL AND ECOLOGICAL TOXICOLOGY: COMPUTATIONAL RISK ASSESSMENT

EMILIO BENFENATI, GIOVANNA AZIMONTI, DOMENICA AUTERI, AND MARCO LODI

Contents

Computational Toxicology: Risk Assessment for Pharmaceutical and Environmental Chemicals, Edited by Sean Ekins

22.1 INTRODUCTION

Assessment of the risk related to the presence of dangerous chemicals in the environment is both important and difficult. However, accurate assessment of the impact of a chemical is a complex task, for there are many factors involved, and chemical substances are in many cases estimated with a large uncertainty. Further, for most of the currently used industrial chemicals there is no knowledge of their environmental and ecotoxicological properties, and even worse, some chemicals are not commercialized but are present in the environment nevertheless as degradation products or impurities of industrial substances. The more widely studied chemicals are well-known contaminants such as dioxin, and pesticides. Pesticides have been addressed within specific regulations (e.g., European Directive 91/414), so this makes them a good example on how to address the problem with defined protocols.

An even bigger problem is the lack of knowledge about the effects of chemical mixtures (see also Chapter 20). The environment is typically contaminated by mixtures, and assessment of the interactions of chemical mixtures in the environment should be a target of scientific interest, but again, the problem is very complex. In most cases because only few data are known, the environmental risk assessment has to depend on many assumptions, with computational models used to replace the missing experimental data. So computational modelers are introducing more and more tools to cover the different scenarios to improve and reduce the approximations of details.

Also useful have been probabilistic tools in obtaining more realistic results. Probabilistic risk assessment may appear less precise as opposed to the current approach, which involves deterministic methods in the calculation of at least some values. However, the practice for the assessor is to set up a scenario and evaluate it using a number of assumptions. Thus typically the experimenter must use good judgment in considering the likeliness of certain events occurring, and to evaluate the probability of these events. The probability of certain situations, pathways, or scenarios has always been at the basis of the evaluation. What is new, now, is the availability of some computerized tools to measure probability.

22.2 METHODS FOR TOXICITY ASSESSMENT

The task of assessing the effect of a chemical substance on the environment is complicated by the complexity of living organisms that may be affected by the chemical substance. To consider toxic effects, a reference species is used. It should be understood that the experiments done are models in a laboratory to understand effects in the outside world at the ecological level. The real challenge is to study the effects produced out in the environment. So a simplified model of real-world phenomena must be also used. The few selected species must represent all the various other species. Because it is not possible to identify the most sensitive species, some features are introduced to take into account this missing species. Typically acute toxicity is the focus because chronic studies are very expensive and require longer experiments. However, this way a large part of the effects can elude suitable identification.

Toxicity is a dynamic phenomenon. It can occur at a point in time after exposure by a process that is often very complex. To measure toxicity at a fixed time, which makes it static rather than dynamic process, the classic protocol is to expose animals to increasing doses of the chemical. This result is usually a sigmoidal curve. In case of lethal toxicity at a given low dose all animals survive. At a high dose, all animals die. Using these points about the effects at intermediate doses, theoretically a dose that kills 50% of the animals is calculated by modeling software. This way the LD_{50} or LC_{50} are found, where LD stands for lethal dose and LC means lethal concentration (in case of a chemical dissolved in water, for aquatic toxicity). It has to be considered that this value is mathematically interpolated and the uncertainty is typical, both in the case of the intra- and (even more so) interlaboratory experiments.

The procedures for performing laboratory experiments have been periodically revised. The process needs to be reviewed periodically to reduce the uncertainties of the models where the definition of certain factors is lacking. In general, older models are considered as less standardized and less reliable. When new versions are introduced, they involve modifications to the previous conceptual scheme of the protocol in order make the protocol closer to a real-world situation or to reduce the impact on laboratory animals by lowering their number per experiment or lessening the cruelty of the study. For instance, there is a tendency to use a lower dose that kills fewer than 50% of the animals.

Standardization of the laboratory experiments is a critical issue, so procedures are changed over time. Naturally this is because laboratory methods improve over time. Nevertheless, standards are milestones that are indicative of current knowledge and agreement on protocol. Standards are used to compare and integrate data, so they represent a fundamental effort to speak the same language. It is important to remember that the actual situation to be understood and protected—environment and life—is much more complex than any standards. Standards only simplify to enable us to set up the scenario and focus on a limited set of conditions. The Organisation for Economic Co-operation and Development (OECD) is promoting the standardization of laboratory activity. There are a number of protocols for assessing environmental and ecotoxicological endpoints, and the reader can find guidelines at the OECD Web site (*http://www.oecd.org*).

22.3 METHODS FOR EXPOSURE

Environmental exposure is usually expressed in terms of environmental concentration in air, water, sediment, or soil. Models can provide estimates for times or locations that are impractical to measure, and these estimates are then extrapolated beyond the range of observation. Because models simplify reality and do not reflect every condition in the real world, critical evaluation of input data is a key action. In exposure assessment a model's output is only as good as the quality of its input variables. Data and models for risk assessment are often developed in a tiered fashion: simple conservative models are used first, followed by more elaborate models that provide more realistic estimates. For example, the case of pesticide can be considered a reference point for modeling an exposure assessment according to Directive 91/414/EEC, a harmonization process of risk assessment methods for pesticide legislation started in the European Union. In Annex VI of this directive, it is clearly stated that member states must use for their estimates a suitable calculation model validated at the community level to calculate, the concentration of the active substance and of relevant metabolites, and degradation and reaction products that can be expected to be present in the groundwater, surface water, and air. This statement has strong implications for the selection of the model to be used in the registration process and for the definition of common environmental criteria to be used in model application.

Over the last 20 years a lot of effort has been put into the development of methods that start with assessment of pesticide exposure in the environment. A large number of environmental outcome models have been developed: CALF [1], CMLS [2], CRACK-NP [3], EPIC [4], EXAMS [5], GENEEC [6], GLEAMS [7], HSPF [8], HYDRUS-1D, and HYDRUS-2D [9, 10], IMPAQT [11], LEACHM/LEACHP [12], MACRO [13,14], MACRO_DB [15], MIKE SHE [16], MARTHE [17], MOPED [18], MOUSE [19], OPUS [20], PEARL [21,22], GEOPEARL [23], PELEP-DSS [24], PELMO [25], PESTLA [26], PESTRAS [27], PLM [28], PRZM [29], RIVWQ [30], RZWQM [31], SIMULAT [32], SLOOT.BOX [33], SOM-3 [34], SWAT [34,35] TOXWA [36,37], TurfPQ [38], VARLEACH [39], and WAVE [40]. However, models and criteria for calculating chemical concentration have been developed mainly for two environmental compartments: surface water and groundwater.

22.3.1 Groundwater

Almost all the evaluation of pesticides leaching into groundwater is performed at 1 m depth. The assumption is that groundwater is unlikely to be affected by pesticides at concentrations exceeding 0.1 μg/L if those concentrations are not encountered at a shallow depth. Little research has been conducted on the fate process of pesticides once they have leached through the soil and below the root zone.

Water flow and contaminant transport in groundwater are difficult to study and investigations of pesticides at the aquifer level are scarce. The tendency has been to isolate the processes that control the chemical composition and to investigate each separately. Studies at the level of the aquifer that have been reported in the literature refer generally to monitoring of seasonal variations of pesticide concentrations in the groundwater [41–43], or in the unsaturated and saturated zones [44]. Recently in a research project funded by the European Commission, PEGASE, mechanistic or semi-empirical tools were developed for the modeling of pesticide contamination in groundwater at various spatial scales. The investigation used a refined screening tool (PESTGW) and 1D root zone models (MACRO and ANSWERS), and additions of pesticide concentrations and crop rotations in integrated models, to predict pesticide concentrations in the soil-unsaturated, zone-saturated, and zone continuum conditions, (MARTHE, TRACE, and POWER) and the coupling of different models (TRACE + 3 DLEWASTE, MACRO + FRAC3DVS, MACRO + MODFLOW, ANSWERS + MODFLOW). The models were upgraded within the framework of the project to allows future deployment of advanced modeling activities, such as automated calibration against field data or sensitivity and uncertainty analyses [45].

In 1993 the FOCUS workgroup (acronym for the FOrum for the Coordination of pesticide fate models and their USe) was formed. The purpose of FOCUS was to develop consensus among the member states, the European Commission, and industry on the role of modeling in the EU review process of active substances. Guidance was first developed for monitoring the leaching of pesticides to groundwater [46–47]. This included a description of relevant models and their strengths and weaknesses. However, of the nine models initially investigated (PRZM, PRZM-2, PELMO, GLEAMS, PESTLA, VARLEACH, LEACHM, MACRO, and PLM), and recommended in the FOCUS guidance document, only four models in updated versions (MACRO, PEARL, successor of PESTLA, PELMO, and PRZM) are currently in use at the EU level. These models must further be applied to the same data sets in order to achieve harmonization of the risk assessment throughout Europe. Hence standard scenarios with regard to soil, weather, and cropping data were needed to increase the consistency of the regulatory evaluation process by minimizing the subjective influence of the model user. Standard scenarios also make the interpretation much easier, and enable the adoption of a consistent scientific process for a stepwise approach to the evaluation of the leaching potential of substances at the EU level (Table 22.1). Therefore the FOCUS workgroup for Groundwater Scenarios was charged in 1997 with developing a set of standard scenarios that could be used to assess potential movement of active substances and metabolites of plant protection products to groundwater as part of the EU process for placing active substances on Annex I. Since this process proceeds at the community level, the standard scenarios had to apply to the whole European Union. The workgroup developed nine realistic worst-case scenarios [47] and appropriate data input files for the models PELMO, PEARL, and PRZM. For MACRO, only the Châteaudun scenario has been parameterized by FOCUS, as at that time there were no reliable pedotransfer functions available (MACRO DB2 has not been released yet).

TABLE 22.1 Overview of the Characteristics of the Nine Leaching Scenarios

Location	Mean Annual Temperature (°C)	Annual Rainfall (mm)	Texture	OM (%)
Châteaudun	11.4	648 + I	Silty clay Loam	2.4
Hamburg	9.2	786	Sandy loam	2.6
Jokioinen	4.3	638	Loamy sand	7.0
Kremsmünster	8.8	900	Loam/silt loam	3.6
Okehampton	10.4	1038	Loam	3.8
Piacenza	13.3	857 + I	Loam	1.7
Porto	14.8	1150	Loam	6.6
Sevilla	18.1	493 + I	Silt loam	1.6
Thiva	16.2	500 + I	Loam	1.3

Sources: Soil texture is based on FAO (1977) and USDA (1975).

Note: I indicates rainfall supplemented by irrigation; OM indicates organic matter (from FOCUS [47]).

Standard scenarios help establish whether "safe" scenarios exist for the supported uses of a substance. Since they are used in a first tier of an assessment, they are designed to represent a realistic worst case. A new FOCUS groundwater Workgroup was organized in 2003 and charged with the responsibility of providing guidance on higher tier leaching assessments and on coordination of risk assessment procedures at European national levels. After considering the types of data that are available for determining the predicted environmental concentration (PEC) in groundwater (PEC_{gw}), the group categorized the risk assessment approaches into four tiers based on the availability of information. Parameter refinements for modeling, scenario refinements (e.g., based on geographical information systems—GIS—data), or higher tier experimental approaches are classified as second tier. Combinations of the modeling, refined parameters, and experimental approaches from tier 2, as well as advanced spatial modeling and "other higher tier modeling approaches" (e.g., 3D aquifer modeling) are classified as third tier, while monitoring of groundwater (with appropriate reality checking) is considered as the highest or fourth tier [48].

22.3.2 Surface Water

A tiered approach is used for pesticide exposure assessment at the surface water level. Depending on the results of the initial risk assessment, more extensive testing relative to the environmental exposure or hazard may be required to define the full environmental risk. The data are generated from such increasingly comprehensive series of studies (higher tiered studies). At each tier a comparison has to take place between the estimated exposure and the estimated hazard; therefore estimation of two separate tiers for both exposure and effects is necessary.

In July 1994 the FOCUS Steering Committee installed a working group on surface water to analyze the usefulness of mathematical models in studying surface water contamination and their role in the registration process [49]. In 1996 another FOCUS working group on surface water scenarios was formed to develop a series of standard agriculturally relevant scenarios for the European Union that can be used with the models identified to fulfill the requirements for calculating the predicted environmental concentration in surface water (PEC_{sw}). Moreover the workgroup defined a stepwise procedure (*tiered approach*) for the exposure assessment of surface water to be used in the registration process in the EU according to Directive 91/414/EEC. The $FOCUS_{sw}$ workgroup procedure [50] consists of four steps, with the first being a very simple and extreme worst-case scenario using first-order kinetics and assuming a loading equivalent to a maximum annual application. The second tier applies surface water loading based on sequential patterns that take into account the degradation of the substance between successive applications. Again, the PEC_{sw} are calculated so that it can be compared to the same and/or different toxicity levels for aquatic organisms. As with the first tier, if use is considered acceptable at this stage, no further risk assessment is required, whereas an unacceptable assessment necessitates further work involving third tier calculation.

In the third step the more sophisticated modeling takes into account realistic worst-case scenarios whereby the amounts enter the surface water via runoff, spray drift, or drainage. The $FOCUS_{sw}$ workgroup considers spray drift, runoff, and drainage as routes of entry but does not take into account such modes as atmospheric deposition, dry deposition, colloid transport, discharge of wastewater, groundwater, and accidents. $FOCUS_{sw}$ includes both runoff of pesticide in water and pesticide adsorbed to soil particles; the pesticide in the water goes to the water layer and the pesticide adsorbed to soil particles is added to the sediment. For step 3, the $FOCUS_{sw}$ workgroup defined the 10 realistic worst-case scenarios that collectively represent all agriculture in the European Union in the assessment of the PEC_{sw} [51]. Realistic worst-case concentrations are calculated in three identified types of small water bodies across the European continent: ditch, stream, and pond. The major characteristics of the 10 scenarios are reported in Table 22.2.

The models chosen in $FOCUS_{sw}$ for estimating the different routes of entry are MACRO for estimating the contribution of drainage, PRZM for the contribution of runoff and erosion, and TOXSWA for the estimation of the final PEC in surface waters. An additional loading is defined as spray drift input. The calculation of the contribution of the spray drift is incorporated in the Graphical User Interface (GUI) for the surface water scenarios called SWASH (Surface WAter Scenario Help). This is a general software shell developed to ensure that the relevant FOCUS scenarios and input are defined consistently for all models [50].

TABLE 22.2 Overview of the 10 Scenarios Defined by $FOCUS_{sw}$

Name	Mean Annual Temperature (°C)	Annual Rainfall (mm)	Topsoil	Organic Matter (%)	Slope (%)	Water Bodies	Weather Station
D1	6.1	556	Silty clay	2.0	0–0.5	Ditch, stream	Lanna
D2	9.7	642	Clay	3.3	0.5–2	Ditch, stream	Brimstone
D3	9.9	747	Sand	2.3	0–0.5	Ditch	Vreedepeel
D4	8.2	659	Loam	1.4	0.5–2	Pond, stream	Skousbo
D5	11.8	651	Loam	2.1	2–4	Pond, stream	La Jailliere
D6	16.7	683	Clay loam	1.2	0–0.5	Ditch	Thiva
R1	10.0	744	Silt loam	1.2	3	Pond, stream	Weiherbach
R2	14.8	1402	Sandy loam	4	20	Stream	Porto
R3	13.6	682	Clay loam	1	10	Stream	Bologna
R4	14.0	756	Sandy clay loam	0.6	5	Stream	Roujan

Source: From Focus (2002).

Note: Six scenarios are called D scenarios (drainage) because, after release of the pesticide, it may enter the neighboring water body via spray drift deposition and water flow through drainage pipes. In the four R scenarios (runoff) pesticide may enter the water body via spray drift deposition and runoff plus erosion.

At step 3, the calculated PEC_{sw} for each scenario are also compared with relevant toxicity data, and a decision made as to whether it is necessary to proceed to the fourth step of exposure estimation. This final (fourth) step includes substance loadings as in the preceding step but takes into account the range of possible uses. The uses are therefore related to the specific and realistic combinations of cropping, soil, weather, field typography, and aquatic bodies adjacent to fields. By its nature, step 4 is conducted case by case for the properties of a compound, its use patterns, and the areas of potential concern as identified in the lower tier assessments.

At the lower tiers, acute and chronic toxicity parameters are determined for the active substance and a representative formulated product, and are then compared to exposure concentrations from FOCUS steps 1, 2, and 3 in an iterative process. Results from lower tier effects assessments are compared in either step 3 or step 4 exposure calculations and similarly results from higher tier effect assessments are compared in either step 3 or step 4 exposure calculations. At higher tiers, all the options for effects and exposure refinement, along with mitigation options, are considered in order to select the most appropriate path for further risk refinement at step 4.

At the highest tier, risk assessment and mitigation must be merged because measures that are used to refine potential exposure assessments are used to identify appropriate mitigation strategies. For this reason a FOCUS working group on Landscape and Mitigation Factors in Ecological Risk Assessment was established in June 2002, to investigate options and feasibilities of including landscape and mitigation factors in higher tier exposure assessments, and to produce a review of the state of the art in landscape and mitigation factors in exposure assessment, as well as to make recommendations for future FOCUS groups on warp to develop this area further. One of the points developed by the group was the possibility for incorporating modeling refinements and mitigation into the exposure assessment of step 4, as classified in three main refinement options. First, relatively simple changes can be made to the existing FOCUS step 3 scenarios by refining input parameters for the chemical or scenario to make them more precisely reflect the potential risks being assessed. Second, mitigation measures can be incorporated into step 3 scenarios (leading to the step 4 calculations). Third, more specific scenarios can be developed to more precisely reflect the environmental and agronomic conditions for use of a plant protection product at a local or regional scale. The locations of such new scenarios would have to follow the procedures adopted by the FOCUS surface water scenarios group.

A wide range of methods and data is available for describing agricultural landscapes that can be deployed to refine the exposure assessments of step 4. The use of GIS allows a quantitative description of the agro-ecosystem landscape, enabling relationships to be explored between cropped land and areas containing nontarget organisms. In FOCUS reports [51,52] a number of technical recommendations have been developed to deal with questions of scale of analysis, site selection, data availability, and setting landscape assessments in a broader regional or even EU context. Currently, in the EU registration process, landscape analyses are provided by a notifier as requiring higher tier studies for aquatic risk assessment. Typical landscape analyses are based on the use of satellite imagery and aerial photographs that can assess the proximity between sources of contamination and the surface water bodies. The percentage of water bodies confirmed to be receiving an exposure as derived from $FOCUS_{sw}$–step 3 is then provided as to tool for pesticide risk assessment. A number of substances have been evaluated at higher tier levels by this approach; an example of this kind of evaluation, performed on an insecticide to be used on citrus fruit, has been discussed in the appendix A4 of the FOCUS report [52] as well as in scientific papers and at conferences [53,54].

The exposure conditions relevant for risk assessment at the edge of a field scenarios are not necessarily the conditions in water bodies draining from agricultural areas since active substances are often used in different preparations for crop types and with application patterns that depend on the soil preparation and crop type. Moreover factors such as local weather conditions or growth stages of plants on individual fields, and time schedules of farmers, can influence contamination. Additionally flow regimes of tributaries are different, and hence the transport velocity of chemical-loaded water coming from different areas of the basin does vary. Therefore the "edge of a field" exposure scenario as used in risk assessment does not in all instances reflect the worst-case scenario [55]. The assumption of a static water body in fact often results in higher dilution than the assumption of a flowing headwater body.

22.3.3 Air Diffusion

In atmospheric pollution the impact of point source (e.g., a chimney stock) or a continuous source in an area (e.g. industrial area or urban motorway) is usually modeled. Different models exist based on different mathematical assumptions. Many, such as AERMOD, CALPUFF, BLP, CALINE3, are developed or accepted for use by the US EPA and more information can be found at US EPA Web site [55]. The current technology allows environmental modeling based on physicomathematical processing of mass flux in the diffusion and dispersion of pollutants that can migrate from emission sources to the environment, both in the air near the ground and in the atmosphere, in general.

In-depth physical and chemical knowledge is of course necessary, since there are a lot of such interactions to model. But from a modeling point of view, the mathematical techniques of representation of the various environmental problems are similar, allowing analysis to proceed in a uniform way with the modeling of the whole ecosystem.

It is important to remember that a model inevitably contains a certain degree of uncertainty. This is due both to the way in which the mathematics are resolved (e.g., the physicochemical interactions) and to the extent of accuracy and completeness of the input data. Many modeling levels exist in applying Gaussian, Lagrangian, or Eulerian algorithms. The choice of the model that is used depends on the questions that need to be solved, and on a complexity variable, such as if a place has flat or hilly geography, if the phenomenon is long or short term, if secondary reactions like photochemical reactions are present, or if the pollutant reactivity is known. Regardless, whatever the model used we must obtain a detailed map with a definition of the polluting concentration or quantity.

22.4 METHODS FOR ASSESSING CONTAMINATION

There are two different methods used to assess contamination: (1) experimental measurements of contamination and (2) prediction of the possible contamination. We considered above some of the detailed predictive methods. For contamination measurements there are chemical methods and bioassays. In the first case a more detailed assessment of the composition of the contaminants is done. The clear identification of the components related to the pollution phenomenon is useful in identifying the remediation or mitigation initiatives to be adopted. However, there are limitations to the chemical approach:

1. Pollutants can escape from the adopted chemical procedure of identification and measurement.
2. Often for a large series of chemicals no suitable environmental and ecotoxicological properties are available.
3. No information is available on the overall effect of the mixture of pollutants.
4. Application of several analytical methods is often necessary because the analytes are numerous, inorganic, and organic. Pollutants include heavy metals, anions, amines, phenols, pesticides, dioxins, and many other classes, requiring different sampling, extraction, and measurement techniques.

Conversely, bioassays of the toxic effect can overcome the limitations of the chemical analysis. However, bioassays also have a number of limitations:

1. No single bioassay is enough to describe the complex potential effects of the pollution phenomena. Thus a battery of bioassays must be used.
2. Most bioassays are sensitive to acute effects but miss the chronic ones.
3. Application of several bioassays is time-consuming, and the resulting variability can be larger than that of chemical analyses.

The best solution is an intelligent integration of different techniques that are complementary so that the reliability for the overall evaluation of the phenomenon can be increased [56].

22.5 ASSESSING MIXTURES OF CHEMICALS

The knowledge described above is often enough to confirm the congruence between environmental conditions and the corresponding limit values, but it does not solve the problem related to the total toxicity of a polluting mixture [57,58]. There are several possible models of chemical interaction (see also Chapter 20). The simple case is that of the additive model, which happens when the component's combined biological effect equals the sum of each agent's effect individually.

A synergistic effect occurs when the combined effect goes beyond the sum of the effects of each component. An antagonist effect occurs when the combined effect is lower than the sum of each component used in the mixture. Beyond the current regulations, if we want to estimate the general environmental health and the effect on human health of pollutants that have been introduced into the en vironment, we have to assess the presence of the pollutants simultaneously. With the uncertainty of the synergistic effects (positive or antagonist), a reasonable solution is to apply the additive methodology; in other words, without an opposite demonstration, several substances' combined effects should be considered as additive.

A total evaluation can be carried out adding mathematically the fractions, with the substance's concentration for the numerator and the concentration's limit for the denominator:

$$\frac{C1}{L1}+\frac{C2}{L2}+\ldots+\frac{Cn}{Ln}.$$

The resulting number is considered an index, and if the result is higher than 1, the limit (index) of the mixture is considered to be exceeded. Obviously at the denominator we can also have other figures, like the warning value or target value. In any case, we will obtain a number that can be interpreted quantitatively as the limit value and/or the objective. This "simple" calculation can be combined in the modeling computation also as post-elaboration so that we have an average evaluation for each environmental impact area to observe and ultimately can obtain a relative toxicological index.

22.6 METHODS FOR RISK ASSESSMENT

According to the US EPA, ecological risk assessment includes three primary phases: problem formulation, analysis, and risk characterization. In problem formulation, goals are evaluated and assessment endpoints are selected together with the conceptual model. Then an analysis plan is developed. During the analysis phase, exposure to stressors and the relationship between stressor levels and ecological effects is evaluated. In the third phase, risk characterization, risk through integration of exposure and stressor-response profiles, is estimated. Because of the diverse expertise required (especially in complex ecological risk assessments), risk assessors and risk managers frequently work in multidisciplinary teams [59].

Considering European legislation, the first drafts of the Technical Guidance Document describing the risk assessment of industrial chemicals (1993–1994), and the first Guidance Documents on the environmental risk assessment of pesticides, including the publication of the technical annex of Directive 91/414/EC (1996), represent some European milestones in the field of environmental risk assessment.

Although environmental risk is defined as the probability of observing/producing adverse environmental/ecological effects, European legislation of environmental risk assessments includes in all cases a low tier assessment based on a deterministic approach: when the predicted exposure is clearly below the toxic concentrations determined in laboratory studies, the environmental risk is supposed to be acceptable [60]. The evaluation in the first step is based on the best available data but in a second step also takes account of potential uncertainties in the data and the range of use conditions that are likely to occur (realistic worst case approach), to determine whether the results could differ significantly. Generally, legislation for chemicals sets specific methods for risk characterization, but in the deterministic approach, the whole assessment is reduced to the acceptability of certain ratios between the expected exposure and the observed toxicity, plus a set of adjustment factors. Low risk is assumed when the exposure level is sufficiently lower than the laboratory toxicity endpoints. The "distance," or ratio between both values, to accept low risk should cover the uncertainty in the assessment. This is defined by an adjustment factor, fixed for low tier assessments through different procedures, such as the use of application factors for deriving ecotoxicological thresholds or setting fixed triggers for the toxicity exposure ratios. Under the current Directive 91/414/EC the principles behind environmental risk assessments are deterministic and are based on single-point estimates of toxicity and exposure.

Even if this kind of approach is based on deterministic elements, it cannot be considered wholly deterministic. In real-world settings both exposure and effects are highly variable in space and time due to chemical use patterns, environmental characteristics, and biological attributes. Probabilistic methods are one of the tools that should be used together with other lines of evidence to improve the understanding of exposure, toxicity, and resulting risk. Moreover landscape analysis may have a role in assessing potential spatial variability in exposure concentrations during validation of probabilistic calculations. Progression to a more probabilistic means of describing exposure and expressing the risks is proposed also by the Scientific Steering Committee [61]. Even though a deterministic approach that involves the definition of a threshold has served the needs of risk managers well in the past, it provides an apparent (often unrealistic) sharp distinction between the levels where there is an effect and that where no effect will occur. Until now, however, probabilistic approaches have gained only limited acceptance, partly due to a lack of guidance on how to implement and evaluate them. For the foreseeable future, deterministic methods are likely to remain the primary tool for the lower tiers of risk assessment. Nevertheless, the necessity of the higher tiers for risk assessment refinement has opened the possibility of using statistical methods for risk assessment, such as probabilistic risk assessment in the evaluation of exposure (Monte Carlo etc.), toxicity (mesocosm studies), and territory (from landscape analysis, satellite images). In this framework the EUFRAM project is expected to provide guidance documents on the different topics of probabilistic risk assessment (see below).

22.7 COMPUTATIONAL EVALUATION OF ECOTOXICOLOGY AND ENVIRONMENTAL PROPERTIES

21.7.1 Predicting Environmental Properties

The properties of a chemical compound are typically predicted from its chemical structure or some simple measured physicochemical properties. The theoretical basis of these models is that the matrix with which the chemical compound interacts is considered fixed, and thus the environmental behavior is related to the chemical structure only. In most cases this assumption, which as in the case of any model is a simplification, works. In the past the relationship between a given physicochemical property and the chemical structure was studied within a certain chemical group of compounds. At the basis of this approach is the assumption that given a certain chemical skeleton, some fragments can modify the properties on the basis of some simple chemical features, mainly of steric and electronic nature. These models are called (quantitative) structure–property relationships ((Q)SPR) [63, 64].

In more recent years there have been attempts to enlarge the chemical domain of applicability of the models. This has resulted in the exploration of many more chemical descriptors, in an attempt to identify chemical parameters that capture the increased complexity of the phenomenon. Thus thousands of descriptors have been proposed as well as thousands of fragments. (The fragments, or fingerprints, arise from the drug discovery experience in which this approach has been widely used.) A problem related to the introduction of so many descriptors is the risk of deriving relationships that are not statistically significant but simply are due to chance correlation. Thus attention has been given to tools to reduce the number of relationships, selecting or compressing their number with chemometric tools. Linear and nonlinear tools have been used, such as principal component analysis (PCA) and genetic algorithm (GA) [65,66].

In a typical QSPR model development, a set of chemicals is used to build up the model. These chemicals have their property values known, and the modeler calculates the chemical descriptors associated with the structure. Then relationships between the descriptors and the property are calculated, with a series of programs. Some programs are currently used to screen chemicals proposed by industry as products to be introduced into the market. Thus the US EPA yearly evaluates about 1500 to 2000 chemicals, and quite often this evaluation is done without appropriate environmental and toxicological properties.

At the base of these programs there is often the calculated $\log P$. This parameter is the the inverse of the logarithm of the partition coefficient between octanol and water (also known as K_{ow}). Such a parameter has been traditionally measured at the equilibrium of the two phases, and it indicates the preference of a chemical for water (the classical polar natural solvent) or a carbon rich, lipid medium. $\text{Log} P$ has been used to mimic the carbon content of soil or the lipid membrane of the organism, for (eco)toxicity studies [64].

A certain variability of this parameter is likely both in the measured values and the calculated values. Typically programs to calculate $\log P$ consider the presence of atoms and fragments of the molecules. Thus the predictive models based on $\log P$, even if they are apparently based on a single parameter, rely on many more parameters, statistically identified within the different programs to calculate $\log P$. There are several $\log P$ programs that are commercially or publicly available. These programs give predictions that are not identical [67]. (See Chapter 9).

From this discussion on this common parameter we now turn to consider some other issues relating to predictive models. Predictive models are typically based on statistical data. They involve a given uncertainty. Their natural variability is also often quite large. In order to compare the uncertainty of the model with that of the natural process, the latter has to be known. All these recommendations may seem obvious, but often they are disregarded or misunderstood. The programs to predict simple environmental properties are quite robust, even though far from perfect. For instance, they have been evaluated in their capability to predict properties of pesticides [68].

22.7.2 Predicting Ecotoxicity

There are many studies on quantitative structure activity relationships (QSAR) for ecotoxicological endpoints [64]. Traditionally these studies started with good results from modeling aquatic toxicity (see Chapter 23). The first studies were based on simple principles. In the model a common skeleton of the chemical compounds was derived, and a few simple substituent groups were considered. Simple linear equations were developed, in which lipophilic properties, and then steric and electrostatic features, were introduced. The principle of these models was that they worked for congeneric substances and the effect was modulated by the substituent groups. Corwin Hansch was a pioneer in these studies. This kind of approach is still at the basis of software developed by the EPA, such as ECOSAR (see below). A fundamental parameter in practically all these models is $\log P$, or $\log K_{ow}$, Octanol mimics the cellular lipid membrane of the fish, and thus the theoretical basis for this model is that if the chemical is preferentially present in octanol than in water, it will be absorbed by the membrane thereby exerting its toxic phenomenon. This principle works well on a number of chemicals for aquatic toxicity, but there are some drawbacks. The toxic mechanism that is supposed is quite simple and does not consider other toxic mechanisms. Thus the QSAR models based on this simple assumption work well for simple chemicals, but there are chemicals that, unfortunately, present toxicity at levels higher than predicted, and this situation is undesirable. Indeed regulators prefer to predict a chemical more toxic than it is. They can accept some false positives but want to avoid false negatives. Many QSAR models have been developed for specific toxicity endpoints, for chemical classes, and for specific algorithms. It is impossible to list them all here. For instance, compounds with acetylcholinesterase activity have been studied [69–71].

In the evolution of QSAR models for ecotoxicity, several pathways have been addressed, to better explore the possibilities. The numerous QSAR models (many thousands) have been in some cases organized within a unified strategy of developing an architecture in which several simple models are present. In other cases global models have been developed by way of more advanced algorithms. Different endpoints have been addressed, even though most of the studies are for aquatic endpoints. Furthermore the discussion has shifted now to addressing specific issues, such as the predictivity and the applicability of the models. Below we will discuss this evolution.

A number of modified submodels have been developed and introduced. In the program ECOSAR, for instance, submodels are used to optimize the predicted toxicity of compounds based on the chemical classes. From the presence of some chemical moieties, several toxic modes of action have been introduced by different authors [72,73]. Indeed, the basis of this approach was not to refer to chemical classes, but instead to biological phenomena related to the toxicity. Thus the chemical fragments are identified as reactive groups involved in a given reactive process. However, there is not a commonly accepted taxonomy of modes of action, and a large percentage of the chemicals do not present a defined mode of action. Studies using this approach on relatively complex chemicals, such as pesticides, show poor results [68]. However, pesticides are very complex, if not difficult, cases, since many functional groups are present along with many toxic features.

The ECOSAR uses simple equations based on $\log P$. Nevertheless, different equations have to be developed for the specific chemical classes, in order to take into account the reactivity component of the toxicity phenomenon using the chemical category and then modulating it with the $\log P$. Many chemical categories are used, and also the $\log P$ value is typically calculated using many parameters. Thus, even though these models look simple, they use tens of parameters. A drawback of these local models is that they have been built from a limited number of compounds, and no proper validation of the individual submodels exists (see below). Furthermore the concurrent presence of several groups is not addressed. Another approach is to develop models that address a large group of chemicals with a single model. In this case many different parameters are used, and the equation are often also complex.

For aquatic toxicity some good examples have been developed. Kaiser and coworkers developed a probabilistic neural network that gave good predictive results in case of fish toxicity [74]. Within the EC-funded project IMAGETOX, some models have been developed for aquatic toxicity, using different tools, linear and nonlinear, with similar results. It is interesting that the results using chemical descriptors based on 2D or 3D structures are similar. Thus there appears to be no advantage to using more complex descriptors [75].

Some general models for other endpoints have been developed. Devillers et al. [76] addressed the acute toxicity of pesticides on honey bee, achieving good results with one hundred pesticides. Some studies exist that use classifiers. Pintore et al. [77] developed a model to classify pesticides, in general, with the lethal dose toward rat. Classifiers are more typical in toxicity studies, and genotoxicity in particular. Use of a classifier can be a simpler approach compared to regression models, since the prediction does not need to be as precise when only prediction of a toxicity class (e.g., high, medium, low) is involved, and this is even simpler if the classifier is a binary one, which is often the case for genotoxicity (toxic/nontoxic).

While the classical QSAR models were the result of studies aimed at describing a given toxicity phenomenon using a limited number of parameters, more recently a focus has been on the possible use of these models to predict the properties of new chemical compounds. In principle, the predictive feature is implicit in the QSAR model, but to really prove that the model is predictive there are other issues to be addressed. In this case, as in several other instances, modification of some premises is involved, and thus a new scenario that requires new criteria. In particular, in order to prove that a model is predictive, some statistical tools have to be used that are capable of showing the predictive features of the model. A second important point is how to identify the limits of the applicability of a given model.

The point of the measurement of predictive power has been discussed over the last few years, and different tools have been proposed for this purpose [78,79]. It is generally agreed that we have to distinguish between the performance of a QSAR model on the training set (see the previous paragraph: this is the set of compounds used to build up the model) and the performance when the model is used to make a prediction for a new chemical. One perspective states that the performance on a new set of compounds has to be measured. This new set of compounds, called a test set, is a series of chemicals that has never been used to build up the model. This fact is important because the model can easily learn the properties associated with a given compound, so the value provided by the model is not a prediction but only a value that the model recalls. In this case the model could also be overfitted, as it follows too closely to the specific properties of the limited number of chemicals used to build the model. The model is not capable of generalizing the properties of the chemicals used to train the model. Thus an external set is proposed as a way to prove the real performance in predicting the toxic property. However, some problems remain, especially if the number of compounds is too small, so other approaches for testing may be preferable [79].

22.7.3 The Demetra Models

Very recently the EC funded project DEMETRA (*http://www.demetra-tox.net*) developed a series of QSAR models for the prediction of toxicity of pesticides toward five endpoints: trout, daphnia, quail (oral and dietary exposure), and bee [80]. This project introduced a number of innovative issues, compared to previous QSAR models. The target of the project was to develop models for pesticides to be used for regulatory purposes in accord with European legislation. A questionnaire was distributed to a great many end-user to identify their needs. The endpoints to be modeled were chosen from among those defined in writing by the end-user, and not by the modeler, in order to make the models as useful as possible. This attention to the needs of the end-users is unique in the use of QSAR for ecotoxicity prediction. Other novel developments are the clear reference to a specific directive and to related guidelines indicated in the regulation. Typically model criteria are only vaguely defined and data from different protocols are used to build up the model. In the case of the DEMETRA models another new development was that the quality of the data was very high. We already identified the specific guidelines used to produce experimental data. Three high-quality databases were used separately (and not merged) to confirm the quality of the data. Indeed, it is quite common that for the same chemical more than one toxicity value is obtained, even with the same experimental protocol. Within DEMETRA if there was variability higher than a factor of four, the pesticide was not used for modeling purposes. This is unusual in QSAR modeling, so it helps to characterize the uncertainty of the toxicity values of the QSAR model. Without the variability of the experimental values being characterized, it would be very difficult to assess the uncertainty of the predicted values, which are based on these values too. Careful checks of the quality of the data must be performed also for the chemical data, as described in detail elsewhere [80].

Another fundamentally new aspect of the DEMETRA models is that they were developed to minimize the false negatives. This is very important for QSAR's regulatory purposes, as we mentioned earlier. Regulators have to avoid errors in the direction of false negatives as much as possible. QSAR have never been assessed in terms of how to reduce false negatives. QSARs are basically statistical measurements whereby the errors are squared so that the errors in one direction or in another (false positives or false negatives) are equivalent. The result is less useful in regulatory models. DEMETRA models have been developed to avoid false negatives as much as possible [80].

Nevertheless, QSAR models are useful within the more general strategy of risk assessment of chemicals. When using QSAR models for regulatory purposes (compared with any model for a specific purpose), one has to be fully aware of the typical environment of the application. We already explained that DEMETRA only uses toxicity values obtained according to defined guidelines. This refers to the input of the QSAR model, and also to the output of the model, the toxicity value, both of which have to be clearly defined. We mentioned the issue of the uncertainty and false negatives, as addressed within DEMETRA. If a QSAR model has a single outlier that has an error of 100 times, and then no explanation is given for that, the regulators will apply the same uncertainty factor of 100 to all predicted values. In a model for fish toxicity another uncertainty factor of 100 is used to account for species sensitivity. Thus overall an uncertainty of 10,000 should be applied for the QSAR predicted values. The simple result is that for almost all cases the pesticide will be predicted as toxic. Thus such a model is of very limited use. This clearly shows that a model for regulatory purposes has to be assessed by each specific use.

DEMETRA includes specific confidence restrictions capable of identifying outliers. The regulator can chose the confidence level for a factor of 50 or 10 and then can apply, or not, some restriction rules [80]. This way models with an uncertainty factor of 10 can be used: this value is very close to the value of the variability of experimental values. Some other novelties introduced by DEMETRA, are more general and aimed at reducing error. Additionally a modularity in the definition of the applicability domain of the QSAR models has been identified, whereas usually a chemical is defined as within or outside the applicability domain, which is a crude approximation.

DEMETRA models are based on hybrid techniques. Each model for trout, daphnia, quail, or bee is composed of a number of submodels. A hybrid model integrates, in an intelligent way, the results of the submodels to achieve better prediction values capable of reducing false negatives. The regulator can see the minimal and maximal values of the individual models, should the regulator decide to use the lowest predicted value. Another innovative feature of DEMETRA is that its models are freely available (*http://www.demetra-tox.net*) for wide use, and the user has only to calculate 2D descriptors.

22.8 ONGOING RESEARCH AND FUTURE PERSPECTIVES

Assessment of environmental and ecotoxicological properties of chemicals is an evolving field of activity. As increasingly large amounts of information are being generated, new models are being introduced, and there is need for better knowledge. In this ever more demanding situation, there is the need for more integrated strategy, capable of taking advantage of all possible tools to produce more accurate assessments. Whereas in the past many computerized models were still being discussed, nowadays the power and the complexity of such computerized models have exceeded all expectations, and they integrate using many different tools. In the long term the objective is to develop an integrated decision system, capable of handling the challenging task of organizing the diverse strategies of regulators, and merging all eventual in vivo, in vitro, and in silico data to obtain a deeper view of chemical properties. Some of the initiatives that are presently ongoing and capable of enlarging the assortment of tools used in assessing the environmental and ecotoxicological properties of chemicals are discussed below.

22.8.1 HAIR

The EC funded HAIR project (*http://www.rivm.nl/stoffen-risico/NL/hair.htm*) is devoted to integrating the different methods for the assessment of pesticides into a combined index. The intention is to go beyond the environmental and ecotoxicological properties of pesticides and to address the human toxicity aspects as well. The ultimate target is to produce an index useful to regulators changed with evaluating the EU policy on pesticides and to reduce the overall impact of the pesticides. This project also considers other chemicals besides pesticides. As in many other examples, the lessons learned from pesticides offer strategic opportunities for other pollutants, since pesticides have been much more studied for a large and detailed series of properties in relation to specific regulations. Interestingly this project includes an objective to integrate and summarize all different risk assessment strategies. So the project offers a unique opportunity to combine a battery of tools that historically were developed independently and themselves present problems relating to their separate uses. The complex architecture proposed by HAIR allows, for instance, aggregation at the geographical level (from small areas to province, region, or country), and/or application of multiple pesticides.

22.8.2 EUFRAM

EUFRAM (*http://www.eufram.com*) represents a project on pesticides that involves a great many organizations—regulatory authorities, government research institutes, agro-chemical companies, consultancy companies, and universities. Its aim is to provide the basic concepts, case studies, principles, and approaches that can help users conduct, report, evaluate, and communicate probabilistic assessments in a uniform way. The defining feature of a probabilistic risk assessment (PRA) is that one or more sources of variability and/or uncertainty in exposure, effects, and the resulting risk can be quantified. The major advantage is that this way the assessment takes more realistic account of the variation in factors that influence risk. PRAs can take the place of or refine worst-case analyses. With a PRA more data are available (but on the other hand, the PRA needs more data). Furthermore the PRA more clearly identifies the major sources of uncertainty to be assessed.

22.8.3 CHEMOMENTUM

Very recently a new EC project, CHEMOMENTUM, was started. The point of interest is information technology and the development of suitable tools for a technology called GRID. GRID refers to an integrated series of hardware and software tools that are geographically distributed and organized in order to make optimal use of the available resources. The end-user does not get involved in technical issues, and the overall architecture and complex integration of tools do not affect the use of the system, which results in a very simplified system. The challenge of the project is to integrate a battery of tools, with the end result being QSAR models that are simple to use. This development is valuable for several reasons. The use of a complex network of computers (in this case very powerful and fast, and some of the best available in Europe) can make for very fast calculations (e.g., *ab initio* calculations). The power of the system should allow more possibility in exploring different QSAR models to achieve better results. The system will surely simplify user operation. Typically QSAR models call for many different steps: going from the 2D structure to 3D, optimizing the conformation manually, calculating the chemical descriptors, developing QSAR models, descriptor selection or compression, and so on. All these steps will be done automatically within CHEMOMENTUM. Another advantage is that the conditions of the QSAR models will be defined and standardized for regulatory purposes. CHEMOMENTUM will automatically make operations currently done manually and avoid the variability associated with this. This way regulators will be sure to get the same type of results from all users.

ACKNOWLEDGMENTS

We acknowledge the EC funded projects DEMETRA (QLRT-2001-00691), HAIR (SSPE-CT-2003-501997), EUFRAM (QLRT-2001-01346) and CHEMOMENTUM (IST-2005-033437).

REFERENCES

1. Nicholls PH. Simulation of the movement of bentazon in soils using the CALF and PRZM models. *J Environ Sci Health A* 1994;A29:1157–66.
2. Nofziger DL, Hornsby AG. A micro-computer base management tool for chemical movement in soil. *Appl Agricul Res* 1986;1:50–56.
3. Armstrong AC, Matthews AM, Portwood AM, Leeds-Harrison PB, Jarvis NJ. CRACK-NP: A pesticide leaching model for cracking clay soils. *Agr Water Manage* 2000;44:183–99.
4. Williams JR, Jones CA, Dyke PT. A modeling approach to determining the relationship between erosion and soil productivity. *Trans ASAE* 1984;27:129–44.
5. Burns LA. *Exposure analysis modeling system (EXAMS): User manual and system documentation.* EPA/600/R-00/081 Sep 2000, Revision G (May 2004).
6. Parker RD, Rieder DD. *The generic expected environmental concentration program, ENEEC. Part B. Users Manual: Tier one screening model for aquatic pesticide exposure.* Washington: Environmental Fate and Effects Division, Office of Pesticide Programs, US Environmental Protection Agency, 1995.
7. Knisel WG, Davis FM, Leonard RA. *GLEAMS Version 2.0, User Manual.* Tifton, GA: USDA-ARS, Southeast Watershed Research Laboratories, 1992.
8. Donigian AS, Imhoff JC, Bicknell BR, Kittle JL. *Application guide for the hydrologic simulation program—FORTRAN.* EPA 600/3–84–066. Athens, GA: US EPA, 1984.
9. Simunek J, Huang K, van Genuchten MT. The HYDRUS code for simulating the one-dimensional movement of water, heat, and multiple solutes in variably-saturated media. Version 6.0, Research report 144, U.S. Salinity Laboratory, USDA, ARS, Riverside, CA, 1998.
10. Simunek J, Sejna M, van Genuchten MT. *The HYDRUS-2D software package for simulating two-dimensional movement of water, heat, and multiple solutes in variably saturated media.* Version 2.0. *IGWMC-TPS-53.* Golden, CO: International Ground Water Modeling Center, Colorado School of Mines, 1999.

11. De Vries DJ. *IMPAQT, a physico-chemical model for simulation of the fate and distribution of micropollutants in aquatic systems.*The Netherlands and TOW-IW T250, Delft Hydraulics, 1987.

12. Hutson JL, Wagenet RJ. *LEACHM: Leaching estimation and chemistry model. A process-based model of water and solute movement, transformations, plant uptake and chemical reactions in the unsaturated zone.* Version 3. Ithaca, NY: Department of Soil, Crop and Atmospheric Sciences. Research series no. 92-3, Cornell University, 1992.

13. Jarvis NJ, Jansson PE, Dik PE, Messing I. Modelling water and solute transport in macroporous soil: I. Model description and sensitivity analysis. *J Soil Sci* 1991;42:59–70.

14. Larsbo M, Jarvis N. MACRO 5.0. A model of water flow and solute transport in macroporous soil: Technical description. *Emergo* 2003:6. Swedish University of Agricultural Sciences, Department of Soil Sciences, Uppsala, 2003.

15. Jarvis NJ, Hollis JM, Nicholls PH, Mayr T, Evans SP. MACRO-DB: A decision-support tool for assessing pesticide fate and mobility in soils. *Environ Modell Softw* 1997;12:251–65.

16. Danish Hydraulic Institute (DHI). *MIKE 11—User guide and technical reference manual* Copenhagen: DHI, 1998.

17. Thiéry D, Golaz C, Gutierrez A, Fialkiewicz W, Darsy C, Mouvet C, Dubus IG. Refinements to the MARTHE model to enable the simulation of the fate of agricultural contaminants from the soil surface to and in groundwater. In: *Saturated and unsaturated zone: Integration of process knowledge into effective models. Proceedings of the COST international workshop,* Rome, 5–7 May 2004. p. 315–20.

18. Klein M. *MOPED model for pesticide drift, User's manual* Berlin: Environmental Protection Agency, Report 126 05 080. (in German).

19. Steenhuis TS, Pacenka S, Porter KS. MOUSE: A management model for evaluating groundwater contamination from diffuse surface sources aided by computer graphics. *Appl Agr Res* 1987;2:277–89.

20. Smith R.E. Opus: An integrated simulation model for transport of nonpoint-source pollutants and the field scale, Vol. 1. Documentation. USDA-ARS, 1992 ARS–98.

21. Boesten JJTI, van der Linden AMA. Modeling the influence of sorption and transformation on pesticide leaching and persistence. *J Environ Qual* 1991;20: 425–35.

22. Tiktak A, van den Berg F, Boesten JJTI, van Kraalingen D, Leistra M, van der Linden AMA. Manual of FOCUS PEARL version 1.1.1. RIVM report 711401 008, Alterra report 28, Bilthoven, Wageningen, The Netherlands, 2000.

23. Tiktak A, de Nie D, van der Linden T, Kruijne R. Modelling the leaching and drainage of pesticides in the Netherlands: The GeoPEARL model. *Agronomie* 2002;22:373–87.

24. PELEP-DSS (2006). <http://newwave.geru.ucl.ac.be/PELEP_DSS>.

25. Klein M. *PELMO: Pesticide leaching model.* Schmallenberg, Germany: Fraunhofer-Institut für Umweltchemie und Ökotoxikologie, 1991.

26. van den Berg F, Boesten JJTI. Pesticide leaching and accumulation model (PESTLA) version 3.4. Description and user's guide. Technical document 43. DLO Winand Staring Centre, Wageningen, Th e Netherlands. 1999.

27. Tiktak A, van der Linden AMA, Swartjes FA. PESTRAS: A one-dimensional model for assessing leaching and accumulation of pesticides in soil. National Institute of Public Health and the Environment. Report 715501003. Bilthoven, The Netherlands. 1994.
28. Nicholls PH, Harris GL, Brockie D. Simulation of pesticide leaching at Vredepeel and Brimstone farm using the macropore model PLM. *Agr Water Manage* 2000;44:307–15.
29. Carsel RF, Mulkey LA, Lorber MN, Baskin LB. The pesticide root zone model (PRZM): A procedure for evaluating pesticide leaching threats to groundwater. *Ecol Model* 1985;30:49–69.
30. Williams WM, Cheplick JM. *RIVWQ users manual.* Leesburg, VA: Waterborne Environmental, Inc., Waterborne Environmental-B Harrison Street, SE 22075, 1993.
31. RZWQM Team-Root Zone Water Quality Model, Version 1.0. Technical documentation. GPSR technical report 2., USDA-ARS-GSPR, Fort Collins, CO, 1992.
32. Aden K, Diekkrüger B. Modeling pesticide dynamics of four different sites using the model system SIMULAT. *Agr Water Manage* 2000;44:337–55.
33. Rasmussen D. Surface water model for pesticides—SLOOT.BOX. Technical report V. National Environmental Research Institute, Roskilde, Denmark, 1995.
34. De Vries DJ, Kroot MPJM. *SOM-3, a simple model for estimating fluxes, concentrations and adaptation times of micropollutants in aquatic systems*: User's manual, T0632. Delft, Delft Hydraulics, Holland, 1989.
34. Arnold JG, Allen PM. A comprehensive surface-groundwater flow model. *J Hydrol* 1992; 142:47–69.
35. Arnold JG, Fohrer N. SWAT2000: Current capabilities and research opportunities in applied watershed modelling. *Hydrol Process* 2005;19:563–72.
36. Adriaanse PI. Exposure assessment of pesticides in field ditches: The TOXSWA model. *J Pestic Sci* 1997;49,210–12.
37. Beltman WHJ, Adriaanse PI. *User's manual TOXSWA 1.2. Simulation of pesticide fate in small surface waters.* SC-DLO Technical Document 54, Wageningen, The Netherlands 1999.
38. Haith DA. TurfPQ, A Pesticide Runoff Model for Turf. *J Environ Qual* 2001;30: 1033–9.
39. Walker A. Evaluation of a simulation model for prediction of herbicide movement and persistence in soil. *Weed Res* 1987;27:143–52.
40. Vanclooster M, Viaene P, Diels J, Christiaens K. WAVE: A mathematical model for simulating water and agrochemicals in the soil and vadose environment. Reference and user's manual (release 2.0). Institute for Land and Water Management, Katholieke Universiteit Leuven, Leuven, Belgium. 1995.
41. Hill BD, Miller JJ, Chang C, Rodvang SJ. Seasonal variation in herbicide levels detected in shallow Alberta groundwater. *J Environ Sci Heath B* 1996;B31: 883–900.
42. Barbash JE, Thelin GP, Kolpin DW, Gilliom RJ. Major herbicides in ground water: results from the national water-quality assessment. *J Environ Qual* 2001; 30:831–45.

43. Cerejeira MJ, Viana P, Batista S, Pereira T, Silva E, Valério MJ, Silva A, Ferreira M, Silva-Fernandes AM. Pesticides in Portuguese surface and ground waters. *Water Res* 2003;37:1055–63.

44. Johnson AC, Besien TJ, Lal Bhardwaj C, Dixon A, Gooddy DC, Haria AH, White C. Penetration of herbicides to groundwater in a unconfined chalk aquifer following normal soil applications. *J Contam Hydrol* 2001;53:101–17.

45. PEGASE. Pesticides in European groundwaters: Detailed study of representative aquifers and simulation of possible evolution scenarios. PEGASE final report. BRGM/RP-52897-FR. Feb 2004.

46. FOCUS. Leaching models and EU registration. Doc 4952/VI/95, 1995. 124p.

47. FOCUS groundwater scenarios in the EU plant protection product review process. Report of the FOCUS Groundwater Scenarios Workgroup. EC document reference Sanco/321/2000.

48. Gottesbüren B. FOCUS groundwater's view on higher tier modelling and experimental approaches for submission on EU and national level with respect to the assessment of leaching to groundwater. Oral presentation at the 4th European Modelling Workshop: Higher tier assessment of the exposure of ground and surface waters: The context of modelling and experiments, Paris, 21–22 Nov 2005. <www.pfmodel.org>.

49. FOCUS. Surface water models and EU registration of plant protection products. European Commission document 6476/VI/96. 1997.

50. FOCUS surface water scenarios in the EU evaluation process under 91/414/EEC. Report of the FOCUS Working Group on Surface Water Scenarios. EC document reference SANCO/4802/2001-rev.1. 2001.

51. FOCUS, landscape and mitigation factors. In: *Aquatic risk assessment. Vol. 1. Extended summary and recommendations*. Report of the FOCUS Working Group on Landscape and Mitigation Factors in Ecological Risk Assessment. EC document reference SANCO/10422/2005, 2005. 133.

52. FOCUS, landscape and mitigation factors. In: *Aquatic risk assessment. Vol. 2. Detailed technical reviews*. Report of the FOCUS Working Group on Landscape and Mitigation Factors in Ecological Risk Assessment. EC document reference SANCO/10422/2005. 434 pp. 2005.

53. Padovani L, Capri E, Trevisan M. Landscape-level approach to assess aquatic exposure via spray drift for pesticides: A case study in a Mediterranean area. *Environ Sci Technol* 2004;38: 3239–46.

54. Carter A, Capri E. Environmental fate of chlorpyrifos: Exposure and effect of chlorpyrifos following use under Southern European conditions. Catania (Italy), 9–10 Apr 2003. *Outlook on Pest Management* 2004; 10/1564/15 Feb.

55. <http://www.epa.gov>.

56. Galassi S, Benfenati E. Fractionation and toxicity evaluation of waste waters. *J Chrom A* 2000;889:149–54.

57. Olmstead AW, LeBlanc GA. Joint action of polycyclic aromatic hydrocarbons: Predictive modeling of sublethal toxicity. *Aquat Toxicol* 2005;75:253–62.

58. Donnelly CK, Lingenfelter R, Cizmas L, Falahatpisheh MH, Qian Y, Tang Y, Garcia S, Ramos K, Tiffany-Castiglioni E, Mumtaz MM. Toxicity assessment of complex mixtures remains a goal. *Environ Toxicol Pharm* 2004;18,:135–41.

59. Lepper P. Use of higher-tier studies conducted in the context of pesticide risk assessment for quality standard setting. Poster presented at the SETAC Europe 16th Annual Meeting 7–11 May 2006, The Hague, The Netherlands.
60. US Environmental Protection Agency. *Guidelines for ecological risk assessment.* EPA/630/R-95/002F. Washington, 1998.
61. SSC. The future of risk assessment in the European Union—The second report on harmonisation of risk assessment procedures adopted by the Scientific Steering Committee, 10–11 Apr 2003.
62. SSC. Report on the ecological risk assessment of chemicals—The second report on harmonisation of risk assessment procedures adopted by the Scientific Steering Committee, 10–11 Apr 2003. App. 4.
63. Katritzky AR, Petrukhin R, Tatham D, Basak S, Benfenati E, Karelson M, Maran U. Interpretation of quantitative structure-property and -activity relationships. *J Chem Inf Comput Sci* 2001;41:679–85.
64. Cronin MTD, Livingstone DJ. *Predicting chemical toxicity and fate.* Boca Raton FL: CRC Press, 2004.
65. Gini GC, Katritzky AR. *Predictive toxicology of chemicals: Experiences and impact of AI tools.* AAAI 1999 Spring Symposium Series. Menlo Park, CA: AAAI Press, 1999.
66. Devillers J. *Genetic algorithms in molecular modeling.* London: Academic Press, 1996.
67. Benfenati E, Gini G, Piclin N, Roncaglioni A, Varì MR. Predicting LOGP of pesticides using different software. *Chemosphere* 2003;53:1155–64.
68. Hansen OC. Quantitative structure-activity relationships (QSAR) and pesticides. Danish Environmental Protection Agency, Pesticides research 94. 2004.
69. Ali HM, Sharaf EHA, Hikal MS. Selectivity, acetylcholinesterase inhibition kinetics and quantitative structure-activity relationships of a series of *N*-(2-oxido-1,3,2-benzodioxa-phosphol-2-yl) amino acid ethyl or diethyl esters. *Pest Biochem Physiol* 2005;83:58–65.
70. Maxwell DM, Brecht KM. Quantitative structure-activity analysis of acetylcholinesterase inhibition by oxono and thiono analogues of organophosphorus compounds. *Chem Res Toxicol* 1992;5:66–71.
71. Lin G, Lai CY, Liao WC. Molecular recognition by acetylcholinesterase at the peripheral anionic site: Structure-activity relationships for inhibitions by aryl carbamates. *Bioorg Med Chem* 1999;7:2683–9.
72. Russom CL, Bradbury SP, Broderius SJ, Hammermeister DE, Drummond A. Predicting modes of toxic action from chemical structure: Acute toxicity in the Fathead minnow (*Pimephales promelas*). *Environ Toxicol Chem* 1997;16:948–67.
73. Vaal M, van der Wal JT, Hoekstra J, Hermens J. Variation in the sensitivity of aquatic species in relation to the classification of environmental pollutants. *Chemosphere* 1997;35:1311–27.
74. Kaiser KLE, Niculescu SP. Using probabilistic neural networks to model the toxicity of chemicals to the fathead minnow (*Pimephales promelas*): A study based on 865 compounds. *Chemosphere* 1999;38:3237–45.
75. Netzeva TI, Aptula AO, Benfenati E, Cronin MTD, Gini G, Lessigiarska I, Maran U, Marjan Vracko M, Schüürmann G. Description of the electronic structure of

organic chemicals using semiempirical and ab initio methods for development of toxicological QSARs. *J Chem Inf Model* 2005;45:106–14.

76. Devillers J, Pham-Delegue MH, Decourtye A, Budzinski H, Cluzeau S, Maurin G. Structure-toxicity modeling of pesticides to honey bees. *SAR QSAR Environ Res* 2002;13:641–8.
77. Pintore M, Piclin N, Benfenati E, Gini G, Chrétien JR. Database mining with adaptive fuzzy partition (AFP): Application to the prediction of pesticide toxicity on rats. *Environ Tox Chem* 2003;22:983–91.
78. Golbraikh A, Shen M, Xiao Z, Xiao YD, Lee KH, Tropsha A. Rational selection of training and test sets for the development of validated QSAR models. *J Comput Aided Mol Des* 2003;17:241–53.
79. Kraker JJ, Hawkins DM, Basak SC, Natarajan R, Mills D. Quantitative structure-activity relationship (QSAR) modeling of juvenile hormone activity: Comparison of validation procedures. *Chemom Intel Lab Sys*, in press, 2006.
80. Benfenati E. *Quantitative structure-activity relationship (QSAR) for regulatory purpose: The pesticide case.* Amsterdam: Elsevier, 2007.

23

APPLICATION OF QSARs IN AQUATIC TOXICOLOGY

JAMES DEVILLERS

Contents

23.1 INTRODUCTION

The 1960s forever changed the way society viewed the effects of human activities on the environment. Although environmental problems were

Computational Toxicology: Risk Assessment for Pharmaceutical and Environmental Chemicals, Edited by Sean Ekins

apparent such as the acute toxicity of a number of xenobiotics on the biota, the increasing resistance of some pests to new pesticides, and the eutrophication of lakes and estuaries, the warning alarm was sounded in 1962 by Rachel Carson with the publication of her seminal book *Silent Spring* [1]. In this book she argued compellingly against organic contaminants, especially pesticides such as DDT, that were being used with little regard for their impact on the environment including human health.

Many changes in the regulation of chemicals, in the development of environmental sciences, and in the evolution of the mentalities have come since the publication of *Silent Spring*. Prominent among the changes in thinking about the adverse effects of chemicals on the biota is the emergence of environmental toxicology as a scientific discipline. It became obvious that the adverse effects of chemicals released into the environment had to be estimated on organisms occupying different trophic levels. In the early years of environmental toxicology, deficits in hazard and risk assessment schemes as well as in representative tests with standardized protocols limited observable progress. Over time the situation has changed, and today there are a number of ecotoxicity tests in use that are ecologically relevant, and some are rapid and cost-effective. This is particularly true in aquatic toxicology where laboratory tests have been developed on bacteria (e.g., *Vibrio fischeri*), algae (e.g., *Chlorella vulgaris*, *Pseudokirchneriella subcapitata*), protozoa (e.g., *Tetrahymena pyriformis*), crustaceans (e.g., *Daphnia magna*, *Ceriodaphnia dubia*, *Artemia salina*), and fish (e.g., *Oncorhynchus mykiss*, *Pimephales promelas*, *Lepomis macrochirus*, *Danio rerio*) for estimating the effects of chemicals on mortality, growth, reproduction, and so on.

Consequently, as Candide (Voltaire) said, “All is for the best in the best of all possible worlds”! Unfortunately, this is not the case. Whereas in 1965 the number of chemical substances registered with the Chemical Abstracts Service (CAS) was 212,000 [2], currently it is about 12 million [3], and the number of commercially available chemicals is continually increasing. With such a large number of xenobiotics, which potentially can be released into the environment, it is impossible to test all of them exhaustively for their adverse effects on living species. The constraints are mainly time and funding.

Fortunately, to overcome the different problems of laboratory testing, modeling approaches are being developed for estimating the toxicity of chemicals. QSARs (quantitative structure-activity relationships) are the fundamental basis of these approaches in environmental toxicology for predicting the toxicity of chemicals from their molecular structure and/or physicochemical properties.

The goal of this chapter is not to review all the QSAR models in use in environmental toxicology. A huge number of different models has already been covered in other chapters of this book and many others are available in the literature. The attempt of this chapter is to limit the focus to the QSARs in aquatic toxicology. This choice is not without good reason. The oceans cover approximately 71% of the planet and the freshwater ecosystems (lakes, pond, rivers, etc.) occupy large spaces on the earth's surface. Although the ecological and economical roles of these aquatic ecosystems are clearly crucial, they are highly polluted by the xenobiotics. Hence there is urgent need to evaluate their ecotoxicity.

The plan of this chapter is as follows: After briefly introducing some basic concepts in environmental QSARs, the characteristic use of QSARs in aquatic toxicology will be discussed and their advantages and limitations will be presented by way of various examples.

23.2 PRINCIPLES FOR THE DEVELOPMENT OF ENVIRONMENTAL QSARs

For deriving a QSAR model, three different types of elements are required (Figure 23.1). First, a measured endpoint for a set of molecules has to be available. Second, chemicals must be described by means of their physicochemical properties or structurally derived parameters. Last, a statistical method must be used for linking the first two elements. These three critical ingredients of the general methodology in the derivation of a QSAR model are briefly discussed in the next sections.

23.2.1 Measured Endpoint

A well-defined endpoint is essential for the establishment of an accurate QSAR. Thus, in deriving an aquatic structure-toxicity model, it is important to select the toxicity data obtained from standardized laboratory tests. For this reason LC_{50} (or EC_{50}) values (i.e., the concentrations that are lethal, or effective, to 50% of the organisms tested) are preferentially used, since they give

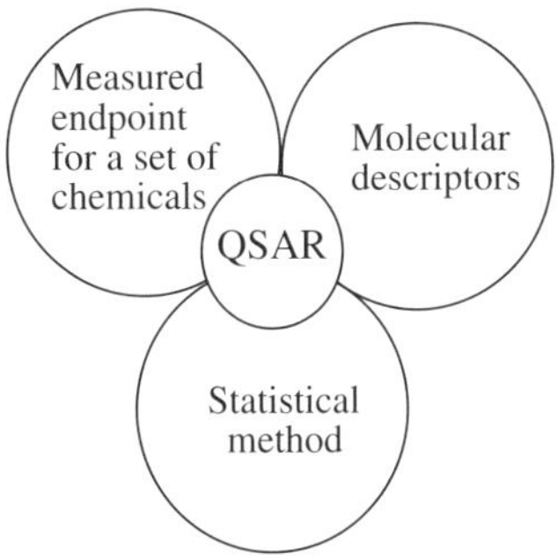

Figure 23.1 Ingredients for deriving a QSAR model.

statistical and toxicological significance values. These toxicity data must be expressed on a molar basis to be structurally comparable. In addition, to derive the models, the data must be converted into a logarithmic scale to avoid statistical problems with classical methods, such as regression analysis. By convention, negative logarithms are preferred to obtain the larger values for the more active chemicals. In general, the biological data should be obtained from the same experimental conditions. One important source of variability in the data upon which some QSARs are developed is their source and age. Thus, if the experimental biological activities were obtained from different laboratories over a long period of time, changes in the protocols (e.g., time of exposure, composition of the media, use of nominal or actual concentrations) and differences in the practical experience for the achievement of the test(s) will result in a variability in the data. Examples of aquatic toxicity tests and their respective variability in the results in the frame of inter-laboratory ring exercises are given in Table 23.1.

Another source of variability in the biological data is that arising from the presence of impurities in the tested chemicals. Ideally, a chemical should be 100% pure and its structure fully known. In practice, a lower purity is acceptable if the impurities do not alter the activity of concern. This judgment can be based on specific knowledge regarding the studied substance or on the toxicological significance of the concentration(s) of the impurity(ies) [4]. The biological data used for deriving a QSAR model are generally retrieved from the open literature. Whereas in the past it was commonplace to first test chemicals and then use one's own data to derive a QSAR model, today this situation is totally unusual. Numerous handbooks are available for obtaining data. They include collections of biological activities for specific taxa [e.g., 5,6] or chemical families [e.g., 7]. Also valuable databases are now available on

TABLE 23.1 Variability of Inter-laboratory Ring Tests

Species	Endpoint	Chemical	% CV*
Fish	Mortality (96-h LC_{50})	Surfactants	20–78
	Growth (28-d NOEC)	Dichloroaniline	57
		LAS	33
D. magna (Crustacean)	Mortality (48-h EC_{50})	Surfactant	26
	Reproduction (21-d NOEC)	Dichloroaniline	53
		$CdCl_2$	80
		Phenol	76
A. salina (Crustacean)	Mortality (24-h LC_{50})	$CuSO_4$	47
S. costatum (Algae)	Growth (72-h EC_{50})	$K_2Cr_2O_7$	44
		3,5-Dichlorophenol	18

Source: Adapted from [4].

*Coefficient of variation.

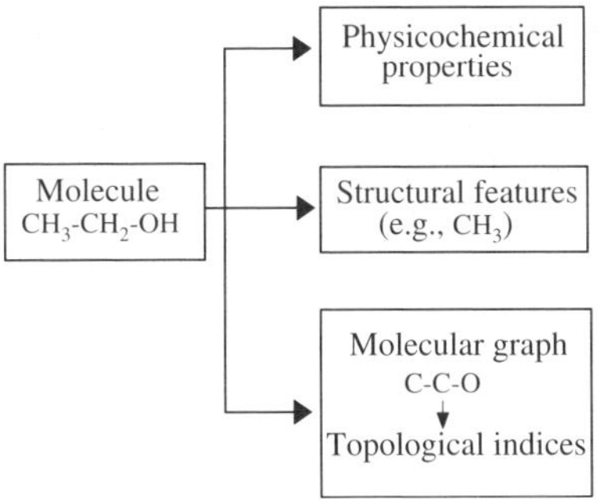

Figure 23.2 Molecular descriptors used in environmental QSAR.

different media (CD-ROM, Internet) to enable rapid access to biological data (for a full overview of these systems, see refs. [8–11]).

23.2.2 Molecular Descriptors

In selecting molecular descriptors upon which to base a QSAR model, one needs to know the role that these parameters play, either the way the chemical behaves or the way the studied endpoint is expressed [4]. While 3D molecular descriptors are now widely used in drug design [12], most QSAR models in environmental sciences are derived from 2D molecular descriptors that are broadly classified into three main types (Figure 23.2).

Descriptors based on measured or calculated physicochemical properties of the molecules have historically been the most widely used in QSAR. Thus it is now well known that the biological responses of the living organisms to xenobiotics (and drugs) are controlled by lipophilic (hydrophobic), electronic, and/or steric properties. Under these conditions the use of these physicochemical properties as molecular descriptors allows one to obtain valuable QSAR models that can be easily interpreted. Because a consensus exists within the QSAR community on the usefulness of these descriptors, it is not necessary in the frame of this chapter to theorize about them nor to catalog QSAR equations showing their interest. This work has been done in numerous valuable articles and books [e.g., 13–15]. Instead it is more useful to regard the problems that undoubtedly arise when these descriptors are used without careful judgment.

No other physicochemical property has attracted as much interest in QSAR as the 1-octanol/water partition coefficient (log *P*). This is because of its direct relationship to membrane permeation and numerous other biological processes (also see Chapter 9). At first sight, the experimental measurement of log *P* seems to be a straightforward exercise. In practice, this is not the case because many factors affect the experimental determination of log *P* such as temperature, pH, and stability to degradation. As a result identical replicate samples can give very different log *P* values, and measurement by different workers using the same methodology (e.g., shake-flask method) can produce marked variations. The problem is worse when both direct and indirect methods are used. Under these conditions the use of experimental log *P* values of different origin for the design of a QSAR model can introduce important biases if a critical analysis of the data has not been previously performed. Similarly, although log *P* has been measured for large collections of organic molecules, the total number of experimental values that can be found in the literature and in the databases will be relatively small compared to the number of existing chemicals. Consequently, since the pioneering work of Fujita and coworkers [16] who considered log *P* to be an additive-constitutive property [17] (i.e., equal to the sum of the log *P* of the parent solute plus a π term representing a substituent), different modeling approaches have been proposed to simulate the lipophilicity of organic molecules (e.g., see [18–23]). Each approach has strengths and weaknesses. For example; numerous log *P* simulators available on the market are based on a simple "additivity" principle or linear relationships, so they cannot correctly encode the effects of an increase in the degree of substitution of a molecule. More specifically, overestimations are often noted with molecules presenting a high degree of substitution. This is clearly shown in Table 23.2. KowWin and PrologP, belonging to the category above, are able to correctly estimate the log *P* value of biphenyl but entirely fail with decachlorobiphenyl. Conversely, ALOGPS and AUTOLOGP, providing a nonlinear estimation of the log *P* values, are well suited for estimating the hydrophobicity of these two chemicals (see also Chapter 9).

Consequently the use of log *P* as molecular descriptor in a QSAR study does not provide the guarantee that the obtained model will be sound. Also problems of availability and lack of accuracy can arise with the other physicochemical descriptors used in QSAR studies.

A suitable description of the molecules can be also simply performed from structural features (i.e., atoms, functional groups, substructures) that are suspected to influence their biological activities. These descriptors are encoded

TABLE 23.2 Variations in the Calculated log *P* Values of Biphenyl and Decachlorobiphenyl

Software	Biphenyl	Decachloro-PCB	Reference
Clog*P* (v. 1.0.0)	4.03	9.20	[18]
KowWin (v. 1.53)	3.76	10.2	[19]
ACD/LogP (v. 1)*	3.98	8.10	[18]
PrologP (v. 5.1)**	3.97	11.4	[18]
KlogP	4.09	9.92	[20]
ALOGPS 2.1	4.02	8.59	[22]
AUTOLOGP (v. 4)	3.81	8.33	[23]
Experimental log *P*	**3.98**	**8.26**	[18]

*Model based on 532 group contributions, 21 carbon atom type contributions, and 2206 intramolecular correction factors [18, p. 116].

**Based on Rekker's method.

in a Boolean manner (0/1) [24,25] or according to their frequency of occurrence in the molecules [26–28] (Figure 23.2). It is also possible to use the whole structure of an organic molecule depicted as a graph, generally without hydrogens, for deriving numerical descriptors termed topological indexes (Figure 23.2). Many algorithms are available in the literature [29–31] for calculating these interesting molecular descriptors, which can be easily computed for all the existing, new, and in development chemicals for a multivariate description of the molecules when they are judiciously combined.

It is important to stress that all the different types of molecular descriptors present advantages and drawbacks [32–37]. However, availability and ease of interpretation are the main criteria to be used in their selection for deriving a QSAR model.

23.2.3 Statistical Methods

The data used in developing environmental QSARs frequently cause problems when being statistically assessed. Problems can arise because, first, some toxicological endpoints are difficult to measure with a sufficient level of accuracy. In addition amalgamation of data from different sources is commonplace, yielding the introduction of noise in the data matrix. Last, speed and availability of computers and the existence of powerful software tools have increased the temptation to derive QSAR models from large collections of calculated descriptors, neglecting the fact that quantity does not mean quality. Indeed problems arise when these parameters are too intercorrelated, not well statistically distributed, or totally meaningless. Techniques exist to reduce the dimensionality of the data matrices and to eliminate the variables of poor interest [38–40], but unfortunately, in practice, they are marginally used.

Statistical methods have to be selected to cope with all these problems. The selection process, however, can be difficult and often requires expertise [41]. The difficulties involved in the selection process are numerous. For example, the partial least squares (PLS) method [42] is widely used in QSAR because this statistical approach is well suited for deriving a model when a large number of molecular descriptors are used or when co-linearity in the descriptors exists. However, because the method is purely linear, PLS cannot deal with nonlinearities and complex interactions between descriptors [4]. Conversely, it is obvious that the artificial neural networks (ANNs) [43] and especially the supervised ANNs, such as the three-layer feedforward neural network trained by the backpropagation algorithm [44], are suited to overcome these problems. ANNs are also able to model highly noisy data [45]. However, their proper use, among other things, requires large data sets [43].

23.2.4 Selection of the Training and Testing Sets

Ideally a QSAR model is based upon a training (learning) set of data. Its accuracy, precision, and applicability are evaluated from an external testing

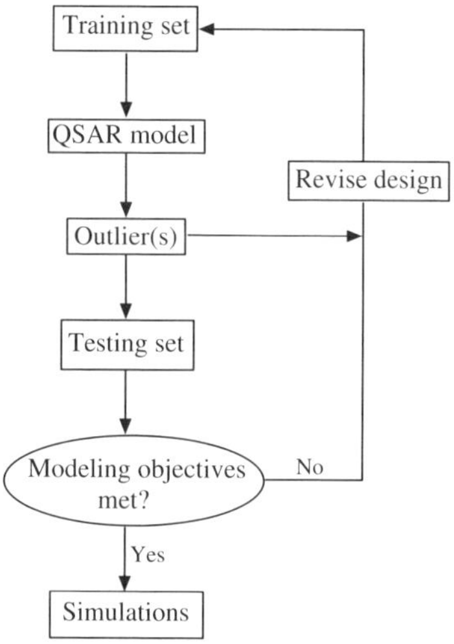

Figure 23.3 The QSAR modeling process.

set (Figure 23.3). It is best to start with a large data set, which is then divided into training and testing sets. The selection of these sets is problem dependent. It is generally performed by means of linear and nonlinear multivariate methods [46–49] to secure the representativeness of the biological activities and chemical structures in both sets. Sometimes, unfortunately, the training set comprises all the chemicals that have been tested because there are so few biological data. In that case, a classical leave-one-out (LOO) cross-validation procedure or a bootstrap [50] is used to try to estimate the performance of the QSAR model. However, it is important to note that these statistical techniques have to be considered as a last resort.

23.2.5 Detection and Analysis of the Outliers

Often in the development of a QSAR the activities of some chemicals belonging to the training set are poorly predicted by the model. These compounds are termed outliers. From a statistical point of view, "an outlier among residuals is one that is far greater than the rest in absolute value and perhaps lies three or four standard deviations or further from the mean of the residuals" [51, p. 152]. From a practical point of view, the outliers must be regarded with interest because they are warnings that there is more to understand about the model before it can be safely used. Moreover an understanding of the cause of the outlier behavior can be of inestimable value in gaining a more fundamental insight into the underlying biochemical processes governing toxic action. The origin of such outlier behavior can be ascribed to one or more of the following [52,53]:

- The outlier is the result of poor training set design.
- Some experimental biological data are inaccurate or wrong. As stressed above, this could be due to differences in the endpoints, experimental conditions, and so on. It could also be due to a simple typo made during a data compilation process or to more pernicious events. For example, the toxicity of chemicals is commonly expressed in ppm (i.e., parts per million) and ppb (i.e., parts per billion) (e.g., see [54]). Unfortunately, ppb in Europe and in North America refer to 10^{12} and 10^{9}, respectively.
- The model requires additional descriptors to correctly encode the studied biological activity or the values of some of the selected descriptors are incorrect.
- The outlier interacts by a different molecular mechanism at its biochemical site of action than the other studied chemicals.
- The outlier yields one or more metabolic or chemical transformation products acting by a different mechanism at its biochemical site of action than the other studied compounds.
- The statistical method could be deficient in finding the functional relationship between the biological activity and the selected molecular descriptors.

After a logical explanation has been obtained for the presence of any outliers, the model is generally refined to increase its predictive power (Figure 23.3). It should be stressed that the elimination of a chemical acting as outlier is only performed when a problem dealing with the biological activity is clearly identified. Otherwise, the strategy consists of the addition of other molecular descriptors and/or the use of another statistical engine. In all cases, refining a QSAR model is a time-consuming process, especially when nonlinear methods (e.g., supervised ANNs) are used to derive the models. In such a case, after each modification, it is necessary to perform different runs to arrive at an optimal solution.

23.2.6 Validation/Verification

The term validation usually refers to a process of showing how accurate and/or precise a model is [4]. In practice, validation of a QSAR model, like that of all the numerical models in the environmental sciences [55], is impossible. A QSAR model is partially verified from the statistical parameters provided by the statistical method used for its design and from the simulation results obtained with the training and testing sets. A way to try to estimate with accuracy the simulation performances of a QSAR model is to split the testing set into an in-sample test set (ISTS) and an out-of-sample test set (OSTS) [56]. The ISTS, including structures widely represented in the training set (e.g., isomers of position), is used for assessing the interpolation performances of the QSAR model while the OSTS, including particular chemicals weakly represented in the training set, is useful for estimating the extrapolation performances of the model [56]. This strategy can be used when the availability of experimental data is not a limiting factor, but in all cases it has to be employed with care. It is well accepted that interpolated data are safer and less likely to be prone to uncertainties than extrapolated data [4].

After a QSAR model has been designed and verified, it can be used to estimate the biological activity of untested chemicals. However, it is important to understand that a QSAR model cannot be better than the data from which it was developed. The next sections attempt to highlight some of the previous discussion points through the analysis of the different QSAR modeling strategies applied in aquatic toxicology.

23.3 CONGENERIC QSAR MODELS IN AQUATIC TOXICOLOGY

Initial investigations in QSAR postulated that xenobiotics belonging to the same chemical class should behave in a toxicologically similar manner [57]. Consequently numerous QSAR models were derived from a specific family of chemicals such as chlorophenols or para-substituted benzenes [58–61]. Unfortunately, it was rapidly demonstrated [62,63] that QSARs based on defined modes of action (MOA) are more relevant that those only strictly based on chemical families. Collections of QSAR models are derived from sets of chemicals acting by the same MOA. "Inert" chemicals are found to act by a nonspecific mode of action related to narcosis. The potency of a chemical to induce narcosis only depends on its lipophilicity encoded by the 1-octanol/water partition coefficient (log *P*). Such a chemical is always as toxic as its lipophilicity indicates. Narcosis-type toxicity is therefore called baseline toxicity or minimum toxicity [64,65]. Compounds acting according to this MOA include numerous industrial chemicals such as aliphatic alcohols, alkanes, and alkyl- and halogenated benzenes. Less "inert" chemicals are slightly more toxic than baseline toxicity and act by a polar narcosis mechanism of toxicity. Broadly speaking, the chemicals in relation with this MOA are usually characterized as possessing hydrogen bond donor acidity (e.g., phenols, anilines) [64,65]. Oxidative phosphorylation uncouplers inhibit the ATP synthesis within the mitochondria. Chemicals belonging to this category are weak acids and phenols, anilines, and pyridines including multiple electronegative groups such as more than one nitro substituents or more than three halogen substituents [57]. For illustrative purposes, QSAR equations dealing with these three MOAs are listed in Table 23.3.

Reactive chemicals enhance toxicity in relation to reactions with chemical structures found in biomolecules (e.g., epoxides) or to metabolism into more toxic compounds [65]. Toxicity of specifically acting chemicals has turned out to be due to their interactions with certain receptors such as inhibition of acetylcholinesterase by organophosphorus compounds [65]. A survey of the

TABLE 23.3 Examples of Baseline Toxicity (BT), Polar Narcosis (PN), and Oxidative Phosphorylation Uncoupling (OPU) QSARs

MOA	Species*	Endpoint	Slope	Intercept	r^2	*n*
BT	*P.r.*	LC_{50}	0.871	1.13	0.98	50
BT	*P.p.*	LC_{50}	0.94	1.25	0.94	60
BT	*C.*sp.	LC_{50}	0.919	0.967	0.97	5
BT	*C.a.*	LC_{50}	0.881	0.989	0.92	5
BT	*T.p.*	IG_{50}	0.929	2.639	0.99	20
PN	*P.r.*	LC_{50}	0.46	3.04	0.82	11
PN	*P.p.*	LC_{50}	0.65	2.29	0.90	39
PN	*T.p.*	IG_{50}	0.574	0.865	0.76	30
OPU	*P.p.*	LC_{50}	0.67	−2.95	0.82	12
OPU	*V.f.*	EC_{50}	0.489	0.126	0.85	16
OPU	*T.p.*	IG_{50}	0.401	0.189	0.82	12

Source: Adapted from [57].

* *P.r.* = *Poecilia reticulata*, *P.p.* = *Pimephales promelas*, *C.* sp. = *Cyprinus* sp., *C.a.* = *Carassius auratus*, *T.p.* = *Tetrahymena pyriformis*, *V.f.* = *Vibrio fischeri*.

literature shows that in aquatic toxicology, most of the congeneric QSAR models are derived from limited learning sets, and their predictive performances are generally not estimated from an external testing set.

Last note that the proper use of these QSAR models designed for molecules with a specific MOA requires that we can assign a molecule to its appropriate class. This task represents a major hurdle in ecological risk assessment, since incorrect mode-of-action-based QSAR selections can result in 10- to 1000-fold errors in toxicity predictions [66]. Solutions have been proposed to overcome this problem [65,67–71], but in practice, none is entirely satisfying. The different modeling strategies proposed to solve this crucial problem are presented in the next section.

23.4 NONCONGENERIC QSAR MODELS IN AQUATIC TOXICOLOGY

With the advent of powerful computers and easy access to them, and the introduction of expert systems, artificial intelligence, and neural networks in QSAR, radically different models designed from noncongeneric large sets of chemicals have been proposed. No attempts are made to design a model that is easily interpretable in terms of MOA. The main objective of the present models is to provide powerful simulators with a wide domain of application for predicting the toxicity of any kind of molecule.

The CASE/M-CASE (computer automated structure evaluation) methodology has been used for modeling different toxicological activities [72–77], including aquatic toxicity endpoints [75–77]. Briefly, the fundamental assumption of the M-CASE methodology is that the observed biological activity of a molecule is always governed by some of its substructures, called biophores. The basic hierarchical algorithm starts off by identifying the statistically most significant substructure among a learning set consisting of active and inactive chemicals. The molecules containing this "top biophore" are then removed from the database and the remaining ones are submitted to a new analysis, yielding the identification of the next biophore. This procedure is repeated until either the activity of all the molecules in the learning set have been accounted for or no additional statistically significant biophore can be found. This way M-CASE splits the original data set into subsets of molecules, each associated with a particular biophore. It is assumed that molecules belonging to the same subset act by the same MOA. For each subset, M-CASE identifies additional parameters, called modulators, which can be structural descriptors or physicochemical properties such as log *P* and water solubility. Once the biophores and modulators are identified, they are used to derive a model based on an additive scheme using local QSARs. The M-CASE program can run with an additional feature called BAIA (baseline activity identification algorithm) [78], allowing it to identify in the first step a baseline activity due to a specific physical attribute of the molecules (e.g., log *P*). All chemicals whose toxicity is not accounted for by BAIA are assumed to have an activity depending on other factors and hence are analyzed by M-CASE.

M-CASE/BAIA was used by Klopman and coworkers for estimating the acute toxicity (LC_{50}) of organic chemicals against the guppy (*Poecilia reticulata*) [75] and the fathead minnow (*Pimephales promelas*) [76] as well as the inhibition of the bioluminescence (EC_{50}) in the marine bacterium *Vibrio fischeri* [77] (Table 23.4). While, undoubtedly, M-CASE/BAIA is original, the

TABLE 23.4 Selected QSAR Models Derived from Large Noncongeneric Data Sets

Species*	*n*	Learning/Testing	Technique**	Reference
P.r.	219	90–10%	M-CASE/BAIA	[75]
P.p.	675	90–10%	M-CASE/BAIA	[76]
V.f.	901	90–10%	M-CASE/BAIA	[77]
V.f.	747	454/150 + 143	BP-ANN	[56]
V.f.	1308	1068/240	BP-ANN	[84]
T.p.	750	600/150	PAAN	[85]
T.p.	1084	1000/84	PAAN	[86]
D.m.	776	700/76	PAAN	[87]
P.p.	865	80–20%	PAAN	[88]
P.p.	886	800/86	PAAN	[89]
P.p.	562	392/170	BP + Fuzzy-ANN	[90]
P.p.	551	LOO + 80–20% + 541/10	CPANN	[91]

* *P.r.* = *Poecilia reticulata*, *P.p.* = *Pimephales promelas*, *V.f.* = *Vibrio fischeri*, *T.p.* = *Tetrahymena pyriformis*, *D.m.* = *Daphnia magna*.

** M-CASE/BAIA (see text). BP-ANN = three-layer feedforward artificial neural network trained by the backpropagation algorithm, PAAN = probabilistic artificial neural network, CPANN = counterpropagation artificial neural network.

design of QSAR models from this approach is not straightforward and the obtained simulation tools present some weaknesses [2]. Thus it is claimed that M-CASE runs automatically. However, before a model can be denied, the chemicals must be classified as inactive, marginally active, or active. In aquatic toxicology there is no rationale to that, and the selected breakpoints remain linked to the size and structural characteristics of the data sets for which observed LC_{50} or EC_{50} values are available. This basically influences the modeling process and the quality of the obtained model. Other criticisms have been published elsewhere [2]. Nevertheless, M-CASE/BAIA represents an original method for modeling large sets of chemicals in aquatic toxicology.

It is now well known that the artificial neural networks (ANNs) are nonlinear tools well suited to find complex relationships among large data sets [43]. Basically an ANN consists of processing elements (i.e., neurons) organized in different oricnted groups (i.e., layers). The arrangement of neurons and their interconnections can have an important impact on the modeling capabilities of the ANNs. Data can flow between the neurons in these layers in different ways. In feedforward networks no loops occur, whereas in recurrent networks feedback connections are found [79,80].

The ANNs are subject to a learning process that can be viewed as a method for updating its architecture and connection weights in order to optimize its efficiency to perform a specific task. The three main learning paradigms are supervised, unsupervised (or self-organized), and reinforcement [79,80]. Each category has many algorithms. Supervised is the most commonly employed learning paradigm to develop classification and prediction applications. The algorithm takes the difference between the correct or desired output and the actual ANN prediction and uses that information to adjust the weights in the network, so that next time the prediction will be closer to the correct answer. Unsupervised learning is used when we want to perform clustering of the input data. ANNs that are trained using this learning process are called self-organizing neural networks [81] because they receive no direction on what the desired output should be. Indeed, when presented with a series of inputs, the outputs self-organize by initially competing to recognize the input information and then cooperating to adjust their connection weights. Over time the network evolves so that each output unit is sensitive to and will recognize inputs from a specific portion of the input space. Reinforcement learning attempts to learn the input–output mapping through trial and error, with a view toward maximizing a performance index called the reinforcement signal. Reinforcement learning is particularly suited to solve difficult temporal (time-dependent) problems [79,80,82,83].

Most of these different learning processes and topologies have been used for designing QSAR models. Selected ANN QSAR models (56,84–91) are presented in Table 23.4.

The main advantages of the supervised ANNs are their ability to find complex relationships between variables and their high flexibility. For example, ANN QSARs have been designed to predict the toxicity of chemicals to the rainbow trout (*Oncorhynchus mykiss*) [92], the bluegill (*Lepomis macrochirus*)

[93], the midge (*Chironomus riparius*) [94], and gammarids (*Gammarus fasciatus*) [95] from a set of variables encoding the structure, physicochemical properties of the molecules, as well as the experimental conditions under which the different tests are performed (e.g., temperature, pH, size of the organisms). A new type of ANN QSAR model has been recently proposed [96] for the common situation in which the toxicity of molecules mainly depends on their log *P*. Briefly, in a first step, a classical regression equation with log *P* is derived. The residuals obtained with this simple linear equation are then modeled from a three-layer feedforward ANN, including different molecular descriptors as input neurons. Finally, results produced by the linear and nonlinear QSAR models are both considered for calculating the toxicity values, which are then compared to the initial toxicity data.

If the ANNs present numerous advantages, they also present some weaknesses. Indeed they are versatile tools, but they require some experience to be correctly used to correctly scale the data, tune the network, and avoid overtraining and overfitting. More complete critical reviews of the ANN models can be found elsewhere [2,97–100].

Support vector machines (SVMs) represent a new class of machine learning algorithms. They are increasingly being used in numerous domains, including QSAR and QSPR (see also Chapter 8). Briefly, SVMs map the *n*-dimensional input space into a higher dimensional space where a linear model can be applied [101]. An SVM QSAR model was derived by Panaye et al. [102] for predicting the toxicity of 203 nitro- and cyanoaromatic chemicals to *Tetrahymena pyriformis*. Different kernel functions were tested, and comparisons were made with radial basis function and general regression ANNs. If the SVMs are powerful statistical devices presenting numerous advantages over the ANNs, it should be noted that they are difficult to tune.

23.5 SPECIFIC QSAR MODELING APPROACHES IN AQUATIC TOXICOLOGY

Most of the QSARs derived in aquatic toxicology allow us to predict the toxicity of organic chemicals while the inorganic compounds also widely contaminate the aquatic ecosystems. Analyzing the list of community high-volume chemicals, Hart [103] observed that among the 1952 entries, 984 were organic chemicals, 255 inorganic compounds, and 713 UCVBs (unknown, variable composition, biologicals). Fortunately, different (Q)SARs have been derived between the physicochemical properties of ions (mostly cations) and the toxicity of metals. These relationships have been recently analyzed by Walker et al. [104].

It is well known that the toxicity of chemicals in a mixture does not necessarily correspond to that predicted from data on pure compounds [105]. Consequently different QSAR approaches have been developed for predicting mixture toxicity (e.g., see [106–110] and also Chapter 8).

Last, even if bioconcentration, the process of accumulation of waterborne chemicals by aquatic organisms through nondietary routes, is a property and not an activity, it is important to stress that there exist a huge number of QSPR models allowing one to predict the accumulation of chemicals in biota. For further information on this important topic, the reader is referred to Devillers et al. [111], Dearden [112], and Dimitrov et al. [113] who review all the existing bioconcentration models and their advantages and limitations.

23.6 USE OF QSARs IN AQUATIC TOXICOLOGY

23.6.1 Regulation of Chemicals

Recent policy developments worldwide have placed increased emphasis on the use of (S)QSAR and QSPR models within various regulatory programs [114–117] for the hazard and risk assessment of chemicals because these models allow one to save time and money and also to reduce the number of laboratory animals used in the tests. Thus QSAR models find applications in the prioritization of existing chemicals for further testing or assessment, the classification and labeling of new substances, and in the risk assessment of new and existing chemicals [117,118]. However, to be safely used, the performance of these models has to be evaluated. Indeed Kaiser and coworkers [119] have divulged that the majority of the linear regression equations proposed in ECOSAR, extensively used by the EPA, were based on fewer than five compounds. In fact many were derived from a single chemical. It is obvious that these particular QSAR models are not statistically valid and their use must be avoided. It is also interesting to stress that these authors [119] indicated that the classification scheme of chemicals to select a QSAR model in ECOSAR was unclear to the user, inconsistent, or faulty. To overcome these problems, different solutions have been proposed, including comparison exercises (e.g., [120]), proposal of rules for selecting QSAR models (e.g., [91,121]), and/or the use of various indices for estimating the quality of the QSAR equations (e.g., [122,123]).

23.6.2 Design of Safer Chemicals

The concept of designing safer chemicals has been defined by Garrett [124] as: "The employment of structure-activity relationships (SAR) and molecular manipulation to achieve the optimum relationship between toxicological effects and the efficacy of intended use." During this process, QSAR models are used to propose new chemical substances that are safe for the environment and efficacious or to modify existing chemicals to reduce their (eco)toxicity, while their target activity or property for which they are currently used is preserved. In this context it is obvious that the selected models must present a high level of accuracy and specificity. QSARs derived from congeneric sets

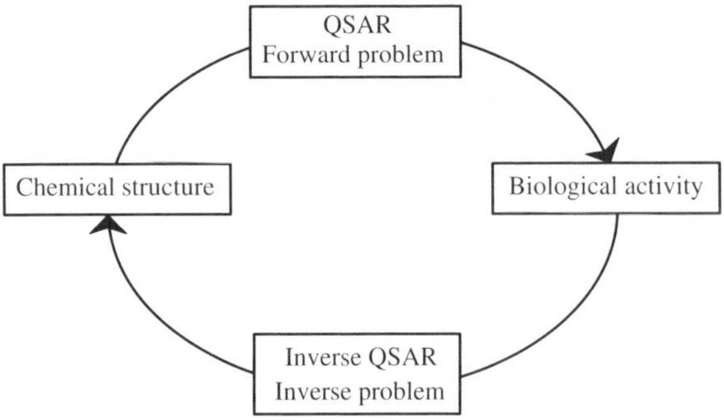

Figure 23.4 The inverse problem in QSAR studies.

of molecules acting according to the same MOA are well suited to fulfill these requirements. With these models it is expected that the effect of a small modification in the structure of a chemical will be shown on the biological activity of interest.

In the classical QSAR modeling process, also termed the forward problem [125], the prediction of the activity of a chemical is made from its chemical structure encoded by one or several molecular descriptors (Figure 23.4). It would be a great advantage if the model could be made to indicate directly the structure that is predicted to possess an activity or a property value in a desired target range. Such a process is called inverse QSAR or the inverse problem [125]. The inverse problem has been approached by a variety of methods (e.g., see [126–128]). Their performance is rather limited. Indeed most suffer from drawbacks due to combinatorial complexity of the search space, design knowledge acquisition difficulties, nonlinear structure-activity or structure-property relationships, and problems in incorporating high-level chemical and biological knowledge. In addition they are generally computationally very expensive.

23.7 CONCLUSIONS

QSAR models represent invaluable tools for simulating the different endpoints necessary to correctly assess the hazard of chemicals in the aquatic ecosystems. Most of the available QSAR models deal with single organic compounds. This does not mean that the mixtures of organic chemicals and the inorganic compounds are not of interest. It rather indicates that much work has to be done in these two areas.

The existing QSAR models fall into two categories: those derived from somewhat limited sets of chemicals acting according to the same MOA and those obtained from large heterogeneous sets of molecules. Even if the latter type of models are still denigrated by most of the developers of QSARs based on specific MOAs, there is a noticeable increase in their use for modeling various toxicological endpoints [129,130] because, correctly designed, they often compete favorably with the classical QSAR models and they represent the unique response when faced with the increasing number of human–made chemicals for which, unfortunately, (eco)toxicological data are missing.

REFERENCES

1. Carson R. *Silent spring.* Boston: Houghton Mifflin, 1962.
2. Devillers J. QSAR modeling of large heterogeneous sets of molecules. *SAR QSAR Environ Res* 2001;12:515–28.
3. Chemical Abstracts Services <http://www.cas.org/cgi-bin/regreport.pl>.
4. Anonymous. *QSARs in the assessment of the environmental fate and effects of chemicals.* Technical report 74. Brussels: ECETOC, 1998.
5. Devillers J, Exbrayat JM. *Ecotoxicity of chemicals to amphibians.* Reading, MA: Gordon and Breach, 1992.
6. Kaiser KLE, Devillers J. *Ecotoxicity of chemicals to* Photobacterium phosphoreum. Reading, MA: Gordon and Breach, 1994.
7. Mayer FL, Ellersieck MR. Manual of acute toxicity: Interpretation and data base for 410 chemicals and 66 species of freshwater animals. US Fish Wildlife Service, Resource Publ. 160, 1986.
8. Voigt, K. Environmental information databases. In: Schleyer PvR, Allinger NL, Clark T, Gasteiger J, Kollman PA, Schaefer HF, Schreiner PR, editors. *The encyclopedia of computational chemistry,* Vol. 2. New York: Wiley, 1998. p. 941–52.
9. Lawson AJ, Jochum CJ. Factual information databases. In: Schleyer PvR, Allinger NL, Clark T, Gasteiger J, Kollman PA, Schaefer HF, Schreiner PR, editors. *The encyclopedia of computational chemistry,* Vol. 2. New York: Wiley, 1998. p. 983–1002.
10. Dassler WL. Using internet search engines and library catalogs to locate toxicology information. *Toxicol* 2001;157:121–39.
11. Guerbet M, Guyodo G. Efficiency of 22 online databases in the search for physicochemical, toxicological and ecotoxicological information on chemicals. *An Occup Hyg* 2002;46:261–8.
12. Kubinyi H. *3D QSAR in drug design: Theory, methods and applications.* Leiden: ESCOM, 1993.
13. Dearden JC. Physico-chemical descriptors. In: Karcher W, Devillers J, editors, *Practical applications of quantitative structure-activity relationships (QSAR) in environmental chemistry and toxicology.* Dordrecht: Kluwer Academic, 1990. p. 25–59.
14. Hansch C, Leo A. *Exploring QSAR: Fundamentals and applications in chemistry and biology.* Washington: American Chemical Society, 1995.
15. Pliska V, Testa B, van de Waterbeemd H. *Lipophilicity in drug action and toxicology.* Weinheim: VCH, 1996.

16. Fujita T, Iwasa J, Hansch C. A new substituent constant, π, derived from partition coefficients. *J Am Chem Soc* 1964;86:5175–80.
17. Leo A. Calculating $\log P_{oct}$ from structures. *Chem Rev* 1993;93:1281–306.
18. Sangster J. *Octanol-water partition coefficients: Fundamentals and physical chemistry.* Wiley Series in Solution Chemistry, Vol. 2. New York: Wiley, 1997.
19. Meylan WM, Howard PH. Atom/fragment contribution method for estimating octanol-water partition coefficients. *J Pharm Sci* 1995;84:83–92.
20. Klopman G, Li JY, Wang S, Dimayuga M. Computer automated $\log P$ calculations based on an extended group contribution approach. *J Chem Inf Comput Sci* 1994;34:752–81.
21. Devillers J, Domine D, Guillon C, Bintein S, Karcher W. Prediction of partition coefficients ($\log P_{oct}$) using autocorrelation descriptors. *SAR QSAR Environ Res* 1997;7:151–72.
22. Tetko IV, Tanchuk VY. Application of associative neural networks for prediction of lipophilicity in ALOGPS 2.1 program. *J Chem Inf Comput Sci* 2002; 42:1136–45.
23. Devillers J. AUTOLOGP™: A computer tool for simulating *n*-octanol-water partition coefficients. *Analusis* 1999;27:23–9.
24. Free SM, Wilson JW. A mathematical contribution to structure-activity studies. *J Med Chem* 1964;7:395–9.
25. Zahradnik P, Foltinova P, Halgas J. QSAR study of the toxicity of benzothiazolium salts against *Euglena gracilis*: The free-Wilson approach. *SAR QSAR Environ Res* 1996;5:51–6.
26. Hall LH, Kier LB, Phipps G. Structure-activity relationship studies on the toxicities of benzene derivatives: I. An additivity model. *Environ Toxicol Chem* 1984;3:355–65.
27. Hall LH, Kier LB. Structure-activity relationship studies on the toxicities of benzene derivatives: II. An analysis of benzene substituent effects on toxicity. *Environ Toxicol Chem* 1986;5:333–7.
28. Devillers J, Zakarya D, Chastrette M, Doré JC. The stochastic regression analysis as a tool in ecotoxicological QSAR studies. *Biomed Environ Sci* 1989;2:385–93.
29. Kier LB, Hall LH. *Molecular connectivity in structure-activity analysis.* Letchworth: Wiley, 1986.
30. Kier LB, Hall LH. *Molecular structure description: The electrotopological state.* New York: Academic Press, 1999.
31. Devillers J, Balaban AT. *Topological indices and related descriptors in QSAR and QSPR.* Amsterdam: Gordon and Breach, 1999.
32. Cronin MTD, Livingstone DJ. Calculation of physicochemical properties. In: Cronin MTD, Livingstone DJ, editors, *Predicting chemical toxicity and fate.* Boca Raton, FL: CRC Press, 2004. p. 31–40.
33. Fisk PR, McLaughlin L, Wildey RJ. Good practice in physicochemical property prediction. In: Cronin MTD, Livingstone DJ, editors, *Predicting chemical toxicity and fate.* Boca Raton, FL: CRC Press, 2004. p. 41–59.
34. Schüürmann G. Quantum chemical descriptors in structure-activity relationships: Calculation, interpretation, and comparison of methods. In: Cronin MTD,

Livingstone DJ, editors, *Predicting chemical toxicity and fate*. Boca Raton, FL: CRC Press, 2004. p. 85–149.

35. Devillers J. No-free-lunch molecular descriptors in QSAR and QSPR. In: Devillers J, Balaban AT, editors, *Topological indices and related descriptors in QSAR and QSPR*. Amsterdam: Gordon and Breach, 1999. p. 1–20.
36. Devillers J. New trends in (Q)SAR modeling with topological indices. *Curr Opin Drug Discov Dev* 2000;3:275–9.
37. Netzeva TI. Whole molecule and atom-based topological descriptors. In: Cronin MTD, Livingstone DJ, editors, *Predicting chemical toxicity and fate*. Boca Raton, FL: CRC Press, 2004. p. 61–83.
38. Devillers J, Thioulouse J, Karcher W. Chemometrical evaluation of multispecies-multichemical data by means of graphical techniques combined with multivariate analyses. *Ecotoxicol Environ Safety* 1993;26:333–45.
39. Domine D, Devillers J, Chastrctte M, Karcher W. Non-linear mapping for structure-activity and structure-property modelling. *J Chemom* 1993;7:227–42.
40. Livingstone DJ. Building QSAR models: A practical guide. In: Cronin MTD, Livingstone DJ, editors, *Predicting chemical toxicity and fate.* Boca Raton, FL: CRC Press, 2004. p. 151–70.
41. Devillers J. Statistical analyses in drug design and environmental chemistry: Basic concepts. In: Coccini T, Giannoni L, Karcher W, Manzo L and R. Roi R, editors, *Quantitative structure/activity relationships (QSAR) in toxicology*. JRC-Ispra: CEC, 1992. p. 27–41.
42. Geladi P, Tosato ML. Multivariate latent variable projection methods: SIMCA and PLS. In: Karcher W, Devillers J, editors. *Practical applications of quantitative structure-activity relationships (QSAR) in environmental chemistry and toxicology.* Dordrecht: Kluwer Academic Press, 1990. p. 171–9.
43. Devillers J. *Neural networks in QSAR and drug design.* London: Academic Press, 1996.
44. Rumelhart DE, Hinton GE, Williams RJ. Learning representations by back-propagating errors. *Nature* 1986;323:533–6.
45. Devillers, J. On the necessity of multivariate statistical tools for modeling biodegradation. In: Ford MG, Greenwood R, Brooks CT, Franke R, editors, *Bioactive compound design: Possibilities for industrial use*. Oxford: BIOS Scientific Publishers, 1996. p. 173–86.
46. Domine D, Devillers J, Chastrette M. A nonlinear map of substituent constants for selecting test series and deriving structure-activity relationships: I. Aromatic series. *J Med Chem* 1994;37:973–80.
47. Domine D, Devillers J, Chastrette M. A nonlinear map of substituent constants for selecting test series and deriving structure-activity relationships: II. Aliphatic series. *J Med Chem* 1994;37:981–7.
48. Domine D, Devillers J, Wienke D, Buydens L. Test series selection from nonlinear neural mapping. *Quant Struct Act-Rel* 1996;15:395–402.
49. Putavy C, Devillers J, Domine D. Genetic selection of aromatic substituents for designing test series. In: Devillers J, editors, *Genetic algorithms in molecular modeling*. London: Academic Press, 1996. p. 243–69.

50. Efron B, Tibshirani RJ. *An introduction to the bootstrap*. New York: Chapman and Hall, 1993.
51. Draper N, Smith H. *Applied regression analysis*, 2 edition. New York: Wiley, 1981.
52. Devillers J, Lipnick RL. Practical applications of regression analysis in environmental QSAR studies. In: Karcher W, Devillers J, editors, *Practical applications of quantitative structure-activity relationships (QSAR) in environmental chemistry and toxicology*. Dordrecht: Kluwer Academic, 1990. p. 129–43.
53. Lipnick RL. Outliers: Their origin and use in the classification of molecular mechanisms of toxicity. In: Hermens JLM, Opperhuizen A, editors, *QSAR in environmental toxicology—IV*. Amsterdam: Elsevier, 1991. p. 131–53.
54. Frear DEH, Boyd JE. Use of *Daphnia magna* for the microbioassay of pesticides: I. Development of standardized techniques for rearing *Daphnia* and preparation of dosage-mortality curves for pesticides. *J Econ Entomol* 1967;60:1228–35.
55. Oreskes N, Shrader-Frechette K, Belitz K. Verification, validation and confirmation of numerical models in the earth science. *Science* 1994;263:641–6.
56. Devillers J, Bintein S, Domine D, Karcher W. A general QSAR model for predicting the toxicity of organic chemicals to luminescent bacteria (Microtox® test). *SAR QSAR Environ Res* 1995;4:29–38.
57. Bradbury SP, Russom CL, Ankley GT, Schultz TW, Walker JD. Overview of data and conceptual approaches for derivation of quantitative structure-activity relationships for ecotoxicological effects of organic chemicals. *Environ Toxicol Chem* 2003;22:1789–98.
58. Ribo JM, Kaiser KLE. Effects of selected chemicals to *Photobacterium phosphoreum* bacteria and their correlations with acute and sublethal effects on other organisms. *Chemosphere* 1983;12:1421–42.
59. Kaiser KLE, Dixon DG, Hodson PV. QSAR on chlorophenols, chlorobenzenes and para-substituted phenols. In: Kaiser KLE, editor, *QSAR in environmental toxicology*. Dordrech: Reidel, 1984. p. 189–206.
60. Devillers J, Chambon P. Toxicité aiguë des chlorophénols sur *Daphnia magna* et *Brachydanio rerio*. *J Fr Hydrol* 1986;17:111–20.
61. Kaiser KLE, Ribo JM, Zaruk BM. Toxicity of para-chloro substituted benzene derivatives in the Microtox™ test. *Water Poll Res J Canada* 1985;20:36–43.
62. Bradbury SP. Fish acute toxicity syndromes: Applications to the development of mechanism-specific QSARs. In: Turner JE, England MW, Schultz TW, Kwaak NJ, editors, *QSAR88, 3rd International Workshop on Quantitative Structure-Activity Relationships in Environmental Toxicology*. 22–26 May 1988, Knoxville, TN. p. 61–70.
63. Bradbury SP, Henry TR, Carlson RW. Fish acute toxicity syndromes in the development of mechanism-specific QSARs. In: Karcher W, Devillers J, editors, *Practical applications of quantitative structure-activity relationships (QSAR) in environmental chemistry and toxicology*. Dordrecht: Kluwer Academic, 1990. p. 295–315.
64. Lipnick RL. Narcosis: Fundamental and baseline toxicity mechanism for nonelectrolyte organic chemicals. In: Karcher W, Devillers J, editors, *Practical applications of quantitative structure-activity relationships (QSAR) in environmental chemistry and toxicology*. Dordrecht: Kluwer Academic, 1990. p. 281–93.

65. Bol J, Verhaar HJM, van Leeuwen CJ, Hermens JLM. *Predictions of the aquatic toxicity of high-production-volume-chemicals: Part A. Introduction and methodology.* The Hague: VROM, 1993.

66. Bradbury SP. Predicting modes of toxic action from chemical structure: An overview. *SAR QSAR Environ Res* 1994;2:89–104.

67. Russom CL, Bradbury SP, Broderius SJ, Hammermeister DE, Drummond RA. Predicting modes of toxic action from chemical structure: Acute toxicity in the Fathead minnow (*Pimephales promelas*). *Environ Toxicol Chem* 1997; 16:948–67.

68. Basak SC, Grunwald GD, Host GE, Niemi GJ, Bradbury SP. A comparative study of molecular similarity, statistical, and neural methods for predicting toxic modes of action. *Environ Toxicol Chem* 1998;17:1056–64.

69. Aptula AO, Netzeva TI, Valkova IV, Cronin MTD, Schultz TW, Kühne R, Schüürmann G. Multivariate discrimination between modes of toxic action of phenols. *Quant Struct-Act Relat* 2002;21:12–22.

70. Spycher S, Pellegrini E, Gasteiger J. Use of structure descriptors to discriminate between modes of toxic action of phenols. *J Chem Inf Model* 2005;45:200–8.

71. Yao XJ, Panaye A, Doucet JP, Chen HF, Zhang RS, Fan BT, Liu MC, Hu ZD. Comparative classification study of toxicity mechanisms using support vector machines and radial basis function neural networks. *Anal Chim Acta* 535; 2005:259–73.

72. Klopman G. Artificial intelligence approach to structure-activity studies: Computer automated structure evaluation of biological activity of organic molecules. *J Am Chem Soc* 1984;106:7315–20.

73. Rosenkranz HS, Cunningham AR, Zhang YP, Klopman G. Applications of the CASE/MULTICASE SAR method to environmental and public health situations. *SAR QSAR Environ Res* 1999;10:263–76.

74. Rosenkranz HS, Cunningham AR, Zhang YP, Claycamp HG, Macina OT, Sussman NB, Grant SG, Klopman G. Development, characterization and application of predictive-toxicology models. *SAR QSAR Environ Res* 1999;10:277–98.

75. Klopman G, Saiakhov R, Rosenkranz HS, Hermens JLM. Multiple computer-automated structure evaluation program study of aquatic toxicity: 1. Guppy. *Environ Toxicol Chem* 1999;18:2497–505.

76. Klopman G, Saiakhov R, Rosenkranz HS. Multiple computer-automated structure evaluation study of aquatic toxicity. 2. Fathead minnow. *Environ Toxicol Chem* 2000;19:441–7.

77. Klopman G, Stuart SE. Multiple computer-automated structure evaluation study of aquatic toxicity: III. *Vibrio fischeri*. *Environ Toxicol Chem* 2003;22:466–72.

78. Klopman G. The MultiCASE program: II. Baseline activity identification algorithm (BAIA). *J Chem Inf Comput Sci* 1998;38:78–81.

79. Wasserman PD. *Neural computing: Theory and practice.* New York: Van Nostrand Reinhold, 1989.

80. Eberhart RC, Dobbins RW. *Neural netwoek PC tools: A practical guide.* San Diego: Academic Press, 1990.

81. Kohonen T. *Self-organizing maps.* Berlin: Springer-Verlag, 1995.

82. Pao YH. *Adaptive pattern recognition and neural networks.* Reading, MA: Addison-Wesley, 1989.
83. Levine DS. *Introduction to neural and cognitive modeling.* Hillsdale, NJ: Lawrence Erlbaum, 1991.
84. Devillers J, Domine D. A noncongeneric model for predicting toxicity of organic molecules to *Vibrio fischeri. SAR QSAR Environ Res* 1999;10:61–70.
85. Niculescu SP, Kaiser KLE Schultz TW. Modeling the toxicity of chemicals to *Tetrahymena pyriformis* using molecular fragment descriptors and probabilistic neural networks. *Arch Environ Contam Toxicol* 2000;39:289–98.
86. Kaiser KLE, Niculescu SP, Schultz TW. Probabilistic neural network modeling for the toxicity of chemicals to *Tetrahymena pyriformis* with molecular fragment descriptors. *SAR QSAR Environ Res* 2002;13:57–67.
87. Kaiser, KLE, Niculescu SP. Modeling acute toxicity of chemicals to *Daphnia magna*: A probabilistic neural network approach. *Environ Toxicol Chem* 2001;20:420–31.
88. Kaiser KLE, Niculescu SP. Using probabilistic neural networks to model the toxicity of chemicals to the fathead minnow (*Pimephales promelas*): A study based on 865 compounds. *Chemosphere* 1999;38:3237–45.
89. Niculescu SP, Atkinson A, Hammond G, Lewis M. Using fragment chemistry data mining and probabilistic neural networks in screening chemicals for acute toxicity to the fathead minnow. *SAR QSAR Environ Res* 2004;15:293–309.
90. Mazzatorta P, Benfenati E, Neagu CD, Gini G. Tuning neural and fuzzy-neural networks for toxicity modeling. *J Chem Inf Comput Sci* 2003;43:513–8.
91. Vracko M, Bandelj V, Barbieri P, Benfenati E, Chaudry Q, Cronin M, Devillers J, Gallegos A, Gini G, Gramatica P, Helma C, Mazzatorta P, Neagu D, Netzeva T, Pavan M, Patlewicz G, Randic M, Tsakovska I, Worth A. Validation of counter propagation neural network models for predictive toxicology according to the OECD principles: A case study. *SAR QSAR Environ Res* 2006;17: 265–84.
92. Devillers J, Flatin J. A general QSAR model for predicting the acute toxicity of pesticides to *Oncorhynchus mykiss. SAR QSAR Environ Res* 2000;11:25–43.
93. Devillers J. A general QSAR model for predicting the acute toxicity of pesticides to *Lepomis macrochirus. SAR QSAR Environ Res* 2001;11:397–417.
94. Devillers J. Prediction of toxicity of organophosphorus insecticides against the midge, *Chironomus riparius*, via a QSAR neural network model integrating environmental variables. *Toxicol Meth* 2000;10:69–79.
95. Devillers J. A QSAR model for predicting the acute toxicity of pesticides to gammarids. In: Leardi R, editor, *Nature-inspired methods in chemometrics: Genetic algorithms and artificial neural networks.* Amsterdam: Elsevier, 2003. p. 323–39.
96. Devillers J. A new strategy for using supervised artificial neural networks in QSAR. *SAR QSAR Environ Res* 2005;16:433–42.
97. Devillers J. Strengths and weaknesses of the backpropagation neural network in QSAR and QSPR studies. In: Devillers J, editor, *Neural networks in QSAR and drug design.* London: Academic Press, 1996. p. 1–46.

98. Kaiser KLE. The use of neural networks in QSARs for acute aquatic toxicological endpoints. *J Mol Struct (Theochem)* 2003;622:85–95.
99. Kaiser KLE. Neural networks for effect prediction in environmental and health issues using large datasets. *QSAR Comb Sci* 2003;22:185–90.
100. Devillers J. Artificial neural network modeling in environmental toxicology. In: Livingstone DJ, editor, *Neural networks: Methods and applications*. Totowa: Humana Press, 2007.
101. Cristianini N, Shawe-Taylor J. *An introduction to support vector machines and other kernel-based learning methods*. Cambridge: Cambridge University Press, 2000.
102. Panaye A, Fan BT, Doucet JP, Yao XJ, Zhang RS, Liu MC, Hu ZD. Quantitative structure-toxicity relationships (QSTRs): A comparative study of various non linear methods: General regression neural network, radial basis function neural network and support vector machine in predicting toxicity of nitro- and cyano-aromatics to *Tetrahymena pyriformis*. *SAR QSAR Environ Res* 2006;17:75–91.
103. Hart JW. The use of data estimation methods by regulatory authorities. In: Hermens JLM, Opperhuizen A, editors, *QSAR in environmental toxicology—IV*. Amsterdam: Elsevier, 1991. p. 629–33.
104. Walker JD, Enache M, Dearden JC. Quantitative cationic-activity relationships for predicting toxicity of metals. *Environ Toxicol Chem* 2003;22:1916–35.
105. Yang RSH. *Toxicology of chemical mixtures: Case studies, mechanisms, and novel approaches*. San Diego: Academic Press, 1994.
106. Hermens J, Canton H, Janssen P, de Jong R. Quantitative structure-activity relationships and toxicity studies of mixtures of chemicals with anaesthetic potency: Acute lethal and sublethal toxicity to *Daphnia magna*. *Aquat Toxicol* 1984;5: 143–54.
107. Xu S, Nirmalakhandan N. Use of QSAR models in predicting joint effects in multi-component mixtures of organic chemicals. *Water Res* 1998;32:2391–8.
108. Tichy M, Cikrt M, Roth Z, Rucki M. QSAR analysis in mixture toxicity assessment. *SAR QSAR Environ Res* 1998;9:155–69.
109. Tichy M, Borek-Dohalsky V, Matousova D, Rucki M, Feltl L, Roth Z. Prediction of acute toxicity of chemicals in mixtures: Worms *Tubifex tubifex* and gas/liquid distribution. *SAR QSAR Environ Res* 2002;13:261–9.
110. Altenburger R, Nendza M, Schüürmann G. Mixture toxicity and its modeling by quantitative structure-activity relationships. *Environ Toxicol Chem* 2003;22:1900–15.
111. Devillers J, Domine D, Bintein S, Karcher W. Comparison of fish bioconcentration models. In: Devillers J, editor, *Comparative QSAR*. Philadelphia; Taylor and Francis, 1998. p. 1–50.
112. Dearden JC. QSAR modeling of bioaccumulation. In: Cronin MTD, Livingstone DJ, editors, *Predicting chemical toxicity and fate*. Boca Raton, FL: CRC Press, 2004. p. 333–55.
113. Dimitrov S, Dimitrova N, Parkerton T, Comber M, Bonnell M, Mekenyan O. Base-line model for identifying the bioaccumulation potential of chemicals. *SAR QSAR Environ Res* 2005;16:531–54.
114. Walker JD, Carlsen L, Hulzebos E, Simon-Hettich B. Global government applications of analogues, SARs and QSARs to predict aquatic toxicity, chemical or

physical properties, environmental fate parameters and health effects of organic chemicals. *SAR QSAR Environ Res* 2002;13:607–16.

115. Worth AP, van Leeuwen CJ, Hartung T. The prospects for using (Q)SARs in a changing political environment: High expectations and a key role for the European Commission's joint research centre. *SAR QSAR Environ Res* 2004;15:331–43.

116. Gerner I, Spielmann H, Hoefer T, Liebsch M, Herzler M. Regulatory use of (Q)SARs in toxicological hazard assessment strategies. *SAR QSAR Environ Res* 2004;15:359–66.

117. Cronin MTD. The use by governmental regulatory agencies of quantitative structure-activity relationships and expert systems to predict toxicity. In: Cronin MTD, Livingstone DJ, editors, *Predicting chemical toxicity and fate*. Boca Raton, FL: CRC Press, 2004. p. 413–27.

118. Russom CL, Breton RL, Walker JD, Bradbury SP. An overview of the use of quantitative structure-activity relationships for ranking and prioritizing large chemical inventories for environmental risk assessments. *Environ Toxicol Chem* 2003;22:1810–21.

119. Kaiser KLE, Dearden JC, Klein W, Schultz TW. A note of caution to users of ECOSAR. *Water Qual Res J Can* 1999;34:179–82.

120. Moore DRJ, Breton RL, MacDonald DB. A comparison of model performance for six quantitative structure-activity relationship packages that predict acute toxicity to fish. *Environ Toxicol Chem* 2003;22:1799–809.

121. OECD. *The principles for establishing the status of development and validation of (quantitative) structure-activity relationships (QSARs).* Paris: OECD, 2004.

122. Schultz TW, Netzeva TI, Cronin MTD. Evaluation of QSARs for ecotoxicity: A method for assigning quality and confidence. *SAR QSAR Environ Res* 2004;15:385–97.

123. Kolossov E, Stanforth R. The quality of QSAR models: Problems and solutions. *SAR QSAR Environ Res* 2007;18:89–100.

124. Garrett RL. Pollution prevention, green chemistry, and the design of safer chemicals. In: DeVito SC, Garrett RL, editors, *Designing safer chemicals: Green chemistry for pollution prevention*. ACS Symposium Series 640. Washington: American Chemical Society, 1996. p. 2–15.

125. Venkatasubramanian V, Sundaram A, Chan K, Caruthers JM. Computer-aided molecular design using neural networks and genetic algorithms. In: Devillers J, editor, *Genetic algorithms in molecular modeling*. London: Academic Press, 1996. p. 271–302.

126. Hall LH, Kier LB. Molecular connectivity Chi indices for database analysis and structure-property modeling. In: Devillers J, Balaban AT, editors, *Topological indices and related descriptors in QSAR and QSPR*. Reading: Gordon and Breach, 1999. p. 307–60.

127. Brüggemann R, Pudenz S, Carlsen L, Sørensen PB, Thomsen M, Mishra RK. The use of Hasse diagrams as a potential approach for inverse QSAR. *SAR QSAR Environ Res* 2001;11:473–87.

128. Weis DC, Foulon JL, LeBorne RC, Visco DP. The signature molecular descriptor: 5. The design of hydrofluoroether foam blowing agents using inverse-QSAR. *Ind Eng Chem Res* 2005;44:8883–91.
129. Devillers J. A decade of research in environmental QSAR. *SAR QSAR Environ Res* 2003;14:1–6.
130. Kaiser KLE. Evolution of the international workshops on quantitative structure-activity relationships (QSARs) in environmental toxicology. *SAR QSAR Environ Res* 2007;18:3–20.

24

DERMATOTOXICOLOGY: COMPUTATIONAL RISK ASSESSMENT

JIM E. RIVIERE

Contents

24.1 INTRODUCTION

Skin is a primary route of exposure for chemicals in environmental and occupational settings, as well as serving as a portal for systemic drug delivery using transdermal patches, or for local therapy of dermatological diseases using topical formulations. Exposure may occur with use of cosmetics and many personal care products. Skin is also a target for toxicity by these same agents, secondary to direct chemical action or as a result of immunological detection with amplification due to prior chemical sensitization. Because of this almost

Computational Toxicology: Risk Assessment for Pharmaceutical and Environmental Chemicals, Edited by Sean Ekins

universal exposure of skin to chemicals and drugs, and because toxicity to skin is manifested in very visible and noticeable reactions, a great deal of attention has been focused on the effects of chemical exposure to this organ.

There are two major types of studies that comprise the field of dermatotoxicology: chemical absorption and irritation/sensitization. By far, computational approaches in this field have been focused on the former due to the ability to generate quantitative data suitable for a computational approach. This is also logical since a chemical must traverse the protective outer layer of the skin in order to gain access to its viable cells and exert a toxicological effect. Thus computational approaches to predict toxicity are confounded by chemical properties that allow absorption to occur, with the best work at this point on models to predict dermal absorption. This will be the focus of the present chapter.

24.2 MODEL SYSTEMS

Assessment of percutaneous absorption for any topically applied drug or chemical, can be classified based either on a model's level of biological complexity (in silico, in vitro, in vivo) or on the specific species studied (human, laboratory rodent, monkey, pig). The goal of the research should also be taken into consideration. Is the work being conducted to study the mechanism of absorption (e.g., identify a specific mathematical model or assess the effect of a vehicle) or to quantitatively predict absorption in humans? Is the study designed to look at a local effect in skin or a systemic effect after absorption? That is, are skin concentrations the relevant metric or is flux of chemical across skin important? Model systems and approaches in use today to assess dermal absorption have recently been extensively reviewed [1].

The primary approach to assess dermal absorption in most computational toxicology studies is the in vitro diffusion cell. In this model, skin sections (full thickness, dermatomed to a specific thickness) are placed in a two-chambered diffusion cell where receptor fluid is placed in a reservoir (static cells) or perfused through a receiving chamber (flow-through cells) to simulate dermal blood flow. The chemical may either be dosed under ambient conditions neat or dissolved in a vehicle (Franz and Bronaugh cells) or in water (side by side diffusion cells), resulting in finite versus infinite dosing conditions, respectively. This is a major variable in selecting the mathematical model to be used to derive the permeability constant. Selection of the receptor fluid (saline, albumin-based media, etc.) is also critical, since absorption is only detected if the penetrating compound is soluble in the receptor fluid. The right receptor fluid is particularly important for hydrophobic penetrants. Many studies of pharmaceutical compounds use saline as the receptor fluid because of the hydrophilic nature of many drugs, a choice that can falsely suggest minimal absorption for lipophilic chemicals, since they are not soluble in the receptor fluid and thus cannot be detected as absorbed. Steady-state flux is measured in these models and permeability calculated using the methods described below. In addition to perfusate composition, the temperature of perfusate is controlled, with pharmaceutical investigations suggesting that studies be conducted at 35°C to mimic the surface temperature of skin. These techniques have been exhaustively reviewed elsewhere [1–3].

The second major approach used to assess dermal absorption is in vivo. This is the primary approach used to assess drug absorption by all routes of administration. It is also the approach used in many toxicology disposition studies where full mass balance is attempted. A chemical is dosed on the surface of an animal, and total excreta (urine, feces, expired air) are collected and analyzed for parent compound or metabolites. Radiolabeled compounds are often employed in these studies. These data are usually expressed as percentage dose absorbed per unit of surface area exposed, and this procedure is well adapted to laboratory rodent models. However, the resulting metrics are not optimal for computational toxicology approaches. Dose may be applied occluded (evaporation of dose prevented) or non-occluded (dose site open to ambient environment). In calculating absorbed dose, all chemical at the dose site must be segregated from other tissues that can reflect absorbed chemical. This usually involves gently washing the nonabsorbed chemical with a soapy solution. When larger animals (e.g., pigs, primates) or humans are studied and total mass balance is not possible (e.g., cannot collect feces and expired air), the fraction of a systemically absorbed compound excreted in the urine must first be determined using parenteral dosing. In some classic studies, this parenteral route correction factor was conducted in monkeys [4] under the assumption that systemic distribution, metabolism, and elimination of these pesticides are similar in human and primate. In pigs, separate parenteral injections have been similarly made to determine fractional excretion by other routes [5].

For many pharmaceutical compounds administered as transdermal drug delivery systems, absorption can be assessed by determining the area under the curve (AUC) of the plasma concentration-time profile, the peak plasma flux, and time of peak flux, much as it is for determining bioavailability from oral and other routes of administration. These are classical metrics of biopharmaceutical bioequivalence studies and are extensively covered in other texts [6,7].

There are several perfused skin preparations with an intact functional microvasculature. The major advantage of such a perfused system is that subsequent systemic influences on absorbed chemical are not present, yet the tissue is fully functional with an intact microcirculation unlike simpler in vitro models. The perfused rabbit ear model, perfused pig ear model, in situ sandwich skin flap in athymic rats, and the hybrid rat-human sandwich flap have been developed [8], but each intuitively has severe limitations. The isolated perfused porcine skin flap (IPPSF) developed in our laboratory is a unique ex vivo skin preparation that has an intact functional cutaneous microcirculation. Predictions from IPPSF studies have correlated well with in vivo absorption

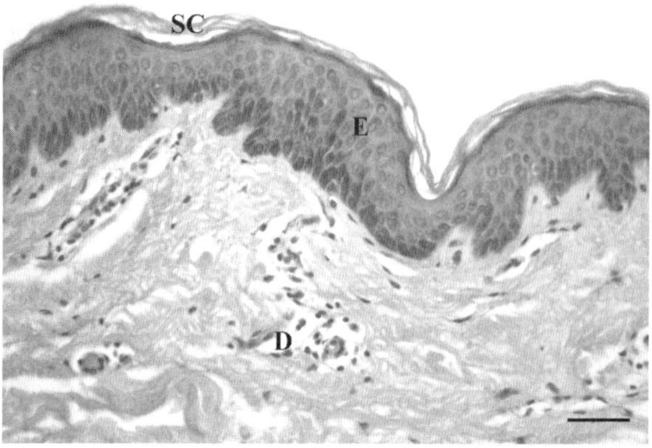

Figure 24.1 Light micrograph of normal human skin. SC: stratum corneum; E: epidermis; D: dermis. —is at 50 μm.

data for several drugs and insecticides [9–11]. IPPSFs are physiologically and biochemically viable and therefore can be used to assess cutaneous toxicity of topically applied chemicals [12]. The latter is most important as cutaneous toxicity as well as dermal absorption of various pesticide formulations can be assessed simultaneously.

24.3 PRINCIPLES OF DERMAL ABSORPTION

The skin is composed of two primary layers, the epidermis, which includes the outermost stratum corneum barrier and underlying viable keratinocytes, and the dermis [13](Figure 24.1). Skin is relatively impermeable to most aqueous solutions and ions, but it may be permeable to more lipophilic compounds. The stratum corneum cell layer is nonviable and considered to be the rate-limiting barrier in drug and chemical percutaneous absorption [13]. It is axiomatic that a topically applied chemical must first traverse the stratum corneum barrier before it is capable of eliciting any toxicological or immunological effect on subsequent cell layers, making absorption both the primary factor in assessing the dermal effects of drugs and chemicals, as well as a confounding factor in all dermal computational toxicology models targeted toward intact skin.

Chemical absorption pathways can hypothetically involve both intercellular and intracellular passive diffusion across the epidermis and dermis and/or transappendageal routes, via hair follicles and sweat pores. Most available research has concentrated on the stratum corneum as the primary barrier to absorption, although the viable epidermis and dermis can also contribute resistance to the percutaneous penetration of specific chemical classes, for example, when the true barrier to absorption is not diffusional but is metabolic. Barrier is thus an operational definition and can be related to either a physical structure (e.g., stratum corneum) or a biological process (e.g., diffusional resistance, metabolism, vascular uptake) that retards absorption of topically applied chemicals.

The accepted hypothesis for dermal absorption is that the dominant pathway for chemicals to traverse the stratum corneum is through the intercellular lipids. Lipophilic compounds diffuse through this lipid milieu, while polar molecules traverse the aqueous region of the intercellular lipids. This intercellular region, described as the mortar in the "brick and mortar" model of the stratum corneum [14], is now considered the most likely path for absorption of lipophilic drugs. Although this model is conceptually simple, the actual physical chemical environment of the intercellular lipids is complex. It is filled with neutral lipids (complex hydrocarbons, free sterols, sterol esters, free fatty acids, and triglycerides) that make up 75% of the total lipids, as well as other polar lipids [15,16]. These intercellular lipids are also inextricably linked to the outer cellular membranes of the corneocytes, making a relatively complex and fluid structure that is often modeled as a simple homogeneous lipid pathway. Successive tape stripping, delipidization techniques, and heat or chemical epidermal separation techniques have been used by investigators to demonstrate the dominant influence that the stratum corneum and the lipid domain holds on penetration of hydrophilic and lipophilic chemicals.

Percutaneous absorption through the intercellular pathway of the stratum corneum is driven by passive diffusion down a concentration gradient described at steady state by Fick's law of diffusion [17–19],

$$Flux = \left[\frac{D \cdot PC \cdot SA}{H}\right](\Delta x),$$

where D is the diffusion coefficient, PC the partition coefficient, SA the applied surface area, H the membrane thickness (or more precisely the convoluted intercellular path length), and Δx the concentration gradient across the membrane. Since in vivo blood or in vitro perfusate concentrations after absorption are negligible compared to applied surface concentration, Δx reduces to the surface concentration (C) available for absorption. It is this relationship that allows the prediction of compound flux across the skin to be correlated to factors predictive of D and PC (e.g., octanol/water partition coefficients). Flux is expressed in terms of applied surface area, often normalized to cm^2. This is an oversimplification of the complex transport processes that occur in dermal diffusion [20,21], but it forms the basis for most computational toxicology modeling efforts in this field.

The term ($D \cdot PC/H$) is compound dependent, and it is termed the permeability coefficient (K_p), reducing the determination of flux to $K_p \cdot \Delta X$ or $K_p \cdot C$, a first-order pharmacokinetic equation ($dx/dt = kX$). Rearrangement of this equation yields the primary method used to experimentally calculate the value of K_p:

$$K_p = \frac{\text{Steady-state flux}}{\text{Concentration}}.$$

It must be stressed that both transdermal flux and K_p are not only chemical dependent but also tightly constrained by the membrane system studied as well as the experimental design of the study used to estimate it (neat compound, vehicle, length of experiment, etc.). The *PC* that is integral to K_p is the *PC* between the surface or applied vehicle and the stratum corneum lipids. Different vehicles will thus result in different *PC*s. Similarly skin from different species may result in different *PC* due to differences in the stratum corneum lipids and intercellular path lengths. From the computational toxicology perspective, this translates into quantitative models whose parameters are very dependent on experimental variables often not appreciated to be significant contributors to the process.

Dermal absorption data are often expressed as percent dose absorbed. This is conceptually correct if one assumes that permeability is unchanged across dose as it represents a first-order pharmacokinetic process. It is also appropriate when comparing experimental treatments (e.g., temperature, vehicle) using the same applied dose. However, in many cases, topical dosing results in applying substantial amounts of chemical compared to what can be absorbed across the skin. In soils, thick layering where most soil is not in contact with skin, and thus not able to reach its surface to partition into, suggests that most of an applied dose is not actually available for absorption. In such studies only a monolayer of soil is in reality in contact with the skin. Unlike in fluid or gel matrices, compound generally does not diffuse from soil layers not in contact with the skin, unless water is added as another vehicle. In these cases, accounting for the applied dose as total dose in a multi-layer system overestimates available dose, which artificially reduces the calculated percent dose absorbed. Similarly it overestimates the surface concentration used to estimate K_p. Caking of heavy dermal formulations results in a similar layering phenomenon.

The dose may also bind to the application device and not be available for absorption. In these cases a large fraction of the dose may not be thermodynamically driving the diffusion process. Finally, when the dose is applied in solution, saturation may result in precipitation of the chemical. Rapid evaporation of a volatile vehicle may precipitate it, thereby decreasing its availability for absorption. All these factors lead to a phenomenon often seen in dermal absorption studies where percent dose absorbed decreases with applied dose. Conducting a study at high applied doses may underestimate absorption of lower applied doses, and vice versa. Alternatively, for some chemicals dosed in volatile vehicles, evaporation increases thermodynamic activity of the penetrating solute through supersaturation, and enhances absorption.

TABLE 24.1 Experimental Variables That Should Be Controlled or Documented When Conducting Dermal Absorption Studies

Species, age and sex of animal
Application site on body or from, where skin obtained for in vitro studies
Dose, vehicle concentration, vehicle volume
Area of skin dosed
Vehicle, impurities, other formulation additives
Occlusion, dosing device
Length of dosing
Length of data collection
Assay method
Experimental endpoints and how calculated (K_p, AUC, etc.)
Additional in vitro systems parameters:
Type of diffusion cell system employed (static, flow-through)
Skin pre-treatment
Skin thickness
Composition of perfusate
Perfusate flow rate, temperature

This brief review illustrates that many experimental variables can effect the determination of K_p, an essential metric in any topical computation toxicology exercise. Table 24.1 tabulates the relevant parameters that should be considered when designing a dermal penetration study or using literature date for a quantitative modeling study.

24.4 DERMATOTOXICITY

Although the focus of this chapter is modeling dermal absorption, it is worth digressing to discuss the types of toxicological reactions that can occur in skin if a penetrating compound has activity against epidermal or dermal constituents. The field of dermatotoxicology has been extensively reviewed elsewhere [22,23] and will only be discussed here in relation to factors that affect computational issues.

If a cytotoxic chemical is capable of traversing the stratum corneum, it may cause toxicity to the skin as a function of its inherent potential to modify cellular function. Complex quantitative structure activity relationship (QSAR) models developed to assess general cytotoxicity may be applicable to define this inherent toxic potential. The clearest approach to assessing chemical-induced damage to skin is to assess what abnormalities occur when the specific anatomical structures discussed above are perturbed after exposure to topical compounds, since this will be the response modeled in a computational toxicology exercise.

This is also a good point to differentiate a chemical's so-called mode from its mechanism of action. This point often causes confusion when consulting literature from classic toxicology with relation to computational modeling sources. The mode of action is a function of the system changes in organ function that occurs as a result of chemical exposure. For example, this could be kidney damage or in our case epidermal necrosis. In contrast, the mechanism of action is the precise biochemical or molecular interaction that results in an effect. For kidneys, it could be irreversibly interrupting renal tubular sodium transport with resulting renal dysfunction. However, a chemical could cause a vascular effect on renal endothelial cells that likewise results in renal dysfunction. In both cases the dysfunction could be a result of a receptor binding or induction of mRNA to produce a new protein. Depending on the endpoint of renal disease selected (e.g., change in blood urea nitrogen—BUN), these two chemicals might be indistinguishable.

In the case of skin, a chemical might cause direct metabolic changes to epidermal keratinocytes and cause a dermal response (e.g., an irritant), or alternatively it may cause the same end result by inducing an immunological reaction (e.g., a sensitizer). In all these cases the chemical has the same "mode of action," that is, damage to skin, but has very different "mechanisms of action" that translate into different dose–response or concentration–response relationships when included in a computational toxicology model. For other chemicals a change in skin function may occur but not to the level of producing a toxicological effect. An example would be a change in proteomic patterns that does not produce irreversible loss of cellular function. In this case, mechanisms of action for two chemicals may be similar, but only one produces dermal toxicity.

Many cutaneous irritants specifically damage the barrier properties of skin that results in an irritation response. These include organic solvents, discussed previously, which extract the intercellular lipid and perturb the skin's barrier as assessed by a marker such as trans-epidermal water loss (TEWL). Some chemicals destroy or digest the stratum corneum and underlying epidermis. These are properly termed corrosives. These compounds cause chemical burns and include strong acids, alkalis, and phenolics. They essentially attack the epidermal barrier and chemically destroy the underlying viable cell layers. These types of reactions are easy to assess using cell-free in vitro models such as the Corrositex® system that detects macromolecular damage to a collagen matrix, which results in a chemical color change in an associated detector system. However, this simple system would not detect a toxin whose mechanism of action, as described below, is to cause epidermal cellular dysfunction from loss of epidermal barrier function and the same end result.

Direct irritation may be defined as an adverse effect of compounds directly applied topically to the skin not involving prior sensitization and thus initiation by an immune mechanism. Several types of adverse effects and chemical interactions may be manifested when a compound is placed upon the skin surface. Irritation is usually assessed by a local inflammatory response characterized by erythema (redness) and/or edema (swelling). Other responses may be present that do not elicit inflammation such as an increase in thickness.

Irritant reactions may be classified by many characteristics including acute, cumulative, traumatic, and pustular. However, two classifications are generally studied by toxicologists. Acute irritation is a local response of the skin usually caused by a single agent that induces a reversible inflammatory response. Cumulative irritation occurs after repeated exposures to the same compound and is the most common type of irritant dermatitis seen. In addition to chemical irritation, ultraviolet light can cause direct toxicity to skin via generation of oxygen-free radicals. Alternatively, UV light can activate certain chemicals (e.g., psoralen, some tetracycline antibiotics) to become cytotoxic. Mechanisms of irritation are very complex and are still being characterized, since they involve interaction of inflammatory cytokines with subsequent involvement of the immune system.

Alteration of epidermal cell function may then initiate other sequelae. If a penetrating compound is capable of interacting with the immune system, the manifestations seen will depend on the type of immunologic response elicited (e.g., cellular versus humoral, acute hypersensitivity). It should be stressed that immune cells (e.g., Langerhans cells, lymphocytes, mast cells) may modulate the reaction or the keratinocytes themselves may initiate the response; again, the same mode but different mechanisms of action. In fact keratinocytes were once thought to produce only keratin and mucopolysaccharides, but recent studies have shown that they can produce growth factors, chemotactic factors, and adhesion molecules. Keratinocytes may act as the key immunocyte in the pathophysiology of allergic contact and irritant contact dermatitis. When skin comes in contact with irritants, there are many pathways that can trigger the production of pro-inflammatory cytokines. Direct irritation of keratinocytes by toxic chemicals may also initiate this cytokine cascade without involvement of the immune system, blurring the distinction between direct and indirect cutaneous irritants. Release of such cytokines is often used as a biomarker of skin irritation in computational toxicology studies.

It must be stressed that the primary mechanism of many topical irritants (e.g., organic solvents, corrosives) is the impairment to the stratum corneum barrier properties discussed earlier. If the stratum corneum barrier is perturbed, a feedback response may be initiated whereby regeneration of the barrier occurs. This reaction is mediated by cytokines (especially TNF-α) originating locally within the epidermis. However, additional responses to these inflammatory mediators may in themselves launch an irritation response mediated by the keratinocytes. Thus, regardless of the initiating mechanism, the sequelae to many irritants is the same, making the definition of unique dermal computational toxicology models difficult.

24.5 LOCAL SKIN VERSUS SYSTEMIC ENDPOINTS

Before a quantitative model is developed, it is important to have some knowledge of precisely what endpoint is being modeled. If a kinetic study is being conducted, is the focus to predict transdermal flux and serum concentration-time profiles, or is the goal to determine how much chemical will reach target sites within the skin? The same experimental model systems may be used for both endpoints; however, different data will have to be collected. For transdermal flux, perfusate is collected and either K_p or AUC determined. If, on the other hand, local skin deposition is the endpoint, then skin samples or biopsies must be collected to assess the amount of chemical retained in skin, not just the amount that traversed skin, which would be estimated if only flux were monitored. As discussed in the methods section above, very lipophilic molecules may not partition into perfusate, so the quantity of chemicals exposed to the skin may be underestimated. Unlike perfusate flux parameters, there are no well-established metrics for dose deposition into skin.

Finally, cell culture studies are often used to assess direct cutaneous toxicity to skin cells such as keratinocytes. As discussed above, this can be used to define if a potential topical chemical that penetrates the skin can cause epidermal cell dysfunction. However, systemically administered chemicals can also distribute to skin and modify keratinocytes cell function. Damage to skin, especially if the mechanism of action is immunological, does not require topical exposure.

24.6 QSAR APPROACHES TO MODEL DERMAL ABSORPTION

Since passive diffusion is the primary driving force behind dermal absorption, physicochemical factors such as molecular weight and structure, lipophilicity, p*K*a, ionization, solubility, partition coefficients, and diffusivity can influence the dermal absorption of various classes of chemicals. In addition penetration of acidic and basic compounds will be influenced by the skin surface, which is weakly acidic (pH 4.2–5.6), since only the uncharged moiety of weak acids and bases is capable of diffusing though the lipid pathway. Several of these factors (e.g., molecular weight, and partition coefficients) have been used to predict absorption of various drug classes [24–26].

The first such relationship widely used to assess chemical absorption is that of Potts and Guy [24]:

$$\log K_p = 0.71 \log PC_{octanol/water} - 0.0061MW - 6.3 \qquad \left(R^2 = 0.67\right),$$

where MW is the molecular weight. This equation was subsequently modified [27] to relate K_p to molecular properties of the penetrants as

$$\log K_p = 0.0256MV - 1.72\sum \alpha_2^H - 3.93\sum \beta_2^H - 4.85 \qquad \left(R^2 = 0.94\right),$$

where MV is molecular volume, $\Sigma\alpha_2^H$ is the hydrogen-bond donor acidity, and $\Sigma\beta_2^H$ is the hydrogen-bond acceptor basicity.

The most promising approach is to further extend this rationale using linear free-energy relationships (LFER) to relate permeability to the physical properties of the penetrant under defined experimental conditions (dose, membrane selection, vehicle). Geinoz and coworkers [28] critically reviewed most such quantitative structure permeability relationships (QSPeR) applied to dermal absorption and should be consulted for a detailed review. Abraham's LFER model is representative of the dermal QSPeR approaches presently available [29]. This model was selected because it is broadly accepted by the scientific community as being descriptive of the key molecular/physiochemical parameters relevant to solute absorption across skin. This basic model can be written as

$$\log k_p = c + a\sum \alpha_2^H + b\sum \beta_2^H + s\pi_2^H + rR_2 + vV_x,$$

where π_2^H is the dipolarity/polarizability, R_2 represents the excess molar refractivity, V_x is the McGowan volume, and the other parameters are as described earlier. The variables c, a, b, s, r, and v are strength coefficients coupling the molecular descriptors to skin permeability in the specific experimental system studied.

Our laboratory has focused significant research on the effects of chemical mixtures on dermal absorption of penetrant compounds. In order to incorporate mixture effects, our laboratory has been exploring use of an additional term operationally called the mixture factor (*MF*), yielding

$$\log k_p = c + mMF + a\sum \alpha_2^H + b\sum \beta_2^H + s\pi_2^H + rR_2 + vV_x.$$

The nature of the *MF* is determined by examining the residual plot (actual – predicted log k_p) generated from the base LFER equation based on molecular descriptors of the permeants, against a function of the physical chemical properties of the mixture/solvents in which they were dosed [1,30]. Figure 24.2 depicts the improvement in prediction of K_p when this approach is used.

The literature on QSPeR is exhaustive and rapidly growing. The limitation of applying these approaches to chemical absorption is the lack of large and comparable databases of chemical dermal absorption, as well as the lack of availability of molecular descriptors for many compounds. As will be discussed below, data suitable for large-scale analyses must be rigorously controlled relative to the species studied, the nature of the experiments (in vitro vs. in vivo), dose, surface area, vehicle, and method of sample collection and analyses.

24.7 PHARMACOKINETIC MODELS

The final area of computational toxicology to be discussed in this chapter applied to dermal absorption is the general area of pharmacokinetic models (see also Chapter 3). These are usually developed as extensions of whole animal-based models using classic compartmental [19,31] or physiological-based [32] approaches. These texts should be consulted for an overview of the

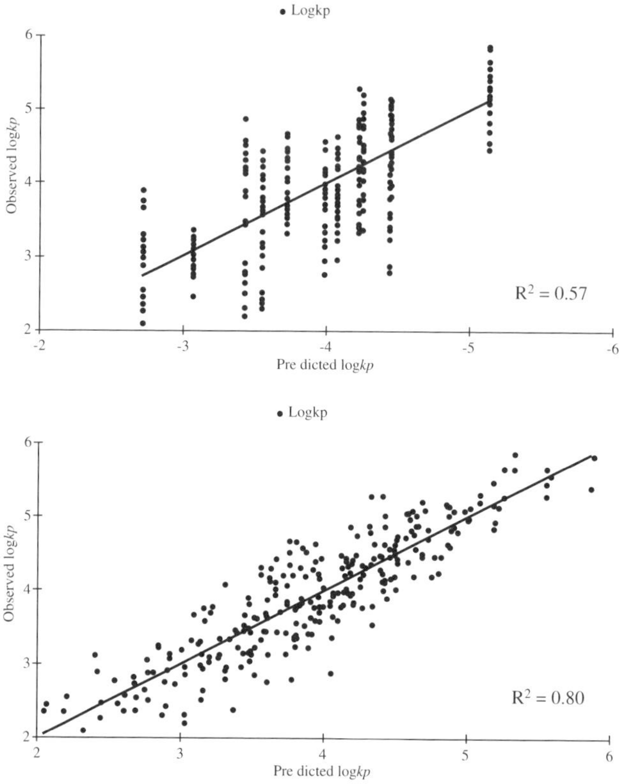

Figure 24.2 QSAR plots showing effects of including a mixture factor (MF) [Refractive Index] on prediction of dermal permeability when dosed in complex chemical mixtures. (*Top plate*) No MF included: $n = 288$, $R^2 = 0.57$, $Q^2 = 0.56$, $s = 0.55$, $F = 77$. (*Bottom plate*) MF included: $n = 288$, $R^2 = 0.80$, $Q^2 = 0.79$, $s = 0.37$, $F = 192$.

basic assumptions inherent in each genre of pharmacokinetic modeling, as these are carried forward when applied to skin. There are an enormous variety of approaches taken to develop dermatopharmacokinetic models. These applications to skin have recently been well reviewed elsewhere [20,33].

The primary purpose for most dermatopharmacokinetic models is to quantitate the linkage between anatomical and physiological properties of skin that play rate-limiting roles in absorption with target sites for which concentration-time profiles are needed. The complexity of the models is a function of the chemical being studied as well as the level of precision (concentration, time frame) required for the prediction. For example, models created to estimate total percent dose absorbed are much simpler than those designed to predict the time and magnitude of peak plasma concentrations in a subject.

Numerous other factors play a role in the structure and complexity of models employed. In general, the stratum corneum is assumed to be a homogeneous membrane into which compound partitions from the surface. Some models attempt to take into account the tortuous nature of the intercellular pathway when defining actual diffusion path length. For volatile compounds, models may specifically include evaporative loss from the surface of the skin or binding to other surface sites (e.g., application device, hair). The next stage of potential complexity is whether metabolism occurs in the epidermis, a process that may dramatically increase model complexity. Drug then enters the dermis or is assumed to directly enter the blood. If a compound is vasoactive, dermal blood flow may be specifically modeled. At all three tissue levels, irreversible or very slow tissue binding may also be included to model the formation of so-called depots in the stratum corneum or fat. Two examples of pharmacokinetic models used in our laboratory [34–36] are depicted in Figures 24.3 and 24.4.

Some dermatopharmacokinetic models attempt to develop linkages between chemical concentrations in a specific compartment with a

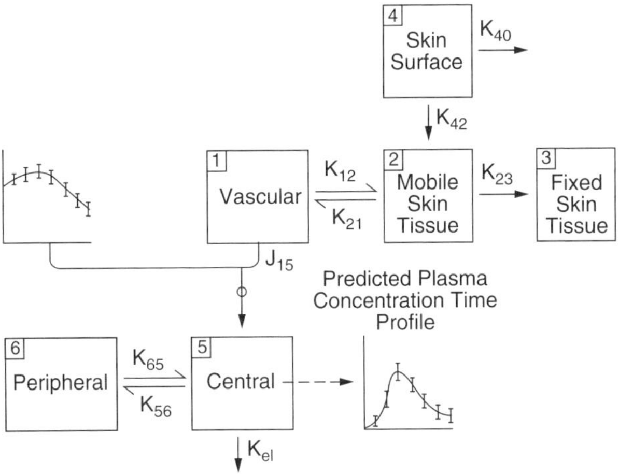

Figure 24.3 Compartmental pharmacokinetic model linking skin absorption determined in an in vitro model to a systemic model to predict plasma concentration time profiles in vivo.

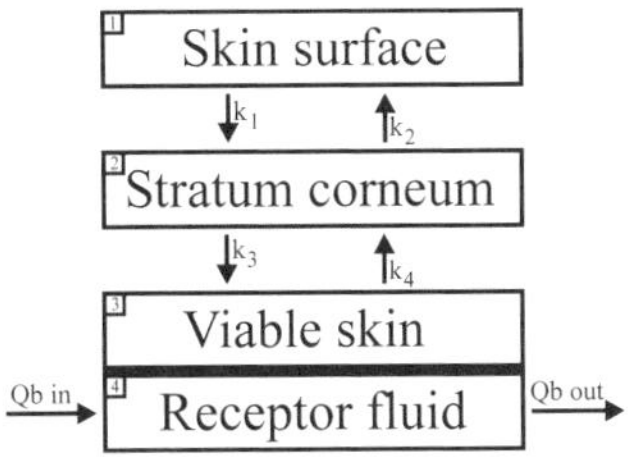

Figure 24.4 Physiological based pharmacokinetic model of chemical absorption across skin.

toxicological effect based on mechanisms of action [33]. This work is in its infancy but holds promise should sufficient data to conduct these modeling exercises become available.

24.8 CONCLUSIONS

The percutaneous absorption and dermatotoxicity of topically applied drugs and chemicals is a major concern in pharmacology, pharmaceutical, and toxicological sciences. As research in these areas continues and regulatory oversight is applied to more classes of substances, efforts have been made to develop predictive models to quantitate chemical exposure to the skin and systemic circulations. Model complexity increases with greater anatomical or physiological details. Before realistic predictive models can be constructed, experimental data must be collected of sufficient quality to make such efforts worthwhile.

REFERENCES

1. Riviere JE. *Dermal absorption models in toxicology and pharmacology.* New York: Taylor and Francis. 2006.
2. Bronaugh RL, Stewart RF. Methods for in vitro percutaneous absorption studies: III. Hydrophobic compounds. *J Pharm Sci* 1984;73:1255–8.
3. Bronaugh RL, Stewart RF. Methods for in vitro percutaneous absorption studies: IV. The flow-through diffusion cell. *J Pharm Sci* 1985;74:64–7.
4. Feldman RJ, Maibach HI. Percutaneous absorption of some pesticides and herbicides. *Tox Appl Pharmacol* 1974;28:126–32.
5. Carver MP, Riviere JE. Percutaneous absorption and excretion of xenobiotics after topical and intravenous administration to pigs. *Fundam Appl Toxicol* 1989;13: 714–22.
6. Chow SC, Liu JP. *Design and analysis of bioavailability and bioequivalence studies*, 2nd edition. New York: Dekker, 2000.
7. Vergnaud JM, Rosca ID. *Assessing bioavailability of drug delivery systems: Mathematical modeling.* New York: Taylor and Francis, 2005.
8. Pershing LK, Krueger GG. New animal models for bioavailability studies. In: Shroot B, Schaefer H, editors, *Pharmacology and the skin: Skin pharmacokinetics.* Basel: S Karger, 1987. p. 57–69.
9. Riviere JE, Bowman KF, Monteiro-Riviere NA, Carver MP, Dix LP. The isolated perfused porcine skin flap (IPPSF): I. A novel in vitro model for percutaneous absorption and cutaneous toxicology studies. *Fundam Appl Toxicol* 1986;7: 444–53.
10. Riviere JE, Monteiro-Riviere NA, Williams PL. Isolated perfused porcine skin flap as an in vitro model for predicting transdermal pharmacokinetics. *Eur J Pharm Biopharm* 1995;41:152–62.

11. Wester RC, Melendres J, Sedik L, Maibach HI. Percutaneous absorption of salicylic acid, theophylline, 2,4-dimethylamine, diethyl hexylphthalic acid, and ρ-aminobenzoic acid in the isolated perfused procine skin flap compared to man. *Toxicol Appl Pharmacol* 1998;151:159–65.
12. Monteiro-Riviere NA. The use of isolated perfused skin in dermatotoxicology. In *Vitro Toxicol* 1993;5:219–33.
13. Monteiro-Riviere NA. Comparative anatomy, physiology, and biochemistry of mammalian skin. In Hobson DW, editor, *Dermal and ocular toxicology: Fundamentals and methods*. Boca Raton, FL: CRC Press, 1991. p. 3–71.
14. Elias PM. Epidermal lipids, membranes, and keritinization. *Int J Derm* 1981;20:1–19.
15. Magee P. Percutaneous absorption: Critical factors in transdermal transport. In: Marzulli FN, Maibach HI, editors, *Dermatoxicology*. New York: Hemisphere Publishing, 1991. p. 1–36.
16. Monteiro-Riviere NA, Inman AO, Mak V, Wertz P, Riviere JE. Effects of selective lipid extraction from different body regions on epidermal barrier function. *Pharm Res* 2001;18:992–8.
17. Roberts MS, Anissimov YG, Gonsalvez RA. Mathematical models. In: Bronaugh RL, Maibach HI, editors, *Percutaneous absorption*, 3rd edition. New York: Taylor and Francis, 1999. p. 3–55.
18. Wester RC, Maibach HI. Cutaneous pharmacokinetics: 10 Steps to percutaneous absorption. *Drug Metab Rev* 1983;14:169–205.
19. Riviere JE. *Comparative pharmacokinetics: Principles, techniques and applications.* Ames: Iowa State Press, 1999.
20. Roberts MS, Anissomov YG. Mathematical Models In: Bronaugh RL, Maibach HI, editors, *Percutaneous absorption*. 4th edition. New York: Taylor and Francis, 2006. p. 1–44.
21. van der Merwe D, Brooks JD, Gehring R, Monteiro-Riviere NA, Baynes RE, Riviere JE. A physiological based pharmacokinetic model of organophosphate dermal absorption. *Toxicol Sci* 2006;89:188–204.
22. Monteiro-Riviere NA, Riviere JE. Skin. In: Marquardt H, Schäfer S, McClellan R, Welsch F, editors, *Toxicology*. New York: Academic Press, 1999. p. 439–57.
23. Zhai H, Maibach HI. *Dermatotoxicology*, 6th edition. Boca Raton, FL: CRC Press, 2004.
24. Potts RO, Guy RH. Predicting skin permeability. Pharm Res 1992;9:663–9.
25. Cleek RL, Bunge AL. A new method for estimating dermal absorption from chemical exposure: 1. General Approach. *Pharm Res* 1993;10:497–506.
26. Bunge AL, Cleek RL. A new method for estimating dermal absorption from chemical exposure: 2. Effect of molecular weight and octanol-water partitioning. *Pharm Res* 1995;12:88–95.
27. Potts RO, Guy RH. A predictive algorithm for skin permeability: The effects of molecular size and hydrogen bond activity. *Pharm Res* 1995;12:1628–33.
28. Geinoz S, Guy RH, Testa B, Carrupt PA. Quantitative structure-permeation relationships (QSPeRs) to predict skin permeation: A critical evaluation. *Pharm Res* 2004;21:83–92.

29. Abraham MH, Martins F. Human skin permeation and partition: General linear free-energy relationship analyses. *J Pharm Sci* 2004;93:1508–23.
30. Riviere JE, Brooks JD. Predicting skin permeability from complex chemical mixtures. *Toxicol Appl Pharmacol* 2005;208:99–110.
31. Gibaldi M, Perrier D. *Pharmacokinetics*, 2nd edition. New York: Dekker, 1982.
32. Reddy MB, Yang RS, Clewell HJ, Andersen ME. *Physiological based pharmacokinetic modeling*. New York: Wiley, 2006.
33. McDougal JN, Zheng Y, Zhang Q, Conolly R. Biologically based pharmacokinetic and pharmacodynamic models of skin. In: Riviere JE, editor, *Dermal absorption models in toxicology and pharmacology*. New York: Taylor and Francis, 2006. p. 89–112.
34. Williams PL, Carver MP, Riviere JE. A physiologically relevant pharmacokinetic model of xenobiotic percutaneous absorption utilizing the isolated perfused porcine skin flap (IPPSF). *J Pharm Sci* 1990;79:305–11.
35. Williams PL, Riviere JE. A model describing transdermal iontophoretic delivery of lidocaine incorporating consideration of cutaneous microvascular state. *J Pharm Sci* 1993;82:1080–4.
36. Riviere JE, Williams PL, Hillman R, Mishky L. Quantitative prediction of transdermal iontophoretic delivery of arbutamine in humans using the in vitro isolated perfused porcine skin flap (IPPSF). *J Pharm Sci* 1992;81:504–7.

PART V

NEW TECHNOLOGIES FOR TOXICOLOGY: FUTURE AND REGULATORY PERSPECTIVES

25

NOVEL CELL CULTURE SYSTEMS: NANO AND MICROTECHNOLOGY FOR TOXICOLOGY

MICHAEL L. SHULER AND HUI XU

Contents

Computational Toxicology: Risk Assessment for Pharmaceutical and Environmental Chemicals, Edited by Sean Ekins

25.1 INTRODUCTION

The integration of modern nano and microfabrication technologies with cell culture has contributed to the development of new devices for toxicological studies and drug evaluations [1,2]. Minimization of cell culture devices lowers the device cost, decreases chemical usage and waste production, allows massively parallel tests, and most importantly, provides opportunities for researchers to control, monitor, and analyze cell responses with few cells, or even at the single-cell or subcellular level [3]. Nano and micro techniques enable manipulations, with nano and micrometer control, on the biochemical composition and topology of the substrate, the medium composition, and the cellular microenvironment [4], allow an improved mimic of the extracellular microenvironment a cell experiences in vivo, and bring in vitro cell responses closer to in vivo cell responses. Miniature cell-based biosensors have been quickly developed for the functional characterization and detection of a wide range of biologically active compounds, including drugs, pathogens, toxicants, and odorants [1].

In this chapter we introduce various techniques for fabricating miniature cell culture devices and cell-based biosensors, provide examples of human and animal cells immobilized on the chip devices, and explain different approaches to pattern multiple types of cells on one device. The application of nano and micro techniques in precise control over the cellular microenvironment is discussed. Selective cell-based biosensors are described later in the chapter. Finally, we conclude that these novel cell culture systems, coupled with predictions from in silico mathematical modeling, can potentially improve predictions of human clinical responses and enable better understanding of toxicological mechanisms.

25.2 NANO AND MICROFABRICATION TECHNIQUES FOR CELL CULTURE

The nano and microfabrication technologies for manufacturing Micro-Electro-Mechanical Systems (MEMS) are now widely used in biological studies which

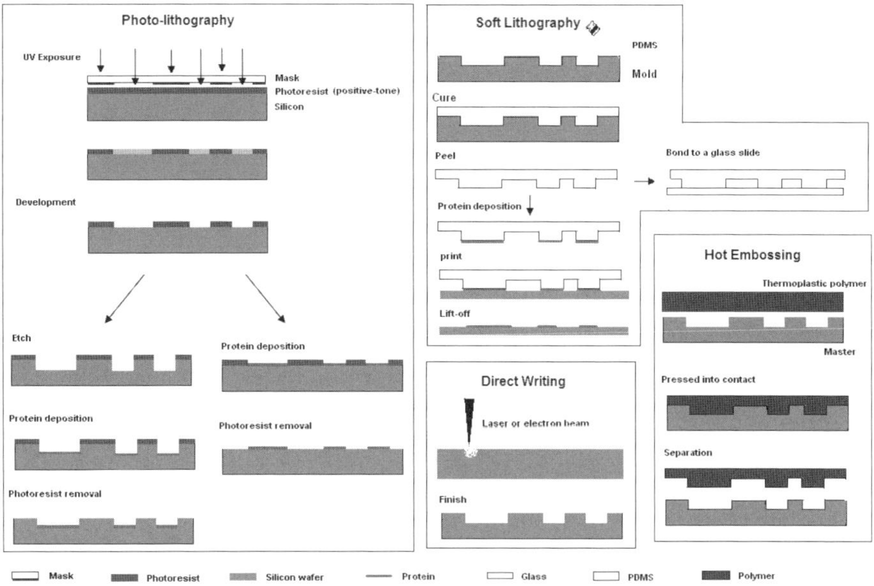

Figure 25.1 Schematics of typical nano and micro fabrication techniques: Photolithography, soft lithography, hot embossing, and direct writing. See color plates.

combine microfluidics and cell handling for microreactors, cell culture, cell stimulation, cell sorting, cell lysis, sample separation, purification, and biochemical analysis [1]. These techniques are capable of fabricating miniature devices for cell culture with specific, pre-designed patterns for housing cells and supporting cell growth, or for biochemical analysis with incorporated optical and electrical components. Such devices are sometimes referred to as Bio-MEMS.

A wide range of fabrication methods have been developed to manufacture miniaturized cell culture devices (Figure 25.1). The examples described here include photolithography, electro-beam lithography, soft lithography, hot embossing, laser micromachining, and electrical discharge machining. Each method has its strengths and weaknesses. Decisions on which to choose should be made based on the requirement for precision, time, cost, volume of production, and the availability of materials and fabrication facilities.

25.2.1 Photo Lithography

Borrowed from the semiconductor industry, photolithography has been well accepted for patterning cells and fabricating micro featured bio-microfluidic devices on silicon and glass substrates [5,6]. The resolution of photolithography is limited by the optical diffraction, with a minimum feature size around 50 nm [5]. The first step is to make photo-masks with the desired patterns. The pattern is generated by direct laser or ultraviolat (UV) light onto a thin photo-resist layer on a chrome (Cr) coated glass or quartz plate (Figure 25.1). The exposed photoresist becomes soluble and is removed in a developer solution. The unprotected Cr is then etched away in an acid bath, leaving the pattern image on the plate. To transfer this pattern onto a silicon or glass substrate, the substrate is spin-coated with photoresist and exposed to ultraviolet light through the mask, which selectively allows light through. The exposed photoresist will be either solubilized (positive-tone photoresist) or crosslinked (negative-tone photoresist), followed by exposure to a developer solution that removes solubilized or uncrosslinked photoresist, and results in a photoresist pattern. This pattern can be used for direct biomolecular (e.g., cell adhesion molecules for cell attachment) deposition [7] or for selective dry or wet etching into the substrates to form chambers for cell placement and fluid handling [8]. The remaining photoresist can then be dissolved in acetone or resist remover or etched off by plasma stripping.

Although a mature technique, the application of photolithography is limited by its requirement for expensive facilities, high processing cost, and limited resolution.

25.2.2 Electro-beam Lithography

Electro-beam (E-beam) lithography has been the primary technique for defining, patterning and connecting experimental structures at the nanoscale [9]. E-beam lithography uses a narrow electron beam to generate patterns. The extremely short quantum mechanical wavelength of high-energy electrons allows the E-beam to overcome the diffraction limit, creating single surface features about 3 to 5 nm in size. This limit increases to 30 to 40 nm for large area surface patterning [5,10,11]. Recently a simple, low-cost E-beam lithography system with a 10 nm resolution at a $50 \times 50\,\mu m$ scan area was presented. It is constructed by insertion of an electrostatic deflector plate system at the electron-beam exit of the column of a scanning electron microscope (SEM) [12].

E-beam lithography allows computer-controlled direct pattern writing onto resist-coated substrates without an expensive physical mask. It has some drawbacks as well such as its high initial capital expense and slow writing speed due to its point-by-point exposure method.

25.2.3 Soft Lithography

Soft lithography is developed for the fabrication of relatively large microstructures (>50 μm) in biological applications [13,14]. It represents a group of surface-patterning techniques, including the micro contact printing, microfluidic patterning, and stencil patterning. One key feature of soft lithography is the use of elastomeric materials, most commonly poly(dimethylsiloxane)

(PDMS), to replicate patterns from a master by molding. PDMS is inexpensive, biocompatible, gas permeable, amenable to surface modification, and optically transparent [15]. Soft lithography enables high-volume production of disposable devices and lowers the cost for frequent design changes common in biological research.

Soft lithography starts with making masters for PDMS molding. The master can be either a silicon master or a SU-8 mold made from photolithography. However, fabricating silicon masters using traditional photolithography is time-consuming and costly. Instead, a SU-8 mold is most frequently used in soft lithography. SU-8 is a transparent negative-tone photoresist originally designed for the MEMS industry. It features a high-aspect ratio, good mechanical strength, excellent chemical resistance, and low cost. The process of making a SU-8 mold is simple: spin coat, expose, and develop. To mold PDMS devices, the PDMS oligomer and crosslinking prepolymer are mixed well and poured onto the SU-8 mold. Negative pressure is applied for removal of air bubbles introduced by mixing. The PDMS precursor mixture is allowed to cure at 65°C overnight, after which the PDMS sheet can be peeled off the mold. This micro featured PDMS slab is then ready for use as a stamp for micro contact printing, or it can be sealed against a substrate for microfluidic applications.

Micro contact printing (uCP) uses a PDMS stamp to pattern chemical molecules, such as proteins or alkanethiolates, onto silicon, glass, gold [16], PDMS [17], and many other plastic substrates. Proteins are deposited or passively absorbed onto the PDMS stamp, then brought into contact with the substrate surface for several minutes. As the stamp is lifted, the protein pattern is left on the surface. Printing of cell adhesive proteins will result in cell attachment only at defined areas. uCP is capable of transferring patterns with features of 1 μm dimensions and with an edge roughness of <100 nm [13]. This resolution is limited by the stamp distortion during its contact with the substrate and by surface diffusion of low molecular weight inks [18].

Microfluidic patterning combines a channel system with laminar flow to deliver chemical molecules in solution or cells in suspension to a restricted area. When a PDMS sheet with features is sealed to a glass slide, water-tight flow pathways are formed. Proteins or cells can be patterned onto the glass slide by forcing the solution flow through these fluidic channels. It should be noted that the laminar flow in microscale channels allows patterning of several different fluids in parallel to each other in one channel [5]. This phenomenon is also used for gradient generation and toxicological assessment, which are explained later in this chapter.

Stencil patterning of cells utilizes a thin piece of PDMS sheet with through holes (stencil) to confine protein solution or cell suspensions within defined boundaries. The stencil works as a physical barrier to avoid contact of cell or protein solution with the substrate except at open areas in the stencil. With multiple openings on the stencil, multiple types of cells can be seeded onto the same piece of substrate for cell co-cultures. Because of its elastic nature and hydrophobicity, a PDMS stencil will stick loosely to a clean smooth surface [1]. This seal is reversible, meaning that the PDMS sheet can be peeled off without destroying either the PDMS or the surface of the substrate. Stencils can be made out of other biocompatible materials as well, but a good seal between the stencil and the substrate is required to avoid cell or chemical cross-contamination.

25.2.4 Hot Embossing

Hot embossing is also called thermal imprinting. This technique is capable of manufacturing features in the thermoplastic polymers with dimensions as small as 10 nm [19]. Hot embossing is capable of manufacturing microfeatures with polymer substrates such as polymethyl methacrylate (PMMA), polycarbonate, cyclo-olefin copolymer (COC) [20], and polyimide [19]. For embossing, a silicon or hard polymer mold is pressed into thermoplastic materials at a temperature higher than the material's glass transition temperature (T_g), forming a relief of its feature in the plastic. The material surface is textured and separated from the mold once the polymer is cooled below the T_g [18]. The major advantages of hot embossing include its very low cost, capability for volume production of disposable devices, and ability to form 3D features that are difficult to produce in silicon [19].

25.2.5 Laser Micromachining

Lasers can be used for both laser ablation and laser polymerization. Laser ablation combines evaporation and melt expulsion to remove metals, ceramics [21], glass, silicon, and plastics materials, while laser polymerization involves building 3D microstructures through laser-induced polymerization of organic-inorganic hybrid biomaterials [22]. Laser ablation can produce surface features in the 5 to 500 μm range with submicron accuracy [23]. Coupled with a microlithographic projection technique, laser micromachining with a metal mask can generate fine microstructures at the micron and submicron level. For a reduction of the mask pattern the laser beam exiting the mask is focused onto the polymer target surface via an optical setup [23,24]. The high-relief topology of the laser lithographed substrates can also be utilized as a master mold for PDMS stamps in microcontact printing and spatial patterning of cell adhesion proteins.

Multiphoton absorption is a nonlinear optical phenomenon where two or more low-energy photons are absorbed instead of a high-energy photon [25]. Since this process requires an intense flux of excitation photons, optical absorption, and consequent laser ablation or polymerization, are tightly confined to the vicinity of focus with a micrometer resolution [26]. Using this technique, Giridhar et al. were able to drill a subsurface tunnel into glass substrate with a high-power femtosecond pulsed laser. Microfluidic pathways and chambers 10 to 100 μm wide and 5 to 50 μm deep on a glass plate were also fabricated.

Multiphoton absorption induced polymerization of organic-inorganic hybrid biomaterials can be used as well to fabricate microscopic 3D topographies, such as barriers and growth lanes, micro-structured medical devices [22]. Polymerization can be done in situ within dissociated neuron cell cultures without compromising cell viability [27]. In a different application, Costantino and coworkers added fluorescent dyes to UV-cured resins and used two-photon optical lithography to make spatial landmarks to quantify molecular transport, cell growth and migration [28].

Laser micromachining is automatic, fast, and cheaper than hot embossing [29]. Because it creates no or very little thermal effects, it allows a very clean cut surface [24]. The direct writing of subsurface features with multi-photon absorption is unique among available nano and micro fabrication methods.

25.2.6 Micro Electrical Discharge Machining

Micro electrical discharge machining (EDM) is primarily used for hard metals and other electrically conductive materials. When two electrodes separated by a dielectric medium come close, the dielectrical medium breaks down and suddenly becomes conductive. Sparks generated between the electrodes release energy and remove metal by melting and evaporation [30]. Murali and Yeo combined micro electro-discharge machining with ultrasound electrode vibration for improved surface quality, and achieved micro features as small as 25 μm on titanium with 20 μm diameter electrodes [30].

25.2.7 Materials for Miniature Cell Culture Devices

Not all materials are suitable for fabricating micro-scale cell culture devices. Below is an incomplete list of requirements and considerations for choosing the material:

- Biocompatibility
- Capability for surface modification that allow cell attachment (for on-chip mono layer cell culture of adherent cells)
- Chemical resistance to cell culture medium or working fluids
- Ease of fabrication or processing
- Durability
- Cost

Silicon and glass are the traditional and well-accepted substrates for micro and nanofabrication. However they are fragile, and the processes for patterning silicon and glass surfaces, such as photolithography and E-beam lithography, are expensive. Many ceramics, such as Titania ceramic (titanium dioxide), are biocompatible implant materials with no cytotoxicity. They can be easily shaped during molding by a microfabricated silicon master [31]. Gold or gold-coated substrates are often used for cell culture in cell-based biosensors for easy detection of electrical signals [32]. Titanium alloy is a material with good biocompatibility but poor machinability [30]. Micro electro-discharge machining is one way to process the titanium alloy without compromising its biocompatibility.

Polymers can be micro-machined in many ways, for example, by soft lithography, hot embossing, and direct laser machining. Flexible plastic substrates used in soft lithography include PDMS, poly(ethylene-dioxythiophene)/polystyrene sulfonate (PEDT/PSS) [33], and polymeric films. Polystyrene, PMMA, polycarbonate, plexiglass, and polyethylene are the primary plastics used for fabricating cell culture devices with hot embossing. Plastics are seldom used in photolithography because of the lack of good contact between mask and substrate. They are generally less mechanically fragile than glass or silicon. Plastics are optically transparent, easy to mold, relatively inexpensive, thus suitable for disposable devices to minimize chemical or biological cross-contamination.

For 3D cell culture, although porous polymers such as polylactic acid (PLA), polyglycolic acid (PGA), and poly (lactic-co-glycolic acid) (PLGA) are commonly used scaffolds in tissue engineering, hydrogels are better accepted for cell culture in micro devices. Hydrogels are three-dimensionally cross-linked macromolecules of hydrophilic polymers. They are soft in texture and easy to handle at microliter scale. Hydrogels used for cell cultures include collagen, matrigel, calcium alginate, agarose, and synthetic polymers such as PEG, polyglycolide, polylactide, and acrylamide derivatives [34,35].

25.3 EXAMPLES OF HUMAN AND ANIMAL CELLS IMMOBILIZED ON A CHIP

Cells can be cultured in monolayer on a solid surface or porous membrane, or embedded into hydrogels or other scaffolds for 3D culture. In general, most adherent cell lines and primary cells can be easily immobilized onto miniature cell culture devices upon substrate surface modifications. Coating with poly-D-lysine, fibronectin, collagen, agarose [35], laminin [36], serum, and other glycoproteins containing the amino acid sequence Arg-Gly-Asp (RGD), or processing with plasma treatment for plastics, are generally sufficient for surface conditioning to support cell attachment and growth on silicon, quartz, glass, PDMS, and many plastic substrates. Suspension cells (cells that normally grow in suspension), unlike adherent cells, are generally cultured in 3D in order to be kept at desired positions.

25.3.1 Monolayer Cell Culture on a Solid Surface

Our laboratory has fabricated several generations of micro cell culture analog (μCCA) devices, also called "animal-on-a-chip" or "body-on-a-chip". In μCCA, mammalian cells are cultured in compartments on the silicon devices, with the cell culture medium circulated among them working as the blood surrogate. Examples of cells or cell lines successfully cultured on the chips are as follows:

Liver Primary rat hepatocytes, H4IIE (rat), HepG2 (human), or HepG2/C3A (human) cells are generally used to study xenobiotic metabolism, toxicology, and liver function.

Mammary carcinoma Breast cancer cells, including MCF-7 (human), T47D (human), and MDA-MB-231(human), have estrogen receptor expression profiles that are different [37]. These cells are used to study the responses of breast tissues to estrogen or pharmaceuticals.

Megakaryoblast MEG-01 (human) is a human megakaryoblastic cell line established from bone marrow [38]. These cells are sensitive to chemical stimulation.

Uterine sarcoma MES-SA is derived from human uterus sarcoma. MES-SA/DX-5 is a multiple drug resistant cell line established from MES-SA cells. These two cell lines are generally used together to study multi-drug resistance and to test modulators to enhance drug delivery across the cell membrane.

Lung cells L2 is a rat lung epithelial cell line used to study chemical toxicity to lung cells and tissues.

Intestine/colon cancer cells The caco-2 intestinal cell monolayer is a good model for testing permeability and oral deliverability of compounds [39].

Pre-adipose cells Differentiated mouse 3T3-L1 cells are capable of accumulating hydrophobic compounds [40] and can be used as an adipose tissue surrogate [41].

There are many other endothelial, epithelial, fibroblast cell lines, and primary cells that have been used for on-chip cultures. Some examples are as follows:

Neuron cells Neuron cells are generally electrically excitable, and their electrophysiological property changes upon physical and chemical stimulation. For this reason neuron cells are widely used as the sensing element in cell-based biosensors [42]. Primary rat pup astrocytes are also used in co-culture with endothelial cells for the in vitro mimic of blood–brain barrier (BBB) [43].

Muscle cells Myocytes are electrically excitable and are popular in cell-based biosensors [44]. Primary myocytes isolated from adult or neonatal rat hearts have been used to study the effect of microtopography on cell functions [45]. Murine skeletal muscle cells have been patterned in line on glass chips to study the differentiation of myoblasts into myotubes [46].

Embryonic stem cells Embryonic stem cells (ESCs) are especially sensitive to their microenvironment and can be used to test micro cell culture devices for their capability to support cell growth. Murine ESC were recently used to test a PDMS cell culture device with continuous perfusion [47].

25.3.2 Cell Culture on Porous Membranes

Epithelial barrier models for the skin [48,49], respiratory tract [50], BBB, and intestine [39] are constructed to study and predict the absorption, penetration, and metabolism of drugs or environmental toxins through these barriers. All the models are physically tight structures, and generally involve cells cultured at the air–liquid interface on porous membrane support, such as a porous polycarbonate filter. The use of a permeable support allows cells to be grown in a polarized state under more natural conditions promote *Cell* differentiation and enhance cell functions.

Ma et al. fabricated an ultra-thin, highly porous silicon nitride membrane using photolithograpy followed by reactive ion etching. The finished membrane chip is 15×15 mm in size, with a grid of porous silicon nitride membrane with 400 nm pores at the center, about 0.8×0.8 mm and 1 μm in thickness [43]. An endothelial cell line, SV-HCEC, and astrocytes isolated from rat pups were cultured separately onto each side of the membrane to mimic the BBB. This porous membrane, about one order of magnitude thinner and at least twice as porous than commercially available membranes, enables better direct interaction between the two types of cells, which is critical for an in vitro BBB mimic. The tailor-made membrane also allows precise control and adjustment of pore size, porosity, and membrane thickness. It could be a better alternative for traditional membranes used in the in vitro model of intestine using Caco-2 and Ht29GlucH mucus-secreting co-culture [51], or in the skin model with human keratinocytes, and in the respiratory tract with Calu-3 immortalized human bronchial epithelial cells [50].

25.3.3 3D On-chip Cell Cultures

Three-dimensional cell cultures allow the formation of 3D tissue-like cell structures, which more accurately represent the natural cellular environment than traditional 2D monolayer cultures.

3D spheroid liver cell culture has attracted attention for its ability to maintain morphological and functional characteristics of primary hepatocytes for extended time [52]. Fukuda et al. fabricated a PMMA chip with 300 μm diameter cylindrical cavities. The centers of these cavities were coated with collagen for cell attachment through micro contact printing, while the other areas were treated with polyethylene glycol (PEG) to prevent cell adhesion. Primary hepatocytes were able to form spheroids in the cavities, with shapes and liver-specific phenotypes similar to those in vivo [52]. Otsuka and colleagues patterned a microarray of ten thousand hepatocyte heterospheroids underlaid with endothelial cells on a 20×20 mm glass substrate through PEG patterning [53]. Hydrogel encapsulation is another popular way for on-chip 3D cell culture. It protects cells from extra hydrodynamic shear stress. Hydrogels used for this purpose include alginate, matrigel, PEG and many others. MCF-7 breast cancer cells have been embedded in a collagen-gel matrix and entrapped in an array of pyramidal-shaped holes in a silicon chip for a chemosensitivity assay involving multi-chemical stimulation [54].

25.4 SIMULTANEOUS ASSESSMENT OF MULTIPLE CELL TYPES THROUGH CELL CO-CULTURE

Traditional toxicity testing involves one single type of cell dosed with one or a mixture of compounds. Since toxicological effects may be obvious in one kind of cell but not in another, identical tests are repeated with different cell types. But only one type is tested each time. This kind of test is labor-intensive and time-consuming. Further, results from such a test might be incomplete and cannot represent in vivo cell responses because of the absence of cell–cell and organ–organ interactions. All the organs have complex multicellular structures. The interactions between these cells are critical for normal organ development and functions. Although organs in the body are physically separated, they are interconnected by the blood circulation, and communicate through secretion, absorption, and metabolism of chemicals, into and from the blood stream. The existence of a second type of cell may alter the cell responses and functions. Indeed it has been shown that co-culture of different cell types, such as primary hepatocytes with non–liver-derived endothelial or fibroblasts, modulate the liver cell phenotype and specific functions [7].

In this section we describe various methods for co-culture of different types of cells from the same organ or from different organs. Co-culture enables simultaneous monitoring on multiple types of cells and take the in vitro culture system closer to a living organism.

25.4.1 Randomly Distributed Cell Co-Culture

Co-culture can be obtained by mixing multiple types of cells before cell seeding (Figure 25.2*a*) or by seeding one type of cell on top of another [55,56], which results in random cell contact and distribution. Alteration of cell seeding density or the ratio of cell population changes the strength of cell–cell interactions. In the mixed co-culture of Caco-2 cells and colorectal cancer

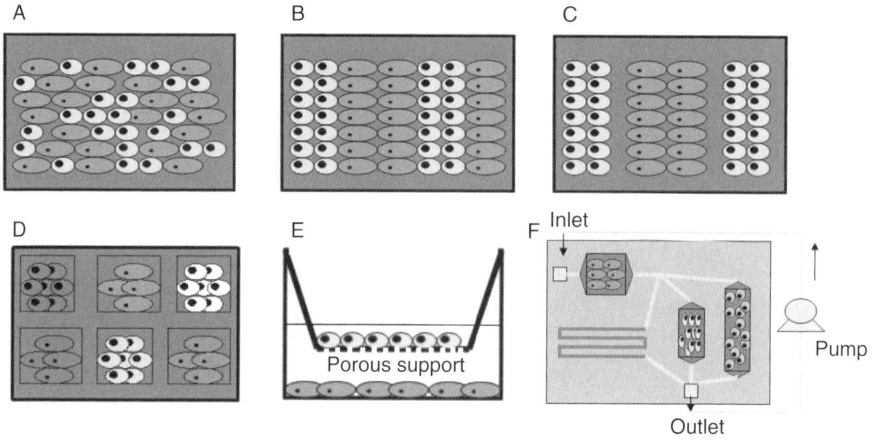

Figure 25.2 Illustration of multiple-type cell co-culture systems: (*a*) Random mixed, (*b*) contact patterning with fuzzy boundaries, (*c*) co-culture without cell contact on a flat surface, (*d*) in chambers, (*e*) the trans-well system, and (*f*) the micro cell culture analog (μCCA) device.

HT29-5M21 cells, seeding ratios influenced the paracellular permeability of the cell monolayer [56]. Adult rat hepatocytes maintained on top of a mesenchymal progenitor cell line C3H/10T1/2 cells [57] or rat 3T3 fibroblast cells [58] maintained better viability and higher ethoxyresorufin *O*-dealkylase (EROD) activity and albumin secretion functions [58].

Randomly distributed cell co-culture is simple in operation, but precise control of cell–cell contact and interactions is difficult. In addition sometimes it can be difficult to differentiate cell type A apart from type B.

25.4.2 Co-Culture with Fuzzy Boundaries

Nano and microfabrication allows organized patterning of two or more types of cells in contact with defined boundaries, which may become less distinct after cell spreading and growth. (Figure 25.2*b*). Bhatia et al. patterned on glass several stripes of liver and fibroblast cells next to each other using photolithography [59]. Primary rat liver cells were patterned based on surface modification with aminosilanes linked to biomolecules, and 3T3-J2 fibroblasts were attached to the remaining unmodified area through nonspecific, serum-mediated attachment. With the same approach, hepatocytes were patterned into round isolated islands while fibroblasts filled the spaces between liver cells [55]. The influence of direct contact on cell functions can be studied by correlating cell function with the distance to the point of direct contact. It was observed that induction of hepatic functions in hepatocytes was increased in the vicinity of fibroblasts and maximal induction of liver-specific functions was correlated with maximal initial heterotypic interaction [55]. It was suggested that heterotypic cell contact is necessary for induction of these functions [60]. This technique also enables control over the strength of cell–cell interaction without changing the numbers of each type of cells. But one limit of this method is that with time motile and mitotic cells will eventually intermix and the relative cell numbers can change [59].

25.4.3 Co-Culture Without Cell Contact

If the subject of study is soluble chemical signals in the culture medium, a no-contact cell co-culture system is preferred to eliminate the potential influence through direct contact. By increasing the distance between patterned cell stripes (Figure 25. 2*c*), Bhatia et al. were able to keep the cells well separated to study how the spacc between co-culture cells affects their interactions [7]. Cells can also be separated by being confined to certain regions by physical barriers such as deep chambers or straight walls. Li et al. developed an integrated discrete multi-organ cell culture system (IdMOC) [61], which is a cell culture plate with relatively large wells subdivided into small subcompartments (Figure 25.2*d*). The IdMOC plates were modified from six-well tissue culture plates with seven inner wells of a diameter of 8 mm, a height of 0.8 mm and a capacity of 100 µl. Cells from multiple organs, including the liver, kidney, lung, central nervous system, blood vessels, and the mammary tissue, were cultured separately in small wells within one big well. For toxicity testing, drug-containing medium is added to flood all inner wells. Thus all cells are exposed to the same medium and communications between them is allowed. But the cells do not physically contact each other.

The transwell (Figure 25.2*e*) or other membrane-based methods also enable isolated cell co-cultures. Membrane filters, with a pore size from 0.1 to 12 µm, are used as cell growth substrates. Generally, one type of cell is seeded onto the membrane, which is immersed into a micro well with a second type of cells cultured on the bottom of the well. Multiwell plate transwell systems are available from manufacturers such as Corning for high-throughput studies.

25.4.4 The Cell Culture Analog: Macro and Micro Designs

Although the cell culture systems described above enable the observation of some cell–cell and "organ–organ" interactions, the gap between in vitro and in vivo testing is still wide. In those systems the fluid to cell volume ratio and the relative size of each "organ" are far from their physiological values. In addition all these cell culture systems are static, but cells in the body experience a more dynamic environment, with blood circulation providing nutrition, carrying away waste, transporting chemical signals generated in one cell to other cells, and exerting necessary mechanical forces for cell growth and differentiation.

The cell culture analog (CCA) system is one approach attempting to narrow this gap. It is developed to address the issues of reactive metabolite formation and tissue–tissue interaction through exchange of metabolites or signal molecules, and to evaluate potential efficacy and toxicity of pharmaceuticals and environmental chemicals. A CCA device can be considered as a simplified and minimized human/animal body, in which mammalian cell cultures are used to represent key functions of specific organs, and cell culture medium is used as a blood surrogate.

The design of the CCA is guided by physiologically based pharmacokinetic (PBPK) modeling, (PBPK is described in Chapter 3 of this book). PBPK is a mathematical model for predicting the overall plasma and tissue kinetic behaviors based on available data on physicochemical properties and specific absorption, distribution, metabolism, and elimination (ADME) processes [62]. A PBPK model describes the body as a set of interconnected compartments, each of which describes either an organ or a tissue. Each compartment represents, from an engineering point of view, a chemical reactor, absorber, or surge tank. Sets of differential equations are derived through the mass balance across various compartments [63]. The CCA is a physical replica of a PBPK model with multiple types of mammalian cells cultured in each compartment instead of a mathematical description of the metabolism or absorption. In its design the fluid fraction feeding each cell culture chamber is the same as the blood fraction received by the corresponding organ in vivo. Additionally the fluid residence time, in each chamber equals its in vivo value. The liquid-to-cell ratio in each cell culture unit is managed to be as close as possible to its physiological value (about 0.5). Shear stress introduced by the flow is calculated and is kept at the physiological level for that type of tissue. Endothelial cells withstand the highest levels (typically < 14 dynes/cm^2), while other cells experience lower shear values (<2 dynes/cm^2) [64].

The CCA was first developed and applied to study the toxicology of naphthalene in rats [65]. It contains three cell reactors representing liver, lung, and the other tissue. A rat hepatoma cell line, H4IIE, and a rat lung cell line L2 were cultured separately in the liver and lung chambers, while the other tissue had no cells and represented the time fluid was retained in other tissues without metabolism or absorption. The first generation CCA was macroscopic in size, sitting on top of a laboratory bench, with cells grown in milk dilution bottles connected with tubing. The medium was pumped into the lung compartment, and then was split into two separate flows into the liver compartment and the other tissue compartment. Fluid exiting from the liver and the other tissue were collected together and then pumped back into the lung for a second circulation loop. Note that the medium was circulated inside a closed loop. This system was challenged with the chemical naphathalene, which is converted into lung-toxic metabolites by liver cells. The macro CCA system was able to demonstrate the potential metabolic toxicity of naphthalene for lung injury, as indicated by L2 cell death [65]. Increasing the number of liver

Figure 25.3 A typical μCCA device filled with red dye for visualization of fluidics pathways. See color plates.

cells and/or inducing cytochrome P450 activity in the liver compartment lead to increased lung cell mortality. These results indicate that the CCA is a potentially useful tool to study the action of compounds with reactive metabolites.

A μCCA is a miniature version of the macroscopic CCA. It achieves more physiologically realistic liquid-to-cell ratios in tissue compartments than the macro CCA. A typical μCCA chip is fabricated using the standard photolithography method [66] in a silicon chip 2.5 × 2.5 cm in size, with small chambers replacing big flasks for cell culture (Figure 25.2*f*). The chambers are coated with poly-D-lysine, collagen, fibronectin, or Matrigel™ for cell attachment. After cell seeding, the μCCA chip is sandwiched between two tailor-made plexiglass pieces (Figure 25.3), and cell culture medium is pumped into the inlet hole drilled into the top plexiglass piece using a peristaltic pump, and it exits the chip from the outlet hole into a medium reservoir. Then the cell culture medium is pumped back into the device through the inlet. The first version of μCCA devices were designed based on the same PBPK model used for the macro CCA, and succeeded in predicting the toxicity of naphthalene on lung cells [8,66]. Later on, a fat chamber was added into the original design to study chemical accumulation in the adipose tissue [8,41]. Differentiated 3T3-L1 cells formed adipocyte-like cells served as a fat surrogate. The μCCA with fat compartment demonstrated that fat protected the lung cells from death and glutathione depletion when challenged with naphthalene or naphthoquinone [41].

The μCCA allows in situ monitoring of the physiological status of cells and chemical bioaccumulation. Cells can be stained with molecular probes to detect cell viability, membrane potential, intracellular calcium, and so forth.

Images are taken with an upright microscope at selected time points, or monitored in real-time with a portable fluorescence cytometric system, which has been specifically designed for µCCA devices [67]. A ruthenium dye-based oxygen sensor was integrated into µCCA for continuous online monitoring of dissolved oxygen content in the circulating medium [66]. In addition a positive displacement micropump has been fabricated for sustaining recirculation of fluid at µl/min through pneumatic pressure [68], holding the potential to replace the bulky peristaltic pumps.

Currently the µCCA devices are being investigated for applications in cancer therapy, selection of drug resistance modulators, environmental toxin detection, and so forth. The CCA technique is being commercialized by the Hurel Corp.

25.5 MANIPULATION OF THE EXTRACELLULAR MICROENVIRONMENT

In vivo, living cells constantly communicate with their surroundings. The interaction between the cells and the extracellular microenvironment regulates cell behavior. With nano and microfabrication techniques, researchers are now able to control cell functions and responses through precise manipulation over the physical and chemical environment around a cell, such as the surface chemical composition and topology of the substrate, the medium composition, and the cellular microenvironment [69].

The importance of substrate surface chemistry has been extensively studied and will not be described in detail here. The cellular environment emphasizes the importance of neighboring cells. Living cells constantly communicate with other cells in their surroundings and this cell–cell interaction partially explains the success of in vitro cell co-cultures in better maintaining the viability and normal functions of primary hepatocytes, which was discussed in Section 25.4. In this section we focus only on the microfluidic control of extracellular environment, and the influence of substrate topography on cell functions.

25.5.1 Microfluidic Control of Extracellular Microenvironment

In miniature cell culture devices, cell culture medium not only maintains cell viability but also enables microfluidic control of the extracellular microenvironment. As is well known, in microfluidic channels with a typical channel size of 50 µm, laminar flow dominates [13]. Fluids flow in parallel to each other without mixing, other than by simple diffusion [70]. Such a flow can be generated and placed across a single cell. Takayama and colleagues fabricated a device with three inlets that allows multiple laminar fluid streams to flow in parallel across a single bovine capillary endothelial cell, to deliver reagents to, and to remove chemicals from cells with subcellular spatial selectivity [3]. This technique, called "partial treatment of cells using laminar flows," made it possible to dislodge limited areas of cell-substrate adhesion enzymatically, to study the subcellular processes of mitochondrial movement and changes in cytoskeletal structure, and to observe cell responses to different toxins within individual cells [3,71].

The characteristic of laminar flow also forms the basis for on-chip generation of stable flow rate or chemical concentration gradients (linear, logarithmical, etc.) [47,72] over a short distance. Effects of flow rate, chemical concentration, and so forth, on cell adhesion, viability, and functions thus can be studied massively in parallel. The spatially and temporally controlled gradient of chemotactic factors provides a robust method to study chemotaxis to investigate migratory cells under a variety of conditions. These gradients are generated using syringe-driven or gravity-driven flow combined with a network of fluidic resistances in channels[47] or on porous membranes [73].

25.5.2 Influence of Substrate Topography on Cell Functions

Topography of a substrate is as important as chemical composition in controlling cell responses and functions [69]. Structural properties of the substrate at molecular, nanometer, and micrometer scales control and influence cell adhesion, spreading, migration, growth, differentiation, and a variety of functions, in a cell-type specific manner [24]. With nano and microfabrication techniques, we are now able to mimic in vivo structures, create new topographic characteristics at the cellular and subcellular level, and guide cells to behave in the way we prefer.

Control of Cell Adhesion and Growth Some surface microstructures may appear to better promote cell adhesion and growth than others. Topographical parameters that could modify cell spreading can be the size, feature, spacing, depth, density, and orientation of the individual surface features, the roughness of the surfaces [74], etc. Astroglial cells attach selectively to 50 micron wide bars fabricated using microcontact printing or photolithography [14]. Smooth transitions in morphology (e.g. minimal microgroove width and depth) appeared to favor osteoblast cell growth [24].

Influence on Cell Orientation and Migration Anisotropic topographic features have been shown to induce many cell types to align and migrate along the direction of the anisotropy [75]. This phenomenon is termed contact guidance. When cultured on a polyimide device with 5 μm deep, 4 μm wide grooves coated with fibronectin and serum albumin, osteoblast-like MC3T3-E1 cells strongly aligned their cell membrane bodies, cell nucleus and focal adhesions with the microgrooves [19]. When seeded on a glass surface micropatterned with grooves, more than 95% of neutrophils moved in the direction of the long axis of ridges/grooves (2 μm width, 3–5 μm in height) regardless of the geometry and chemistry [69]. In addition the rate of cell movement was strongly dependent on the topographical microgeometry of the ridges, with the highest rate achieved at a ridge of 5 μm in height and 10 μm in spacing, about 10 times faster than on a smooth glass surface.

Influence on Cell Functions Schmidt and Von Recum tested seven different microtextured silicone surfaces for their effect on macrophage mitochondrial metabolic activity [74]. Cells on some textures were metabolically more active than cells on the other textures. The ability of macrophages to respond to phorbol 12-myristate 13-acetate (PMA) stimulations was also varied by microstructures, possibly due to a structural alteration of protein kinase C [74]. In another study, myocytes plated on microtextured surfaces display an increase in myofibrillar height [45], which elicits microenvironmental remodeling of proteins that mechanically attach the cell to its surroundings. But vinculin, a focal adhesion protein, was found to decrease in expression [45].

25.6 MINIATURE CELL-BASED BIOSENSORS

Cell-based biosensors incorporate living biological cells as the sensing element to convert immediate environmental perturbations, in to processible signals capable of being processed [76,77]. As the minimum functional and integrating communicable unit of living systems, cells transduce and transmit a variety of signals upon physical and chemical stimulation [78]. These physiological signals, including gene/protein expression, substance production or depletion, electrical signals, change of cell shape, and cell migration, are useful for obtaining information on the chemical or physical stimuli.

Traditional methods for pathogen identification and chemical detection include whole animal tests, high-performance liquid chromatography (HPLC), mass spectrometry (MS), and molecular biology methods based on antibody binding, enzyme activity, or nucleic acid sequence, such as immunoassays and polymerase chain reaction (PCR). Cell-based biosensors have certain advantages over these traditional assays. Assays such as immunoassays or PCR techniques are only for detection of a specific compound, which is generally well known and extensively studied. In contrast, living cells generally respond to a broad range of types of stimulation. Thus cell-based biosensors are versatile assays for detection of unknown agents [79], compounds with unknown toxicological mechanisms, or unpredictable effects such as the synergistic effect of chemical mixtures [77]. In addition biosensors with mammalian cells, especially human cells, have a distinct advantage of responding in a manner that offers insight into the physiological effect of an analyte, on humans [76].

An increasing number of studies of cell-based biosensors are combining cell culture with nano/micro techniques in their designs and fabrications. The advantages of miniaturized sensing devices are that they are portable, materials and reagent saving, suitable for field applications, and capable of being operated in a high-throughput mode. Further, as sensors can be positioned directly at the vicinity of the cell surface, sensitivity is improved because of a more efficient transport of analytes from the cell surface to the device [78].

A useful cell-based biosensor should have at least three components: living cells as a signal generator upon stimulation, a cell culture component supporting cell growth, and an electrical/optical detection component for signal collection. In addition an assay that connects cell responses to a measurable parameter is necessary if the response itself is difficult to measure. For electrically excitable cells, their electrical activities are indicators for cell status. But for cells with no obvious electrical activity, sometimes a reporter is introduced to translate the cell response to electrical or optical signals.

25.6.1 Cell-Based Biosensors with Electrically Excitable Cells

Electrically excitable cells, such as the neurons and cardiomyocytes, are popular candidates for cell-based sensors because of their intrinsic electrophysiological characteristics. Their electrical activities reflect the physiological status of the cells, and report the dynamics of cell death, receptor-ligand interactions, alterations in metabolism, and generic membrane perforation processes [80]. Changes in action potential patterns appear before morphological changes and cell damage occur, which provides sensitivity and reversibility [80]. This sensitivitity could be further enhanced by receptor up-regulation [42]. Strategies of cell response include substance-dependent changes in spontaneous native activity patterns, in network oscillations and in pathological membrane currents [80]. Such biological systems can be used for rapid detection and identification of novel pharmacological substances, toxic agents, certain odorants, and a great variety of chemical substances.

Microelectrode arrays are used to stimulate electrical activity in individual cells and record extracellular signals from neurons and cardiomyocytes. Gilchrist and colleagues built a microelectrode array with electrodes on glass substrates fabricated using the standard photolithography method. The arrays were mounted in 40-pin, ceramic dual-in-line packages. A cell chamber was formed by drilling an 8 mm hole in the center of a polystyrene cell culture dish mounted to a package [44,81]. HL-1 mouse atrial myocytes, which fire spontaneous action potentials, were seeded into the cell chamber coated with fibronectin and gelatin. Nifedipine at concentration between 10 and 100 nM decreased the beat rate up to 23% [44]. Maher et al. fabricated a silicon micromachined device upon which cultured mammalian neurons can be continuously and individually stimulated and monitored [82]. The neurochip is based upon a 4×4 array of metal electrodes, each of which has a caged well structure designed to hold a single mature cell body.

25.6.2 Cell-Based Biosensors with Electrically Non-excitable Cells

Non-excitable cells with no obvious electrical activity play the same important role as excitable cells in cell-based biosensors. The changes in electrical signals, such as cell impedance, and nonelectrical parameters such as cell morphology, proliferation, metabolism, cell viability, pH, and extracellular analyte concentrations, can be measured upon chemical exposure and physical stimuli. Cells can also be genetically engineered to express reporters or biomarkers, such as the green fluorescence protein (GFP), upon specific stimulation.

The morphological and functional status of cells can be assessed by evaluating their electrical impedance. Giaever and Keese [32] described an electrical cell-substrate impedance-sensing (ECIS) device that was capable of monitoring morphological changes of adherent cells quantitatively in real time. This method measures changes in impedance of small gold-film electrodes (with a $10^{-4}\,cm^2$ surface area) deposited on the bottom of a culture dish used as growth substrate. When mammalian cells attach, spread, change shape or move across the electrode, the measured impedance changes with great sensitivity. ECIS was used to study quantitatively the attachment and spreading of epithelial MDCK cells on different protein coatings, and the influence of divalent cations on cell spreading kinetics [83]. DePaola et al. utilized a similar device with an additional flow chamber for quantitative evaluation of flow-induced dynamic changes in cultured cells [84]. Bovine aortic endothelial cells grown to confluence on thin film gold electrodes were exposed to fluid shear stress of 10 dynes/cm^2. At the onset of flow, the monolayer electrical resistance sharply increased in about 15 minutes, then decreased with a prolonged exposure to flow. The observed changes in endothelial impedance were reversible upon flow removal. The dynamic change of impedance with flow indicated the morphological and/or functional changes in the cell layer [84]. The ECIS method is capable of detecting vertical motion of cells on the order of 1 nm, much lower than the resolution of an optical microscope [32]. This noninvasive, highly-sensitive, continuous, and real-time detection method can be used to study the nature and mechanisms of cell morphological variations.

A microphysiometer aims to determine the changes in extracellular analyte concentrations [85]. Such a device for measuring extracellular pH, known as Cytosensor Microphysiometer®, which was described first in 1992 [86], is commercially available from Molecular Devices®. This device is made on top of a silicon chip with cells retained in a disk-shaped region 50 µm high and 6 mm in diameter between two track-etched microporous polycarbonate membranes. Cell culture medium flows tangentially across the surface of the upper membrane. The lower membrane contacts the sensing surface, which is a silicon chip coated with silicon nitride. As energy metabolism in living cells is tightly coupled to cellular ATP usage, any event that perturbs cellular ATP levels will result in a change in acid excretion [87]. The system uses a light addressable potentiometric sensor (LAPS) to measure small changes in extracellular acidification. Recently modified electrodes were incorporated into the Cytosensor Microphysiometer® for simultaneous measurement of changes in extracellular glucose, lactose, and oxygen concentrations [85,88]. Glucose and lactate are measured indirectly at platinum electrodes by amperometric oxidation of hydrogen peroxide, and oxygen is measured at a platinum electrode coated with a Nafion film [85]. This system was tested with mouse fibroblast cells (A9 L HD2 S.C.18) and Chinese hamster overy (CHO) cells challenged with 2,4-dinitrophenol and antimycin A.

Gene transfer, or genetic transformation, is a common technique in molecular biology that introduces foreign genes into cells. By inserting a reporter gene of a measurable biomarker after a promoter whose activation is well controlled by a specific stimulation, expression of this reporter can be used to quantify this simulation. Genes used for this purpose include those coding for enzymes, such as luciferase, β-lactamase, and β-galactosidase, and for GFP. Enzyme-based assays generally require cell lysis and addition of enzyme substrates. Recently developments on cell membrane-permeable substrates for luciferase and β-lactamase have partially eliminated the need for cell lysis [89], and allows direct detection of responses in intact cells. GFP is a powerful tool for cell-based assays owing to its intrinsic fluorescence that allows nearly real-time analysis of molecular events in living cells. GFP assays are less expensive and faster than enzyme-based assays because no substrate or additional washing steps are needed [90]. Different GFP variants have been developed for different colors (blue, cyan, and yellow), brighter fluorescence (enhanced GFP, EGFP) and optimal properties for both high-throughput and high-content screening [90]. As GFP is relatively stable in mammalian cells (EGFP has a half-life of more than 24 hours), short half-life, or destabilized EGFP (dEGFP) is preferred in transcription assays for dynamic studies [91].

A pathogen identification biosensor was constructed using B lymphocytes engineered to express cytosolic aequorin, a calcium-sensitive bioluminescent protein, as well as membrane-bound antibodies specific for pathogens of interest. Cross-linking of the antibodies by even low levels of the appropriate pathogen elevated intracellular calcium concentrations within seconds, causing the aequorin to emit light [92]. In another study, when transfected with a DNA construct with the promoter and regulatory regions of the endothelial-leukocyte adhesion molecule (ELAM) gene (a gene that responds to inflammation signals such as interleukin-1 (IL-1)) adjacent to the coding sequence for β-lactamase, human umbilical vein endothelial ECV304 cells become fluorescent upon subsequent exposure to IL-1 [79].

Lee et al. built a cell-based array [93] which used 20 recombinant bioluminescent bacteria having different promoters fused with the bacterial Lux genes to detect and classify environmental toxicity. About 2 μl of the cell-agar mixture was deposited into the wells of a cell chip made of a 1 mm thick acryl plate with 96 1.2 mm diameter wells. The chip was immersed into a solution containing toxic chemicals for 30 minutes and the bioluminescence from the cell arrays was measured with a CCD camera. These arrays are fast, portable, and economical high-throughput biosensor systems for detecting environmental toxicities [93]. Although this assay uses bacteria, minor modification could make it compatible with mammalian cell cultures.

Caution should be taken when applying these transcription based assays. First, the biomarkers introduced into cells may alter cell response. Second, if fluorescence biomarkers such as GFP are used, the analyte of interest should be examined for autofluoresce to determine the feasibility of using a cellular fluorescence assay for the resolution of small effects [76]. Furthermore the process of transcription and translation can introduce a time lag between exposure and observable responses.

25.6.3 Some General Considerations for Using Cell-Based Biosensors

Cell-based biosensors offer opportunities to detect, screen, and characterize toxicological properties of toxins, environmental chemicals, and pharmaceuticals with high throughput, excellent sensitivity, and low cost. However, the use of cells requires special attention.

First, cells with different various, or if maintained differently, may respond differently. Actually individual variance of cells from the same cell culture dish is often observed. Inconsistent results between laboratories or two sets of individual experiments may happen. One way to alleviate this issue is to use cloned cells all derived from one single cell to improve stability, reproducibility and consistency. Secondly, because many cell-based biosensors use only a few cells, the signal may be weak despite the higher efficiency of signal collection at the cellular level. Noise from the environment may interfere with cellular signals. Gilchrist and colleagues noticed that in a system with stable temperature within ±0.2°C, pH within ±0.05 units, and no significant change in osmolarity, the overall beat rate variation with myocytes was as large as ±4.7%. The magnitude of the noise has a large impact on the detection capability of a cell-based system [44] and should be considered in device design and system evaluation.

Other challenges remain in using living cells as the sensing element. Variables such as cell density and cell interaction can significantly affect the sensor properties [2]. Additionally long-term cell storage and use of cells in the field by personnel with minimal technical background is a practical limitation for widespread adoption of this technology. The hurdle remains that cells themselves are not as stable as other sensing elements, such as antibodies or nucleic acid probes, and have a limited operational lifetime. However, Bloom et al. have demonstrated the concept of air-dry stabilization of mammalian cells with subsequent recovery following rehydration. Human 293H cells transfected with a cyanobacterial gene encoding sucrose phosphate synthase were desiccated and rehydrated with exogenous addition of an extracellular anhydrophile glycan [94]. This technique may potentially be used to provide stabilized mammalian cells for mammalian cell-based biosensors with extended shelf lives.

25.7 CONCLUDING REMARKS: HOW WILL THESE DATA BE USED FOR MODELING EXPOSURE, TOXICOKINETICS, AND COMPUTATIONAL MODELS

Computational modeling is a powerful tool to predict toxicity of drugs and environmental toxins. However, all the in silico models, from the chemical structure-related QSAR method to the systemic PBPK models, would benefit from a second system to improve and validate their predictions. The accuracy of PBPK modeling, for example, depends on precise description of physiological mechanisms and kinetic parameters applied to the model. The PBPK method has primary limitations that it can only predict responses based on assumed mechanisms, without considerations on secondary and unexpected effects. Incomplete understanding of the biological mechanism and inappropriate simplification of the model can easily introduce errors into the PBPK predictions. In addition values of parameters required for the model are often unavailable, especially those for new drugs and environmental toxins. Thus a second validation system is critical to complement computational simulations and to provide a rational basis to improve mathematical models.

Whole animal models and traditional cell culture models have been the primary mode of validation for this purpose. However, animal models are subject to significant expense, lengthy experiments, ethical issues, and it is often difficult to extrapolate results from animals directly to effects on human beings. Results from traditional in vitro cell culture are of limited usefulness due to the difficulty sustaining in vivo functions of primary cells as an isolated, static system [95].

In this chapter recently developed cell culture systems based on nano and microfabrications, including cell-based biosensors, were presented. These systems are shrinking toxicological tests onto small silicon or plastic chips, resulting in reduced reagent consumption, waste production and space needs. Such devices also enable parallel toxicological tests and high-throughput processing. Combining cell culture with microfluidics allows simultaneous tests with multiple-dose treatments and with multiple-types of cells. Substrate surface modifications at the nano/micro scale enable defined cell patterning, 3D cell culture, and simulation of in vivo cellular microenvironment, which helps maintain cell function and produce more authentic experimental results.

Using the CCA system as an example, with the cell culture medium circulated through all the culture chambers as the blood surrogate, a CCA allows the circulation of chemical information among pseudo-organs, to mimic dose dynamics, chemical metabolism, and interchange between compartments. The CCA system aims to reproduce the scenario of animal studies, and enable better understanding of ADME and toxicology of compounds. This system is intended to work in conjugation with a PBPK model to test and refine mechanistic hypotheses. With CCA as a complete replica of PBPK, the predicted responses and the measured CCA responses should match. A disagreement between CCA results and PBPK predictions indicates an incomplete understanding of the underlying molecular mechanisms. Further, validated PBPK models based on molecular mechanisms provide a basis for prediction of human responses to chemical or drug mixtures. A close interaction of the CCA with the PBPK serves as a basis to probe mechanisms for toxicity and for use in cross-species extrapolation.

We expect that these novel cell culture systems, coupled with predictions from in silico and mathematical modeling, will provide new insights into the toxicity of environmental and pharmaceutical chemicals, expand our understanding of toxicological mechanisms, and improve prediction of human clinical responses.

ACKNOWLEDGMENTS

This work was supported in part by the Cornell NanoBiotechnology Center (NBTC) and New York State Office of Science, Technology and Academic Research (NYSTAR). NBTC is supported by National Science Foundation (NSF), NYSTAR, and cooperate sponsors.

REFERENCES

1. Park TH, Shuler ML. Integration of cell culture and microfabrication technology. *Biotechnol Prog* 2003;19:243–53.
2. El-Ali J, Sorger PK, Jensen KF. Cells on chips. *Nature* 2006;442:403–11.
3. Takayama S, Ostuni E, LeDuc P, Naruse K, Ingber DE, Whitesides GM. Subcellular positioning of small molecules. *Nature* 2001;411:1016.
4. Li N, Tourovskaia A, Folch A. Biology on a chip: Microfabrication for studying the behavior of cultured cells. *Crit Rev Biomed Eng* 2003;31:423–88.
5. Madou MJ. *Fundamentals of microfabrication*. Boca Raton, FL.: CRC Press, 1997.
6. Britland S, Perez-Arnaud E, Clark P, McGinn B, Connolly P, Moores G. Micropatterning proteins and synthetic peptides on solid supports: A novel application for microelectronics fabrication technology. *Biotechnol Prog* 1992;8:155–60.
7. Bhatia SN, Balis UJ, Yarmush ML, Toner M. Effect of cell–cell interactions in preservation of cellular phenotype: Cocultivation of hepatocytes and nonparenchymal cells. *FASEB J* 1999;13:1883–900.
8. Viravaidya K, Sin A, Shuler ML. Development of a microscale cell culture analog to probe naphthalene toxicity. *Biotechnol Prog* 2004;20:316–23.
9. Tseng S, Tsai C. Fabrication of individual aligned carbon nanotube for scanning probe microscope. *J Phys Conf Ser* 2005;10:186–9.

10. Vieu C, Carcenac F, Pepin A. Electron beam lithography: Resolution limits and applications *Appl Surf Sci* 2000;164:111–7(7).
11. Norman JJ, Desai TA. Methods for fabrication of nanoscale topography for tissue engineering scaffolds. *An Biomed Eng* 2006;34:89–101.
12. Molhave K, Madsen DN, Boggild P. A simple electron-beam lithography system. *Ultramicroscopy* 2005;102:215–9.
13. Whitesides GM, Ostuni E, Takayama S, Jiang X, Ingber DE. Soft lithography in biology and biochemistry. *An Rev Biomed Eng* 2001;3:335–73.
14. Kane RS, Takayama S, Ostuni E, Ingber DE, Whitesides GM. Patterning proteins and cells using soft lithography. *Biomaterials* 1999;20:2363–76.
15. McDonald JC, Duffy DC, Anderson JR, Chiu DT, Wu H, Schueller OJ, et al. Fabrication of microfluidic systems in poly(dimethylsiloxane). *Electrophoresis* 2000;21:27–40.
16. Chen CS, Mrksich M, Huang S, Whitesides GM, Ingber DE. Micropatterned surfaces for control of cell shape, position, and function. *Biotechnol Prog* 1998;14:356–63.
17. De Silva MN, Desai R, Odde DJ. Micro-patterning of animal cells on PDMS substrates in the presence of serum without use of adhesion inhibitors. *Biomed Microdevices* 2004;6:219–22.
18. Truskett VN, Watts MP. Trends in imprint lithography for biological applications. *Trends Biotechnol* 2006;24:312–7.
19. Charest JL, Bryant LE, Garcia AJ, King WP. Hot embossing for micropatterned cell substrates. *Biomaterials* 2004;25:4767–75.
20. Becker H, Gartner C. Polymer based micro-reactors. *J Biotechnol* 2001;82:89–99.
21. Hao L, Lawrence J, Chian KS. Osteoblast cell adhesion on a laser modified zirconia based bioceramic. *J Mater Sci Mater Med* 2005;16:719–26.
22. Doraiswamy A, Jin C, Narayan RJ, Mageswaran P, Mente P, Modi R, et al. Two photon induced polymerization of organic-inorganic hybrid biomaterials for microstructured medical devices. *Acta Biomater* 2006;2:267–75.
23. Kearsley A. Laser micromachining. *Med Device Technol* 2003;14:18–9.
24. Duncan AC, Weisbuch F, Rouais F, Lazare S, Baquey C. Laser microfabricated model surfaces for controlled cell growth. *Biosens Bioelectron* 2002;17:413–26.
25. Zipfel WR, Williams RM, Webb WW. Nonlinear magic: Multiphoton microscopy in the biosciences. *Nat Biotechnol* 2003;21:1369–77.
26. Giridhar MS, Seong K, Schulzgen A, Khulbe P, Peyghambarian N, Mansuripur M. Femtosecond pulsed laser micromachining of glass substrates with application to microfluidic devices. *Appl Opt* 2004;43:4584–9.
27. Kaehr B, Allen R, Javier DJ, Currie J, Shear JB. Guiding neuronal development with in situ microfabrication. *Proc Natl Acad Sci USA* 2004;101:16104–8.
28. Costantino S, Heinze KG, Martinez OE, De Koninck P, Wiseman PW. Two-photon fluorescent microlithography for live-cell imaging. *Microsc Res Tech* 2005;68:272–6.
29. Klank H, Kutter JP, Geschke O. CO(2)-laser micromachining and back-end processing for rapid production of PMMA-based microfluidic systems. *Lab Chip* 2002;2:242–6.

30. Murali M, Yeo SH. Rapid biocompatible micro device fabrication by micro electro-discharge machining. *Biomed Microdevices* 2004;6:41–5.
31. Petronis S, Eckert KL, Gold J, Wintermantel E. Microstructuring ceramic scaffolds for hepatocyte cell culture. *J Mater Sci Mater Med* 2001;12:523–8.
32. Giaever I, Keese CR. A morphological biosensor for mammalian cells. *Nature* 1993;366:591–2.
33. Cosseddu P, Bonfiglio A. Soft lithography fabrication of all-organic bottom-contact and top-contact field effect transistors. *Appl Phys Lett* 2006;88:23506.1–23506.3.
34. Woerly S, Plant GW, Harvey AR. Cultured rat neuronal and glial cells entrapped within hydrogel polymer matrices: A potential tool for neural tissue replacement. *Neurosci Lett* 1996;205:197–201.
35. Dillon GP, Yu X, Sridharan A, Ranieri JP, Bellamkonda RV. The influence of physical structure and charge on neurite extension in a 3D hydrogel scaffold. *J Biomater Sci Polym Ed* 1998;9:1049–69.
36. Lam MT, Sim S, Zhu X, Takayama S. The effect of continuous wavy micropatterns on silicone substrates on the alignment of skeletal muscle myoblasts and myotubes. *Biomaterials* 2006;27:4340–7.
37. Bhat HK, Vadgama JV. Role of estrogen receptor in the regulation of estrogen induced amino acid transport of system A in breast cancer and other receptor positive tumor cells. *Int J Mol Med* 2002;9:271–9.
38. Ogura M, Morishima Y, Ohno R, Kato Y, Hirabayashi N, Nagura H, et al. Establishment of a novel human megakaryoblastic leukemia cell line, MEG-01, with positive Philadelphia chromosome. *Blood* 1985;66:1384–92.
39. Artursson P, Palm K, Luthman K. Caco-2 monolayers in experimental and theoretical predictions of drug transport. *Adv Drug Deliv Rev* 2001;46:27–43.
40. Viravaidya K, Shuler ML. Prediction of naphthalene bioaccumulation using an adipocyte cell line model. *Biotechnol Prog* 2002;18:174–81.
41. Viravaidya K, Shuler ML. Incorporation of 3T3-L1 cells to mimic bioaccumulation in a microscale cell culture analog device for toxicity studies. *Biotechnol Prog* 2004;20:590–7.
42. Gross GW, Rhoades BK, Azzazy HM, Wu MC. The use of neuronal networks on multielectrode arrays as biosensors. *Biosens Bioelectron* 1995;10:553–67.
43. Ma SH, Lepak LA, Hussain RJ, Shain W, Shuler ML. An endothelial and astrocyte co-culture model of the blood–brain barrier utilizing an ultra-thin, nanofabricated silicon nitride membrane. *Lab Chip* 2005;5:74–85.
44. Gilchrist KH, Giovangrandi L, Whittington RH, Kovacs GT. Sensitivity of cell-based biosensors to environmental variables. *Biosens Bioelectron* 2005;20:1397–406.
45. Motlagh D, Senyo SE, Desai TA, Russell B. Microtextured substrata alter gene expression, protein localization and the shape of cardiac myocytes. *Biomaterials* 2003;24:2463–76.
46. Tourovskaia A, Figueroa-Masot X, Folch A. Differentiation-on-a-chip: A microfluidic platform for long-term cell culture studies. *Lab Chip* 2005;5:14–9.
47. Kim L, Vahey MD, Lee HY, Voldman J. Microfluidic arrays for logarithmically perfused embryonic stem cell culture. *Lab Chip* 2006;6:394–406.

48. Poumay Y, Dupont F, Marcoux S, Leclercq-Smekens M, Herin M, Coquette A. A simple reconstructed human epidermis: Preparation of the culture model and utilization in in vitro studies. *Arch Dermatol Res* 2004;296:203–11.

49. Downing BR, Cornwell K, Toner M, Pins GD. The influence of microtextured basal lamina analog topography on keratinocyte function and epidermal organization. *J Biomed Mater Res* A 2005;72:47–56.

50. Florea BI, Cassara ML, Junginger HE, Borchard G. Drug transport and metabolism characteristics of the human airway epithelial cell line Calu-3. *J Control Release* 2003;87:131–8.

51. Walter E, Janich S, Roessler BJ, Hilfinger JM, Amidon GL. HT29-MTX/Caco-2 cocultures as an in vitro model for the intestinal epithelium: In vitro–in vivo correlation with permeability data from rats and humans. *J Pharm Sci* 1996;85:1070–6.

52. Fukuda J, Sakai Y, Nakazawa K. Novel hepatocyte culture system developed using microfabrication and collagen/polyethylene glycol microcontact printing. *Biomaterials* 2006;27:1061–70.

53. Otsuka H, Hirano A, Nagasaki Y, Okano T, Horiike Y, Kataoka K. Two-dimensional multiarray formation of hepatocyte spheroids on a microfabricated PEG-brush surface. *Chembiochem* 2004;5:850–5.

54. Torisawa YS, Shiku H, Yasukawa T, Nishizawa M, Matsue T. Multi-channel 3-D cell culture device integrated on a silicon chip for anticancer drug sensitivity test. *Biomaterials* 2005;26:2165–72.

55. Bhatia SN, Balis UJ, Yarmush ML, Toner M. Microfabrication of hepatocyte/fibroblast co-cultures: Role of homotypic cell interactions. *Biotechnol Prog* 1998;14:378–87.

56. Nollevaux G, Deville C, El Moualij B, Zorzi W, Deloyer P, Schneider YJ, et al. Development of a serum-free co-culture of human intestinal epithelium cell-lines (Caco-2/HT29-5M21). *BMC Cell Biol* 2006;7:20.

57. Langenbach R, Malick L, Tompa A, Kuszynski C, Freed H, Huberman E. Maintenance of adult rat hepatocytes on C3H/10T1/2 cells. *Cancer Res* 1979;39:3509–14.

58. Bhandari RN, Riccalton LA, Lewis AL, Fry JR, Hammond AH, Tendler SJ, et al. Liver tissue engineering: A role for co-culture systems in modifying hepatocyte function and viability. *Tissue Eng* 2001;7:345–57.

59. Bhatia SN, Yarmush ML, Toner M. Controlling cell interactions by micropatterning in co-cultures: Hepatocytes and 3T3 fibroblasts. *J Biomed Mater Res* 1997;34:189–99.

60. Bhatia SN, Balis UJ, Yarmush ML, Toner M. Probing heterotypic cell interactions: Hepatocyte function in microfabricated co-cultures. *J Biomater Sci Polym Ed* 1998;9:1137–60.

61. Li AP, Bode C, Sakai Y. A novel in vitro system, the integrated discrete multiple organ cell culture (IdMOC) system, for the evaluation of human drug toxicity: Comparative cytotoxicity of tamoxifen towards normal human cells from five major organs and MCF-7 adenocarcinoma breast cancer cells. *Chem Biol Interact* 2004;150:129–36.

62. Poulin P, Theil FP. Prediction of pharmacokinetics prior to in vivo studies: II. Generic physiologically based pharmacokinetic models of drug disposition. *J Pharm Sci* 2002;91:1358–70.

63. Gerlowski LE, Jain RK. Physiologically based pharmacokinetic modeling: Principles and applications. *J Pharm Sci* 1983;72:1103–27.

64. Powers MJ, Domansky K, Kaazempur-Mofrad MR, Kalezi A, Capitano A, Upadhyaya A, et al. A microfabricated array bioreactor for perfused 3D liver culture. *Biotechnol Bioeng* 2002;78:257–69.

65. Sweeney L, Shuler ML, Babish JG, Ghanem A. A cell culture analogue of rodent physiology: Application to naphthalene toxicology. *Toxicol in Vitro* 1995;9: 307–16.

66. Sin A, Chin KC, Jamil MF, Kostov Y, Rao G, Shuler ML. The design and fabrication of three-chamber microscale cell culture analog devices with integrated dissolved oxygen sensors. *Biotechnol Prog* 2004;20:338–45.

67. Tatosian DA, Shuler ML, Kim D. Portable in situ fluorescence cytometry of microscale cell-based assays. *Opt Lett* 2005;30:1689–91.

68. Sin A, Reardon CF, Shuler ML. A self-priming microfluidic diaphragm pump capable of recirculation fabricated by combining soft lithography and traditional machining. *Biotechnol Bioeng* 2004;85:359–63.

69. Tan J, Saltzman WM. Topographical control of human neutrophil motility on micropatterned materials with various surface chemistry. *Biomaterials* 2002;23: 3215–25.

70. Kenis PJ, Ismagilov RF, Whitesides GM. Microfabrication inside capillaries using multiphase laminar flow patterning. *Science* 1999;285:83–5.

71. Takayama S, Ostuni E, LeDuc P, Naruse K, Ingber DE, Whitesides GM. Selective chemical treatment of cellular microdomains using multiple laminar streams. *Chem Biol* 2003;10:123–30.

72. Li N, Baskaran H, Dertinger SK, Whitesides GM, Van de Water L, Toner M. Neutrophil chemotaxis in linear and complex gradients of interleukin-8 formed in a microfabricated device. *Nat Biotechnol* 2002;20:826–30.

73. Diao J, Young L, Kim S, Fogarty EA, Heilman SM, Zhou P, et al. A three-channel microfluidic device for generating static linear gradients and its application to the quantitative analysis of bacterial chemotaxis. *Lab Chip* 2006;6:381–8.

74. Schmidt JA, von Recum AF. Macrophage response to microtextured silicone. *Biomaterials* 1992;13:1059–69.

75. Teixeira AI, Abrams GA, Bertics PJ, Murphy CJ, Nealey PF. Epithelial contact guidance on well-defined micro- and nanostructured substrates. *J Cell Sci* 2003;116:1881–92.

76. Pancrazio JJ, Whelan JP, Borkholder DA, Ma W, Stenger DA. Development and application of cell-based biosensors. *An Biomed Eng* 1999;27:697–711.

77. Stenger DA, Gross GW, Keefer EW, Shaffer KM, Andreadis JD, Ma W, et al. Detection of physiologically active compounds using cell-based biosensors. *Trends Biotechnol* 2001;19:304–9.

78. Haruyama T. Micro- and nanobiotechnology for biosensing cellular responses. *Adv Drug Deliv Rev* 2003;55:393–401.

79. Durick K, Negulescu P. Cellular biosensors for drug discovery. *Biosens Bioelectron* 2001;16:587–92.
80. Gross GW, Harsch A, Rhoades BK, Gopel W. Odor, drug and toxin analysis with neuronal networks in vitro: Extracellular array recording of network responses. *Biosens Bioelectron* 1997;12:373–93.
81. Gilchrist KH, Barker VN, Fletcher LE, DeBusschere BD, Ghanouni P, Giovangrandi L, et al. General purpose, field-portable cell-based biosensor platform. *Biosens Bioelectron* 2001;16:557–64.
82. Maher MP, Pine J, Wright J, Tai YC. The neurochip: A new multielectrode device for stimulating and recording from cultured neurons. *J Neurosci Meth* 1999;87: 45–56.
83. Wegener J, Keese CR, Giaever I. Electric cell-substrate impedance sensing (ECIS) as a noninvasive means to monitor the kinetics of cell spreading to artificial surfaces. *Exp Cell Res* 2000;259:158–66.
84. DePaola N, Phelps JE, Florez L, Keese CR, Minnear FL, Giaever I, et al. Electrical impedance of cultured endothelium under fluid flow. *An Biomed Eng* 2001; 29:648–56.
85. Eklund SE, Taylor D, Kozlov E, Prokop A, Cliffel DE. A microphysiometer for simultaneous measurement of changes in extracellular glucose, lactate, oxygen, and acidification rate. *Anal Chem* 2004;76:519–27.
86. McConnell HM, Owicki JC, Parce JW, Miller DL, Baxter GT, Wada HG, et al. The cytosensor microphysiometer: Biological applications of silicon technology. *Science* 1992;257:1906–12.
87. Hafner F. Cytosensor Microphysiometer: Technology and recent applications. *Biosens Bioelectron* 2000;15:149–58.
88. Eklund SE, Kozlov E, Taylor DE, Baudenbacher F, Cliffel DE. Real-time cell dynamics with a multianalyte physiometer. *Meth Mol Biol* 2005;303:209–23.
89. Zlokarnik G, Negulescu PA, Knapp TE, Mere L, Burres N, Feng L, et al. Quantitation of transcription and clonal selection of single living cells with beta-lactamase as reporter. *Science* 1998;279:84–8.
90. Kain SR. Green fluorescent protein (GFP): Applications in cell-based assays for drug discovery. *Drug Discov Today* 1999;4:304–12.
91. Kain SR, Ganguly S. Uses of fusion genes in mammalian transfection: Overview of genetic reporter systems. *Curr Protocols Mol Biol* 1995:9.6.1–9.6.12.
92. Rider TH, Petrovick MS, Nargi FE, Harper JD, Schwoebel ED, Mathews RH, et al. A B cell-based sensor for rapid identification of pathogens. *Science* 2003;301:213–5.
93. Lee JH, Mitchell RJ, Kim BC, Cullen DC, Gu MB. A cell array biosensor for environmental toxicity analysis. *Biosens Bioelectron* 2005;21:500–7.
94. Bloom FR, Price P, Lao G, Xia JL, Crowe JH, Battista JR, et al. Engineering mammalian cells for solid-state sensor applications. *Biosens Bioelectron* 2001; 16:603–8.
95. Shuler ML, Ghanem A, Quick D, Wong MC, Miller P. A self-regulating cell culture analog device to mimic animal and human toxicological responses. *Biotechnol Bioeng* 1996;52:45–60.

26

FUTURE OF COMPUTATIONAL TOXICOLOGY: BROAD APPLICATION INTO HUMAN DISEASE AND THERAPEUTICS

Dale E. Johnson, Amie D. Rodgers, and Sucha Sudarsanam

Contents

Computational Toxicology: Risk Assessment for Pharmaceutical and Environmental Chemicals, Edited by Sean Ekins

26.1 INTRODUCTION

Computational toxicology applications have made a major contribution in increasing the reliability of risk assessments for environmental chemicals, food additives, and pharmaceuticals at the regulatory agency level. These efforts have become an essential part of chemical risk assessment in the environment particularly when regulators are faced with assessing possible hazards from tens of thousands of chemicals where only a few will have been evaluated with extensive toxicology testing. However, predictive applications have made less of an impact in drug discovery and development within the pharmaceutical industry. The two most important pharmaceutical applications envisioned over the last decade have been (1) impacting the attrition rate where computational models would be used as filters during early drug design, highlighting structural features that may lead to toxicity liabilities in individual compounds as well as chemical libraries, and (2) predicting toxicities induced by individual compounds in in vitro and in vivo toxicology models prior to extensive preclinical testing or as part of the lead identification process leading to the selection of development candidates. The more elusive goal of predicting a potential adverse event in humans has not been fully realized [1].

Commercially available computational toxicology tools have been applied in pharmaceutical research and development primarily for predicting mutagenicity and carcinogenicity outcomes [2–6]. Computational models for predicting the potential of a compound (by chemical structure) to induce adverse events in humans are highlighted by drug-induced QT prolongation where predictive structure-activity relationship models have been constructed from hERG screening [7] and a hERG pharmacophore model constructed based on IC_{50} data [8] (see also Chapter 13). This has been possible because predictive animal models and screening programs were established that led to the construction of more extensive databases. Other potential toxicities hold the same promise when predictive screening assays can be developed to generate relevant data sets. However, to date this approach has had limited utility with predicting other toxicities in humans induced by pharmaceutical compounds.

There are signs that computational toxicology will change significantly in the next five years because advances seen in the medical sciences will drive an increasing demand for innovative solutions for drug safety [1]. There is a general consensus that predictive simulations (computational models) of both preclinical (cellular and animal models) and human clinical trial outcomes will have enormous value in reducing time, cost, and failures in the process of discovering and developing new drugs [9]. Moreover the goal of having computational data management and simulation tools enable work at the hit-to-lead discovery stage with information directly applicable to specific patient susceptibilities in targeted therapeutic areas would, in essence, transform the way drug discovery and development is currently conducted [10–13]. In this chapter we describe reasons why computational predictive models for human adverse drug reactions have been difficult to develop and how new approaches for biomarker discovery and quantization may provide to be a rational approach in solving this long-standing dilemma.

26.2 THE FUTURE: LINKED TO MOLECULAR TOXICOLOGY AND SYSTEMS BIOLOGY

The future of computational toxicology is consistent with, and in fact mirrors, predictions for the future of molecular toxicology, in general. Both seek advances in the understanding of mechanisms at the interface of chemistry and biology to answer important biological questions pertaining to risks chemicals pose to the environment and public health. Molecular toxicology will provide the endpoints for computational modeling while computational toxicology will provide hypotheses for molecular toxicology inquiry. The field of computational toxicology, once only defined as a process or tool used to integrate chemical and biological data to predict an endpoint or outcome, is following close behind advances in other technologies, informatics, and information access to head in a direction more relevant to human disease and adverse outcomes [1]. The link to advances in molecular toxicology is essential because our concepts of toxicity endpoints are expanding rapidly to molecular insights into disease and health. What once was considered the final endpoint of toxicity evaluation (e.g., histopathologic findings) is now considered the phenotypic anchoring of complex genomic and proteomic information [14] giving added insight into mechanisms of toxicity.

In an editorial in 1999, Stevens and Marnett [15] described the field of molecular toxicology as one that seeks to understand mechanisms of toxicity in precise molecular terms; expanding current knowledge about both chemical biological mediators and downstream effectors that lead to the complex biology associated with disease. They stated that understanding mechanisms of toxicity would have translational potential in both applied and basic science, but noted that the inability to predict adverse drug reactions still remained a major impediment to development of new pharmaceuticals. The hope was that adverse responses could be identified early in the drug discovery and development process, allowing more accurate risk assessment and increasing the efficiency of compound selection. It was thought that understanding patterns of gene expression and biological responses to toxicity would hold unlimited potential for rapid "biological profiling" of new chemical entities and that defining mechanisms of toxicity would provide a unique window on the most basic of the life processes.

In 2006 Marnett wrote about the exciting times in the present field of toxicology, stating that "advances in analytical chemistry enable investigators to identify and quantify toxic agents and their metabolites at ever lower concentrations, sometimes into the zeptomole range." He added that "structural biology provides high-resolution images of ligand-protein, protein-protein, and DNA-protein complexes that provide visions of the molecular basis of biological function." He offered that "transcriptional and proteomic profiling enable investigators to integrate global cellular responses to chemical and biological stress" [16]. What was considered to be a potential advance in 1999 was considered a reality in 2006.

From a practical aspect, the rapid advance of new technologies over this relatively short time period has created a new challenge for toxicologists: making sense of enormous data sets from microarray studies. For computational toxicologists, attempting to merge these new molecular endpoints into existing datasets and analytical procedures is an enormous challenge. This has led to the development of new tools and approaches in order to organize and interpret these complex data. On the computational side, these informatics advances have pushed the window of revealing "the most basic of life processes" closer to the present timeframe than envisioned in 1999. A current emerging strategy, called functional genomics, a term distinguished from but connected to phenotypic anchoring of genomic information, is designed to expand the scope of biological investigation from studying single genes or proteins to studying all genes or proteins at once in a systematic fashion [17]. The term functional genomics describes a process whose goal is to discover the biological function of specific genes and to uncover how sets of genes and their products work together in health and disease [18,19]. Yu et al. [20] have recently described a systematic approach to quantitate the degree by which functional gene systems change by dose or on a temporal basis, linking the original gene expression data with functional gene category results.

The future goals of computational toxicology will involve predictive algorithms; the difference between 1999 and today is in the complexity of biological information—*the toxicity endpoint*—and our more advanced information on chemical entities; how they are distributed, metabolized, concentrated, in biological compartments; and polymorphisms that define genetic differences in individuals that affect these metabolism, distribution, and toxicity characteristics. Technological advances will also lead to a greater understanding of how findings from cellular and animal models can be correlated to human studies to define relevant biomarkers for the toxicity-inducing mechanisms. Quantitative biomarkers will be the future endpoints that allow predictive computational algorithms to be constructed and new screening systems to be created [1].

Key challenges that still remain in the computational assessment of risk for therapeutics are (1) the formidable task of predicting adverse outcomes of drugs in specific human populations, (2) understanding risk factors that make some individuals more susceptible to toxicity and incorporating these risk factors into computational models [1], and (3) the process of generating or extrapolating new biological data at therapeutic or physiologically relevant exposure levels [21]. The basis for risk and safety assessment of chemical compounds traditionally drew on mechanisms of toxicity of single compounds on living organisms at high levels of exposure, and such data currently populate the majority of databases used in computational risk assessment. In some instances, high doses can "swamp out" subtle changes in biological pathways, masking relevant fluxes that occur in response to external stimuli.

26.3 ASSESSING THE NEEDS OF INDUSTRY AND THE HEALTH CARE SYSTEM

26.3.1 Solving the Unexpected Human Adverse Drug Reaction Dilemma

Integrating toxicity data from animals, cellular systems, and human subjects has proved to be a difficult and formidable task, and it remains a stumbling block in creating relevant computational models for human adverse drug reactions (ADRs). This is particularly true when model systems have not been (or cannot be) developed to generate screening data for database construction. A reasonable goal in the future would be to use human data exclusively to create predictive ADR models. Most toxicological data developed in animals or cellular systems involves high-dose endpoint evaluation; whereas adverse drug reactions in humans occur at therapeutic doses, generally in low incidences. Factors that influence both the incidence and severity of adverse drug reactions include the variability of the human response, particularly when the disease context of the treated individual is considered. Most common diseases entail extremely complex patterns of pathogenesis that involve the regulation of hundreds of genes and their protein products [22]. These underlying diseases and stressors become the intrinsic risk factors that interact on a patient-to-patient basis with complex environmental factors and individual genetic variability. For example, it is becoming increasingly apparent that activation of the innate immune system, which occurs as a result of bacterial or viral infections and/or necrotic cell death, may predispose individuals to a more severe form of drug-induced liver injury from therapeutic levels of a variety of drugs. It is also well known that preexisting liver disease and coexisting illnesses may have a greater effect on the ability of the patient to recover from an initial liver injury thereby increasing the severity of the adverse outcome [23–26].

26.3.2 Risk Factors That Affect the Incidence of Drug-Induced Hepatotoxicity

We use drug-induced hepatotoxicity as an example to highlight the complexities of correlating preclinical and clinical data and introducing human risk factors into computational models. Drug-induced hepatotoxicity is a well-studied drug-induced toxicity (if not the most studied), yet it continues to be the primary cause of drug development failures and market withdrawal of approved drugs, and is a major contributor to the high cost burden that adverse drug reactions impose on the health care system. In addition drug-related hepatotoxicity is the leading cause of acute liver failure in the United States among patients referred for liver transplantation. The main cause of these cases is toxicity from intentional or unintentional overdoses of acetaminophen. However, it is still unknown whether the actual incidence of drug-induced hepatotoxicity is underestimated; the diagnosis requires ruling out other causes of liver injury including hepatitis A, B, or C infection, alcoholic or autoimmune hepatitis, biliary tract disorders, and hemodynamic problems. In several cases symptoms such as fatigue, anorexia, nausea, discomfort, in the right upper quadrant, and dark urine may precede distinct biochemical abnormalities. Drugs associated with human liver injury have been classified according to "signature" biochemical and clinical symptomology. For instance, hepatocellular or cytolytic injury as seen with isoniazid or troglitazone involves marked elevations of aminotransferase levels that precede increases in total bilirubin and modest elevations in alkaline phosphatase levels. Cholestatic injury as seen with amoxicillin-clavulanic acid or chlorpromazine is characterized by increased levels of alkaline phosphatase that precede or are relatively more prominent than increases in aminotransferases. Hypersensitivity or immunologic injury as with phenytoin, nitrofurantoin, or halothane is often delayed or more severe after repeated exposure and can be associated with fever, rash, or eosinophilia. Mitochondrial injury as with valproic acid or high-dose parenteral tetracycline involves microvesicular steatosis on biopsy, lactic acidosis, and modest elevations of aminotransferase levels [27].

The Drug-Induced Liver Injury Network (DILIN) (*http//dilin.dcri.duke.edu/*) has been established to conduct research into the causes of drug-induced liver disease (DILI). DILIN is a nonprofit endeavor sponsored by the National Institute of Diabetes and Digestive and Kidney Diseases (NIDDK) of the US National Institutes of Health. DILIN hopes to be able to discover why some people have these unwanted liver reactions and others do not.

Two studies are being conducted by DILIN. In the retrospective study *http://dilin.dcri.duke.edu*, DILIN will establish a registry of people who experienced a liver injury due to taking certain prescribed medications since 1994. The four drugs are isoniazid (INH), phenytoin (Dilantin), clavulanic acid/amoxicillin (Augmentin), and valproic acid (Depakote). They were chosen because they have a characteristic, or "signature," clinical presentation, making them good candidates for scientific research. In the prospective study *http://dilin.dcri.duke.edu*, patients who recently experienced an adverse liver reaction after taking any drug or herbal medicine will be studied. DILIN will follow these patients over time to find out what happens to them as a result of their injury. Information and samples (blood, biopsy) will be available for future study. It is anticipated that these studies and potentially the samples may provide insight into both genetic and nongenetic causes of hepatotoxicity in certain patients.

26.3.3 Drug-Induced Hepatotoxicity and Underlying Inflammatory Conditions

The current level of knowledge of mechanisms responsible for hepatotoxicity in humans and on the role of contributing risk factors such as underlying inflammatory conditions is still unclear despite the large amount of research devoted to these topics over the last several years. Studies in normal animal models are useful for detecting dose-related, predictable hepatotoxicity but generally not useful in detecting unpredictable hepatotoxicity in humans. In certain cases, combining these two types of data into hepatotoxicity databases without establishing stringent meta analysis criteria leads to an over prediction type of error. Information on inflammatory disease and its association with hepatotoxicity is predominately based on histopathology samples from animal studies or biopsies from humans. Biopsies from patients with microsteatosis and cholestasis show minimal inflammation; whereas hepatocellular injury either via necrosis or an immunological reaction show marked inflammation [27]. In the future the more rational way of establishing connections will be through detailed pathway analyses.

Drug-induced hepatotoxicity and underlying inflammatory conditions are certainly linked, but the extent to which this represents a risk factor for idiosyncratic hepatotoxicity is unknown. Consider the response to microorganism infections as an example. The innate immune system utilizes germline-encoded pattern recognition receptors (PRRs), such as toll-like receptors (TLRs), to recognize pathogen-associated molecular patterns (PAMPs) that are common in microbial pathogens but rare in the host. Examples of PAMPs include CpG motifs found in the DNA of some intracellular pathogens or bacterial cell-wall components. The first mammalian TLR, TLR4, was identified in 1997. TLRs are generally expressed where pathogens are first encountered, bind PAMPs during an infection, and signal for activation and maturation of adaptive immune cells. For example, both hepatitis C (HCV) [24] and lipopolysachharide (LPS or endotoxin) [25] use TLR4 to initiate the inflammatory cascade.

Dysfunction in TLRs, and the resulting inadequate or inappropriate immune response, has been implicated in infectious disease, cancer, and in disorders that can follow bacterial infections, such as sepsis, peridontitis, cardiac ischaemia, and cerebral palsy. Inappropriate TLR stimulation can lead to inflammation or autoimmunity [28]. LPS is the most widely studied TLR agonist, and several data sets are available, including those where LPS treatment preceded chemically induced hepatotoxicity in rats showing that LPS pretreatment does exacerbate hepatotoxicity with certain compounds [29–31]. Low-dose LPS exposure stimulates mononuclear phagocyte function among other host responses as a defense against invading microorganisms; moderate LPS exposure can evoke tissue injury through neutrophil activation and intravascular coagulation [32], while high-dose LPS exposure initiates a cascade of inflammatory events that results in cell death, tissue injury, and organ failure [23].

Polymorphonuclear leukocytes (neutrophils) accumulate in liver vasculature as a response to both infection and tissue trauma. Their primary function is to eliminate invading organisms and/or to remove dead and dying cells [26]. In the liver excessive inflammatory response also initiates additional tissue damage. This has been shown by intravenous infusions of certain interleukins such as IL-2 where monocytic infiltration occurs in several tissues. In the liver, when doses exceed therapeutic doses, the inflammatory response is marked and tissue damage occurs [33]. In drug-induced hepatotoxicity, it is well known that necrotic cell death causes an inflammatory response with neutrophil infiltration in the liver via activation of the complement cascade triggering Kupffer cell activation [34]. An inflammatory mediator specifically released by necrotic cells is high mobility group box 1 (HMGB1), a chromatin-bound nuclear factor. HMGB1 can bind to TLR4 on Kupffer cells and induce the generation of pro-inflammatory cytokines [35]. Underlying "pro-inflammatory" conditions, such as microbial diseases and xenobiotic use that could initiate minor, often subclinical, inflammatory cascades may indeed be a risk factor for drug-induced hepatotoxicity that exists in certain segments of the population. This is currently unknown and exemplifies the difficulty of including certain types of risk factors in computational models. However, it will become increasingly important to gain a full understanding of the complexities of chemical and biological interactions that lead to hepatotoxicity, since new therapeutics and vaccine adjuvants are being developed that specifically interfere with cell signaling pathways and/or modulate the immune system, thereby altering intrinsic risk factors in certain patient populations.

26.3.4 ADR Risk Factors and Other Underlying Disease Processes in Patients

The following examples are used to further illustrate the potential difficulty of predicting human adverse drug reactions without incorporating risk factors. Approximately three years after approval Rezulin (troglitazone) was withdrawn from the market based on numerous reports of liver failure associated with its use. Research and opinions on the mechanism of hepatotoxicity have stimulated several scientific discussions and numerous presentations at scientific symposia. Possible mechanisms include direct hepatotoxicity via necrosis, apoptosis, mitochondrial injury, bile salt retention, and covalent binding to macromolecules. Several proposed mechanisms, including the formation of electrophilic metabolites (all with convincing data), have not fully explained why certain individuals were so susceptible to liver toxicity. There is evidence that host mechanisms or risk factors in patients with diabetes (the disease being treated) may cause patients to be predisposed to liver toxicity. Patients with a history of cholestasis and ongoing bile salt retention could be more susceptible to developing salt-mediated apoptosis and hepatotoxicity while undergoing therapeutic treatments [36].

Human variability can also play a major role in the development and severity of ADRs and certain pharmacogenetic analyses have highlighted individual susceptibilities of patients developing safety issues [37]. During a phase III trial with the antirestenosis drug tranilast, approximately 4% of patients (out of 11,500) developed hyperbilirubinemia. All treated patients had a clinically insignificant increase in bilirubin while on tranilast, but patients with two 7-repeat alleles, a DNA repeat polymorphism in the UDP-glucuronosyl transferase 1A1 (UGT1A1) gene, developed hyperbilirubinemia [37]. With another drug, irinotecan, a topoisomerase inhibitor used in the treatment of colon and rectal cancer, this same polymorphism, which is associated with decreased glucuronidation, was associated with a higher risk of developing neutropenia after treatment. Patients with the polymorphism (which can be present in approximately 10% of certain populations) experienced higher circulating levels of a reactive metabolite [13].

Certain adverse reactions can also be associated with individuals who are carriers of rare disease genes, patients with compromised immune systems, and patients with certain chronic diseases. Peripheral neuropathy from anticancer drugs such as vincristine can be accelerated in specific patients with Charcot-Marie tooth syndrome [38]. Park et al. [39] has discussed potential mechanisms that make HIV-positive patients more susceptible to enhanced drug toxicity, and Graham et al. [40] showed that the risk of developing rhabdomyolysis from statin-fibrin combinations increased significantly in older patients with diabetes mellitus.

All of this suggests that ADR predictions may require a more complex multifactorial approach, and this represents a challenge to any existing or newly introduced computational toxicology system. It may not be possible to link chemistry with toxicity outcome in regard to ADRs without some consideration for additional susceptibility factors.

26.4 GAP ANALYSIS ON CURRENT APPROACHES

26.4.1 The Pharmaceutical or Therapeutic Perspective

From a pharmaceutical or therapeutic perspective, gaps that exist within current computational systems and approaches include the following:

1. Ability to predict adverse outcomes of drugs in specific human populations, discussed in detail above.
2. Understanding risk factors that make some individuals more susceptible to toxicity and incorporating these risk factors into computational models. This knowledge would be highly beneficial early in the R&D process so that appropriate biomarkers or diagnostics can be established.

3. Generating or extrapolating new biological data at therapeutic or physiologically relevant exposure levels for more relevant computational endpoints. This becomes important in reducing error associated with false positive predictions based on high-dose extrapolations.
4. Understanding the potential pharmacological target distribution in tissues other than those intended for therapeutic intervention—as they may represent specific points for toxicity. Knowledge of this a priori would allow for more targeted screening and early estimates of therapeutic index.

A key question that frequently emerges in pharmaceutical applications is, Can mechanism- and chemistry-based toxicity be distinguished very early in the R&D process? This is important because chemistry-based toxicity can be resolved via chemical optimization, whereas mechanism-based toxicity can indicate the wrong target, wrong chemical scaffold, or insufficient Therapeutic Index because of target distribution regardless of scaffold. Based on the future directions that computational approaches are likely headed, this is exactly the type of question that will be answered by the process of functional genomics described above: essentially the discovery of the biological function of specific genes and the understanding of how sets of genes and their products work together in health and disease. Off-target toxicity is often termed "unexpected" primarily because of a lack of knowledge of the consequences of inhibiting, activating, or modulating a biological function (including downstream events) in an unwanted or unexpected tissue. Computational models and pathway analysis will be expected to aid in these discoveries and applications. For example, Bugrim, et al. [41] suggest that side effects of pyrazinamide could have been predicted from pathway analysis a priori, and Apic et al. [42] suggest the blue vision side effect of sildenafil citrate is due to binding to phosphodiesterase-6 in the eye though the intended target is phosphodiesterase-5 in smooth muscle. Rajasethupathy et al. [43] point out that rofecoxib exerts its intended anti-inflammatory action by inhibiting PGE2 activation downstream of COX-2. The potential increase in cardiovascular side effects could possibly be linked to the inhibition of protacyclin, a platelet inhibitor. These examples were developed retrospectively. Data and tools are available to conduct similar prospective analyses, and these activities will fall within the purview of computational toxicologists now and in future.

Targeted cancer therapies, such as tyrosine kinase inhibitors, represent a different type of off-target toxicity potential [44]. With these compounds, often displaying multiple kinase inhibiting profiles, toxicity is thought to be mechanism-based, as the same targets may exist and function in normal tissues. Therefore a higher distribution differential to tumors in relation to plasma becomes advantageous for anti-tumor activity. However, high distribution as a property also can direct more compound to tissues, in general. Moreover tumor tissue often expresses raised levels of efflux transporters such as P-gp and MDR-1, compared to healthy tissues [13]. Recently cardiotoxicity was discovered as an unexpected side effect of imatinib mesylate (Gleevec), a small molecule inhibitor of the fusion protein Bcr-Abl, the causal agent in chronic myelogenous leukemia [45]. These studies indicate that imatinib-induced inhibition of c-Abl is also associated with cardiomyocyte toxicity and that c-Abl has a previously unknown survival function in cardiomyocytes. Predicting and discovering potential off-target toxicity such as in the imatinib example could become a computational requirement during early stages of research on compounds that interfere with signal transduction pathways existing in several tissues.

26.4.2 Current and Developing Approaches to Predict or Model ADRs

Johnson and Rodgers [1] have described the human ADR network based on broad mechanisms of action, separating ADRs into mechanism- and chemistry-based categories.

- **Mechanism-related** (on-target) The pharmacology on the desired target is exaggerated, or (off-target) the right target is located in undesired tissues, or the parent compound interacting with an undesired target is exerting a different mechanism.
- **Chemistry-related** The compound or metabolite/s interacts (e.g., covalently binds) to an undesired target.

When the toxicity is dose-responsive (either as chemistry- or mechanism-based) the toxicity can be correlated with the exposure of the toxic chemical species, in the right compartment for the sufficient time to cause toxicity. When the toxicity is not dose-responsive, an inquiry ensues to discover what makes an affected individual more susceptible. This is the case in idiosyncratic toxicity in humans.

On-target mechanism-based toxicity (exaggerated pharmacology) is generally associated with exposure response and individual susceptibilities as root causative factors. Computational approaches currently used include pharmacokinetic/pharmacodynamic modeling (PK/PD) (see Chapter 3) and QSAR modeling (see Chapters 8–13). Approaches under development include pharmacogenetic analyses and diagnostic determinations of individual differences in metabolism, transport, and clearance that affect the exposure at the site of action.

Off-target mechanism-based toxicity, or unintended adverse effects, also involve exposure response and individual variations, but at an unintended site of action. Root causes can include lack of specificity of the compound and undefined distribution or sequestration in an off-target compartment. PK/PD modeling, and in some instances depending on the type of data base that can be constructed, QSAR modeling are used for analyses. Approaches under development are detailed databases on target distribution throughout the body (the so-called body map of global gene expression in normal tissues) and pharmaceutical-based PB/PK models. Application of practical, user-friendly

PB/PK modeling to pharmaceutical and biotechnology research holds great promise to predict off-target effects of compounds where therapeutic targets are known to be expressed in normal tissues.

Chemistry-based toxicity also involves exposure response, individual variability, and altered clearance. However, the proximal toxicant is typically a reactive chemical species. Several current systems are used for this purpose, such as mutagenicity (see Chapter 14) and carcinogenicity software programs, fragment-based chemistry and biology predictors, adduct determinations, and metabolite predictors.

26.5 COMPUTATIONAL APPROACHES TO TOXICITY BIOMARKERS

26.5.1 Biomarkers in Drug Discovery and Development

The concept of using biomarkers to monitor normal biological processes, disease progression, efficacy, and safety of therapeutic intervention has been steadily gaining ground. The Biomarkers Definitions Working Group [46] formally defines biomarker as an "indicator of normal biologic processes, pathogenic processes, or pharmacologic responses to a therapeutic intervention." This definition of a biomarker is very broad and not restricted to molecular biomarkers. Biomarkers are starting to find use in drug development and treatment; for example, Gleevec, which targets c-Abl kinase, has been approved to treat chronic myeloid leukemia (CML) for Philadelphia chromosome positive (Ph+) patients. In this case the chromosomal abnormality resulting from translocation and fusion of c-Abl and BCR served as a biomarker for both discovery and clinical development stages. Frank and Hargreaves [47] review biomarker definitions and provide examples of molecular- and image-based biomarkers for various therapeutic areas. Because biomarker discovery often involves genome-wide profiling of relevant samples using high-throughput technologies such as gene expression profiling via microarrays, spectrometry-based proteomics and metabolomic technologies, computational approaches play a major role in biomarker discovery. An example of a computational model involving 21 genes that predict recurrence of breast cancer is given by Paik et al. [48].

26.5.2 Systems Biology Approach

Despite the few success stories outlined above, discovery of predictive biomarkers remains a difficult problem. The difficulty can be attributed to several factors, among them the lack of well-annotated relevant clinical specimens for profiling and the fact that static sampling of biological processes fails to model dynamic events. The realization that studying and modeling individual components fail to predict behavior of biological systems as a whole has given rise to the concept of systems biology. Although there are several definitions of systems biology, it can be simply described as a method used to model interactions among various components. The need for systems biology has also been felt in toxicology. In a review, Waters and Fostel [17] persuasively argue for a systems approach to toxicology, as we attempt in this chapter.

26.5.3 Future Directions in the Discovery of Toxicity Biomarkers through Computational Approaches

Toxicogenomics There has been a steady increase in publications probing markers of toxicity using genome-wide expression profiling via microarrays [49]. This area of research, generally called toxicogenomics, has produced large amounts of datasets that are publicly available (see also Chapter 5). Toxicogenomics publications typically deal with toxic effects of specific chemicals either known to be toxins or drugs that cause toxicity at high doses. Often a rodent or in vitro model is used for the study and differential gene expression of the treatment is measured. The end result is typically a list of genes that are differentially expressed by applying well-accepted statistical analyses to the datasets. This approach has shed light on mechanism of action of the toxic agent in the form of affected genes and has produced signatures of toxicity. While these studies have been valuable, extracting biological insight from toxicogenomics datasets remain a challenge, and new computational approaches are needed.

Meta-Analysis One such approach is to take advantage of the fact that there have been multiple studies addressing specific mechanisms such as hepatotoxicity due to different toxic agents. In the meta-analysis approach all datasets are collected and normalized by such variables as experimental conditions, species used in the experiment, microarray platform variability, and gene expression normalization. Because the approach collectively analyzes data from different analyses, it is called a meta-analysis, as is well documented in the literature. A meta-analysis for cancer gene expression data has been performed by Rhodes et al. [50], and it has revealed some common molecular signatures of cancer, which were validated on independent datasets. To perform the meta-analysis, all datasets must be made available as well as the common experimental protocols that were used. Standards for describing toxicogenomics datasets are still evolving, and future publications may adhere to these standards. Also efforts are underway to integrate toxicogenomics datasets with other types of data, such as sequence homology across species, chemicals, protein–protein interactions, and biomedical literature [51]. Meta-analysis of toxicogenomics datasets will likely prove their value as more datasets become available and standard protocols are established.

Ontology-Based Analysis To analyze a list of significant genes such as those resulting from a toxicogenomics experiment, a well-defined ontology is used to map genes to ontology terms. In particular, use of gene ontology (GO) to analyze toxicogenomics data has proved to be useful. GO consists of three ontologies: biological processes, molecular functions, and cellular localizations. Terms in ontology are arranged hierarchically and are part of a controlled vocabulary. For example, glucose catabolism is part of the biological process ontology but is specifically a subterm of hexose catabolism. In addition a term can have synonyms and be associated with more than one ontology. The GO terms make it easier to associate genes with function because the genes products have been annotated.

In the gene ontology based approach, the first step is to determine a list of differentially expressed genes using a suitable statistical method. The list of significant genes is then mapped to GO terms and compared with GO terms for all genes in the array, revealing GO terms that are significantly over- or underrepresented. These "differential GO terms" describe biological processes that are significantly active due to experimental perturbation. This approach was initially proposed for the analysis of genes differentially expressed in cancer, and it has been coded into GoMiner (*http://discover.nci.nih.gov/gominer/*). An example of the use of GO terms analysis in toxicogenomics is presented by Currie et al. [52] who apply the ontology-based technique to elucidate the mechanism of action of nongenotoxic carcinogen diethylhexylphthalate in rodents. Yu et al. [20] have proposed a method called GO-quant that uses GO terms in a quantitative manner in toxicogenomics analysis.

Ontology-based approaches allow a list of genes to be analyzed in a meaningful way such as the ones obtained from toxicogenomics experiments. GO is constantly undergoing revisions while cell, disease, and pathway ontologies are also emerging. In toxicology also SysToxOntology has been proposed [53] for standardizing and analyzing toxicogenomics information.

26.5.4 Pathway Analysis as a Discovery Tool

While ontology-based methods have been shown to be useful to infer processes represented by a list of significant genes in toxicogenomics analysis, these methods do not take into account known functional interactions among genes described in pathways. Canonical pathways such as metabolic or signal transduction pathways represent consensus views of biological processes and offer another method to analyze a list of significant genes. In this approach, one projects a list of genes onto canonical pathways highlighting genes that are under- or over expressed in the experiment. Such an approach has been applied by Mao et al. [54] using KEGG pathways to annotate clusters arising from analysis of gene expression data.

Pathway analysis of gene expression data need not be restricted to only canonical pathways. Generally, the interactions described in the literature among gene products, proteins, and small molecules can be extracted from the literature and visualized as networks. Natural language processing (NLP) tools have been applied to process biomedical literature and extract meaningful interactions. There are also commercially available tools such as those offered by GeneGo, Ingenuity, PathART, and Pathway Studio for both manually curated and NLP-based interactions among genes. Networks of interactions resulting from text mining approaches can be characterized as "literature networks," and their overlapping pathways are included among canonical pathways. However, overlapping pathways cannot be equated with canonical pathways, so errors can result from automated text-mining methods. If the literature were not unduly biased toward certain genes (see Hoffman and Valencia [55] for genes that are overrepresented in literature), the literature network would offer an automated way of inferring pathways for list of genes, and occasionally would reveal insight into overlapping pathways not detected by analyzing only canonical pathways (Figure 26.1). Ekins [49] reviews methods that use the literature network approach to analyze toxicogenomics data. This method is still in its infancy in toxicogenomics, but there are applications elsewhere demonstrating its utility [56,57].

We see we articipate future development of a more multi-factorial approach to determine the most relevant biomarkers of drug-induced toxicity. Certainly "signatures" of differentially expressed genes are helpful, but the discovery of the more complex interaction of genes and proteins in response to various perturbations will hold the key to biomarker discovery and utilization in the future.

26.5.5 Inferring Human Endpoints from Animal Data

Toxicogenomics studies are usually carried out in rodents, and methods such as the ones outlined above could be used to infer pathway mechanisms and further functional insights. But translating inferences from animal data to mechanism of action in humans remains a difficult problem to solve.

The availability of complete or nearly complete genomes of human, rodents, and dog enables assigning gene orthology and offers one way to transfer functional information between species. Generally, the orthology assignment is not trivial due to the split of paralogues before and after speciation. Remm and Sonnhammer [58] have proposed an algorithm that takes into account so-called in-paralogues (paralogues that arise after the species split) and out-paralogues (paralogues that arise before the species split). Because of the high degree of sequence conservation, orthology assignment can be straightforward in some cases. However, the orthology assignment does not imply that the gene functions identically in two species. There may be cases where the rodent orthology of a human gene does not exist. For example, IL8, which is one of the major mediators in inflammatory response in humans, has no known orthologue in rodents.

Another approach used to infer human functional significance from animal data is to focus on conserved pathways between species. There is experimental evidence to show that innate immunity is conserved across nematodes,

(a)

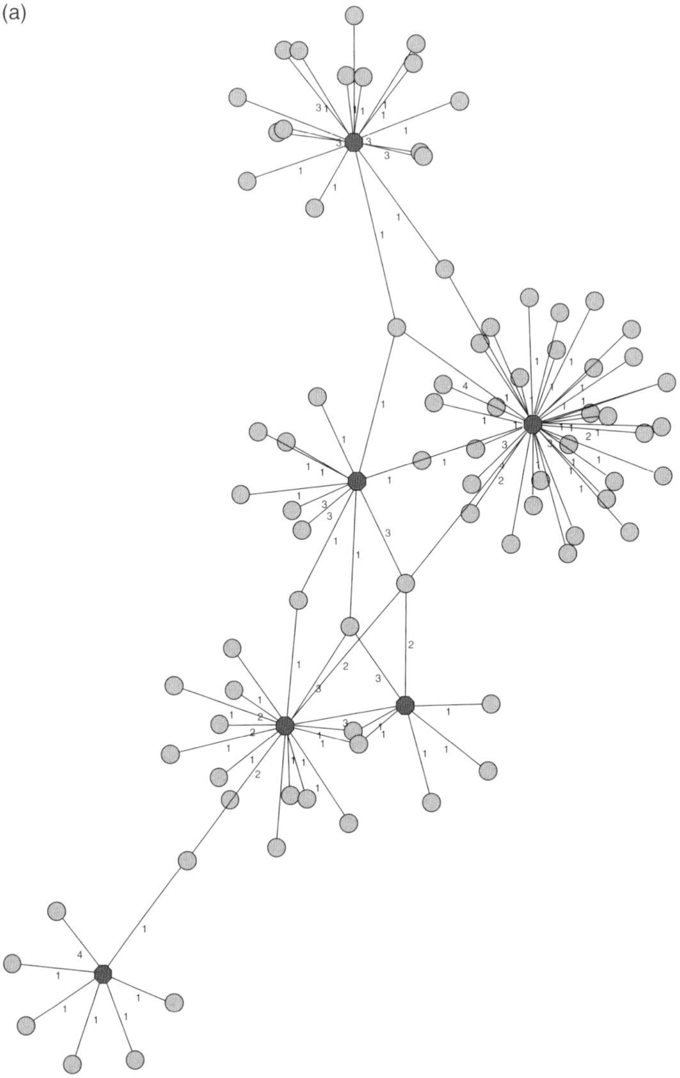

Figure 26.1 An example of literature based network generation. (*a*) Putative human orthologs of signaling response genes (from Table 4 of Currie et al. [52]) are shown in red. Associated genes from literature are sown in cyan. If two genes occur in the same abstract they are joined together. Images were generated using LitNet (Personal Software, S. Sudarsanam), which uses graph theoretical methods to query PubMed for co-occurring genes. (*b*) Details are shown of an enlargement of a portion of the network. Numbers on the connecting line shows number of abstracts that link two genes together.

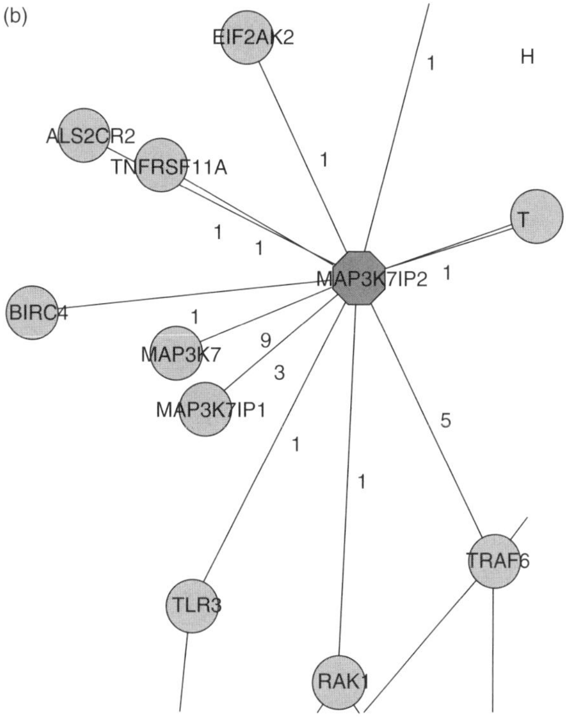

Figure 26.1 (*Continued*)

arthropods, and vertebrates [59], resulting in conserved pathways. It stands to reason that conserved pathways imply conserved functions across species. Kelley et al. [60] outline a network alignment strategy to detect pathways conserved between yeast and bacteria. In a toxicogenomics context this might be accomplished in the following way: First, a list of genes is converted into a network of interacting genes using a pathway or literature network approach, as described above, for human and the other species in which the experiment was carried out (Figure 26.1). Second, the networks are aligned and a score representing degree of conservation is obtained. Depending on the degree of network conservation, one could transfer inference from animal to human with confidence.

26.6 TOOLBOX FOR THE COMPUTATIONAL TOXICOLOGIST

The computational toxicologist of the near future in the pharmaceutical or biotechnology sector will be asked to use in silico tools to (1) provide predictions on chemical structures on specific endpoints (e.g., mutagenicity, carcinogenicity, QTc prolongation), (2) provide information on potential toxicophores

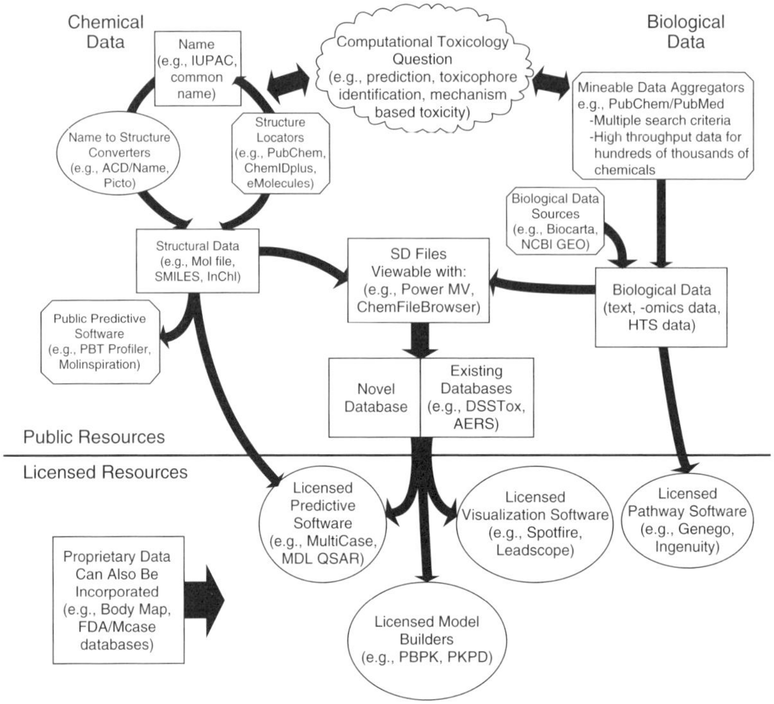

Figure 26.2 Public and commercial resources for the computational toxicology toolbox.

existing within structures for chemical synthetic alternatives, (3) predict metabolic pathway information of compounds and predict metabolites in various species and the toxicological consequence of forming these metabolites, (4) predict information on target-based toxicity (mechanism-based) and off-target potential in various tissues, (5) predict potential drug and drug–drug interactions leading to toxicity, and (6) predict the potential for human adverse events in specific clinical trial designs and in various patient populations. Resources will come from publicly available sources as well as commercially available applications. Figure 26.2 shows some of these applications with examples of software programs separated into public domain and commercial sources.

Equipping the computational toxicologist with the necessary tools begins with access to data as datasets incorporating both chemical and biological information are fundamental to computational toxicology. Chemical data are fairly well standardized and publicly available, and moreover easily translatable from name to structure and back using publicly available tools (ACD/Name, Picto, PubChem, ChemIDplus). Certain biological endpoints, such as mutagenicity and carcinogenicity, have previously been extracted from standardized publicly available sources and compiled into databases (*http://www.epa.gov/ncct/dsstox/DataFiles.html*). However, other biological endpoints, specifically toxicogenomic and proteomic data, have yet to be standardized completely but are now starting to be compiled publicly. Current projects include PubChem, which seeks to compile high-throughput data for public use (*http://www.ncbi.nlm.nih.gov/entrez/query.fcgi?db=pcassay*), and NIEHS's Chemical Effects in Biological Systems (CEBS) database, which is projected to be a database for mining genomics, proteomics, and metabolomics data (*http://cebs.niehs.nih.gov/*). Also there is the Comparative Toxicogenomics Database (CTD; *http://ctd.mdibl.org/*) that contains scientifically reviewed and curated information on chemicals, relevant genes, and proteins, and their interactions in vertebrates and invertebrates. CTD provides sequence, reference, species, microarray, and general toxicology information and so is a unique centralized resource for toxicogenomic research. The database includes capabilities that enable visualization of cross-species comparisons of gene and protein sequences. ArrayTrack *http://edkb.fda.gov/webstart/arraytrack* is a freely available product from the US FDA National Center for Toxicological Research (NCTR). It functions as an integrated platform for managing, analyzing, and interpreting microarray gene expression data. The FDA is also working on databases utilizing ADR data from both the Spontaneous Reporting System (SRS) and the Adverse Event Reporting System (AERS) *http://fda.gov/Cder/Offices/OPS_IO/adverse_effect_database.htm*.

Standardization of data and visualization of datasets are paramount for maximal use of data. A computational toxicologist must be able to create databases "on the fly," adding and subtracting data when necessary to form relevant datasets for answering very specific questions that may arise during various phases of projects. This is especially important in systems biology where the number of potential connections is large. The computational toxicologist must be able to pare away extraneous data to uncover relevant gene interactions at a local level, while still having the capability of searching for other potential gene targets on a global scale. Since many targets of pharmaceutical intervention are expressed in other areas of the body as well as the intended tissue or organ, the analysis for potential toxicity has to be extended throughout the body. Local datasets have the advantage of providing a detailed map of one target organ or cell type, although global datasets are still necessary to predict potential off-target toxicity.

It is unlikely that any one program will ever become a complete computational toxicology toolbox. Rather, it is more likely that certain predictive software will be used for specific applications, while more general applications will be used to create relevant datasets for multiple uses. Systems biology pathway software, in concert with metabolite prediction, pharmacokinetics, and ADME software will be critical in reducing cost and time necessary for the development of new pharmaceuticals [49].

26.7 INITIATIVES FOR COMPUTATIONAL TOXICOLOGY STUDY AND TRAINING

Computational toxicology is an emerging field that draws on several scientific disciplines:

- Molecular toxicology
- Computational chemistry, where physical-chemical mathematical modeling at the molecular level includes quantum chemistry, force fields, molecular mechanics, molecular simulations, molecular modeling, molecular design, and cheminformatics
- Computational biology or bioinformatics including development of molecular biology databases and analysis of data
- Systems biology, where mathematical modeling and reasoning is applied to the understanding of biological systems and the explanation of biological phenomena

Most people working in the field today have graduate degrees in a field related with one or two of the disciplines noted above. There are several new graduate programs in the field, ranging from individual courses to concentrated majors. The following summary is not an exhaustive listing of programs but serves to highlight certain areas of academic interest. The University of Pittsburgh, the first dedicated computational toxicology graduate program and the only academic program on the Society of Toxicology Web site listing computational toxicology as an area of concentration, emphasizes a strong toxicology core and development of computer programming skills so that students get to understand existing analytical systems and create new software programs. Colorado State University has a core graduate program, the Quantitative and Computational Toxicology Research Group, with a strong program in mathematical modeling and a concentration in physiologically-based pharmacokinetic modeling. Other graduate programs with research opportunities in computational toxicology include the University of North Carolina Chapel Hill, Michigan State University, the Massachusetts Institute of Technology, Rutgers University, and Imperial College London, to name a few. The academic scope is typically multidisciplinary with relation to the systematic analysis and modeling of complex biological phenomena. The University of California at Berkeley has included computational toxicology course work as a requirement in its undergraduate Molecular Toxicology major. Students have the option of independent study and undergraduate honors research in computational toxicology with a strong concentration in therapeutic applications. The ideal background for undergraduate study is course work in organic chemistry, biology, biochemistry, molecular toxicology including toxicogenomics, pharmacology, physiology, and statistics. At the Berkeley undergraduate level there are no computer pre-requisites, but it is noted that all students entering the curriculum have sufficient baseline computer skills. The program at Berkeley includes placement into internships (pre- and postdegree) currently at US FDA in the Informatics and Computational Safety Analysis Staff (ICSAS) group and at local San Francisco Bay area biotechnology companies. Postdoctoral training in computational toxicology is available at a number of universities as well as programs sponsored and administered through the US EPA. Almost all academic programs or research activities have an environmental focus, primarily because of the essential applications and funding in chemical risk assessment. The wave of the future appears to be in a collaborative-consortium direction. For example, the Environmental Bioinformatics Computational Toxicology Center is a collaborative effort between Robert Wood Johnson Medical School, University of Medicine and Dentistry of NJ; Rutgers, the State University of NJ, Princeton University; and US FDA Center for Toxicoinformatics, NCTR. Investigators interact and collaborate through these institutions and US EPA centers and laboratories on specific areas of interest according to an operating agreement and plan.

26.8 SUMMARY

This chapter has highlighted several areas where the field of computational toxicology appears to be heading. While predicting the potential of a compound to induce specific ADRs continues to be a formidable task, the field of computational biology is taking a direction more relevant to human disease and adverse outcomes. The elusive goal of predicting potential human ADRs from new drugs will involve defining better biomarkers to quantify toxicity endpoints. There is an important need to determine the distribution of potential targets in all tissues and gain an understanding of how disease risk factors and individual susceptibilities affect human ADRs. Such an overall systems approach will be necessary if relevant advances are to be made in computational toxicology.

REFERENCES

1. Johnson DE, Rodgers AD. Computational toxicology: Heading toward more relevance in drug discovery and development. *Curr Opin Drug Disc Devel* 2006;9(1):29–37.
2. Greene N. Computer systems for the prediction of toxicity: An update. *Adv Drug Deliv Rev* 2002;54:417–31.
3. Matthews E, Contrera J. A new highly specific method for predicting the carcinogenic potential of pharmaceuticals in rodents using enhanced MCASE QSAR-ES software. *Regul Toxicol Pharmacol* 1998;28(3):242–2.
4. Snyder RD, Pearl GS, Mandakas G, Choy WN, Goodsaid F, Rosenblum IY. Assessment of the sensitivity of the computational programs DEREK, TOPKAT, and MCASE in the prediction of the genotoxicity of pharmaceutical molecules. *Environ Mol Mutagen* 2004;43(3):143–58.

5. White AC, Mueller RA, Gallavan RH, Aaron S, Wilson AGE. A multiple in silico program approach for the prediction of mutagenicity from chemical structure. *Mutat Res* 2003;539:77–89.
6. Cariello NF, Wilson JD, Britt BH, Wedd DJ, Burlinson B, Gombar V. Comparison of the computer programs DEREK and TOPKAT to predict bacterial mutagenicity. *Mutagenesis* 2002;17(4):321–9.
7. Roche O, Trube G, Zuegge J, Pflimlin P, Alanine A, Schneider G. A virtual screening method for prediction of the hERG potassium channel liability of compound libraries. *Chembiochem* 2002;3:455–9.
8. Ekins S, Boulanger B, Swaan PW, Hupcey MA. Towards a new age of virtual ADME/TOX and multidimensional drug discovery. *J Compuct Aided Drug Des* 2002;16:381–401.
9. Johnson DE, Smith DA, Park BK. Linking toxicity and chemistry: Think globally, but act locally? *Curr Opin Drug Disc Devel* 2004;7(1):33–5.
10. Johnson D. The discovery-development interface has become the new interfacial phenomenon. *Drug Discov Today* 1999;4(12):535–6.
11. Johnson DE, Wolfgang GHI. Predicting human safety: screening and computational approaches. *Drug Discov Today* 2000;5(10):445–54.
12. Ekins S, Crumb WJ, Sarazan RD, Wikel JH, Wrighton SA. Three-dimensional quantitative structure-activity relationship for inhibition of human ether-a-go-go-related gene potassium channel. *Pharmacol Exp Ther* 2002;301:427–34.
13. Johnson DE, Smith DA, Park BK. Safety/toxicity threshold concepts in drug discovery and development. *Curr Opin Drug Disc Devel* 2005;8(1):24–6.
14. Paules R. Phenotypic anchoring: Linking cause and effect. *Environ Health Perspect* 2003;111:A338–9.
15. Stevens JL, Marnett LJ. Defining molecular toxicology: A perspective. *Chem Res Toxicol* 1999;12(9):747–8.
16. Marnett LJ. The future of toxicology. *Chem Res Toxicol* 2006;19(5):609.
17. Waters MD, Fostel JM. Toxicogenomics and systems toxicology: Aims and prospects. *Nature Rev* 2004;5:936–48.
18. Hieter P, Boguski M. Functional genomics: It's all how you read it. *Science* 1997;278:601–2.
19. Cunningham ML. Putting the fun into functional toxicogenomics. *Tox Sci* 2006; 92(2):347–8.
20. Yu X, Griffith WC, Hanspers K, Dillman JF, Ong H, Vredevoogd MA, Faustman EM. A system-based approach to interpret dose- and time-dependent microarray data: Quantitative integration of gene ontology analysis for risk assessment. *Toxicol Sci* 2006;92(2):560–77.
21. Rietjens IM, Alink GM. Future of toxicology: Low-dose toxicology and risk–benefit analysis. *Chem Res Toxicol.* 2006;19(8):977–81.
22. Ionannidis J. Materializing research promises: Opportunities, priorities and conflicts in translational medicine. *J Tanslational Med* 2004;2:5.
23. Pestka J, Zhou H-R. Toll-like receptor priming sensitizes macrophages to proinflammatory cytokine gene induction by deoxynivalenol and other toxicants. *Tox Sci* 2006;92(2):445–55.

24. Machida K, Cheng KT, Sung VM, Levine AM, Foung, S and Lai MM: Hepatitis C virus induces toll-like receptor 4 expression, leading to enhanced production of beta interferon and interleukin-6. *J Virol* 2006;80(2):866–74.
25. Qi H-Y, Shelhamer JH. Toll-like receptor 4 signaling regulates cytosolic phospholipase A2 activation and lipid generation in lipopolysaccharide-stimulated macrophages. *J Biol Chem* 2005;280(47):38969–75.
26. Jaeschke H. Mechanisms of liver injury: II. Mechanisms of neutrophil-induced liver cell injury during hepatic ischemia-reperfusion and other acute inflammatory conditions. *Am J Physiol Gastrointest Liver Physiol* 2006;290:1083–88.
27. Navarro VJ, Senior JR. Drug-related hepatotoxicity. *N Engl J Med* 2006;354:731–9.
28. Hoffman ES, Smith RET, Renaud RC. TLR-targeted therapeutics. *Nature Rev Drug Discov* 2005;4:879–80.
29. Luyendyk J, Lehman-McKeeman L, Nelson D, Bhaskaran V, Reilly T, Car B, Cantor G, Maddox J, Ganey P, Roth R. Unique gene expression and hepatocellular injury in the lipopolysaccharide-ranitidine drug Idiosyncrasy rat model: Comparison with famotidine. *Toxicol Sci* 2006;90:569–85.
30. Luyendyk J, Lehman-McKeeman L, Nelson D, Bhaskaran V, Reilly T, Car B, Cantor G, Deng X, Maddox J, Ganey P, Roth R. Coagulation dependent gene expression and liver injury in rats given lipopolysaccharide with ranitidine but not with famotidine. *J Pharmacol Exp Ther* 2006;317:635–43.
31. Waring J, Liguori M, Luyendyk J, Maddox J, Ganey P, Stachlewitz R, North C, Blomme E, Roth R. Microarray analysis of lipopolysaccharide potentiation of trovafloxacin-induced liver injury in rats suggests a role for proinflammatory chemokines and neutrophils. *JPET* 2006;316:1080–7.
32. Pearson JM, Bailie MB, Fink GD, Roth RA. Neither platelet activating factor nor leukotrienes are ccritical mediators of liver injury after lipoolysaccharide administration. *Toxicology* 1997;121(5):181–9.
33. Wolfgang GH, McCabe RD, Johnson DE. Toxicity of subcutaneously administered recombinant human interleukin-2 in rats. *Toxicol Sci* 1998;42(1):57–63.
34. Jaeschke H. Molecular mechanisms of hepatic ischemia-reperfusion injury and preconditioning. *Am J Physiol Gastrointest Liver Physiol* 2003;284:G15–G26.
35. Tsung A, Hoffman RA, Izuishi K, Critchlow ND, Nakao A, Chan MH, Lotze MT, Geller DA, Billiar TR. Hepatic ischemia/reperfusion injury involves functional TLR4 signaling in nonparenchymal cells. *J Immunol* 2005;175(11):7661–8.
36. Smith MT. Mechanisms of Troglitizone Hepatotoxicity. *Chem Res Toxicol* 2003;16(16):679–87.
37. Roses AD. Pharmacogenetics and drug development: the path to safer and more effective drugs. *Nature Rev Genet* 2004;5:645–65.
38. Griffiths JD, Stark RJ, Ding JC, Cooper IA. Vincristine neurotoxicty in Charcot-Marie-tooth syndrome. *Med J Aust* 1985;143:305–6.
39. Park BK, Pirmohamed -M, Kitteringham NR. Role of drug disposition in drug hypersensitivity: A chemical, molecular, and clinical perspective. *Chem Res Toxicol* 1998;11(9):969–88.
40. Graham DJ, Staffa JA, Shatin D, Andrade SE, Schech SD, LaGrenade L, Gurwitz JH, Chan KA, Goodman MJ, Platt R. Incidence of hospitalized rhabdomyolysis in patients treated with lipid-lowering drugs. *JAMA* 2004;292(21):2585–90.

41. Bugrim A, Nikolskaya T, Nikolsky Y. Early prediction of drug metabolism and toxicity: Systems biology approach and modeling. *Drug Discov Today* 2004;9(3): 127–35.

42. Apic G, Ignjatovic T, Boyer S, Russell RB. Illuminating drug discovery with biological pathways. *FEBS Lett* 2005;579(8):1872–7.

43. Rajasethupathy P, Vayttaden SJ, Bhalla US. Systems modeling: A pathway to drug discovery. *Curr Opin Chem Biol* 2005;9:400–6.

44. Rishton G. Failure and success in modern drug discovery: guiding principles in the establishment of high probability of success drug discovery organizations. *Med Chem* 2005;1:519–27.

45. Kerkela R, Grazette L, Yacobi R, Iliescu C, Patten R, Beahm C, Walters B, Shevtsov S, Pesant S, Clubb F, Rosenzweig A, Salomon R, Van Etten R, Alroy J, Durand JB, Force T. Cardiotoxicity of the cancer therapeutic agent imatinib mesylate. *Nature Med* 2006;12(8):908–916.

46. Biomarkers Definitions Working Group, Bethesda, Md. Biomarkers and surrogate endpoints: Preferred definitions and conceptual framework. *Clin Pharmacol Ther* 2001;69:89–95.

47. Frank R, Hargreaves R. Clinical biomarkers in drug discovery and development. *Nat Rev Drug Discov* 2003;2(7):566–80.

48. Paik S, Shak S, Tang G, Kim C, Baker J, Cronin M, Baehner FL, Walker MG, Watson D, Park T, Hiller W, Fisher ER, Wickerham DL, Bryant J, Wolmark N. A multigene assay to predict recurrence of tamoxifen-treated, node-negative breast cancer. *N Engl J Med* 2004;351(27):2817–26.

49. Ekins S. Systems-ADME/Tox: Resources and network approaches. *J Pharmacol Toxicol Meth* 2006;53:38–66.

50. Rhodes DR, Yu J, Shanker K, Deshpande N, Varambally R, Ghosh D, Barrette T, Pandey A, Chinnaiyan AM. Large-scale meta-analysis of cancer microarray data identifies common transcriptional profiles of neoplastic transformation and progression. *PNAS* 2004;101(25):9309–14.

51. Mattingly CJ, Rosenstein MC, Davis AP, Colby GT, Forrest JN, Boyer JL. The comparative toxicogenomics database: A cross-species resource for building chemical–gene interaction networks. *Toxicol Sci* 2006;92(2):587–95.

52. Currie RA, Bombail V, Oliver JD, Moore DJ, Lim FL, Gwilliam V, Kimber I, Chipman K, Moggs JG, Orphanides G. Gene ontology mapping as an unbiased method for identifying molecular pathways and processes affected by toxicant exposure: Application to acute effects caused by the rodent non-genotoxic carcinogen diethylhexylphthalate. *Toxicol Sci* 2005;86(2):453–69.

53. Xirasagar S, Gustafson SF, Huang C, Pan Q, Fostel J, Boyer P, Merrick BA, Tomer KB, Chan DD, Yost KJ, Choi D, Xiao N, Stasiewicz S, Bushel P, Waters MD. Chemical effects in biological systems (CEBS) object model for toxicology data, SysTox-OM: Design and application. *Bioinformatics* 2006;22(7):874–82.

54. Mao X, Cai T, Olyarchuk JG, Wei L. Automated genome annotation and pathway identification using the KEGG orthology (KO) as a controlled vocabulary. *Bioinformatics* 2005;21(19):3787–93.

55. Hoffmann R, Valencia A. Life cycles of successful genes. *Trends Genet* 2003; 19(2):79–81.

56. Calvano SE, Xiao W, Richards DR, Felciano RM, Baker HV, Cho RJ, Chen RO, Brownstein BH, Cobb JP, Tschoeke SK, Miller-Graziano C, Moldawer LL, Mindrinos MN, Davis RW, Tompkins RG, Lowry SF. A network-based analysis of systemic inflammation in humans. *Nature* 2005;437(7061):1032–7.
57. Bredel M, Bredel C, Juric D, Harsh GR, Vogel H, Recht LD, Sikic BI. Functional network analysis reveals extended gliomagenesis pathway maps and three novel MYC-interacting genes in human gliomas. *Cancer Res* 2005;65(19):8679–89.
58. Remm M, Storm CEV, Sonnhammer ELL. Automatic clustering of orthologs and in-paralogs from pairwise species comparisons. *J Mol Biol* 2001;314(5):1041–52.
59. Kim DH, Ausubel FM. Evolutionary perspectives on innate immunity from the study of *Caenorhabditis elegans*. *Curr Opin Immunol* 2005;17(1):4–10.
60. Kelley BP, Sharan R, Karp RM, Sittler T, Root DE, Stockwell BR, Ideker T. Conserved pathways within bacteria and yeast as revealed by global protein network alignment. *PNAS* 2003;100(20):11394–99.

27

COMPUTATIONAL TOOLS FOR REGULATORY NEEDS

ARIANNA BASSAN AND ANDREW P. WORTH

Contents

Computational Toxicology: Risk Assessment for Pharmaceutical and Environmental Chemicals, Edited by Sean Ekins

27.1 INTRODUCTION

In the regulatory framework there is a growing need for in silico methods that can be used to gain information about environmental fate, and ecological and health effects of chemicals. Computer-aided toxicity prediction mainly makes use of the relationship between chemical structure and biological activity to compute (eco)toxicity and fate of chemicals (e.g., physicochemical properties, toxicological activity, distribution, and fate), thus generating nontesting data about the effects of the chemicals on humans and the environment. The different techniques that are used to derive nontesting information include (quantitative) structure–activity relationship models, expert systems, and read-across/category approaches. These nontesting methods are based on the idea that the biological activity of a chemical is intrinsic to its nature and so can be directly inferred from its molecular structure and the properties of similar compounds whose activities are known [1].

In principle, nontesting methods can be applied at different stages in the development and registration of chemicals, from in-house research and development to the compilation of dossiers on chemical safety for submission to regulatory authorities. In practice, the ways in which these approaches are used depends on the requirements of the specific legislation and the possibilities offered by regulatory authorities.

This chapter focuses on the future use of nontesting methods in the regulatory assessment of chemicals. A brief explanation is provided of the main types of nontesting methods, and reference is made to the use of these methods in the European Union, as foreseen by the REACH legislation. The uptake of nontesting methods in the European Union is an example of a trend across many countries within the Organisation for Economic Cooperation and Development (OECD). The need to use nontesting methods has led to the development and implementation of integrated testing strategies (ITS) based as far as possible on the use of nontesting data [2]. The use of nontesting methods within such strategies implies the need for computational tools to facilitate the entire workflow. In our view, the main functionalities of ITS should be incorporated into a decision support system (DSS) to facilitate the implementation of this workflow. Such a DSS could be useful in any regulatory context.

27.2 IN SILICO TOXICITY PREDICTION TECHNIQUES

A wide spectrum of in silico techniques can be used to generate nontesting data for filling in data gaps needed in hazard assessment. The meanings of the different terms used to indicate the estimation methods employed within various regulatory frameworks worldwide are given below.

27.2.1 SAR

Structure–activity relationships (SARs) are theoretical models that can be used to predict, in a qualitative manner, the physicochemical, biological (e.g., toxicological), and environmental fate properties of molecules from knowledge of their chemical structures. More specifically, a SAR is a qualitative relationship (i.e., association) between a molecular (sub)structure and the presence or absence of a given biological activity, or the capacity to modulate a biological activity imparted by another substructure. The term substructure refers to an atom, or group of adjacently connected atoms, in a molecule. A substructure associated with the presence of a biological activity is sometimes called a structural alert. A SAR can also be based on the ensemble of steric and electronic features considered necessary to ensure intermolecular interaction with a specific biological target molecule, which results in the manifestation of a specific biological effect. In this case the SAR is sometimes called a 3D SAR or pharmacophore.

27.2.2 QSAR

A quantitative structure-activity relationship (QSAR) is a measurable relationship between a biological activity (e.g., toxicity) and one or more molecular descriptors that are used to predict the activity. A molecular descriptor is a structural or physicochemical property of a molecule, or part of a molecule, which specifies a particular characteristic of the molecule and is used as an independent variable in a QSAR.

27.2.3 (Q)SAR

The term (Q)SAR makes generic reference to a relationship that can be qualitative or quantitative in nature, between structural information and biological activity.

27.2.4 Expert System

This is a very broadly used term for any formal system, generally computer-based, that enables a user to obtain rational predictions about the properties or biological activity of chemicals [3,4]. A combination of different SARs and

QSARs (mainly computer-based combinations) can give rise to an expert system. One or more databases may also be integrated in the system.

27.2.5 Read-Across or Analogue Approach

In the read-across or analogue approach, endpoint information for one chemical is used to make a prediction of the endpoint for another chemical, which is considered to be similar in some way. In principle, read-across can be used to assess physicochemical properties, environmental fate, and (eco)toxicity effects, and it may be performed in a qualitative or quantitative manner. A one-to-one read-across is an ad hoc comparison based on the similarity between two chemicals. Read-across carried out between three or more chemicals can lead to the formulation of generalizations about the group, and eventually to establish that a common substructure can be associated with a SAR. Although the distinction between SAR and read-across may appear vague, the term SAR usually refers to an approach that has been subjected to some degree of statistical validation, and thus to a more formalized approach than read-across.

27.2.6 Qualitative Read-Across

In qualitative read-across the potential of a chemical to exhibit a property is inferred from the established potential of one or more analogues. The analogue approach could simply be regarded as the use of SAR. The process involves: (1) the identification of a chemical substructure that is common to two substances (which are considered to be analogues) and (2) the assumption that the presence (or absence) of a property/activity for a substance can be inferred from the presence (or absence) of the same property/activity for the analogous substance.

27.2.7 Quantitative Read-Across

In quantitative read-across the numerical value of a property (or potency of an endpoint) of a chemical is inferred from the quantitative data of one or more analogues. Finding the analogue involves: (1) the identification of a chemical substructure that is common to two substances (which are considered to be analogues) and (2) the assumption that the known value of a property for one substance can be used to estimate the unknown value of the same property for another substance. When applying quantitative read-across, there are four general ways of estimating the missing data point: (1) by using the endpoint value of a source chemical (e.g., the value for the closest analogue), (2) by processing the endpoint values of two or more source chemicals (e.g., by averaging, by taking the most representative value), (3) by taking the most conservative value of two or more analogues, and (4) by using the analogue data to develop a "mini-QSAR" that can be used to provide an appropriate scaling (or proportionality factor) for the chemical of interest.

27.2.8 Category Approach

A chemical category is a group of chemicals whose physicochemical properties, ecological effects, environmental fate, and human health effects are likely to be similar or follow a regular pattern as a result of structural similarity. The similarities may be based on: (1) a common functional group (e.g., aldehyde, epoxide, ester, metal ion); (2) the likelihood of common precursors and/or breakdown products, via physical or biological processes, that result in structurally similar chemicals (e.g., a metabolic series); or (3) an incremental and constant change across the category (e.g., a chain-length category). Read-across can be used to estimate values (qualitatively or quantitatively) for one member of a category from known information on the same property or biological activity for other category members [5,6].

27.2.9 Grouping

Normally grouping involves collecting chemicals and establishing if they can form the basis for read-across or category because of some measure of commonality.

27.3 PREDICTING TOXICITY FOR THE REGULATORY ASSESSMENT OF CHEMICALS

National and international agencies have a number of reasons to encourage the use of in silico methods [7]. First of all, computational methods are faster and cheaper compared to empirical testing methods, and their use results in considerable savings of time and money during the assessment of chemical hazard. Second, the rising public concern for animal welfare has led to a strong commitment by companies and governmental agencies to reduce animal testing by means of an increased use of computational toxicology as well as validated alternative tests such as in vitro methods. Laws in many countries (including European Commission Directive 86/609/EEC on the protection of animals used for experimental and other scientific purposes [8]) require, wherever possible, the use of alternative methods in place of animals to reduce the number of sacrificed animals and diminish the amount of distress or pain suffered by animals.

It is possible to identify a variety of applications for computer-aided toxicology in regulatory frameworks [9,10]:

- Prioritization of existing chemicals for further testing and assessment: Chemicals of higher concern are tested before chemicals of lower concern.
- Hazard assessment of chemicals: Experimental testing is replaced by filling in data gaps needed for hazard identification, potency estimation and classification and labeling.

- Risk characterization of chemicals: Experimental testing is replaced by filling in data gaps needed to derive dose metrics or assessment factors that are combined with exposure information in risk assessment.

The specific use of (Q)SARs and expert systems by different regulatory agencies worldwide (e.g., European Union, United States, Canada) is described in a number of excellent reviews, where the specific regulatory applications are discussed in detail [9,11–14].

In silico methods are expected to play an increasing important role in toxicity prediction for hazard and risk assessment when the new chemicals legislation proposed by the European Commission on 29 October 2003 [16] comes into force on 1st June 2007. The new EU regulatory framework known as REACH (Registration, Evaluation and Authorisation of CHemicals) aims at improving the protection of human health and the environment through the better and earlier identification of the properties of chemical substances. As a consequence of the new regulatory system, additional environmental and toxicological data on approximately 30,000 chemicals has to be acquired. To limit the cost and the number of animals used for testing, REACH explicitly encourages the use of computer-aided methods such as (Q)SAR methods and category/read-across approaches for filling in this enormous knowledge gap of chemical information. In order to be used in place of experimental data, REACH requires that the in silico methods meet certain conditions. For example, in the case of (Q)SARs, these requirements include: (1) the model has to be valid (see below), (2) the substance has to fall within the applicability domain of the (Q)SAR model, and (3) the applied method has to be provided with adequate and reliable documentation.

The use of in silico methods, and in particular, (Q)SARs, as valuable components of the regulatory assessment strategy is hampered by two major factors. First, model estimations can be properly interpreted only by specialists with specific expertise in the field of computational toxicology, and this factor certainly limits the widespread use of in silico approaches for regulatory purposes. Second, for a method to be accepted in a regulatory framework, its scientific validity has to be established in accordance with internationally agreed validation principles. At present there is a lack of information on the scientific validity of many models, which then limits their regulatory acceptance [17].

Under REACH, (Q)SAR methods may be used as alternatives to testing only if their validity has been ascertained by reference to the OECD (Organisation for Economic Cooperation and Development) Principles for (Q)SAR validation. In November 2004 in the context of the 37th Joint Meeting of the Chemicals Committee and the Working Party on Chemicals, Pesticides, and Biotechnology, the OECD member countries and the European Commission adopted the OECD principles for the validation, for regulatory purposes, of (Q)SAR models. The agreed OECD principles are as follows:

To facilitate the consideration of a (Q)SAR model for regulatory purposes, it should be associated with the following information:

1. *A defined endpoint.*
2. *An unambiguous algorithm.*
3. *A defined domain of applicability.*
4. *Appropriate measures of goodness-of-fit, robustness and predictivity.*
5. *A mechanistic interpretation, if possible.*

The intent of principle 1 is to ensure clarity in the endpoint being predicted by a given model, since a given endpoint could be determined by different experimental protocols and under different experimental conditions. It is therefore important to identify the experimental system that is being modeled by the (Q)SAR. The term endpoint refers to any physicochemical, biological, or environmental effect that can be measured and therefore modeled.

The (Q)SAR estimate of an endpoint is the result of applying a (mathematical) algorithm to a set of parameters that describe the chemical structure. The intent of principle 2 is to ensure transparency in the model algorithm that generates predictions of an endpoint from information on chemical structure and/or physicochemical properties. The issue of reproducibility of the predictions is covered by this principle.

Principle 3 defines an applicability domain that refers to the response and chemical structure space in which the model makes predictions with a given reliability. Ideally the applicability domain should express the structural, physicochemical, and response space of the model. The chemical structure space can be expressed by information on physicochemical properties and/or structural fragments. The response can be any physicochemical, biological, or environmental effect that is being predicted.

Principle 4 points to performing statistical validation to establish the performance of the model. Goodness-of-fit and robustness refer to the internal model performance while predictivity refers to the external model validation.

According to principle 5, a (Q)SAR should be associated with a mechanistic interpretation, if possible. The mechanistic interpretation refers to the assignment of physicochemical or biological meaning to the descriptors, and to the assignment of a plausible relationship between the descriptors and the modeled endpoint.

27.4 SUPPORTING THE USE OF COMPUTATIONAL TOXICOLOGY IN REGULATORY FRAMEWORKS

The regulatory assessment of chemicals involves one or more of the following procedures: (1) hazard assessment (which includes hazard identification and dose–response characterization), possibly leading to classification and labeling; (2) exposure assessment; (3) risk assessment based on hazard and exposure assessments; and (4) the identification of persistant, bioaccumulative, and toxic (PBT) chemicals according to formal PBT criteria.

To promote the use of in silico methods for regulatory purposes, a well-defined and transparent framework should be established that provides support to nonspecialists in choosing, handling, and applying valid in silico methods. A clear workflow assisting nonspecialists all the way through the generation of reliable nontesting data would certainly facilitate the exploitation of computational toxicology for regulatory purposes. More specifically the workflow should aid the following processes:

- Retrieving existing physicochemical properties and (eco)toxicological information for a given chemical.
- Selecting relevant in silico approaches for predicting individual toxic endpoints.
- Providing information on the reliability of the estimates so that the user can make informed judgments on their adequacy for the purpose.
- Predicting the effects of a given chemical on humans and the environment.
- Exploiting the capabilities of various in silico methodologies.
- Integrating the results obtained from different in silico approaches such as (Q)SAR and read-across.
- Supporting the compilation of robust summaries that document in a transparent way the use of in silico methods for specific regulatory purposes.

Relevance refers to the appropriateness of the method (or of the nontesting data derived from the application of the method) for a particular regulatory purpose, while reliability refers to the performance (predictivity and reproducibility) of the method. Adequacy defines the usefulness (suitability) of the method (or of the nontesting data derived from the application of the method) to fulfill specific information requirements. High reliability and high relevance are likely to result in high adequacy of the approach, but adequacy is also dependent on the context. In other words, the availability of other information and the regulatory consequences of the decision will affect the judgment of adequacy.

An important feature of the workflow is that the entire process can be appropriately documented, for example, by compiling available reporting formats that are being implemented in various regulatory frameworks. These are templates developed to provide a standard framework for summarizing and structuring key information about the application of in silico methods and about the predictions calculated with these methods. In a risk assessment context the reporting formats serve the purpose of supplying the necessary information elements for interpreting and evaluating nontesting data derived from the use of relevant and reliable in silico approaches. The reporting formats are not meant to limit the use of (Q)SAR approaches or impose what methods should be used. Different types of reporting formats may exist. Those documenting the method itself provide robust summaries of the methodology (i.e., description of the algorithm and of its development and validation). Other reporting formats are intended to accommodate information on the actual generation of an estimate for a given chemical and with a specific method. Some national and international governmental bodies are working to optimize formats for summarizing the reasoning that leads to a specific regulatory decision (e.g., skin irritant or not a skin irritant) for a specific substance and endpoint. This reasoning makes use of information derived from different sources such as: (1) (Q)SAR predictions, which are thoroughly described in the corresponding prediction reporting formats linked in turn with the corresponding model reporting formats; (2) conclusions obtained by applying a read-across or category approach, also documented by appropriate reporting formats; and (3) in vivo and in vitro data (if available).

Eventually the workflow should be automated as much as possible into a computer-based decision support system (DSS) that aids the process of decision making when generating and using nontesting data for regulatory purposes. The DSS may prove invaluable as a means of promoting the regulatory use of in silico methods.

27.5 DECISION SUPPORT SYSTEM (DSS) FOR NONTESTING STRATEGIES

In general, a DSS is an interactive, flexible, and adaptable computer-based information system especially developed for supporting decision making in semistructured and unstructured situations by providing an organized set of tools. The following general properties should characterize a DSS, some of which have been noted by other authors [18,19]:

- It should facilitate decision processes and lead to improved decisions.
- It should support rather than automate decision making.
- It should provide an easy-to-use interface.
- It should allow for the decision maker's own insight.
- It should allow for data retrieval and storage, data analysis, identification of alternatives, and choice among alternatives.
- It should be able to respond quickly to the changing needs of decision makers.

In summary, a DSS should incorporate a body of knowledge that can be retrieved on an ad hoc basis in various customized ways as well as in standardized reports [20]. Moreover a subset of stored knowledge can be selected for deriving new knowledge. Importantly, a DSS must be designed to interact directly with the user who will be provided with the effective and relevant tools supporting operation flows and decision making.

A DSS that can support the implementation of nontesting strategies in a regulatory framework, such as REACH, should have the following specific characteristics:

- It should assist the end-user in collecting (eco)toxicological and chemical information (including biokinetics and environmental fate).
- It should provide guidance and support for selecting relevant in silico methods and for generating reliable and nontesting data.
- It should provide all the necessary supporting information needed to document the nontesting strategy, to help the user make informed judgments on the adequacy of the nontesting data.
- It should have an open-source software application to guarantee maximum transparency not only of the methodologies used by the system but also of the information stored therein.

The choice for an open-source DSS is first of all driven by the belief that in this way the DSS would be able to respond quickly to any progress and change in the scientific field as well as in the regulatory framework. Moreover the availability of the source code together with the right to modify it is likely to stimulate new improvements and further implementations. The open-source issue is a delicate subject that also takes into account that the DSS should be available to all the stakeholders and should be developed and managed by an organization that is independent of national and commercial interests such as the economic interests associated with the use of specific software.

It is possible to categorize the building blocks of the DSS into two main groups:

- Inventories (i.e., databases) storing information on chemicals, molecular structures, models, predictions, and experimental data.
- Tools for generating estimates, providing information on their reliability, and documenting their use.

Inventories are basically structured knowledge repositories. They can be independent systems or alternatively can integrate data from other information systems. Information Technology (IT) tools are software applications that assist the user in specific operations. Some inventories and IT tools that can be considered as essential components of the DSS are described in generic terms below, with emphasis on their scientific features rather than to specific software applications (e.g., commercially available software).

27.6 INVENTORY OF CHEMICALS

An important building block of a DSS for generating nontesting data for a given chemical is, of course, an inventory of chemicals that stores a variety of identification data such as names, CAS numbers, EC numbers, molecular weight, structural formulas, and molecular structures (2D and 3D information). Calculated descriptors can be accommodated in this inventory as well.

Standards on how to store chemical structures should ensure compatibility among the different components of the DSS and thus allow for an easy and smooth communication between them. Ideally a quality control should be performed on the correspondence of name to CAS number and chemical structure for all the chemicals stored in the inventory. This procedure is a rather demanding and time-consuming and probably will limit the number of chemicals that can be stored in the database. It is therefore proposed that the quality-checked chemicals be tagged with a special label. The structures can be stored in 2D formats using MDL's "mol file" connectivity table format [21], or SMILES [22] or InChI [23] codes, which are text-based representations of chemical structures [24]. SMILES (Simplified Molecular Input Line Entry System) is already widely used in many software applications. It is reported that there exists a specific canonicalization algorithm that generates one unique SMILES among all valid possibilities [25]. The IUPAC International Chemical Identifier (InChI) was launched in April 2005 and is expected to play a central role as a useful, unique, and public-domain chemical representation.

The chemicals stored in the inventory can be searched by exact structure, substructure, or similarity [26]. Similarity searching aims at retrieving compounds that are similar to a query compound by one or more measures of similarity. A set of structural features of the target molecule is compared with those of each chemical in the database, generating a similarity measure by a chosen metric such as the Tanimoto coefficient [27]. More details about chemical similarity are given below in relation to the chemical similarity tool.

27.7 INVENTORY OF DATA SOURCES

One of the first steps in the risk assessment process involves the collection of available information on the physicochemical properties, ecological effects, environmental fate, and health effects for a given chemical. In a nontesting strategy, data are also essential to make predictions by means of the read-across approach. Data can be retrieved from books, Internet-based resources (free and commercial resources), or commercial databases. This is a vast and rapidly developing field, so the reader is referred elsewhere for discussions of data sources [28–31].

It is recognized that a single database of experimental data will always contain a limited amount of information, and this information may become obsolete within a short period. The retrieval of data for a given chemical can be optimized by structuring available data sources in a list that can be consulted (queried), for example, by endpoint. Ultimately the efforts needed to collect existing information on a given chemical can be reduced by designing an information system tool that is capable of interfacing (communicating) with different online databases and that allows for the retrieval of the entire set of available information for the chemical of interest in a single run. This tool would simply be an ad hoc interface that integrates a number of freeware and commercial databases. Ideally the entire set of databases integrated in the inventory of data sources could be queried by structure, that is, exact structure, substructure, or similarity.

27.8 INVENTORY OF (Q)SAR MODELS

An assessment of the compliance of a (Q)SAR model with the OECD principles for (Q)SAR validation provides the basis for determining whether the model and the nontesting data it generates can possibly replace the need for testing. According to the discussions occurring at the OECD level, it is expected that the acceptance of (Q)SARs as nontesting alternative sources of data in making decisions will be based on the validity and transparency of a specific (Q)SAR model within a specific regulatory context [32]. Therefore a transparent description of a (Q)SAR model is the foundation for its acceptance. In this respect, reporting formats are being developed to provide a standard framework for compiling robust summaries of (Q)SAR models and their corresponding validation studies. The structure of this format has been designed to include the essential information that can be used to evaluate the compliance of the (Q)SAR model with the OECD principles [33]. The (Q)SAR model reporting formats, which should be regarded as a communication tool, are intended to enable a more efficient exchange of information between industry and regulators.

A database of (Q)SAR models should serve two purposes: it should provide easy and direct access to (Q)SAR models to be used as an alternative to testing methods, and it should also provide structured key information about the models themselves and the corresponding validation studies. In line with the structure of the (Q)SAR model reporting formats, the (Q)SAR model database should include the following kinds of information grouped in different sections:

- *Information about the source of the model.* This section includes information about the model developers (i.e., contact details), together with references to the original peer-reviewed papers describing the development of the model. There is also a field to identify the author(s) and creation date of the format.
- *Description of the type of model.* This section specifies the type of model (e.g., SAR, regression-based QSAR, expert system, battery of (Q)SARs) and defines the endpoint and the dependent variable being modeled, reporting also information (if available) on the quality of the data used

to develop the model. In this respect the Klimisch approach for the evaluation of the quality of experimental toxicological and ecotoxicological data can be employed [34]. This piece of information is useful for evaluating the reliability of the predictions made with the model. This section also stores valuable information about the algorithm, the goodness-of-fit statistics, the applicability domain, and the mechanistic basis of the model.

- *Information about the development of the model.* This section reports the details of the development of the model, specifying how the descriptors have been generated and selected and listing the chemicals included in the training set. It follows that the database can be queried to determine whether a chemical of interest is present in the training set/s of one or more models.
- *Information about the validation of the model.* This section describes how the model has been validated and reports the statistics obtained by techniques such as leave-one-out cross validation, leave-many-out cross validation, Y-scrambling, and external validation. The chemicals forming the test set are also listed in this section.

To avoid duplication of effort, this database of robust summaries of (Q)SAR models should be made freely available and considered as a reference source for the use of (Q)SAR approaches in a regulatory context.

27.9 INVENTORY OF (Q)SAR PREDICTIONS

(Q)SARs that are adequately documented in the inventory of (Q)SAR models can be used to generate predictions for chemicals of interest such as the chemicals within a regulatory inventory (e.g., the EU Inventory). The predictions can then be stored in a database of (Q)SAR predictions so that each prediction is linked to a robust summary describing the corresponding model used to generate it. As for the case of the inventory of (Q)SAR models, the inventory of (Q)SAR predictions should provide all the information to compile a robust report documenting the generation of the prediction in a transparent and complete way. The following types of information should be included in this database:

- *Information about the substance.* General information about the substance (e.g., chemical name, SMILES, structure, CAS).
- *Information about the model.* Model used to generate the prediction, with reference to its broader description in the (Q)SAR model database.
- *Information about the prediction.* Actual value of the prediction stored in the database, together with the description of how the algorithm was

applied to the chemical in question. The reliability of the prediction should be reported by means of well-defined criteria. It would also be desirable to provide an indication of the possible regulatory applicability of the prediction.

Querying this database would then give the prediction for a given chemical and a given endpoint, together with the link to the reporting format for the model used to generate the prediction.

27.10 INVENTORY OF EXISTING CHEMICAL CATEGORIES

This inventory would be useful to apply category/read-across approaches, since if a compound is a member of an existing category, available data for the other members of the category can be used to perform read-across in a straightforward way. The database of categories should include all the information necessary to adequately document the category and its boundary, and to properly document the predictions made using the category concept. Preliminary guidance on the implementation of category approaches has been developed by the OECD [5], and is being extended for REACH purposes by the European Commission. Since experience in applying these approaches is still evolving in various regulatory frameworks, the user requirements for this inventory are not easy to define at present.

27.11 SIMILARITY TOOL

Toxicity predictions rely on the similarity principle, which states that similar compounds have similar biological activities [35], and on the neighborhood principle, which states that molecules located in the same region of the descriptor space show similar biological activities [36]. Therefore the ability to quantify chemical similarity (i.e., the measure of how similar a chemical is to another) can be useful when predicting the unknown toxicity of a given chemical on the basis of the known toxicity of another similar compound. As the outcomes of appropriate measures of chemical similarity are calculated by means of a similarity tool, chemicals can be classified into categories facilitating the use of a read-across approach [5]. A similarity tool should also allow one to investigate the chemical space occupied by the training data and assess the applicability domain of the model.

It is clear that chemical similarity depends on the molecular features that are compared, and this means that the numerical representations chosen to describe the chemical (i.e., molecular descriptors) affect the outcome of the similarity quantification. A molecular descriptor is defined as "the final result of a logic and mathematical procedure that transforms chemical information encoded within a symbolic representation of a molecule into a useful number." [37]. The thousands of descriptors that can be derived from various theories and approaches can be classified in:

- 0D (zero-dimensional) descriptors such as molecular weight and atomic composition indexes.
- *1D* (mono-dimensional) descriptors derived by counting structural fragments in the molecule (molecular fingerprints).
- 2D (two-dimensional) descriptors derived from algorithms applied to a topological representation (molecular graph).
- 3D (three-dimensional) derived from 3D chemical structures that have to have previously undergone geometry optimization.

Quantification of similarity involves three different processes [26]:

- Representing the chemical in the descriptor space (i.e., defining the molecular descriptors that characterize the chemical in the descriptor space).
- Selecting and weighting the relevant descriptors for the comparison (i.e., relevant descriptors to the endpoint of interest).
- Comparing quantitatively the selected descriptors by calculating a similarity coefficient.

There are many different types of similarity indexes, including the association coefficients (e.g., Tanimoto coefficient [27], Jaccard coefficient [38], Hodgkin-Richards coefficient [39,40]), the correlation coefficients or cosine-like indexes, and the distance coefficients or dissimilarity indexes (e.g., Hamming distance) [26].

The Tanimoto index is the most common similarity index implemented in a number of structure searchable interfaces, where one compound is compared to another on the basis of fingerprints. The structure (most commonly, 2D structure) of a molecule is encoded as a pattern of bits set within a bit string (fingerprint): if a particular fragment is present at least once, then a corresponding bit is set in the bit string.

The similarity tool should encode different similarity indexes, and since the similarity indexes make use of molecular descriptors, the tool should also be able to derive the numerical representation from the given structure or to upload it from an external file. As for all the tools included in the DSS, the use of the chemical similarity tool should lead to results (e.g., category formation) that can be documented by the user in a transparent and comprehensive way.

27.12 APPLICABILITY DOMAIN TOOL

A way to gain information on the reliability of a prediction generated by the DSS is by means of an applicability domain tool capable of estimating the applicability domain from the training set using a suitable statistical approach. Once the applicability domain has been established, it is then possible to assess whether the compound of interest falls within defined boundaries.

Establishing the applicability domain of a (Q)SAR model and determining whether a given chemical falls within the applicability domain is a crucial step in generating a reliable estimate, as discussed in the third OECD validation principle ("defined domain of applicability"). This principle is based on the assumption that a model is capable of making reliable predictions only within the structural, physicochemical and response space. If the chemical is not an outlier in terms of the structural domain, the descriptor values, or the response value, it is likely that the prediction is reliable. However, falling within the applicability domain does not automatically imply full reliability of the prediction as in the case of a chemical acting with a different mechanism than that underpinning the (Q)SAR model.

The chemical space occupied by the training data set is the basis for describing the applicability domain of the model and thus estimating the reliability of the estimates. Training sets can be analyzed directly by structural similarity analysis or in the model descriptor space where chemicals are represented as points in a multivariate space. The similarity approach, based on the chemical similarity concept described above, relies on the assumption that a (Q)SAR prediction is reliable if the compound is similar to the compounds in the training set. Assessment of the applicability domain by means of this approach would certainly benefit from the use of the similarity tool described above. Regarding the analysis of the descriptor space, it is possible to identify four major types of statistical methods to characterize the interpolation space (i.e., its coverage) defined by the descriptors: range based, distance based, geometrical, and probability density distribution based [41].

27.13 OTHER TOOLS

27.13.1 Reliability Scoring Tool

When relevant (Q)SAR models are selected to generate predictions, it would be useful to develop an automated scoring system that quantifies the reliability of the prediction for a given compound. Some parameters that might affect the reliability score are: (1) falling within the applicability domain of the model, (2) having the same mechanism of action, (3) finding similar compounds in the training set, (4) reliability of the experimental data used to develop the model (e.g., Klimisch code [34]), and (5) performance of the model.

27.13.2 Classification Schemes

There exist various classification schemes that enable the derivation of certain chemical information, such as the mode of (eco)toxic action, on the basis of structural characteristics alone. An example is the Verhaar classification scheme for aquatic toxicology [42], which provides information on the mode of action of the chemical (this scheme is already encoded in a freeware application called Toxtree [43]). Using various rules the chemicals are distributed among five groups: (1) inert chemicals (narcotics), (2) less inert chemicals (polar narcotics), (3) reactive chemicals, (4) specifically acting chemicals (acetylcholine esterase (AChE) inhibitors, compounds interacting with Na channel regulating receptors), and (5) chemicals with unknown mode of action (compounds not classified in the first four classes). Application of the Verhaar classification scheme aids the choice of the (Q)SAR model for making reliable predictions of aquatic toxicity for a given substance. Another classification scheme for aquatic toxicity is the scheme developed by Russom et al. [45] and implemented in the ASTER program.

Another example is the Cramer decision tree, which has been used for structuring flavoring substances and food contact materials according to levels of concern [43,44]. Relying primarily on chemical structure and an estimate of total human intake to establish priorities for testing, it classifies the chemical in one of three classes: class 1 contains substances of simple chemical structure with known metabolic pathways and innocuous end products, which suggest a low order of oral toxicity; class 2 contains substances that are intermediate; class 3 contains substances with a chemical structures that permit no strong initial impression of safety and may even suggest a significant toxicity. The classification procedure utilizes recognized pathways for metabolic deactivation and activation, data on toxicity, and the presence of a substance as component of traditional foods or as an endogenous metabolite.

The Cramer classification scheme can be used to make a threshold of toxicological concern (TTC) estimation. TTC is a concept that aims to establish a level of exposure for all chemicals below which there would be no appreciable risk to human health; the threshold is based on a statistical analysis of the toxicological data from a broad range of different and/or structurally related chemicals and on the extrapolation of the underlying animal data to a no-effect dose considered to represent a negligible risk to human health.

27.13.3 Estimation Tools Based on (Q)SARs and Expert Systems

Nonproprietary algorithms that are stored in the (Q)SAR model database could be automated in a software tool that computes directly the prediction for a given chemical. Many of these algorithms are published in the scientific literature, and computer-based freeware tools are increasingly being developed to make the algorithms usable and readily available. For further information, the reader is referred to various reviews (e.g., [46]).

27.14 STEPWISE APPROACH FOR THE GENERATION AND USE OF NONTESTING DATA

A schematic workflow comprising various steps (Figure 27.1) is proposed for the generation and use of nontesting data in any regulatory context. The proposed schema can be used as a starting point for implementing more elaborate nontesting strategies that take into account specific regulatory needs (e.g., the needs of industry and government authorities under REACH). At all stages the stepwise approach depends also on expert judgment, which may then result in slight modifications to the proposed workflow. Depending on the specific needs and goals of the user, a single step may prove to be sufficient for the purpose. The proposed workflow implicitly makes use of the different tools and databases described above. Therefore the DSS can be used to facilitate the various steps of this workflow.

In the *starting step*, information about the substance under consideration is collected (regulations, purity/impurity profile, composition), and the information gaps are identified after comparing the information requirements under the given regulatory framework and the available (eco)toxicological and fate information acquired by browsing different data sources. As the substance under consideration may comprise one or multiple compounds, a specific compound on which it is useful to apply in silico approaches is selected (parent compound). Subsequently (Q)SAR and read-across approaches will be applied to the parent compound to make predictions of a given (toxic) endpoint. If the parent compound is known by CAS or EC number or by name, it is essential to derive its structure (e.g., in the form of the SMILES code) to be used in the prediction generation process. A working matrix is generated to store all the information collected in this phase and in the following steps of the workflow.

After the initial step of data collection, a preliminary analysis (*step 1*) of the toxicity/fate/uptake of the parent compound is performed using information about abiotic and biotic reactions. Biotic reactions refer to the chemical reactions occurring in living organisms (e.g., plants, animals, bacteria, fungi) during the biotransformation process of foreign compounds (xenobiotics). Different enzymes catalyze a wide variety of reactions for facilitating the elimination of xenobiotics. Abiotic reactions are those that change the chemical composition of the compound, but that do not involve the participation of

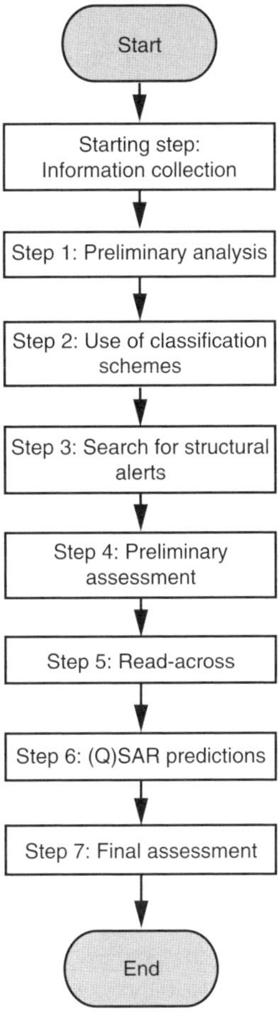

Figure 27.1 Schematic workflow proposed to provide guidance on the generation and use of in silico methods in a regulatory context.

any living organism. Examples of abiotic reactions are photochemical reactions, oxidation, reduction, and hydrolysis. Knowledge of these reactions is needed when assessing the environmental distribution and fate of chemicals. The stability/reactivity of the parent compound may be further estimated by analyzing fragments and descriptors like molecular orbital energies (HOMO, LUMO). Some considerations that should help the preliminary analysis are as follows:

- How size (molecular weight), shape and lipophilicity ($\log K_{ow}$) affect uptake and toxicity.

- Whether ionisation can take place at the relevant pH.
- What chemical reactivity (i.e., what type of reactions) is expected for the parent compound.
- Which metabolites and reaction products (e.g., hydrolysis products) are generated.

Additional compounds (metabolites and other reaction products) for which it is useful to make predictions can be chosen. The (Q)SAR and read-across approaches will be applied to each of these compounds.

In the following process (*step 2)*, further information on the likely biological activity of the compound may be obtained through classification schemes (where available) for the endpoint of interest. For example, classification schemes by Verhaar [42] and Russom [45] can be used when assessing the mode of action for acute fish toxicity. The classification scheme developed by Cramer et al. [44] is useful for evaluating the likely systemic toxicity of a compound.

In *step 3* structural alerts (where available) for the endpoint of interest are searched. Both 2D and 3D structural alerts can be used, although 2D SAR approaches are likely to be more readily used because they are more intuitive and easier to automate.

In *step 4* a preliminary assessment of the likely uptake/toxicity/fate profile of the parent compound is performed using the outcomes of *steps* 1 through 3. This preliminary assessment involves an analysis of the uptake/toxicity/fate profile of all the query compounds, and the following issues should be considered:

- The outcome of the preliminary analysis done in step 1 (physicochemical properties, metabolites, reaction products).
- Information collected by applying the appropriate classification schemes that may serve to classify the mode of action of the compound in question. This information is useful when evaluating which (Q)SAR models should be applied to the compound in question.
- The presence of structural alerts.

This evaluation step should also lead to the identification of a clearer nontesting strategy that can be applied in subsequent steps.

Step 5 is aimed at filling data gaps for all the query compounds using a read-across (or analog) approach, where the endpoint information for one chemical is used to make a prediction of the endpoint for another chemical, which is considered to be "similar" in some way. It follows that read-across first requires the identification of similar compounds, either by means of a traditional (nonformalized) approach or by means of a computer-based analog search engine. Analogues can be sought from the database/s within the DSS and/or external databases. If a compound is a member of an existing category, available data for the other members of the category can be used to perform read-across. If there is no existing category, this step can include the development of a new category.

In *step 6* relevant (Q)SARs are selected and (Q)SAR predictions are generated for the query compounds, together with the reporting formats for the models and the predictions. The reliability of the predictions also has to be established. When more than one relevant and reliable prediction is available a judgment of the relative reliability of the different predictions might be necessary if the predictions do not lead to the same conclusions. This might lead the user to select a "representative" value or to weight the available data in some way, or possibly use a consensus approach.

In the final step (*step 7*) an overall analysis of the outcome of steps 1 through 6 for the endpoint of interest is performed. The toxicity of the parent compound is assessed via the information obtained for all the query compounds (metabolites, reaction products, analogues). The overall assessment should make use of all the available information (testing and nontesting data). A formal logical approach, such as decision theory, that evaluates all the possible options and uses probabilistic calculations to identify the best decision could be used to support the overall assessment process. This implies that additional decision analysis tools will need to be integrated into the DSS.

27.15 CONCLUDING REMARKS

In this chapter we described the main types of nontesting approaches being increasingly developed around the world for use in the regulatory assessment of chemicals. We considered the REACH legislation as a particular example of a new legislative framework that explicitly incorporates the need to use nontesting approaches, in the interests of cost-effectiveness and animal welfare. We showed how a DSS would contain the main functionalities needed to support the implementation of ITS based on nontesting data. The European Commission through the European Chemicals Bureau (ECB) is developing such a system in collaboration with a range of stakeholders, including industry, governmental authorities, and the OECD. While the particular interest of the ECB is to develop a DSS suitable for REACH, it is intended that the system be developed so that it can be used in any regulatory framework. Different frameworks are likely to use nontesting data in different ways. More generally applicable computational tools are needed to provide these data, and to serve the sciences on which they are based. In our view, the DSS should be designed according to the principles of user-friendliness, transparency, and adaptability. A DSS incorporating these functionalities should also be freely available to all stakeholders, and especially to industry and the regulatory authorities. This will ensure that the companies submitting regulatory dossiers and the regulatory bodies checking/approving them have access to the same information.

The availability of publicly accessible tools should encourage a coordinated use of nontesting data by different authorities across countries.

REFERENCES

1. Schultz TW, Cronin MTD, Walker JD, Aptula AO. Quantitative structure-activity relationships (QSARs) in toxicology: A historical perspective. *J Mol Struct (Theochem)* 2003;622:1–22.
2. Worth AP. The tiered approach to toxicity assessment based on the integrated use of alternative (non-animal) tests. In: Cronin MTD, Livingstone DJ, editors, *Predicting chemical toxicity and fate*. Boca Raton, FL: CRC Press, 2004. p. 391–412.
3. Dearden J. In silico prediction of drug toxicity. *J Comput Aided Mol De* 2003;17: 119–27.
4. Dearden JC, Barratt MD, Benigni R, Bristol DW, Combes RD, Cronin MTD, Judson PM, Payne MP, Richard AM, Tichy M, Worth AP, Yourick JJ. The development and validation of expert systems for predicting toxicity. The report and recommendations of an ECVAM/ECB workshop (ECVAM workshop 24). *ATLA* 1997;25:223–52.
5. Rosenkranz HS, Cunningham AR. Chemical categories for health hazard identification: A feasibility study. *Regul Toxicol Pharm* 2001;33:313–18.
6. OECD. Section 3.2 of the OECD Manual for Investigation of HPV Chemicals: Chapter of guidance document on the formation and use of chemical categories. 2005. Available at: <http://www.oecd.org/document/7/0,2340,en_2649_34379_1947463_1_1_1_1,00.html>.
7. Cronin MTD. Predicting chemical toxicity and fate in human and the environment—an introduction. In: Cronin MTD, Livingstone DJ, editors, *Predicting chemical toxicity and fate*. Boca Raton, FL: CRC Press, 2004. p. 3–13.
8. European Commission. Council Directive 86/609/EEC of 24 November 1986 on the approximation of laws, regulations and administrative provisions of the member states regarding the protection of animals used for experimental and other scientific purposes. *Off J Eur Commun* 1986;L358:1–29, 18 Nov 1986.
9. Cronin MTD. The use by governmental regulatory agencies of quantitative structure-activity relationships and expert systems to predict toxicity. In: Cronin MTD, Livingstone DJ, editors, *Predicting chemical toxicity and fate*. Boca Raton, FL: CRC Press, 2004. p. 414–27.
10. Worth AP, Bassan A, de Bruijn J, Gallegos Saliner A, Netzeva TI, Patlewicz G, Pavan M, Tsakovska I, Eisenreich S. The role of the European Chemicals Bureau in promoting the regulatory use of (Q)SAR methods. *SAR QSAR Environ Res* 2007;18:111–25.
11. Walker JD, Carlsen L, Hulzebos E, Simon-Hettich B. Global government applications of analogues, SARs, and QSARs to predict aquatic toxicity, chemicals or physical properties, environmental fate parameters and health effects of organic chemicals. *SAR QSAR Environ Res* 2002;13:607–16.
12. Cronin MTD, Walker JD, Jaworska JS, Comber MHI, Watts CD, Worth AP. Use of QSARs in international decision-making frameworks to predict ecologic effects

and environmental fate of chemical substances. *Environ Health Perspect* 2003;111: 1376–90.

13. Cronin MTD, Walker JD, Jaworska JS, Comber MHI, Watts CD, Worth AP. Use of QSARs in international decision-making frameworks to predict health effects of chemical substances. *Environ Health Perspect* 2003;111:1391–401.
14. Walker JD. Applications of QSAR in toxicology: A US government perspective. *J Mol Struct (Theochem)* 2003;622:167–84.
15. OECD. *Report on the regulatory uses and applications in OECD member countries of (quantitative) structure-activity relationship [(Q)SAR] models in the assessment of new and existing chemicals*. Paris: Organisation of Economic Cooperation and Development, 2006. Available at: <http://www.oecd.org/document/30/0,2340,en_2649_34365_1916638_1_1_1_1,00.html>.
16. European Commission. Proposal for a Regulation of the European Parliament and of the Council concerning the registration, evaluation, authorisation and restriction of chemicals (REACH), establishing a European Chemicals Agency and amending Directive 1999/45/EC and regulation (EC) {on persistent organic pollutants}. Proposal for a Directive of the European Parliament and of the Council amending Council Directive 67/548/EEC in order to adapt it to Regulation (EC) of the European Parliament and of the Council concerning the registration, evaluation, authorisation and restriction of chemicals. Brussels, Belgium. 29 Oct 2003. Available at: <http://ec.europa.eu/environment/chemicals/reach/reach_intro.htm>.
17. Worth AP, Cronin MTD, Van Leeuwen, CJ. A framework for promoting the acceptance and regulatory use of (Quantitative) structure activity relationships. In: Cronin MTD, Livingstone DJ, editors, *Predicting chemical toxicity and fate*. Boca Raton, FL: CRC Press, 2004. p. 429–40.
18. Turban E, Aronson RH. *Decision support systems and intelligent systems*, 5th edition. Upper Saddle River, NJ: Prentice Hall, 1997.
19. Alter S. *Decision support systems: Current practice and continuing challenges*. Reading, MA: Addison-Wesley, 1980.
20. Holsapple CW, Whinston AB. *Decision support systems: A knowledge based approach*. Minneapolis: West, 1996.
21. Elsevier MDL CTFile Formats (MDL mol and SD file format) downloadable documentation on <http://www.mdli.com/downloads/public/ctfile/ctfile.jsp>.
22. Weininger D. SMILES, a chemical language and information system: 1. Introduction to methodology and encoding rules. *J Chem Inf Comput Sci* 1988;28: 31–6.
23. IUPAC Web site describing the IUPAC Chemical Identifier (InChI): <http://www.iupac.org/inchi>.
24. Richard AM, Swirsky GL, Nicklaus MC. Chemical structure indexing of toxicity data on the Internet: Moving toward a flat world. *Curr Opin Drug Discov Devel* 2006;9:314–25.
25. Weininger D, Weininger A, Weininger JL. SMILES: 2. Algorithm for generation of unique SMI LES notation. *J Chem Inf Comput Sci* 1989;29:97–101.
26. Gallegos Saliner A. Mini-review on chemical similarity and prediction of toxicity. *Curr Comput Aided Drug Des* 2006;2:105–22.

27. Flower DR. On the properties of bit string-based measures of chemical similarity. *J Chem Inf Comput Sci* 1998;38:379–86.

28. Kaiser KLE. Toxicity data sources. In: Cronin MTD, Livingstone DJ, editors, *Predicting chemical toxicity and fate*. Boca Raton, FL: CRC Press, 2004. p. 174–29.

29. The QSAR and Modelling Society maintains an extensive list of databases that can be accessed through their Web site: <http://www.qsar.org>.

30. The European Chemicals Bureau publishes a list of databases on the Web dedicated to computational toxicology: <http://ecb.jrc.it/QSAR/information_sources/information_databases.php>.

31. Yang C, Benz RD, Cheeseman MA. Landscape of current toxicity databases and database standards, *Curr Opin Drug Discov Devel* 2006;9:124–33.

32. Worth AP, Bassan A, Gallegos A, Netzeva TI, Patlewicz G, Pavan M, Tsakovska I, Vracko M. The characterisation of (quantitative) structure-activity relationships: Preliminary guidance. 2005;JRC report EUR 21866 EN. European Chemicals Bureau, Ispra (Italy). Available at: <http://ecb.jrc.it>.

33. QSAR Model Reporting Formats can be downloaded from European Chemicals Bureau Web site: <http://ecb.jrc.it/QSAR>.

34. Klimisch HJ, Andreae M, Tillmann U. Systematic approach for evaluating the quality of experimental toxicological and ecotoxicological data. *Regul Toxicol Pharm* 1997;25:1–5.

35. Johnson M, Maggiora GM. *Concepts and applications of molecular similarity*. New York: Wiley, 1990.

36. Walters WP, Stahl MT, Murcko MA. Virtual screening—An overview. *Drug Discov Today* 1998;3:160–78.

37. Todeschini R, Consonni V. *Handbook of molecular descriptors*. Weinheim: Wiley-VCH, 2000.

38. Jaccard P. Étude comparative de la distribution florale dans une portion des Alpes et des Jura. *Bull. Soc Vaudoise Sci Nat* 1901;37:547–79.

39. Hodgkin EE, Richards GW. Molecular similarity based on electrostatic potential and electric field. *Int J Quant Chem Quant Biol Symp* 1987;14:105–10.

40. Good AC, Hodgkin EE, Richards GW. Utilization of Gaussian functions for the rapid evaluation of molecular similarity. *J Chem Inf Comput Sci* 1992;32:188–91.

41. Jaworska J, Nikolova-Jeliazkova N, Aldenberg T. QSAR applicability domain estimation by projection of the training set in descriptor space: a review. *ATLA* 2005;33:445–59.

42. Verhaar HJM, van Leeuwen CJ, Hermens JLM. Classifying environmental pollutants: Structure-activity relationships for prediction of aquatic toxicity. *Chemosphere* 1992;25:471–91.

43. Toxtree. Software developed by Dr Nina Jeliazkova (Ideaconsult Ltd; contact nina@acad.bg) on behalf of the European Chemicals Bureau (ECB). Copyright European Communities (2005). Available at: <http://ecb.jrc.it/QSAR>.

44. Cramer GM, Ford RA, Hall RL. Estimation of toxic hazard—A decision tree approach. *J Cosmet Toxicol* 1978;16:255–76.

45. Russom CL, Anderson EB, Greenwood BE, Pilli A. ASTER: An integration of AQUIRE data base and the QSAR system for use in ecological risk assessments. *Sci Total Environ* 1991;109–10:667–70.
46. Worth AP, Netzeva T, Patlewicz GY. Predicting toxicological and ecotoxicological endpoints. In: Van Leeuwen VJ, Vermeire T, editors, *Risk assessment of chemicals: An introduction*, 2nd edition. Berlin: Springer-Verlag, 2007.

INDEX

Wiley Series on Technologies for the Pharmaceutical Industry
Sean Ekins, Series Editor

Computational Toxicology: Risk Assessment for Pharmaceutical and Environmental Chemicals

Edited by Sean Ekins